D0892103

Aspects of Organic Chemistry

Structure

Distribution
VCH, P.O. Box 101161, D-69451 Weinheim (Federal Republic of Germany)
Switzerland: VCH, P.O. Box, CH-4020 Basel (Switzerland)
United Kingdom and Ireland: VCH (UK) Ltd., 8 Wellington Court, Wellington Street, Cambridge CB1 1HZ (England)
USA and Canada: VCH, Suite 909, 220 East 23rd Street, New York, NY 10010-4606 (USA)

ISBN 3-906390-15-2

Gerhard Quinkert
Ernst Egert
Christian Griesinger

Aspects of Organic Chemistry

Structure

Verlag Helvetica Chimica Acta, Basel

Weinheim · New York · Basel
Cambridge · Tokyo

Professor Dr. Gerhard Quinkert
Professor Dr. Ernst Egert
Professor Dr. Christian Griesinger
Institut für Organische Chemie
Marie-Curie-Straße 11
D-60439 Frankfurt am Main

Translated by Andrew Beard

Published jointly by
VHCA, Verlag Helvetica Chimica Acta, Basel (Switzerland)
VCH Verlagsgesellschaft mbH, Weinheim (Federal Republic of Germany)
VCH Publishers, Inc., New York, NY (USA)

Editorial Directors: Dr. M. Volkan Kisakürek, Dr. Birgit Gröne
Production Manager: Jakob Schüpfer
Layout: Erich Brossog
Cover Design: Bruckmann & Partner, Basel

Library of Congress Card No. applied for.

A CIP catalogue record for this book is available from the British Library.

Die Deutsche Bibliothek - CIP-Einheitsaufnahme

Quinkert, Gerhard:
Aspects of organic chemistry / Gerhard Quinkert ; Ernst Egert ; Christian Griesinger. - Basel : Verl. Helvetica Chimica Acta ; Weinheim ; New York ; Basel; Cambridge ; Tokyo : VCH.
Literaturangaben. - Dt. Ausg. u.d.T.: Quinkert, Gerhard: Aspekte der organischen Chemie
NE: Egert, Ernst:; Griesinger, Christian:
Structure. - 1996
ISBN 3-906390-15-2

Printed on acid-free paper.

Printing: Birkhäuser+GBC AG, CH-4153 Reinach BL
Printed in Switzerland

Foreword

Immanuel Kant, writing in 1786, was doubtful that 'chymistrie', an experimental field of study, could ever develop into a science. The 'chymistrie' he meant, however, could only have been alchemy, and it is now known that alchemists, to preserve the Hermetic mysteries, consciously invented elaborate metaphor to produce a deliberately confused terminology, 'to avert the curiosity of the uninitiated'. Chemistry has a completely different view of language as a vehicle of communication, as this volume makes clear in various places.

Like earlier adherants of romantic natural philosophy, today's supporters of postmodernism and poststructuralism not only make no secret of their aversion to the methodical procedures of natural science, but also, in many cases, have developed aggressive ideologies and antiindustrial political stances against it. (Organic) chemistry and the chemical industry have been so closely connected with one another for a century that, in the public mind, the word 'chemistry' is often seen as synonymous with 'chemical industry'. Without wishing to imply any stand-off between the two fields, which exist in a fruitfully symbiotic relationship with one another, the term 'organic chemistry', when used here, refers exclusively to the empirically based scientific discipline.

In 16 chapters, this volume lays the foundations for the study of the structure of the molecules and supermolecules of organic chemistry. The first eight chapters form a single unit, in their order and in their content. In these, the first topic is the categorization of the enormous diversity of molecules into ensembles of distinct individuals, all possessing the same general or individual molecular formula. Representatives of such ensembles, of the same individual molecular formula, known as 'isomers', contain the same atoms in the same numerical rations, but in different sequences (constitution) or spatial arrangement (configuration). Molecules may be described and classified with the aid of structure models: irrespective of whether these molecules have already been prepared or exist solely in the mind of the chemist. The *structure model of the classical organic chemistry*, making use of a chemical model managing with concepts borrowed from classical mechanics, says nothing about the nature of the chemical bond: the general construction principles it assumes are essentially the result of experimental observation. Individual atoms of a molecule are bound to one another. If the bonds involved are single bonds, then the atom groups connected with one another were originally assumed to rotate freely with respect to one another about the connecting bond. Contrary to the assumption of 'free rotation', certain idealized arrangements of atoms with characteristic torsion-angle combinations (conformations) are energetically favored. These play a dominant role in conformational analysis, a discipline devoted to solving the relationships between the conformation of a molecule and its associated properties.

Conformational analysis is also the central theme when relating the molecules and supermolecules of micro-, macro-, and supramolecular chemistry with one another. The 'basic rules' are introduced in *Chapt. 6*, using molecules of low mass, while *Chapt. 7* demonstrates how they also apply to molecules of high mass. *Chapt. 7* is the central focus of this volume, and not only because of its size. In the past, organic chemists, entirely out of keeping with the original sense of 'organic chemistry', have effectively avoided involvement with biomacromolecules, leaving it to interested parties of whatever discipline to occupy themselves with this lively – in the truest sense of the word – field of chemistry. The authors are particularly concerned to assist in bringing these separately living 'relatives', which have meanwhile developed in markedly different ways, closer together.

The first seven chapters deal with solving structural problems with the aid of the structure model of the classical organic chemistry. In certain cases, though, experimentally ascertained properties of molecules contradict the assumed conformation, configuration, or constitution. This is where the structure model of the classical organic chemistry hits the limits of its range of validity. In its place, there came a physical model characterized by the use of wave mechanics and placing its main emphasis on the chemical bond: the (qualitative) MO model. *Chapt. 8* demonstrates how, using the MO model, it is possible to solve problems which the structure model of the classical organic chemistry cannot explain.

While the first eight chapters, taken as a whole, offer a view over the enormous diversity of organic chemical molecules and supermolecules, enabling their complex interrelationships to be perceived and grasped, the final eight, more like essays in their style, pursue broader or deeper themes relating to molecular structure already mentioned earlier. The molecules and supermolecules here were selected with the aim of helping to traverse the divide between chemistry and biology – to make this a routine occurrence. *Chapt. 16*, especially, deals with bridging the gap between chemistry and biology.

Each of the authors has – more than once – given an introductory course of lectures about 'structural organic chemistry' (provided for all third-semester chemistry students), based on the contents of this book, at the University of Frankfurt on Main. They were acting on the assumption that their audience, which, parallel to the lecture course, was occupied with practical experimental work in the laboratory and the solving of chemical problems in seminars, was making reference to one or other of the current text-books about 'organic chemistry'. Furthermore, the authors have borne in mind that, as well as providing *tuition for all*, the university must not lose a chance of *cultivating the ability to self-education for some*. Selected aspects of organic chemistry, complete with detailed references to further literature at the end of each chapter, were developed with this in mind. The author names given in the further literature sections have not been included in the author index.

Organic chemistry uses a codified iconography to describe the structure of molecules, and nomenclature rules to name them. *Chapt. 10* contains detailed information of how molecules can be represented through formulae and names.

This text has a complex history, which – to give due thanks to all those involved – will be outlined briefly. It began with the lecture program and its associated draft script. The content of the lecture program was documented in the notes of numerous students over many semesters. Of the written-up notes, which were not without influence on the content and form of later variants of the lecture course, those of *Peter Eckes* – because of their quality and their widespread dissemination at the time – should be highlighted. The manuscript which finally saw the light of day has matured over the years. Without the superlative word-processing skills, patience, and judiciousness of *Anneliese Dlabal*, it would never have seen completion. A whole army of efficient helpers was involved in the production of the preliminary illustrations, using the program ChemDraw. *Petra Schmalz*, *Claudia Schierloh*, *Astrid Döring*, *Dietmar Reichert*, *Klaus Urbahns*, and *Birgit Gröne* were especially invaluable here. From the collected notes to the manuscript is a long way. From the manuscript to the completed book is no shorter. It is thanks to *M. Volkan Kısakürek* that this final, demanding stage could be executed at all. His feel for language, knowledge of chemical nomenclature, and esthetic principles have left their imprint on nearly every page.

As well as the authors, *Joachim W. Engels*, *Dieter Rehm*, and *Wolf-Dieter Stohrer* were active participants in the development of the teaching program. Their contributions are interwoven into the text. The same also applies to the many contributions of the Visiting Professors (of the *Rolf Sammet* Foundation and the *Degussa* Foundation at the *Johann Wolfgang Goethe* Universität, Frankfurt am Main): *Peter Dervan*, *Albert Eschenmoser*, *Meir Lahav*, *Jean-Marie Lehn*, *K. Barry Sharpless*, *W. Clark Still*, and *Rolf K. Thauer*. We are indebted to *Albrecht Berkessel*, *Steffen Glaser*, *Michael W. Göbel*, *Clemens Jochum*, *Reiner Luckenbach*, *Michael Reggelin*, *Wolfgang R. Roth*, *Hans-Günther Schmalz*, *Wolfgang Schubert*, and *Hans Ulrich Stilz* for constructive commentaries on individual chapters.

With the best will in the world, authors frequently cannot recall who it was who led them to this thought or to that picture. However, they can usually name one or two individuals who had a lasting influence on their patterns of thought. For one of us (*G.Q.*), who essentially was forced to self-education, *Edgar Heilbronner* was responsible, early on, for sharpening the view of the role of models, while *Albert Eschenmoser*, over the decades, has given the directions and set the standard.

Frankfurt am Main, January, 1996

Gerhard Quinkert
Ernst Egert
Christian Griesinger

Contents

1 Introduction to Organic Chemistry

1.1. Outline of the General Aims of '*Aspects of Organic Chemistry*'

If the tasks of a philosopher are assumed to include the exposure and the clarification of unspoken assumptions then this text will also be suited for profitable study in a philosophical seminar. The authors have therefore been careful to ensure that the reader is made aware of many of the assumptions which underlie the terms employed and the correct understanding of their implicit meanings, so that the text is easily understandable and can be of use to the reader who is not yet very experienced in chemistry. An exception, however, has been made in the case of this section. Here a broad brush has been used, to outline the themes covered through the whole series, including those parts not yet published, using concepts which have not been painstakingly introduced and with allusions requiring a level of comprehension that only an experienced chemist will possess. *Aspects of Organic Chemistry*, however, should be a useful aid to both the beginner and the advanced chemist.

The awarding of the *Nobel* Prize for Chemistry to *Donald J. Cram, Jean-Marie Lehn*, and *Charles J. Pedersen* in 1987 made a wide public aware that molecular chemistry (with its molecules) has extended its reach into the field of supramolecular chemistry (with its supermolecules). In *Aspects of Organic Chemistry*, Volume 1: *Structure*, the foundations are laid for the structural description of molecules and supermolecules.

For the first hundred years of organic chemistry, the central field of interest was the molecular structure of individual compounds. Nowadays, interest is primarily focused on the molecular function which a compound can exhibit because of its structure-related properties. This change in emphasis reflects the dramatic widening of horizons ushered in by *James D. Watson* and *Francis H. C. Crick* (*Nobel* Prize for Physiology or Medicine 1962) after their elucidation of the structure of DNA (**d**eoxyribo**n**ucleic **a**cid) in 1953. Each of the two single strands of DNA, which, in the supermolecule, are coupled together in the famous double-helix structure, serves, when unwound, as a template for the production of an exact DNA copy, which may be passed on to a daughter cell upon cell division.

Because of the exceptional degree of attention recently given to the catalysis of stereochemically asymmetric reactions (through the work of

K. Barry Sharpless and *Ryoji Noyori*, for example), it has become easier to define conditions which must be fulfilled before a consensual definition of a fully developed chemical reaction can be agreed upon. A chemical reaction may be regarded as fully developed if it leads to the desired product in high chemical yield and with a high and predictable degree of regio- and stereoselectivity, with not too narrow a substrate specificity, using substoichiometric quantities of effective catalysts and which can easily be scaled up. In *Aspects of Organic Chemistry*, Volume 2: *Reactivity*, the foundations are laid of the description and the application of chemical reactions. Particular attention is given to the degree to which the processes presented are fully developed, so as to sharpen the perspective for further potential advances.

At the center of chemistry stands synthesis. Since *Adolf von Baeyer* (*Nobel* Prize for Chemistry 1905), chemical synthesis has been understood as meaning a system of methods, with the help of which structurally complex chemical compounds may be built up from simple ones. The ultimate which has so far been achieved in synthetic organic chemistry is the two total syntheses of vitamin B_{12}, which was made into reality by a total of 130 doctoral students and post-doctoral researchers under the guidance of *Robert B. Woodward* (Harvard University) and/or *Albert Eschenmoser* (Eidgenössische Technische Hochschule-Zürich, Switzerland). *Woodward* (*Nobel* Prize for Chemistry 1965) demonstrated how multistep syntheses of structurally complex target compounds may be planned consistently and carried out flexibly. *Eschenmoser* showed how, through the invention of new reactions in the course of a synthesis project, the original strategy may be retained and the synthesis be used as an important source of new discoveries about the reactivity of chemical compounds. *Elias J. Corey* received the *Nobel* Prize for Chemistry in 1990 for his development of the theory and methodology of multistep organic synthesis including computer-assisted analysis of complex synthetic problems.

The attention of the synthetic chemist is, on one hand, directed towards natural products, and, on the other, towards those target compounds which have not arisen during the course of naturalistic evolution, but which promise to exhibit particular functions as a result of their chemical, physical, or biological properties (as a reagent or catalyst, as a material or as a biologically active substance). In *Aspects of Organic Chemistry*, Volume 3: *Synthesis*, the current state of knowledge of the synthesis of structurally complex target compounds is disclosed.

The spectrum of synthetic methods expressly includes both those of chemical synthesis and those of biological synthesis. Under the heading of chemical synthesis procedures should be understood not only transformations using reagents or catalysts which chemists have developed, but also those in which enzymes, antibodies, and even dead cells with their natural genomes find application. Living cells possess an extremely finely tuned synthesis program, used for the numerous synthetic steps taking place inside them, in their DNA. This synthesis program, and the synthesis procedure governed by it, may be reprogrammed using genetic engineering techniques (through the introduction of *in vitro* newly combined (recombinant) nucleic-acid molecules). Synthetic steps which exploit the synthetic capabilities of living cells with original or recombinant DNA in an intentional manner are classified under biological synthesis.

After the successful 'physicalization' of chemistry in the first half of this century, it is high time to break down the barriers between chemistry and the neighboring discipline of biology: at cultivating the interdisciplinarily thinking chemist who can identify and improve upon biologically active structures and make the ensuing target compounds accessible through chemical or biological synthesis. *Aspects of Organic Chemistry*, Volume 4: *Methods of Structure Determination*, will demonstrate how X-ray crystal-structure analysis, NMR spectroscopy and computer simulation may be applied to the determination of the structure of a biologically interesting active substance and how it may be modified in specific ways after consideration of a suitable structure-activity relationship. A point of concern raised as early as 1907 by *Emil Fischer* (*Nobel* Prize for Chemistry 1902) in his *Faraday Lecture* of that year ('*Synthetical Chemistry in its Relation to Biology*') is thus addressed: '*In its early youth, organic chemistry was so closely connected with biology. I do consider it not only possible but desirable, that the close connexion of chemistry with biology ... should be reestablished, as the great chemical secrets of life are only to be unveiled by cooperative work*'.

1.2. The Compounds of Organic Chemistry

Chemistry is the empirically based science of the properties and the transformation of substances. Properties, in general, can run through a continuous scale of 'values'. Substances are classifiable in a discrete manner, continuous progression there lead merely to mixtures. This differentiation between mixtures and discrete substances – chemical compounds, as the chemist calls them – stands at the beginning of every journey into chemistry. Every property of a mixture belongs to a continuum between the properties of the pure substances. The properties of a chemical compound, on the other hand, are novel, and characterize a discrete substance. The discrete nature of the pure substance is a fundamental chemical phenomenon, and can easily be recognized in everyday life. It reveals the fundamental structure of physical reality and is not in any way trivial. *Carl Friedrich von Weizsäcker* has claimed that presumably no conceptualization would be possible, were there no phenomena of discreteness, and therefore no discontinuity in nature.

1.2.1. Delimitations and Categorizations

The name does it suggest: organic chemistry is devoted to the study of the building blocks of life. The differentiation of organic and inorganic compounds is not as simple as it might appear at first glance, though. In the course of two centuries, all manner of boundaries have been drawn.

Originally, it was assumed that it would not be possible to synthesize compounds of organic chemistry outside of living cells. A special vital force, the so-called *vis vitalis*, was held to be responsible for their coming into being. Because of this belief, organic compounds such as vegetal oxalic acid or animalic urea were regarded separately. *Fig. 1.1* shows how *Friedrich Wöhler* was able to identify oxalic acid (1825) and urea (1828) as product components, after he had introduced cyanogen into an aqueous ammonia solution.

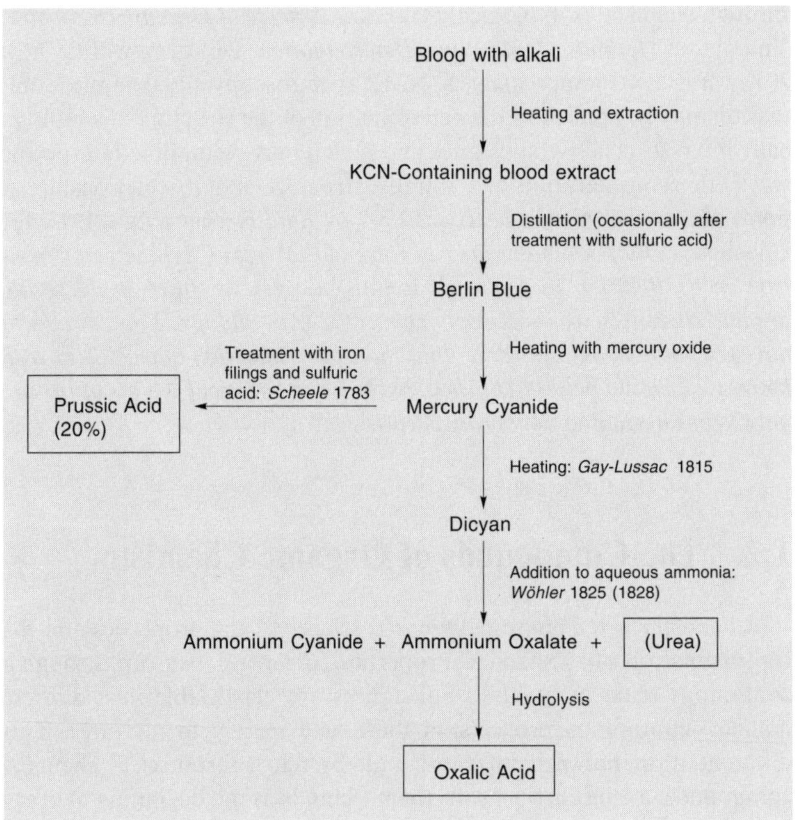

Fig. 1.1. *First preparation of prussic (hydrocyanic) acid, oxalic acid, and urea in a laboratory*

This finding has subsequently been the subject of many different interpretations, which have given rise to the legend that *vis vitalis* had by then already been rendered obsolete, solely as a result of *Wöhler*'s observation. A careful historical analysis, however, does not endorse such sweeping conclusions. It was necessary in any case, though, to find a new borderline between inorganic and organic compounds, since compounds synthesized by organs or organisms could now also be prepared by chemists in the laboratory.

Another definition describes all carbon compounds as organic compounds. This classification, however, also fails to produce a satisfactory result, as conventionally the elemental modifications of carbon itself, its oxides, carbonates, and carbides are regarded as inorganic.

An often-used definition classifies all hydrocarbons and their derivatives as organic compounds. A delimitation of this type is persuasive, viewed from the perspective of a systematic development of the nomenclature of chemical compounds. However, the fact that the definition is entirely artificial should not be overlooked; it does not correspond to biological, nor to potential prebiotic chemistry, and so contributes little towards an understanding of naturalistic evolution.

In such a situation, one may ask what science behold as public knowledge understands by organic chemistry. A handbook of this area of study should presumably be an authoritative guide here. The *Beilstein Institute for Literature of Organic Chemistry*, situated in Frankfurt am Main, produces the *Beilstein Handbook of Organic Chemistry* (hereafter referred to simply as *Beilstein*). This work is quite literally an encyclopedia of known organic compounds. However, *Beilstein* does not concern itself with macromolecular carbon compounds, and does so only in a limited way with organometallic compounds. Comprehensive information about organometallic compounds is to be found in the *Gmelin Handbook of Inorganic and Organometallic Chemistry*, published by the *Gmelin Institute for Inorganic Chemistry and Related Areas*, also situated in Frankfurt am Main.

Even though *Beilstein* is not optimally suited to provide a definition of what is meant by organic chemistry, it does draw attention to another fact: that sources of information exist relating to previously acquired knowledge in the field of organic chemistry, wherever the exact borderline of that discipline may lie in particular instances.

1.2.2. Documentation and Retrieval of Chemical Knowledge

Communication is an integral part of the scientific process. It may be done orally, to a limited public, as a lecture at conferences or at special, more or less regularly occurring scientific meetings. If a scientific publication is to be more widely disseminated, and at the same time be recorded in perpetuity, it may be presented in written form in scientific books or periodicals. To document the progress of science on an ongoing basis, handbooks are produced, recording and citing original literature in particular fields.

Pessimists have imagined Doomsday scenarios, in which the rising deluge of published information bursts the dams of information retrieval capabilities. They prophesy that in the foreseeable future, ever more authors, ever more frequently, will publish the same research findings two or three times over. Since even the most conscientious researcher is no longer in the position of being able to take account of every relevant publication of importance in reference organs, let alone in the original literature, it would simply not be possible to afford the wasted human effort and capital outlay needed to oversee such runaway overproduction. Fortunately, this vision conveys more the tone of nightmare than the inevitable reality of science of today or tomorrow.

The revolutionary changes undergone in publishing, from the production of a manuscript by the author on a personal computer to the storage and retrieval of information in data bases, ensure that the technical prerequisites for an intelligent solution to the flood of published knowledge are at hand. The important data bases which concern the field of organic chemistry are primarily *Beilstein*-ONLINE, *CAS*-ONLINE and SCI-SEARCH. These data bases and their associated printed media are comprehensively described in *Chapt. 9*.

1.2.3. The 'Literature' Problem

If the current degree of completeness of, for example, *Beilstein* is considered, it will be observed that at the present time, only publications dating from 1959 and earlier have been completely registered, and literature from 1960 to 1979 has, up until now, only been processed in the field of heterocyclic compounds. How is such a backlog to be explained?

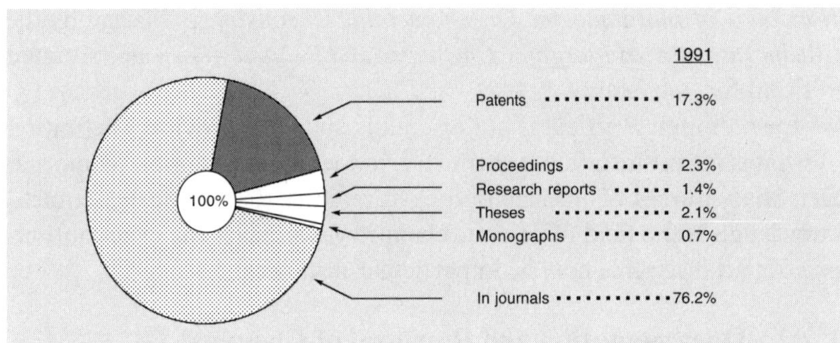

Fig. 1.2. *Classification by type of publication of material excerpted by* CAS *for 1991*

The answer to this question is relatively simple, looking at the gigantic number of publications. The *Beilstein* Institute has set itself the task of completely reviewing, checking, and critically assessing all the literature for a given time-period; a goal which cannot be attained swiftly. At present, more than 500,000 publications in this area of chemistry appear annually: one publication a minute, round the clock, every day of the year, Sundays and public holidays included. Because of this flood of information, whose end is not in sight, the total sum of knowledge of

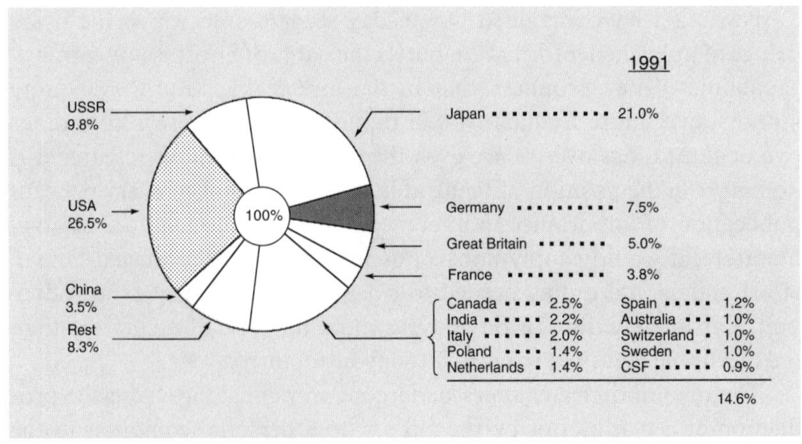

Fig. 1.3. *Classification by national origin of material excerpted by* CAS *for 1991*

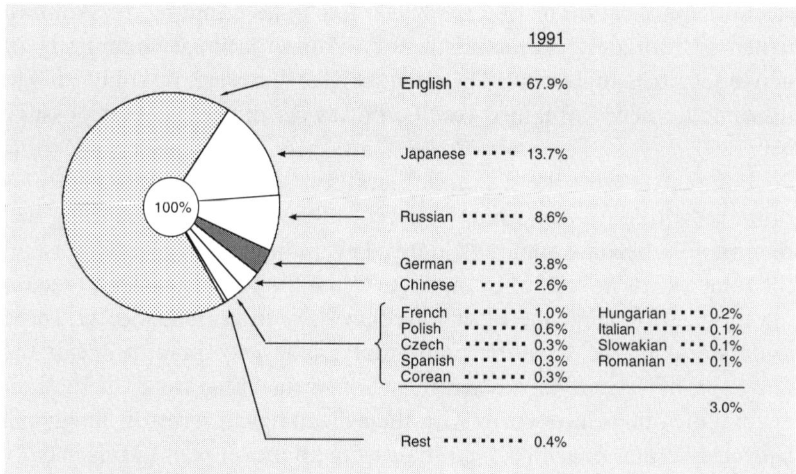

Fig. 1.4. *Classification by language of publication of material excerpted by* CAS *for 1991*

chemistry multiplies in ever shorter time-periods. *Beilstein* lists about one million organic compounds from its own area of specialization reported in the period before 1960. From 1960 to 1979, 2.5 million organic compounds were added to their number. In 1990, the total figure was somewhere around 4.6 million compounds. The number (at present more than 10 million in total) of *Chemical Abstracts Service* (*CAS*) listed registry numbers (*CA-RN*) is sadly not identical with any given number of chemical compounds: if a particular *CAS* examiner is not certain whether a chemical compound described in a publication already has a *CA-RN* ascribed to it or not, then, as a rule, a new *CA-RN* will be assigned.

Classification by type of publication, national origin, and language of publication of material excerpted by *CAS* in a given period provides an interesting insight into the research of different countries and linguistic groups. *Figs. 1.2–1.4* give the results.

1991 was chosen as the reference year, because the criteria used – national origin and language of publication – would give a distorted picture for a later year, following the political changes in Eastern Europe.

1.3. Conclusion

Communication is an essential component of scientific activity. A person who was aware of experimentally determined answers and profound interpretations relating to original scientific questions and retained them for him- or herself would perhaps be a great intellectual. That person would not, however, act as a scientist. The scientist has a duty to make known new discoveries and illuminating results (although not with excessive haste). Firstly, because experimental results, hypotheses, and theories are tempered in the fire of informed criticism, establishing an evolutionary selection procedure. Furthermore, because every scientific publication or lecture plays a role in determining the standard which is ultimately characteristic for the international community of active scientists of its time. The scientist will, therefore, not only wish to disseminate newly obtained results, but to do this in the best possible way.

Familiarity with the scientific literature, scientific lecture presentation and with data bases simply cannot come too soon. One thing will then rapidly become apparent: normal everyday language makes use of the same words as scientific language. Often they have a different meaning in scientific language to the one they have in normal speech. To be understood in the scientific community, it is necessary to speak the language of science in a scientific way: translating words with their normal meanings into words with their meanings in scientific language. *Gaston Bachelard* has observed that, were attention paid to this usually unconscious translation process, one would notice that, in the language of science, many expressions sit inside quotation marks.

One more word: much has been written and spoken, particularly since the *Rede Lecture* of 1959, in which *Charles P. Snow* coined the '*metaphor of the two cultures*', of the chasm separating natural scientists from adherents of the arts and humanities, of mutual incomprehension between members of the literary and the scientific intelligentsia, of two mutually alien world views; one naturalistic, one culturalistic. We do not in any way underestimate our stated intention of conquering existing prejudices and letting nothing arise in their place. However, we believe that those discoveries and observations which chemists, by their endeavors to analyze and responsibly alter existing reality, have collated and recorded in writing, constitute a significant part of human cultural achievement. We have not forgotten though, that man is also part of nature. To overstress the naturalistic/culturalistic alternative could do more to conserve the two-culture mentality, rather than reformulate it for dialogue.

Further Reading

To Sect. 1.1.

D. J. Cram, *Nobel Lecture: The Design of Molecular Hosts, Guests, and Their Complexes, Angew. Chem. Int. Ed.* **1988,** *27,* 1009.

J.-M. Lehn, *Nobel Lecture: Supramolecular Chemistry – Scope and Perspectives – Molecules, Supermolecules, and Molecular Devices, Angew. Chem. Int. Ed.* **1988,** *27,* 89.

C. J. Pedersen, *Nobel Lecture: The Discovery of Crown Ethers, Angew. Chem. Int. Ed.* **1988,** *27,* 1021.

K. B. Sharpless, *Discovery of the Titanium-Catalyzed Asymmetric Epoxidation – A Personal Account, Proceedings of the Robert A. Welch Foundation Conferences on Chemical Research XXVI:* Stereospecificity in Chemistry and Biochemistry, Houston, Texas, 1984.

R. Noyori, *Chemical Multiplication of Chirality: Science and Application, Chem. Soc. Revs.* **1989,** *18,* 187.

J. D. Watson, *Nobel Lecture: The Involvement of RNA in the Synthesis of Proteins, Science* **1963,** *140,* 17.

F. H. C. Crick, *Nobel Lecture: On the Genetic Code, Science* **1963,** *139,* 461.

J. D. Watson, F. H. C. Crick, *Molecular Structure of Nucleic Acids. A Structure for Deoxyribose Nucleic Acid, Nature* **1953,** *171,* 737;
– *Genetic Implications of the Structure of Deoxyribonucleic Acid, Nature* **1953,** *171,* 964.

A. Baeyer, *Über die chemische Synthese,* Verlag der K. B. Akademie, München, 1878.

R. B. Woodward, *Synthesis* in *Perspectives in Organic Chemistry* (Ed. A. Todd), Interscience Publ., New York, 1956.

A. Eschenmoser, C. E. Wintner, *Natural Product Synthesis and Vitamin B_{12}, Science* **1977,** *196,* 1410.

E. J. Corey, *Nobel Lecture: The Logic of Chemical Synthesis: Multistep Synthesis of Complex Carbogenic Molecules, Angew. Chem. Int. Ed.* **1991,** *30,* 455.

E. Fischer, *Nobel Lecture: Synthesis in the Purine and Sugar Group* in *Nobel Lectures Chemistry 1901–1921,* p. 21, Elsevier Publ. Comp., Amsterdam, 1966.
– *Faraday Lecture: Synthetical Chemistry in its Relation to Biology, J. Chem. Soc.* **1907,** 1749.

To Sect. 1.2.

C. F. von Weizsäcker, *Aufbau der Physik,* Deutscher Taschenbuch Verlag, München, 1985.

G. Quinkert, *Spuren der Chemie im Weltbild unserer Zeit* in *Chemie und Geisteswissenschaften* (Ed. J. Mittelstraß, G. Stock), Akademie Verlag, Berlin, 1992.

1.2.1.

F. Wöhler, *Über Cyan-Verbindungen, Ann. Phys. Chem.* **1825,** *3,* 177;
– *Über künstliche Bildung des Harnstoffs, Ann. Phys. Chem.* **1828,** *12,* 253.

H. Bauer, *Die ersten organisch-chemischen Synthesen, Naturwissenschaften* **1980,** *67,* 1.

J. Weyer, *150 Jahre Harnstoffsynthese, Nachr. Chem. Tech. Lab.* **1978,** *26,* 564.

1.2.2.

M. Mücke, *Die chemische Literatur,* Verlag Chemie, Weinheim, 1982.

Y. Wolman, *Chemical Information – A Practical Guide to Utilization* (2nd Edn.), John Wiley & Sons, New York, 1988.

B. Fabian, *Buch, Bibliothek und geisteswissenschaftliche Forschung,* Vandenhoeck & Ruprecht, Göttingen, 1983.

To Sect. 1.3.

G. Bachelard, *Die Bildung des wissenschaftlichen Geistes,* Suhrkamp Verlag, Frankfurt am Main, 1984.

J. Ben-David, *The Scientist's Role in Society,* Prentice-Hall Inc., Englewood Cliffs, New Jersey, 1971.

C. P. Snow, *The Two Cultures and A Second Look,* Cambridge University Press, Cambridge, 1980.

H. Kreuzer (Ed.), *Die zwei Kulturen – Literarische und naturwissenschaftliche Intelligenz – C. P. Snows These in der Diskussion,* dtv/Klett-Cotta, München, 1987.

W. H. Brock, *C. P. Snow – Novelist or Scientist? Chem. Br.* **1988,** 345.

P. Janich, *Grenzen der Naturwissenschaft – Erkennen als Handeln,* C. H. Beck, München, 1992.

2 The Structure Model of the Classical Organic Chemistry

2.1. The Concept of the Molecule

The discrete nature of chemical compounds was postulated at the beginning of the nineteenth century in the *laws of constant and multiple proportions*. These laws are based upon the premise that the large number of chemical compounds are composed of a small number of chemical elements.

According to the law of constant proportions, the weights of the different elements combined in a given chemical compound are fixed in a constant ratio. In the chemical compound water, the elements hydrogen and oxygen are combined together: to every weight-unit of hydrogen there are eight weight-units of oxygen.

The law of constant proportions, also called the fundamental principle of stoichiometry, was augmented by the law of multiple proportions, which states that the ratios of the weights of two or more chemical elements, which can combine to give two or more distinct chemical compounds, is always a relationship of simple integers. Like water, hydrogen peroxide contains the elements hydrogen and oxygen. In this case, sixteen weight-units of oxygen are present for every one of hydrogen. The ratios of 1:8 in the case of water, and 1:16 in that of hydrogen peroxide, reveals that the proportions of oxygen combined with a given proportion of hydrogen in the two chemical compounds, water and hydrogen peroxide, are in a 1:2 ratio.

As a simple interpretation of the laws of constant and multiple proportions, the atom theory immediately presented itself.

According to the atom theory, a chemical element, such as hydrogen or oxygen, is characterized by its consisting of a particular sort of atoms; a chemical compound, such as water or hydrogen peroxide, by its consisting of a particular sort of molecules. In these molecules, the postulated atoms of the elements concerned, determined by chemical analysis, are chemically combined with one another. The discrete nature of the molecule explains the discrete nature of the chemical compound and makes clear the difference between mixtures and compounds; in mixtures, the components (elements or compounds) are present in arbitrary proportions; in chemical compounds, however, they are not. In order to calculate how many H-atoms are combined with how many O-atoms in a molecule of water or hydrogen peroxide, it is necessary to know the masses of the individual atoms and those of the respective molecules. These values have been determined by a multitude of physical methods, as a result of which it is known that, in the water molecule, two H-atoms are combined with one O-atom, and, in the hydrogen-peroxide molecule, two H-atoms are combined with two O-atoms. Furthermore, it is now known how big these molecules are, and that 1 l of water contains more than 10^{25} water molecules.

On the foundation of the atomic composition of material reality, chemists have erected the grandiose edifice of their structure theory, using the answers which they have obtained to questions concerning which atoms in the molecules are directly connected to each other.

2.2. Molecular Models

The molecular structure of chemical compounds is the basis of chemistry and biology. What does the term structure mean in this context? It is the objective of this volume to answer this question.

The *International Union of Pure and Applied Chemistry (IUPAC)* has refrained from defining the concept of structure: '*The term structure may be used in connexion with any aspect of the organization of matter*'. This is regrettable, especially as contemporary linguistic usage employs the word as a generic term at the vertex of a pyramid of concepts, including – in anticipation of future themes – the subsidiary concepts of *constitution*, and even more so of *configuration* and *conformation*. In older literature, however, the term structure is used synonymously with constitution.

Structural description of a chemical compound normally proceeds through utilization of a particular model – a geometric figure, for example – assigned to the molecule under consideration. The purpose of such a model is always to reduce a complex and confusing set of factors to an idealized, easily understandable representation. The static stereomodel, which we will start with here, consists of a rigid arrangement of points, connected by straight lines. The points are depictions of atomic nucleus positions, the straight lines represent chemical bonds. The description of molecules by formulae (and names) in the context of the structure model of the classical organic chemistry will be dealt with in *Chapt. 10* of this volume.

From the perspective of the chemist, it is typical to formulate structure models which:
- are tailored for entire classes of chemical compounds,
- describe the discrete particular case of a given chemical compound (water or hydrogen peroxide, for example) as a special case of the relevant class of compounds (hydroxy compounds). The chemist cultivates a certain 'chemical intuition', with the aid of which cross links are established between members of a class of compounds (or a group of reactions). Presumed analogies may then be formulated as questions and, where possible, experimentally answered.

The structure model of organic chemistry was developed in a time when nothing was yet known about electrons and their significance for chemical bonding. It says nothing about the physical nature of the chemical bond and is restricted to the description of the number of bound atoms or atom groups (ligands) and their arrangement relative to each other and to common central atoms.

For the present, it is didactically sensible to refrain from discussing the physical foundation of the structure model of organic chemistry. It is in fact possible to uncover the stereostructure of organic chemical compounds purely phenomenologically. Experience, that the significance of the three-dimensional structure of organic compounds is encountered in many areas and numerous statements of chemistry, supports the assertion that *chemistry, in essence, is always stereochemistry.*

Only when questions arise which may no longer be satisfactorily answered on the basis of the structure model of the classical organic chemistry does it become necessary to construct and apply another model (*Chapt. 8*).

To determine the structure of a molecule in the sense of the structure model of the classical organic chemistry requires that the
- nature,
- number,
- connectivity, and
- spatial orientation

of the involved atoms be established. The necessary procedure is outlined in *Table 2.1.* Practical implementation, from top to bottom, of the steps described affords a gradually discriminating information about the chemical compound examined.

Table 2.1. *Procedure for the Determination of the Structure of a Molecule*

Conceptual Basis	Experimental Methods	Information	
		Expression	**Content**
Substances with different physical and chemical properties are different in their physical composition and/or their chemical structures	Various separation techniques (crystallization, sublimation, zone melting, distillation, extraction, and various forms of chromatography)	Pure substance	Uniform chemical composition
Atomic structure of matter, validity of laws of stoichiometry	Qualitative and quantitative analysis including determination of molecular mass	Molecular formula	Nature and number of atoms per molecule
Model representation of organic compounds	Chemical and/or physical techniques (nowadays mainly spectroscopic)	Constitutional formula	Nature, number, and connectivity of the atoms of each molecule
Model representation of the three-dimensional structure of organic compounds	Chemical and/or physical techniques (nowadays mainly diffraction or nuclear magnetic resonance based)	Stereoformula	Nature, number, connectivity, and three-dimensional structure of the atoms of each molecule

2.3. Molecular Formula

Having obtained a pure substance through the application of various separation methods, the next step is to ascertain the molecular formula of the compound examined. Specialized analytical laboratories exist to

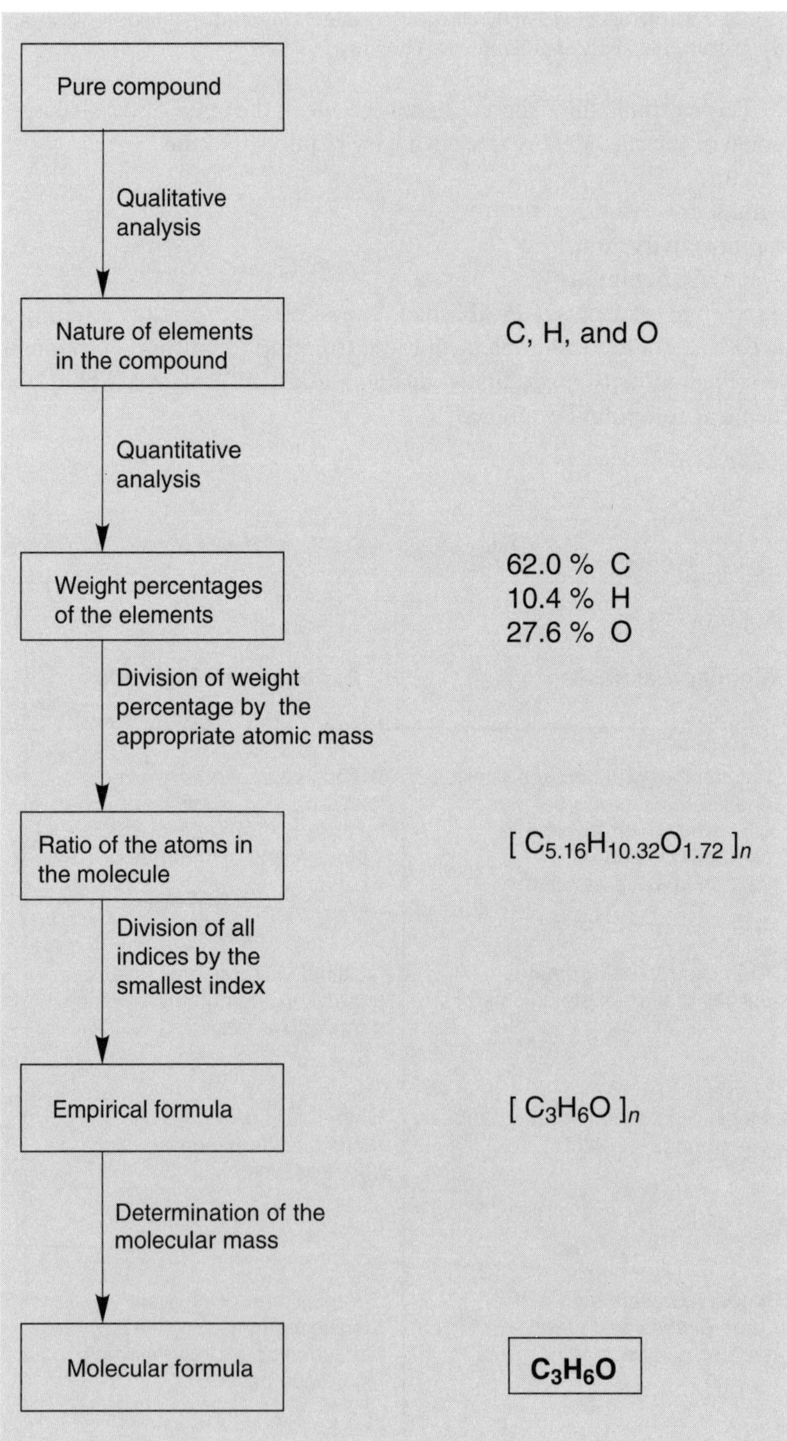

Fig. 2.1. *Experimental determination of molecular formula*

perform this task, which relies on qualitative and quantitative analysis and the determination of the molecular mass (*Fig. 2.1*). The molecular formula thus acquired (*e.g.* C_3H_6O) leaves open the question of the manner in which the given (in this case ten) atoms are connected to each other. In order to make a statement about the connectivity (or sequence) of the atoms in a molecule, a conceptual model of the constitution of organic compounds is required.

2.4. Constitutional Isomerism

The constitutional model of organic chemistry was developed by *Johann J. Loschmidt*, *F. August Kekulé von Stradonitz*, *Archibald C. Couper*, and *Alexander M. Butlerov*. While *Kekulé* (and probably also *Loschmidt*) were responsible for the complete design, *Couper* introduced the straight-line bond symbol (to allow for the different valencies of particular atoms), and *Butlerov* formulated the requirement that each organic compound should be represented by only one formula symbol (implying that each formula symbol would represent only one compound). The model uses the following *construction guidelines* for the connectivity of atoms within a molecule: Each atom may form a total number of bonds corresponding to its particular valency. These bonds are symbolically represented as straight lines. So carbon is tetravalent (four-bonded), for example, while nitrogen is trivalent, oxygen bivalent, and hydrogen monovalent. C-Atoms, within their valency, can participate in single, double, or triple bonds, and are assembled in branched or unbranched chains and/or rings. Because of the numerous possible ways in which they may combine, a large number of constitutional formulae may be constructed according to the constitutional model of the classical organic chemistry, even for a molecule containing only relatively few atoms.

Different compounds which possess the same molecular formula are called (after *Jöns Jacob Berzelius*) either *isomeric compounds*, or simply *isomers*. The molecular formula C_3H_6O may correspond to nine different constitutional isomers, if the construction guidelines specified above are applied. Their constitutional formulae and names are given in *Table 2.2*.

The total number of possible constitutional isomers of a molecular formula is strongly dependent on the size of the molecule, and can reach astronomically high values even for medium-sized molecules. Some examples from the homologous series of alkanes (with the general molecular formula C_nH_{2n+2}) are given on the next page.

It is not only the large number of possible constitutional isomers that is notable here, but also the small number of them that, frequently, are actually known. For example, in the case of $C_{25}H_{52}$, of the total of 36,797,588 possible constitutional isomers, only 34 were to be found in *CAS* ONLINE at the beginning of 1990.

For a long time, graph theory has addressed itself to the combinatorial problem of determining how many possible constitutional isomers

Molecular formula	Number of possible constitutions
CH_4	1
C_2H_6	1
C_3H_8	1
C_4H_{10}	2
C_5H_{12}	3
C_6H_{14}	5
C_7H_{16}	9
C_8H_{18}	18
C_9H_{20}	35
$C_{10}H_{22}$	75
$C_{15}H_{32}$	347
$C_{20}H_{42}$	366319
$C_{25}H_{52}$	36797588

Table 2.2. *The Molecular Formula C_3H_6O: Constitutional Formulae and Names of the Complete Set of Constitutional Isomers*

Molecular formula **C_3H_6O**

Constitutional formulae

$H_2C{=}C(OH){-}CH_3$	$H_2C{=}CH{-}CH_2{-}OH$	CH_2 / $H_2C{-}CH{-}OH$ (ring)
$H_3C{-}CH_2{-}CH{=}O$	$H_2C{=}CH{-}O{-}CH_3$	O / $H_2C{-}CH{-}CH_3$ (ring)
$H_3C{-}C({=}O){-}CH_3$	$H_3C{-}CH{=}CH{-}OH$	$O{-}CH_2$ / $H_2C{-}CH_2$ (ring)

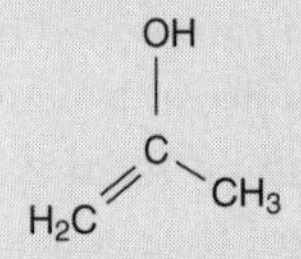

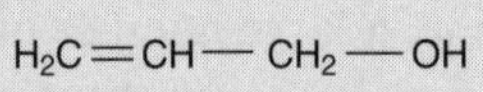

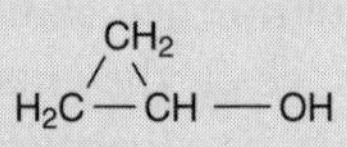

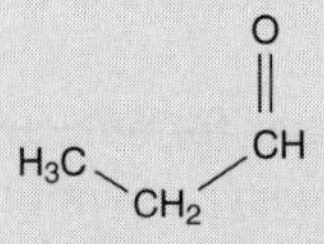

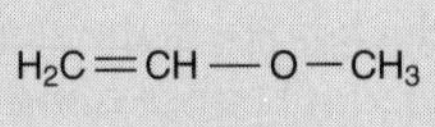

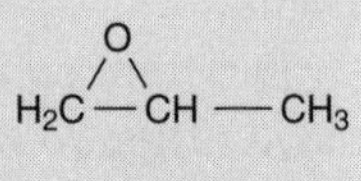

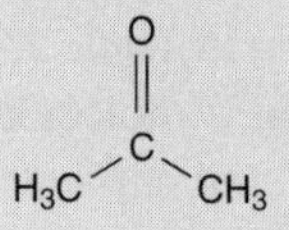

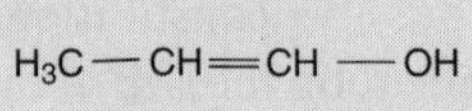

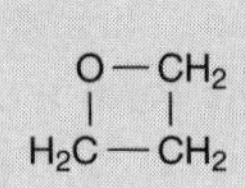

Names

Prop-1-en-2-ol	Prop-2-en-1-ol	Cyclopropanol
Propionaldehyde	Methyl vinyl ether	Methyloxirane
Acetone	Prop-1-en-1-ol	Oxetane

can exist for a particular molecular formula. Graph theory makes use of intuitively comprehensible and aesthetically pleasing diagrams. These diagrams, which are also known as 'trees', are used for the description of constitutional isomers (*Chapt. 10.1*). A further possibility for describing the constitution of organic molecules affords so-called matrices. Electronic data-processing especially makes use of these (*Table 10.9*).

2.5. Stereoisomerism

Now that the nature, number, and connectivity of the atoms of a molecule have been described, with the help of molecular formulae based on constitutional models, the next step must be to consider the spatial arrangement of the atoms within a molecule. Classical organic chemistry's model of the stereostructure of organic compounds is based on the *construction guideline* that the four bonds of a C-atom are directed towards the vertices a, b, c, and d of a tetrahedron (*Fig. 2.2*).

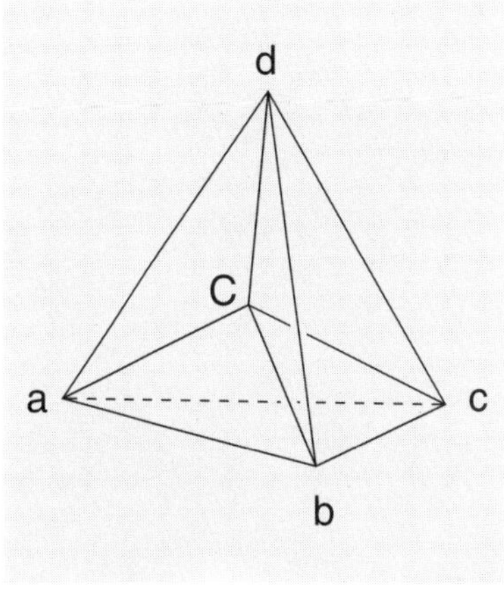

Fig. 2.2. *Tetrahedron model of methane after* van't Hoff

Independently of each other, in 1874, *Jacobus H. van't Hoff* and *Joseph Achille Le Bel* recognized that the tetrahedron model allows the prediction and interpretation of further isomers which are not constitutional isomers.

2.5.1. Stereoisomerism in Acyclic Compounds

2.5.1.1. Methane and Its Derivatives

For the unsubstituted methane molecule, two pieces of new information arise from the transition from constitutional formula to stereoformula:

- all four bonds are equivalent,
- the figure depicted by the stereoformula may be superimposed upon its mirror image by pure translation and rotation; as a result of this, no stereoisomers exist in this case.

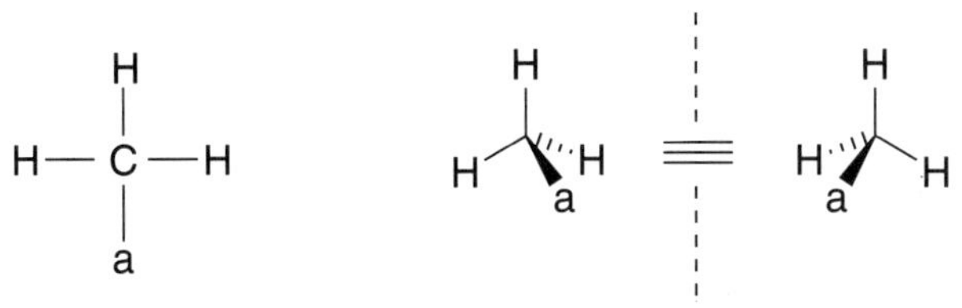

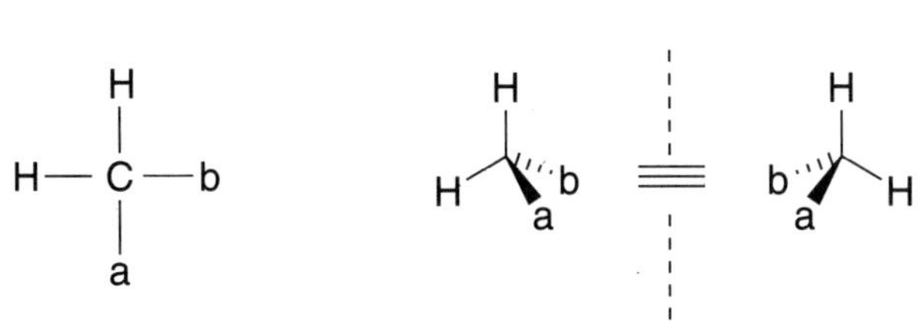

Neither are stereoisomers obtained in the case of mono- or disubstituted methane derivatives: each constitutional formula may be superimposed upon its mirror image.

For tri- or tetrasubstituted methane derivatives with four different ligands, the situation is different: now two stereoformulae may both fit one and the same constitutional formula. For precision, these stereoformulae are denoted as *configurational formulae.* A configurational formula describes the respective arrangement of the atoms of a molecule in space (without, however, consideration of the various alterations in atomic location which arise from rotation about single bonds).

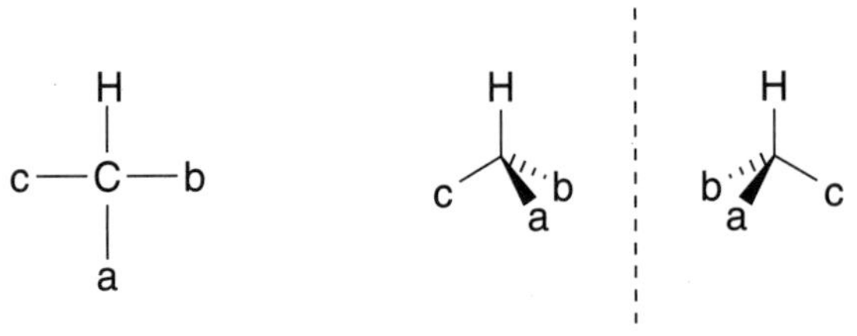

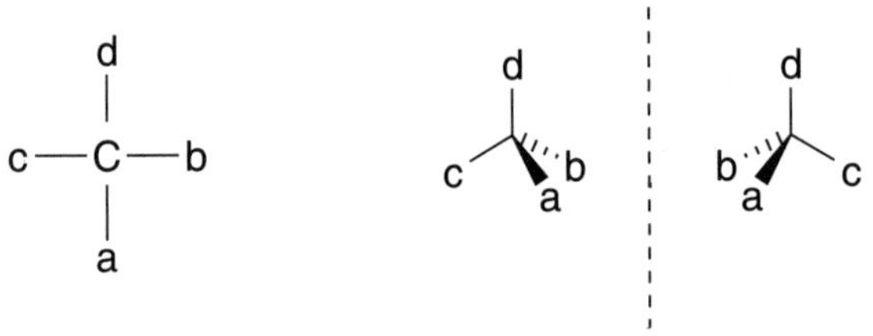

The corresponding stereoisomers, related to each other as mutual mirror images, and which cannot be superimposed upon one another by pure translation and rotation, are called *enantiomers* (*reflection isomers*). Enantiomers differ from each other through their *absolute configuration* at the stereogenic center (or, more generally, at those stereogenic centers) present in the molecule (*vide infra*). The property of an object not being superimposable upon its mirror image through the above-mentioned operations is known as *chirality* (or *handedness*). Chemical compounds or their models, which display this property, are called *chiral* (or *handed*). How the two individual components of an enantiomer pair are specified, using notation of their absolute configuration, is described in *Chapt. 10.2.1.*

The tetrahedron model illustrates that enantiomerism already occurs when four different ligands are grouped around a central atom. It is not essential that the stereogenic center should be a C-atom. Constitutionally equivalent ligands may be configurationally identical or distinct. The consequences of this fact are discussed in *Chapt. 3.* A central atom with enantiomerism-determining ligand orientation of this type is described in the literature as a *chiral center* (or a *center of chirality*) or as a *stereogenic center*. In this sense, all stereoisomerism-generating structural units are referred to as *stereogenic units*. As is still to be seen, the stereogenic center only represents one kind of stereogenic unit.

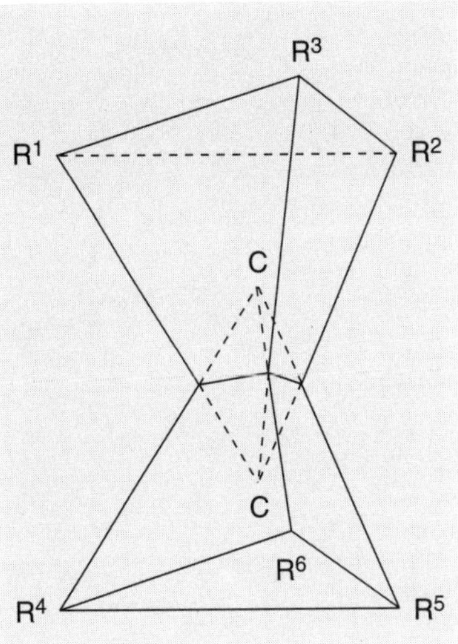

Fig. 2.3. *Tetrahedron model of ethane after* van't Hoff

2.5.1.2. Ethane and Its Derivatives

For the next homologue above methane, new aspects concerning discretness of chemical compounds appear. For ethane and its derivatives, each C-atom finds itself in the center of one tetrahedron and at one of the vertices of the other, respectively (*Fig. 2.3*). Just as valid as the orientation depicted in *Fig. 2.3,* however, is every other figure which may be imagined following rotation of either tetrahedron about the C–C axis. To circumvent the possibility of an infinite degree of isomerism, *van't Hoff* introduced the *concept of free rotation*. In the case of unsubstituted ethane, under normal conditions, no stereoisomers have actually been observed: quite different from those compounds in which free rotation about a C–C bond must obviously be hindered on steric grounds.

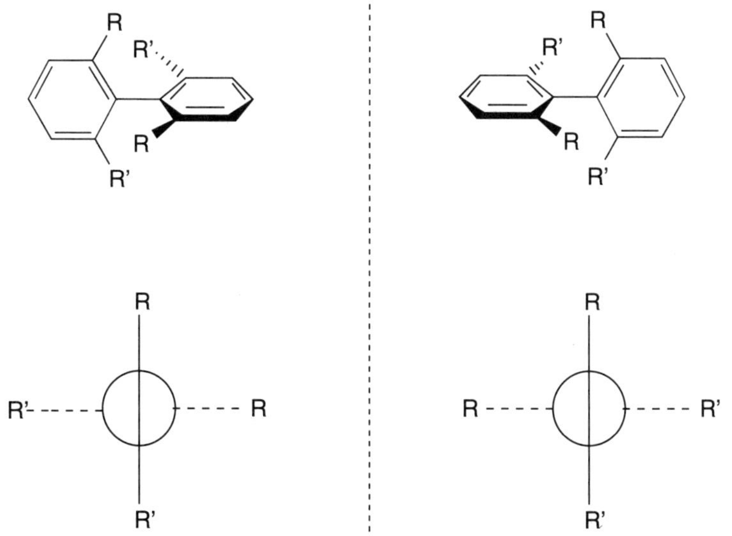

Those stereoisomers which can be interconverted by rotation about a C–C bond are called *conformational isomers* (*conformers*), as opposed to those stereoisomers which may not be interconverted through rotation about a C–C bond, and which are called *configurational isomers*. *Chapt. 10* describes the conventional use of *constitutional*, *configurational*, and *conformational formulae*.

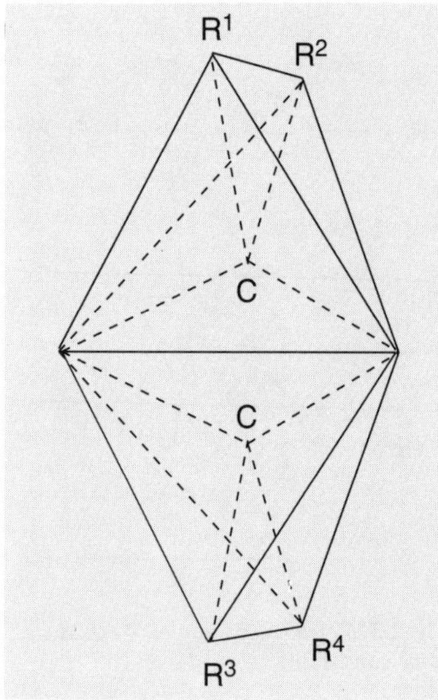

Fig. 2.4. *Tetrahedron model of ethene after* van't Hoff

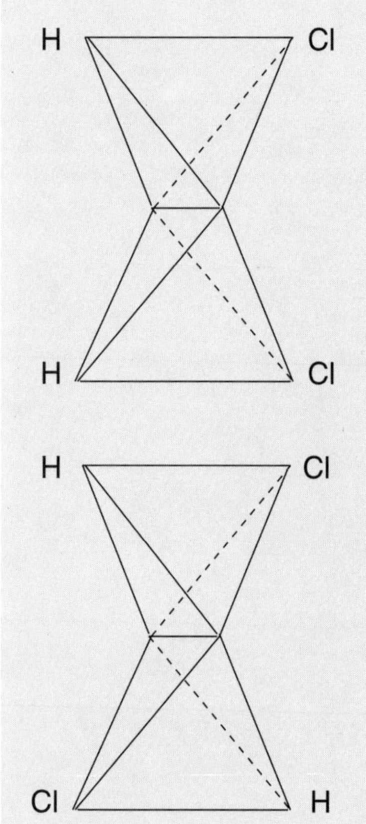

Fig. 2.5. *Tetrahedron model of both 1,2-dichloroethene stereoisomers after* van't Hoff

2.5.1.3. Ethene and Its Derivatives

Ethene (ethylene) is the simplest example of an alkene (hydrocarbon with a C=C bond of general molecular formula C_nH_{2n}). In the *van't Hoff* model of ethene, the two tetrahedra have one edge in common (*Fig. 2.4*).

The C=C bond may be regarded as a two-membered ring, as a result of which the following verifiable consequences may be drawn:

- free rotation is abolished about a C=C bond;
- the central atoms of the four ligands which are connected directly to the C-atoms participating in the C=C bond lie in a plane;
- two stereoisomers (*configurational isomers*) should be considered (*Fig. 2.5*) as soon as the members of both pairs of ligands which are bound to the same central atom of a C=C bond (the *geminal* ligands) differ from each other, irrespective of whether the *vicinal* ligand pairs (those located on the two adjacent atoms constituting the C=C bond) are identical or distinct from one another.

A *stereogenic double bond* is spoken of in these cases. Having become acquainted with the *stereogenic center* (*Sect. 2.5.1.1*), we now encounter a second *stereogenic structural unit* in the stereogenic C=C bond.

In the case of unsubstituted ethene, and its monosubstituted and 1,1-disubstituted derivatives, configurational isomers do not arise. 1,2-Disubstitution or corresponding polysubstitution, however, leads in all cases to two configurational isomers (*cf. Chapt. 10.2.2.*):

```
a     a    |    a     H
 \   /     |     \   /
  C=C      |      C=C
 /   \     |     /   \
H     H    |    H     a
           |
a     c    |    a     d
 \   /     |     \   /
  C=C      |      C=C
 /   \     |     /   \
b     d    |    b     c
```

As the resulting configurational isomers do not exist in a relationship of *enantiomers* (reflection isomers), this case is one of *diastereoisomers*.

2.5.1.4. Ethyne and Its Derivatives

The consistent application of the tetrahedron model to ethyne (acetylene) and its derivatives predicts a linear structure. The two tetrahedra are connected at three vertices (*Fig. 2.6*). For ethyne and its derivatives, as expected, no stereoisomerism is observed.

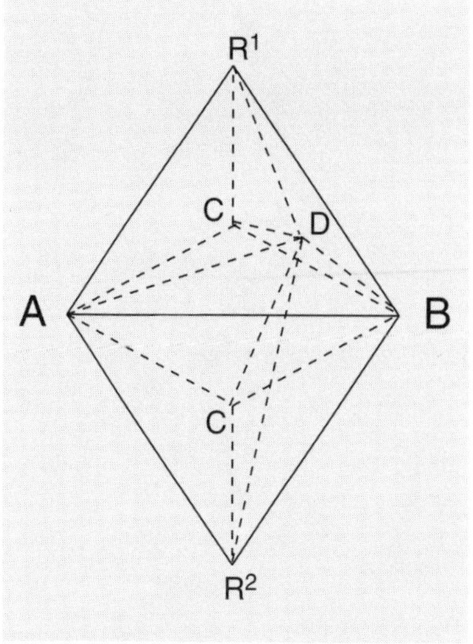

Fig. 2.6. *Tetrahedron model of ethyne after* van't Hoff

2.5.2. Stereoisomerism in Cyclic Compounds

Cyclic compounds may be classified according to

- number of centers in the ring, into those with *small* rings (3 and 4), *normal* rings (5 to 7), *medium* rings (8 to 11), and *large* rings (> 11);
- type of atoms in the rings, into *isocyclic* compounds (only one type of atom; *carbocyclic* in the case of carbon) or *heterocyclic* compounds;
- topological complexity, into *monocyclic*, *bicyclic*, or *polycyclic* ring systems. Bicyclic compounds already allow for further classification according to whether only one center is common to both rings (*spiro* compounds), or two adjacent centers (*fused* ring systems) or two non-adjacent centers (*bridged* ring systems).

In *Sect. 2.5.1.3*, the C=C bond was interpreted as a two-membered ring system. This has the advantage of permitting the division of three-dimensional space into two half-spaces, using the plane of the two-membered ring as a reference plane. The four ligands of a C=C bond are distributed pairwise between the two half-spaces, so that possible configurational isomers, for the case of a stereogenic C=C bond, may be unambiguously specified. A similar relationship applies in the case of the three-membered ring. Here the plane of the three-membered ring plays the role of reference plane for the ligands under consideration. Rings with more than three centers are, as a rule, not planar. In these cases, a best-fit plane is usually superimposed on the ring system, and this functions as a reference plane for relative ligand orientations.

2.6. The Number of Stereoisomers

2.6.1. The Formal Maximum Number of Stereoisomers

In the previous sections, it was established that a methane derivative with one stereogenic center led to two stereoisomers. For the example of a polysubstituted ethane derivative, we saw that for a system with two stereogenic centers, four stereoisomers came into being. Furthermore, the study with ethene derivatives revealed that for each stereogenic C=C bond, two stereoisomers may be observed. These results may be summarized in a single formula for the determination of the formal maximum number of possible stereoisomers (*Table 2.3*).

Table 2.3. *Determination of the Formal Maximum Number of Stereoisomers and Corresponding Notations*

$N^{formal}_{max.} = 2^n$ n = Number of stereogenic centers

$N^{formal}_{max.} = 2^m$ m = Number of stereogenic double bonds

Number of stereogenic units	Notation of possible combinations	Formal maximum number of stereoisomers
1 stereogenic center	*R* *S*	2
2 stereogenic centers	*R,R* *R,S* *S,R* *S,S*	4
3 stereogenic centers	*R,R,R* *R,R,S* *R,S,R* *S,R,R* *R,S,S* *S,R,S* *S,S,R* *S,S,S*	8
1 stereogenic C=C bond	*E* *Z*	2
2 stereogenic C=C bonds	*E,E* *E,Z* *Z,E* *Z,Z*	4
3 stereogenic C=C bonds	*E,E,E* *E,E,Z* *E,Z,E* *Z,E,E* *E,Z,Z* *Z,E,Z* *Z,Z,E* *Z,Z,Z*	8

$N^{formal}_{max.} = 2^{n+m}$

The example marked by the molecular formula C_3H_6O, which has accompanied us on the way from the pure compound, *via* the molecular formula to the number of possible constitutional isomers (*Table 2.2*), should now be completed through consideration of the number of possible stereoisomers which may fit it (*Table 2.4*). Among the constitutional formulae corresponding to C_3H_6O one is to be found with a stereogenic center and another one with a stereogenic C=C bond, so that all four stereoisomers (one pair of enantiomers and one pair of diastereoisomers) and a total of eleven different chemical entities are to be expected together.

Table 2.4. *Constitutional Isomers and Stereoisomers Derived from the Molecular Formula C_3H_6O*

Molecular formula C_3H_6O

Constitutional formulae

$H_2C{=}C(OH){-}CH_3$	$H_2C{=}CH{-}CH_2{-}OH$	$H_2C{-}CH{-}OH$ (cyclopropane ring with CH_2)
$H_3C{-}CH_2{-}CH{=}O$	$H_2C{=}CH{-}O{-}CH_3$	$H_2C{-}CH{-}CH_3$ (epoxide ring with O) [boxed]
$H_3C{-}C(=O){-}CH_3$	$H_3C{-}CH{=}CH{-}OH$ [boxed]	$O{-}CH_2$ / $H_2C{-}CH_2$ (oxetane)

Configurational formulae

Diastereoisomers	H, H_3C C=C H, OH	H, H_3C C=C OH, H
Enantiomers	H, H C–C (O) H, CH_3	H, H C–C (O) CH_3, H

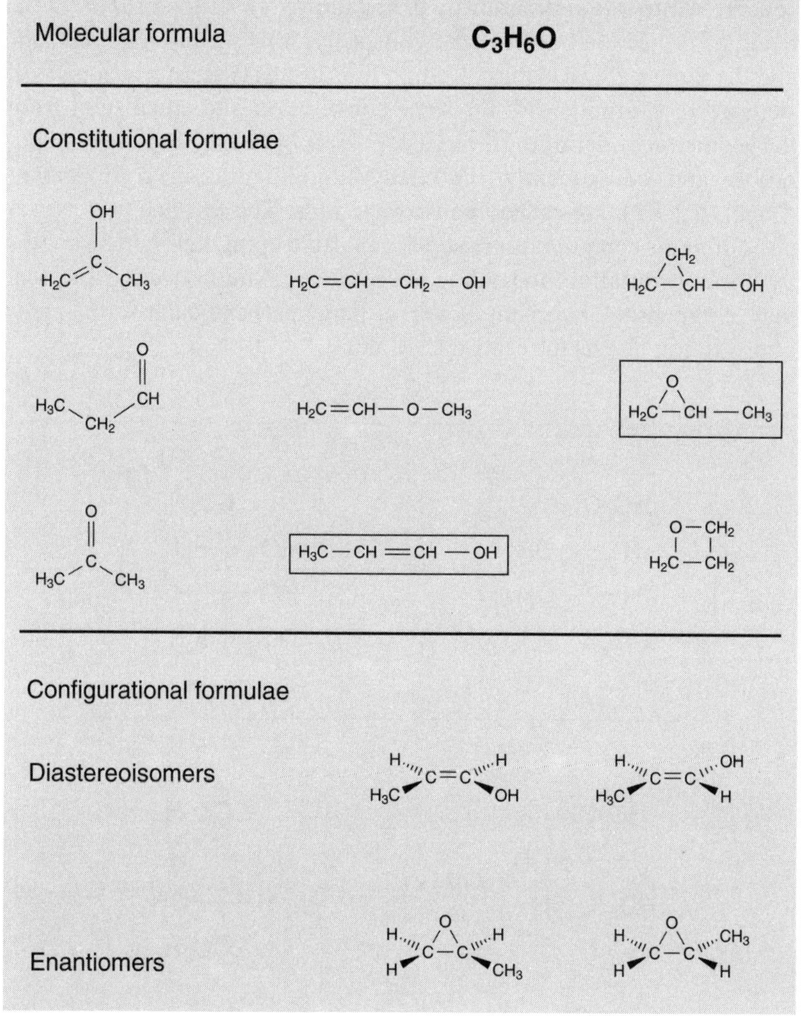

2.6.2. Limitations of the Formal Maximum Number of Stereoisomers

2.6.2.1. Limitations on Grounds of Constitutional Symmetry

The actual number of possible stereoisomers does not agree with the formal maximum number in every instance. If constitutional symmetry is present, for example, fewer stereoisomers may exist than the calculation of the formal maximum number would lead one to expect. This explains the case of tartaric acid (=2,3-dihydroxybutanedioic acid), for which only three stereoisomers are known. For the constitutional formula of tartaric acid, like a constitutionally asymmetrical butanedioic-acid derivative, four stereoformulae (as *Fischer* projections; *cf. Chapt. 10.2.1*) may, for now, be depicted. By closer consideration, though, it may be demonstrated that the two formulae (a) and (b) may be converted into one another through an (allowed) rotation of 180° (in the plane of projection), so that the two projections merely represent one and the same stereostructure. Compounds of this type, in which pairs of stereogenic C-atoms, with the same constitution and equidistant from the geometric center of the molecule, possess opposite absolute configurations and, consequently, the relative *u*-configuration (*u* for *unlike*; *Chapt. 10.2.1.2*), are called *meso*-compounds. The formula representations of *meso*-compounds are superimposable upon their mirror images by linear translation and (allowed) rotation. Alternatively expressed, they 'compensate' internally: a mirror plane perpendicular to the plane of projection bisects the central C–C bond.

(a) HO_2C; H–C–OH; H–C–OH; HO_2C ≡ (b) CO_2H; HO–C–H; HO–C–H; CO_2H

(c) HO_2C; H–C–OH; HO–C–H; HO_2C (d) CO_2H; HO–C–H; H–C–OH; CO_2H

Constitution: 2,3-Dihydroxybutanedioic acid (tartaric acid)

Relative configuration: *u* (*meso*) or *l* (*rac*)

Absolute configuration: (*R,S*) ≡ (*S,R*) or (*R,R*) and (*S,S*)

The *Fischer* projections (c) and (d) cannot be superimposed upon one another by any combination of linear translation or rotation in the plane of projection. However, permitted 180° rotation around the geometric center of the molecule superimposes both projection (c) and projection (d) upon themselves. In contrast with *meso*-tartaric acid, the relative configuration of the two enantiomers is assigned the descriptor *l* (*l* for *like*). The absolute configuration is (*R*,*R*) in one case and (*S*,*S*) in the other (*cf. Chapt. 10.2.1*).

1,2-Dimethylcyclopropane provides an example of how a *meso*-compound can also occur in a set of stereoisomeric cyclic compounds. Instead of the expected four stereoisomers, there are in this case only three, as the geometric figure of the *cis*-isomer (a), with the relative *u*-configuration, is identical to its own mirror image (*cf. Chapt. 10.2.3*).

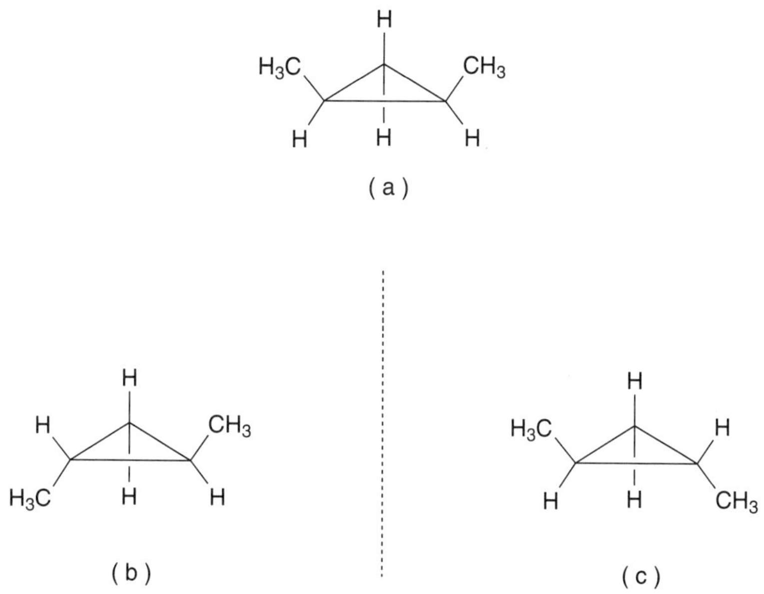

H H_3C CH_3 H H H

(a)

H H CH_3 H_3C H H

(b)

H H_3C H H H CH_3

(c)

Constitution: 1,2-Dimethylcyclopropane

Relative configuration: *u* (*meso* or *cis*) or *l* (*rac* or *trans*)

Absolute configuration: (*S*,*R*) ≡ (*R*,*S*) or (*R*,*R*) and (*S*,*S*)

For the two *trans*-isomers (b) and (c), with the relative *l*-configuration, two enantiomers of opposite absolute configuration, (*R*,*R*) or (*S*,*S*), can be specified.

The phenomenon that the total number of stereoisomers may be decreased on grounds of constitutional symmetry is also observed in the case of stereogenic C=C bonds. For hexa-2,4-diene, with two (conjugated) C=C bonds, four stereoisomers would formally be expected. It may be seen, however, that the (*E*,*Z*)-formulation is identical to its (*Z*,*E*)-counterpart. Consequently, in this instance, there are also only three stereoisomers in total. As these may not be superimposed on each other through rotation about C–C bonds, they are configurational isomers.

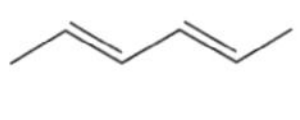

(*E*,*E*) 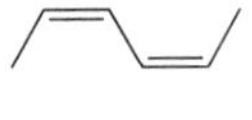 (*Z*,*Z*)

(*Z*,*E*) ≡ (*E*,*Z*)

The class of carotenoids offers some more complex examples. It contains representatives which are formally derived from the acyclic hydrocarbon φ,φ-carotene ($C_{40}H_{56}$), and which are constitutionally modified by hydrogenation, dehydrogenation, cyclization, oxidation, or any combination of these processes. Any constitution almost always corresponds to a great number of associated configurations. This number usually is not only dependent upon whether the compound under discussion is symmetrical or not, and how many stereogenic C=C bonds are present, but is also influenced by whether the number of stereogenic C=C bonds is odd or even.

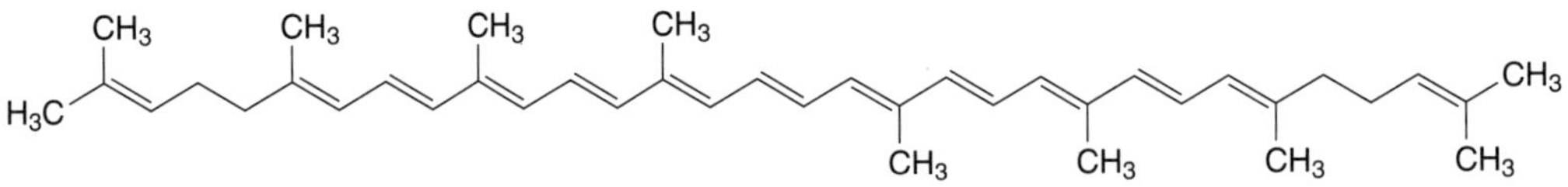

φ,φ-Carotene

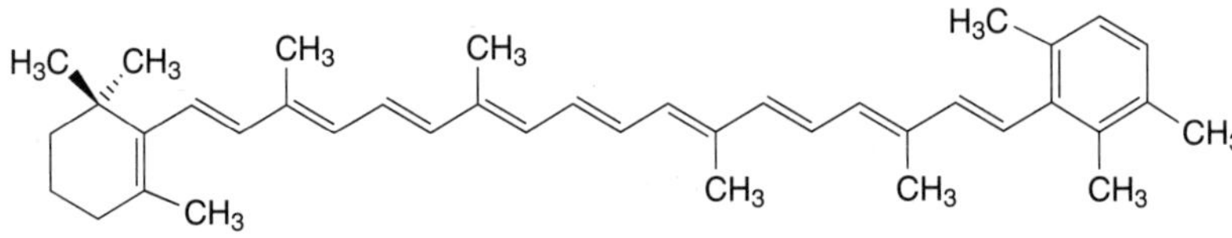

β,φ-Carotene

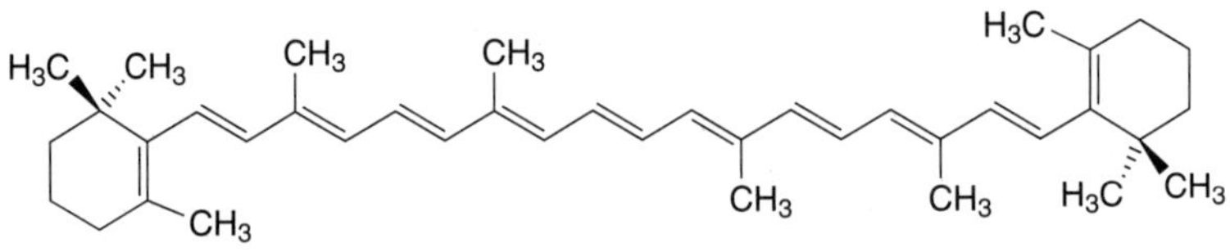

β,β-Carotene

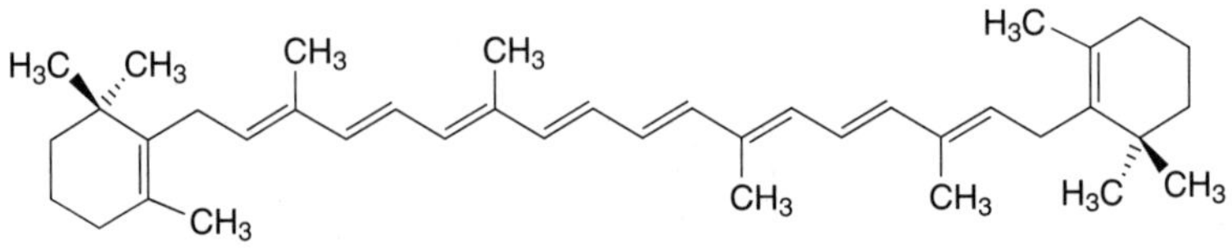

7,7′-Dihydro-β,β-carotene

For carotenoids which are constitutionally asymmetrical, and which exhibit n stereogenic C=C bonds the number of configurational isomers N is given by the expression:

$$N^{real}_{max.} = N^{formal}_{max.} = 2^{n}$$

So β,φ-carotene is one of 512 ($N = 2^9$) possible configurational isomers. For carotenoids which are constitutionally symmetrical, and which are provided with an odd number of stereogenic C=C bonds, N amounts to:

$$N^{real}_{max.} = 2^{(n-1)/2} \times (2^{(n-1)/2} + 1)$$

So naturally-occuring β,β-carotene is one among 272 ($N = 16 \times 17$) possible configurational isomers. For constitutionally symmetrical carotenoids with an even number of stereogenic C=C bonds, N adds up to:

$$N^{real}_{max.} = 2^{(n/2)-1} \times (2^{n/2} + 1)$$

So 7,7′-dihydro-β,β-carotene is one of 136 ($N = 8 \times 17$) possible configurational isomers.

2.6.2.2. Limitations on Grounds of Molecular Strain

A further cause of the limitation of the formal maximum number of stereoisomers is molecular strain, which is particularly conspicuous in the cases of, for example, bridged systems and cycloalkenes (cyclic hydrocarbons with one C=C bond of general molecular formula C_nH_{2n-2}).

So, in the case of the bicyclo[2.2.1]heptane derivative camphor (= bornan-2-one), only two stereoisomers (one pair of enantiomers) are observed, although, because of the two stereogenic centers (bridgehead atoms C(1) and C(4)), four stereoisomers would be expected formally. For the bicyclic system of camphor, however, only those configurations can exist in which those ligands (not comprising a part of the bicyclic skeleton) attached to the bridgehead atoms are located, because of molecular strain, at the outside of the molecule.

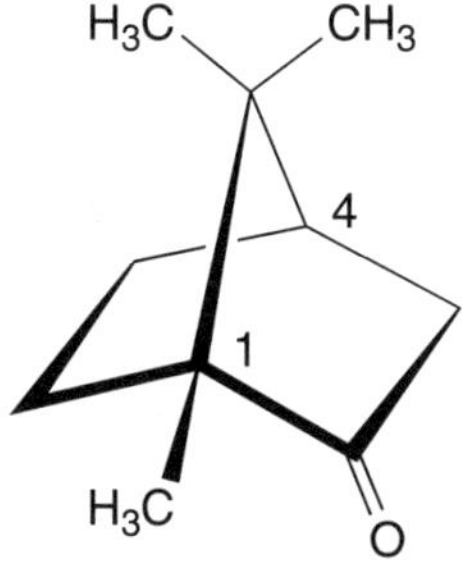

(1*R*,4*R*)-Camphor

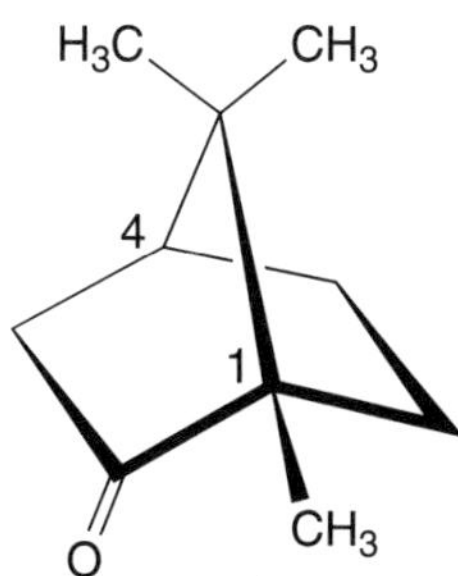

(1*S*,4*S*)-Camphor

This strain-determined limitation lapses, naturally, when one or more of the rings of a bicyclic system attains a certain size, so that, especially, H-atoms situated on a bridgehead atom may be accommodated at the inside of the molecule. In the case of bicyclo[6.5.1]tetradecane, for example, four stereoisomers have indeed proved possible to prepare. The *out,out*- and *in,in*-stereoisomers are diastereoisomers. The *out,in*- and *in,out*-stereoisomers are enantiomers and occur in equimolar quantities in a mixture of the two enantiomers. There is even experimental evidence that the two diastereoisomers may interconvert through *homeomorphic isomerization*. In a homeomorphic isomerization, the polymethylene segment of one of the rings moves through the remaining ring, so that the whole molecule is turned 'inside out'. A homeomorphic isomerization of the *out,in*- and the *in,out*-stereoisomers could not, however, be observed in the case of bicyclo[6.5.1]tetradecane.

out, out ⇌ *in, in*

out, in | *in, out*

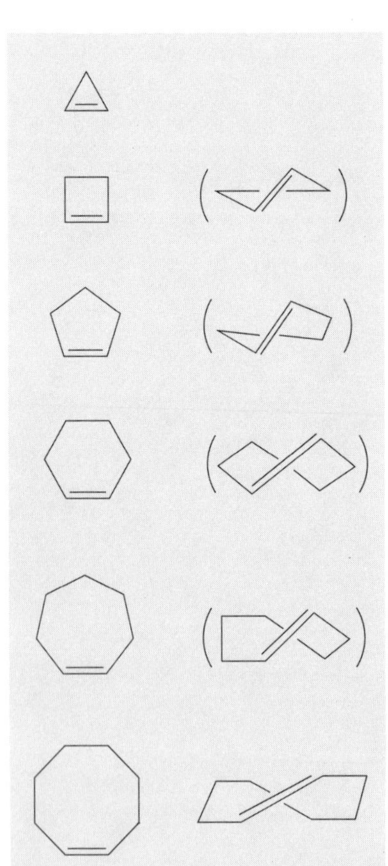

Fig. 2.7. *Cycloalkenes C_3–C_8* ((*E*)-isomers unstable at room temperature in brackets)

(*E*/*Z*)-Isomerism at C=C bonds is also limited in cyclic systems. For rings with fewer than eight members, only the (*Z*)-isomer can be isolated (*Fig. 2.7*), as the respective (*E*)-isomers give rise to intolerable molecular strain.

Cases in which the formal maximum number of stereoisomers are not obtained, due to grounds of molecular symmetry or of molecular strain, are encountered again and again. For the up-and-coming stereochemist, working out the appropriate real maximum number of stereoisomers for a given constitutional formula with stereogenic centers and stereogenic C=C bonds represents the best fitness training.

2.6.3. Extension of the Formal Maximum Number of Stereoisomers

2.6.3.1. Extension in the Case of 2,6,2′,6′-Tetrasubstituted 1,1′-Biphenyl Derivatives

For 2,6,2′,6′-tetrasubstituted 1,1′-biphenyl derivatives, there exist two stereoisomers (enantiomers; *Fig. 2.8*), although this type of compound possesses no stereogenic centers or C=C bonds. The formal maximum number of stereoisomers (2^{n+m}) is, therefore, exceeded, as a case of stereoisomerism arises, which was previously considered only as a phenomenon (*Sect. 2.5.1.2*), rather than as a quantifiable attribute of a compound. This new type of stereoisomerism is observed:

- when bulky ligands in the 2,6- and the 2′,6′-positions of a biphenyl system hinder free rotation about the C–C bond;
- when the ligands in the 2,6- and 2′,6′-positions of the phenyl residues differ from each other.

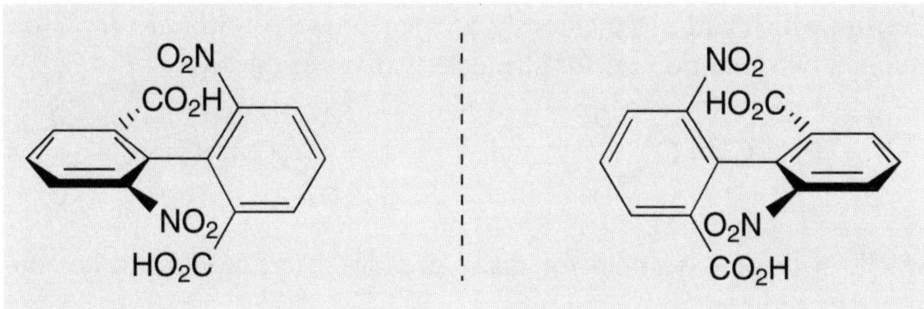

Fig. 2.8. *Enantiomers of 6,6′-dinitro-1,1′-biphenyl-2,2′-dicarboxylic acid*

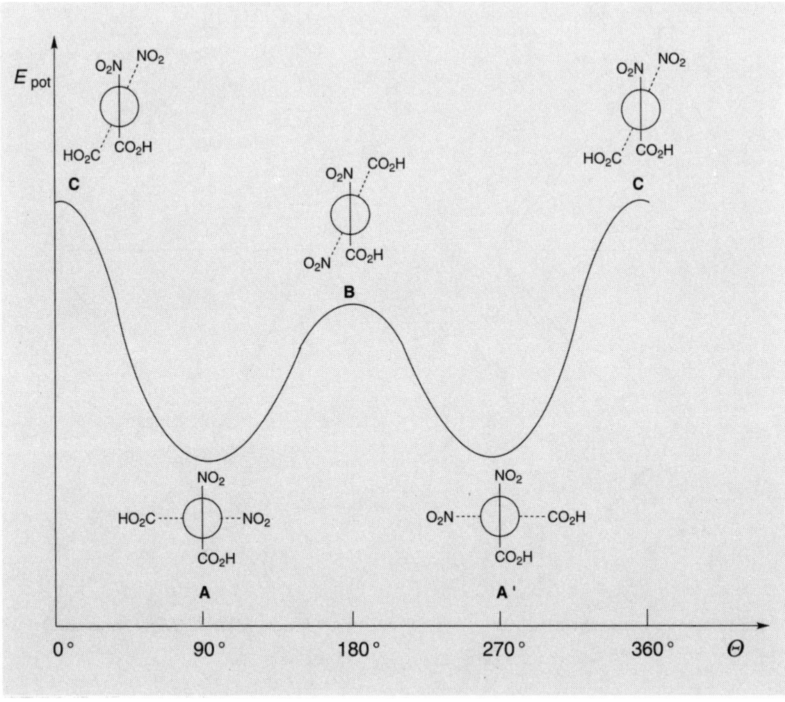

Fig. 2.9. *Energy profile of 6,6′-dinitro-1,1′-biphenyl-2,2′-dicarboxylic acid with discrete conformations corresponding to the extreme points*

With the help of an *energy profile* (*Fig. 2.9*) – a diagram showing the potential energy of a molecular system as a function of a structural parameter (here the torsion angle) – possible conformations and their relative energies may be clearly represented. The diagram allocates particular conformations (in *Newman* projection; *cf. Chapt. 10.3.1*) to the extreme points in the curve. The conformations of minimum energy (**A** and **A′**) are enantiomorphic conformers (*cf. Fig. 2.17*). Structures **B** and **C** are conformations of maximum energy, through which state one enantiomer would have to pass in transition to the other enantiomer, were free rotation possible.

2.6.3.2. Extension in the Case of Allenes, Alkylidene-cycloalkanes, and Spiro Compounds

The tetrahedron model was not only suited to give the correct number of configurational isomers in the case of chemical compounds, with the same constitution and the same number of stereogenic centers and/or stereogenic C=C bonds, but, on top of that, foresaw new types of isomerism, which had not previously been encountered. So *van't Hoff* formulated two enantiomers for allene derivatives of the type

$(a)(b)C{=}C{=}C(a)(b)$ | $(a)(b)C{=}C{=}C(a)(b)$

which, in those days (and for many decades after), could not be confirmed experimentally.

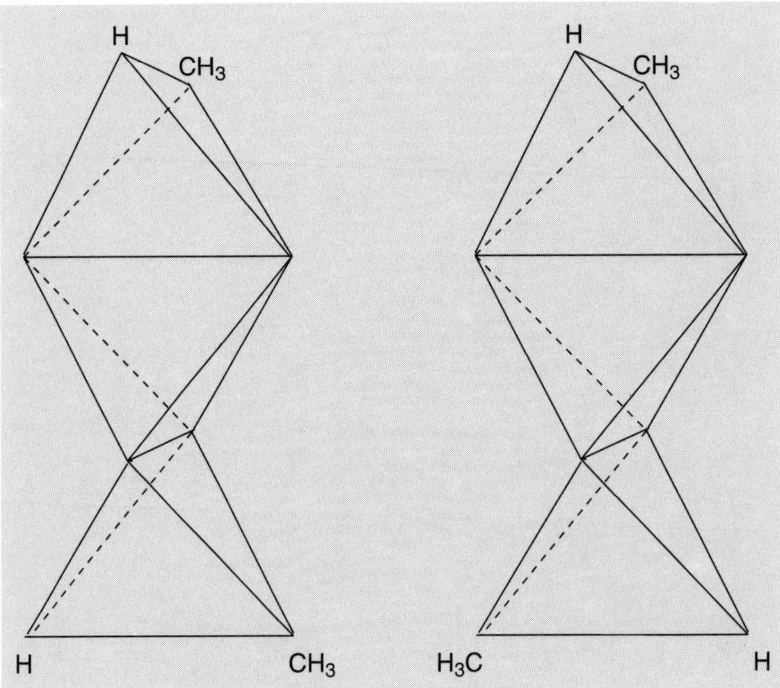

Fig. 2.10. *Tetrahedron model of both penta-2,3-diene enantiomers after* van't Hoff

The *van't Hoff* diagram of *Fig. 2.10* reveals how, in the allene structure, the two two-membered rings (C=C bonds) are situated perpendicularly to each other. The same applies to the two planes, in each of which is situated a terminal stereogenic C=C bond atom and its two ligands.

If, in a thought experiment, a tetrahedron (as a model for a methane derivative) is stretched, so that, instead of the central atom, three linearly ordered centers are created, then a formation (as a model for an allene derivative; *Fig. 2.11*) is obtained, which corresponds to the relative orientation of the two terminal ligand pairs of an allene structure. While, for methane derivatives, enantiomers only occur when all four ligands are different (*Sect. 2.5.1.1*), allene derivatives merely require that each terminal center bears two different ligands, irrespective of whether the two pairs differ from each other or not.

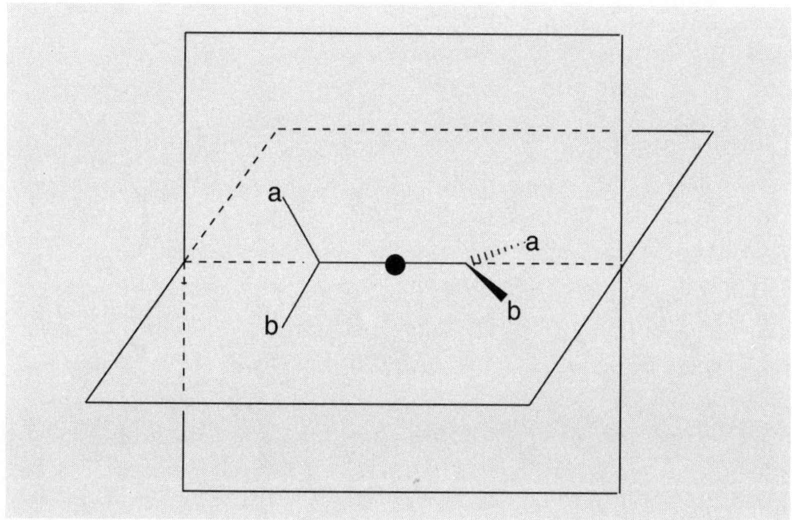

Fig. 2.11. *Illustration of a thought experiment in which an allene derivative is obtained from a methane derivative*

For a cumulene with three (or, generally, any other odd number of) cumulated C=C bonds (in contrast to the situation with an allene, or any other cumulene with even number of C=C bonds), both terminal ligand pairs lie in the same plane. Therefore, it follows that, instead of a pair of enantiomers, two diastereoisomers ((*E*/*Z*)-configurational isomers) may be observed, similarly to the case of isolated C=C bonds (*Sect. 2.5.1.3*).

```
a          a          a          b
 \        /            \        /
  C=C=C=C               C=C=C=C
 /        \            /        \
b          b          b          a

    (Z)                   (E)
```

Allenes, alkylidene-cycloalkanes, and spiro compounds possess an axis of chirality as their stereogenic unit. *Fig. 2.12* summarizes a whole series of structures of various chirality types: structures **1**–**5** are charac-

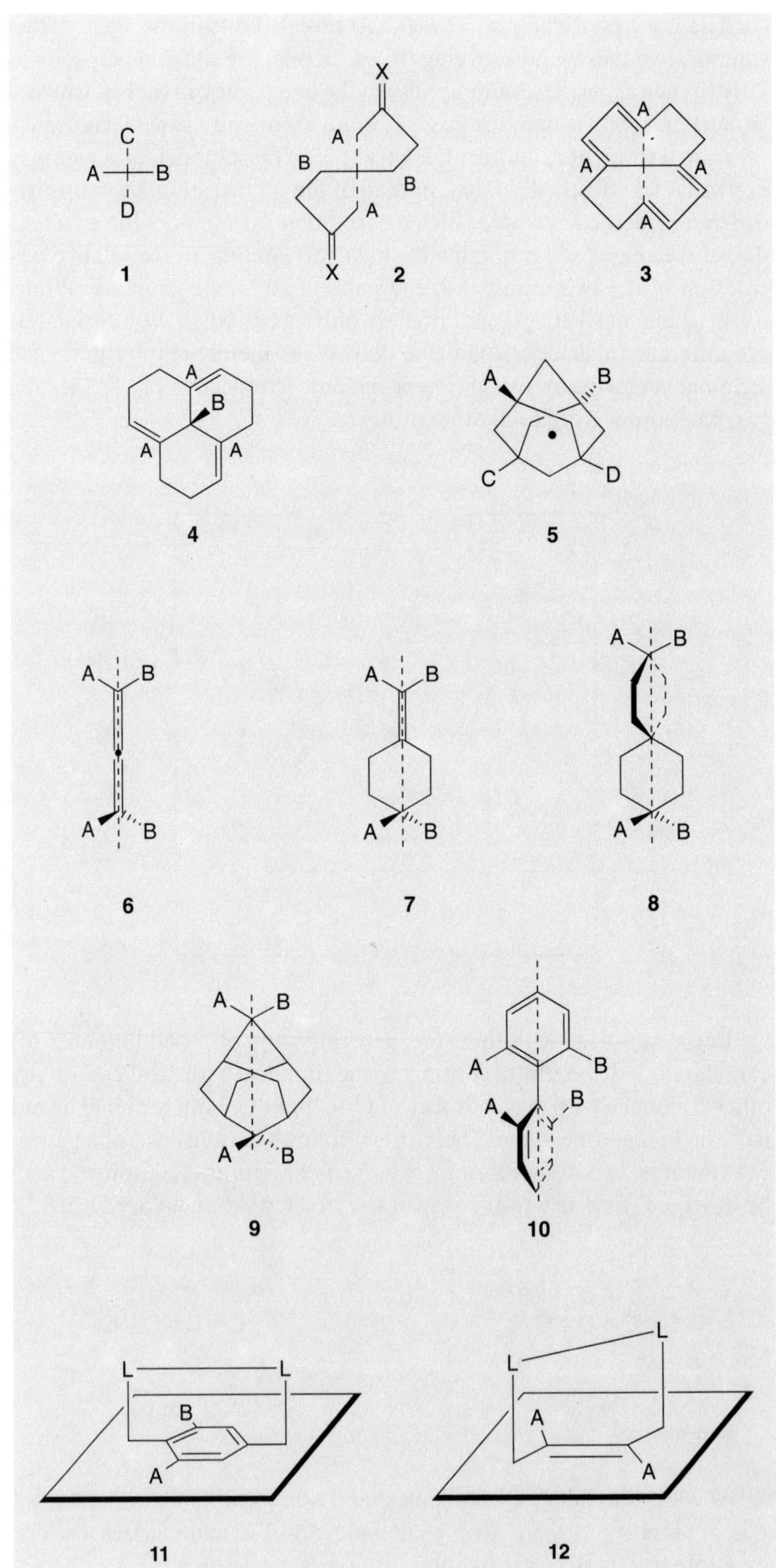

Fig. 2.12. *A selection of chiral structures with a* chirality center (**1–5**), chirality axis (**6–10**), *or* chirality plane (**11** and **12**) *in common*

terized by a *center of chirality*, structures **6**–**10** by an *axis of chirality*, and structures **11** and **12** by a *plane of chirality*.

How enantiomers of allenes, alkylidene-cycloalkanes, and spiro compounds are specified is described in *Chapt. 10.2.4*.

It should not go unmentioned that some of the stereoisomers in *Fig. 2.12* must be interpreted as *conformational isomers* under the definition '*those stereoisomers which interconvert by rotation about a C–C bond are called conformational isomers*'. On the other hand, they may be characterized by their specific sense of chirality (by their absolute configuration), and so could, therefore, be described as *configurational isomers*. Here it becomes obvious that the differentiation of stereoisomers into configurational or conformational isomers is not 'computer-pure', in contrast to the delineation between enantiomers and diastereoisomers.

2.6.4. Is the Entire Set of Causes of Stereoisomerism Known?

Vladimir Prelog answered this question with 'no'. He has extended the frontiers of the seemingly closed field of stereochemical statics. Not only has he always emphasized that fundamental stereostructural concepts and ideas should be based on group theory, but has experimentally closed previously unidentified gaps in stereochemical knowledge.

In the search for previously unrecognized manifestations of isomerism, it would be most wise to make use of structures in which repetitive structural units may be discerned. It can happen in such cases that isomers come into being which simply do not exist in the repeating units themselves. An example of *constitutional isomerism* will show what is meant by this. It concerns the four different uroporphyrinogens of types **I**–**IV** (*Fig. 2.13*), to which a monopyrrole precursor may be ascribed. *Fig. 2.14* illustrates the statistical distribution of the various types and the conspicuous preference for uroporphyrinogen **III** upon repeated coupling and final cyclization.

A new kind of stereoisomerism, which is only observed, when a compound of greater structural complexity is examined, is encountered in the case of an antibiotic of the macrotetrolide class, nonactin. This compound is an *ionophor*, which manages the transport of cations through biological membranes with remarkable specificity. Four C_{10}-hydroxycarboxylic-acid residues (nonactinic acid; configurational formulae **1** and **2**) are joined together *via* ester linkages and closed into a 32-membered ring. The constitutional formula of nonactin (**3**) exhibits 16 (denoted with a star) stereogenic centers.

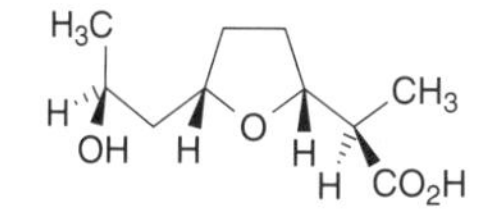

1

● (–)-(2*R*, 3*R*, 6*S*, 8*S*)-Nonactinic acid

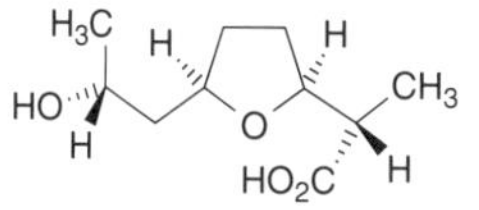

2

○ (+)-(2*S*, 3*S*, 6*R*, 8*R*)-Nonactinic acid

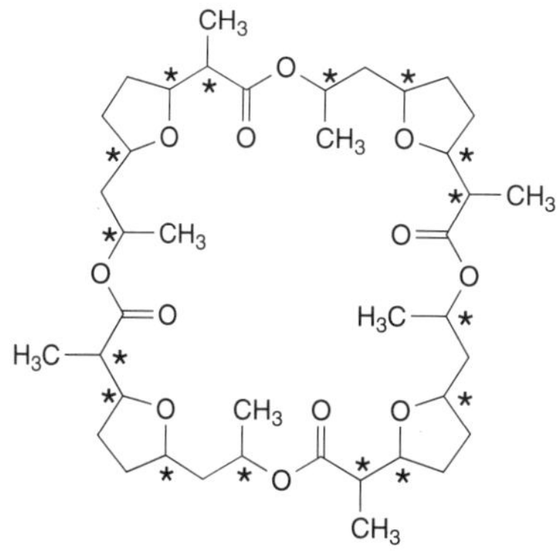

3

The formal maximum number of possible stereoisomers is therefore 65,536. In reality, however, only six stereoisomers with the regular head-to-tail sequence need to be considered (**1**–**6** in *Fig. 2.15*). As naturally occuring nonactin is optically inactive, and a racemic mixture may be excluded on the grounds of the crystal structure, only the two schematically simplified formula representations of two *meso*-compounds (**5** and **6** in *Fig. 2.15*) remain for the final verdict for determination of the configuration. As X-ray crystal-structure analysis shows, the antibiotic is correctly represented by the symbol **6** (*Fig. 2.15*): the enantiomorphic nonactinic-acid residues each occur twice,

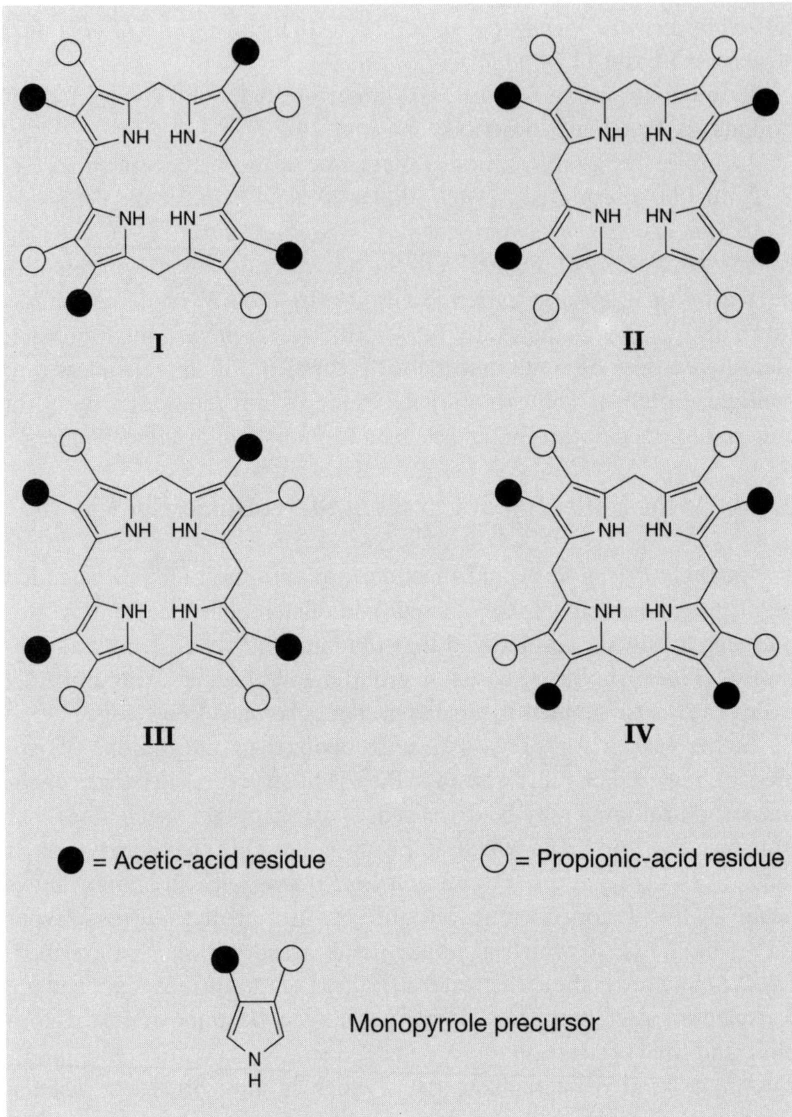

Fig. 2.13. *Uroporphyrinogens* **I–IV**

alternately arranged. The molecule shows S_4 symmetry (**7** in *Fig. 2.15*), and nonactin, despite the 16 stereogenic centers present, is consequently achiral. The meaning of S_4 symmetry will be described in *Chapt. 11*. From the crystal structure, it may further be deduced that the molecule possesses a topology which reminds you of the seam of a tennis ball.

Another type of isomerism (*cyclostereoisomerism*) occurs when two enantiomorphic subunits (*e.g.* (*R*)- and (*S*)-alanine; *Fig. 2.16*) assemble in the regular head-to-tail sequence to form a higher-molecular cyclic system out of six lower-molecular structural subunits. In the set of four distribution patterns containing equal numbers of both enantiomorphic alanine residues, two *cycloenantiomers* (**2** and **3** in *Fig. 2.16*) can be observed. As these exist in a mirror-image relationship, they must also, naturally, reflect each other's *ring sense* (clockwise or anticlockwise orientation of the head-to-tail sequence), and thus reverse it.

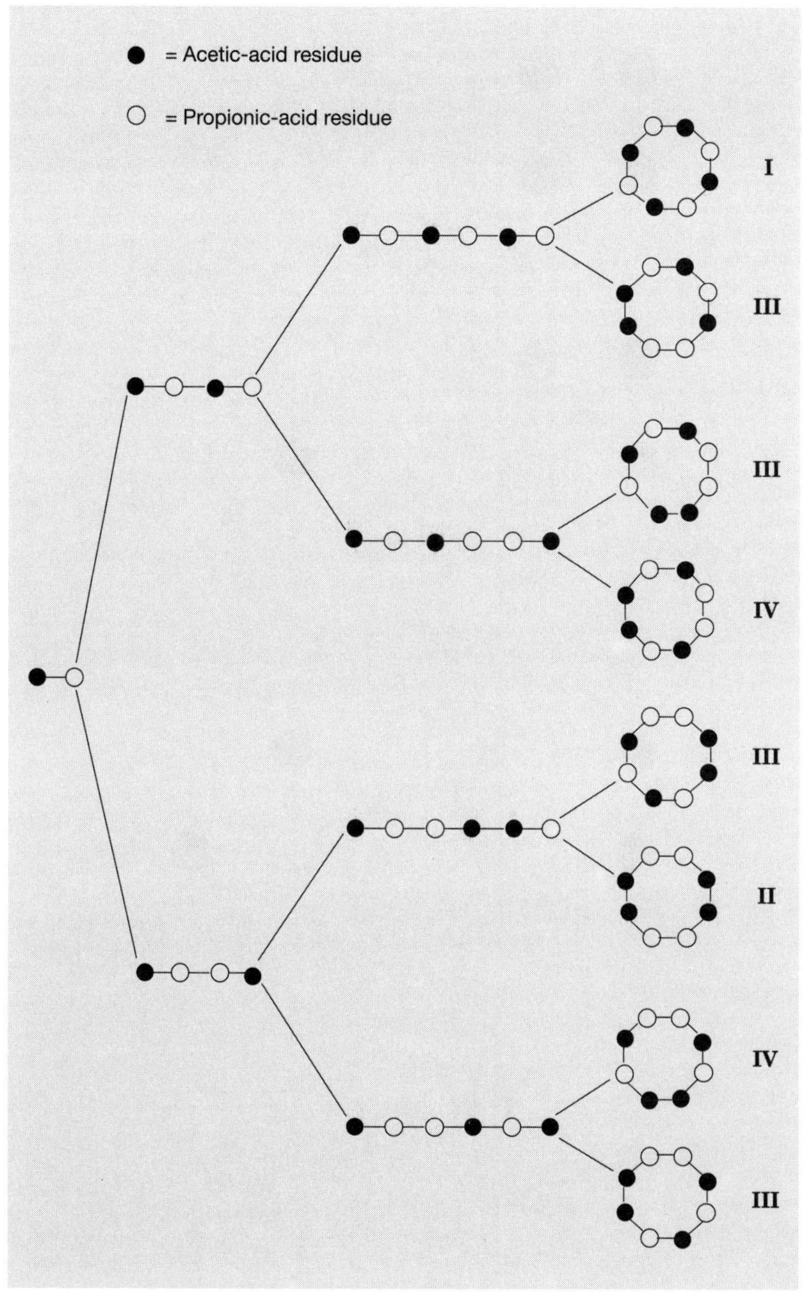

Fig. 2.14. *Statistical distribution for the assembly of uroporphyrinogens* **I–IV** *from the (symbolically simplified) monopyrrole precursor and formaldehyde*

Turning to cyclooctapeptides constructed in the head-to-tail sequence and containing equivalent numbers of the enantiomorphic alanine subunits, two pairs of *cycloenantiomers* (**6** and **7**, and **11** and **12** in *Fig. 2.16*) are encountered.

For cyclodecapeptides constructed along these lines, there already exist besides five pairs of enantiomers (**18** and **19**, **20** and **21**, **26** and **27**, **28** and **29**, **31** and **32**) six pairs of cycloenantiomers (**16** and **17**, **22** and **23**, **24** and **25**, **33** and **34**, **36** and **37**, and **38** and **39** in *Fig. 2.16*). Additionally, there are four pairs of *cyclodiastereoisomers* which do not exist in a

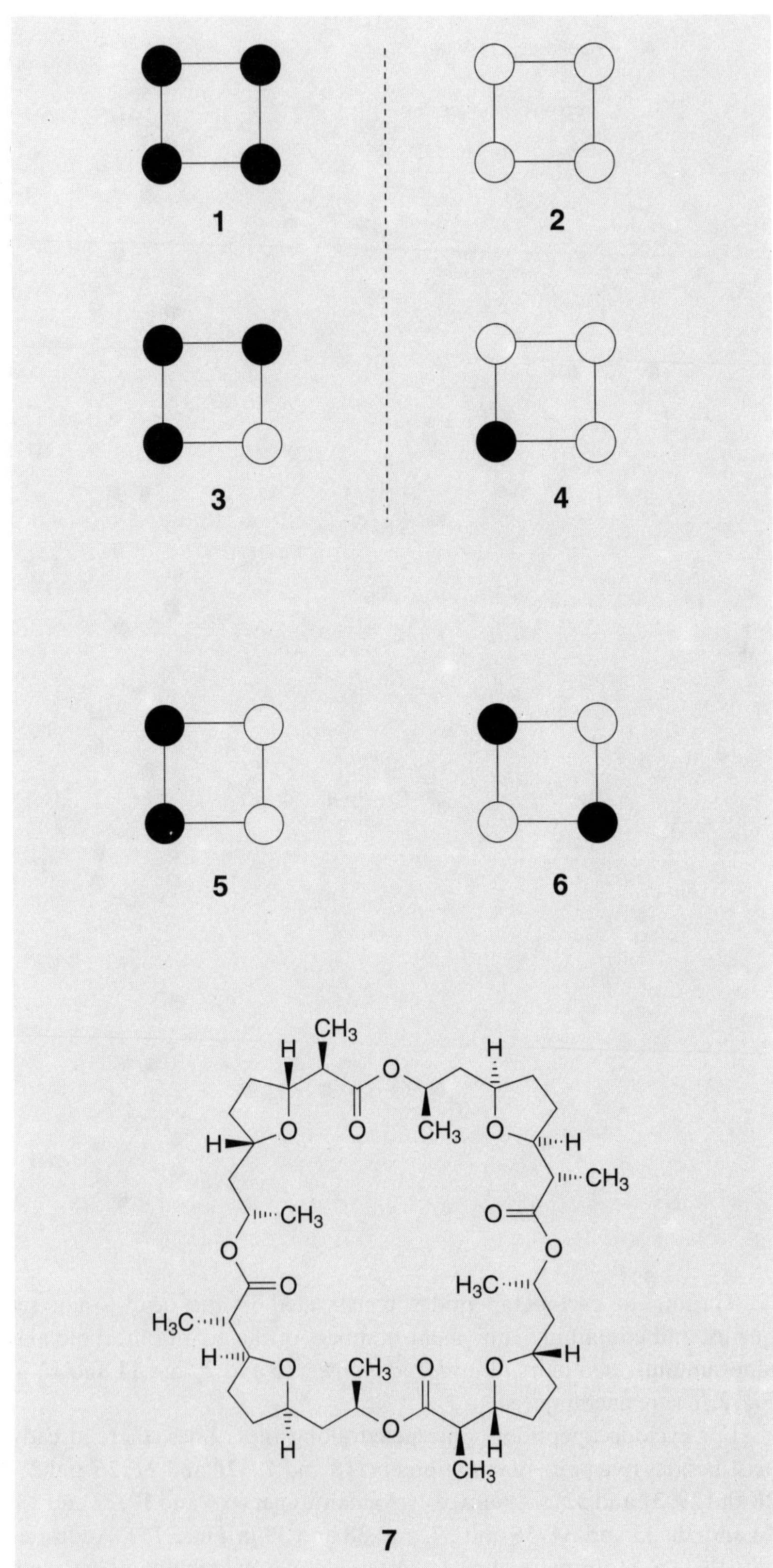

Fig. 2.15. *Representation of racemic mixtures* (from **1** and **2**, or **3** and **4**) *and* meso-*compounds* (**5** or **6**) *containing equal numbers of enantiomorphic nonactinic-acid residues, together with the configurational formula* (**7**) *of nonactin*

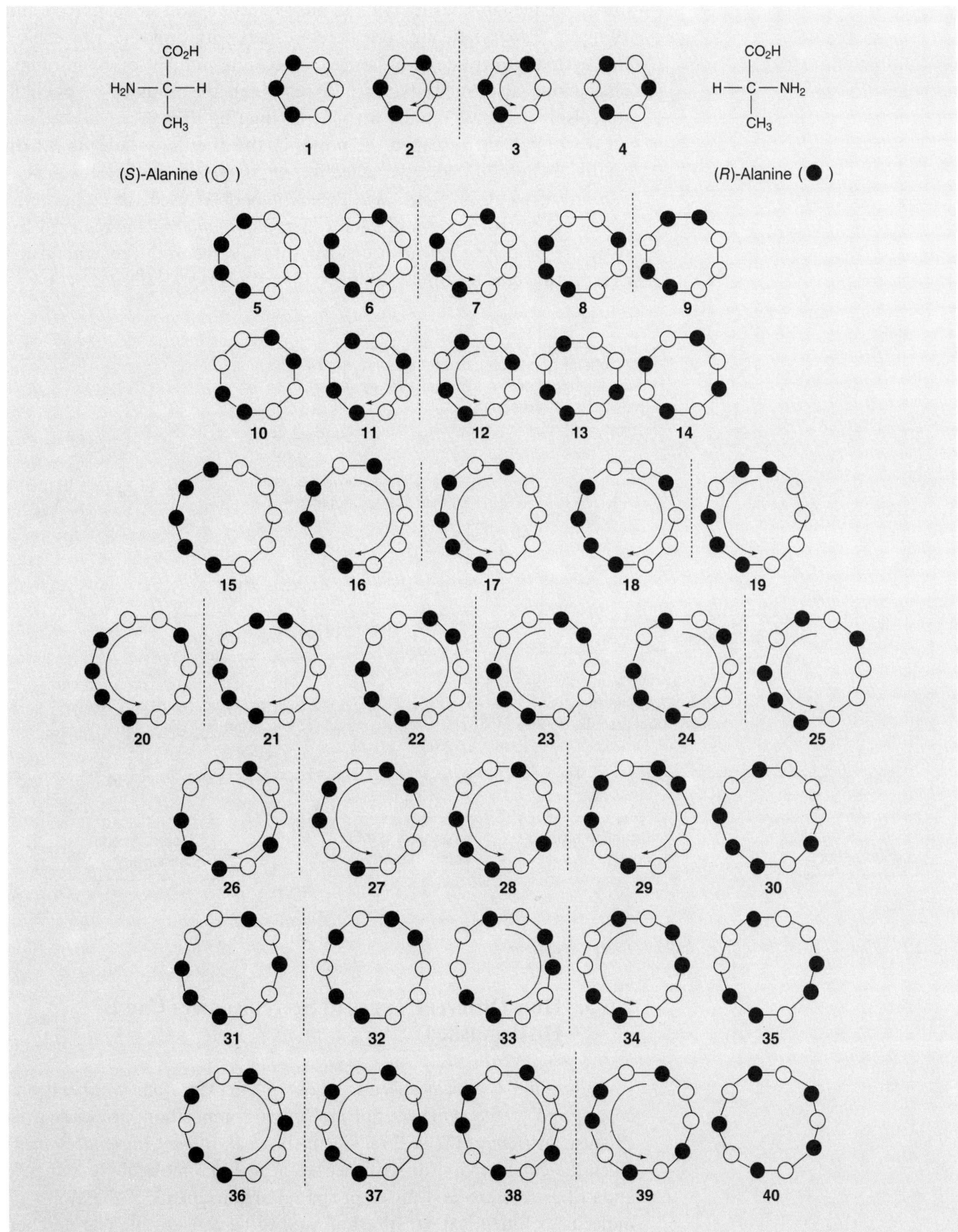

Fig. 2.16. *The complete set of representations of cyclopeptides with six, eight, or ten building units of alanine enantiomers in head-to-tail sequence, with the same number of (*R*)- and (*S*)-enantiomers*

mirror-image relationship (**18** and **20**, **19** and **21**, **26** and **28**, and **27** and **29** in *Fig. 2.16*): these are cyclostereoisomers with opposite 'ring sense'. For cyclooligopeptides consisting of the same number of enantiomorphic amino-acid subunits (such as alanine), the number of possible stereoisomers grows rapidly with increasing ring size.

It will not have escaped the notice of the attentive reader that here (and elsewhere), besides the adjective *enantiomeric*, the adjective *enantiomorphic* has been used. While *enantiomeric* is used for compounds which are mirror images of each other, *enantiomorphic* is reserved for structure models (*e.g.* transition structures), structural fragments (such as repeating constitutional subunits), or entire crystals (*Chapt. 3.1*).

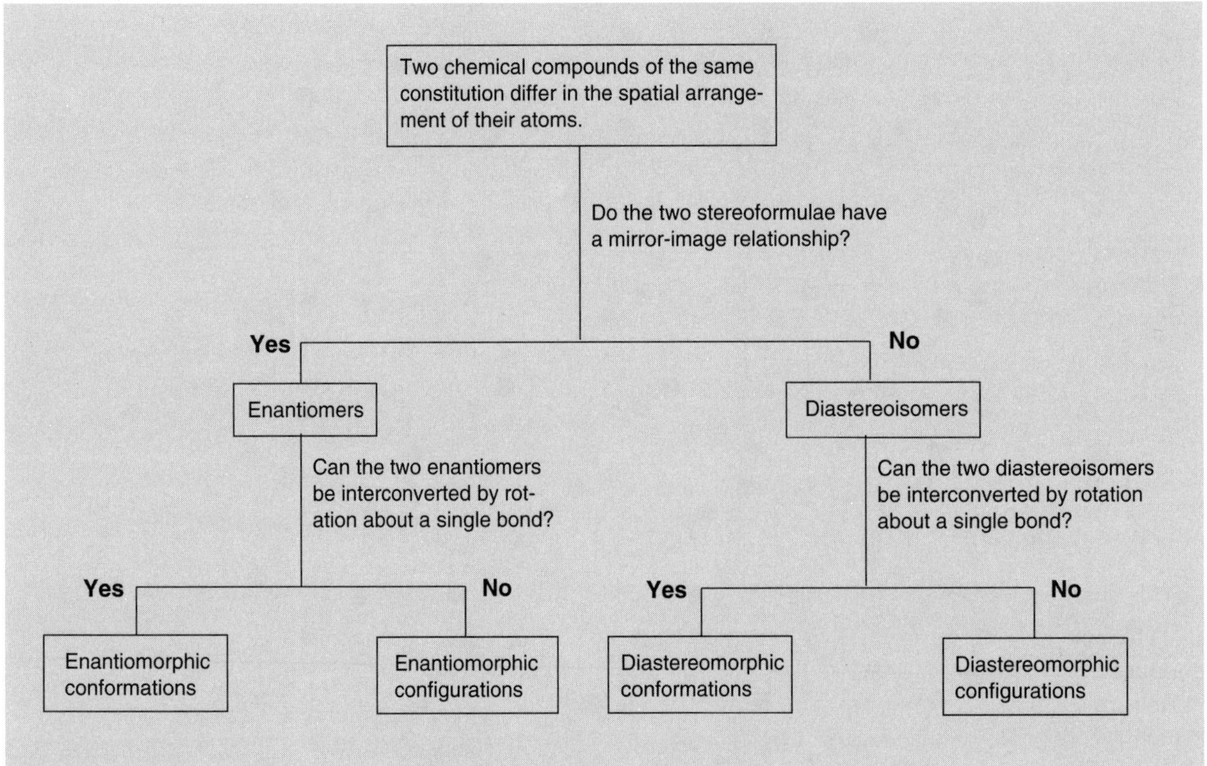

Fig. 2.17. *Flow diagram for differentiation between stereoisomers*

2.6.5. How Different Types of Stereoisomers Can Be Distinguished

Individual chemical species, whose particular atom combinations correspond to the same overall molecular composition, are known to chemists as *isomers* (*Sect. 2.4*). Chemists speak of *constitutional isomers* when the connectivity (or sequence) of atoms in a molecule is changed, anda of *stereoisomers* if different spatial arrangements of the atoms of molecules of identical constitution need to be considered. The number and stereostructure of possible stereoisomers (enantiomers or diastereoisomers) can be derived in a clear manner from the tetrahedron model.

Up-and-coming chemists must have heightened awareness, though, of the limitations which may be traced back either *a priori* to constitutional symmetry, or *a posteriori* to intolerable molecular strain.

Fig. 2.17 shows a way, which helps to differentiate programmatically between enantiomers and diastereoisomers, and pragmatically between conformational isomers (conformers) and configurational isomers.

2.7. Conclusion

Since the middle of the last century, the development of organic chemistry represents the most remarkable use of logical reasoning of a non-quantitative type that has ever taken place. Atoms (most importantly the elements C, H, O, N, P, S) may be assembled, according to their particular valencies, in the manner of a module-based construction system, into molecules containing branched or unbranched chains and/or rings with single, double, and/or triple bonds. The simple construction guidelines of the constitutional model of the classical organic chemistry stand in dramatic contrast to the enormous complexity of possible constitutional isomers. In the meantime, chemists have recognized that constitutional formulae, representing in a simple way the connectivity of participating atoms, are nothing else but graphs or multigraphs. Therefore, graph theory is excellently suited for the correct formulation of the number and individual type of atom combinations of possible constitutional isomers. Today, computers are assigned the task of depicting the various atom combinations under examination, from which they may be simply counted off. The necessary algorithms must, of course, take account of the construction guidelines of molecular architecture as demonstrated here.

The application of graph theory to the determination of the real number of chemical compounds with an identical molecular formula which can exist, with particular respect to avoiding multiple counting which arises as a consequence of molecular symmetry (*Chapt. 3*), is a contribution of *mathematical chemistry*. In contrast to mathematical chemistry, which sets out to mathematicize chemistry without the diversion through physics, and seeks to make apparent the logical structure of chemistry, one does speak of *physical chemistry*, if ideas and outcomes of physics are applied to chemical compounds (and their reactions). If the adoption of thoughts and results from theoretical physics plays a role, then one speaks of *theoretical chemistry*.

The tetrahedron model was not only well suited for the correct interpretation of those isomers whose existence could be explained by stereogenic centers (enantiomers or diastereoisomers) or stereogenic C=C bonds (diastereoisomers), but was additionally able to predict (for allenes, alkylidene-cycloalkanes, and spiro compounds) other cases of stereoisomerism, which could not be experimentally verified for decades afterwards. It may be assumed that, in the future, further cases of constitutional isomerism or stereoisomerism, which today are still unknown, will be found.

Interest in constitutional isomerism and stereoisomerism is so great because, with every individual case examined, the structure theory of the classical organic chemistry undergoes a falsification test. The numerical agreement between predicted and experimentally verified isomers has been confirmed again and again. Occasionally, exceptions were stumbled upon. This led, for certain very particular classes of compounds, to the structure theory of the classical organic chemistry being slightly modified or, eventually, even given up (*cf. Chapt. 8*).

In this chapter, the structure of chemical compounds stood in the foreground. It seems appropriate to emphasize at this point that it is, however, not the structure, but the effect that the structure exerts upon one's mind that plays the major role in chemical science. (Organic) chemists differ from other scientists in that they use a pictographic language, and are close to architects in their mentality. This becomes particularly clear when observing synthetic chemists in their planning and execution of a synthesis (*Aspects of Organic Chemistry*; Volume 3: *Synthesis*). The emerging richness of expression of the chemical pictographic language and the clarity of the symbols used (on recognition of their conventional specification) will follow in *Chapt. 10* of this volume.

Further Reading

To Sect. 2.2.

D. H. R. Barton, *Stereochemistry* in *Perspectives in Organic Chemistry* (Ed. A. Todd), Interscience Publ., New York, 1956.

H. Hartmann, *Die Bedeutung Quantentheoretischer Modelle für die Chemie,* Franz Steiner Verlag, Wiesbaden, 1965.

To Sect. 2.4.

W. J. Wiswesser, *Johann Josef Loschmidt (1821–1895): A Forgotten Genius, Aldrichim. Acta* **1989**, *22*, 17.

O. T. Benfey, *Kekulé Centennial*, American Chemical Society, Washington, DC, 1966.

J. H. Wotiz (Ed.), *The Kekulé Riddle*, Cache River Press, Vienna, Il., 1993.

To Sect. 2.5.

IUPAC-Rules on Organic Stereochemistry, Pure Appl. Chem. **1976,** *45*, 11.

J. H. van't Hoff, *Die Lagerung der Atome im Raume,* 3. Aufl., F. Vieweg und Sohn, Braunschweig, 1908.

O. B. Ramsay, *van't Hoff – Le Bel Centennial,* American Chemical Society, Washington, DC, 1975;
– *Nobel Prize Topics in Chemistry: Stereochemistry*, Heyden & Son, London, 1981.

K. Freudenberg (Ed.), *Stereochemie*, Franz Deuticke, Leipzig, 1932.

E. L. Eliel, *Stereochemistry of Carbon Compounds,* McGraw-Hill, New York, 1962.

B. Testa, *Principles of Organic Stereochemistry,* Marcel Dekker, New York, 1979.

J. Dale, *Stereochemie und Konformationsanalyse,* Verlag Chemie, Weinheim, 1978.

Topics in Stereochemistry, Interscience, Vol. 1–20, 1967–1991.

K. Mislow, *Introduction to Stereochemistry,* W. A. Benjamin, New York, 1966.

To Sect. 2.6.

2.6.2.1.

D. Seebach, V. Prelog, *The Unambiguous Specification of the Steric Course of Asymmetric Syntheses, Angew. Chem. Int. Ed.* **1982,** *21,* 654.

IUPAC Nomenclature of Carotenoids, Pure Appl. Chem. **1975**, *41,* 407.

2.6.2.2.

C.H. Park, H.E. Simmons, *Bicyclo[8.8.8]hexacosane. Out,In-Isomerism, J. Am. Chem. Soc.* **1972,** *94,* 7184.

M. Saunders, *Stochastic Search for the Conformations of Bicyclic Hydrocarbons, J. Comput. Chem.* **1989**, *10,* 203.

M. Saunders, N. Krause, *The Use of Stochastic Search in Looking for Homeomorphic Isomerism: Synthesis and Properties of Bicyclo[6.5.1]tetradecane, J. Am. Chem. Soc.* **1990**, *112,* 1791.

R.W. Alder, *Intrabridgehead Chemistry, Tetrahedron* **1990**, *46,* 683.

2.6.4.

D. Mauzerall, *The Condensation of Porphobilinogen to Uroporphyrinogen, J. Am. Chem. Soc.* **1960**, *82,* 2605.

W. Keller-Schierlein, H. Gerlach, *Makrotetrolide, Progress in the Chemistry of Organic Natural Products* **1968,** *26*, 161.

M. Dobler, *The Crystal Structure of Nonactin, Helv. Chim. Acta* **1972**, *55,* 1371.

M. Dobler, J.D. Dunitz, B.T. Kilbourn, *Die Struktur des KNCS-Komplexes von Nonactin, Helv. Chim. Acta* **1969**, *52,* 2573.

H. Gerlach, K. Oertle, A. Thalmann, S. Servi, *Synthese des Nonactins, Helv. Chim. Acta* **1975**, *58,* 2036.

V. Prelog, H. Gerlach, *Cycloenantiomerie und Cyclodiastereomerie, Helv. Chim. Acta* **1964**, *47,* 2288.

H. Gerlach, J.A. Owtschinnikow, V. Prelog, *Über cycloenantiomere cyclo-Hexaalanyle und ein cycloenantiomeres cyclo-Diglycyl-tetraalanyl, Helv. Chim. Acta* **1964**, *47,* 2294.

To Sect. 2.7.

R. Hoffmann, P. Laszlo, *Representation in Chemistry, Angew. Chem. Int. Ed.* **1991,** *30,* 1.

J.G. Nourse, R.E. Carhart, D.H. Smith, C. Djerassi, *Exhaustive Generation of Stereoisomers for Structure Elucidation, J. Am. Chem. Soc.* **1979**, *101,* 1216.

I. Ugi, J. Bauer, K. Bley, A. Dengler, A. Dietz, E. Fontain, B. Gruber, R. Herges, M. Knauer, K. Reitsam, N. Stein, *Computer-assisted solution of chemical problems: a new discipline in chemistry, Angew. Chem. Int. Ed.* **1993,** *32,* 201.

3 Optical Activity, Chirality, and Symmetry of Molecules

3.1. Fundamentals

3.1.1. The Physical Phenomenon of Optical Activity

With the aid of linearly polarized light, it is possible to divide organic molecules into two groups. Members of one group, in solution, rotate the plane of linearly polarized light (*Fig. 3.1*): these are defined as optically active. Representatives of the other group do not rotate the plane of linearly polarized light: these are optically inactive.

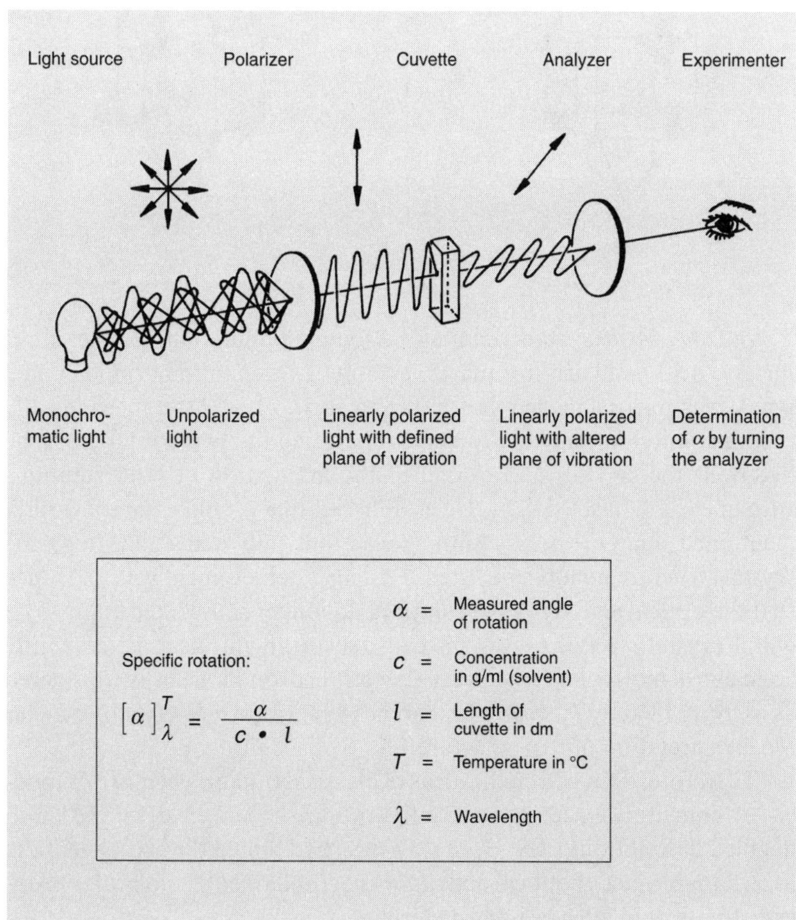

Fig. 3.1. *Rotation of the plane of vibration of linearly polarized light by a solution of a levorotatory substance in a polarimeter and definition of the specific rotation*

The experimental investigation of how molecules interact with linearly polarized light (the determination of the specific rotation $[\alpha]_{\lambda}^{T}$) is an important analytical method. Using it, two enantiomers may be differentiated by their optical rotatory activity: under identical measurement conditions (identical molecular concentration, identical solvents, temperatures, and light wavelengths), one enantiomer will rotate the plane of linearly polarized light by a particular angle to the left (anticlockwise; symbolized as (−)); the other enantiomer will rotate the plane of linearly polarized light by the same angle to the right (clockwise; symbolized as (+)).

It was already known at the time of *Louis Pasteur* that two enantiomorphic types of quartz crystal exist (*Fig. 3.2*). It was also known that the two enantiomorphic quartz crystal types are optically active. Sections of quartz crystal of equivalent depth all rotate linearly polarized light by the same angle. In quartz, the optical activity is tied to the chiral crystal structure. The monomeric SiO_2 unit, on the contrary, is achiral.

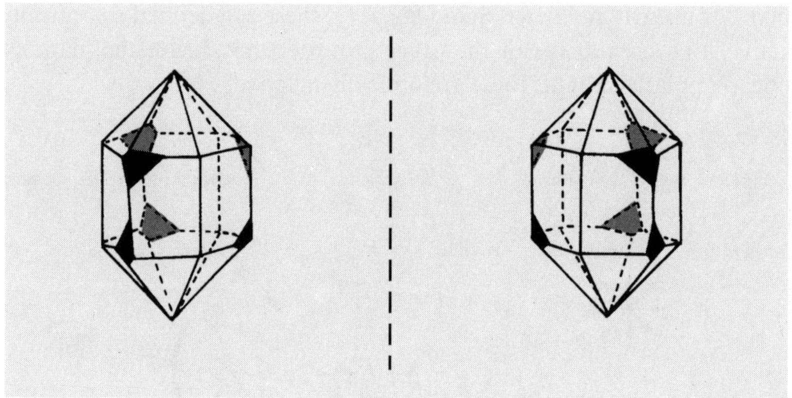

Fig. 3.2. *Enantiomorphic quartz crystals*

In 1848, *Pasteur* observed that the sodium ammonium double salt of tartaric acid, similarly to quartz, exhibits two enantiomorphic crystal types, of a mirror-image relationship to each other. After mechanically (with magnifying glass and forceps) separating the two crystal types, he dissolved the crystals and examined the interaction of both solutions with linearly polarized light. He established that – unlike the previously mentioned quartz – both solutions were optically active, although the crystal structure no longer existed. *Pasteur*'s achievement was to deduce that the optical activity of the solution, no longer caused, naturally, by a chiral crystal structure, was instead due to an intrinsic, permanently fixed chiral molecular structure. (The tetrahedron model was considered as early as 1860 by *Pasteur* as *a*, and in 1874 by *van't Hoff* and *Le Bel* as *the* interpretation of that structure.)

Therefore, even before the ideas of classical organic chemistry's models of constitution, let alone configuration, had been developed into detailed descriptions (*vide supra*), *Pasteur* had linked the experimentally ascertainable fact of optical activity with a fundamental molecular property, that would later be named chirality.

3.1.2. Chirality of Molecules

'*Chirality in Chemistry*' was the theme of the Stockholm *Nobel* lecture of *V. Prelog* in 1975. At the same time, it is also the leitmotiv of a devotion to chemical studies spanning over more than 132 semesters (*studium chymiae nec nisi cum morte finitur*). The concept of *chirality* originated from *Lord Kelvin* and was introduced into chemistry by *Kurt Mislow*.

An object (be it a (model of a) molecule or be it a *Henry Moore* sculpture) is chiral, if it does not permit itself to be superimposed upon its mirror image by linear translation or rotation. We have already encountered this phenomenon, and always when two enantiomers were being discussed.

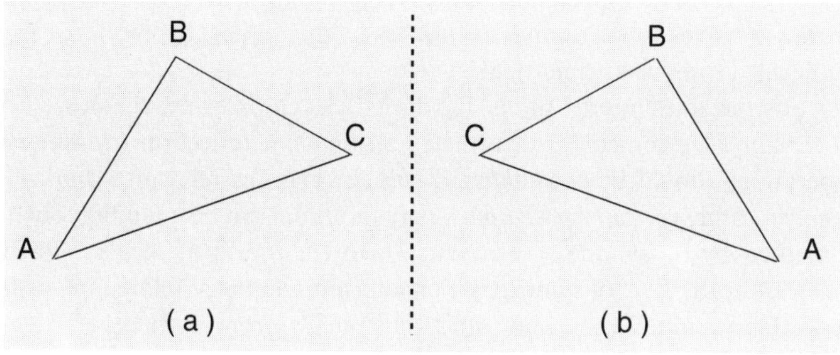

Fig. 3.3. *Enantiomorphic triangles*

Like the triangle in two-dimensional space, the tetrahedron represents the simplest figure, the simplex, in three-dimensional space. In three-dimensional space, a tetrahedron with four distinguishable apexes (model for a methane derivative with four different ligands; *cf. Chapt. 2.5.1.1*) is the simplest chiral figure. In two-dimensional space (*'Flatland'*), a triangle with three different apexes is the simplest chiral figure (*Fig. 3.3*). No combination of linear translation and in-plane rotation can bring the two triangles (a) and (b) into superposition. Only reflection across a line located outside the figures (here the dashed line in *Fig. 3.3*) can convert the two triangles into one another. Although our interest in chirality lies in three-dimensional structures, considerations of a two-dimensional space are very helpful: in *'Flatland'* it may be demonstrated on relatively simple figures how chiral and achiral objects may be distinguished. The upper-case letters of the Latin alphabet, for example, are simple planar figures of which some (A, B, C, D, E ...) are two-dimensionally achiral, and others (F, G, J, L, N ...) are two-dimensionally chiral.

achiral : ABCDEHIKMOTUVWXY

chiral : FGJLNPQRSZ

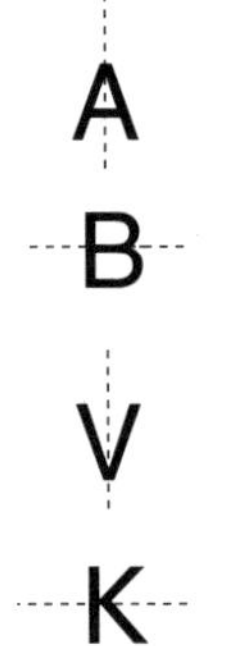

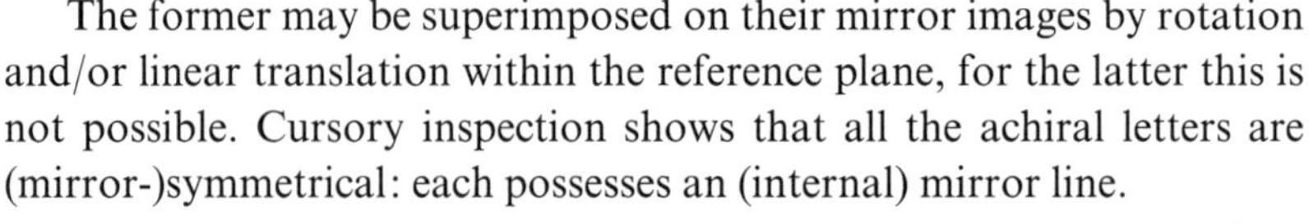

The former may be superimposed on their mirror images by rotation and/or linear translation within the reference plane, for the latter this is not possible. Cursory inspection shows that all the achiral letters are (mirror-)symmetrical: each possesses an (internal) mirror line.

3.1.3. Symmetry of Molecules

Obviously, symmetry determines whether or not a figure is chiral. We must now assign a clearly defined meaning to the concept of symmetry, previously applied in a somewhat 'loose way' (*cf. Chapt. 2.6.2.1*). The symmetry of a molecule may be determined from the behavior of its model in particular symmetry operations. For *cis*-1,2-dimethylcyclopropane (*cf. Chapt. 2.6.2.1*), for example, a plane may be laid through the static formula representation, dividing the three-dimensional figure into two halves which are mirror images of each other: the model of the molecule is mirror-symmetrical.

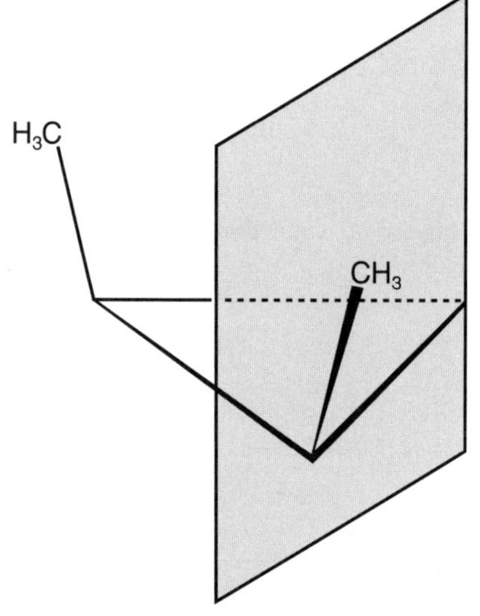

For the static model of *cis*-1,2-dimethylcyclopropane, therefore, in the plane of symmetry σ (symmetry element), a reflection (symmetry operation) should be considered. *Table 3.1* lists the relevant *symmetry elements* and *symmetry operations*. Any molecule can be unambiguously classified by its symmetry elements. Two (or more) molecules which possess the same set of symmetry elements are symmetry-related, regardless of how much they may differ chemically from each other: they belong to the same symmetry point group. Further details are to be found in *Chapt. 11*.

Table 3.1. *Types of Symmetry Operations with Corresponding Symmetry Elements in Three-Dimensional Space*

Symbol	Symmetry operation	Symmetry element
E	Identity operation	Identity element
C	Rotation	Rotation axis
σ	Reflection	Mirror plane
i	Inversion	Inversion center
S	Rotary reflection	Rotary-reflection axis

The relationship between the two pairs of properties – symmetrical or asymmetrical, and chiral or achiral – should now be demonstrated for the various symmetry point groups. There are three combinations:

– a molecule is *asymmetrical* (point group C_1) and *chiral* (*cf.* examples in *Fig. 3.4*);

- a molecule is *symmetrical* (for point groups C_n, D_n, T, O, and I) and *chiral* (*cf.* examples in *Fig. 3.5*);
- a molecule is *symmetrical* (for all point groups except those given above) and *achiral* (*cf.* examples in *Fig. 3.6*).

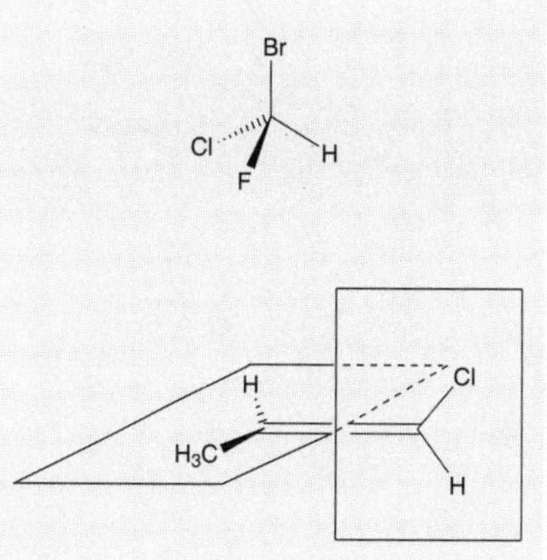

Fig. 3.4. *Examples of* asymmetrical *and* chiral *molecules*

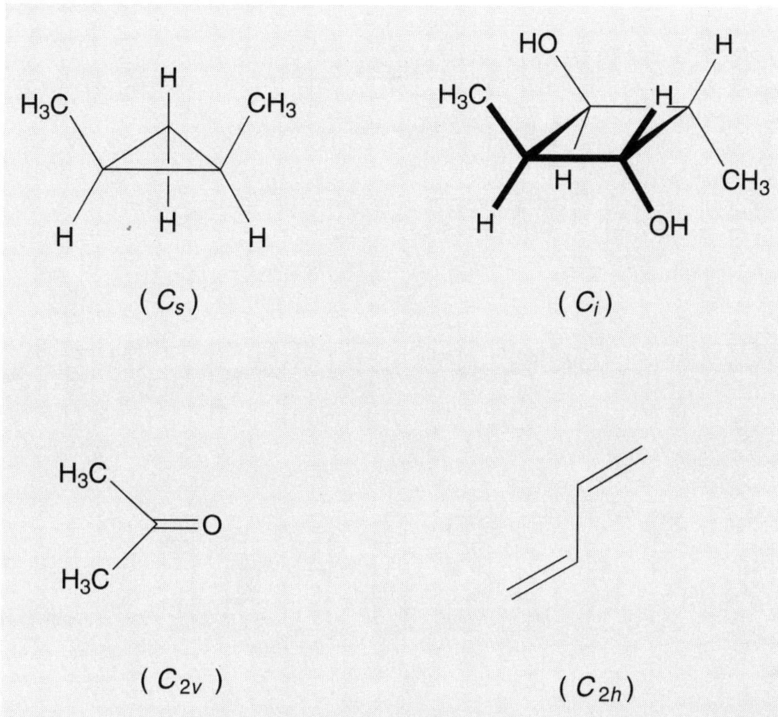

Fig. 3.6. *Examples of* symmetrical *and* achiral *molecules*

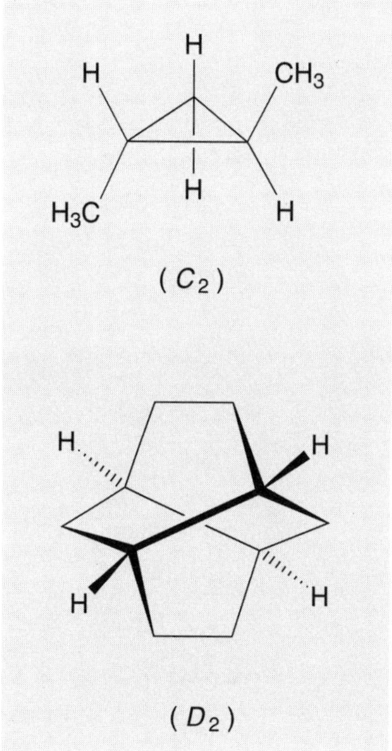

Fig. 3.5. *Examples of* symmetrical *and* chiral *molecules*

It is, therefore, possible to differentiate between chiral and achiral molecules in two ways, using (consciously or not) symmetry operations:
- by reflection across an extramolecular mirror plane,
- by determination of the relevant symmetry point group.

3.2. The *Prelog* Catalog of Regular Tetrahedra

The permutation of two-dimensional achiral symbols (A, B, C, D), denoting achiral ligands, and of two-dimensional chiral symbols (F, G, J, L), denoting chiral ligands, in the form of *Young* distribution diagrams gives 30 different combinations (*Table 3.2*).

Nine of these distribution diagrams possess four different ligands. These represent those tetrahedra, which, by exchanging the positions of two ligands, can be converted into an enantiomorphic or diastereomorphic tetrahedron. This, then gives a grand total of 39 possible different combinations (*Table 3.3*), which can in turn be arranged into various subclasses of compounds with a tetrahedral C-atom as the center atom: the *Prelog* catalog of regular tetrahedra.

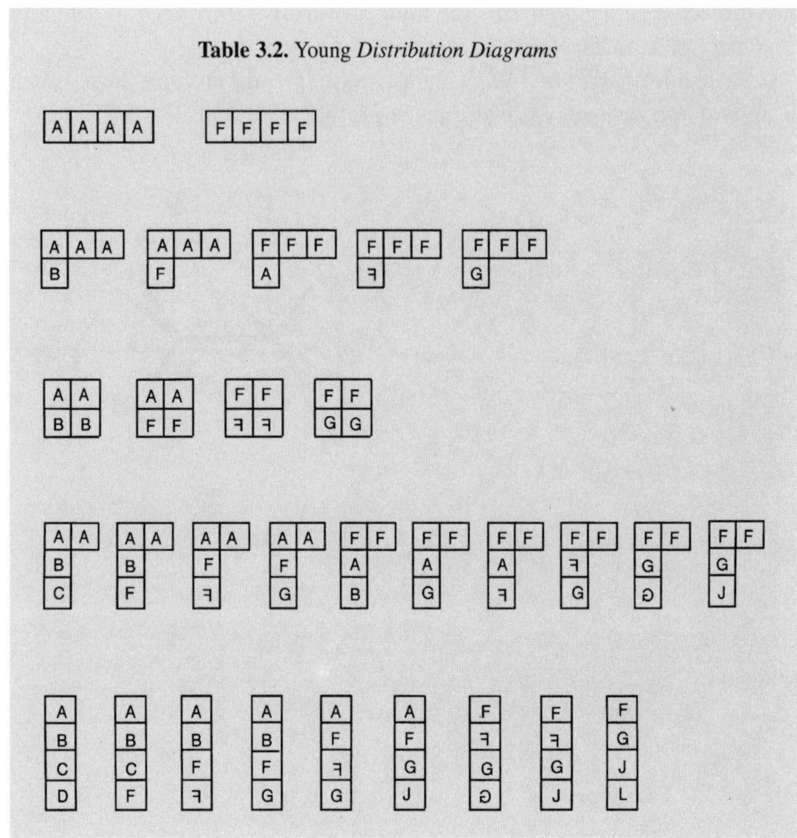

Table 3.2. Young *Distribution Diagrams*

OH
H–C–H
CH$_3$

Ethanol

OH
H–C–D
CH$_3$

(R)-Configuration

OH
D–C–H
CH$_3$

(S)-Configuration

OH
H_S–C–H_R
CH$_3$

Differentiation between two H-atoms at the prochirality center

3.2.1. The Chirality Center (The Asymmetrical C-Atom)

In the *Prelog* catalog of regular tetrahedra, cases 24 and 25 represent models for the C-atom of coordination number 4 as *chirality center*.

3.2.2. The Prochirality Center

Cases 5 and 6 (*Table 3.3*) are models for the C-atom of coordination number 4 as *prochirality center*. A prochirality center is a central atom in the middle of a regular tetrahedron, with two constitutionally identical and two constitutionally or configurationally non-identical ligands at the apexes: X(AABC) or X(AAFꟻ). The structure fragment X(AA) may also describe a two-membered ring consisting of X and A. In each case, this tetrahedron possesses C_s symmetry. Ethanol, therefore, has a prochirality center.

That H-atom of the CH$_2$ group, which, upon substitution with D (deuterium), would give rise to an (R)-chirality center, is specified as H_R; the other H-atom, consequently, as H_S.

Achiral citric acid (= 3-carboxy-3-hydroxypentanedioic acid) has a prochirality center at C(3). The unambiguous designation of the two HOOC–CH$_2$ ligands is given in the formula shown on the next page.

Table 3.3. *The* Prelog *Catalog of Regular Tetrahedra*
(d = diastereomorphic, e = enantiomorphic)

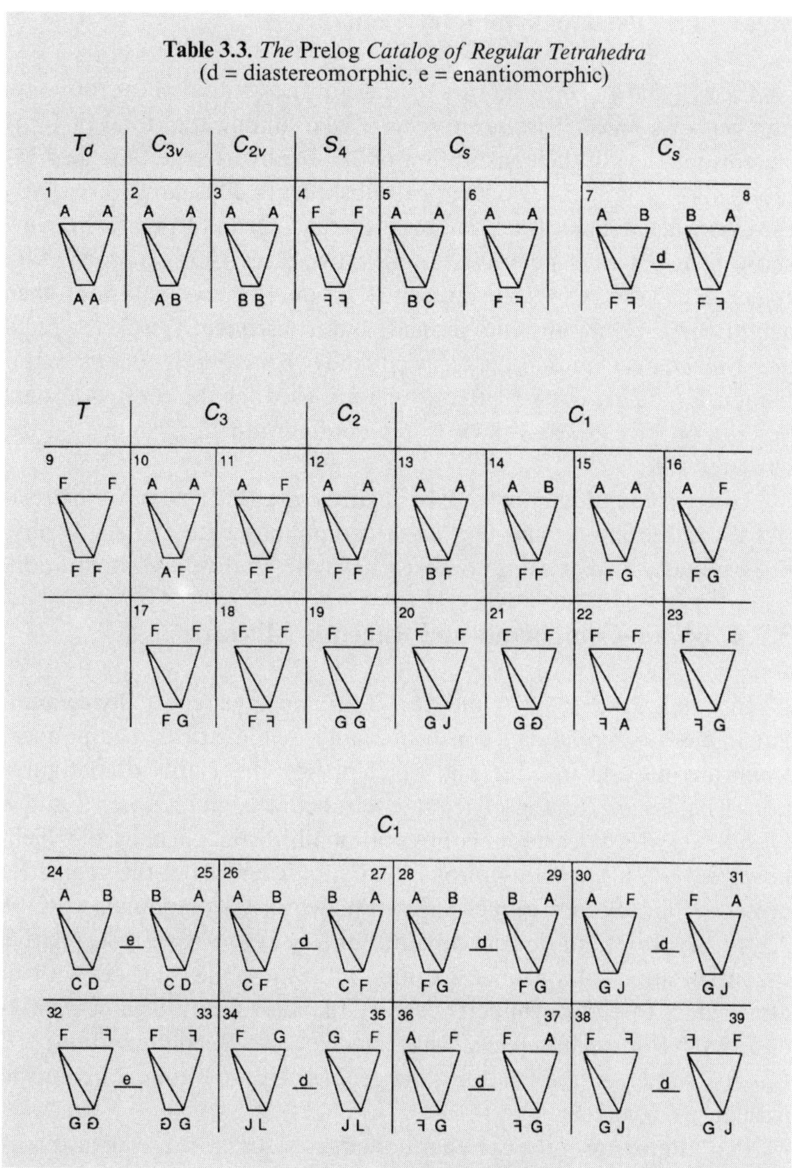

This notation is arrived at by assigning the descriptor *pro-R* to that HOOC–CH_2 ligand of the pair, which would result in an (R)-configurated compound, were this ligand given a higher priority than the other HOOC–CH_2 ligand, according to the sequence rules of *Cahn*, *Ingold*, and *Prelog* (*CIP*), which are described in *Chapt. 10.2.1*. The other ligand is assigned *pro-S*. *Prochirality centers*, *prochirality axes*, and *prochirality planes* may be defined in a manner analogous to that one by which chiral compounds may be designated by chirality centers, chirality axes, and chirality planes (*Fig. 2.12*).

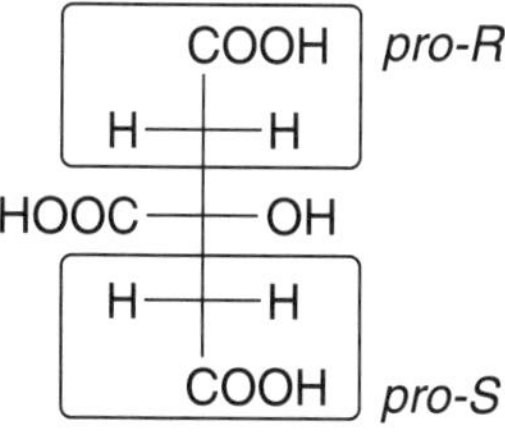

Citric acid

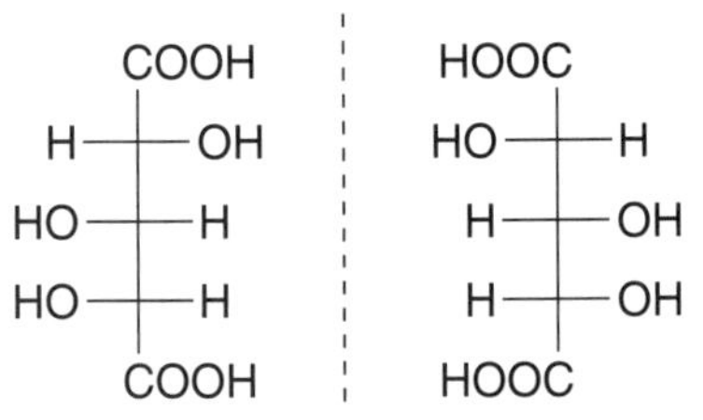

(2*R*,4*R*) (2*S*,4*S*)

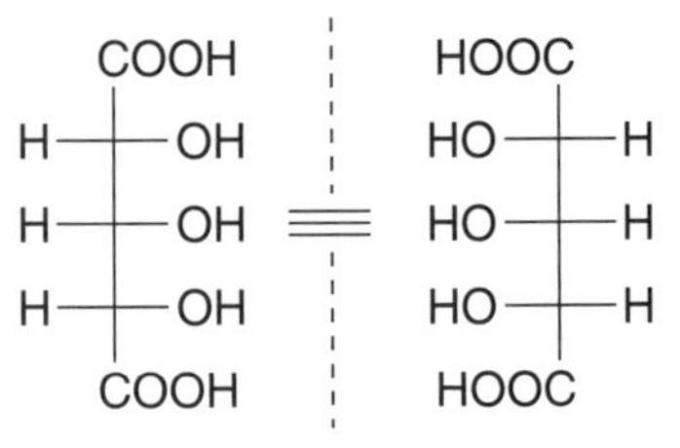

(2*R*,3*r*,4*S*) (2*S*,3*r*,4*R*)

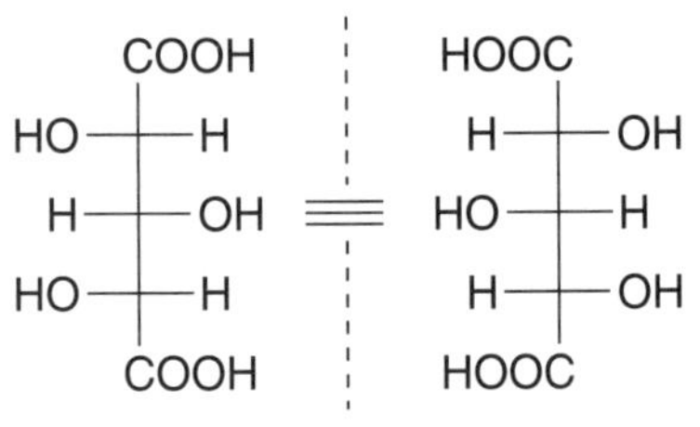

(2*S*,3*s*,4*R*) (2*R*,3*s*,4*S*)

2,3,4-Trihydroxyglutaric acid

3.2.3. The Pseudoasymmetry Center

Cases 7 and 8 (*Table 3.3*) are models for the C-atom of coordination number 4 as *pseudoasymmetry center*. Two among the (total of four) stereoisomers with the constitution of 2,3,4-trihydroxyglutaric acid exhibit – like citric acid – two identical substituents. This, however, is only the case when the absolute configurations at C(2) and C(4) are the same showing in (*R*)- or (*S*)-chirality. In these two cases, C(3) is a prochirality center. If C(2) and C(4) have opposite ((*R*) and (*S*)) configurations, then achiral *meso*-compounds are present, and, consequently, C(3) is a *pseudoasymmetry center*. In order to specify the two *meso*-compounds unambiguously, the lower-case letters *r* or *s* are used for the configurational description of C(3), as shown in the configurational formulae on the left-hand side.

The two enantiomers of 2,3,4-trihydroxyglutaric acid have the relative *l*-configuration. The two *meso*-compounds with the 2,3,4-trihydroxyglutaric-acid constitution each have the relative *u*-configuration.

3.2.4. *meso*-Compounds and Racemic Mixtures

In *Chapt. 2.6.2.1*, *2.6.4*, and *Sect. 3.2.3*, we have repeatedly encountered *meso*-compounds, constitutionally symmetrical compounds, which are optically inactive, and which possess the relative *u*-configuration. The *Prelog* catalog of regular tetrahedra contains cases 7 and 8 (*Table 3.3*) with at least four configurationally distinct ligands, of which, however, two are enantiomorphic and equidistant from the geometric center of the molecule. In such a case (the two *meso*-compounds with the 2,3,4-trihydroxyglutaric-acid constitution, for example, with the relative *u*-configuration), the geometric center of the molecule coincides with an atom center (pseudoasymmetry center). In *meso*-tartaric acid or *cis*-1,2-dimethylcyclopropane (each with the relative *u*-configuration; *cf. Chapt. 2.6.2.1*), on the contrary, the geometric center does not coincide with an atom center.

If a 1:1 mixture of the two enantiomers of tartaric acid is taken and its optical activity examined, no rotation of the plane of linearly polarized light is observed. The so-called *racemic mixture* is optically inactive, since the degrees of rotation of the two enantiomers (same magnitude, opposite sign) compensate externally (descriptor: (±)).

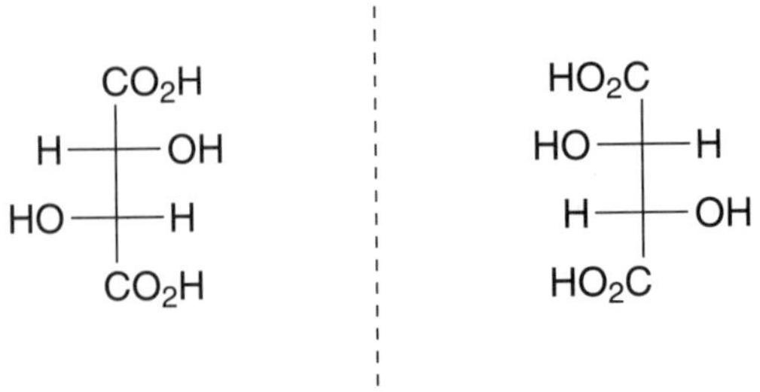

(+)-(*R*,*R*)-Tartaric acid (–)-(*S*,*S*)-Tartaric acid

The *IUPAC* rules of the stereochemistry of organic compounds prescribe a 1:1 correspondence between chirality and optical activity. According to them, all molecules of optically active compounds are chiral, and all chiral molecules are optically active. What, however, is the situation regarding racemic mixtures, which contain identical numbers of chiral molecules with opposite senses of chirality? In order to avoid unnecessary talking at cross-purposes, we say that all chiral compounds which are not present as racemic mixtures, and which, consequently, are optically active, are *chiral, non-racemic*.

3.3. Racemization

3.3.1. Fundamentals

In certain cases it may be observed that an optically active compound, under certain conditions (at elevated temperatures, in the presence of acid or base, or on exposure to light, for example), racemizes, meaning that it turns into an optically inactive mixture of its enantiomers. As early as 1904, *Alfred Werner* formulated a theory of racemization for optically active compounds with one stereogenic center. He assumed that the bonds to the four ligands of the tetrahedrally coordinated central atom were set into powerful vibration, so that a square-planar transition structure (with D_{4h} symmetry, ignoring non-identical ligands) could be reached and passed through (*Fig. 3.7*). This transition structure according to the theory was in equilibrium with the two enantiomers **A** and **A′**. From here, the low-energy structures **A** and **A′** were attainable with identical degrees of probability. Today, it is known that the inversion of a tetrahedrally coordinated center *via* a planar transition structure is unlikely: the energy of the planar structure would be extremely high.

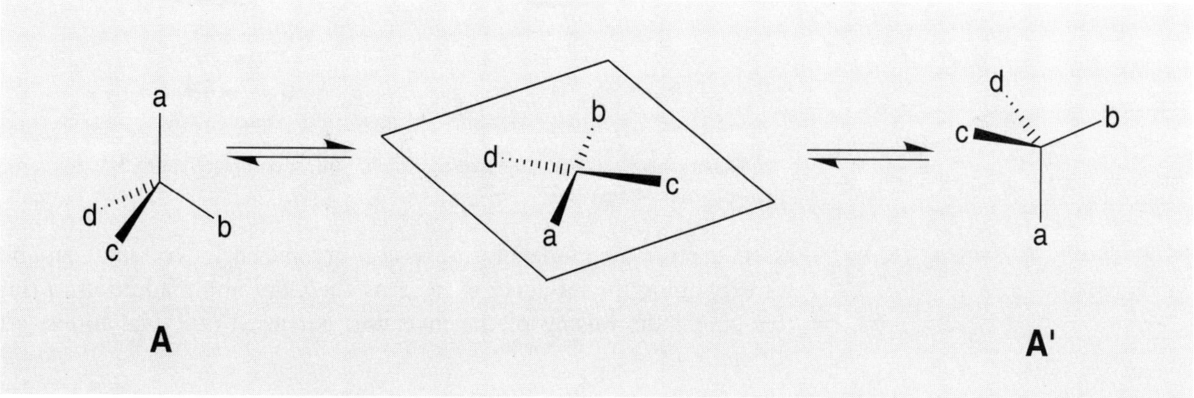

Fig. 3.7. Alfred Werner*'s model of racemization*

The energy of a C_s-symmetrical transition structure is predicted to be comparatively low. It is not so low, however, that optically active compounds with a stereogenic C-atom are, in general, easily racemized. Easily occurring racemizations are the exception, not the rule, and must each be explained individually.

For compounds of general formula X(ABC), the question in this context concerns geometry. If they exhibited a pyramidal structure, then they would be chiral. If they were permanently or transiently structured in a planar manner, then they would be achiral. In 1944, *Prelog* succeeded in separating racemic *Tröger*'s base (*rac*-**1**; *cf. Chapt. 10.2.6*) into its two enantiomers. These latter are configurationally stable for reasons of molecular architecture.

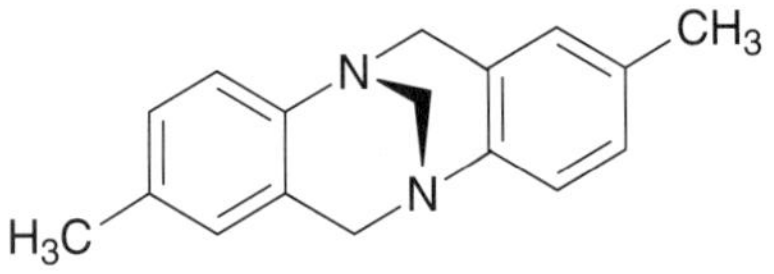

In nineteen-twenties and -thirties *Jacob Meisenheimer* repeatedly attempted to achieve resolution of enantiomers in the case of amines which were configurationally not 'frozen'. Going as far as the imaginative interpretation of the structure model of the classical organic chemistry could take him, he attempted an enantiomer separation of (±)-2-methylaziridine (*rac*-**2**).

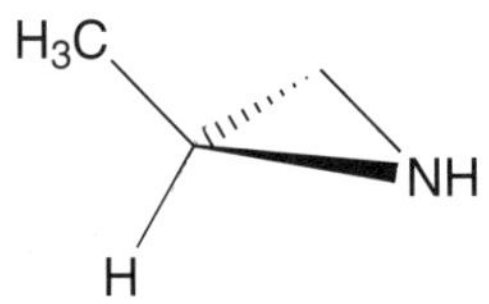

1

2

Should pyramidal amines be capable of easily passing through the planar transition structure, then the angle ∢ CNC should widen (from somewhat under 109° to practically 120°) in the process. Any structural influence which would inhibit such a widening of the angle would cause the inversion barrier to be raised. A strained three-membered ring should be especially effective. This goal was not obtained, however. Consequently, it was even more astonishing, when, at the end of the nineteen-sixties, it was observed that, in *N*-halo-aziridines, the isomerization barrier reaches a level sufficient for the separation of 'invertomers'.

A. Eschenmoser did indeed succeed in separating the two diastereoisomers **3** and **4** of 7-chloro-7-azabicyclo[4.1.0]heptane from each other.

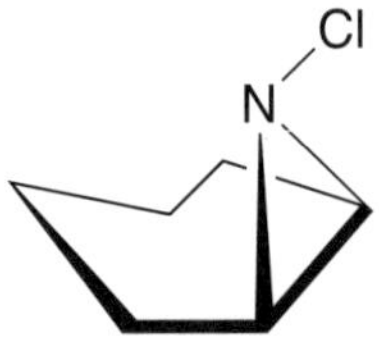

3 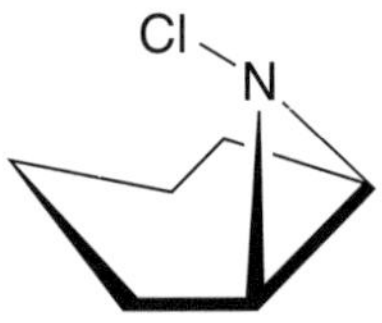 **4**

Diastereoisomer **4** is converted at 29.5 °C almost completely into **3** within a day (half-life time: 4.5 h).

The structure model of the classical organic chemistry is not capable of explaining the influence exerted by the Cl-atom, in addition to ring strain, on the raising of the inversion barrier. For explanations, see *Chapt. 8.4.2*.

3.3.2. Examples

Four examples should explain how racemizations do, in fact, take place: as long as particular experimental conditions and structural requirements are fulfilled.

3.3.2.1. 6,6′-Difluoro-1,1′-biphenyl-2,2′-dicarboxylic Acid

If one of the two enantiomers of 6,6′-difluorodiphenic acid (= 6,6′-difluoro-1,1′-biphenyl-2,2′-dicarboxylic acid) are heated to 140 °C, a steady decrease in the optical activity occurs, until, after a period of time, complete racemization has taken place (*Fig. 3.8*). The reason for this behavior lies in the exceeding of the energy barrier, which, at lower temperatures, hinders free rotation about the central C–C bond. *Fig. 2.9* makes clear that the bulky substituents in the *ortho*-positions must pairwise slip past each other, if racemization is to take place. After the two enantiomers have reached equivalent populations, the thermal equilibrium does not stop; however, the overall composition does not change any further.

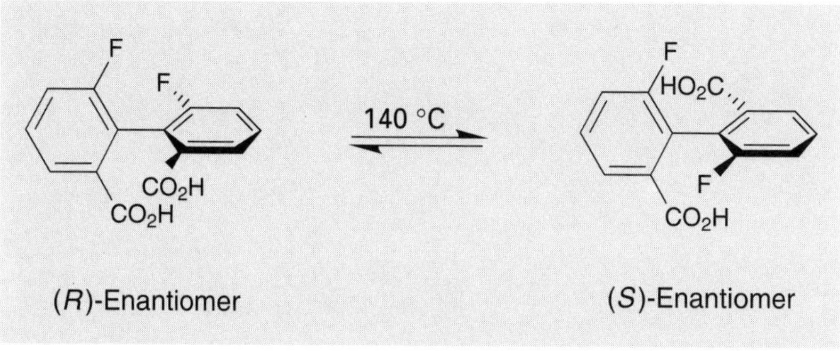

Fig. 3.8. *Racemization of 6,6′-difluoro-1,1′-biphenyl-2,2′-dicarboxylic acid*

3.3.2.2. Acyl Derivatives of *N*-Acyl-α-amino Acids

In the synthesis of proteins (biomacromolecules with repeating units –NH–CH(R)–CO–, R = variable side chain; *Chapt. 10.5.4.2*), acyl derivatives of *N*-acyl-α-amino acids (**A**) play a notable role. Under the influence of bases, they cyclize to oxazolones (azlactones, **B**). The latter are converted in the presence of base

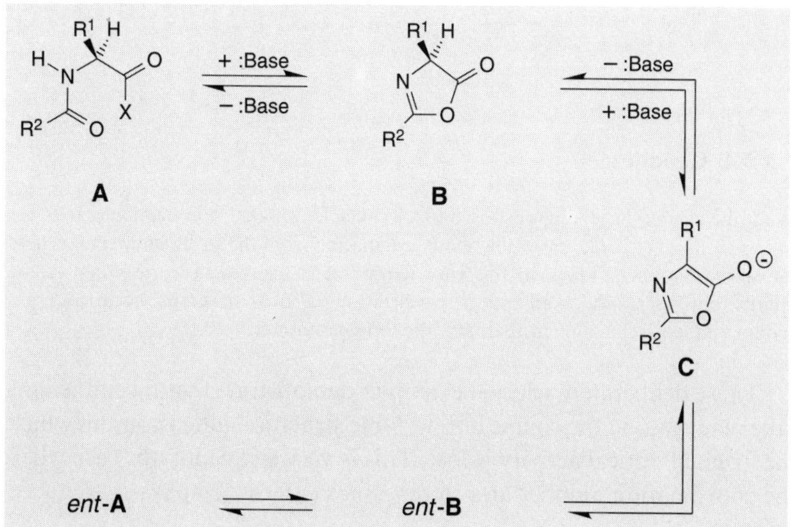

Fig. 3.9. *Racemization of derivatives of* N-*acyl-α-amino acids*

(or acid) into oxazoles (**C**). When the oxazoles are finally converted back into *N*-acyl-α-amino acids, not only **A** (X = OH), but also, with an identical degree of probability, *ent*-**A** (X = OH; *Fig. 3.9*) is produced. (For the unambiguous labeling of enantiomers through omission or utilization of the prefix *ent*, see *Chapt. 10.2.6.*)

3.3.2.3. Usnic Acid

Usnic acid, found in lichens, occurs in nature mostly in optically active form. It racemizes on exposure to light. The cyclohexadienone ring is temporarily opened, giving rise to an achiral diketene, that then rapidly thermally recyclizes. Upon recyclization, the (*R*)- and (*S*)-enantiomers of usnic acid are equally likely to be produced (*Fig. 3.10*).

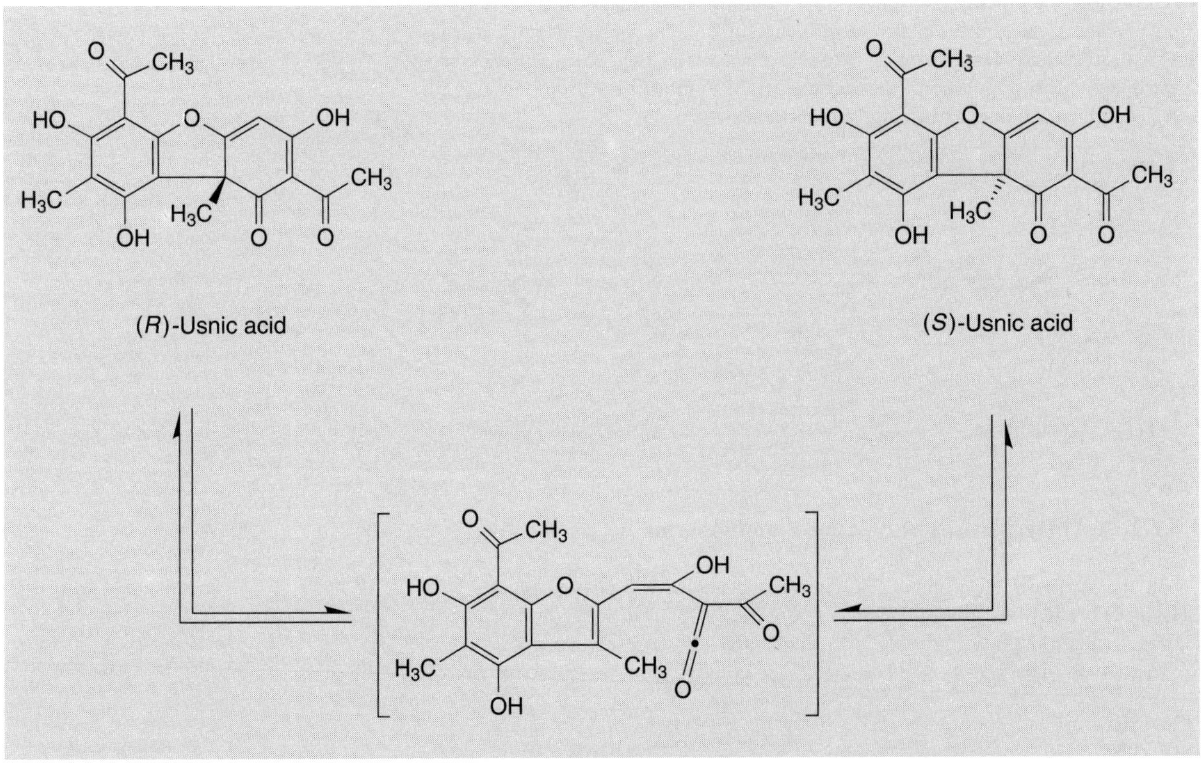

Fig. 3.10. *Racemization of usnic acid*

3.3.2.4. Geodin

If the antibiotic geodin is exposed to a weak acid or base, racemization occurs. As can be seen in *Fig. 3.11*, one of the bonds originating from the stereogenic center (here the spiro C-atom) is also, in this case, temporarily cleaved. The resulting ketene intermediate is achiral, and, upon recyclization, affords naturally occurring (+)-geodin and, with the same probability, the (−)-enantiomer.

These deliberately selected examples demonstrate that racemizations take place during the course of reversible structural alterations, in which the original optical activity is lost. This is very frequently the case when the coordination number at a stereogenic center is temporarily reduced by one.

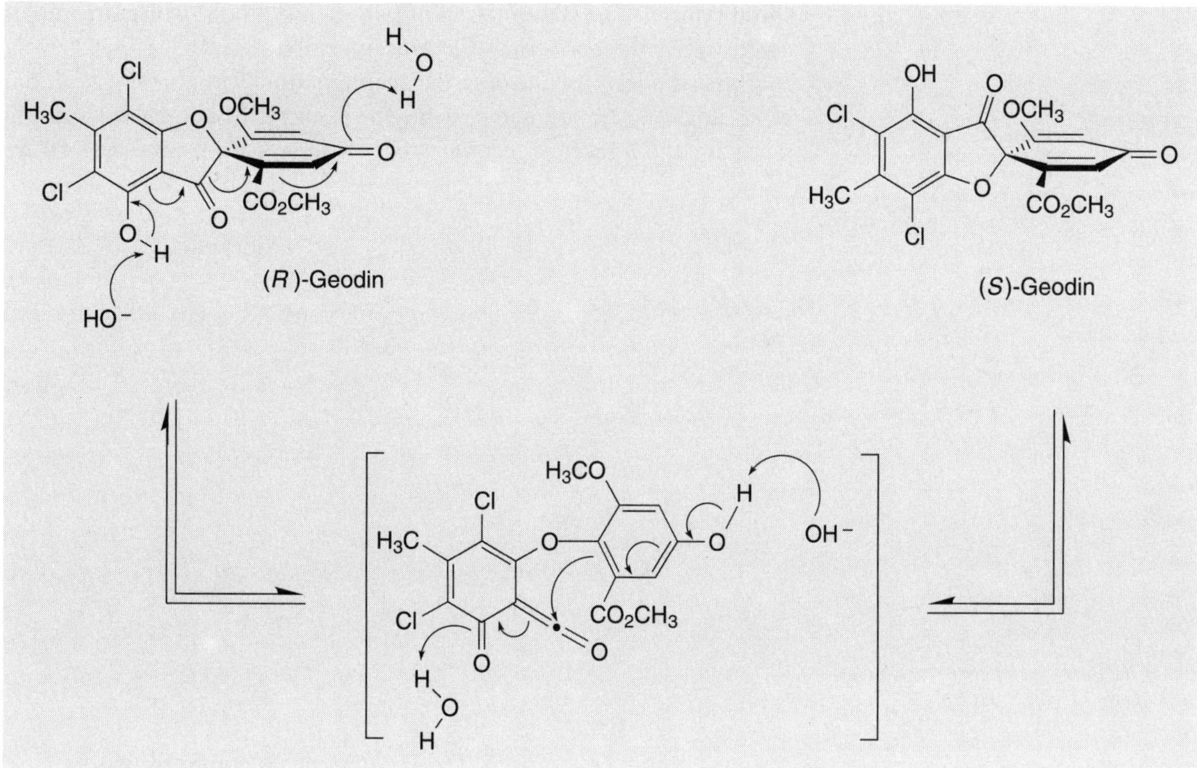

Fig. 3.11. *Racemization of geodin*

3.4. Separation of Enantiomers

3.4.1. Fundamentals

In previous sections (*3.3.2.1–3.3.2.4*), examples were shown of how an enantiomerically pure compound, under particular conditions, can be converted into a mixture of the two enantiomers (a racemic mixture, if amounts are equal). We had already encountered a racemic mixture in the equilibrated composition of *out/in* enantiomers in *Chapt. 2.6.2.2*, without going into the matter *expressis verbis*. Racemic mixtures are optically inactive, although consisting of chiral molecules (*Sect. 3.2.4*). If a racemic mixture is to hand, the question arises of whether, and if so, how, this mixture may be separated into its two enantiomers. At this point it should be noted that diastereoisomers, which, to a greater or lesser extent, differ from each other in their chemical and physical properties, are, in principle (if not always in practice), easy to separate. Enantiomers, in contrast, demand special methods for their separation. To approach the problem of such a separation of enantiomers (also known as resolution of a racemate), we shall again make use of a two-dimensional model (*cf. Sect. 3.1.2*; 'Flatland'). In this model (*Fig. 3.12*), a three-dimensional chiral compound is represented by a two-dimensional

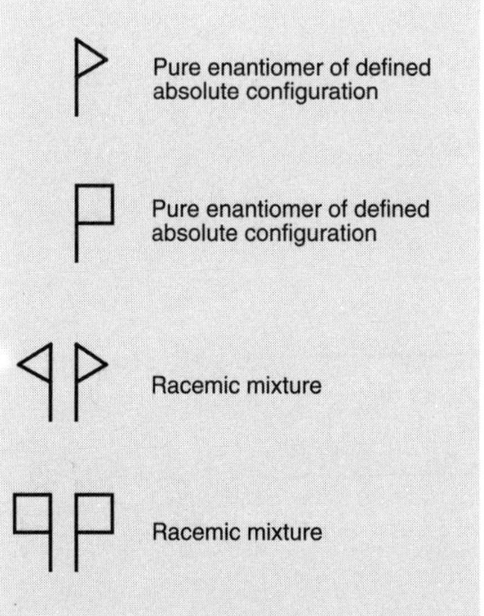

Fig. 3.12. *Two-dimensional symbols, after* Seebach, *for the representation of enantiomerically pure compounds and their racemic mixtures*

chiral symbol. The combination of two such symbols, which are mirror images of each other, stands for a racemic mixture.

If this model is used to describe an interaction (formation of a salt out of an acid and a base component, for example), between the starting racemate and a racemic *auxiliary reagent* (*Fig. 3.13*), a product mixture

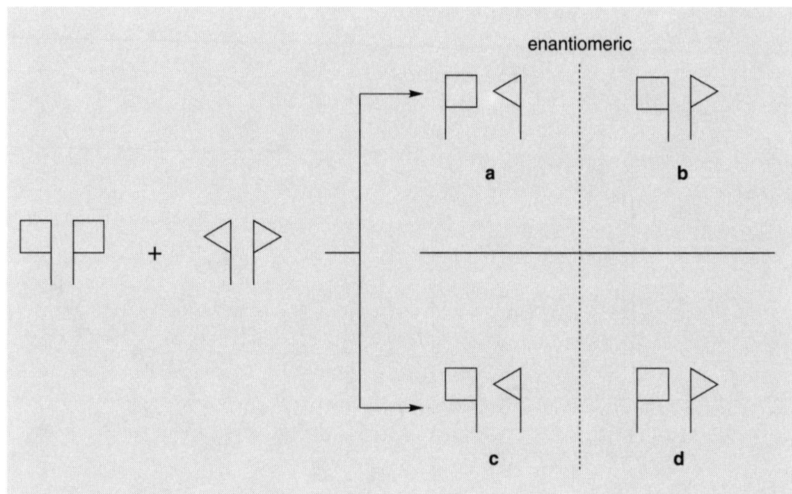

Fig. 3.13. *Two-dimensional representation of the interaction between the molecules of two racemic mixtures*

of four components results. These components comprise two pairs of enantiomers (**a** and **b**, and **c** and **d**). A separation of the racemic mixtures so obtained (**a** and **b** from **c** and **d**) would not constitute a resolution of the starting racemate: consequently, nothing would be gained.

If the number of combinations is reduced, and a racemic mixture is treated with an enantiomerically pure auxiliary reagent (*Fig. 3.14*), a mixture of two product components is obtained. These two product components are now not enantiomers, but diastereoisomers. If these

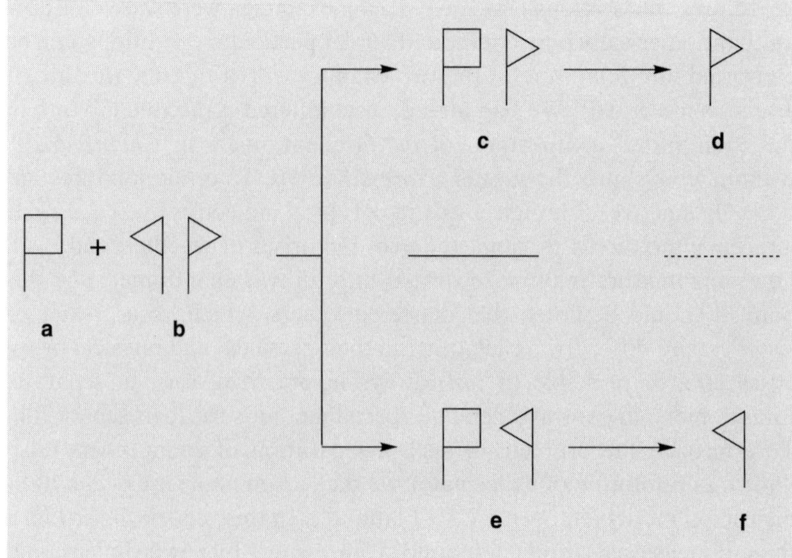

Fig. 3.14. *Two-dimensional representation of the interaction between the molecules of one enantiomerically pure compound and those of a racemic mixture*

diastereoisomeric components are separated, then the two enantiomers of the racemic starting material **b** are separated with them, in the form of the product components **c** and **e**. Subsequent splitting-off of the chiral auxiliary reagent **a** enables the pure enantiomers **d** and **f** to be isolated.

Before practical examples of enantiomer separation (racemate resolution) are presented, it should be stressed that statements about *relative* or *absolute* configurations must be strictly distinguished from each other. *Chapt. 10.2.6* refers to the unambiguous application of configurational formulae.

3.4.2. Examples

The examination of a two-dimensionally chiral model in the previous section has led us to an important finding: the separation of enantiomers always demands some form of interaction between the two components of a racemic mixture and a chiral, non-racemic *auxiliary component*. The following examples provide an insight into what is meant by 'interaction' in these cases.

3.4.2.1. Participation of a Chiral Auxiliary

The racemic mixture of the keto carboxylic acid *rac*-**A** (*Fig. 3.15*) can be converted into a mixture of two diastereoisomeric salts, by treatment with the chiral base (+)-(1*R*)-1-phenylethylamine. These salts may be separated from each other by fractional crystallization. The use of a stronger acid than **A** can liberate the pure enantiomers **A** and *ent*-**A** from the separated salts.

The resolution of racemic mixtures into their enantiomers *via* fractional crystallization of diastereoisomeric salts is a special case of *molecular recognition*. In one of the two salt diastereoisomers, the packing

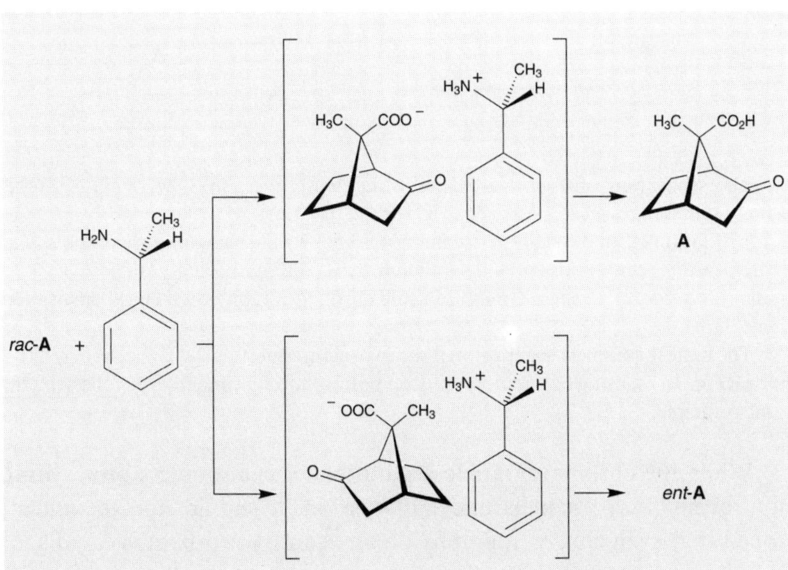

Fig. 3.15. *Resolution of the carboxylic acid mixture* rac-**A** *with the assistance of a single amine enantiomer* via *diastereoisomeric salts*

density, and, therefore, the total sum of attractive interactions, is greater – or to put it in another way, the potential energy is less – than in the other salt diastereoisomer. In the case of brucine, for example, one enantiomer of the chiral guest molecule fits much more tightly than the other one into the chiral hollow environment between the layers of the chiral host molecule.

If the fractional crystallization of diastereoisomeric compounds, with the final goal of isolation of the pure enantiomers, is under discussion, it should be pointed out that a single crystallization of an enantiomerically enriched mixture frequently yields a crystalline product which contains over 98% of one of the two enantiomers. In order to predict a result of this type, though, it would be absolutely essential to know the phase diagram of the binary enantiomer mixture (*Aspects of Organic Chemistry*, Volume 4: *Methods of Structure Determination*).

3.4.2.2. Participation of a Chiral Adsorbent

Another possible means of enantiomer separation is chromatography on a chiral, non-racemic adsorption medium. In accordance with the principle explained above (*Sect. 3.4.1*), the two enantiomers enter into (diastereomorphic) interactions of different magnitudes with the chiral stationary phase. This leads to different retention times on (for example) the column, so that the enantiomers may finally be trapped in separated fractions. The racemic mixture of *Tröger*'s base (*rac*-**B**; *Fig. 3.16*) was chromatographically separated in this way on D-lactose hydrate.

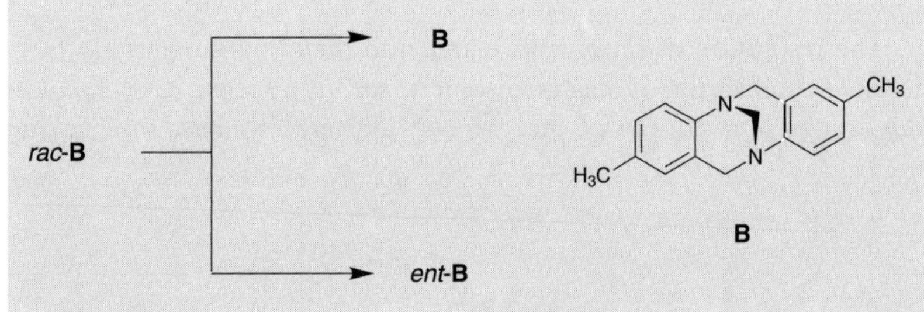

Fig. 3.16. *Partial enantiomer separation using chromatography:* Tröger*'s base (rac-*B*) on a chiral adsorbent*

The separation of *Tröger*'s base into its enantiomers in 1944 by *V. Prelog* is doubly significant: success was achieved for the first time in

- the isolation of an optically active amine (*Sect. 3.3.1*), which, because of molecular architecture, cannot invert, and so, therefore, is configurationally stable;
- the resolution of a racemic mixture using chromatography on a chiral adsorption medium.

The almost complete separation of the two enantiomers of the biphenyl derivative *rac*-**C** (*Fig. 3.17*), using chromatography on potato starch, was achieved following the same principle.

While the chromatographic separation of enantiomers on a chiral, non-racemic adsorbent is under discussion, it should not go unmentioned that even chromatography on an achiral adsorbent is capable of enriching one enantiomer. Since, in principle, molecular interactions

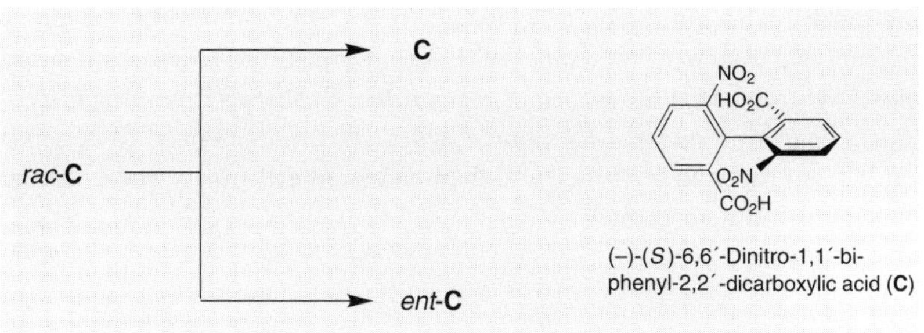

Fig. 3.17. *Enantiomer separation using chromatography: 6,6′-dinitro-1,1′-biphenyl-2,2′-dicarboxylic acid* (*rac*-**C**) *on a chiral adsorbent*

between E and E, or between *ent*-E and *ent*-E (*homochiral*), are energetically different in magnitude from interactions between E and *ent*-E (*heterochiral*), chromatographic discrimination between the two components of the non-racemic mixture is possible. This, once more, constitutes a special case of *molecular recognition*, in which an adsorbed molecule E enters into an energetically favored association with E (or with *ent*-E). As homochiral and heterochiral associations are the norm in concentrated solutions, it is essential to be permanently vigilant for any changes in enantiomeric excess (*ee*) of non-racemic enantiomeric mixtures which may occur during a reaction, its workup, or purification by chromatography, crystallization, sublimation, or extraction.

With such considerations in mind, it should not be surprising that the spectra of single enantiomers in highly concentrated solutions can differ from those ones of the corresponding racemic mixtures.

3.5. Conclusion

With their ideas about the constitution of organic compounds, *J.J. Loschmidt*, *F.A. Kekulé*, *A.C. Couper*, and *A.M. Butlerov* laid the ground for structural chemistry. Stimulated by the work of *L. Pasteur* and *J. Wislicenus*, *J.H. van't Hoff* and *J.A. Le Bel* extended the edifice of knowledge of structural chemistry (in the truest sense of the word) into a new dimension, and, by their considerations of the details of configuration, founded stereochemistry. In the modern era, *V. Prelog* has anchored stereochemical statics solidly in the mathematical discipline of group theory, thereby opening this seemingly worked-out area for new developments. Knowledge of chirality goes back to *L. Pasteur* and was cultivated especially by *V. Prelog* and *K. Mislow*.

During the course of naturalistic evolution, chiral, non-racemic compounds have come into existence (*how*?). The fact that enantiomers have also been subjected to natural selection has been beneficial, as demonstrated by the differentiated molecular functions of the selected enantiomers. In the same way that a non-handed (achiral) mitten allows a lesser dexterity to the right hand placed within it than does a right-handed glove, two molecules must also fit together chirally, if they are to interact with each other in the optimal manner. Molecules which fit together in this way are said to be complementary, or to act complemenmentarily. The ability of a molecule to recognize its complementary partner in a system of molecules of different types, and to associate more closely with it, is called *molecular recognition*. Problems involving molecular recognition occur, for example, when enantiomers need to be separated from racemic mixtures, and their enantiomeric purity and absolute configuration determined.

Further Reading

General

S. F. Mason, *Molecular optical activity and the chiral discriminations*, Cambridge University Press, Cambridge, 1992.

V. Prelog, *Nobel Lecture: Chirality in Chemistry, Science* **1976**, *193*, 17.

To Sect. 3.1.

3.1.2.

V. Prelog, *My 132 Semesters of Chemistry Studies*, American Chemical Society, Washington, DC, 1991;
– *Problems in Chemical Topology, Chem. Br.* **1968**, *4*, 382;
– *Das Asymmetrische Atom, Chiralität und Pseudoasymmetrie, Konikl. Nederl. Akad. Wetenschap.* **1968**, *71B*, 108.

R. S. Cahn, C. Ingold, V. Prelog, *Specification of Molecular Chirality, Angew. Chem. Int. Ed.* **1966,** *5*, 385.

V. Prelog, G. Helmchen, *Basic Principles of the CIP-System and Proposals for a Revision, Angew. Chem. Int. Ed.* **1982,** *21*, 567;
– *Pseudoasymmetrie in der organischen Chemie, Helv. Chim. Acta* **1972,** *55*, 2581.

D. Seebach, V. Prelog, *The Unambiguous Specification of the Steric Course of Asymmetric Syntheses, Angew. Chem. Int. Ed.* **1982,** *21*, 654.

E. A. Abott, *Flatland*, Dover, 6th Edn., New York, 1968.

To Sect. 3.3.

3.3.1.

A. Werner, *Lehrbuch der Stereochemie*, G. Fischer, Jena, 1904.

M. S. Gordon, M. W. Schmidt, *Does Methane Invert through Square Planar? J. Am. Chem. Soc.* **1993**, *115*, 7486.

D. Felix, A. Eschenmoser, *Slow inversion at pyramidal nitrogen. Isolation of diastereomeric 7-chloro-7-azabicyclo[4.1.0]heptanes at room temperature, Angew. Chem. Int. Ed.* **1968,** *7*, 224.

3.3.2.

E. E. Turner, *Configuration and Steric Effects in Conjugated Systems* in *Steric Effects in Conjugated Systems* (Ed. G. W. Gray), p. 1, Butterworths Scientific Publ., London, 1958.

G. Stork, *The Racemization of Usnic Acid, Chem. Ind. (London)* **1955,** 915.

D. H. R. Barton, A. I. Scott, *The Constitution of Geodin and Erdin, J. Chem. Soc.* **1958,** 1767.

D. H. R. Barton, G. Quinkert, *Photochemical Cleavage of Cyclohexadienones, J. Chem. Soc.* **1960,** 1.

To Sect. 3.4.

3.4.1.

J. Jacques, A. Collet, S. W. Wilen, *Enantiomers, Racemates, Resolutions,* John Wiley & Sons, New York, 1981.

D. Seebach, E. Hungerbühler, *Synthesis of Enantiomerically Pure Compounds (EPC-Synthesis)* in *Modern Synthetic Methods,* Vol. 2 (Ed. R. Scheffold), Salle + Sauerländer, Frankfurt am Main, 1980.

3.4.2.

G. Quinkert, U. Schwartz, H. Stark, W.-D. Weber, F. Adam, H. Baier, G. Frank, G. Dürner, *Asymmetrische Totalsynthese von 19-Nor-Steroiden mit photochemischer Schlüsselreaktion: Enantiomerenreine Zielverbindungen, Liebigs Ann. Chem.* **1982,** 1999.

R. O. Gould, M. D. Walkinshaw, *Molecular Recognition in Model Crystal Complexes: The Resolution of D and L Amino Acids, J. Am. Chem. Soc.* **1984**, *106*, 7840.

V. Prelog, P. Wieland, *Über die Spaltung der Tröger'schen Base in optische Antipoden, ein Beitrag zur Stereochemie des dreiwertigen Stickstoffs, Helv. Chim. Acta* **1944**, *27*, 1127.

K. R. Lindner, A. Mannschreck, *Separation of enantiomers by high performance liquid chromatography on triacetylcellulose, J. Chromatogr.* **1980**, *193*, 308.

Y. Okamoto, I. Okamoto, H. Yuki, *Chromatographic Resolution of Enantiomers Having Aromatic Group by Optically Active Poly(triphenylmethyl methacrylate), Chem. Lett.* **1981**, 835.

H. Hess, G. Burger, H. Musso, *Complete separation of enantiomers by chromatography on potato starch, Angew. Chem. Int. Ed.* **1978,** *17,* 612.

W.-L. Tsai, K. Hermann, B. Rohde, A. Dreiding, *Enantiomer-Differentiation Induced by an Enantiomeric Excess during Chromatography with Achiral Phases, Helv. Chim. Acta* **1985**, *68*, 2238.

T. Williams, R. G. Pitcher, P. Bommer, J. Gutzwiller, M. Uskoković, *Diastereomeric Solute-Solute Interactions of Enantiomers in Achiral Solvents. Nonequivalence of the Nuclear Magnetic Resonance Spectra of Racemic and Optically Active Dihydroquinine, J. Am. Chem. Soc.* **1969**, *91*, 1871.

R. Noyori, M. Kitamura, *Enantioselective Addition of Organometallic Reagents to Carbonyl Compounds: Chirality Transfer, Multiplication, and Amplification, Angew. Chem. Int. Ed.* **1991,** *30,* 49.

To Sect. 3.5.

J. Wislicenus, *Über die räumliche Anordnung der Atome in organischen Molekülen, Abhandl. sächs. Ges. Wiss.* **1887**, *14*, 1.

4 Topicity

4.1. Considering Atoms or Atom Groups within a Molecule

4.1.1. Basic Principles and Definitions

Our previous examination (*Chapt. 2* and *3*) of constitution and stereostructure of organic molecules led to a technique for describing whole molecules and comparing them with one another. So as to determine whether stereoisomers must be taken into consideration for a particular constitution, we have compared stereoformulae with one another. When these were found to be non-equivalent, not depicting the very same molecule, then the presence of either enantiomers or diastereoisomers could be deduced. While trying to grasp mentally the structure of a molecule (or a model of it), we more or less deliberately moved constitutionally equivalent components of a molecule around relative to each other, in order to detect symmetry properties. In a manner wholly analogous to this intercomparison of two molecules (or, better, models of them) in order to determine whether we are dealing with equivalent or non-equivalent (enantiomorphic or diastereomorphic) objects, it is easy to see that constitutionally equivalent ligands (atoms or atom groups) exist in one of three possible relationships to each other: they may be *homotopic*, *enantiotopic*, or *diastereotopic*. They are homotopic if partial substitution (either real or 'on paper') of one or the other ligand leads to equivalent products. For example, the two H-atoms in the CH_2 group (or the two CO_2H groups) of malonic acid are *homotopic* (*Fig. 4.1*).

Two atoms or atom groups of a molecule are called enantiotopic, if partial substitution of one or the other ligand leads to two enantiomeric products. For example, the two mono-deuterated substitution products of bromochloromethane are enantiomers, and so the two H-atoms of the starting compound are *enantiotopic* (*Fig. 4.2*).

Two atoms or atom groups of a molecule are called diastereotopic, if partial substitution of one or the other ligand leads to two diastereoisomeric products. For example, the two stereoformulae of the mono-deuterated compound with the constitution of 1,2,3,4-tetrahydro-*N*,1,4-triphenylnaphthalene-2,3-dicarboximide represent diastereoisomers and accordingly the two H-atoms of the starting compound are *diastereotopic* (*Fig. 4.3*)

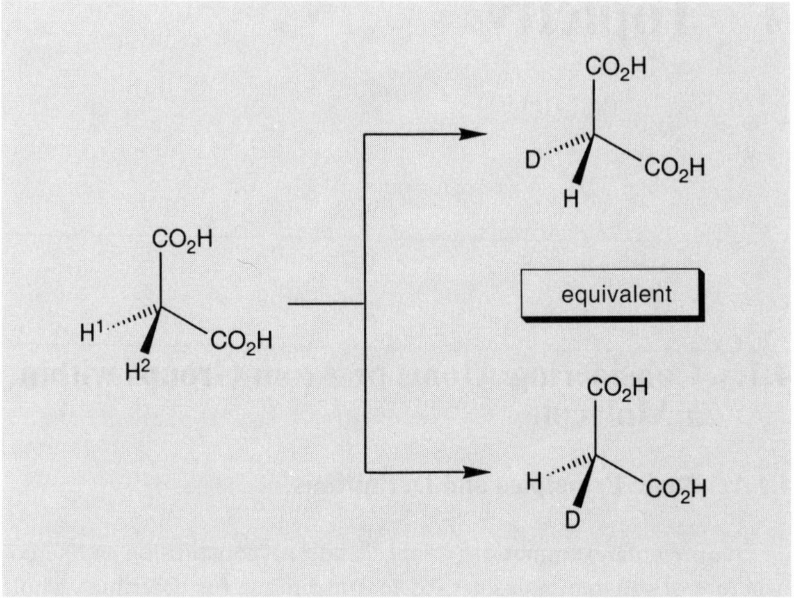

Fig. 4.1. *Substitution of either H^1 or H^2 with D, leading to* equivalent *substitution products.* The H-atoms of the CH_2 group are *homotopic.*

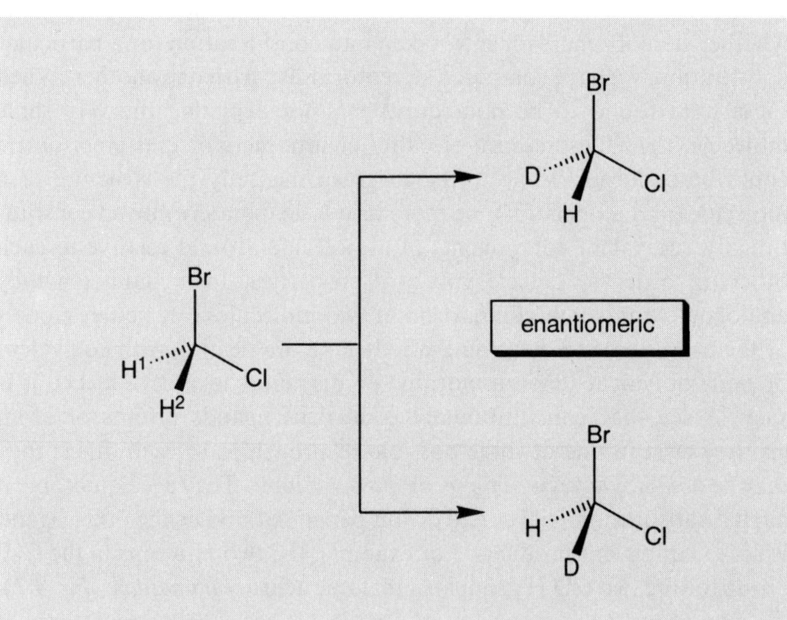

Fig. 4.2. *Substitution of either H^1 or H^2 with D, leading to* enantiomeric *substitution products.* The H-atoms of the CH_2 group are *enantiotopic.*

As we shall see later (*Sect. 4.1.3*), the last example has been taken from a set of a total of ten possible stereoisomers. For the moment, it is sufficient merely to use another two stereoformulae among the whole set, in order to illustrate the different forms of topical relationships between atoms and atom groups of one and the same molecule.

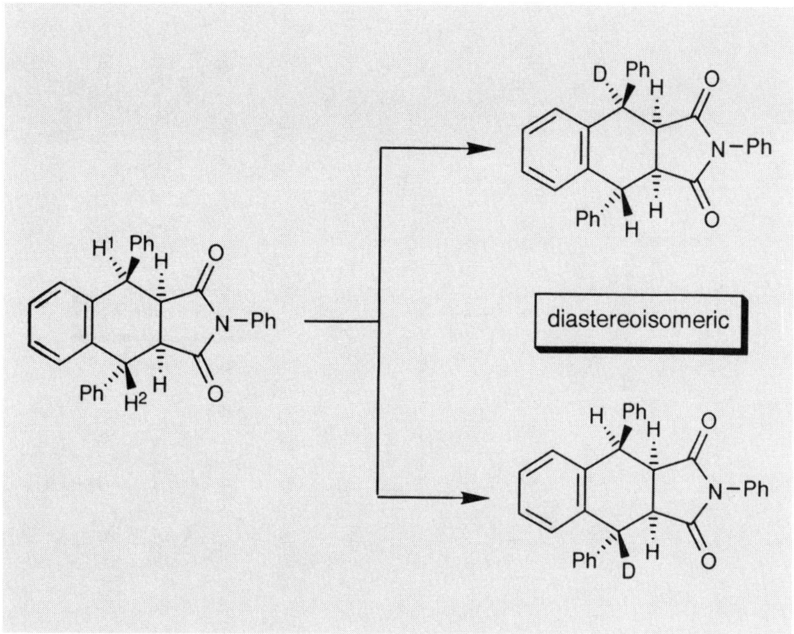

Fig. 4.3. *Substitution of either H^1 or H^2 with D, leading to* diastereoisomeric *substitution products.* The H-atoms indicated (the same applies to the Ph groups, and also to the bridgehead H-atoms) are *diastereotopic.*

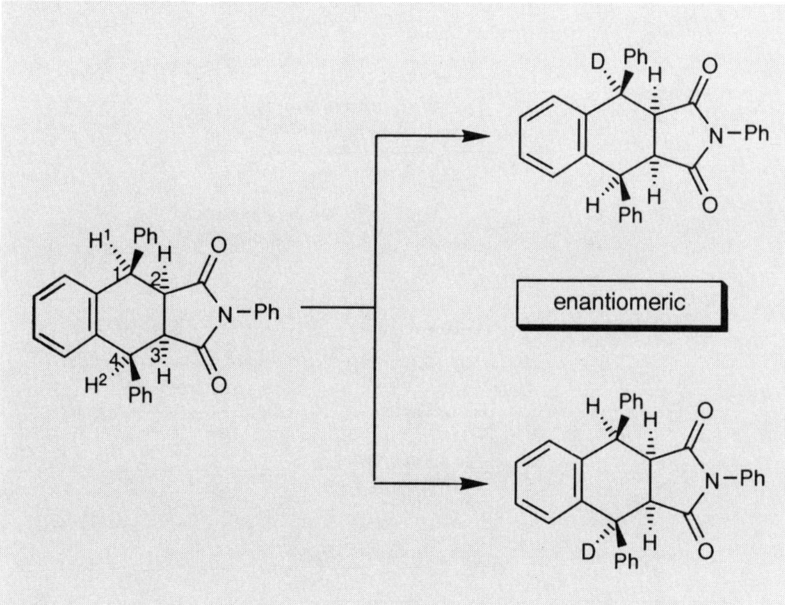

Fig. 4.4. *Substitution of either H^1 or H^2 with D, leading to* enantiomeric *substitution products.* The H-atoms indicated are *enantiotopic.*

In the above example (*Fig. 4.4*), the stereoformula on the left depicts an achiral *meso*-compound with enantiotopic H-atoms (or Ph groups) in the cyclohexene ring and with enantiotopic H-atoms in the bridgehead positions. In the following example (*Fig. 4.5*), the stereoformula on the left depicts a chiral compound of symmetry point group C_2. The numbered H-atoms (and the Ph groups) in the cyclohexene ring and the H-atoms in the bridgehead positions are *homotopic.*

Fig. 4.6 shows how the relative topical relationship of two ligands in the same molecule may be unambiguously determined.

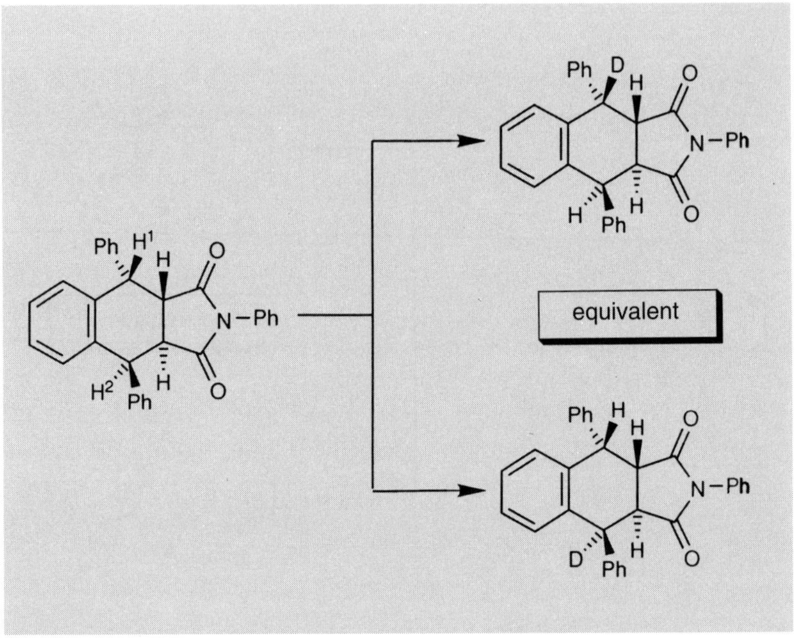

equivalent

Fig. 4.5. *Substitution of either H^1 or H^2 with D, leading to* equivalent *substitution products*. The H-atoms indicated are *homotopic*.

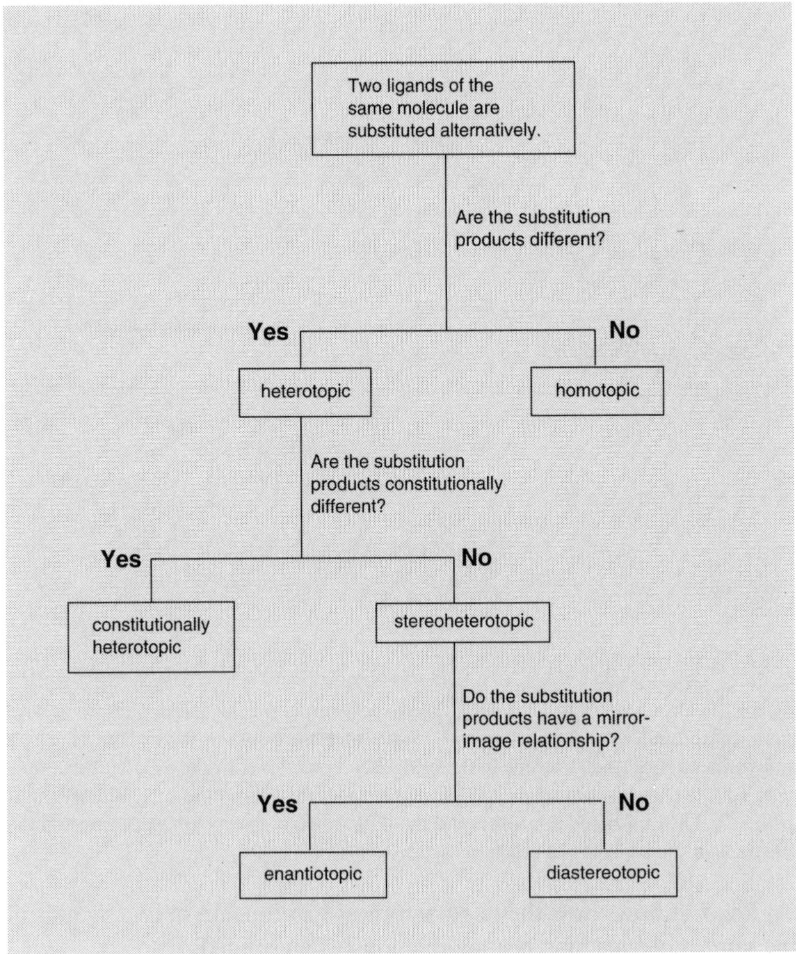

Fig. 4.6. *Differentiation of topical relationships between atoms and atom groups of an individual molecule*

4.1.2. NMR-Spectroscopic Symmetry Analysis

It will not have escaped the notice of the attentive reader that the division of constitutionally equivalent atoms of a molecule into *homotopic*, *enantiotopic*, or *diastereotopic* categories permits conclusions to be drawn about the symmetry of the compound concerned. As can easily be seen on inspection of the undeuterated compounds shown in *Figs. 4.1–4.5*, the relationship is as follows:

- *homotopic* atoms are transposed onto one another by rotation about a C_n axis ($1 < n < \infty$);
- *enantiotopic* atoms are transposed onto one another by rotary reflection across an S_n axis ($S_1 \equiv \sigma$, $S_2 \equiv i$);
- *diastereotopic* atoms cannot be transposed onto one another by any symmetry operation (except the trivial identity operation).

It would undoubtedly be very desirable to have an experimental method at our disposal, with the help of which it would be possible to distinguish between *homotopic*, *enantiotopic*, and *diastereotopic* atoms, and thereby between molecules which are asymmetrical or symmetrical, and, in the latter case, chiral or achiral.

NMR Spectroscopy is capable of doing this (see *Chapt. 13.2.4*). Two constitutionally equivalent atoms of a given molecule either absorb at the same frequency, or they do not. In the former case, the two atoms are designated as being *isochronous*; in the latter case, they are designated as being *anisochronous*. Now:

- two *homotopic* atoms or atom groups are *isochronous* under all conditions;
- two *diastereotopic* atoms or atom groups can only be *isochronous* just by chance;
- two *enantiotopic* atoms or atom groups are *isochronous* under all conditions when in an achiral environment. However, when in a chiral environment, they can only just by chance be *isochronous*.

4.1.3. Examples from Practice

Table 4.1 contains the complete set of possible configurations subsumed under the constitution of 1,2,3,4-tetrahydro-*N*,1,4-triphenylnaphthalene-2,3-dicarboximide, including the three configurations which have already been mentioned. As the middle ring has four stereogenic centers and belongs to a constitutional type of the abba substitution pattern, the formal maximum number of eight relative (16 absolute) configurations is reduced to six:

$$\begin{array}{cc} \textit{syn/cis/syn} & \textit{syn/trans/syn} \\ \left\{\begin{array}{c}\textit{syn/cis/anti}\\ \textit{anti/cis/syn}\end{array}\right\} & \left\{\begin{array}{c}\textit{syn/trans/anti}\\ \textit{anti/trans/syn}\end{array}\right\} \\ \textit{anti/cis/anti} & \textit{anti/trans/anti} \end{array}$$

The relative configurations given in brackets are equivalent, so one of them must in each case be discounted. Furthermore, as the *syn/cis/syn*- and the *anti/cis/anti*-configurations both denote *meso*-compounds, consideration of absolute configurations merely requires that the remaining four configurations be doubled in number. In total then, there are only ten different configurations, or to put it in another way, two *meso*-compounds and four pairs of enantiomers.

The *syn* and *anti* descriptors here give the relative configurations of a to b and/or of b to a. They are analogous to the *cis* and *trans* descriptors; the latter pair denotes the relative configuration of b to b.

Table 4.1. *Determination of the Configuration of Each of the Stereoisomers with the Constitution of 1,2,3,4-Tetrahydro-*N*,1,4-triphenylnaphthalene-2,3-dicarboximide*

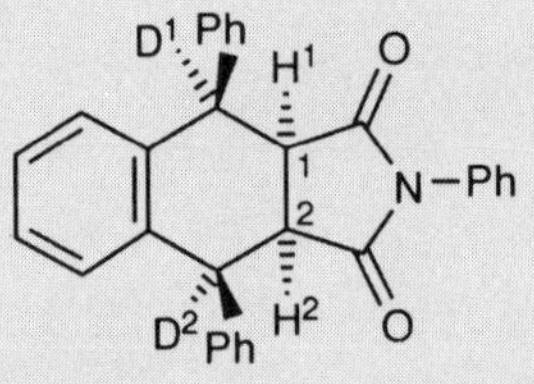

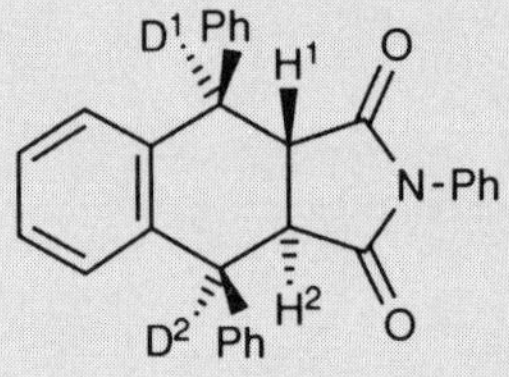

	1	2	*rac*-3	*rac*-4	*rac*-5	*rac*-6
Symmetry point group	C_s	C_s	C_1	C_1	C_2	C_2
Chirality	achiral	achiral	chiral	chiral	chiral	chiral
Relative configuration	*syn/cis/syn*	*anti/cis/anti*	*syn/cis/anti*	*anti/trans/syn*	*anti/trans/anti*	*syn/trans/syn*
Topical relationship (H^1 and H^2)	enantiotopic	enantiotopic	diastereotopic	diastereotopic	homotopic	homotopic
NMR Signals of H^1 and H^2 – in achiral solvents	isochronous	isochronous	anisochronous	anisochronous	isochronous	isochronous
– in chiral solvents	anisochronous	anisochronous	anisochronous	anisochronous	isochronous	isochronous

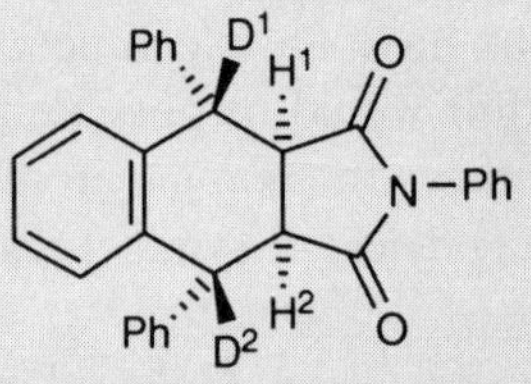

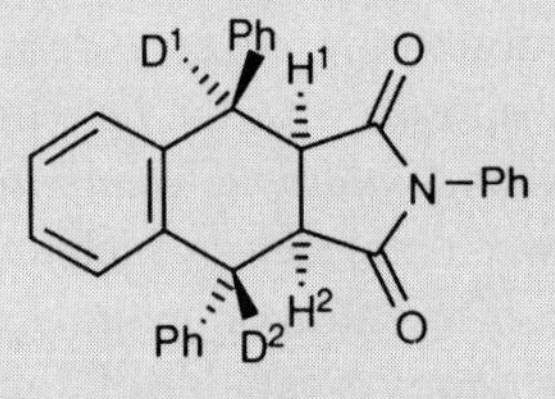

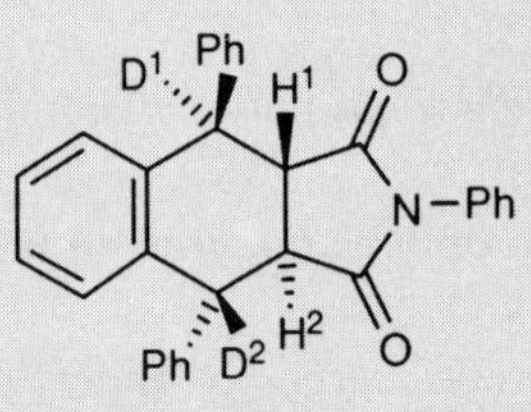

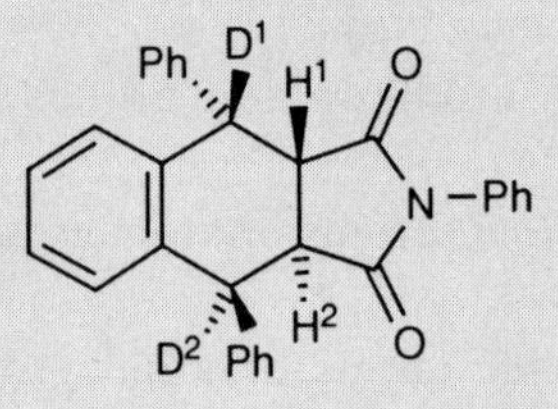

Disregarding the benzene ring protons (which absorb in the low-field region, away from our region of interest), there are two types of protons which warrant our attention: the two bridgehead protons and the two protons on the center ring. Protons in constitutionally different environments should absorb at different frequencies in a ^{1}H-NMR spectrum, and it, therefore, ought to be easy to distinguish between these two types. Since a ^{1}H-NMR spectrum does not only provide information, in the form of individual signals, about protons in different chemical environments, but also, through the fine structure, about interactions (coupling through chemical bonds) between protons of one type and other, non-equivalent protons in the near vicinity, it is possible that an undesirable complex partial spectrum may be obtained for the bridgehead protons. The spectrum may be appreciably simplified, permitting a clear statement about the symmetry of the particular structure, if the protons in the center ring may in some way be easily replaced with deuterons. Then there are only two possible outcomes:

- either the two bridgehead protons absorb at the same resonance frequency, in which case they are *isochronous*. If this is not purely accidental, they must be either *homotopic* or *enantiotopic*, and, therefore, either C_n symmetry ($1 < n < \infty$) or S_n symmetry must be present, resulting in an A_2 *singlet* in the ^{1}H-NMR spectrum;
- or the two bridgehead protons absorb at different resonance frequencies, in which case they are *anisochronous* and therefore *diastereotopic*, implying that the structure of the compound must be asymmetrical, resulting in an *AB double-doublet* in the ^{1}H-NMR spectrum.

Table 4.1 lists the relative configurations of the various isomers, gives their symmetry and chirality character, and indicates the topicity and the isochronicity of the bridgehead protons. It can be seen that, using NMR spectroscopy, it is possible to differentiate between:

- in achiral solvents, isomers with C_s or C_2 symmetry (with A_2 *singlets* for H^1 (H–C(2)) and H^2 (H–C(3))) from those with C_1 symmetry (with *AB double-doublets* for H^1 and H^2),
- in chiral solvents, isomers with C_2 symmetry (with A_2 *singlets*) from those with C_s or C_1 symmetry (with *AB double-doublets*).

Instead of using chiral solvents (or chiral shift reagents), use can also be made of a different experimental criterion: the resolution of racemic mixtures into their corresponding enantiomers (*Chapt. 3.4*). The mixture *rac*-**3** has indeed been resolved into its enantiomers in this way. *rac*-**4** also displays an *AB double-doublet* in its ^{1}H-NMR spectrum in achiral solvents, like *rac*-**3**. That the resolved racemate is *rac*-**3**, and not *rac*-**4**, can be confirmed by the base-catalyzed isomerization of **1** into **2** *via rac*-**4**. With this additional information, it is possible to unambiguously assign to *rac*-**3** and *rac*-**4** their respective relative configurations *syn/cis/anti* and *anti/trans/syn*. Further information is necessary in order to similarly assign the correct configurations to the two *meso*-compounds **1** and **2**. The same also applies for deciding which of the two still unassigned relative configurations (**5** and **6**) should be given to each of the remaining C_2-symmetrical isomers.

In the case described above, a combination of the fine structure of relevant proton signals in the NMR spectrum and the resolution of a racemic mixture into its enantiomers proved sufficient for our purposes (at least for assigning of individual isomers to one of the three symmetrically differentiated isomeric pairs). In the following example, concerning the two stereoisomers with the constitution of 1,3-diphenylindan-2-one (*Table 4.2*), the fine structure of clearly assigned proton signals is used in conjunction with observations of the number of product components obtained on reduction, achieving an unambiguous (due to symmetry principles) configurational assignment. The ketone which, upon reduction (with $LiAlH_4$), affords two diastereoisomeric alcohols, can only be assigned the stereoformula **7**; **7**, **9a**, and **10a** belonging to the class of C_s-symmetrical compounds. Consequently, C_2 symmetry and configurational formula **8** must be ascribed to the remaining ketone. Its single reduction product (*rac*-**11a**) must consist of a racemic mixture with C_1 symmetry in both enantiomers. As *Table 4.2* shows, this assignment is supported by the fine structure of the signals for H—C(1) and H—C(3) in the ketones and for H—C(1), H—C(2), and H—C(3) in the acetates of the reduction products.

It might be asserted that the correlation of the number of components in the alcoholic reduction products with the symmetry of the starting ketone is, in itself, so simple and certain, that the use of NMR spectroscopy here would be redundant. This objection is perfectly correct (disallowing the fact that NMR-spectroscopic characterization should be carried out in any case). However, in some cases, the trivial obstruction can occur, that the total number of expected components in the reaction products is not in fact observed experimentally. That this is not as far-fetched as it might seem is demonstrated in the next example, concerning the two stereoisomers with the constitution of 2,4-dimethyl-2,4-diphenylcyclobutane-1,3-dione (*Table 4.3*).

There are two dimers of methyl phenyl ketene, whose melting points differ by almost 100°C. One is represented by stereoformula **12**, the other by **15**, but which configuration belongs to which ketone? As **12** is of symmetry point group C_{2h} and **15** of symmetry group C_{2v}, the signals of the isochronous Me protons in the corresponding ^{1}H-NMR spectra appear as a *singlet* in either case, and so do not permit any unambiguous assignment. In principle, an unambiguous configurational determination is possible by an analysis of suitable follow-up products, be it by verification of the maximum number of isomers obtained and/or by NMR-spectroscopic differentiation between protons with constitutionally equivalent environments, but with different topicity. Conversion of stereoisomers **12** and **15** of 2,4-dimethyl-2,4-diphenylcyclobutane-1,3-dione into the corresponding diols (using excess $LiAlH_4$) might possibly solve the problem by itself: from the C_{2h}-symmetrical dimer **12**, there could only be a maximum of two diols (**13** with C_s, **14** with C_i symmetry) produced, and from the C_{2v}-symmetrical dimer **15**, a maximum of three

Table 4.2. *Determination of the Configuration of Each of the Stereoisomers with the Constitution of 1,3-Diphenylindan-2-one Using Their Reduction Products*

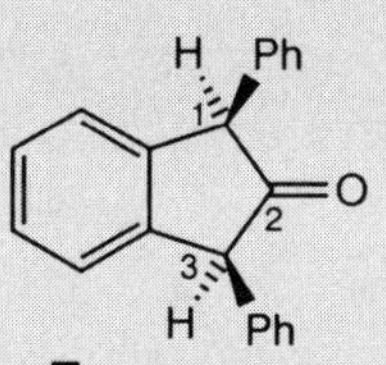

a X = H; **b** X = $COCH_3$

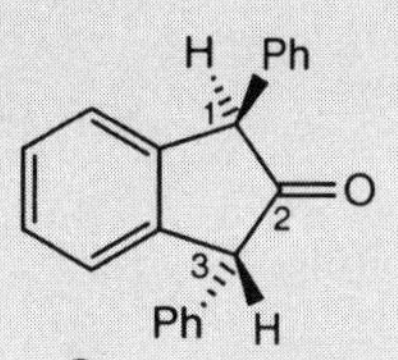

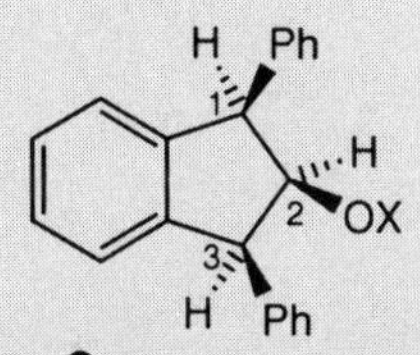

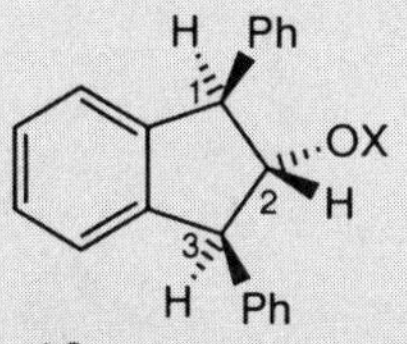

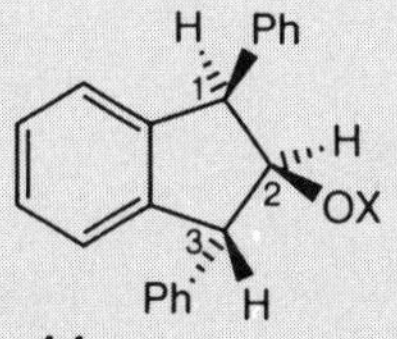

7 → 9, 10

rac-**8** → *rac*-**11**

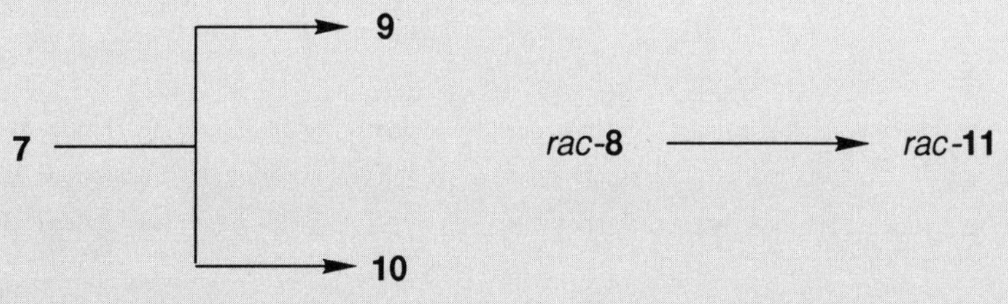

	7	**8**	**9**	**10**	**11**
Symmetry point group	C_s	C_2	C_s	C_s	C_1
Chirality	achiral	chiral	achiral	achiral	chiral
Topical relationship (H–C(1) and H–C(3))	enantiotopic	homotopic	enantiotopic	enantiotopic	diastereotopic
Fine structure of the signals of H–C(1) and H–C(3)	*s*	*s*	*d*	*d*	2 *d*
Fine structure of the signals of H–C(2) of the acetates	–	–	*t*	*t*	*dd*

diols (**16** and **17**, both with C_{2v}, and **18** with C_s symmetry). However, as both the higher-melting and the lower-melting diketones afforded only two diol components, the premise for certain configurational assignment deduced exclusively from the number of resulting diols was invalidated.

Table 4.3. *Determination of the Configuration of Each of the Stereoisomers with the Constitution of 2,4-Dimethyl-2,4-diphenylcyclobutane-1,3-dione Using Their Reduction Products*

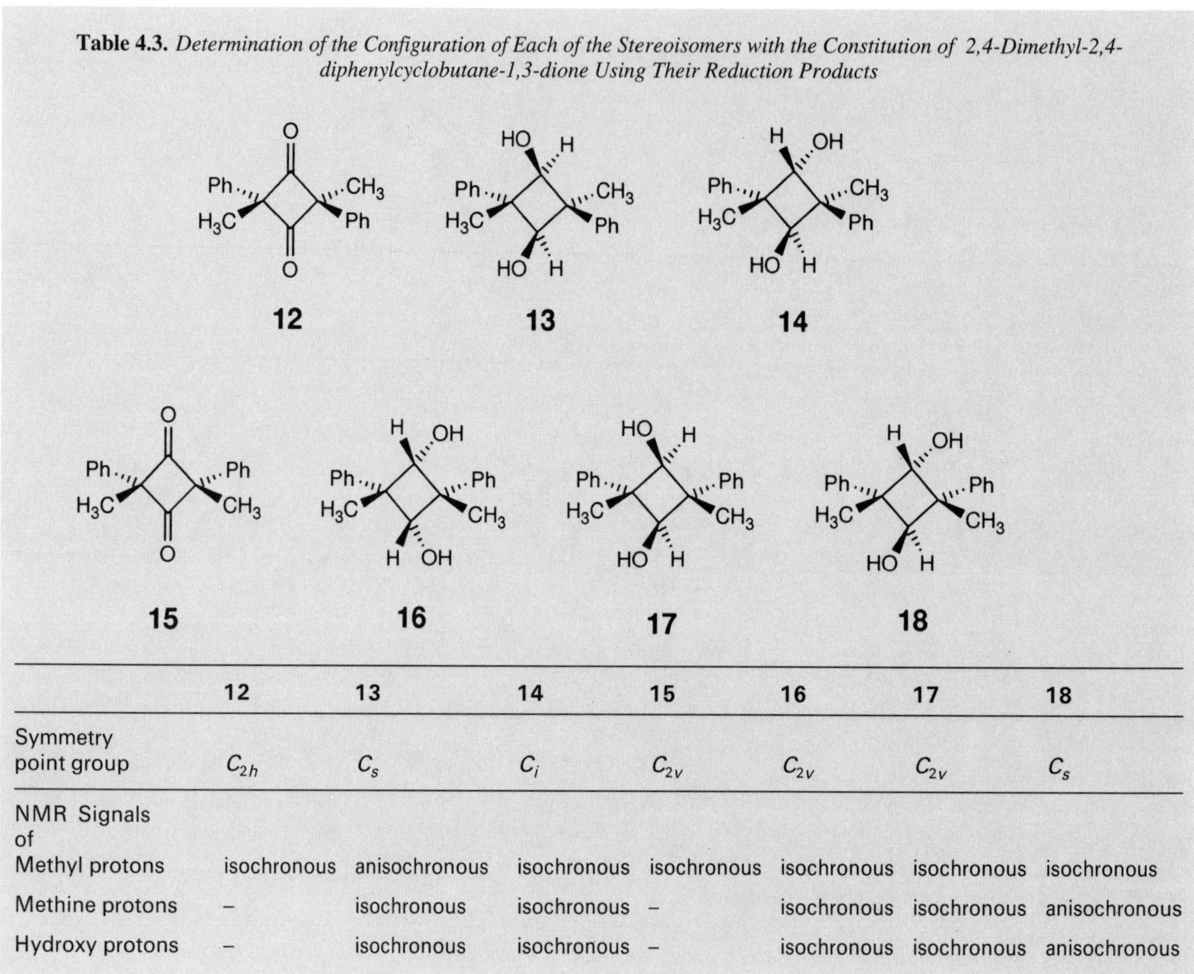

	12	13	14	15	16	17	18
Symmetry point group	C_{2h}	C_s	C_i	C_{2v}	C_{2v}	C_{2v}	C_s
NMR Signals of							
Methyl protons	isochronous	anisochronous	isochronous	isochronous	isochronous	isochronous	isochronous
Methine protons	–	isochronous	isochronous	–	isochronous	isochronous	anisochronous
Hydroxy protons	–	isochronous	isochronous	–	isochronous	isochronous	anisochronous

A second approach to the configurational determination of **12** and **15** is based on the symmetry point groups to which the derived diols belong. In both diol pairs (*i.e.*, **13** and **14**, and the pair which may be **16** and **17**, or **16** and **18**, or **17** and **18**), there exists the possibility of obtaining one diol (**13** derived from **12**, and **18** from **15**) each with C_s symmetry. As the plane of symmetry in **13** is differently oriented in space from that in **18**, these diols would be unambiguously distinguishable both from each other and from the remaining (**14**, **16**, and **17**) diols. Of the total of five diols (**13**, **14**, **16**, **17**, and **18**), only **13** exhibits anisochronous Me protons and only **18** anisochronous OH and CH protons.

As one of the two diols (namely **13**) derived from the higher-melting dione exhibits two different *singlets* for Me protons, and one of the two diols (namely **18**) derived from the lower-melting dione shows two separated signals both for OH and for CH protons, it must follow that the cyclobutane-1,3-dione derivative of m.p. 162.5 °C can only possess C_{2h} symmetry, and that one of m.p. 69.5–70 °C can only possess C_{2v} symmetry. Therefore, these two diketones are correctly represented by the stereoformulae **12** and **15**, respectively.

This method of structural assignment is also of epistemological interest, beyond its practical utility: it cannot be untrue, because it is founded upon consideration of symmetry principles.

The *Prelog* catalog of regular tetrahedra with any possible combination of ligands (*Chapt. 3.2*) acquainted us with models in which a C-atom of coordination number 4 acts as a prochirality center. Bromochloromethane is a concrete example of an achiral compound with a prochirality center. Its two H-atoms are enantiotopic and, in an achiral solvent system, the protons are isochronous: a *singlet* is observed in the ^{1}H-NMR spectrum. Viewed from an NMR-spectroscopic perspective, achiral compounds with a prochirality center and diastereotopic CH_2 protons are more interesting. One such example is citric acid. It was mentioned earlier (*Chapt. 3.2.2*) that it possesses a prochirality center in C(3) (with two enantiotopic HOOC–CH_2 ligands). Regarding the protons of one of these CH_2 groups more closely, it can be ascertained that they are diastereotopic. An *AB double-doublet* would be expected in the ^{1}H-NMR spectrum, and this is indeed what is observed. An analogous situation is found in the case of 1,1-dibenzylphthalan (=1,1-dibenzyl-1,3-dihydroisobenzofuran). In acetaldehyde diethyl acetal, the CH_2 protons of the two enantiotopic EtO groups are diastereotopic and an *ABX_3 multiplet* is observed in the ^{1}H-NMR spectrum.

Citric acid
(after H/D exchange)

1,1-Dibenzylphthalan

Acetaldehyde diethyl acetal

4.2. Considering the Two Faces of a Molecule with Centers of Coordination Number 3

Molecules with double bonds have an upper and a lower face. As '*upper*' and '*lower*' are arbitrary terms, it is necessary to have unambiguous assignments, in all cases where the two faces are not equivalent. These assignments are derived by application of the *sequence rules* to the relevant reference atom of coordination number 3 (the C-atom of a C=O group, the reference atoms C(1) and C(2) of a C=C group). Details of the specification of two-dimensional chiral molecules, molecule fragments, or their models (*Chapt. 3.1.2*) can be found in *Chapt. 10.2.7*.

The two faces of formaldehyde cannot be distinguished from each other: they are equivalent (homotopic). The two faces of acetaldehyde are mirror images of each other: they are enantiotopic.

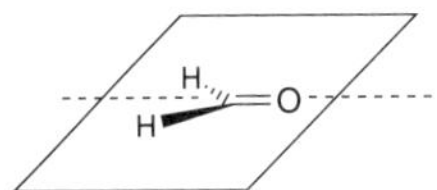

Of the ketones **7** and **8** of *Table 4.2*, the *meso*-compound **7** of *u*-configuration, exhibits two non-equivalent faces that do not show a mirror-image relationship to each other. On the *Re*-side (*Si*-side), the sequence proceeds clockwise (anticlockwise) from O, *via* the (*R*)-configurated C-atom, to the (*S*)-configurated C-atom. The enantiomers **8** and *ent*-**8** both exhibit an *l*-configuration; their two faces are equivalent.

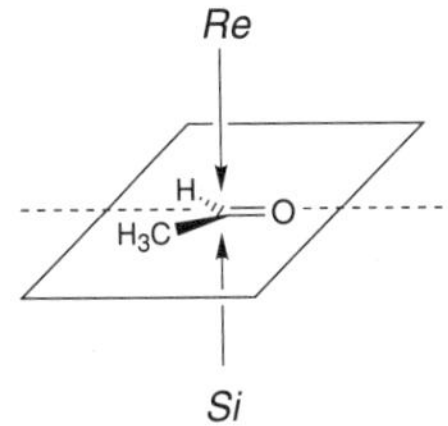

Of the two stereoisomers **12** and **15** with 2,4-dimethyl-2,4-diphenylcyclobutane-1,3-dione constitution, **12**, of symmetry point group C_{2h} and *l*-configuration ((2*R**,4*R**)), presents two equivalent faces; a situation which neither requires nor allows a specification. (*R** and *S** are descriptors for relative configuration; for further details see *Chapt. 10.2.1.2.*) Stereoisomer **15**, of symmetry point group C_{2v} and *u*-configuration ((2*R**,4*S**) or, equally, (2*S**,4*R**)), has two non-equivalent faces

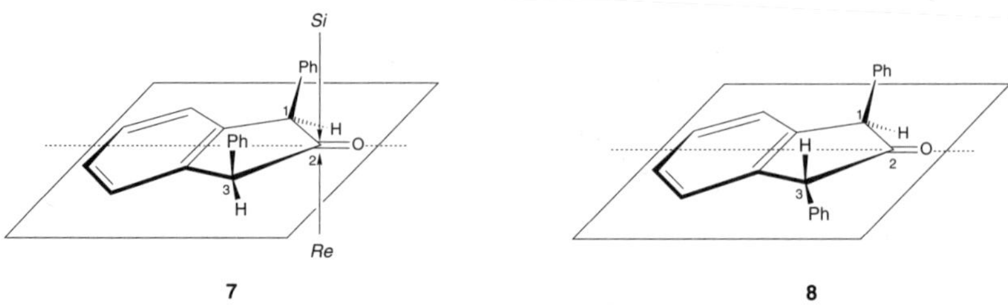

which are not mirror images of each other either, and which, therefore, are diastereotopic. To work out their specification, the relative configuration at C(2) and C(4) is first established, thus enabling the *Si*-face to be specified for the C(1)=O group (with O, C(2), and C(4) anticlockwise) and the *Re*-face (with O, C(2), and C(4) clockwise) for the C(3)=O group. For greater clarity, the view onto the previously unmentioned side, after rotation of 180° of the plane, is also shown.

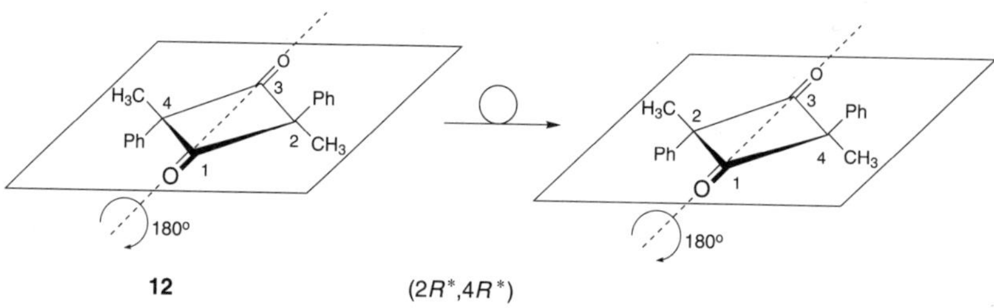

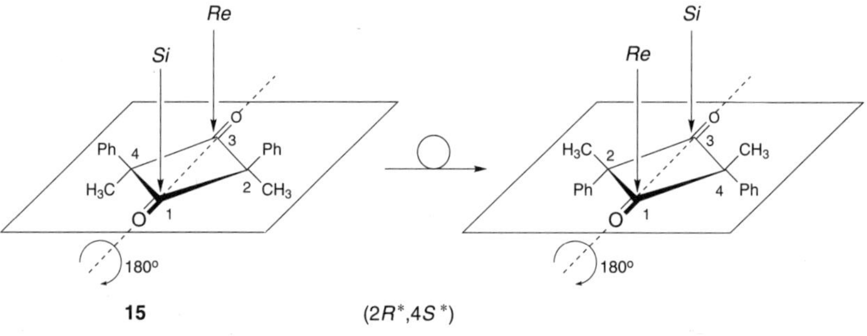

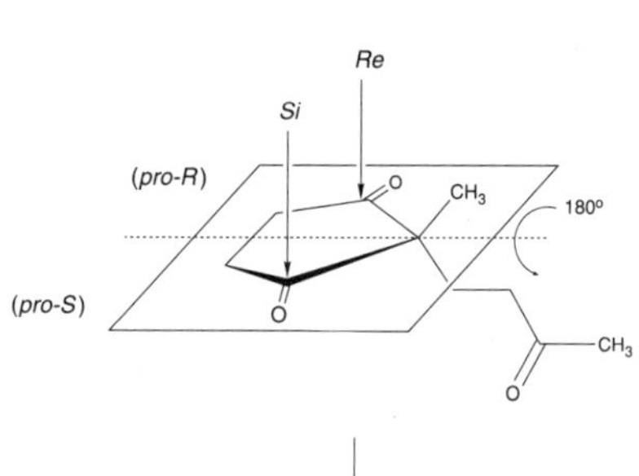

2-Methyl-2-(3-oxobutyl)cyclopentane-1,3-dione, the formula of which is shown on the left, is achiral. The two C=O groups are enantiotopic and, bearing in mind that C(2) is a prochirality center, are specified in the usual way (*Chapt. 3.2.2*). The face of the molecule that is denoted *Si* relative to the (*pro-S*)-carbonyl group is the *Re*-face relative to the (*pro-R*)-carbonyl group. For ease of understanding, the plane here has also been rotated by 180°, showing the previously unmentioned face.

Fig. 4.7 shows how the relationship of the two faces of a two-dimensional molecular model may be determined.

4.3. Conclusion

The *concept of isomerism* (*J. J. Berzelius*) defines the relationship between different molecules of the same molecular formula. Intercomparison of two molecules of the same constitution has one of three

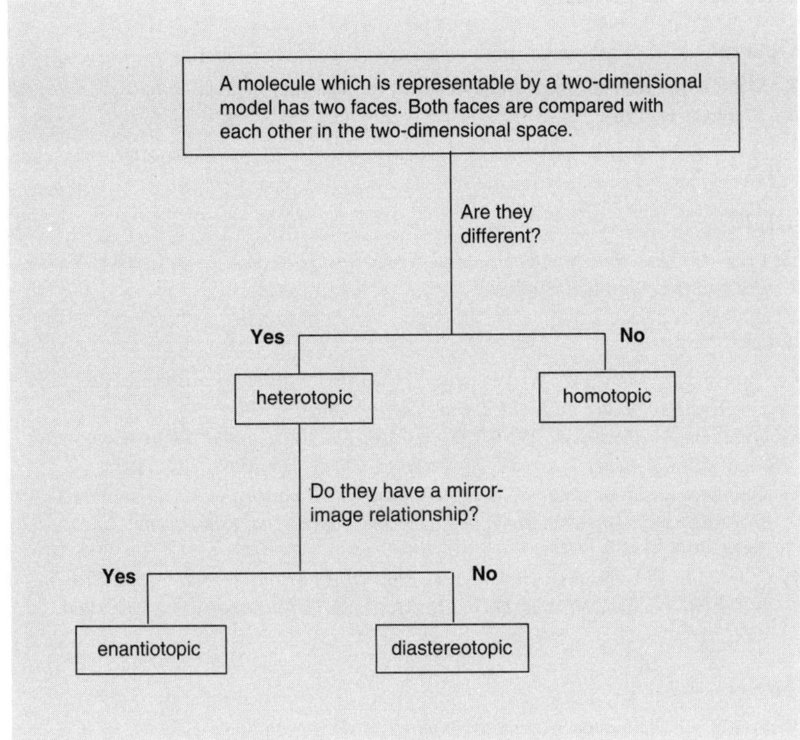

Fig. 4.7. *Differentiation of topical relationships of the two faces of a molecule represented by a two-dimensional model*

possible outcomes: either the two molecules are equivalent or they are not; in the latter case, they then are either enantiomeric or diastereoisomeric (*cf. Fig. 2.17*).

The *concept of topicity* (*K. Mislow*) describes the relationship which exists between two different atoms or atom groups, which are of identical atomic mass and located within one and the same molecule. If these atoms or atom groups are alternately substituted (either physically or intellectually), the same three outcomes are once more encountered: if the atom groups are homotopic, then the substitution products are equivalent. Otherwise, the substitution products are enantiomers (diastereoisomers), if the atom groups are enantiotopic (diastereotopic).

Classifying particular atom groups of the same molecule by topicity criteria turns out to be opportune, when it can be used in the interpretation of NMR spectra. Consideration of topicity is not only applicable to atoms and atom groups, but also to molecular faces: of which it is desirable to know whether they are equivalent or, if not, whether they are enantiotopic or diastereotopic. This may apply to either the two faces of a molecule (with one or more double bonds), or to two surfaces of a crystal, which may be 'recognized' by another molecule in a more or a less selective manner. This type of 'recognition' plays an important role in processes such as chemical reactions (*Aspects of Organic Chemistry*, Volume 2: *Reactivity*) and crystal growth (*Chapt. 12.2.2*).

Further Reading

General

K. Mislow, M. Raban, *Stereoisomeric Relationships of Groups in Molecules, Topics in Stereochem.* **1967,** *1,* 1.

To Sect. 4.1.

4.1.2.

H. Friebolin, *Ein- und zweidimensionale NMR-Spektroskopie,* 2. Aufl., VCH Verlagsgesellschaft, Weinheim, 1992.

4.1.3.

G. Quinkert, K. Opitz, W.-W. Wiersdorff, M. Finke, *Stereospezifische Adduktbildung bei Benzocyclobutenen, Liebigs Ann. Chem.* **1966,** *693,* 44.

G. Quinkert, H.-P. Lorenz, W.-W. Wiersdorff, *Darstellung und Konfigurationszuordnung symmetrischer Indanon-(2)-Derivate, Chem. Ber.* **1969,** *102,* 1597.

G. Quinkert, P. Jacobs, *Die Stereospezifität der lichtinduzierten Cyclobutanon/Tetrahydrofuryliden-Isomerisierung und ihre mechanistische Konsequenz, Chem. Ber.* **1974,** *107,* 2473.

G.C. Brumli, R.L. Baumgarten, A.I. Kosak, *Effect of the Aromatic Ring Current on NMR Resonance Splitting Owing to Molecular Dissymmetry, Nature* **1964,** *201,* 388.

To Sect. 4.2.

V. Prelog, G. Helmchen, *Pseudoasymmetrie in der organischen Chemie, Helv. Chim. Acta* **1972,** *55,* 2581.

5 Configurational Analysis (Demonstrated on Carbohydrates)

5.1. Fundamentals

The most abundant organic compound on this planet, either in its free form, or, still more frequently, in bound form, is glucose: a simple sugar (monosaccharide) from the compound class of carbohydrates.

CHO
H—C—OH
HO—C—H
H—C—OH
H—C—OH
CH_2OH

The class-name carbohydrate derives from the empirical formula, which describes representatives of this compound class as 'hydrates of carbon'.

$$C_n(H_2O)_n$$

A distinction is made in the monosaccharide group between *aldoses* (possessing an aldehyde functionality) and *ketoses* (possessing a ketone functionality). Stereochemical considerations will here be confined to the aldoses, as they exhibit a more diverse stereochemistry than the ketoses. Aldoses can be described using the general constitutional formula given on the right.

CHO
|
$(CH(OH))_m$
|
CH_2OH

As can be seen, the functional groups consist of an aldehyde functionality, a primary OH group, and a variable number of secondary OH groups. Each of these secondary OH groups is bound to a chirality center, so that the number of possible stereoisomers may be calculated from the number of these secondary OH functionalities (*Chapt. 2.6.1*). For the simplest aldose with one chirality center, accordingly there exist two enantiomers.

The aldotrioses have the trivial name of glyceraldehyde (= *glycero*-triose; systematic name: 2,3-dihydroxypropanal; *Fig. 5.1*). Experimentally, the two enantiomers of glyceraldehyde may be distinguished from each other by measurement of their optical rotation (*Chapt. 3.1.1*): one enantiomer rotates the plane of linearly polarized light to the right, while the other affords a rotation of identical magnitude to the left. An unambiguous assignment of absolute configuration can be made (using a suitable crystalline derivative) with the aid of X-ray crystal-structure analysis (*Chapt. 12.1*).

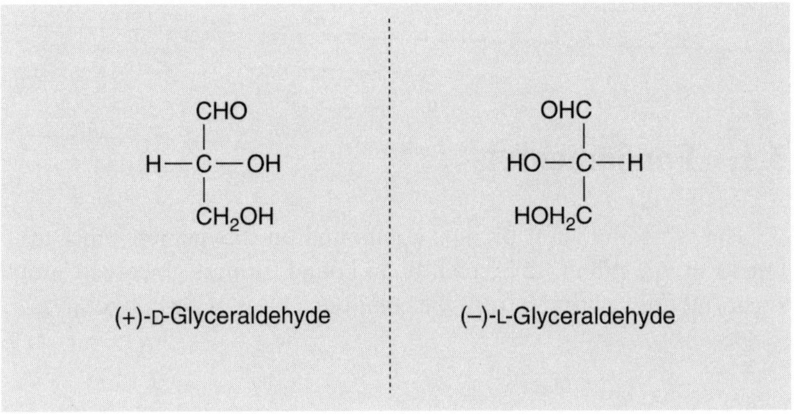

CHO | H—C—OH | CH_2OH

(+)-D-Glyceraldehyde

OHC | HO—C—H | HOH_2C

(−)-L-Glyceraldehyde

Fig. 5.1. Fischer *projection of the two enantiomeric aldotrioses*

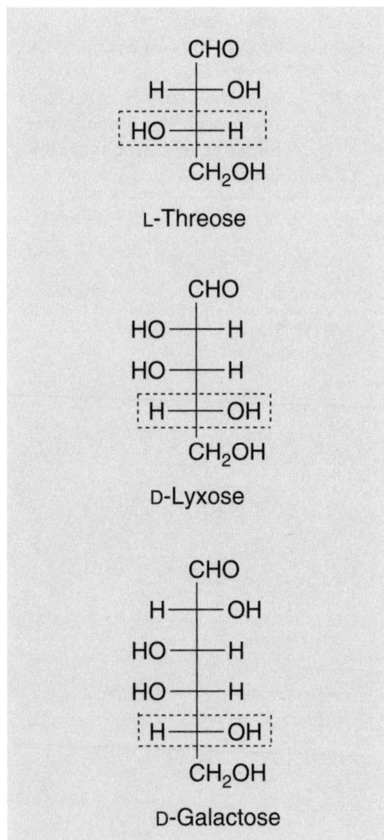

Fig. 5.2. *Determination of* D- *or* L-*assignment by means of the stereogenic center (serving as the reference center) furthest away from the CHO group*

At the end of the last century, before X-ray crystal-structure analysis had been discovered (*cf. Chapt. 10.2.1*), *Emil Fischer* arbitrarily assigned projection formulae to the enantiomers of glyceraldehyde. The enantiomer rotating linearly polarized light to the right was assigned the projection formula in which the OH group connected to the chirality center is situated on the right-hand side of the main C-atom chain. This (+)-enantiomer was given the descriptor D (from *dexter* = right). The enantiomer rotating light to the left was consequently assigned the projection formula in which the secondary OH group is situated on the left. The letter L (from *laevus* = left) was thus assigned as a prefix for (−)-glyceraldehyde. For longer-chain aldoses, *Fischer* prescribed that the secondary OH group furthest away from the aldehyde functionality should be the one used as the reference center for assigning of the D/L-notation (*Fig. 5.2*).

The absolute configuration at each chirality center can also be given using the later developed (*R*/*S*)-convention of the *CIP* system (*Chapt. 10.2.1*). (+)-D-Glyceraldehyde thus receives the absolute configuration (*R*) for its reference center, and (−)-L-*glyceraldehyde* (*S*).

For the characterization of a sugar enantiomer by the sign of its specific rotation [α] (*Chapt. 3.1.1*), it is necessary to ensure that the conditions for the initial value of so-called mutarotation (*Chapt. 6.6.2*) apply.

5.2. The Family Tree of Aldoses

The higher aldoses, with four (tetroses), five (pentoses), six (hexoses), *etc.* C-atoms, and so on, may all be derived, in thought experiments, from one of the two enantiomers of glyceraldehyde. Addition of a further CH(OH) unit, between the aldehyde functionality of (+)-D-glyceraldehyde and its neighboring secondary OH functionality, affords two stereoisomeric aldotetroses, *i.e.*, (−)-D-erythrose ((−)-D-*erythro*-tetrose) and (−)-D-threose ((−)-D-*threo*-tetrose) (*Fig. 5.3*). Two stereoisomers (like (−)-D-threose and (−)-D-erythrose), which possess more than one stereogenic center, but which only differ from each other in the absolute configuration of one stereogenic center, are defined as *epimers*. For every epimer, there exists its enantiomeric compound ((+)-L-erythrose and (+)-L-threose, respectively; *Fig. 5.3*), which is derived from (−)-L-glyceraldehyde. Four stereoisomeric aldotetroses arise, therefore, from the two enantiomeric aldotrioses.

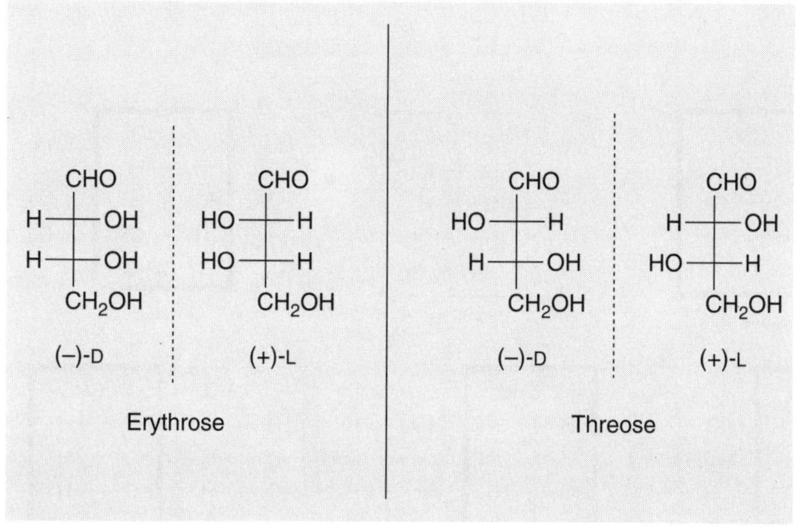

Fig. 5.3. Fischer *projections of the stereoisomeric aldotetroses*

A continuation of the thought experiment shows that each of the four tetroses, upon 'formal insertion' of an additional CH(OH) unit at the designated site, affords two epimeric aldopentoses, so that a total of eight aldopentoses (four enantiomer pairs) results.

Fig. 5.4 summarizes the family tree of Aldoses of the D-series, which is derived from (+)-D-glyceraldehyde and is explained in greater detail in *Sect. 5.3*.

For the question of the total number of stereoisomeric aldoses, it must be taken into account that, for each of the illustrated D-aldoses, there exists an enantiomeric (not illustrated) L-aldose. As well as the ecliptical *Fischer* projections of monosaccharides, the staggered 'zig-zag' stereoformulae (*Fig. 5.5*) also, particularly in more modern literature, find application for the representation of relative configuration, and, if the sense of chirality at one of the stereogenic centers is known, for the absolute configuration. In the case of the monosaccharides, however, the *Fischer* projection does present advantages, as this form of specification allows for particularly easy recognition of the systematic build-up of the compound class.

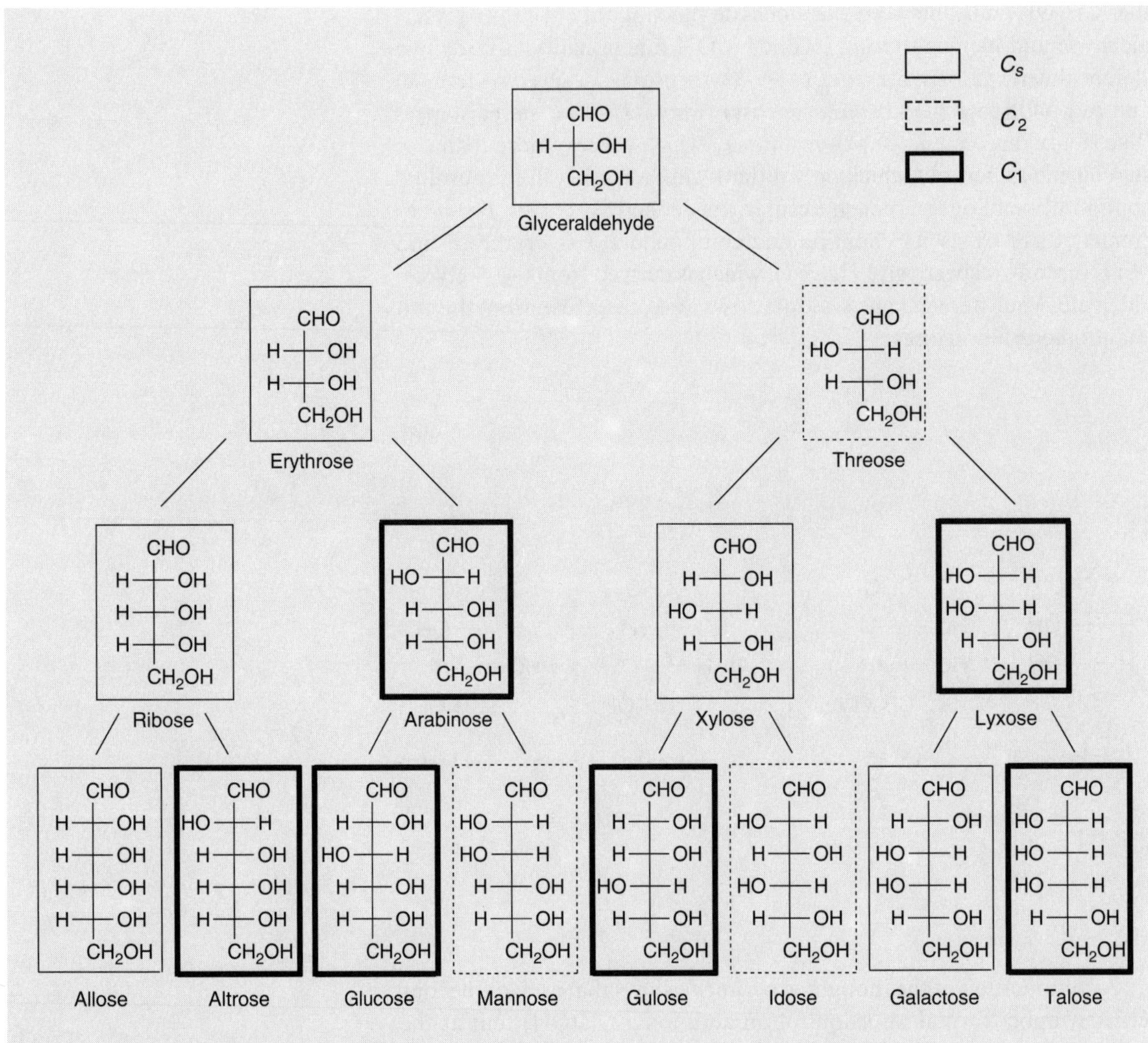

Fig. 5.4. *The family tree of aldoses derived from (+)-glyceraldehyde*. The *Fischer* projections of the corresponding aldaric acids are, variously, chiral and asymmetrical (C_1), chiral and symmetrical (C_2), or achiral and symmetrical (C_s).

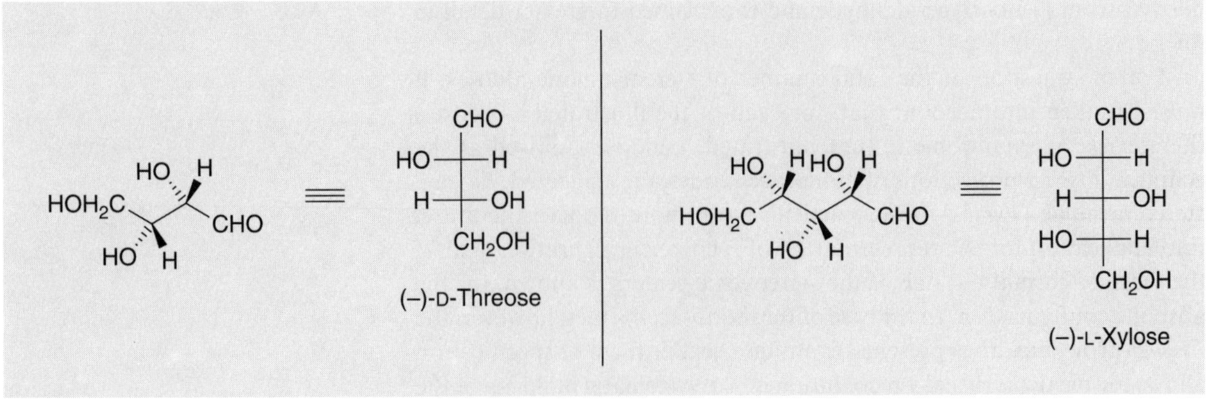

(–)-D-Threose

(–)-L-Xylose

Fig. 5.5. *Examples of 'zig-zag' stereoformulae for monosaccharides*

5.3. Identification of Aldoses

As the number of stereoisomers of the various aldoses (trioses, tetroses, *etc.*) is now known, the question posed on consideration of the aldose family tree of *Fig. 5.4* is how the diastereoisomeric aldoses derived from, say, D-glyceraldehyde can be identified. Nowadays, identification is possible with the aid of physical methods. Before these were available, however, *Fischer* had developed a procedure, impressive in its elegant use of logic, by which the relative configuration of aldoses could be determined. For this, it is necessary to know the lower aldose from which the structure under consideration could be derived by introducing a CH(OH) unit. For higher aldoses built up in such a manner, it follows that all the stereogenic centers bar the newly introduced one will be identical to those of the lower aldose precursor, and so only the configuration at the new stereogenic center will be unknown. The next step is to transform the aldose into an aldaric acid, by partial oxidation of the two terminal (CHO and CH_2OH) groups into COOH functionalities. If one of the aldaric acids obtained from two aldoses with a common precursor (and hence only differing in the absolute configuration of the newly introduced stereogenic center) is optically active, while the other is optically inactive (a *meso*-compound), then the two aldoses can be unambiguously distinguished from each other. The configuration of the newly introduced stereogenic center can thus be deduced by this experimental differentiation, and so the epimeric aldoses can be identified. The heart of *Fischer*'s argumentation is the correlation between an aldaric acid with n stereogenic centers and an aldose also with n, or two aldoses with $n + 1$ stereogenic centers. Consequently, we must next turn our attention to the aldaric acids.

5.3.1. Stereoisomerism of Aldaric Acids

The term aldaric acid denotes the constitutionally symmetrical polyhydroxy diacids which are accessible through partial oxidation of the functional end groups of the corresponding aldoses. The general molecular formula HOOC–$(CHOH)_n$–COOH applies to them. Their symmetry, both constitutional and configurational, dictates that the actual number of their stereoisomers is less than the formal maximum number. In fact, instead of the formula $N = 2^n$, which applies for the aldoses, the aldaric acids are constrained to $N = 2^{(n/2)-1}(2^{n/2} + 1)$ for n = even, and $N = 2^{n-1}$ for n = odd. *Fig. 5.6* shows *Fischer* projections for the aldaric acids in the range of n = 2–5 with special attention given to the symmetry operations (rotation about a C_2 axis or reflection across a σ plane) which may be performed on them.

As may be inferred from the names of the given aldaric acids, two (or more) stereodescriptors are used for the naming of acyclic monosaccharides with more than four consecutive stereogenic centers. Hence, D-*glycero*-D-glucaric acid should be understood to mean 6-D-*glycero*-2,3,4,5-D-glucaric acid.

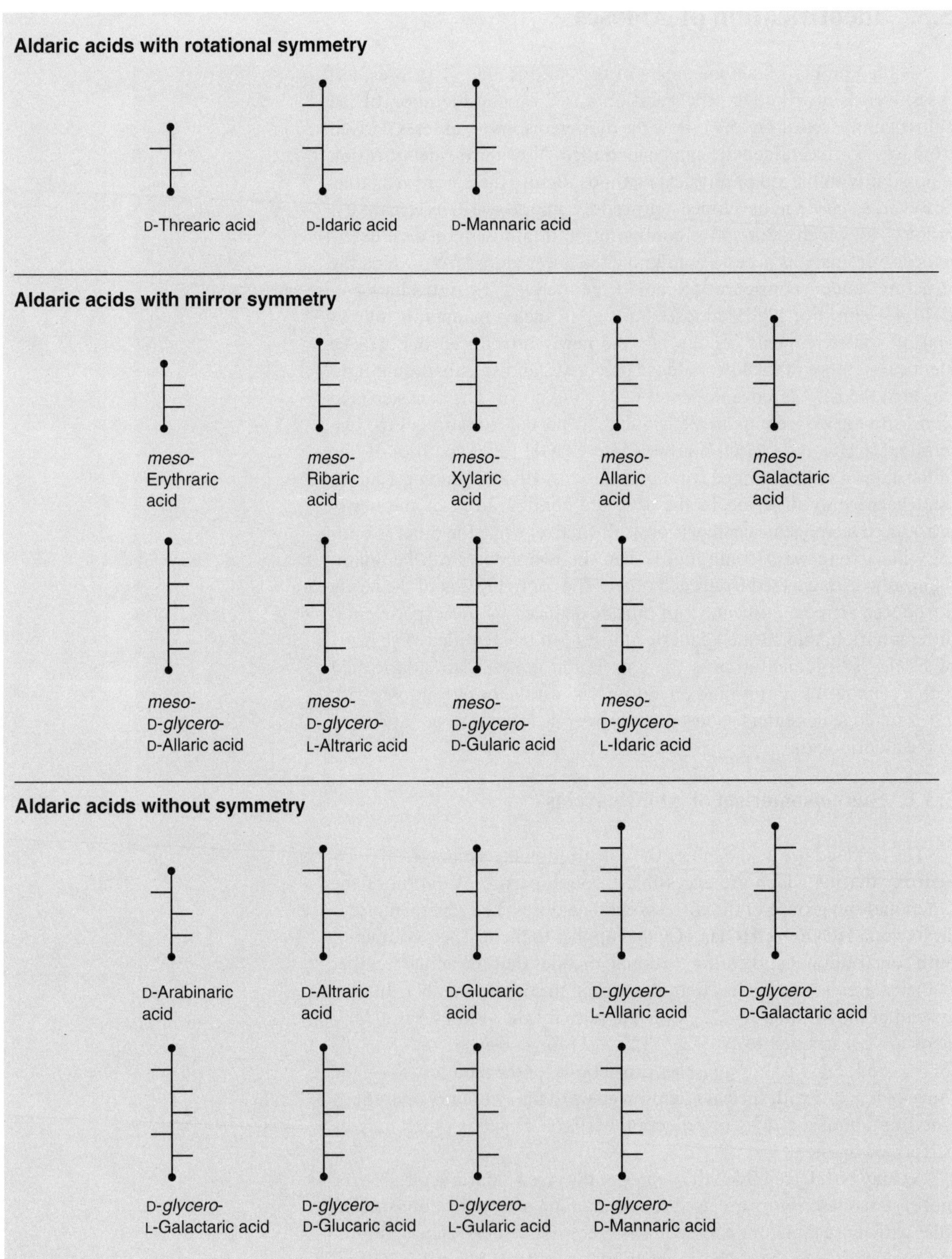

Fig. 5.6. Fischer *projections for aldaric acids in the range of* n = *2–5, considering symmetry operations carried out on them*

5.3.2. Unambiguous Configurational Assignment for Aldotetroses, -pentoses, and -hexoses

The procedure can be explained using the example of the aldoses derived from D-glyceraldehyde. The two D-aldotetroses are both optically active, with their structural formulae belonging to the symmetry point group C_1. Partial oxidation leads to an optically active diacid in one case, and to a *meso*-compound in the other (*Fig. 5.7*). Only D-erythrose, whose *Fischer* projection exhibits both secondary OH functionalities on the same side of the main chain, would be capable of leading to a *meso*-compound: consequently, the optically active oxidation product can only have originated from the threose.

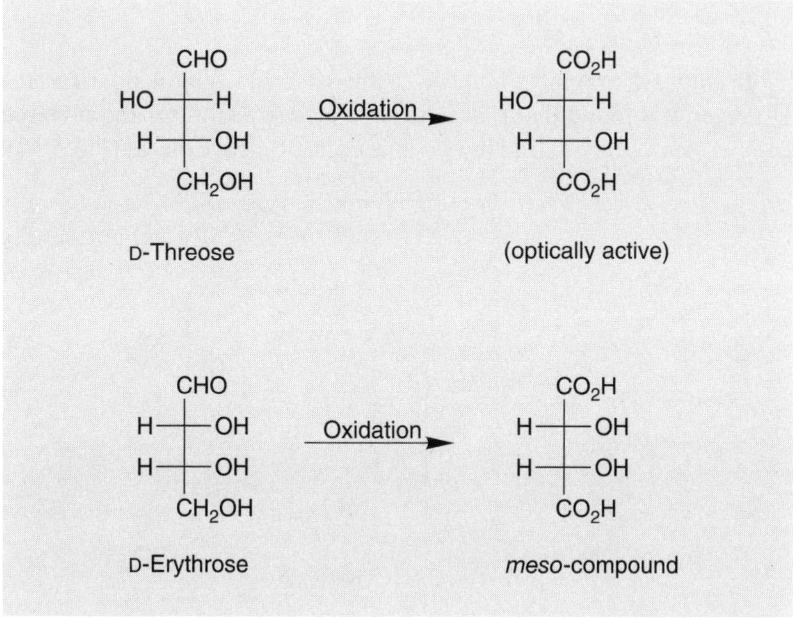

CHO
HO
H
H
OH
CH2OH
Oxidation
CO2H
HO
H
H
OH
CO2H
D-Threose
(optically active)
CHO
H
OH
H
OH
CH2OH
Oxidation
CO2H
H
OH
H
OH
CO2H
D-Erythrose
meso-compound

Fig. 5.7. *Partial oxidation of* D-*aldotetroses*

In the case of the aldopentoses, the oxidation products have the constitution of 2,3,4-trihydroxyglutaric acid (*cf. Fig. 5.6* for the symmetry of the oxidation products). Thus, the oxidation of D-ribose (D-*ribo*-pentose) leads to an optically inactive *meso*-compound (*Fig. 5.8*).

CHO
H
OH
H
OH
H
OH
CH2OH
Oxidation
CO2H
H
OH
H
OH
H
OH
CO2H
(–)-D-Ribose
meso-compound

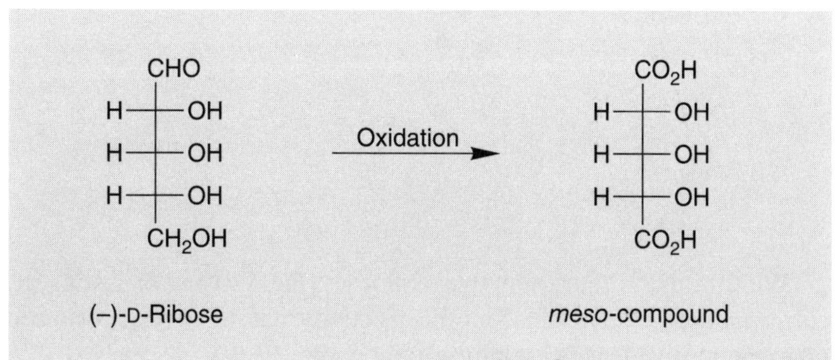

Fig. 5.8. *Partial oxidation of (–)-*D-*ribose*

The diacid originating from D-xylose (D-*xylo*-pentose) also has C_s symmetry, and so shows no optical activity (*Fig. 5.9*).

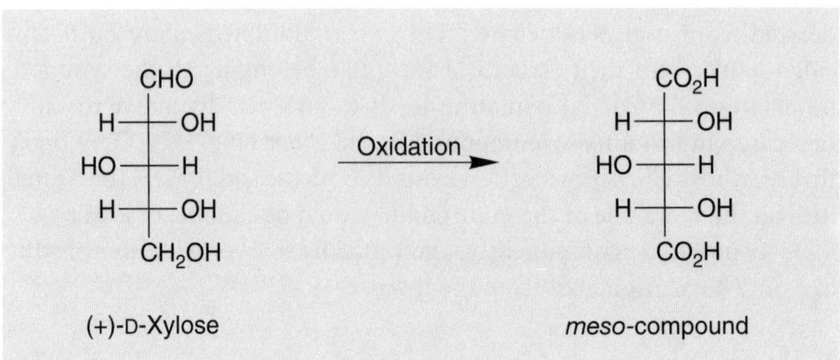

Fig. 5.9. *Partial oxidation of (+)-D-xylose*

In contrast, arabinose and lyxose are oxidized to chiral aldaric acids. However, it is found that D-arabinose (D-*arabino*-pentose) and D-lyxose (D-*lyxo*-pentose) both lead to the same optically active diacid (*Fig. 5.10*).

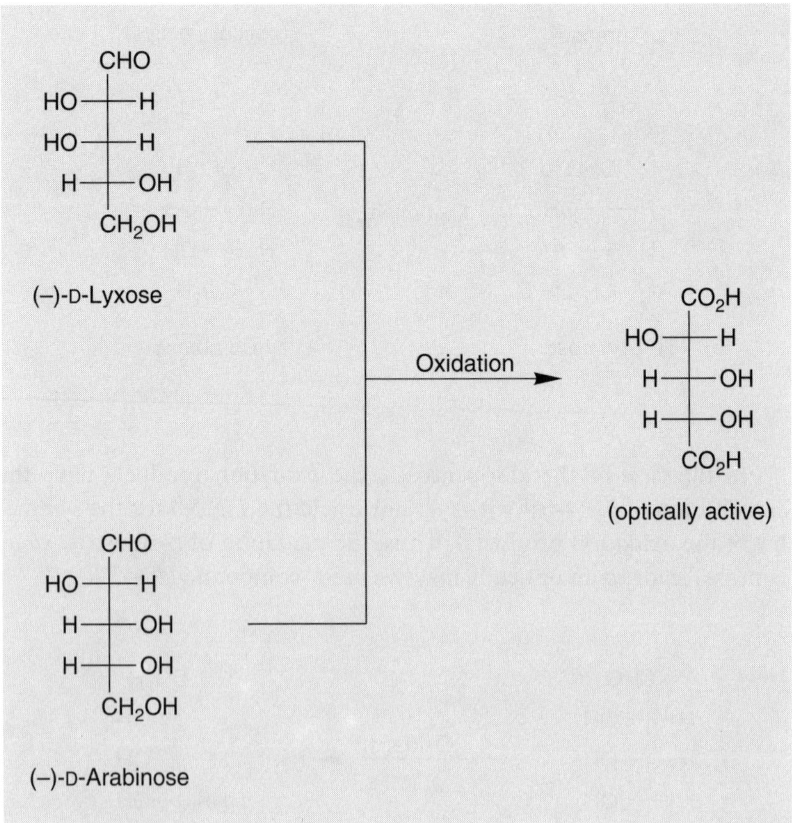

Fig. 5.10. *Partial oxidation of (−)-D-lyxose and (−)-D-arabinose*

With the aid of the family tree of aldoses (*Fig. 5.4*) and by testing the corresponding aldaric acids for optical activity, it is, however, possible to unambiguously identify arabinose and lyxose.

Differentiating between the various aldohexoses (eight diastereoisomers) is a more complex problem. Formally, there should be eight corresponding aldaric acids with the 2,3,4,5-tetrahydroxyadipic-acid constitution. Constitutional symmetry, though, implies that this number is reduced to six: two optically inactive *meso*-compounds and four optically active stereoisomers.

The two *meso*-compounds arise from the partial oxidation of D-allose (D-*allo*-hexose) and D-galactose (D-*galacto*-hexose) (*Fig. 5.11*).

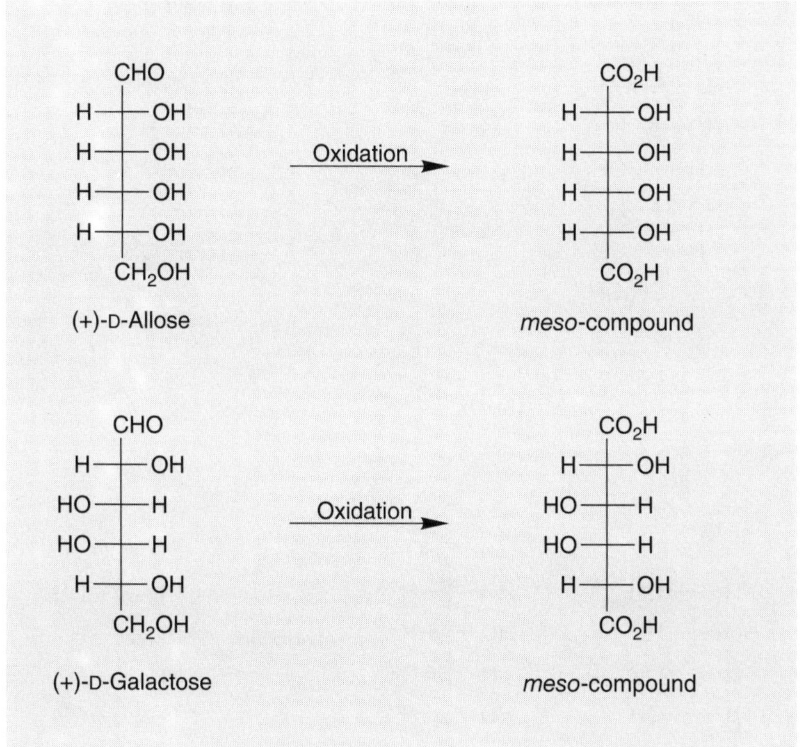

Fig. 5.11. *Partial oxidation of (+)-D-allose and (+)-D-galactose*

From D-mannose (D-*manno*-hexose) and D-idose (D-*ido*-hexose), two different diacids are obtained, which, moreover, are distinct from all of the other oxidation products (*Fig. 5.12*).

From D-altrose (D-*altro*-hexose) and D-talose (D-*talo*-hexose), there is obtained one and the same optically active aldaric acid. A similar situation arises in the cases of D-glucose (D-*gluco*-hexose) and L-gulose (L-*gulo*-hexose) (*Fig. 5.13*).

Obviously, the relationships laid out in *Figs. 5.11–5.13* do not in themselves lead to unambiguous results for differentiating between the D-aldohexoses (*cf.* also the symmetry of the oxidation products, *Fig. 5.6*). Only answers to three questions (QH1 to QH3, in *Fig. 5.14*), relating to the mirror symmetry of the associated pentaric acids, hexaric acids, and heptaric acids, allow an unambiguous identification.

While obtaining answers to three questions permits configurational assignments to be made for the individual aldohexoses, unambiguous

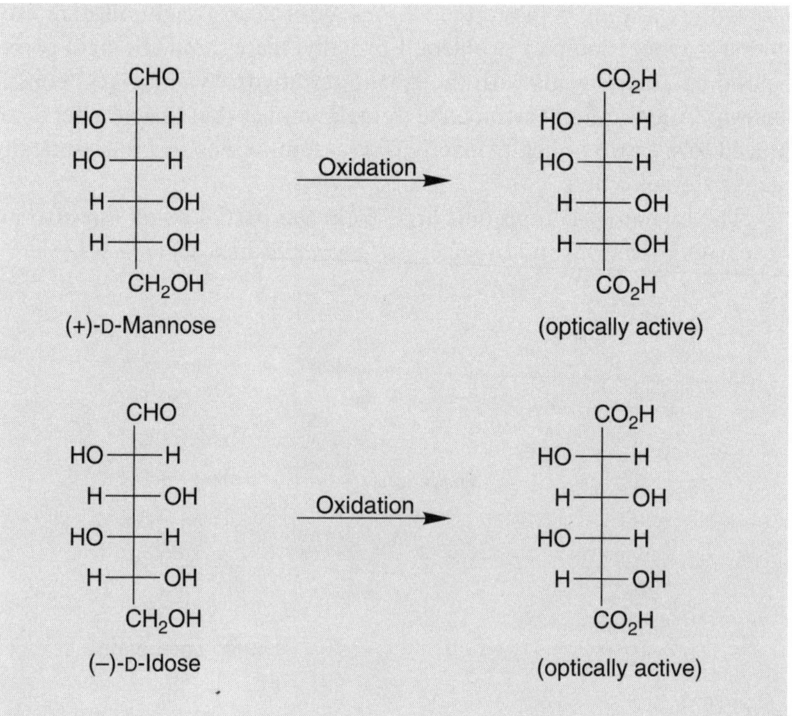

Fig. 5.12. *Partial oxidation of (+)-D-mannose and (−)-D-idose*

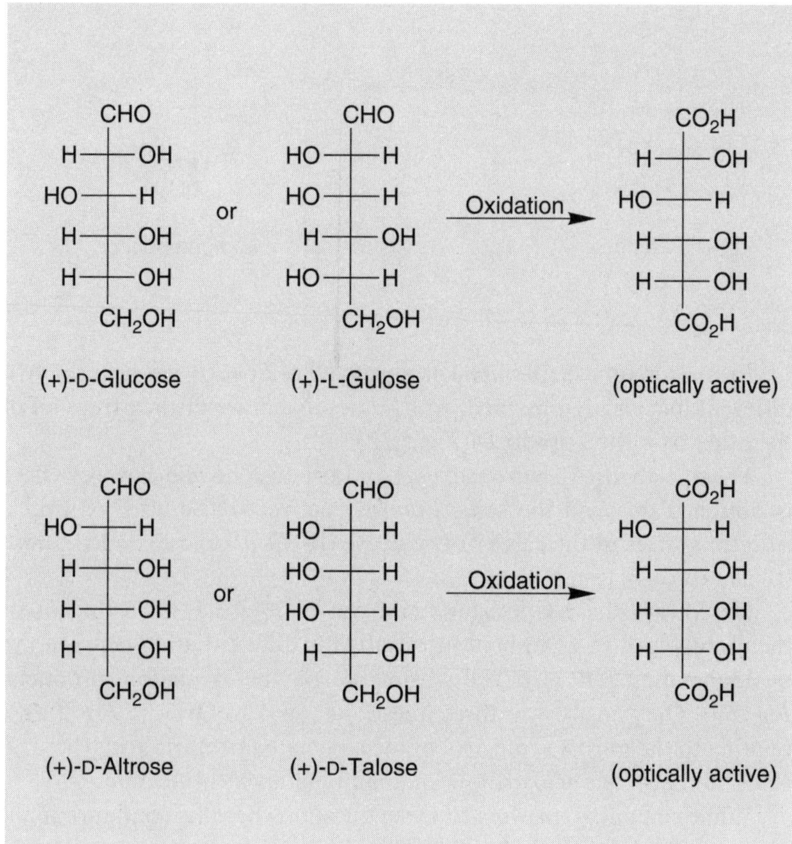

Fig. 5.13. *Partial oxidation of (+)-D-glucose or (+)-L-gulose, and (+)-D-altrose or (+)-D-talose*

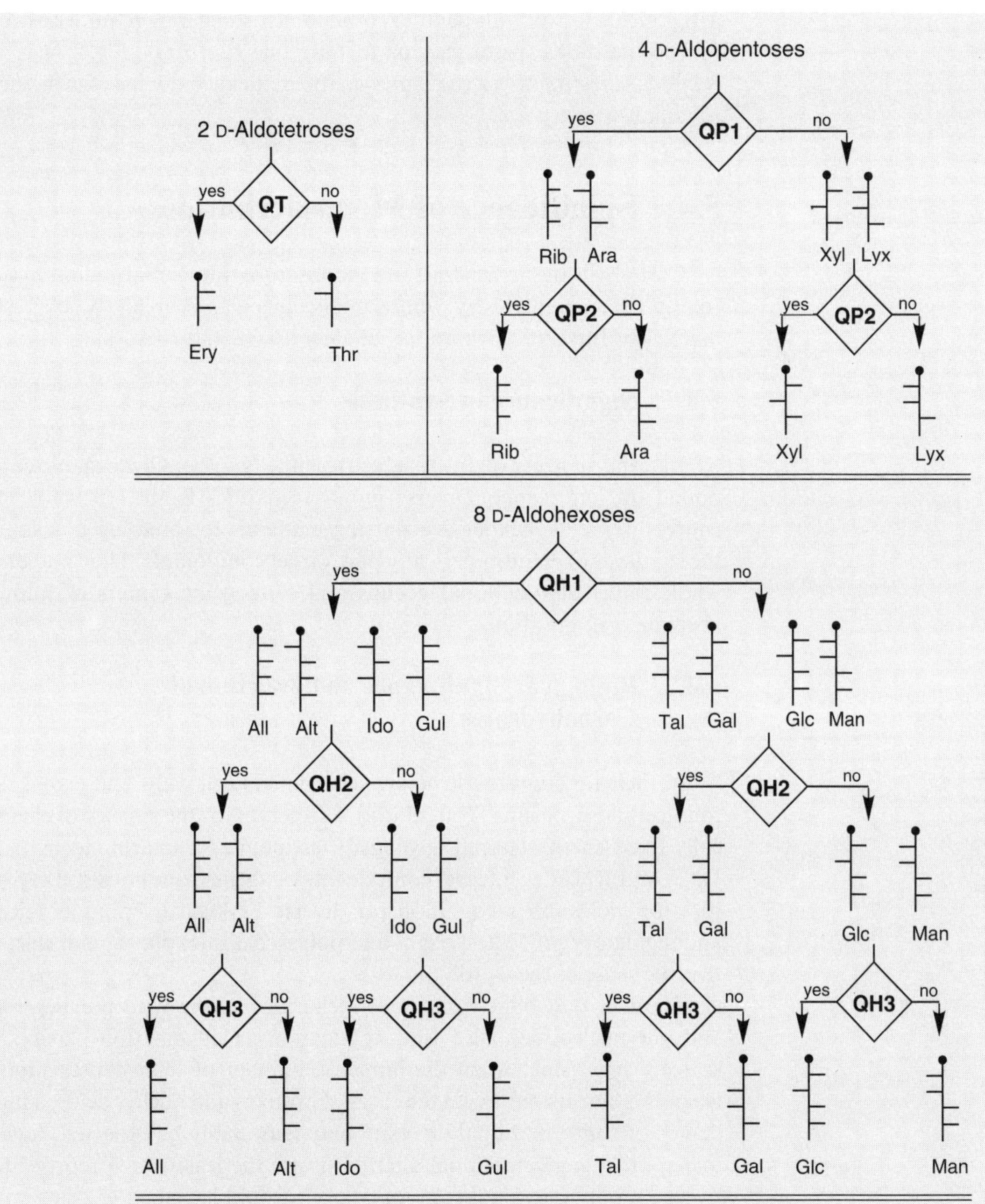

QT: Is the resulting tartaric acid mirror-symmetrical (optically inactive)?
QP1: Is the tartaric acid obtained on descending from the pentose to the tetrose series mirror-symmetrical (optically inactive)?
QP2: Is the associated pentaric acid mirror-symmetrical (optically inactive)?
QH1: Is the pentaric acid obtained on descending from the hexose to the pentose series mirror-symmetrical (optically inactive)?
QH2: Is one of the two C(2) epimeric hexaric acids mirror-symmetrical (optically inactive)?
QH3: Is one of the two C(2) epimeric heptaric acids obtained on ascending from the hexose to the heptose series mirror-symmetrical (optically inactive)?

Fig. 5.14. *Identification of the individual aldoses in the range of* n = *2–4*

assignments for the aldopentoses or aldotetroses merely require answers to two questions (pentoses), or to only one (tetroses). *Fig. 5.14* is an impressive testimony to the logico-mathematical character of structural organic chemistry.

5.4. Significance of Monosaccharides

Carbohydrates in general, and monosaccharides in particular, play a role in organic chemistry whose importance can scarcely be overestimated (for further information, see *Chapt. 6.6.2* and *10.5.3*).

5.4.1. Significance in Synthesis

In terms of sheer quantity, glucose is the Number One organic compound on our planet. Diverse monosaccharides from 'renewable resources' serve as inexpensive starting materials for building blocks for the synthesis of enantiomerically pure target compounds. They possess a whole range of functional groups and stereogenic centers of known absolute configuration.

5.4.2. Biological Significance and Function of Carbohydrates

In their functions as structural elements in cell walls and exo-skeletons (cellulose, pectins, agar, chitin), as reserves for the storage of chemically fixed energy (starch, glycogen), as mobile structural elements in DNA and RNA, as integral components of agents with powerful capacities for molecular recognition for diverse biological, immunological, and regulatory processes, oligo- and polysaccharides play a vital role, in the true sense of the word.

The fact that oligo- and polysaccharides, along with polypeptides and polynucleotides, also find application as information carriers is scarcely surprising, when the immense number of their constitutional isomers is considered. As in the cases of proteins and nucleic acids, a high degree of conformational diversity may reasonably be expected. Easily manageable conformational alterations are the least which representatives of a polyfunctional compound class would have had to bring with them, in order to have been selected and maintained as macro- or supramolecular functional units, during the course of naturalistic evolution. The structural complexity made the non-specialized organic chemists ignore those biomolecules for a long time. With the increasing number of well documented X-ray crystal-structure analyses and with the advances in multidimensional NMR-spectroscopic methods (*Aspects of Organic Chemistry*, Volume 4: *Methods of Structure Determination*), it has become possible to open up the field, in the past isolated on the edge of organic chemistry, of conformational analysis of *glycobiological* molecules.

A therapeutic strategy of the foreseeable future is directed at the search for *antiadhesive* drugs. Antiadhesive molecules should be taken to mean unbound oligosaccharides which, preferentially to oligosaccharides bound to the surface of a cell, can be recognized by the receptors of pathogenic invaders. In this manner, the adhesion of pathogens to human cells, and hence the initial stage of an infection, may be inhibited.

5.5. Conclusion

The previous sections have demonstrated the way in which both the total number of stereoisomeric aldoses, and also the absolute configuration of each individual stereoisomer, can be determined. Monosaccharides, each with its complete set of enantiomerically pure diastereoisomers, proved to be the foundation stones for the build-up of stereochemistry. In the historical development of organic chemistry, *Emil Fischer*'s work on the configurational analysis of the aldoses plays an important role, as the tetrahedron concept of *van't Hoff* and *Le Bel* was not generally accepted until *Fischer*'s successful application of it. The observation that, in the case of glucose (and its derivatives), for example, the expected number of stereoisomers turns out to be smaller than that detected in reality will considerably enhance our view and knowledge of the structure of monosaccharides (*Chapt. 6.6.2*).

There are numerous further grounds for affording monosaccharides a special status in organic chemistry. Monosaccharides are excellently suited as a source of chiral, non-racemic building blocks for the synthesis of enantiomerically pure target compounds, for example. Apart from that, monosaccharides are interesting because of their biological origins and their biological functions.

Chapt. 12 informs about the direct '*Determination of Absolute Configuration*'. Ever since the absolute configuration could be established experimentally, decades-long uncertainty has been cleared away. Chemists nowadays are obliged to use names and formulae for chemical compounds sufficiently precisely that the absolute (or relative, when that is adequate) configuration should be easily and unambiguously recognized.

Further Reading

To Sect. 5.2.

K. Freudenberg, *Emil Fischer and his contribution to carbohydrate chemistry, Adv. Carbohydrate Chem.* **1966,** *21,* 1.

To Sect. 5.3.

R. Bentley, J. L. Popp, *Configurations of Glucose and Other Aldoses, J. Chem. Educ.* **1987,** *64,* 15.

To Sect. 5.4.

5.4.1.

A. Vasella, *Chiral Building Blocks in Enantiomer Synthesis-Ex Sugars* in *Modern Synthetic Methods*, Vol. 2 (Ed. R. Scheffold), Salle + Sauerländer, Frankfurt am Main, 1980.

B. Fraser-Reid, *Progeny of 2,3-unsaturated sugars. They little resemble Grandfather Glucose, Acc. Chem. Res.* **1975,** *8,* 192.

B. Fraser-Reid, R. C. Anderson, *Carbohydrate Derivatives in the Asymmetric Synthesis of Natural Products, Progress in the Chemistry of Organic Natural Products* **1980,** *39,* 1.

S. Hanessian, *Approaches to the Total Synthesis of Natural Products Using 'Chiral Templates' Derived from Carbohydrates, Acc. Chem. Res.* **1979,** *12,* 159;
– *Total Synthesis of Natural Products: The 'Chiron' Approach,* Pergamon Press, Oxford, 1983.

5.4.2.

S. Borman, *Glycotechnology Drugs Begin To Emerge from the Lab, Chem. Eng. News* **1993,** *June 28,* 27.

6 Conformational Analysis (Demonstrated on Steroids)

6.1. The Significance of Steroids

It is not difficult to name more than fifteen *Nobel* Prize winners from 1927 to 1990, who have been preoccupied with the isolation, structure determination, biosynthesis, biological activity at the molecular level, or total chemical synthesis of steroids.

Thanks to their rigid structure, steroids display a definite relationship between their configuration and their conformation. They serve as a prime historical example of conformational analysis (see *Fig. 6.1*).

Chapt. 10.5.1 summarizes the essential points which must be followed for the formulaic description of the structure of a steroid molecule.

6.1.1. The Significance of Steroids in Chemical Synthesis

More than any other class of compounds, steroids have served as 'test objects' for investigations into reaction mechanisms or for the evaluation of new reagents. In countless partial syntheses, inexpensive steroids have been converted into steroids with desirable biological properties. Today, as in the past, steroids are prominent target compounds for total synthesis.

6.1.2. The Biological Significance of Steroids

Cholesterol has attracted the attention, and will continue to do so, of generations of *chemists* (*structure and total synthesis*), *biochemists* (*biosynthesis of cholesterol, and of steroidal hormones from cholesterol*), *cell biologists* (*cholesterol as a lipoidal component of eukariotic plasma membranes, as the precursor for bile acids and steroid hormones, in the receptor-mediated endocytosis of LD-lipoproteins*), and *physicians* (*heart attacks and strokes because of elevated levels of blood cholesterol (arteriosclerotic plaques in the bloodflow) and their avoidance through receptor regulation (diet and medication)*).

Intense research activity is additionally directed at steroid hormones (*estrogens, gestagens, androgens, glucocorticoids, mineralocorticoids, vitamin D_3 derivatives*). The accent is currently on the steroid hormone/

receptor supermolecules and their interaction with specific areas of the DNA supermolecule, with the consequence of altered gene expression.

Among biologically significant steroids, synthetic agonists and antagonists (*anti-hormones*) have occupied the foreground. Their use as hormonal contraceptives (*(−)-norethindrone*, *(−)-norgestrel*, *(−)-gestoden*) for human birth-control, or for non-surgical treatment of male sexual deviation (*(+)-cyproteron acetate*) implies that scientific and medicinal considerations alone are insufficient criteria for evaluation of such research. The chemist of today is, more than ever, confronted with Mankind's problems, which go far beyond his or her immediate neighborhood and lifetime.

Fig. 6.1. *The stamp issued in honor of* Sir Derek (Harold Richard) Barton *for his contribution to the development of conformational analysis and its application in chemistry, especially that of steroids*

6.2. The Significance of Conformational Analysis for Steroid Chemistry

The concept of *conformation* was coined in 1929 by *Sir Walter (Norman) Haworth* in the discussion of carbohydrates. *Haworth* was not able, however, to elucidate the appropriate conformations for sugars. The expression 'conformational analysis' originates from *William S. Johnson* (1951). However, by this time there had already appeared 32 year old *Barton*'s epoch-making publication '*The Conformation of the Steroid Nucleus*' in *Experientia*, and its significance for the development of steroid chemistry was already apparent. In the same year as *Barton*'s work was published, *Edward C. Kendall, Tadeus Reichstein*, and *Philip S. Hench* received the *Nobel* Prize for their investigations into the hormones of the adrenal cortex. These so-called *corticosteroids* had enabled patients, previously confined, in great pain, to wheelchairs, to move freely and painlessly, provided that the appropriate corticosteroids were regularly administered. In numerous laboratories, in industry and in academia, intensive efforts were understandably made to achieve the synthesis of these beneficent natural products and their derivatives. To have success in this, though, it was essential to exploit chemical reactions available which could achieve precisely defined structural alteration during the course of a synthesis. The courses of chemical reactions in those days, however, were difficult to interpret, and hardly to be predicted. This applied even to steroids, which, as the X-ray crystal-structure determination of cholesteryl iodide (*Fig. 6.2*) had shown, are flat and rigid in their construction. If a 'best fit' plane is used dividing three-dimensional space into two half-spaces, the two angular Me groups at C(10) and C(13) are both to be found in the same 'front' half-space (described as β by *Louis F. Fieser*). The 'back' half-space (described by *Fieser* as α; *Fig. 6.3*) contains no angular Me groups. The difficulty of planning synthetic steps before the above-mentioned *Experientia* paper is shown in a publication by *Fieser* on the subject of '*Steric Course of Reactions of Steroids*', which immediately preceded that of *Barton* in the same journal.

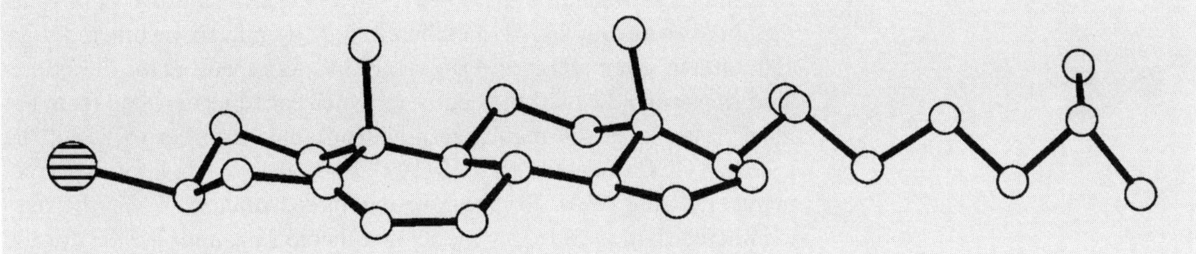

Fig. 6.2. *Molecular structure of cholesteryl iodide in crystalline state after* C. H. Carlisle *and* D. Crowfoot

Fieser made the convincing assumption that the molecule of a reagent would approach that one of a steroid more frequently from the (usually) less hindered α-side. He then considered each C-atom of the ring system individually, and examined whether two (empirically assumed) hindering effects – an *intraradial* and an *extra-radial* – would work with or against one another. The results obtained, however, are not especially convincing and offer, at best, an *a posteriori* explanation.

Barton likewise started out from the assumption that a reagent molecule would approach a steroid molecule from the (usually) less hindered α-side. He then considered the bonds, not constituting a part of the ring system, attached to the cyclohexane rings present and divided these into two (topologically determined) bond types: into *axial* (originally, ambiguously called *polar*) and *equatorial* bonds (*Fig. 6.3*). The results are convincing and allow *a priori* predictions: for example, that a ligand in an axial orientation at a given C-atom is subjected to greater steric hindrance than the same ligand in an equatorial position.

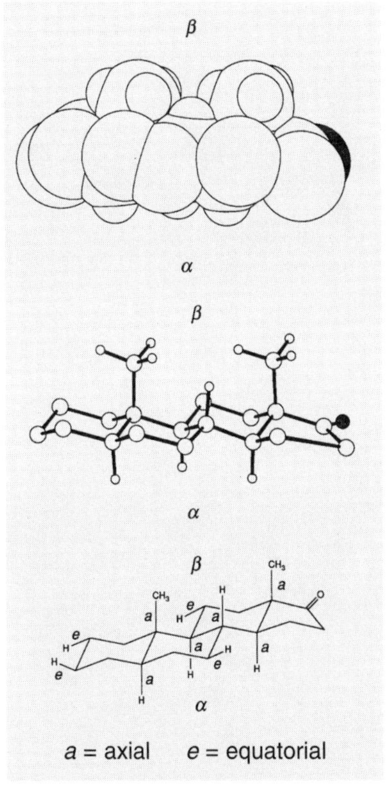

Fig. 6.3. *Various representations of the 5α-androstan-17-one skeleton*

In *Barton*'s *Experientia* contribution, there are already hints of rules, with the help of which the products of various types of reactions would be made predictable. These rules will be considered more closely in *Aspects of Organic Chemistry*, Volume 2: *Reactivity*. For now, it suffices to say that synthetic chemists with interests in corticosteroids (and other natural products) were greatly encouraged by the possibilities of conformational analysis so that, soon, no synthetic target was believed unattainable.

6.3. Stereostructure of Cyclohexane and Its Derivatives

6.3.1. The Special Position of Cyclohexane in the Homologous Series of Saturated Carbocycles: *Baeyer* Strain

To close an open chain into a ring is to drastically reduce the number of its possible conformations. This topologically dictated limitation is even more strict if the ring system is symmetrical. In the expectation that, here, less complicated relationships would be encountered than with open-chain compounds, special interest has always been brought to bear on cyclic compounds.

As early as 1885, *Adolf von Baeyer* had put forward a theory, in which ring size and potential energy of cycloalkanes could be correlated with one another. He set out from the assumption that idealized cycloalkane molecules all adopted the geometry of the relevant regular polygon, and that the CCC valence angle in three-membered rings was 60°, in four-membered rings 90°, in five-membered rings 108°, in six-membered rings 120°, and in seven-membered rings 128° 34′. As a measure for the associated ring strain, he used the degree to which each line deviated from the ideal situation of the regular tetrahedron angle in order to attain the specified CCC valence angle: 1/2 (109° 28′ − 60°) = 24° 44′ for the three-membered ring, 9° 44′ for the four-membered ring, 0° 44′ for the five-membered ring, − 5° 16′ for the six-membered ring, and − 9° 33′ for the seven-membered ring.

Baeyer's strain theory only bears scientific fruit once the theoretically determined ring strain (*Baeyer* strain) is compared with experimentally determined values, and the theory thereby refuted in principle. For the individual cycloalkanes, a specific energy value can be obtained from the heats of combustion. For cycloalkanes of the general molecular formula $(CH_2)_n$, these experimentally determined energy values may be easily normalized and tabulated relative to the cycloalkane of least energy (*Table 6.1*).

Table 6.1. *Relationship Between the Valence-Angle Deviations in the Planar Ring, the Relative Strain of the Molecule, and a CH_2 Group Increment*

CH_2 Groups per ring	3	4	5	6	7
Valence-angle deviation in the planar ring	24° 44'	9° 44'	0° 44'	− 5° 16'	− 9° 33'
Relative strain of the molecule [kJ/mol]	116.0	109.0	25.1	0.0	26.4
Relative strain per CH_2 group [kJ/mol]	38.6	27.2	5.0	0.0	3.8

According to this, in the homologous series of cycloalkanes, cyclohexane, and not cyclopentane, is the least strained compound. Cycloalkanes with large rings (*cf. Chapt. 2.5.2*) attain comparable values.

This contradiction can be resolved as soon as preconceived ideas of the planar geometry of the idealized ring compounds are abandoned. The three-membered ring has no choice in the matter and must be planar. Four- and five-membered rings may just as well be planar, to keep the valence-angle deviation as low as possible. From the six-membered ring upwards, however, non-planar models with no *Baeyer* strain may easily be constructed. The fact that cyclohexane is free of *Baeyer* strain can actually be used as an argument in itself for the non-planar construction of this system.

6.3.2. The Favored Chair Conformation of Cyclohexane

Hermann Sachse (1890) and *Ernst W.M. Mohr* (1918) had persuasively argued that, for cyclohexane, there exists more than one conformation (as we say today), which exhibits no *Baeyer* strain. These conformations are described perceptibly now as the rigid *chair* conformation and the flexible *boat* conformation (*Fig. 6.4*).

The assumption of strain-free chair and boat conformations, however, remained practically without consequences. *Giulio Natta* and *Mario Farina*, in an introduction to stereochemical problems, have pointed out that this fact delayed the systematic development of chemistry in virtually all areas. If this is correct, then it is worth fathoming the collective psychology which hindered the adoption of *Sachse*'s consideration.

The most important reason is the lack of agreement between the number of isomers actually observed and that one expected, should cyclohexane and its derivatives occur in more than one conformation: although *Sachse* had emphatically stressed that, because of rapid conformational alteration, the formal maximum number of isomers would not come into being, a situation already known in the case of ethane (*Chapt. 2.5.1.2*). This explanation was regarded by *Sachse*'s contemporaries as an unnecessary complication: rather than lay to rest *von Baeyer*'s planar molecule models for cycloalkanes in some transitory, impermanent condition of averages, people stayed with the *status quo ante*. The tried and tested criterion of using an experimentally assessible number of isomers as the basis for decisions about structure models was, in this instance, ossified into dogma.

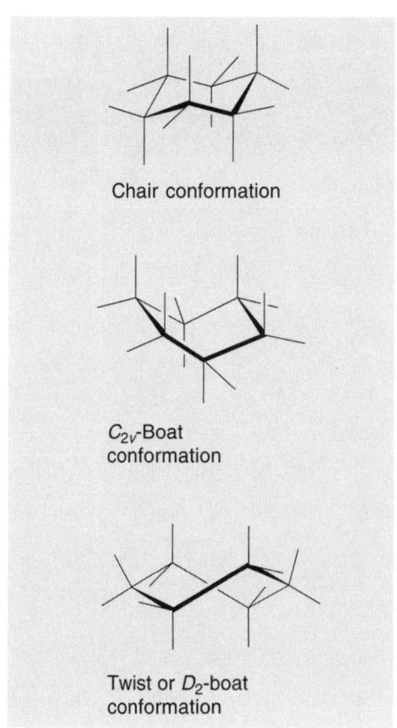

Fig. 6.4. *Chair, C_{2v}-boat, and D_2-boat conformations drawn in perspective*

Odd Hassel, who, together with *Derek Barton*, received the *Nobel* Prize for Chemistry in 1969, had, in the intervening period, used electron diffraction to disclose that cyclohexane (and its derivatives), when possible, exists in the chair conformation, as X-ray crystal-structure analysis had already shown for the steroid ring system (*cf. Fig. 6.2*). It was *Hassel* who uncovered that one half of the twelve C–H bonds in the chair conformation of cyclohexane were arranged parallel to the 3-fold rotation axis, and that the other six C–H bonds were close to the imaginary 'equator' plane of the ring system. Accordingly, the former group of C–H bonds were later named *axial* (abbreviated to *a*), and the latter one as *equatorial* (abbreviated as *e*, *Fig. 6.5*). To comprehend why the chair conformation is energetically preferred to the boat conformation, it seems appropriate first to discuss the energetically favored conformations of simple representatives from the homologous series of alkanes.

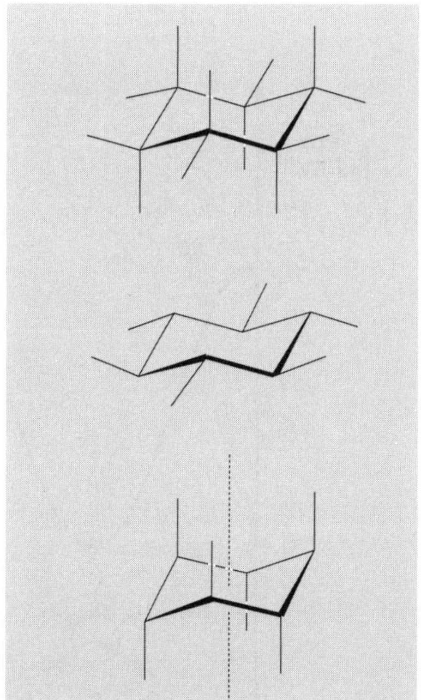

Fig. 6.5. *Different representations* (from top to bottom) *of cyclohexane, of cyclohexane with only equatorial C–H bonds shown, and of cyclohexane with only axial C–H bonds shown*

6.3.3. *Pitzer* Strain

In the application of the tetrahedron theory to compounds with C–C bonds, *van't Hoff* had expressly allowed free rotation, in order to avoid *the possibility of infinite isomerism*, as, without this assumption, *a whole* *host of isomers would be imaginable, even for a compound like C_2H_6* (*Chapt. 2.5.1.2*). However, in the case of 2,6,2′,6′-tetrasubstituted 1,1′-biphenyl derivatives, additional isomers were discovered: free rotation here is obviously limited (*Chapt. 2.6.3.1*). In the meantime, it has become known that the rotational barrier between two stereoisomers must amount to at least 100 kJ/mol, in order for the isomers to have a separate existence, at room temperature, of a few hours. From this dependence of lifetime on the height of the isomerization barrier, it follows that a molecule can sometimes be present in one conformation, sometimes in another, and does not have to be fixed. It is easy to imagine cases in which the rotational barrier is less than 100 kJ/mol, and rotational isomerism at room temperature is not easy to detect.

This barrier must be considered whenever the energy of a molecule, which consists of *translational energy*, *rotational energy*, and *vibrational energy*, is to be estimated. *Kenneth S. Pitzer* had emphasized in the thirties that the theoretically calculated enthalpy for ethane only agreed with the experimentally determined one, if, in addition to the rotation of the whole molecule (*external rotation*), the hindered rotation of the two Me groups against each other (*internal rotation*) was also taken into account. He obtained a sine curve (*Fig. 6.6*) for the inner rotational energy of ethane, in which the peak amplitudes totaled *ca.* 12 kJ/mol. This value is characteristic of the barrier to internal rotation of ethane. While much lower than the previously mentioned 100 kJ/mol – the value required for two stereoisomers to have a brief, separate existence at room temperature – it is appreciably higher than the 2.5 kJ/mol

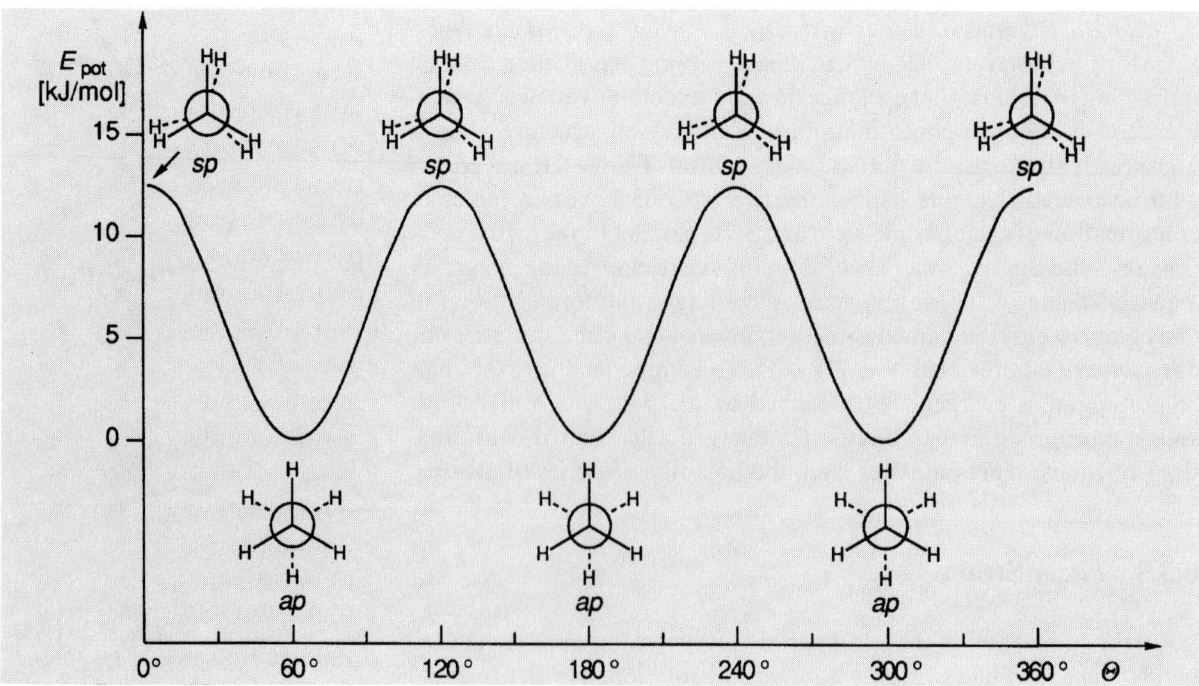

Fig. 6.6. *Energy profile of ethane*

which corresponds to the product $R \cdot T$ – the mean thermal energy available, at room temperature, for the C–C bond rotation. Ethane, therefore, is a molecule in which, under standard conditions, the Me groups, although hindered, are able to rotate relative to one another.

For the two Me groups to undergo a complete rotation, the barrier must be traversed three times. The extremes in the potential curve of ethane (*Fig. 6.6*) correspond to the illustrated conformations. Each minimum corresponds to the *staggered* conformation, and each maximum to the *eclipsed* conformation. The H-atoms of the two Me groups are *at maximum distance* in the staggered and *in closest proximity* in the eclipsed conformation. The assignment made in *Fig. 6.6* was carried out using IR spectroscopy. Each structure with eclipsed C–H bonds will, as a consequence, possess a certain degree of additional strain. This so-called torsion strain is known as *Pitzer* strain, distinguishing it from *Baeyer* strain caused by valence-angle deformation. The existence of *Pitzer* strain cannot be predicted from the structure model of the classical organic chemistry.

6.3.4. Estimation of Energy Differences Using the *Turner* Increment Procedure: *Newman* Strain

Richard B. Turner has developed a method which permits an illustrative determination of the energetic order of the various conformations of cyclohexane and its methyl derivatives. He examined which conformations of butane, which serve as a set of conformational modules, how often could be fitted into each of the conformations under comparison of a cyclohexane structure. If this number was multiplied by the energy value characteristic for the corresponding conformational module ($E_{ap} = 0$; $E_{sc} = 3.6$; $E_{sp} = 26.4$ kJ/mol), and the products obtained summed up, then energy values suitable for comparison were obtained. The specifications of the conformations assigned to the extremes of the butane energy profile are to be found in *Chapt. 10.3.2*.

Turner's method, applied to the chair conformation of unsubstituted cyclohexane (*Fig. 6.7*), allows the *sc*-conformational module ($E_{sc} = 3.6$ kJ/mol) of butane to be fitted into the chair six times. A relative potential energy of 6×3.6 kJ/mol, therefore, results for the chair conformation: 21.6 kJ/mol. Similarly, four *sc*- ($E_{sc} = 3.6$ kJ/mol) and two *sp*-conformational modules ($E_{sp} = 26.4$ kJ/mol) comprise the boat conformation. The resulting relative potential energy of the boat conformation is, therefore, 4×3.6 kJ/mol $+ 2 \times 26.4$ kJ/mol $= E_{boat} = 67.2$ kJ/mol. Comparison of the chair and boat potential energies shows that the chair conformation is favored by 45.6 kJ/mol.

Turner's increment procedure is a rough approximation. Two terminal H-atoms of butane in the *sc*-conformational module, which approach each other closely, must be replaced by two C-atoms which are bound to each other in the cyclohexane chair conformation. That these approximations nevertheless afford useful results in many cases is due to the cancelling out of shortcomings in many subtractions.

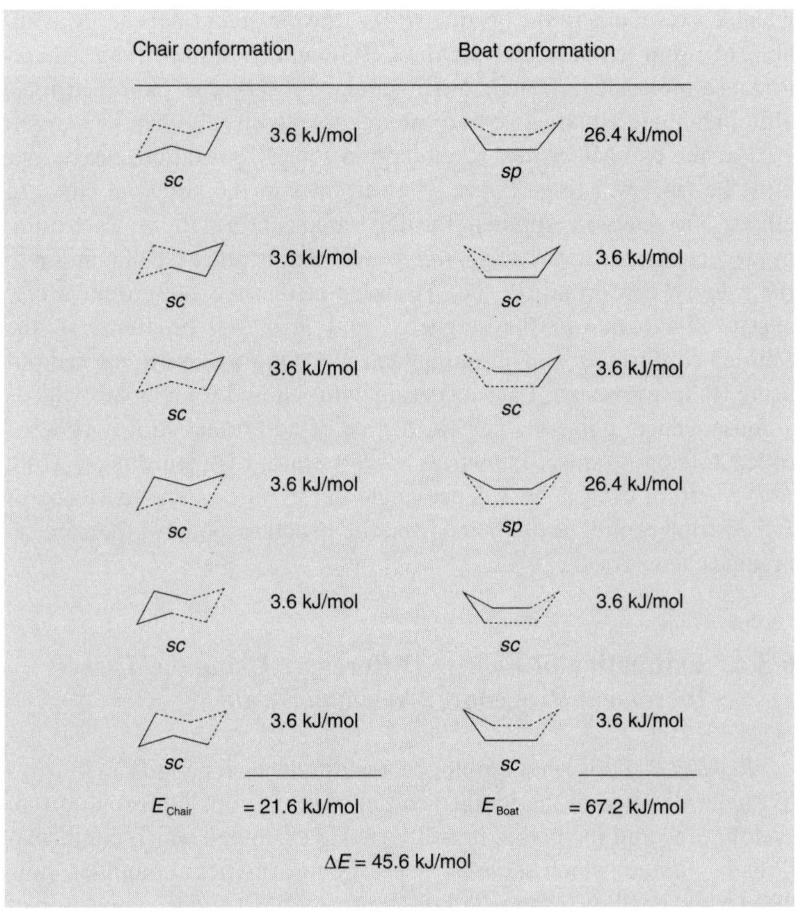

Fig. 6.7. *Estimation of the potential energy of cyclohexane in the chair or in the boat conformation*

The simplest representative of the Me-substituted cyclohexanes, the monosubstituted derivative, can exist in two chair conformations, which differ in the orientation (axial or equatorial) of the Me group (*Fig. 6.8*). Using *Turner*'s method, six *sc*- (E_{sc} = 3.6 kJ/mol) and two *ap*- (E_{ap} = 0

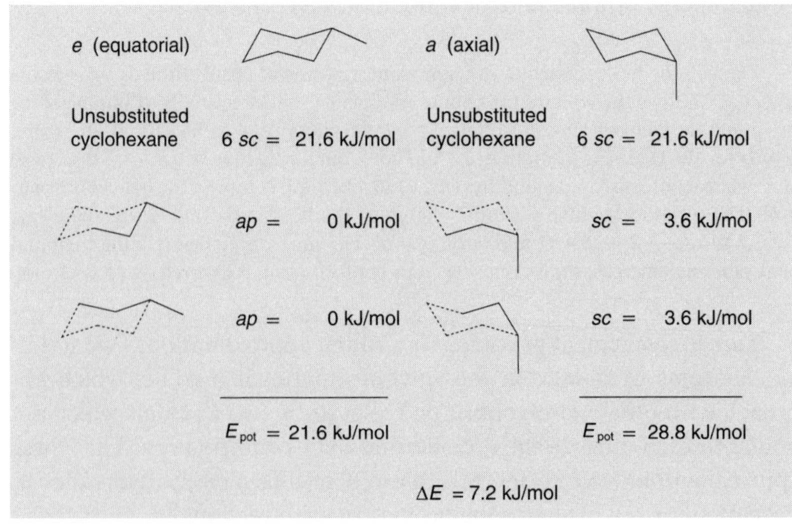

Fig. 6.8. *Estimation of the potential energy of methylcyclohexane in the chair conformation with an equatorial or an axial Me group*

kJ/mol) butane conformational modules may be fitted into the conformation with an equatorial Me group. For this conformer, an energy value of 21.6 kJ/mol ($6 \times 3.6 + 2 \times 0$ kJ/mol) is, therefore, obtained. The conformer with an axial Me group can be put together out of eight *sc*-conformational modules of butane, and so has a potential energy of 8×3.6 kJ/mol = 28.8 kJ/mol. The conformer with the equatorial Me group is thus about 7.2 kJ/mol lower in energy than the axially substituted conformer.

In 1,2-disubstituted cyclohexane derivatives, the two substituents may be *cis*- or *trans*-oriented. A planar configurational formula of *cis*-1,2-dimethylcyclohexane depicts an achiral *meso*-compound. The non-planar conformational formulae show that two chiral chair conformations are present, which exist as mirror images of each other (*Table 6.2*). *Turner*'s method allows the same butane conformational modules, in the same proportions (nine *sc*- and two *ap*-conformational modules in each case), to be fitted into both conformational formulae. The two conformers of *cis*-1,2-dimethylcyclohexane are energetically equivalent, and exist as a racemic mixture, whose components continually interconvert *via* ring inversion (see *Sect. 6.3.6*).

trans-1,2-Dimethylcyclohexane (*Table 6.2*) is already chiral in the planar formula. The two conformers (with equatorial or with axial Me groups) are diastereoisomers. Estimation of the potential energy for both conformations shows the conformer with equatorial Me groups, with E_{pot} = 25.2 kJ/mol (seven butane *sc*- and four *ap*-conformational modules), to be the lower-energy compound, overwhelmingly predominating in the equilibrium between the two conformers. The conformer with axial Me groups has a potential energy of 36.0 kJ/mol (ten *sc*- and one *ap*-conformational module(s) of butane).

For 1,3- and 1,4-dimethylcyclohexane, similarly, a *cis*- or *trans*-orientation of the Me groups is possible. *cis*-1,3-Dimethylcyclohexane is a *meso*-compound, and hence achiral. This can be seen in the planar configurational formula just as easily as in the two conformational formulae, which are interconvertable through ring inversion (*Table 6.2*). Of the latter, one conformer exhibits two equatorially oriented Me groups. *Turner*'s method results in further four butane *ap*-conformational modules being added to the six *sc*-conformational modules already present in the unsubstituted compound. This does not alter the relative potential-energy content, however, provided that all that is being considered is comparisons with other dimethylcyclohexanes. The comparison with the other chair conformation, in which the two Me groups are axially oriented, is particularly relevant. *Turner*'s method leads to an increase to a total of ten in the number of butane *sc*-conformational modules which fit, therefore, increasing the relative potential-energy content to 36.0 kJ/mol (*Table 6.2*). However, the *repulsive 1,5-interaction* has not yet been taken into account. This concerns the fact that the two axial Me groups in 1,3-positions are located on the same side of the molecule, when in the undistorted chair conformation, and that their *van der Waals* radii overlap: the potential energy would, therefore, be further increased by a considerable *van der Waals* strain.

Table 6.2. *Information on the Different Dimethyl-Substituted Cyclohexane Derivatives. Left-hand column:* number of configurational isomers (achiral: no configurational isomers; chiral: two enantiomers). *Center column:* orientation of the two Me groups (*e:* equatorial, *a:* axial) and indication of the relationship between the inverse-chair conformations (identical, enantiomeric, or diastereoisomeric). *Right-hand column:* estimation of the potential energy using *R. B. Turner*'s method.

Planar formulae for representation of constitution and configuration	Conformational formulae	Estimated energy after *R. B. Turner*
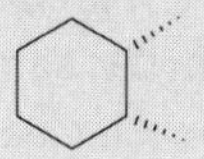 achiral	*a,e* ⇌ *e,a* 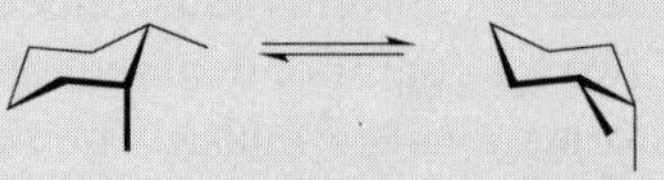 enantiomeric	*a,e* = *e,a* 9 *sc* = 32.4 kJ/mol 2 *ap* = 0 kJ/mol E_{pot} = 32.4 kJ/mol
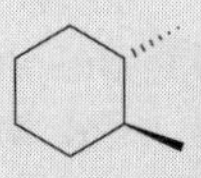 chiral	*e,e* ⇌ *a,a* 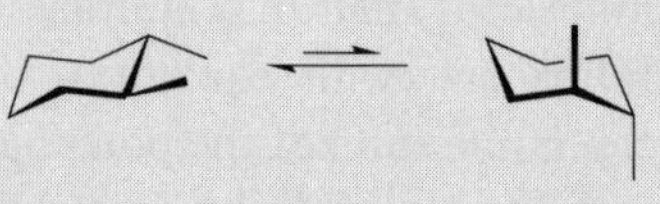 diastereoisomeric	*e,e* 7 *sc* = 25.2 kJ/mol 4 *ap* = 0 kJ/mol E_{pot} = 25.2 kJ/mol *a,a* 10 *sc* = 36.0 kJ/mol 1 *ap* = 0 kJ/mol E_{pot} = 36.0 kJ/mol
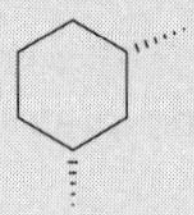 achiral	*e,e* ⇌ *a,a* 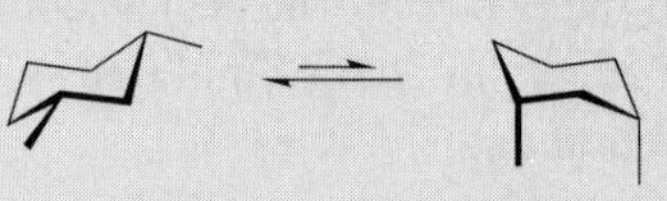 diastereoisomeric	*e,e* 6 *sc* = 21.6 kJ/mol 4 *ap* = 0 kJ/mol E_{pot} = 21.6 kJ/mol *a,a* 10 *sc* = 36.0 kJ/mol E_{pot} = 36.0 kJ/mol
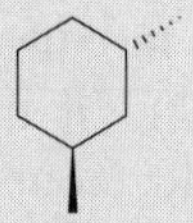 chiral	*a,e* ⇌ *e,a* 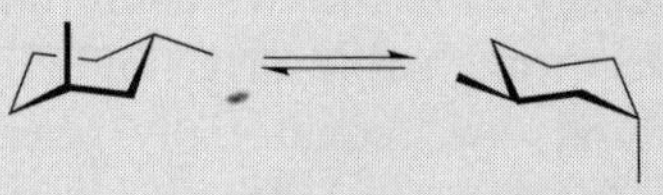 equivalent	*a,e* = *e,a* 8 *sc* = 28.8 kJ/mol 2 *ap* = 0 kJ/mol E_{pot} = 28.8 kJ/mol
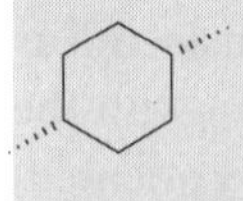 achiral	*a,e* ⇌ *e,a*  equivalent	*a,e* = *e,a* 8 *sc* = 28.8 kJ/mol 2 *ap* = 0 kJ/mol E_{pot} = 28.8 kJ/mol
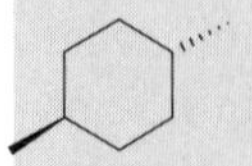 achiral	*e,e* ⇌ *a,a* 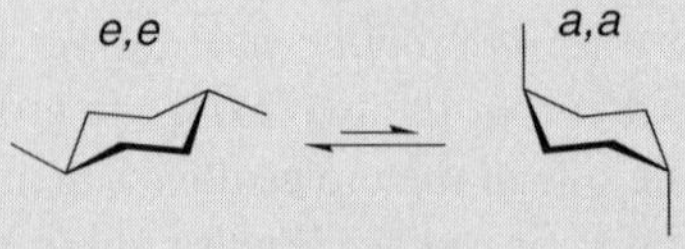 diastereoisomeric	*e,e* 6 *sc* = 21.6 kJ/mol 4 *ap* = 0 kJ/mol E_{pot} = 21.6 kJ/mol *a,a* 10 *sc* = 36.0 kJ/mol E_{pot} = 36.0 kJ/mol

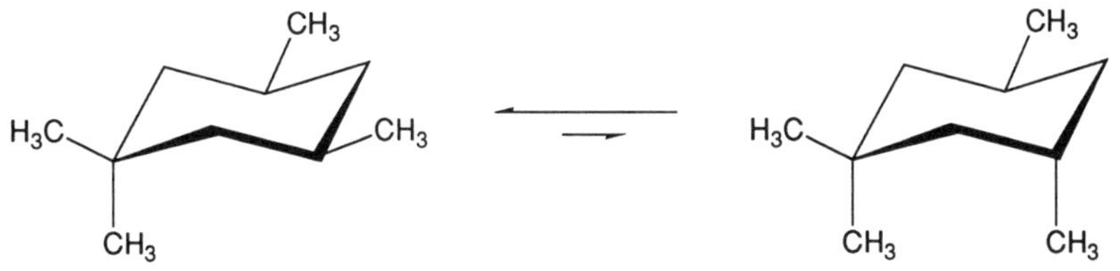

The additional strain energy for a cyclohexane derivative with 1,3-diaxial Me/Me interactions has been determined, with the aid of the equilibrium between *cis*- and *trans*-1,1,3,5-tetramethylcyclohexane (which occurs at 520 to 631 K in the presence of Pd/C), as 15.5 kJ/mol.

This value represents the *van der Waals* strain of an idealized cyclohexane ring (without *Baeyer* and *Pitzer* strain). The potential-energy difference of the two conformations of *cis*-1,3-dimethylcyclohexane would, therefore, increase to $36.0 + 15.5 - 21.6 = 29.9$ kJ/mol. In reality, the potential energy would be decreased by a 'compromise' between the three different (assumed) causes of strain. So, in this case, the two Me groups may distance themselves from each other, causing a flattening of the six-membered ring and leading to a certain additional degree of *Baeyer* and/or *Pitzer* strain. Under normal conditions, however, it may be assumed that, in the equilibrium of the two diastereoisomorphic conformations, the one with equatorially oriented Me groups will be present almost exclusively (which would mean that the conformation with axial Me groups is '*forbidden*').

The rule that conformations with 1,5-repulsion between two non-H-atoms may be discounted – assuming that no plausible exception based on constitutional grounds can be made – was formulated as early as 1950 by *Melvin S. Newman* in his '*Rule of Six*'. From now on, we will only occasionally speak of *1,5-repulsion between two non-H-atoms*, using instead – following a proposal made by *Albert Eschenmoser* – the term *Newman* strain.

trans-1,3-Dimethylcyclohexane is chiral, regardless of whether the planar configurational formula or the chair conformational formulae are considered. The two conformational formulae, with one equatorial and one axial Me group, which interconvert by ring inversion, indicate identical conformations. In contrast to *cis*-1,2-dimethylcyclohexane, a separation into enantiomers is possible for *trans*-1,3-dimethylcyclohexane.

At high temperatures, in the presence of hydrogenation catalysts, *cis*- and *trans*-1,3-dimethylcyclohexane can be brought into equilibrium, in which the *cis*-isomer predominates. This result is in agreement with that of *Turner*'s approach: the 1*e*,3*e*-conformation (*cis*) is about 7.2 kJ/mol lower in energy than the 1*e*,3*a*-conformation (*trans*) (*Table 6.2*).

cis-1,4-Dimethylcyclohexane is achiral; as both the planar configurational formula and also the chair conformational formula show, reflection across a mirror plane (in which the substituted centers are located) is possible. Two conformations can interconvert by ring inversion; on closer consideration these are shown to be identical. We have already encountered (*Chapt. 2.6.2.2*) such a case of *degenerate isomerization* or *topomerization* (*Chapt. 2.6.2.2*). According to *Turner*'s method, two *ap*- and eight *sc*-conformational modules of butane (E_{pot} = 28.8 kJ/mol; *Table 6.2*) may be fitted into the conformation with one equatorial and one axial Me group.

trans-1,4-Dimethylcyclohexane is achiral, in analogy with its *cis*-isomer. Its two conformations, which can interconvert by ring inversion, are diastereoisomers and energetically distinct. The dominant conformer in the equilibrium has two equatorial Me groups and, with a total of four *ap*- and six *sc*-conformational modules of butane, totals up to a relative potential energy of 21.6 kJ/mol (*Table 6.2*). The higher-energy conformer contains ten butane *sc*-conformational modules: the relative total potential energy increases to 36.0 kJ/mol. It may clearly be predicted that the *trans*-1,4-dimethylcyclohexane is lower in energy than the *cis*-isomer, and that the 1*e*,4*e*-conformer is lower in energy than the 1*a*,4*a*-conformer.

On comparing the energy of each of the *cis*- and *trans*-configurational isomers of the 1,2-, 1,3- or 1,4-disubstituted cyclohexanes described here, it can be established that, in general, the most energetically favored configuration is that in which one conformation exists in which both Me groups are in an equatorial orientation. In the case of 1,2-dimethylcyclohexane, it is the *trans*-isomer, for 1,3-dimethylcyclohexane the *cis*-isomer, and for 1,4-dimethylcyclohexane once more the *trans*-isomer.

From ethane and butane we have learned that the staggered conformational type is energetically favored over the eclipsed, and the *ap*-conformation over the *sc*-conformation. From cyclohexane, we know that the chair conformation is energetically favored over the boat conformation. For various dimethylcyclohexane isomers, those with the greatest possible numbers of equatorial Me groups are the lowest in energy. In the case of *cis*-1,3-dimethylcyclohexane, we hit upon an instance in which *Newman* strain caused such a reduction in conformational space that, under normal conditions, only one of the two chair conformations should be considered. We will repeatedly encounter *Newman* strain from now on. The introduction of additional 1,5-repulsions into cyclic systems leads to cases which, structurally, are essentially 'frozen'. *cis*-1-Ethyl-2-isopropylcyclohexane is one such example. No conformation comes into question, apart from the one shown.

H
CH_3
CH_3
H
H
H
H_3C
H

To illustrate further stereostructural phenomena involving complex cyclohexane derivatives, we shall consider fused and bridged bi- or polycyclic systems (see *Chapt. 2.5.2*).

6.3.5. Fused and Bridged Ring Systems

6.3.5.1. Decalins

Decalin (= decahydronaphthalene, bicyclo[4.4.0]decane) lies at the beginning of a series of fused polycyclic ring systems consisting purely of saturated six-membered rings. The two six-membered rings may be either *trans*- or *cis*-fused. A substantial simplification may also be made in this case, by regarding the ring system as being planar only in order to predict the correct number of stereoisomers, which are persistent and separately detectable: there are only two, *cis*- and *trans*-decalin.

If *trans*-decalin is regarded as a 1,2-disubstituted cyclohexane derivative, it can easily be seen from the conformational formula with two fused chairs (*Table 6.3*) that the bridgehead H-atoms are both axially and the adjacent C-atoms of the fused ring are each equatorially oriented. As axial bonds are converted into equatorial and equatorial into axial upon ring inversion of a chair conformation (*Sect. 6.3.6*), the two six-membered rings of *trans*-decalin cannot invert: the fused ring is too small to undergo the necessary conformational alterations. *trans*-Decalin is rigid and conformationally inflexible. At most, either of the two cyclohexane rings could assume the higher-energy boat conformation: the tendency to do this will be slight, however. *trans*-Decalin belongs to the symmetry point group C_{2h} and is, therefore, achiral. The planar configurational formula, the conformational formula with two fused chairs, and the *Newman* projection for *trans*-decalin are given in the upper row of *Table 6.3*.

Table 6.3. *Representation of* cis- *and* trans-*Decalin. Left-hand column:* planar formulae with the real number of configurational isomers. *Center column:* conformational formulae for the achiral *trans*-decalin and for the equilibrium of the enantiomers of *cis*-decalin. *Right-hand column:* torsion angles.

Planar formulae for representation of constitution and relative configuration	Conformational formulae	*Newman* Projection
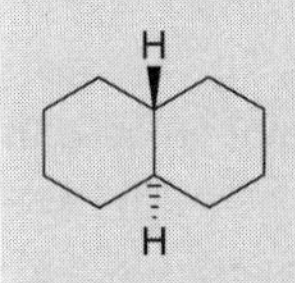 H, H *trans*-Decalin	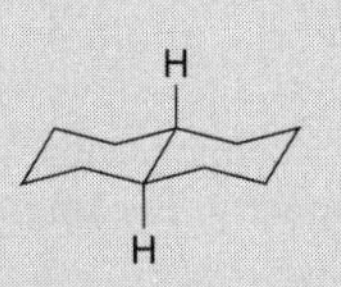 H, H	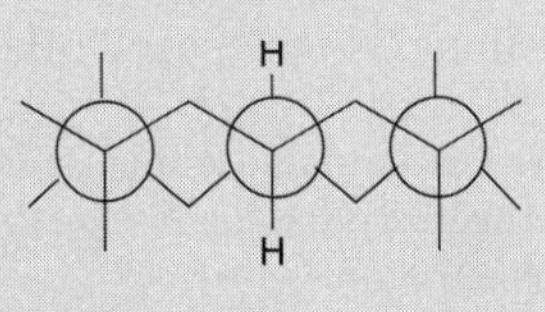 H, H
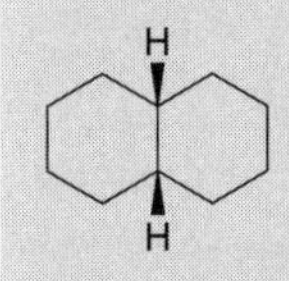 H, H *cis*-Decalin	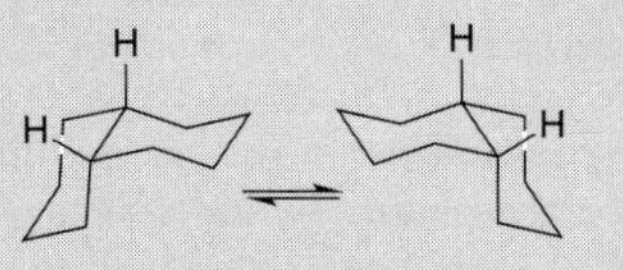 H, H ⇌ H, H	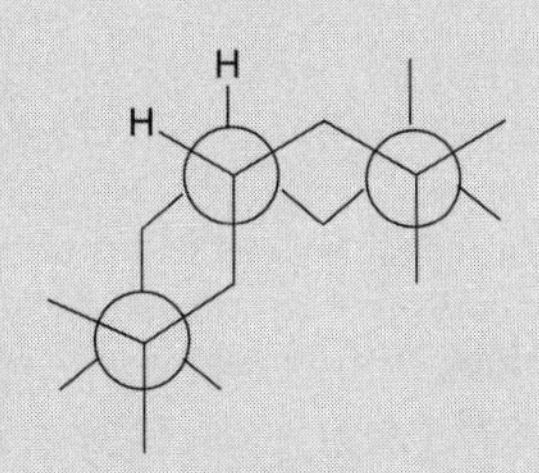 H, H

In *cis*-decalin, one of the two bridgehead H-atoms is equatorially oriented, the other axially. Hence, one of the adjacent C-atoms in the fused ring must be axially oriented, the other equatorially. This has the consequence that an inversion of both chair-form rings is possible, through a cooperative rotation about the C–C bonds. From the planar formula, one would suspect that, for *cis*-decalin, one would be dealing with a *meso*-compound. The individual conformational formulae, however, do not permit the discernment of any mirror symmetry. For *cis*-decalin, therefore, there exist two enantiomorphic conformations. These cannot be separated from each other, as, under normal conditions, they rapidly and continually interconvert. That *cis*-decalin (under normal conditions) cannot be optically active can be deduced both from the configurational formula and from the conformational formulae. The cause of the optical inactivity is different in the two cases, however: the configurational formula suggests internal compensation of a *meso*-compound, the dynamic system of enantiomorphic conformers exhibits external compensation. The lower row of *Table 6.3* gives the static configurational formula, the dynamic relationship between the two conformers, and the *Newman* projection for one of these two conformers.

Comparison of the relative energy values of *cis*- and *trans*-decalin using *Turner*'s method gives a potential energy of 54.0 kJ/mol (15 *sc*- and three *ap*-conformational modules of butane) for the *cis*-isomer. The *trans*-isomer, with $E_{pot} = 43.2$ kJ/mol (twelve *sc*- and six *ap*-conformational modules of butane), is significantly lower in energy and predominates by far if a *cis*/*trans*-equilibrium can be established.

6.3.5.2. Perhydroanthracenes (= Tricyclo[8.4.0.0^{3,8}]tetradecanes)

Fusion of a further saturated six-membered ring onto decalin leads to perhydroanthracene (linear) or perhydrophenanthrene (angular). In both cases, the stereodescriptors *cis* or *trans* are used to describe the nature of the fusion of the two terminal rings onto the center ring. The stereodescriptors *syn* or *anti* give the relative orientation of the two terminal rings to each other (*Tables 6.4* and *6.5*).

The formal maximum number of 16 configurational isomers is not attained in perhydroanthracene, for reasons of symmetry. Looking at the eight formal relative configurations,

trans/*syn*/*trans*	*trans*/*anti*/*trans*
trans/*syn*/*cis*	*trans*/*anti*/*cis*
cis/*syn*/*trans*	*cis*/*anti*/*trans*
cis/*syn*/*cis*	*cis*/*anti*/*cis*

it can easily be seen that the descriptors *trans*/*syn*/*cis*, *cis*/*syn*/*trans*, *trans*/*anti*/*cis*, and *cis*/*anti*/*trans* describe identical arrangements. Of the remaining five relative configurations,

trans/*syn*/*trans*
cis/*syn*/*cis*
cis/*anti*/*cis*
cis/*anti*/*trans*
trans/*anti*/*trans*

the first three describe *meso*-structures, so that here no further configurational description is necessary or possible. The last two each describe a pair of enantiomers, so that it is necessary to assign an absolute configuration for an unambiguous configurational description. Here, however, only the potential-energy values are of interest, and these do not exhibit any differences between enantiomers.

Table 6.4 shows the relative configurations of perhydroanthracene, ordered by potential-energy content. This order corresponds to a large extent to the predictions based on previous discussion, provided that ascertainment is only made of

- how many equatorial or axial C–C bonds radiating from the center ring are included in the whole C-skeleton,
- the nature of the interactions through 1,3-diaxial bonds to ligands at the bridgehead atoms.

The *trans/syn/trans*-configuration is without doubt the one of lowest energy. Four equatorial C–C bonds radiate from the center ring, which has a chair conformation, and the 1,3-diaxial interactions all involve H/H interactions. Following this comes the *cis/anti/trans*-configuration, with three equatorial C–C bonds, one axial C–C bond, one diaxial

Table 6.4. *The Five Relative Configurations of Perhydroanthracene. Left-hand column:* notation; listing of the C–C bonds radiating from the center ring and description of the 1,5-arrangement of axial ligands. *Center column:* conformational formulae. *Right-hand column:* relative potential energy calculated by molecular mechanics (force field: MM2 (85)).

Description of relative configuration	Conformational formulae	Relative energy using MM2 (85) [kJ/mol]
trans/syn/trans 4 *e* C–C 2 diaxial H/H	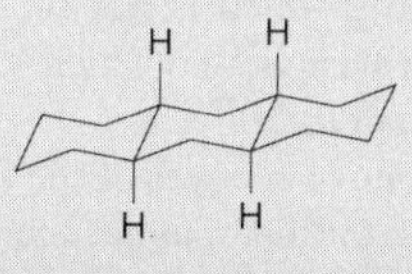	0.0
cis/anti/trans 3 *e* (1 *a*) C–C 1 diaxial H/H 1 diaxial C/H	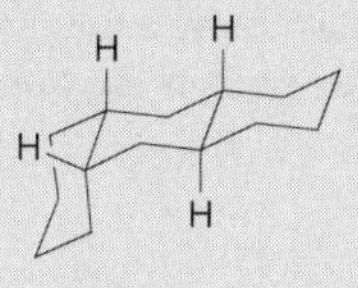	11.3
cis/anti/cis 2 *e* (2 *a*) C–C 2 diaxial C/H	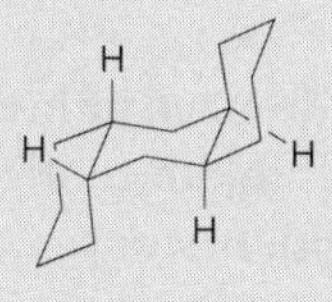	23.4
trans/anti/trans	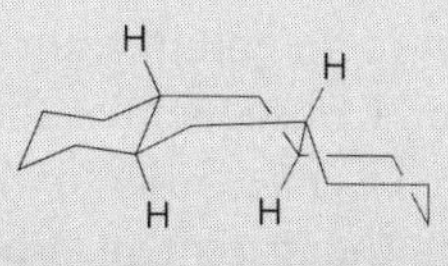	25.5 (Twist-boat)
cis/syn/cis 2 *e* (2 *a*) C–C 1 diaxial H/H 1 diaxial C/C	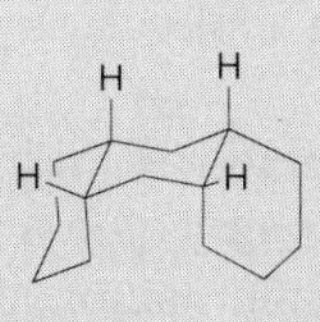	31.8

interaction of the H/H type and one of the C/H type. In the next *cis/anti/cis*-configuration, two equatorial and two axial C–C bonds radiate from the center ring in the chair conformation, and the diaxial interactions are both of the C/H type.

Without question, the *cis/syn/cis*-configuration takes the last place: there are only two equatorial and two axial C–C bonds radiating from the center ring in chair conformation, but the 1,3-diaxial interactions include one of the C/C type, as well as one of H/H type. It is less clear which position the *trans/anti/trans*-configuration comes in. This configuration is noteworthy, in that its central six-membered ring cannot adopt a chair conformation. This offers the opportunity to experimentally find out something about the potential energy of a boat conformation relative to a chair conformation. Namely, if the potential energies of each of the *trans/syn/trans*- and the *trans/anti/trans*-isomers are determined by combustion, then the difference is the degree to which the *trans/anti/trans*-isomer is higher in energy than the *trans/syn/trans*-isomer, and hence the energy difference between the chair and the boat conformations of cyclohexane. The difference, determined in this manner, amounts to 23 kJ/mol.

Using a *Dreiding* model of the boat conformation of cyclohexane, it is not difficult to see that the untwisted boat (with C_{2v} symmetry) can easily, without having to alter valence angles, be converted into the twist-boat (with D_2 symmetry). The *Pitzer* strain in the latter, and therefore the total strain, is lower than in the former.

The *trans/anti/trans*-configuration belongs in the vicinity of the *cis/anti/cis*-configuration in the energetic ordering, and should be a little higher in energy. Precise observation (of *Dreiding* models, for example) shows that the two configurational isomers do not only differ from each other in the conformation of the center ring, but that, in the stereoisomer with the boat conformation, the terminal rings introduce a small, additional *Pitzer* strain, which is not present in the configurational isomer consisting purely of chair-form rings, and which will not be eliminated by subtraction of the corresponding increment total.

In the meantime, the order of energies of the five perhydroanthracene configurations has been calculated with the aid of molecular mechanics, which will be dealt with in greater detail in *Aspects of Organic Chemistry*, Volume 4: *Methods of Structure Determination*. *Table 6.4* lists the relative amounts by which the other four configurational isomers are higher in energy than the *trans/syn/trans*-configuration, which is defined as the zero level.

6.3.5.3. Perhydrophenanthrenes (= Tricyclo[8.4.0.0^{2,7}]tetradecanes)

Among the ten perhydrophenanthrene stereoisomers, which differ from each other in their absolute configurations, are found one *meso*-perhydrophenanthrene (with *trans/syn/trans*-fusion), and one case (with *cis/syn/cis*-fusion) in which, under normal conditions, two enantiomorphic conformations exist in an equilibrium sufficiently rapid to exclude any chance of separation. Consequently, there are a total of six relative configurations (*Table 6.5*). The order of their potential energies can largely be obtained using the increment procedure already described. It has been confirmed using molecular-mechanics methods (right-hand column of *Table 6.5*).

As in the perhydroanthracene series, the configuration in which the center ring has the boat conformation is higher in energy than would be expected from the boat conformation alone. The terminal rings cause additional *Pitzer* strain here as well.

Table 6.5. *The Six Relative Configurations of Perhydrophenanthrene.* For further details, see the legend of *Table 6.4.*

Description of relative configuration	Conformational formulae	Relative energy using MM2 (85) [kJ/mol]
trans/anti/trans 4 *e* C–C 2 diaxial H/H	H H H H	0.0
cis/syn/trans 3 *e* (1 *a*) C–C 1 diaxial H/H 1 diaxial C/H	H H H H	10.5
cis/anti/trans 3 *e* (1 *a*) C–C 1 diaxial H/H 1 diaxial C/H	H H H H	10.9
cis/anti/cis 2 *e* (2 a) C–C 2 diaxial C/H	H H H H	23.0
cis/syn/cis 2 *e* (2 *a*) C–C 1 diaxial H/H 1 diaxial C/C	H H H H	31.4
trans/syn/trans	H H H H	35.1 (Twist-boat)

6.3.5.4. Adamantane, Twistane, Bicyclo[2.2.2]octane

The bridged systems adamantane (= tricyclo[3.3.1.1^{3,7}]decane), twistane (= tricyclo[4.4.0.0^{3,8}]decane), and bicyclo[2.2.2]octane are of interest, because, in each of these rigid structures, different cyclohexane conformations are 'frozen', owing to constitutional (molecular strain) factors.

Hence, the (D_{3d}-symmetrical) chair conformation of cyclohexane may be discerned four times in adamantane. Twistane conceals five (D_2-symmetrical) twist-boat conformations of cyclohexane in its structure. Three (C_{2v}-symmetrical) boat conformations of cyclohexane fit into

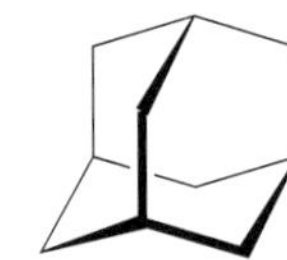
Adamantane

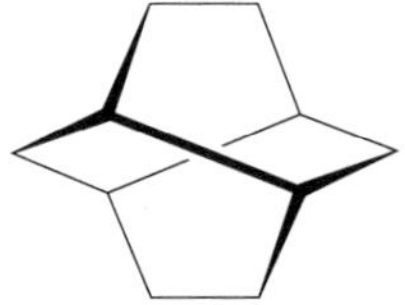
Twistane

Bicyclo[2.2.2]octane

bicyclo[2.2.2]octane, assuming that D_{3h}, rather than D_3 symmetry, is present.

The above-mentioned compounds provide examples of how organic chemists can experimentally verify structural hypotheses by the deliberate design and synthesis of suitable model compounds.

6.3.6. The Energy Profile of Cyclohexane

The agreement between the predicted and the experimentally determined number of isomers was, for a hundred years, the determining criterion of how the structure of organic compounds was viewed (*Chapt. 2.7*). This agreement is possible for cyclohexane and its derivatives, as long as a planar ring (with D_{6h} symmetry; valence angle δ = 120°, torsion angle τ = 0°) is assumed. A planar, saturated six-membered ring, however, exhibits *Baeyer* and also high *Pitzer* strain. It would not be compatible with the fact that cyclohexane is the least strained compound in the homologous series of cycloalkanes (*Sect. 6.3.1*). In the idealized model (*Sect. 6.3.2*) of cyclohexane with a chair-like arrangement (D_{3d} symmetry; δ = 109° 28′, τ = ± 60°), neither *Baeyer* nor *Pitzer* strain arises. In reality, the CCC valence angles come out at 111.4° and the torsion angles at ± 54.5°. As a consequence, the cyclohexane ring, compared to, for example, a *Dreiding* model, is somewhat flattened, and the axial ligands are distorted a little away from the molecule. The six axial H-atoms, however, remain homotopic relative to each other, as do the six equatorial H-atoms. Axial and equatorial H-atoms are, in contrast, diastereotopic. From the perspective of the chair-like cyclohexane ring, two stereoisomers should arise for a monosubstituted derivative: one with an axial, one with an equatorial substituent. As two such stereoisomers could never be detected separately, however, it has been assumed that, under normal conditions, ring inversion occurs.

Through ring inversion, every axial substituent becomes equatorial, every equatorial substituent becomes axial. This ring inversion proceeds through a comparatively high barrier ($\Delta G^{\neq}$ = 45.2 kJ/mol) from the chair conformation (in the global minimum) to the twist-boat conformation (in a local minimum) and finally to the inverse chair conformation (*Fig. 6.9*).

The twist-boat conformation (D_2 symmetry; $\delta \approx$ 109° 28′, $\tau \approx$ ± 33° or ∓ 70°) is in an equilibrium with the chair conformation in such a low proportion (< 0.1 %) that, at room temperature, experimental evidence of its existence is difficult to obtain. At higher temperatures, however, the concentration of the higher-energy twist-boat conformation should reach a not inconsiderable level (*ca.* 30% at 800 °C). *Frank A. L. Anet* and *Orville L. Chapman* measured IR spectra of two samples of cyclohexane at 10 K. One sample complied with the normal conformational mixture of cyclohexane. The other sample contained cyclohexane that had previously been heated to 1070 K and which had then been quenched under high vacuum at 20 K. From the difference in the IR spectra, the

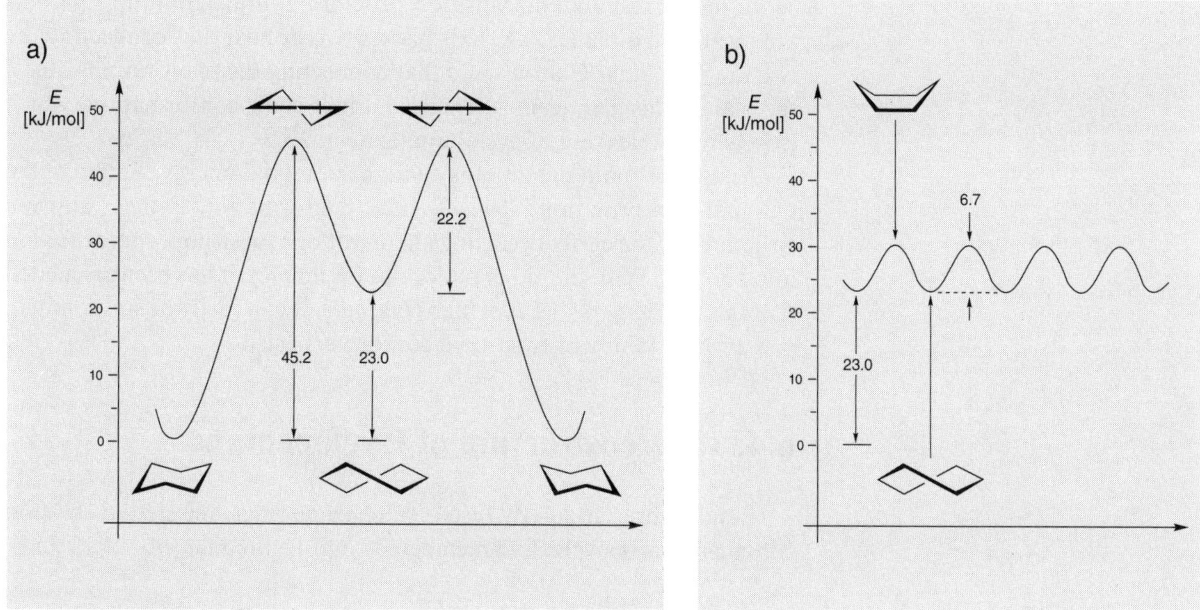

Fig. 6.9. *Energy profile of cyclohexane:* a) *chair, half-chair, and* D_2*-boat conformations,* b) D_2*- and* C_{2v}*-boat conformations*

proportion of the twist-boat conformation at 1070 K could be deduced, and, from the speed with which the characteristic bands of the higher-energy cyclohexane conformer disappeared, it was possible to calculate the energy barrier for the interconversion from twist-boat to chair conformation as 22.2 kJ/mol. Were the height of the energy barrier (45.2 kJ/mol according to *Fig. 6.9*) for the interconversion from chair to twist-boat conformation known, then it would be possible to calculate, by subtraction, the energy difference between chair and twist-boat conformations of cyclohexane (23.0 kJ/mol according to *Fig. 6.9*). *Fig. 6.9* contains the full set of data. But how was it possible to measure the energy barrier between the chair and the twist-boat conformation?

Only a method of measurement can be considered in whose 'exposure time' the change from chair conformation to inverse chair conformation takes place. It has been demonstrated that the time scale of NMR spectroscopy corresponds to that of the dynamic behavior of cyclohexane. Dynamic NMR spectroscopy is discussed in *Chapt. 13.*

The axial and equatorial protons of cyclohexane may be distinguished from each other using NMR spectroscopy at low temperatures, when the ring inversion has been slowed down relative to the NMR timescale. With NMR spectroscopy, it can not only be established that the protons in cyclohexane occupy two distinct positions, but also that their positions are continually alternating. Were it possible to measure the rate of this exchange, then it would be possible to determine the energy barrier for the ring inversion, which is the cause of the continual alternation in position of the individual protons. NMR Spectroscopy has indeed been used to obtain a value of 45.2 kJ/mol for this barrier.

In the corresponding transition structure, four neighboring C-atoms are located in a plane. A C_2 axis bisects two bonds: that connecting the two middle planar C-atoms and that connecting the two C-atoms out of the plane. This arrangement is called a half-chair conformation (which closely resembles that of cyclohexene; *Sect. 6.5.1*).

And what about the topology and the potential-energy content of the C_{2v}-boat conformation? The C_{2v}-boat conformation is the transition structure for the easily occurring change of one twist-boat conformation into another. With the aid of molecular mechanics, it has been calculated that this barrier is 6.7 kJ/mol high (*Fig. 6.9*). It can easily be surmounted, and a whole family of twist-boat conformers exists.

6.4. Stereostructure of Cyclopentane

Besides three angularly fused cyclohexane rings, the steroid skeleton contains a carbocyclic five-membered ring. In the majority of steroids,

Table 6.6. *Conformations of Cyclopentane. Left-hand column:* molecular models for (from top to bottom) planar, envelope, and half-chair conformations. *Center column:* symmetry elements respective to the plane of perspective (trace of the plane of symmetry perpendicular to the plane of perspective: —; two-fold rotational axis in the plane of perspective: ·····; if the trace of the plane of symmetry and rotational axis coincide, the sign — is used). *Right-hand column:* torsion angles with indication of sign in each case.

Molecular models of characteristic cyclopentane conformations	Symbols with symmetry elements	Symbols with torsion angles
(planar)		0, 0, 0, 0, 0
(envelope)		28.6 (+), 0, 28.6 (−), 46.1 (+), 46.1 (−)
(half-chair)		15.1 (−), 15.1 (−), 39.4 (+), 48.1 (−), 39.4 (+)

the five-membered ring *D* and the six-membered ring *C* are *trans*-fused to one another, as a result of which a considerable strain is introduced into the entire molecule. The conformation of ring *D* is non-planar. X-Ray crystal-structure analyses of numerous steroids have revealed that ring *D* adopts a conformation somewhere between the C_s and the C_2 conformations of cyclopentane (*Table 6.6*; for further details, see *Chapt. 11.1.2*). In cyclopentane itself, the five-membered ring is somewhat flatter than in substituted cyclopentanes, and, hence, than in steroids.

A planar cyclopentane ring would be almost free of *Baeyer* strain (CCC valence angle $\delta = 108°$), but would, however, exhibit considerable *Pitzer* strain (many eclipsed bonds). By way of compromise, the valence angles become significantly smaller than 108°. Using molecular mechanics, computer-assisted conformational analyses and subsequent energy minimizations have been carried out. According to these, the potential energy for the C_2 and the C_s conformations is the same, and attains its highest value for the planar conformation. Everything suggests that the local deviation from planarity wanders about the ring. It can be imagined starting from four C-atoms located in a plane, the fifth lying some 50 pm above or below it. It is not always the same ring center that protrudes out of the plane, but the unevenness wanders with the vibrating C-atoms through the ring (*pseudorotation*). As the pseudorotation cycle is not subject to any barrier greater than the vibrational energy, cyclopentane is freely mobile, and, in reality, a whole ensemble of conformers is present (*Fig. 6.10*).

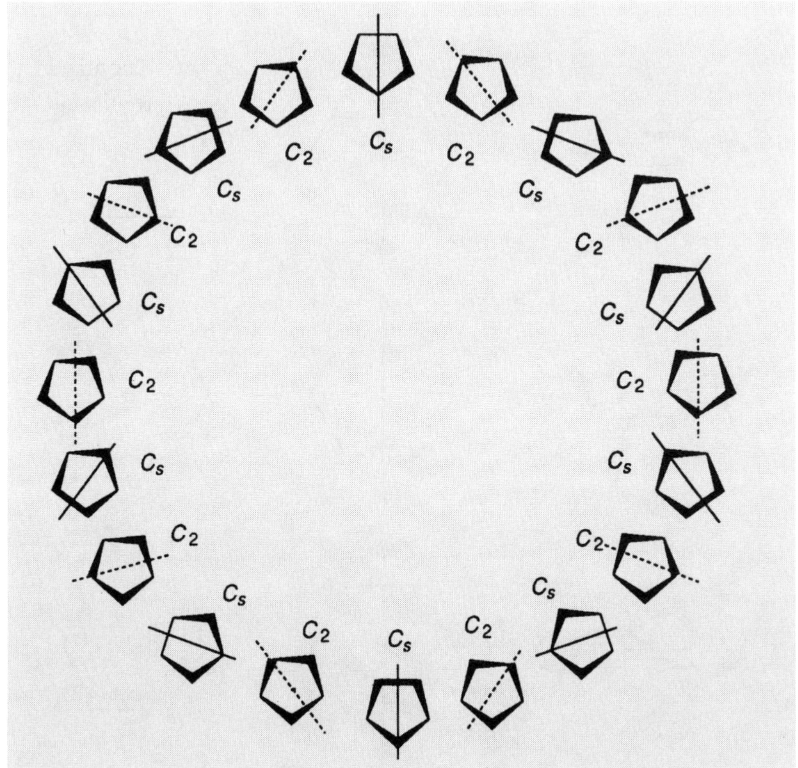

Fig. 6.10. *Pseudorotation cycle of cyclopentane*

6.5. Stereostructure of Selected Cycloalkenes

6.5.1. Cyclohexene

Cyclohexene is optically inactive. Under normal conditions, an equilibration (racemization) *via* ring inversion occurs between the two chiral half-chair conformations with C_2 symmetry. In the half-chair conformation, equatorial (*e*) and axial (*a*) bonds are only found at the centers C(4) and C(5). The bonds at C(3) and C(6) are described as *pseudoequatorial* (*e′*) and *pseudoaxial* (*a′*).

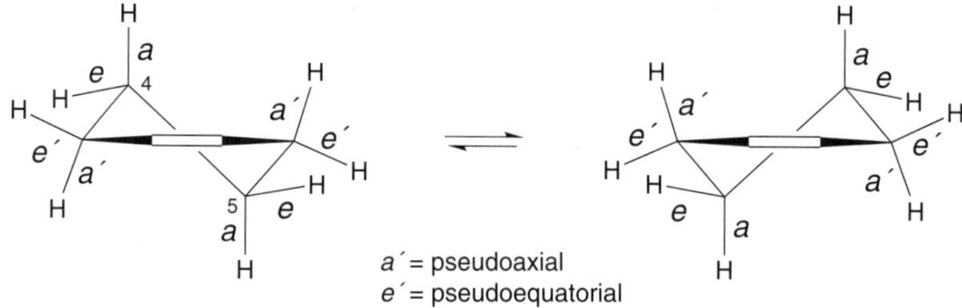

Table 6.7. *Conformations of Cyclohexene. Left-hand column:* molecular models for (from top to bottom) planar, boat, and half-chair conformations. For further details, see the legend of *Table 6.6.*

Molecular models of characteristic cyclohexene conformations	Symbols with symmetry elements	Symbols with torsion angles
		0, 0, 0, 0, 0, 0
		0, 37, 37, 39, 39, 0; − + + −
		61, 44, 44, 15, 15, 0; + − − + +

The barrier has been calculated (25.5 kJ/mol) by molecular mechanics (see *Aspects of Organic Chemistry*, Volume 4: *Methods of Structure Determination*) and measured by NMR spectroscopy (22.6 kJ/mol). Opinions are divided as to whether the boat conformation (*Table 6.7*) corresponds to a potential-energy maximum (transition structure) in the energy profile or to a local minimum. In the latter case, a 'sofa' conformation would also require consideration. The planar cyclohexene structure does not merit consideration even as a transition structure.

6.5.2. The Configurational Isomers of Cyclooctene

In the homologous series of cycloalkenes, the first case in which (Z)-and (E)-configurational isomers can both exist side by side under normal conditions is encountered with eight-membered rings. The (Z)-

Table 6.8. *Conformers with Different Cyclooctene Configurations. Left-hand column:* molecular models (from top to bottom) for asymmetric (Z)-cyclooctene, the C_2-symmetrical crown conformation of (E)-cyclooctene, and the C_2-symmetrical chair conformation of (E)-cyclooctene. For further details, see the legend of *Table 6.6.*

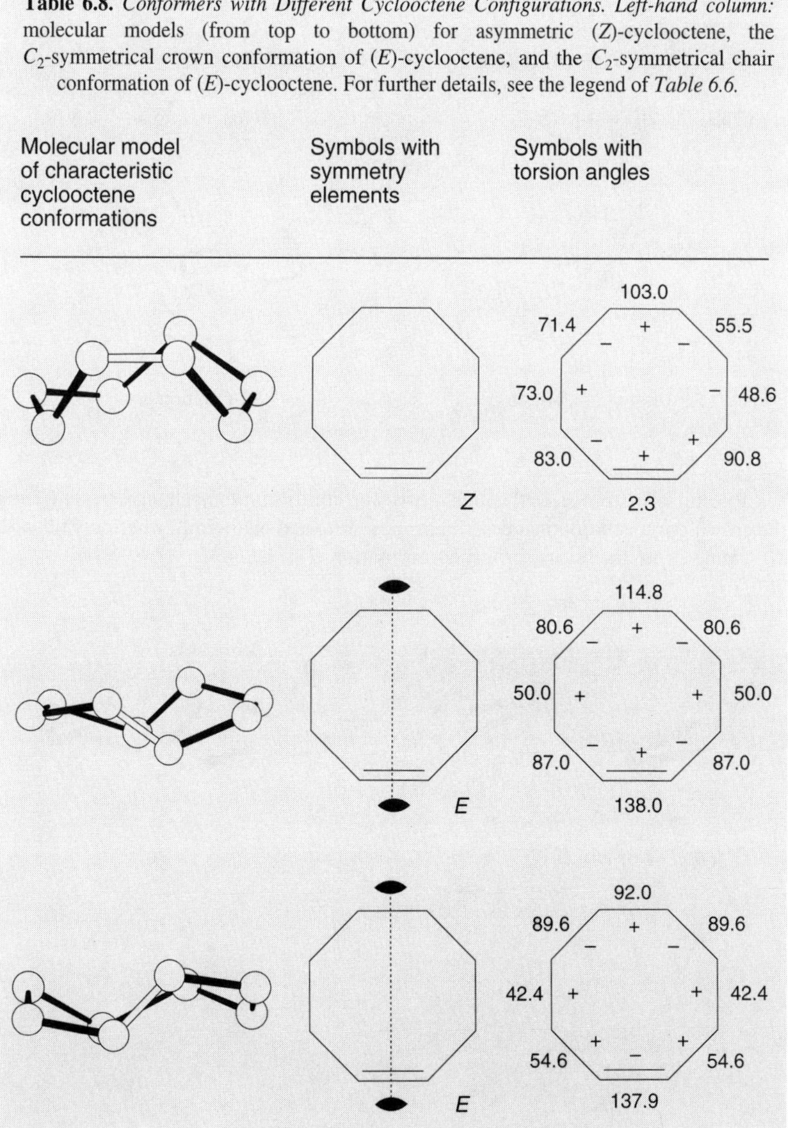

isomer occurs preferentially in asymmetrical conformations, which interconvert easily. One such is illustrated in *Table 6.8*. It is lower in energy than the (*E*)-isomer, its hydrogenation enthalpy undercutting that of the (*E*)-isomer by almost 39 kJ/mol.

The lowest-energy conformation of the (*E*)-isomer has a deformed C=C bond (with a C–C=C–C torsion angle of 138°), shows a dipole moment (0.8 D), and is C_2-symmetrical (*Table 6.8*). As the racemization barrier is very high (almost 150 kJ/mol), it is possible to separate the racemic mixture (*via* diastereoisomeric platinum complexes) into its enantiomers (*Fig. 6.11*).

$K_2[PtCl_4] + C_2H_4 \longrightarrow K[PtCl_3(C_2H_4)] + KCl$

$K[PtCl_3(C_2H_4)] + \mathbf{1} \longrightarrow [PtCl_2(C_2H_4)(\mathbf{1})] + KCl$

$[PtCl_2(C_2H_4)(\mathbf{1})] + \textit{rac-}\mathbf{2} \longrightarrow [PtCl_2(\mathbf{1})(\mathbf{2})] + [PtCl_2(\mathbf{1})(\textit{ent-}\mathbf{2})]$

$[PtCl_2(\mathbf{1})(\mathbf{2})] + [PtCl_2(\mathbf{1})(\textit{ent-}\mathbf{2})] \longrightarrow [PtCl_2(\mathbf{1})(\mathbf{2})]$

$[PtCl_2(\mathbf{1})(\mathbf{2})] + 4\,KCN \longrightarrow K_2[Pt(CN)_4] + 2\,KCl + \mathbf{1} + \mathbf{2}$

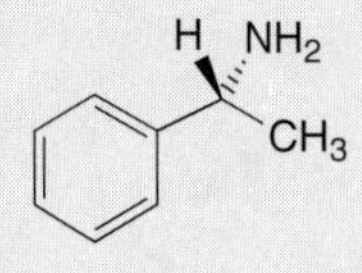

H NH$_2$ CH$_3$

1

(+)-(1*R*)-Phenylethylamine

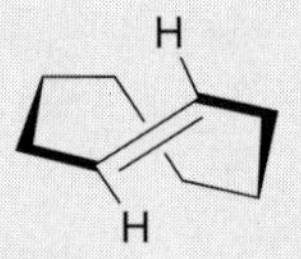

H H

2

(–)-(*R*,*E*)-Cyclooctene

Fig. 6.11. *Separation of enantiomers of (±)-(*E*)-Cyclooctene* (*rac*-**2**)

Racemization proceeds through a two-step conformational change, passing from the crown conformation just described, *via* a distorted chair conformation (still with C_2 symmetry) to the inverse crown conformation (*Fig. 6.12*).

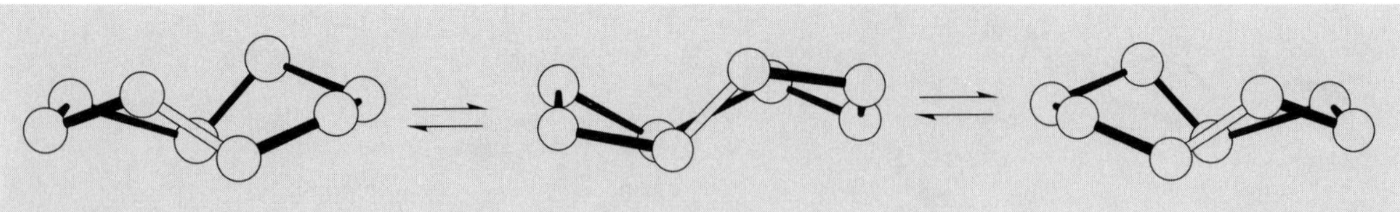

Fig. 6.12. *Racemization of (–)- or (+)-(*E*)-Cyclooctene* (**2** or *ent*-**2**)

6.6. On the Conformations of Heterocyclic Systems

6.6.1. General

The standard work '*Conformational Analysis*' by *E.L. Eliel et al.* puts it shortly and succinctly: '*The replacement of a methylene group in an alicyclic ring by a heteroatom usually does not have a profound effect on the conformational properties of the system. Certainly there are quantitative differences between the homocyclic and the heterocyclic rings, but in general the similarities, rather than the differences, are the striking features.*'

6.6.2. Conformational Analysis of Monosaccharides

The structural interpretation of (+)-D-glucose as an open-chain pentahydroxy aldehyde (*Chapt. 5*) does not fit a series of observations. It has not *one* pentaacetate, as would be expected, but *two* diastereoisomeric pentaacetates, for example. These observations are best explained by the assumption that an addition of the OH group at C(5) to the C(1)=O bond takes place in glucose, forming a cyclic pyranose. It follows directly that two configurational isomers (*anomers*) of glucopyranose must exist, as, upon the cyclo-isomerization of the seco-isomeric aldohexose, an additional stereogenic center (the so-called *anomeric center*) comes into existence at C(1):

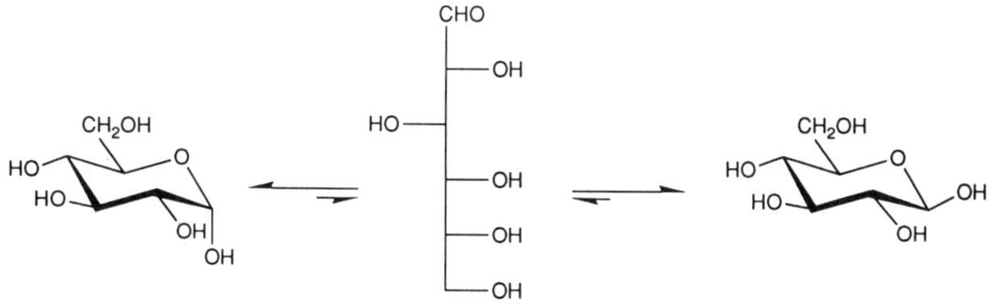

α-D-Glucopyranose D-Glucose β-D-Glucopyranose

For the conformational analysis of heterocyclic compounds, we will, for now, make use of the known facts about related carbocycles. For discussion of the tetrahydropyran ring, the cyclohexane ring (*Sect. 6.3.2*) will serve as a basis, and, for discussion of the tetrahydrofuran ring, the cyclopentane ring (*Sect. 6.4*). Naturally, the replacement of a CH_2 group by an O-atom will result in a certain degree of 'perturbation', which might be found by close study. For the time being, however, it is sufficient to be aware that the tetrahydropyran ring will also exhibit its lowest possible level of potential energy in the chair conformation. In the case of a pyranose, the predominant chair conformation in the equilibrium will be the one with the greatest number of equatorially oriented substituents at C(2) to C(5). For the two D-glucopyranose *anomers*, it is possible for the entire set of substituents on these centers to adopt an equatorial arrangement. Crystal-structure analysis by X-ray diffraction has revealed that the OH group at C(1) is axially oriented in the so-called α-anomer, and equatorially oriented in the β-anomer (see *Chapt. 10.5.3.1)*. Surprisingly, α-D-glucopyranose constitutes still 36% of the equilibrium mixture in aqueous solution at 40 °C: a fact which cannot be understood in terms of the structure model of the classical organic chemistry. The observed proportion of the stereoisomer (*anomer*) with an axial OH group (or, quite generally, an electronegative substituent) at the *anomeric center* C(1) in pyranoses, which is higher than expected, is caused by what is called the *anomeric effect* (*Chapt. 8.1.2*). *Table 6.9* lists the observed equilibrium compositions of some aldoses at 40 °C in

Table 6.9. *Percentages of Anomeric Pyranoses and Furanoses for Diverse Aldoses*

Aldose	α-Pyranose	β-Pyranose	α-Furanose	β-Furanose
Ribose	20	56	6	18
Arabinose	63	34	---------- 3 ----------	
Xylose	33	67	-------- < 1 ----------	
Lyxose	71	29	-------- < 1 ----------	
Allose	18	70	5	7
Altrose	27	40	20	13
Glucose	36	64	-------- < 1 ----------	
Mannose	67	33	-------- < 1 ----------	
Gulose	< 22	> 78	-------- < 1 ----------	
Idose	31	37	16	16
Galactose	27	73	-------- < 1 ----------	
Talose	40	29	20	11

aqueous solution. To obtain these figures, the individual components are converted into their corresponding trimethylsilyl ethers and analyzed by gas chromatography.

It may easily be established experimentally that an equilibrium process is operating in all cases, by commencing from the relevant individual anomers. After a period of time, the same mixture will have arisen from each of the starting components. If optical rotation is used as the experimental criterion for this equilibration, then, for D-glucose in aqueous solution, the value $[\alpha]_D^{20} = +53$ is reached, irrespective of whether α-D-glucopyranose ($[\alpha]_D^{20} = +113$) or β-D-glucopyranose ($[\alpha]_D^{20} = +19$) was used to begin with. The phenomenon whereby two different solutions of optically active compounds, after a period of time, show one and the same degree of optical rotation (which is different from zero, hence excluding a racemization), and the solutions both contain mixtures of the two starting compounds, is called *mutarotation*.

Monosaccharides usually exist in solution as a mixture of diverse constitutional isomers (seco-isomeric polyhydroxy aldehyde, cyclo-isomeric pyranose, furanose, or septanose; the two last-named with five- or seven-membered rings, arising from addition of the OH group at either C(4) or C(6) to the C=O bond) and configurational isomers (anomeric furanoses, pyranoses, or septanoses). The six-membered-ring constitutional isomers predominate as a rule (*Table 6.9*), although even a minor variation in experimental conditions, such as a change in solvent, can lead to exceptions.

Subordinate components, or even those present in proportions below the level of detection, can nevertheless be removed from the equilibrium by the use of specific reagents and their existence evidenced by means of their irreversibly generated ultimate products, arising out of the perpetually disturbed equilibrium.

Chapt. 10.5.3 provides more details of the description of carbohydrates with pyranose or furanose rings.

6.7. Conclusion

Constitution, configuration, and conformation describe the structure of a chemical compound in varying degrees of resolution. For the description of *constitution*, attention is paid to the kind and connectivity of the atoms present in a molecule. Atom-specific valencies and the ability of the atoms to arrange themselves in open chains and in rings, hereby participating in single, double, or triple bonds, are the points of interest. For the description of the *absolute configuration* of organic chemical compounds – a term synonymous with the *sense of chirality* – an indication of the way in which the four apexes of a tetrahedron are ordered in a Cartesian coordinate system suffices. For the description of the *relative configuration*, it is necessary to state how the ligands at two (or more) stereogenic centers, in two- or three-dimensional stereomodels, relate to each other: whether they are *like* or *unlike* (*Chapt. 10.2.1.2*). For the description of *conformation*, one concentrates primarily on the different torsion angles. When attempting to identify discrete cases of idealized conformation, it is necessary, as far as constitution permits, to consider staggered conformational moieties only, among them to favor those ones with minimum *Newman* strain, and prefer *ap*- to *sc*- to all the other conformational modules of butane.

Whilst endeavoring to structurally describe discrete substances, it is advisable to bear in mind that putatively uniform substances may behave as mixtures, especially if discrete conformers are present. Preoccupation with different conformations becomes acute, when the reactivity of a compound is under scrutiny (see *Aspects of Organic Chemistry*, Volume 2: *Reactivity*): where the conformation of the reacting compound in the ground state has to be compared with that in the transition state. *Barton*'s founding of *dynamic stereochemistry* – stimulated by the works of *Pitzer* and *Hassel* – was a result of such preoccupations. Ever after, the interest of the chemist was no longer confined merely to the most energetically favored conformation of a chemical compound, if chemical reactivity or biological activity were asked for. This shift of emphasis, from molecular statics to molecular dynamics, is characteristic of the transformation which chemistry has undergone, from preoccupation with problems of molecular structure to concentration on problems of molecular function.

By developing the rules of conformational analysis using as his example the then highly topical steroids, *Barton* created the prerequisite conditions for the total synthesis of steroid hormones, highly sought after because of their biological functions, to be achieved. Whole synthetic chemistry was soon to profit from the clarified perspective on the relationship between conformation and chemical reactivity. No wonder that the leading synthetic chemists were well aware of how large a role conformational analysis was soon going to play in the subsequent blossoming of the synthesis of biologically active substances. To those who are interested in the development of ideas, it is always conspicuous how important it is to introduce new ideas and concepts at the right time, and using the right examples.

Results from molecular mechanics have been worked into *Chapt. 6* in various places; so as not to leave hanging in the air stability orders which were otherwise left incomplete. Conformational analysis today involves the use of powerful computers and multicolored graphics to search for diverse minima, representing conformations, in a potential hypersurface, and to display results as three-dimensional images (*Aspects of Organic Chemistry*, Volume 4: *Methods of Structure Determination*). The conformation analyst uses interactive computer programs for so-called *molecular modelling*, these programs being capable of immediately displaying the effects of proposed structure alterations on the screen.

In the age of computer-assisted conformational analysis, an understanding of conformation as it was worked out by methods such as *Turner*'s should not, however, be underestimated. In his everyday activity, the chemist wants to apply what he has perceived. He will have perceived what has been comprehended as combination of the various conformation-determining factors on the qualitative level. The rules deduced in a given class of compounds will be of value in other compound classes as well.

Molecular structure remained beyond comprehension, as long as no accompanying 3D model, initially hand-made and afterwards computer-generated, was available. This is especially true for the molecules of macromolecular and the supermolecules of supramolecular chemistry.

Further Reading

To Sect. 6.1.

R. Witzmann, *Schlüssel des Lebens – Die Steuerung biologischer Vorgänge durch Steroide*, Molden, Wien, 1977.

I. McKinley, *Synthesis of a Postal Issue, Chem. Br.* **1977,** *13,* 51.

6.1.1.

D. N. Kirk, M. P. Hartshorn, *Steroid Reaction Mechanisms,* Elsevier Publ. Comp., Amsterdam, 1968.

J. Fried, J. A. Edwards (Ed.), *Organic Reactions in Steroid Chemistry,* Vol. 1 and 2, Van Nostrand Reinhold Comp., New York, 1972.

A. A. Akhrem, Y. A. Titov, *Total Steroid Synthesis,* Plenum Press, New York, 1970.

R. T. Blickenstaff, A. C. Ghosh, G. C. Wolf, *Total Synthesis of Steroids,* Academic Press, New York, 1974.

6.1.2.

M. S. Brown, J. L. Goldstein, *Nobel Lecture: A Receptor-Mediated Pathway for Cholesterol Homeostasis, Angew. Chem. Int. Ed.* **1986,** *25,* 583.

M. Sluyser (Ed.), *Interaction of Steroid Hormone Receptors with DNA,* VCH Verlagsgesellschaft, Weinheim, 1985.

V. K. Moudgil (Ed.), *Recent Advances in Steroid Hormone Action,* Walter de Gruyter, Berlin, 1987.

J. Elks, G. H. Phillips, *Discovery of a Family of Potent Topical Anti-inflammatory Agents* in *Medicinal Chemistry* (Ed. S. M. Roberts, B. J. Price), Academic Press, London, 1985.

F. J. Zeelen, *Steroid Contraceptives* in *Medicinal Chemistry* (Ed. S. M. Roberts, B. J. Price), Academic Press, London, 1985.

F. J. Zeelen, *Medicinal Chemistry of Steroids,* Elsevier, Amsterdam, 1990.

E. E. Baulieu, *Contragestion and Other Clinical Applications of RU 486, an Antiprogesterone at the Receptor, Science* **1989,** *245,* 1351.

E. E. Baulieu, P. A. Kelly, *Hormones,* Hermann, Paris, 1990.

F. Neumann, R. Wiechert, *Die Geschichte von Cyproteronacetat – Ungewöhnliche Wege bei der Entwicklung eines Arzneimittels,* Medizinisch Pharmazeutische Studiengesellschaft e. V., Mainz, 1984.

W. Frobenius, *Ein Siegeszug mit Hindernissen,* Schriftenreihe des Scheringianums, Berlin, 1989.

To Sect. 6.2.

D. H. R. Barton, *The Conformation of the Steroid Nucleus, Experientia* **1950,** *6,* 316;
– *The Principles of Conformational Analysis, Science* **1970,** *169,* 539;
– *The Stereochemistry of cyclo-Hexane Derivatives, J. Chem. Soc.* **1953,** 1027;
– *Some Recollections of Gap Jumping,* American Chemical Society, Washington, DC, 1991.

D. H. R. Barton, G. A. Morrison, *Conformational Analysis of Steroids and Related Natural Products, Fortschr. Chem. Organ. Naturst.* **1961,** *19,* 165.

L. F. Fieser, *Steric Course of Reactions of Steroids, Experientia* **1950,** *6,* 312.

To Sect. 6.3.

K. Freudenberg (Ed.), *Stereochemie – Eine Zusammenfassung der Ergebnisse, Grundlagen und Probleme,* Franz Deuticke, Leipzig, 1933.

E. L. Eliel, S. H. Wilen, *Stereochemistry of Organic Compounds,* John Wiley, New York, 1994.

E. L. Eliel, N. L. Allinger, S. J. Angyal, G. A. Morrison, *Conformational Analysis,* Interscience Publ., New York, 1965.

O. B. Ramsay, *Stereochemistry* in the Series *Nobel Prize Topics in Chemistry,* Heyden, London, 1981.

J. Dale, *Stereochemie und Konformationsanalyse,* Verlag Chemie, Weinheim, 1978.

B. Testa, *Principles of Organic Stereochemistry,* Marcel Dekker, New York, 1979.

6.3.1.

A. Wassermann, *Spannungstheorie und physikalische Eigenschaften ringförmiger Verbindungen* in *Stereochemie* (Ed. K. Freudenberg), Franz Deuticke, Leipzig, 1933.

A. J. Ihde, *The Development of Strain Theory* in *Kekulè Centennial* (Ed. O. T. Benfey), *Advances in Chemistry* Series 61, American Chemical Society, Washington, DC, 1966.

6.3.2.

H. Sachse, *Über die Konfigurationen der Polymethylenringe, Z. Physik. Chem.* **1892,** *10,* 203;

– *Über die geometrischen Isomerien der Hexamethylenderivate, Ber. Dtsch. Chem. Ges.* **1890,** *23,* 1363.

E. Mohr, *Die Baeyersche Spannungstheorie und die Struktur des Diamanten, J. Prakt. Chem.* **1918,** *98,* 315.

G. Natta, M. Farina, *Struktur und Verhalten von Molekülen im Raum,* Verlag Chemie, Weinheim, 1976.

D. H. R. Barton, *How to Win a Nobel Prize: an Account of the Early History in Stereochemistry of Organic and Bioorganic Transformations* (Ed. W. Bartmann, K. B. Sharpless), VCH Verlagsgesellschaft, Weinheim, 1987.

O. Hassel, *The Cyclohexane Problem, Topics in Stereochem.* **1971,** *6,* 11.

J. D. Dunitz, *The Two Forms of Cyclohexane, J. Chem. Educ.* **1970,** *47,* 488.

6.3.3.

W. G. Dauben, K. S. Pitzer, *Conformation Analysis* in *Steric Effects in Organic Chemistry* (Ed. M. S. Newman), Chapman & Hall, New York, 1956.

R. M. Pitzer, *The Barrier to Internal Rotation in Ethane, Acc. Chem. Res.* **1983,** *16,* 207.

6.3.4.

R. B. Turner, *Energy Differences in the cis- and trans-Decalins, J. Am. Chem. Soc.* **1952,** *74,* 2118.

E. Juaristi, *The Attractive and Repulsive Gauche Effects, J. Chem. Educ.* **1979,** *56,* 438.

N. L. Allinger, M. A. Miller, *The 1,3-Diaxial Methyl-Methyl-Interaction, J. Am. Chem. Soc.* **1961,** *83,* 2145.

M. S. Newman, *Some Observations Concerning Steric Factors, J. Am. Chem. Soc.* **1950,** *72,* 4783.

6.3.5.

W. S. Johnson, *The Relative Stability of Stereoisomeric Forms of Fused Ring Systems, Experientia* **1951,** *7,* 315;

– *Energy Relationships of Fused Ring Systems, J. Am. Chem. Soc.* **1953,** *75,* 1498.

N. L. Allinger, B. J. Gorde, I. J. Tyminski, M. T. Wuesthoff, *The Perhydrophenanthrens, J. Org. Chem.* **1971,** *36,* 739.

W. S. Johnson, V. J. Bauer, J. L. Margrave, M. A. Frisch, L. H. Dreger, W. N. Hubbard, *The Energy Difference between the Chair and Boat Forms of Cyclohexane. The Twist Conformation of Cyclohexane, J. Am. Chem. Soc.* **1961,** *83,* 606.

J. L. Margrave, M. A. Frisch, R. G. Bantista, R. L. Clarke, W. S. Johnson, *Further Studies on the Energy Difference between the Chair and Twist Forms of Cyclohexane, J. Am. Chem. Soc.* **1963,** *85,* 546.

O. Ermer, J. D. Dunitz, *Zur Konformation des Bicyclo[2.2.2]octan-Systems, Helv. Chim. Acta* **1969,** *52,* 1861.

6.3.6.

M. Squillacote, R. S. Sheridan, O. L. Chapman, F. A. L. Anet, *Spectroscopic Detection of the Twist-Boat Conformation of Cyclohexane. A Direct Measurement of the Free-Energy Difference between the Chair and the Twist-Boat, J. Am. Chem. Soc.* **1975,** *97,* 3244.

F. A. L. Anet, A. J. R. Bourn, *Nuclear Magnetic Resonance Line-Shape and Double-Resonance Studies of Ring Inversion in Cyclohexane-d_{11}, J. Am. Chem. Soc.* **1967,** *89,* 760.

F. R. Jensen, D. S. Noyce, C. H. Sederholm, A. J. Berlin, *The Rate of the Chair-Chair Interconversion of Cyclohexane, J. Am. Chem. Soc.* **1962,** *84,* 386.

S. L. Spassor, D. L. Griffith, E. S. Glazer, K. Nagarijan, J. D. Roberts, *Conformational Properties of Substituted 1,1-Difluorocyclohexanes, J. Am. Chem. Soc.* **1967,** *89,* 88.

R. Bucourt, *The Torsion Angle Concept in Conformational Analysis, Topics in Stereochem.* **1974,** *8,* 159.

To Sect. 6.4.

W. L. Duax, C. M. Weeks, D. C. Rohrer, *Crystal Structures of Steroids, Topics in Stereochem.* **1976,** *9,* 271.

C. Altona, H. J. Geise, C. Romers, *Geometry and Conformation of Ring D in Some Steroids from X-Ray Structure Determinations, Tetrahedron* **1968,** *24,* 13.

C. Altona, *Geometry of Five-Membered Rings* in *Conformational Analysis* (Ed. G. Chiurdoglu), Academic Press, New York, 1971.

J. E. Kilpatrick, K. S. Pitzer, R. Spitzer, *The Thermodynamics and Molecular Structure of Cyclopentane, J. Am. Chem. Soc.* **1947,** *69,* 2483.

To Sect. 6.5.

6.5.1.

D. H. R. Barton, R. C. Cookson, W. Klyne, C. W. Shoppee, *The Conformation of Cyclohexene, Chem. Ind. (London)* **1954,** 21.

6.5.2.

R. B. Turner, W. R. Meador, R. E. Winkler, *Heats of Hydrogenation. Hydrogenation of Some* cis- *and* trans-*Cycloolefins, J. Am. Chem. Soc.* **1957,** *79,* 4133.

A. C. Cope, C. R. Ganellin, H. W. Johnson, jr., T. V. Van Auken, H. J. S. Winkler, *Resolution of* trans-*Cyclooctene, J. Am. Chem. Soc.* **1963,** *85,* 3276.

P. C. Manor, D. P. Shoemaker, A. S. Parkes, *The Conformation and Absolute Configuration of (−)-*trans-*Cyclooctene, J. Am. Chem. Soc.* **1970,** *92,* 5260.

A. C. Cope, B. A. Pawson, *Kinetics of Racemization of (+ or −)-*trans-*Cyclooctene, J. Am. Chem. Soc.* **1965,** *87,* 3649.

O. Ermer, S. Lifson, *Vibrations, Conformations, and Heats of Hydrogenation of Nonconjugated Olefins, J. Am. Chem. Soc.* **1973,** *94,* 4121;

O. Ermer, *Conformation of* trans-*cyclooctene, Angew. Chem. Int. Ed.* **1974,** *13,* 604; – *Aspekte von Kraftfeldrechnungen,* W. Baur Verlag, München, 1981.

M. Traetteberg, *The Molecular Structure of* trans-*Cyclooctene, Acta Chem. Scand.* **1975,** *29B,* 29.

To Sect. 6.6.

6.6.1.

E. L. Eliel, N. L. Allinger, S. J. Angyal, G. A. Morrison, *Conformational Analysis,* Interscience Publ., New York, 1965.

C. Romers, C. Altona, H. R. Buys, E. Havinga, *Geometry and Conformational Properties of Some Five- and Six-Membered Heterocyclic Compounds Containing Oxygen or Sulfur, Topics in Stereochem.* **1969,** *4,* 39.

6.6.2.

C. Altona, M. Sundaralingam, *Conformational Analysis of the Sugar Ring in Nucleosides and Nucleotides. A New Description Using the Concept of Pseudorotation, J. Am. Chem. Soc.* **1972,** *94,* 8205.

To Sect. 6.7.

V. Prelog, *Conformational Analysis – Scope and Present Limitations, Pure Appl. Chem.* **1971,** *25,* 465.

M. Saunders, R. M. Jarret, *A New Method for Molecular Mechanics, J. Comput. Chem.* **1986,** *7,* 578.

M. Saunders, *Stochastic Exploration of Molecular Mechanics Energy Surfaces. Hunting for the Global Minimum, J. Am. Chem. Soc.* **1987,** *109,* 3150.

V. L. Shannon, H. L. Strauss, R. G. Snyder, C. A. Elliger, W. L. Mattice, *Conformation of the Cycloalkanes $C_{14}H_{28}$, $C_{16}H_{32}$, and $C_{22}H_{44}$ in the Liquid and High-Temperature Crystalline Phases by Vibrational Spectroscopy, J. Am. Chem. Soc.* **1989,** *111,* 1947.

J. B. Hendrickson, *Machine Computation of The Common Rings, J. Am. Chem. Soc.* **1961,** *83,* 4537.

N. L. Allinger, M. A. Miller, F. A. VanCatledge, J. A. Hirsch, *The Calculation of the Conformational Structures of Hydrocarbons by the Westheimer-Hendrickson-Wiberg Method, J. Am. Chem. Soc.* **1967,** *89,* 4345.

J. M. A. Baas, B. van de Graaf, D. Tavernier, P. Vanhee, *Empirical Force Field Calculations. Conformational Analysis of* cis-*Decalin, J. Am. Chem. Soc.* **1981,** *103,* 5014.

7 Macromolecular and Supramolecular Chemistry

7.1. Definitions

An important criterion for the classification of chemical compounds is the relative molecular mass. Compounds with a relative molecular mass above about 10,000 (= 10 kDa) are ascribed to *macromolecular chemistry*. Consequent reasoning would require that compounds with a relative molecular mass lower than 10 kDa be assigned to *micromolecular chemistry*. The limit of 10 kDa was proposed by the founder of macromolecular chemistry, *Hermann Staudinger*. It is arbitrary, though takes account of the empirical finding that the so-called macromolecular properties, such as the high viscosity of dilute solutions, become pronounced beyond this limit.

Molecular chemistry is concerned with molecules, of any size, whose atoms are bound together by covalent bonds. *Supramolecular chemistry* is concerned with supermolecules, whose molecules are bound together by non-covalent forces.

The terms *macromolecules*, *plastics*, or *polymers* are quite often used synonymously. By a polymer (oligomer) is meant a molecule which can be described structurally by many (some) repetitions of constitutional repeating units (*vide infra*). It is not possible, without additional information, to infer which monomer was used in the preparation (polymerization) of a polymer from its structure and systematic name. Macromolecular compounds may be expediently subdivided into technical polymers (plastics) and biopolymers.

Technical polymers, like biopolymers, were more or less fenced off by organic chemists in the past (*Chapt. 1.2.1*). However, it is our wish to encourage the increasing awareness that molecular size may no longer be assumed as a criterion for demarcation of scientific disciplines. In the question of molecular structure, we see a unifying interest for previously separate fields of study and put our faith in conformational analysis to cross the border.

The affix *mer* or *meric* is used as a constituent in nouns (relating to the smallest structural unit of *monomers*, *dimers*, *oligomers*, *30mers*, *polymers*, *diversomers*) or adjectives (*isomeric*, *enantiomeric*, *anomeric*, allowing for structural distinctions of a given structural unit).

In idealized chain polymers, there exists a structural regularity, which also applies, in enhanced mode, for the description of an idealized crystal: structural repeating units, recurring with periodic regularity by means of translation, are present in both cases. Linear polymers with their one-dimensional order stand at the beginning of a systematics which ends with crystals and their three-dimensional order. Crystal symmetry is discussed in *Chapt. 11*.

7.2. Technical Polymers

7.2.1. Constitution and Configuration of Poly(methylene) and Poly(propylene)

Polymerization products result from spontaneous or catalyzed polymerizations of the same monomer unit or of different ones. Origin-derived trivial names such as, for example, *polyethylene* for a polymer indicate the monomer used: *ethylene*. The systematic name *poly(methylene)*, in contrast, is based on the *bivalent constitutional repeating unit* ($-CH_2-$).

If polymers consist of only one kind of *constitutional repeating unit*, then they are described as *regular* polymers; if, however, they contain two or more different *constitutional repeating units*, then one speaks of *irregular* polymers (*Fig. 7.1*). The term 'regular' has already been used, without further comment, in *Chapt. 2.6.4*, for a constitutional repeating unit occurring several times in a chemical compound.

For *poly(propylene)* of regular constitution, the *constitutional repeating unit* is the 1-methylethane-1,2-diyl fragment ($-CH(CH_3)-CH_2-$), identical here with the *constitutional base unit*. According to convention, the bivalent constitutional repeating unit begins at the atom of highest priority. Regular ordering is attained, if the connectivity of the constitutional base unit is uniform (*head-tail*). In poly(propylene) of irregular constitution, the constitutional base units are strung together in irregular sequence by means of head-head, tail-tail, or head-tail connections. For stereostructural description of polymers of poly(propylene) type by their

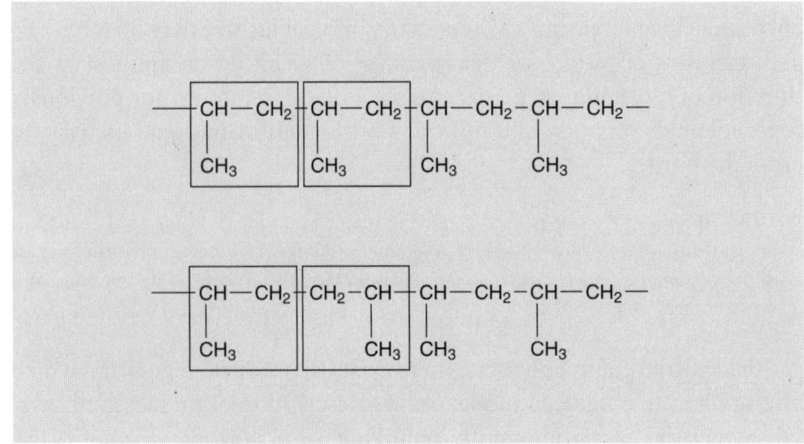

Fig. 7.1. *Fragments of poly(propylene) with* regular (above) *and with* irregular (below) *constitution*

configurational repeating units, it should be born in mind that the latter may be identical with the *configurational base units*, but that they do not have to be.

It is, therefore, appropriate to distinguish between the *configurational base unit* and the *configurational repeating unit*.

H
|
— C— CH_2—
|
CH_3

and

CH_3
|
— C— CH_2—
|
H

For linear polymers it is customary, instead of the original *Fischer* projection with a vertically oriented chain, to use a modified *Fischer* projection with a horizontal chain, in which horizontal lines are directed in front of the plane of projection and vertical lines behind it. Zig-zag representation of the chain is also very frequently employed. *Fig. 7.2* shows how to translate the *Fischer* projection into the zig-zag representation and *vice versa*.

For poly(propylene) of regular constitution, there exist configurationally regular (*tactic*) and configurationally irregular (*atactic*) variants. Tactic poly(propylene) may be *isotactic* or *syndiotactic*, depending

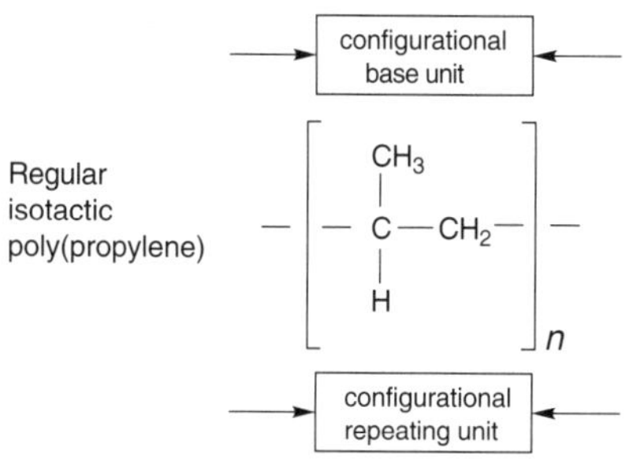

The configurational repeating unit is identical to the configurational base unit.

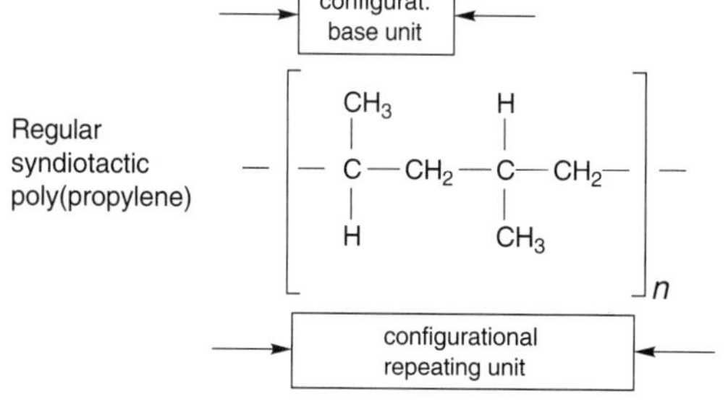

The configurational repeating unit consists of two enantiomorphic configurational base units.

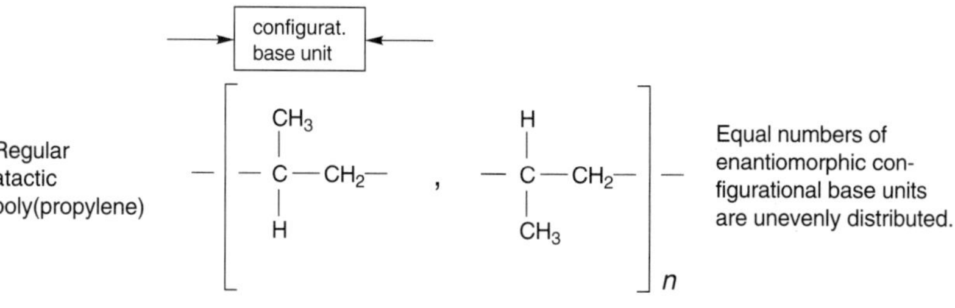

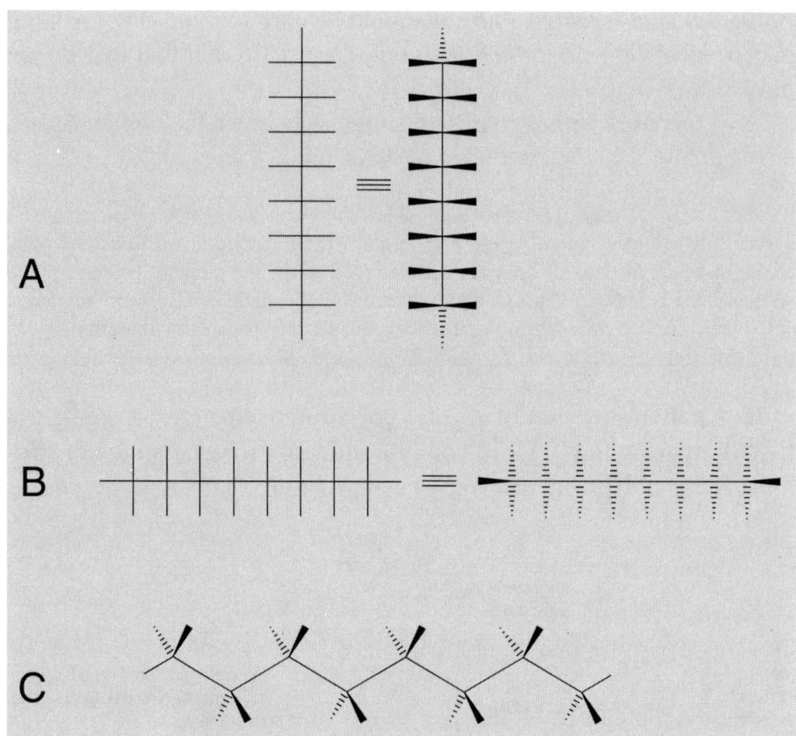

Fig. 7.2. Fischer *projection* (A), *modified* Fischer *projection* (B), *and zig-zag representation* (C)

on whether or not the configurational base unit and the configurational repeating unit are identical.

The *isotactic* polymer (*Fig. 7.3*) is characterized by the recurrence of only one and the same configurational base unit in a constitutionally regular pattern. In isotactic poly(propylene) (R = Me), the configurational base unit and the configurational repeating unit are identical. The Me groups are all to be found on the same side of the chain.

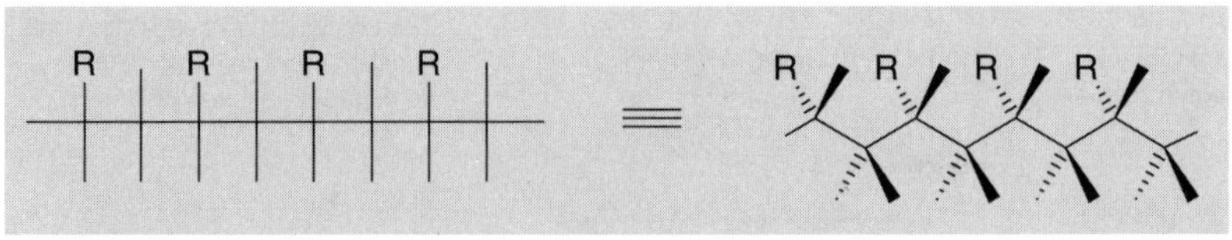

Fig. 7.3. *Modified* Fischer *projection and zig-zag representation of an isotactic polymer*

The *syndiotactic* polymer (*Fig. 7.4*) is characterized by the constitutionally regular alternation of two base units, distinct by virtue of their enantiomorphic configurations. Here the configurational base unit and the configurational repeating unit are not identical. In syndiotactic poly(propylene) (R = Me), the Me groups are located on alternating sides of the chain.

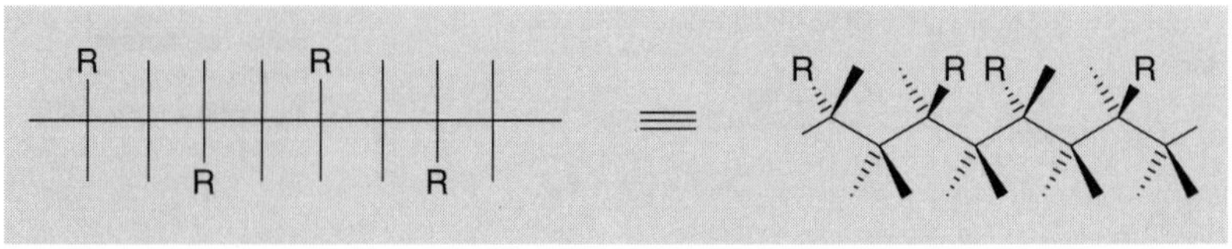

Fig. 7.4. *Modified* Fischer *projection and zig-zag representation of a syndiotactic polymer*

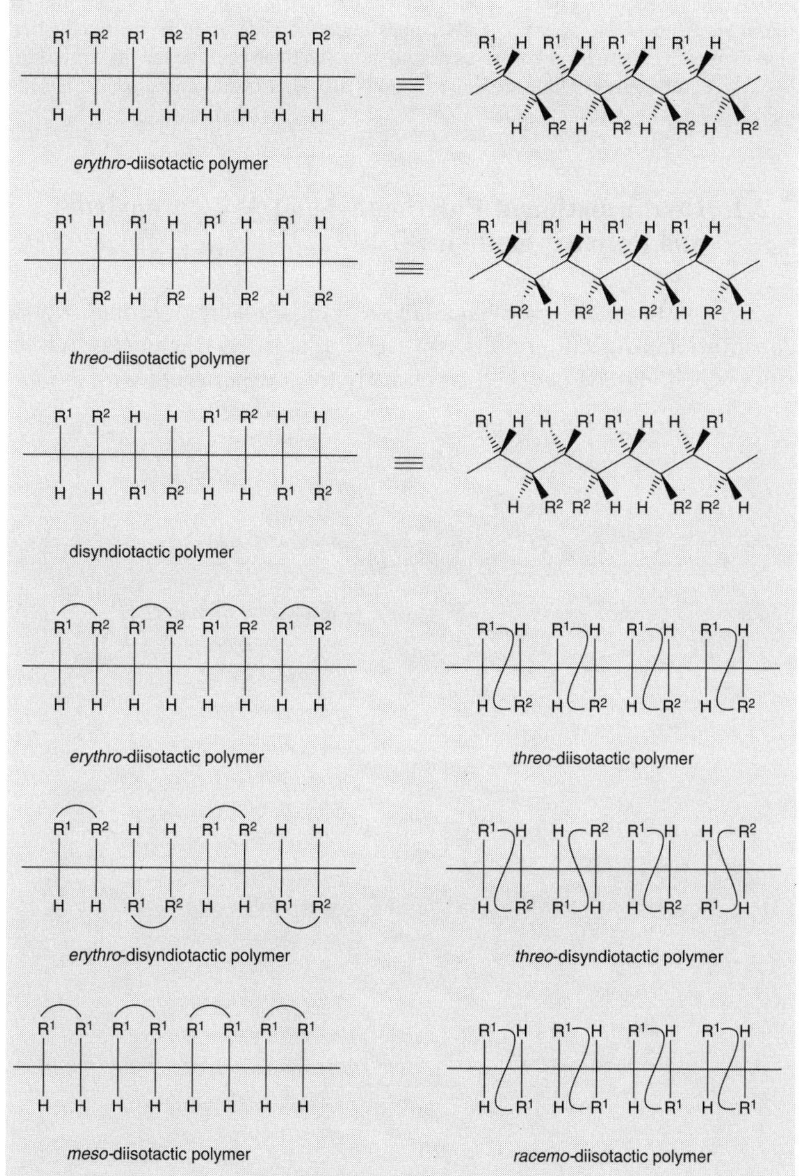

Fig. 7.5. *Stereochemical specifications of polymers*

As a rule, the two portions of the main chain extending outwards from any one constitutional repeating unit are not identical. This has the consequence that a C-atom in the backbone of a polymer, connected to two additional, non-identical ligands, represents a stereogenic center. Unless expressly stated otherwise, the formula representations for configurational base units or repeating units give only the relative configuration.

Polymers may be specified in greater detail regarding their relative configuration. *Fig. 7.5* shows ditactic polymers, in which each configurational base unit contains two (distinct) stereogenic centers. The lower six representations show ring-connected polymers: in instances of *erythro*-diisotactic and *threo*-diisotactic polymers, the chirality centers should (when known) be specified with (R) or (S).

If the base units in a polymer occur in their possible configurations in equal numbers, but irregularly distributed, then the polymer is *atactic*.

In many polymers, tacticity (the regular ordering of configurational base units in the main chain of the polymer) only comes under consideration in relatively short chain segments, a situation known as partial tacticity. Such chain segments containing two, three, four, *etc.* configurational base units are described as *diads*, *triads*, *tetrads*, *etc.* (*Fig. 7.6*).

7.2.2. Conformation of Poly(methylene), Poly(propylene), and Poly(oxymethylene)

Systematic conformational analysis of those macromolecules, whose constitutional repeating units consist of alkanediyl fragments, follows rules which may be deduced empirically from experiences with *ethane*,

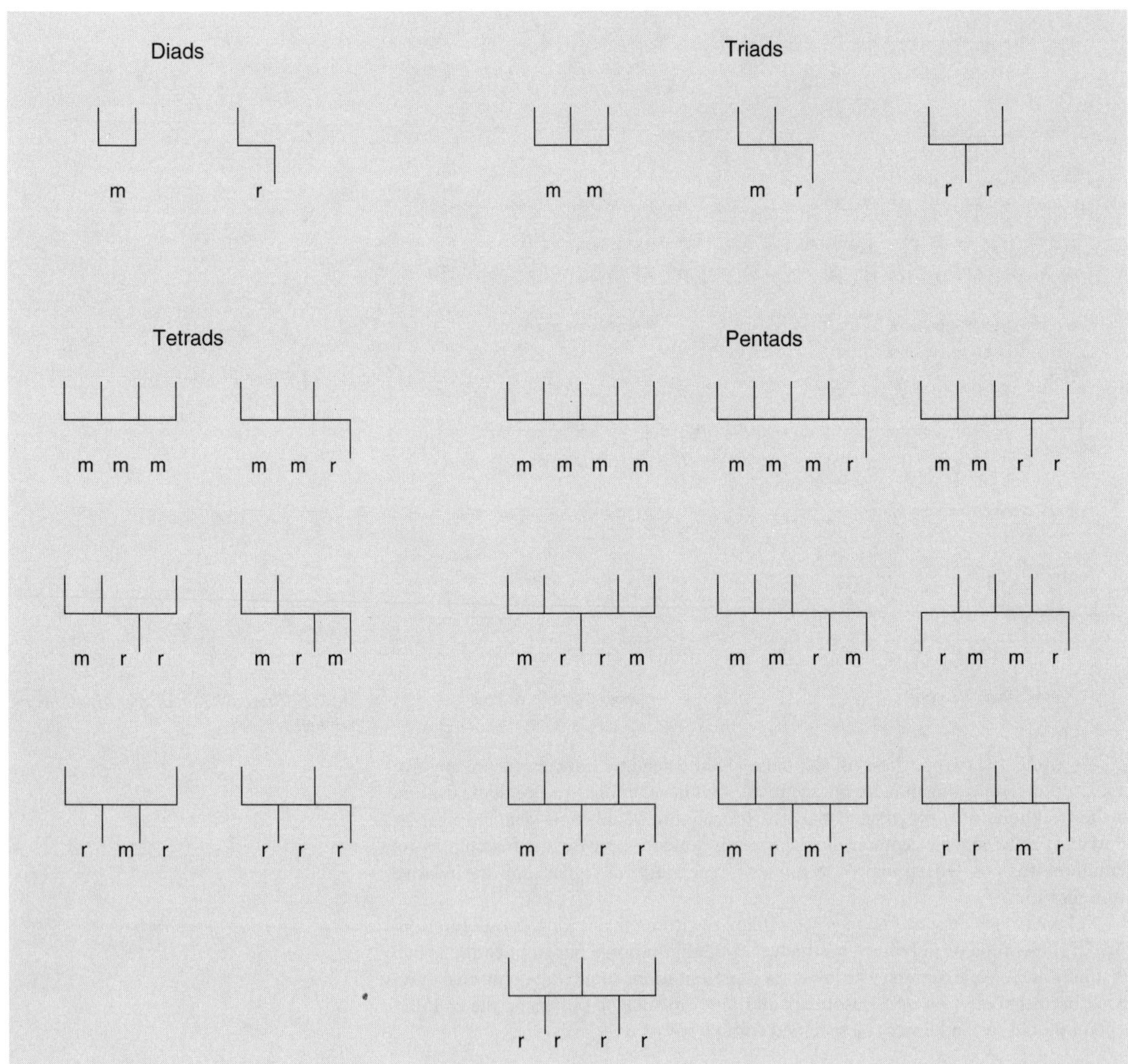

Fig. 7.6. *Symbolic representation of chain segments with two, three, four, or five configurational base units.* Letter m (from *meso*) or r (from *racemo*) depicts the isotactic or syndiotactic polymer segment, respectively.

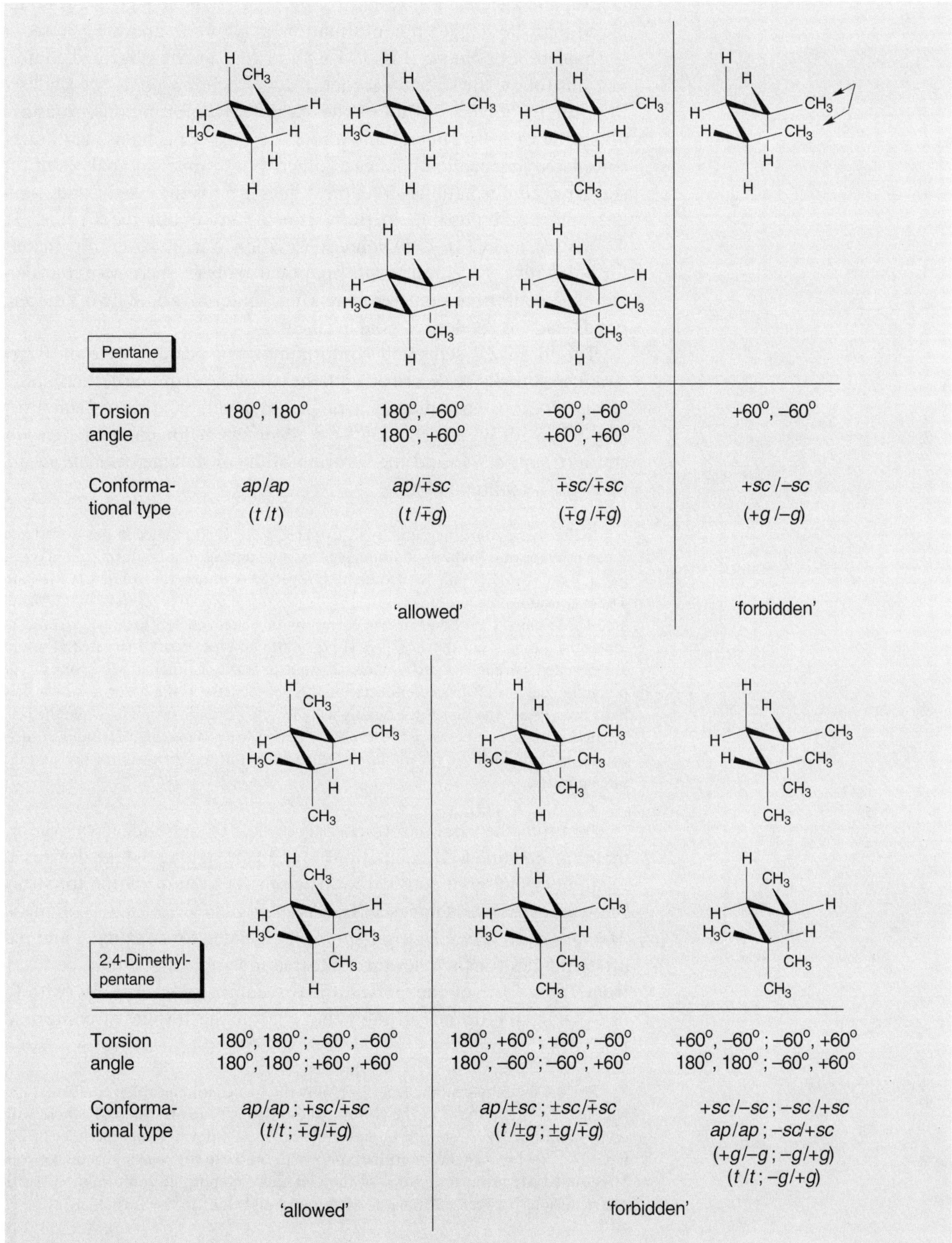

Fig. 7.7. *Systematic conformational analysis for pentane and 2,4-dimethylpentane* (*t* : *trans*, *g* : *gauche*)

butane, and *pentane*. For *ethane* (*Chapt. 2.5.1.2* and *6.3.3*), it was established that the staggered conformation is 'allowed', and the eclipsed is 'forbidden'. For *butane* (*Chapt. 6.3.4* and *10.3.2*), the staggered conformations follow the stability sequence *ap* > *sc*. For *pentane*, we shall see that, for the analysis of all of the staggered conformations, rotations about the C(2)–C(3) bond and about the C(3)–C(4) bond may not be considered independently of each other. For *hexane*, an analysis of the staggered conformations along these lines affords the insight that, similarly, it is not possible to draw conclusions about the C(2)–C(3), C(3)–C(4), and C(4)–C(5) bonds in isolation, but, however, fortunately for qualitative, systematic conformational analysis, successive consideration of conformational sections after rotations about two adjacent bonds leads to the desired goal in all cases.

The idealized, staggered conformations of pentane, free of *Pitzer* strain, may easily be described with the aid of the diamond-lattice matrix (*Chapt. 10.3.3*). They may be arranged in a suitable order of stabilities (*Fig. 7.7*) with the assistance of the requirement for minimal *Newman* strain (*Chapt. 6.3.4*) and the favoring of the *ap*-butane over the *sc*-butane conformational module.

In the *ap*/*ap*-conformational type, two H,H pairs, each situated at positions 1 and 3, can be made out. In the enantiomorphic *ap*/*−sc*- and *ap*/*+sc*-conformational types, one 1,3-situated H,H pair is encountered, together with one 1,3-situated H,Me pair. The enantiomorphic *−sc*/*−sc*- and *+sc*/*+sc*- conformational types have two 1,3-situated H,Me pairs, their bonds to the corresponding chain centers arranged parallel to each other. Finally, in the *+sc*/*−sc*- conformational type, two 1,3-situated H-atoms are oriented parallel to each other, as well as two 1,3-situated Me groups. The potential energies of these conformational types increase in the order in which they have been listed: the first three qualify as 'allowed' on steric grounds, while the last, because of repulsion between the two 1,3-situated Me groups (high *Newman* strain), counts as 'forbidden': in the diamond lattice, two H-atoms compete for one and the same position.

As a rule, the true conformers may deviate from the idealized conformations. *Ab initio* MO calculations today can attain such high degrees of reliability as to merit consideration as sources of information for structure-energy relationships (see *Aspects of Organic Chemistry*, Volume 4: *Methods of Structure Determination*). Calculated torsion angles and relative energies for the relevant conformations of pentane may be taken from *Table 7.1*. Not unexpected, the two conformers listed last (with C_s or C_1 symmetry) deviate strongly enough from the 'forbidden' conformation with C_s symmetry.

For 2,4-dimethylpentane (*Fig. 7.7*), only the two enantiomorphic conformations with two 1,3-orientations of the H-atom and the Me group are 'allowed': those with *ap*/*ap*- and *+sc*/*+sc*-, or with *ap*/*ap*- and *−sc*/*−sc*-conformational modules of butane. The other two enantiomorphic conformations (both with *one* repulsion between 1,3-situated Me groups) and one of the two diastereomorphic conformations (with *two* repulsions between 1,3-situated Me groups) all belong to the 'forbidden' ones.

Table 7.1. *Torsion Angles and Calculated Relative Energies for the Conformations of Pentane*

Idealized [°]	Calculated [°]	Calc. [a]) rel. energies [kJ/mol]
180/180	180/180	0.00
180/±60	177/±68.76	4.19
±60/±60	±63.77/±63.40	7.87
±60/∓60	———	21.60
	±78.13/∓77.14	17.21
	±63.23/∓94.61	16.24

[a]) K. B. Wiberg, M. A. Murcko, *Energies of Alkane Rotamers. An Examination of Gauche Interactions, J. Am. Chem. Soc.* **1988**, *110*, 8029.

The comparison of pentane with 2,4-dimethylpentane underlines the reduction of conformational space caused by repulsion between 1,3-situated non-H-atoms (*Newman* strain; *cf. cis*-1-ethyl-2-isopropylcyclohexane in *Chapt. 6.3.4*). This reduction is the cause of conformational preorganization, important, for example, for the function of acyclic ionophores, the cyclization tendency of acyclic precursors of macrocyclic, synthetic target compounds, and the generation of regular secondary structural elements (*Chapt. 10.5.4.2*) through formation of H-bonds (*Chapt. 15*).

Drastic restrictions in molecular mobility are attained by constitutional restraint caused by ring formation. If two imaginary Me groups, oriented parallel to each other, are replaced by the CH_2 groups of a propane-1,3-diyl fragment (*Fig. 7.8*), then the 'allowed' conformations of the acyclic system lose their validity as a starting basis. From the four 'forbidden' conformations of 2,4-dimethylpentane, one arrives at the two 'allowed' conformations of *trans*-1,3-dimethylcyclohexane (*Table 6.2*), as well as at the 'allowed' conformation of the configurationally isomeric *cis*-1,3-dimethylcyclohexane with equatorially oriented Me groups.

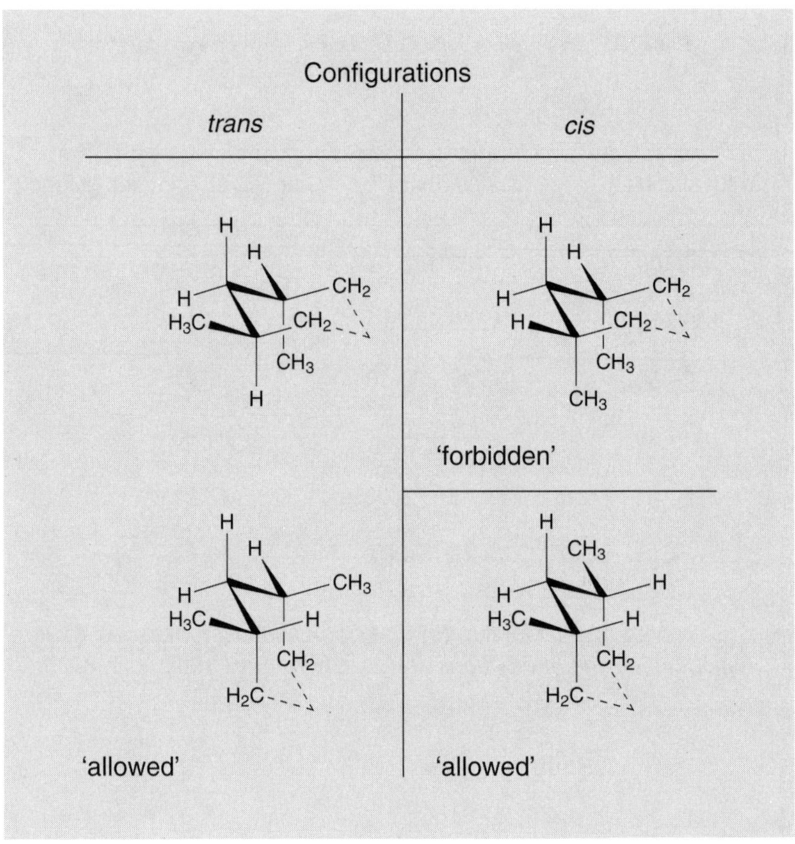

Fig. 7.8. *Conformers of pentane bridged between C(2) and C(4) by a propane-1,3-diyl fragment*

Of the ten conformational types of hexane (*Fig. 7.9*) together with their corresponding torsion-angle combinations, all but the last three listed are 'allowed'. Of the three 'forbidden' conformational types, the first two cases each have one repulsion between 1,3-situated Me or CH_2 groups, while, in the last case, there are even two repulsions of that kind.

The $\pm sc/\mp sc$-conformational type with two 1,3-situated non-H-atoms parallel to each other is always cited in the chemical literature as a conformational type to be avoided whenever possible (high *Newman* strain).

If the *ap*-oriented butane-1,4-diyl fragment ($-CH_2-CH_2-CH_2-CH_2-$) is selected as a conformational repeating unit, one arrives at the idealized zig-zag chain of poly(methylene) (*Fig. 7.11, a*). The idealized, linear zig-zag chain is a special case, as, here, two enantiomorphic, degenerate helices coincide. Every deviation from the idealized, linear conformation results in the zig-zag chain adopting a helical arrangement. For the precise description of a helix, it is necessary to know the parameters given in *Fig. 7.10*.

In the idealized case of poly(methylene), the minimum distance between the H-atoms in 1,2- or 1,3-positions amounts to 250 pm, and so is larger than the sum of their *van der Waals* radii (for H: $2 \cdot 120$ pm = 240 pm). Zig-zag chains with a distance of 255 pm between the neighboring repeating units have indeed been observed (*Fig. 7.11, a*).

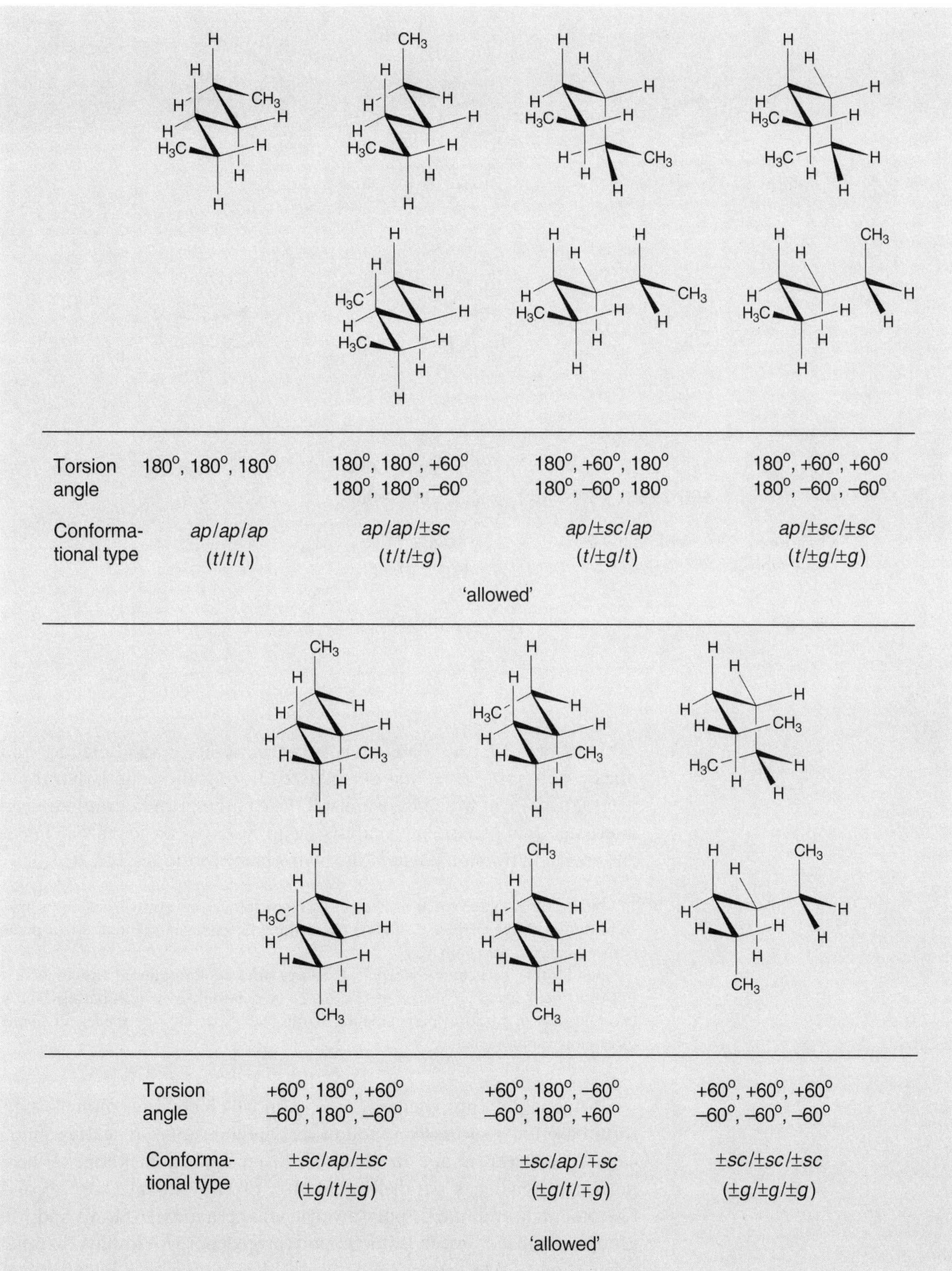

Fig. 7.9. *Systematic conformational analysis for hexane.* The conformations put together in vertical pairs are each enantiomorphic.

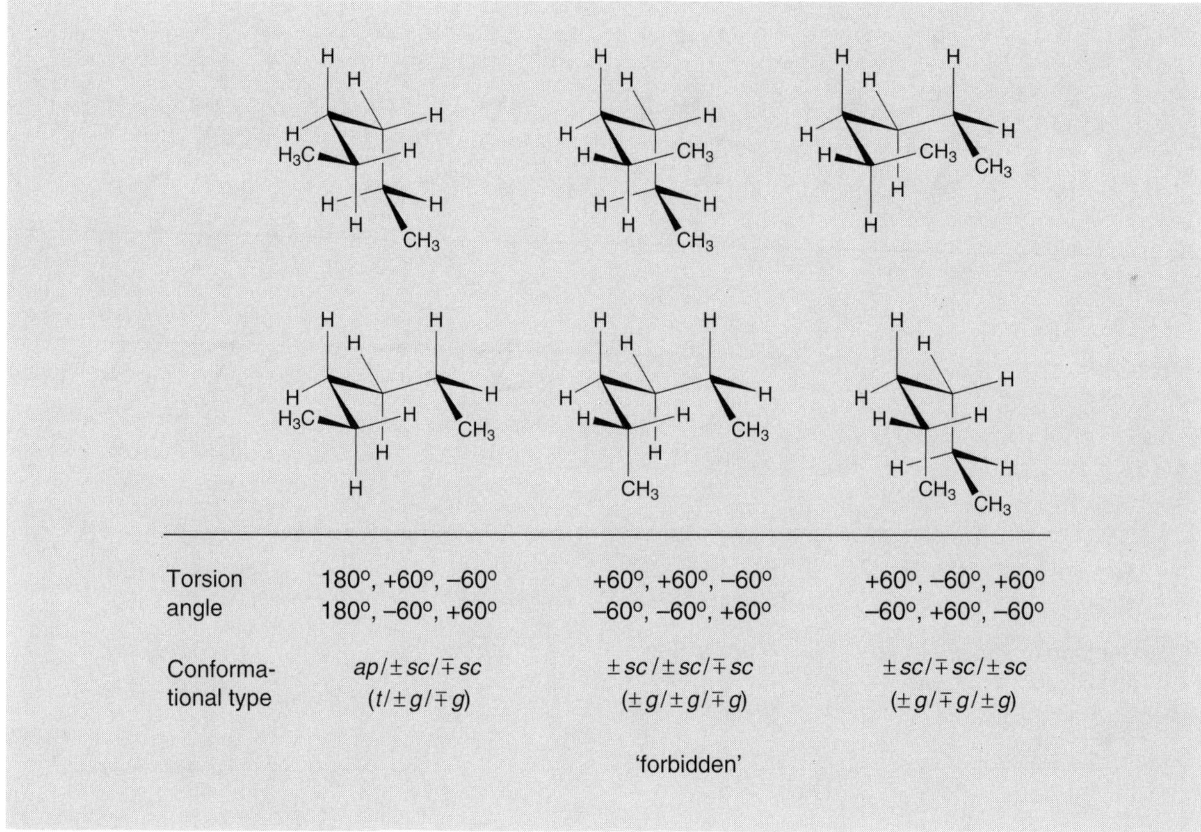

Fig. 7.9. (*cont.*)

In *Sect. 7.2.1*, the distinction between isotactic, syndiotactic, and atactic poly(propylene) was introduced. For syndiotactic poly(propylene) (*Fig. 7.4*), in which base units of the enantiomorphic configuration alternate, conformational analysis again leads to an idealized, linear zig-zag chain (torsion angle of the main chain: 180°; *Fig. 7.11, b*).

In the three-dimensional structure, it is possible to discern a translation axis, coinciding with the chain axis, a glide plane along the chain axis, and a mirror plane perpendicular to the chain axis.

Besides this, two enantiomorphic 4_1-helices (four configurational base units in a full turn) must also be considered (*Fig. 7.12*). A two-fold screw axis parallel and a two-fold axis of rotation perpendicular to the helix axis may be made out in the three-dimensional structure.

Isotactic poly(propylene) (*Fig. 7.3*), in which configurationally uniform base units successively follow as configurational repeating units, cannot, however, adopt an idealized, linear zig-zag backbone. Where poly(methylene) has parallel orientation of H-atoms at every second C-atom, and syndiotactic poly(propylene) alternating H-atoms and Me groups, idealized, linear isotactic poly(propylene) would have parallel orientation of Me groups, and so would be prone to a high level of *Newman* strain.

Instead of this, two enantiomorphic 3_1-helices must be considered. All the bonds in the idealized, helical main chain are staggered, alternating between *ap*- and +*sc*-, or between *ap*- and −*sc*-conformational fragments (*Fig. 7.13*). A three-fold screw axis parallel to the helix axis is recognizable in the three-dimensional structure.

The symmetry elements employed in *Figs. 7.11–7.13* – *translation vector*, *screw axis*, *glide plane* – play a role in the specification of space groups. Space groups and symmetry point groups are dealt with in *Chapt. 11*.

p

h

p: Pitch = *n* · *h*

n: Number of constitutional repeating units per helix turn

h: Distance of repeating units along the helix axis

For *n* = 2, the helix degenerates into a linear, zig-zag pattern. For *p* = 0, the helix degenerates into a circle.

Fig. 7.10. *Schematic representation of a helix section*

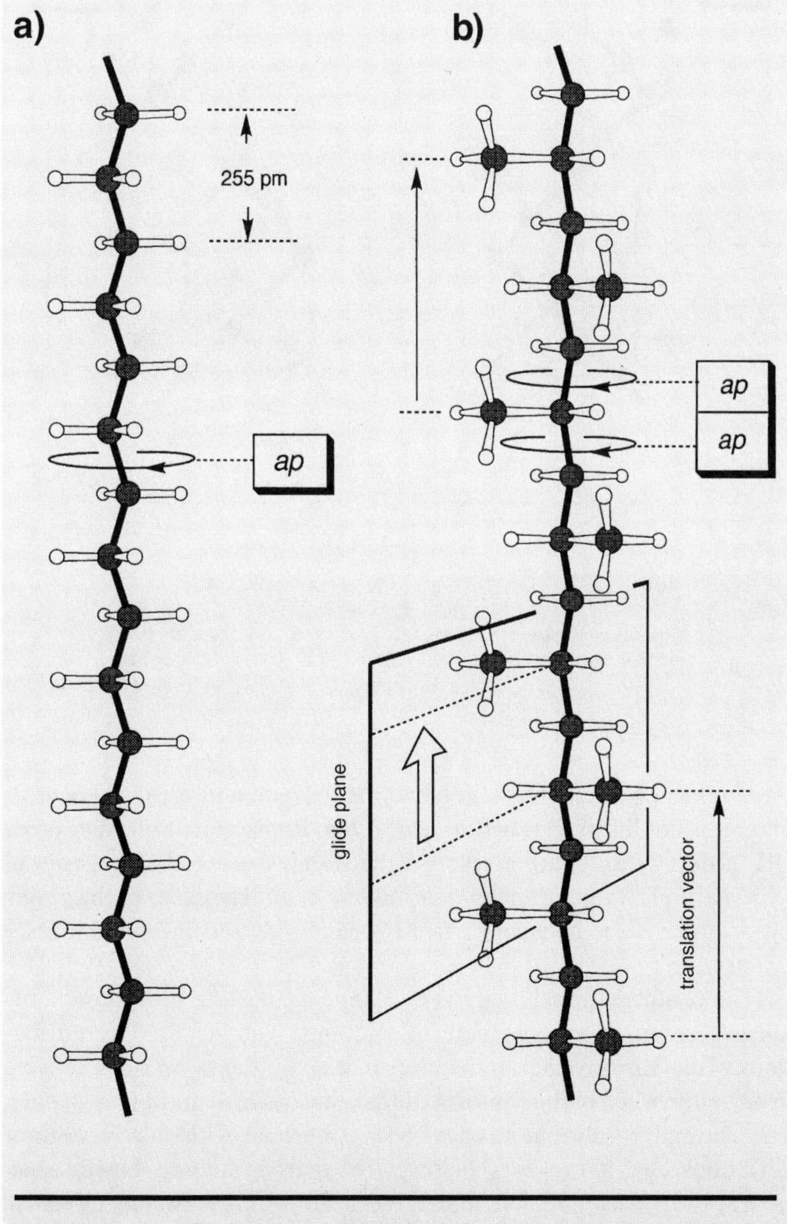

The torsion angles of the idealized, linear backbone are each 180°.

Fig. 7.11. *Sections of poly(methylene)* (a) *and syndiotactic poly(propylene)* (b), *each with idealized, linear chain*

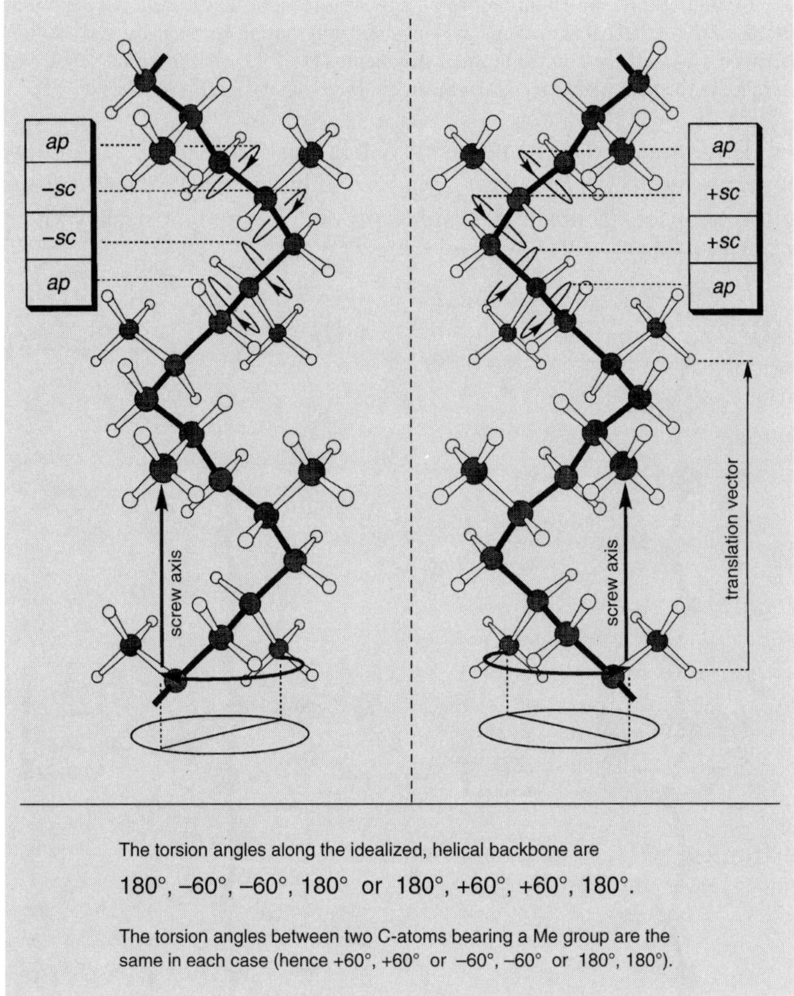

Fig. 7.12. *Sections of syndiotactic poly(propylene) with helical* 4_1*-conformation*

Unlike poly(methylene), poly(oxymethylene) in its idealized conformation is not linear, but helical (*Fig. 7.14*). Replacement of every other CH_2 group by an O-atom reverses the stability order characteristic of hydrocarbons; rather than *ap* > *sc*, now *sc* > *ap*. The cause of this reversal of order is the *anomeric effect* (*Chapt. 8.1.2*; for interpretation see *Chapt. 8.4.3*).

The conformational analysis of alkane chains has, up to now, only taken account of intracatenate interactions. The idealized conformations of poly(propylene) are subject to exactly the same rules as were firmly established in the conformational analysis of pentane (*vide supra*). The 'allowed' conformations have been confirmed, with only very minor deviations, by X-ray crystal-structure analysis and/or NMR spectroscopy. In *Chapt. 13.2.4*, under the heading *Topicity and Chemical Shift*, it is explained how topicity relationships discernible through NMR spectroscopy permit examination of the tacticity of regular polymers.

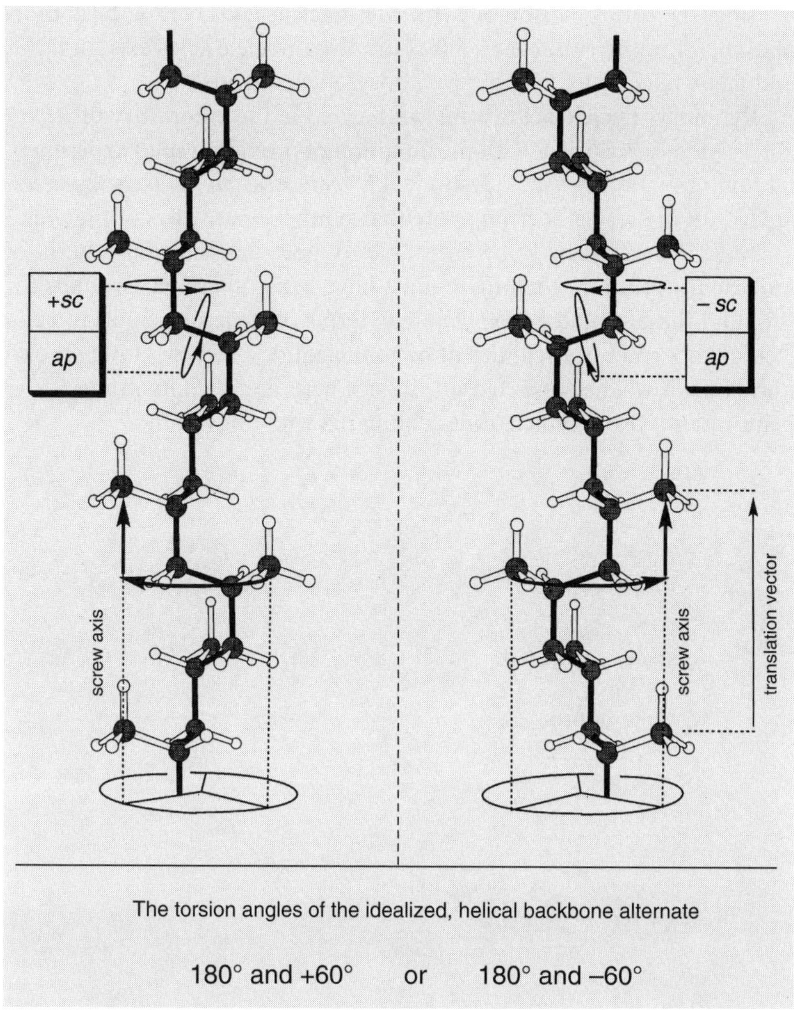

Fig. 7.13. *Sections of isotactic poly(propylene) with helical* 3_1*-conformation*

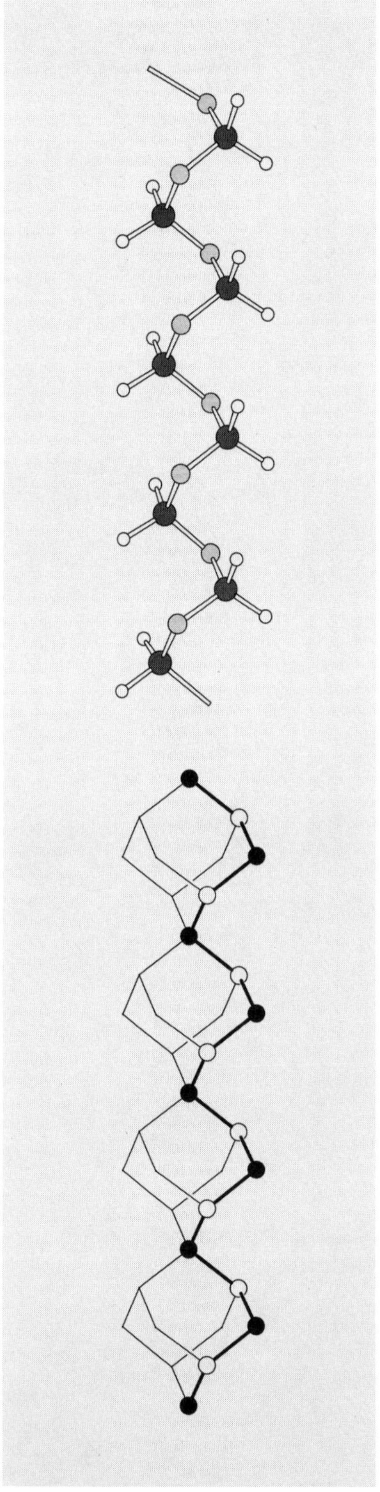

Fig. 7.14. *Idealized, helical conformation of poly(oxymethylene)*

7.3. Stereostructural Aspects of Biopolymers

Knowledge of the molecular structure of biopolymers and the supramolecular structure of functional units of biology is a basic prerequisite for comprehending the phenomena occurring in the living cell. We will, therefore, put *nucleic acids* (their molecular single strands first, then their supramolecular double strands), *polysaccharides*, and *proteins* 'under the lens' of conformational analysis.

7.3.1. Nucleic Acids

The saturated six-membered ring is conformationally more rigid than the corresponding five-membered ring. Hence, didactically, it was more appropriate to place the conformational analysis of cyclohexane before that of cyclopentane (*Chapt. 6.3* and *6.4*), and it is instrumental, for the topological description of carbohydrates (*Chapt. 10.5.3*), to discuss pyranoses before furanoses. It would, therefore, be desirable to consider the conformation of pyranosyl-nucleic acids before the conformation of furanosyl-nucleic acids. Furanosyl-nucleic acids exist in DNA and RNA (*Chapt. 10.5.2*), but pyranosyl-nucleic acids?

Pyranosyl-nucleic acids were synthesized in the laboratory of *Albert Eschenmoser*, in Zürich, with the intention of providing solid experimental findings to answer the question '*Why pentose- and not hexose-nucleic acids?*', of interest concerning potential synthetic pathways of the prebiotic era. If hexoses had not been able to meet the selection criteria of molecular evolution, then there must have been functional grounds for this, and these should be explicable in terms of structure. So as to get a feel for the structural realities of oligonucleotides, hexose- and pentose-nucleic acids will be presented side by side here, and pentopyranosyl- and pentofuranosyl-oligonucleotides compared with one another.

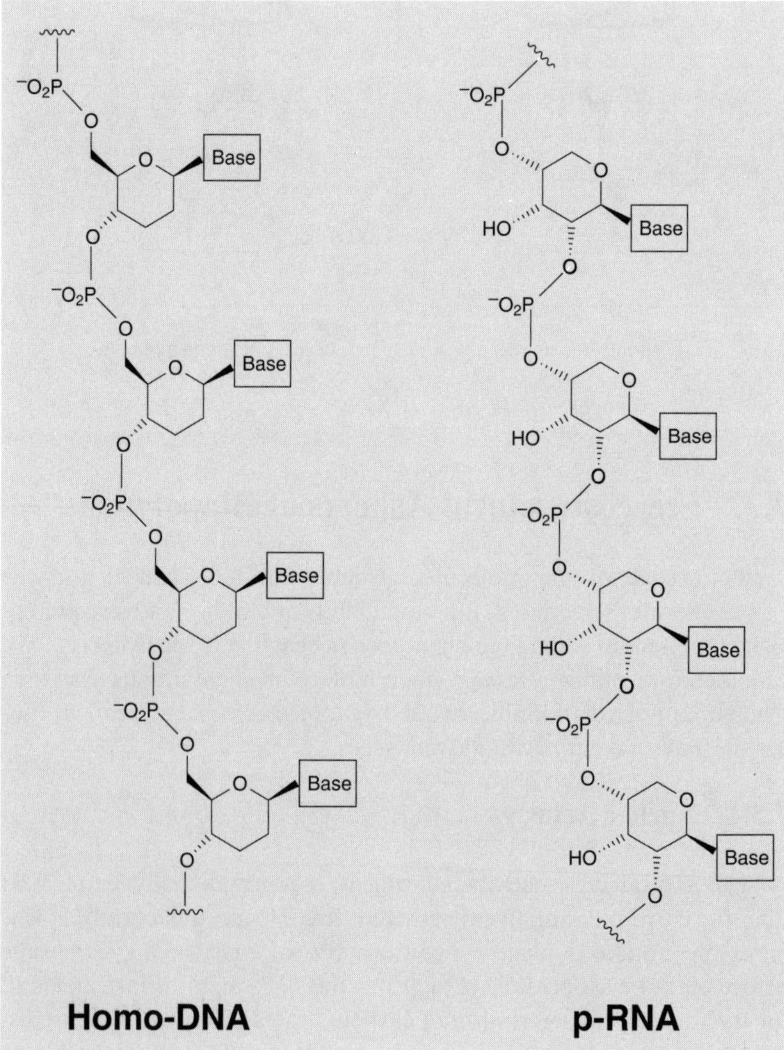

Fig. 7.15. *Sections of hexopyranosyl (4′ → 6′)- and pentopyranosyl (2′ → 4′)-oligonucleotides*

A delineation of how nucleic-acid single strands are built up, by the formal arrangement of constitutional, configurational, or conformational repeating units, is given in *Chapt. 10.5.2*.

7.3.1.1. Systematic Conformational Analysis of Pyranosyl-Oligonucleotide Single Strands

The two non-natural pyranosyl-oligonucleotides *homo-DNA* and *pyranosyl-RNA* (*p-RNA*) (*Fig. 7.15*) are virtually ideal demonstration objects for systematic conformational analysis.

Homo-DNA and p-RNA were synthesized in the Zürich laboratory in order to study the relationship of the pairing behavior of oligonucleotide single strands to the ring size of the participating furanose- or pyranose-sugar moieties, in the hope of finding clues relating to the origin of RNA.

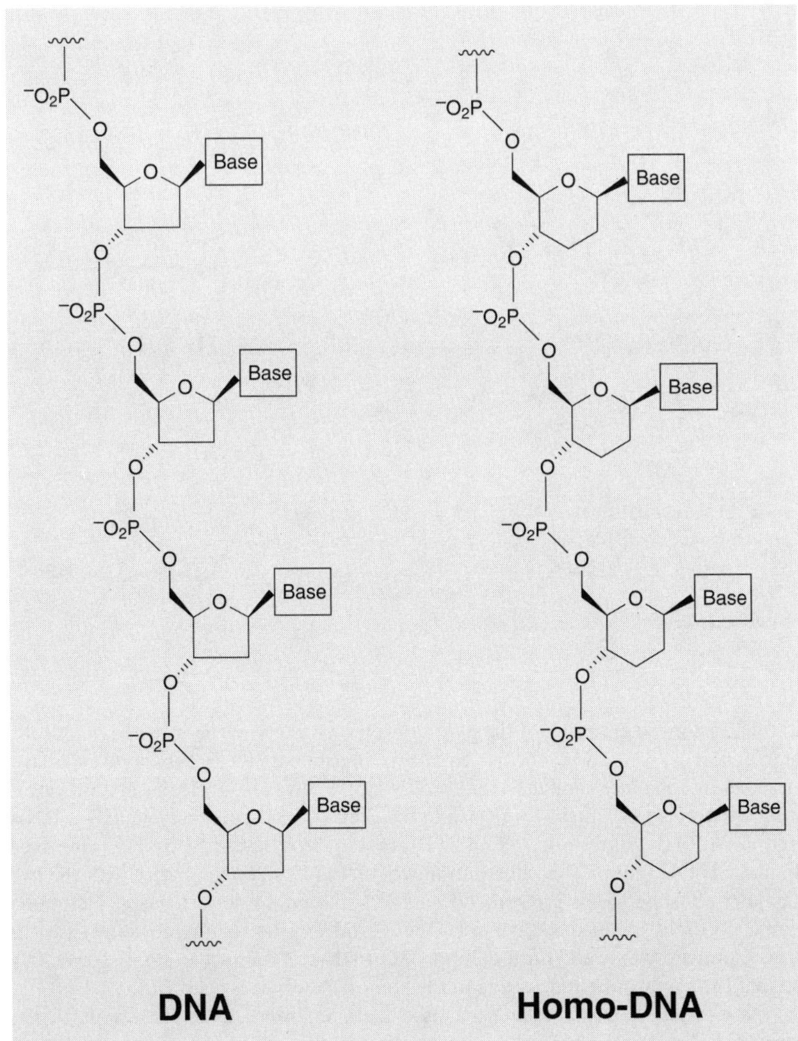

Fig. 7.16. *Sections of homologous pentofuranosyl (3′ → 5′)- and hexopyranosyl (4′ → 6′)-oligonucleotides*

7.3.1.1.1. Homo-DNA

The constitutional and configurational nature of a homo-DNA single strand, in comparison to that one of a DNA single strand, is shown in *Fig. 7.16*.

Fig. 7.17 shows phosphorylated sections of nucleotides, with a furanose and pyranose ring shown side by side in each case: the sugar building blocks of DNA (**A**) and homo-DNA (**B**), those of RNA (**C**) and homo-RNA (**D**), and, finally, those with an allofuranose (**E**) and an allopyranose (**F**) ring.

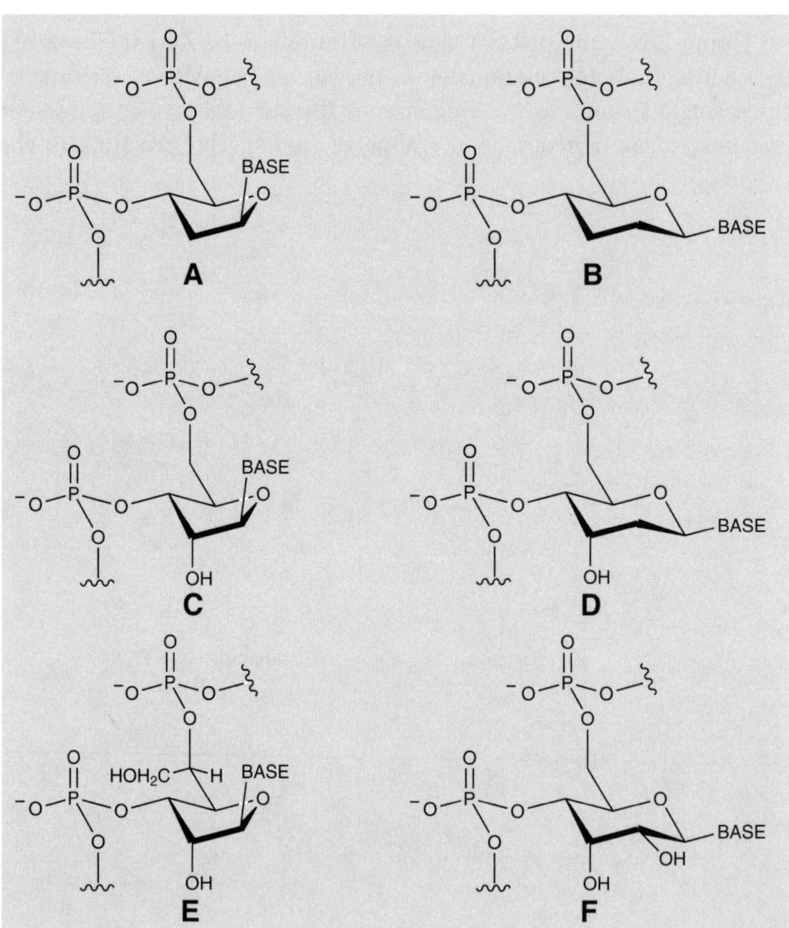

Fig. 7.17. *Sections of nucleotide single strands, each with a furanose or a pyranose ring as the sugar component*

The six-membered ring of the pyranose variants exists in the thermodynamically stable chair conformation. The two equatorial phosphodiester side chains are spatially oriented in a manner similar to that in DNA or RNA. The nucleobase, however, is also equatorially oriented with regard to the chair-formed pyranoses, unlike in DNA and RNA. In the allofuranose (**E**), one H-atom in the exocyclic 5-position is replaced by the CH_2OH group. This substitution would lead to steric hindrance with the two flanking phosphodiester groups (*Newman* strain), and hence to **E** being thermodynamically disadvantaged relative to **F**. The 2,3-dideoxy-*β*-D-allopyranose (**B**) has one CH_2 group more than the sugar component of DNA, 2-deoxy-*β*-D-ribofuranose (**A**), so that the naming oligonucleotide of the homo-DNA type is justified.

Non-standard terms have been used here, in order to emphasize particular structural details already in the nomenclature. Rather than talk of homo-RNA, it would also be possible, to emphasize the difference between **D** and **F**, to refer to 2-deoxy-β-D-allopyranose, or, in accordance with *IUPAC* rules, 2-deoxy-β-D-*ribo*-hexopyranose. *Table 7.2* offers a little glossary, which also includes the established *IUPAC* symbols.

Table 7.2. *Nomenclature of Pyranoses after* A. Eschenmoser *and after* IUPAC

Chemical formula	Nomenclature after *Eschenmoser*	*IUPAC* Nomenclature	*IUPAC* Symbol
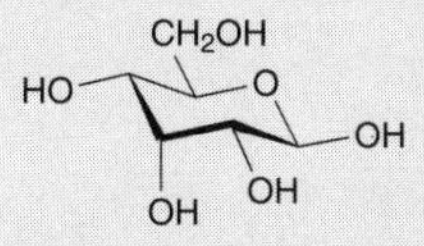 CH_2OH, HO, O, OH, OH, OH	β-D-Allopyranose	β-D-Allopyranose	All*p*
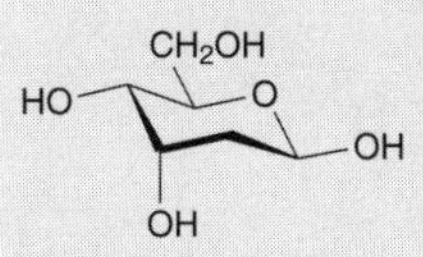 CH_2OH, HO, O, OH, OH	2-Deoxy-β-D-{allo, altro} pyranose	2-Deoxy-β-D-*ribo*-hexopyranose	d[2]All*p*
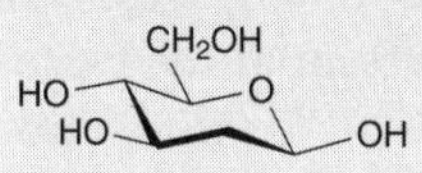 CH_2OH, HO, HO, O, OH	2-Deoxy-β-D-{gluco, manno} pyranose	2-Deoxy-β-D-*arabino*-hexopyranose	d[2]Glc*p*
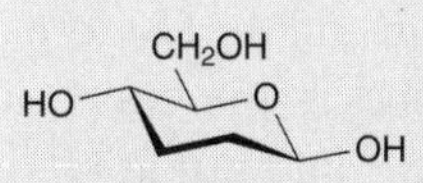 CH_2OH, HO, O, OH	2,3-Dideoxy-β-D-{allo, altro, gluco, manno} pyranose	2,3-Dideoxy-β-D-*erythro*-hexopyranose	d[2]d[3]Glc*p*

Idealized conformations of a homo-DNA strand, free of *Pitzer* strain, may be graphically described with the help of the diamond-lattice matrix (*Chapt. 10.3.3*). This method ignores the specific nature of the atom centers and different bond lengths, while assuming tetrahedral coordination, not only for C- and P-, but also for the O-centers. For this last instance, it means that additional phantom ligands must be introduced, in order to reach the coordination number 4. This fitting into the diamond-lattice matrix makes it clear that idealized conformations of the polynucleotide have torsion angles of ±60° or 180°, and that the pyranose ring adopts a chair conformation with all three substituents in equatorial orientations (*Fig. 7.18*).

It is possible to specify the idealized, staggered conformations of a homo-DNA strand with the aid of the conformational types of pentane (*Fig. 7.7*). However, to pick out the favored backbone section(s) between two adjacent sugar rings from the large number of possible staggered conformations, additional selection criteria are required.

The steric requirement for minimum *Newman* strain has already been taken into account as a selection criterion in previous instances of sys-

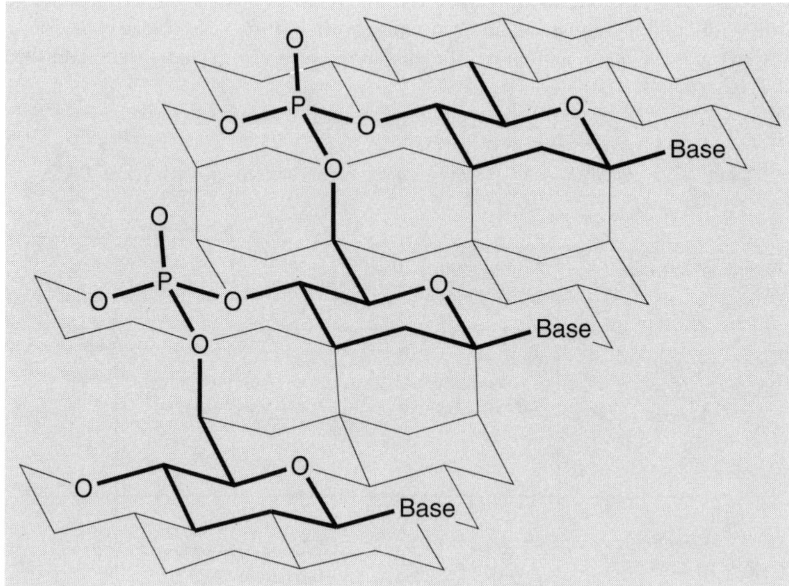

Fig. 7.18. *Diamond-lattice fitting of an idealized homo-DNA trinucleotide conformation, obtained as one of four favored conformations from systematic conformational analysis (Fig. 7.20) and the only one remaining after consideration of conformational repetition*

tematic conformational analysis (*Sect. 7.2.2*) and is also valid here: $\pm sc/\mp sc$-conformational types are 'forbidden'. A further criterion concerns the relative orientation of the two alkoxy groups present at the P-atom of the phosphodiester group. Among the 'allowed' conformations (*Fig. 7.19*), the favored ones are those in which the two alkoxy groups are oriented so that the bond between a phantom ligand and the adjacent O-atom is oriented *ap* to the bond between the P-atom and the O-atom of the other alkoxy group. This conformational type is present twice for $+sc/+sc$ and $-sc/-sc$, once for $ap/+sc$ or $ap/-sc$, and not at all for ap/ap. The special conformational effect, in a linear chain, arising from the alternate replacement of CH_2 groups by O-atoms, taken into account in the stability sequence of *Fig. 7.19*, finds no explanation in the context of the structure model of the classical organic chemistry: it is going to be interpreted in *Chapt. 8.4.4*.

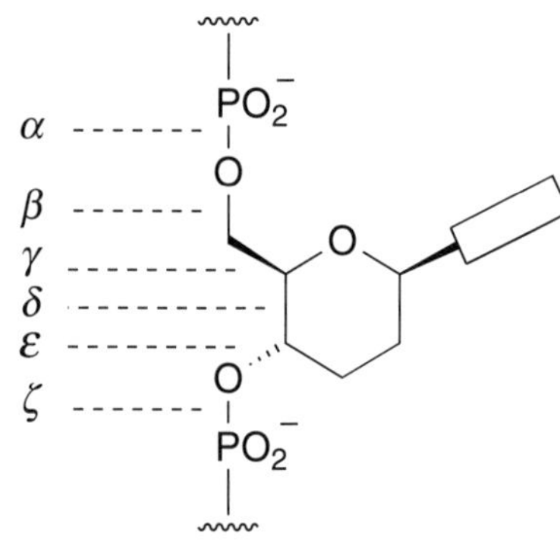

A total of six torsion angles α–ζ along the homo-DNA backbone must be considered in selecting the favored conformations of a homo-DNA single strand. The description by formulae and conventional specification of repeating units relevant to nucleic acids are given in *Chapt. 10.5.2*. One of the six bonds concerned, the endocyclic C–C bond in the pyranose ring, unlike the other five bonds, has, intrinsically, only two (60°, 180°) options in its torsion angle δ, rather than three (60°, −60°, 180°). The number of possible conformations, then, is obtained from the product of $2 \cdot 3^5$ (one bond with two options, five bonds each with three options). This $2 \cdot 3^5 = 486$ formally possible conformations reduces to 243 at the first step, as only the idealized pyranose chair conformation ($\delta = 60°$), with three equatorial substituents, is 'allowed' (*Fig. 7.20*). The inverse chair conformation is 'forbidden', because two of the three axial substituents would cause *Newman* strain in that instance.

The remaining five torsion angles are considered as demonstrated in *Fig. 7.20*. The analysis at any given option branch is always broken off as soon as a partial conformation fails to meet the selection criteria above.

Thus, for example, the value −60° is not permissible for the torsion angle γ, as the corresponding partial conformation exhibits a 1,5-repulsion. Options in which a 1,5-repulsion is present are marked with * in *Fig. 7.20*. For ascertaining the torsion angles α and ζ, the stability order in *Fig. 7.19* must be taken into account: options which do not lead to the energetically favored *+sc*/*+sc*- or *−sc*/*−sc*-conformations of the phosphodiester fragment are signified by **. For the torsion angle ε, there is no case in which the secondary bound phosphate group would be free of *Newman* strain in the idealized model, in any of the four remaining option branches. The value $\varepsilon = 180°$ represents the conformation of least destabilizing 1,5-repulsion in all four branches.

Fig. 7.21 depicts *Newman* projections to demonstrate that high *Newman* strain always arises when – as in pentane (*Fig. 7.7*) – in a chain of five atoms (1–5) the torsion angles between the atoms 1–4 and 2–5 assume the values +60°, −60° or −60°, +60°. Hence, for instance, the torsion angle $\varepsilon = -60°$ is excluded because of 1,5-repulsion between P and C(6′), as the P—O(4′)—C(4′)—C(5′) torsion angle is −60° and the O(4′)—C(4′)—C(5′)—C(6′) torsion angle +60°.

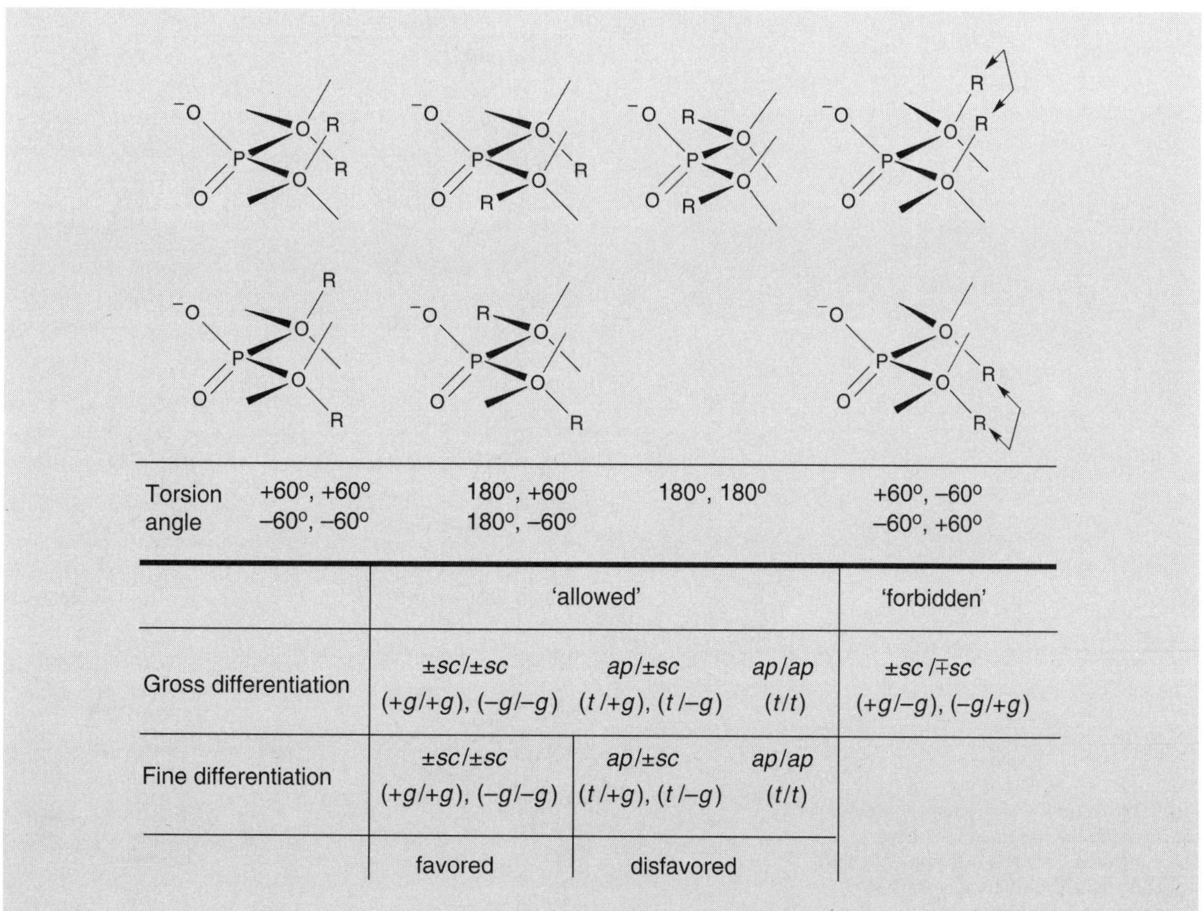

Torsion angle	+60°, +60° / −60°, −60°	180°, +60° / 180°, −60°	180°, 180°	+60°, −60° / −60°, +60°

	‘allowed’			‘forbidden’
Gross differentiation	±*sc*/±*sc* (+*g*/+*g*), (−*g*/−*g*)	*ap*/±*sc* (*t*/+*g*), (*t*/−*g*)	*ap*/*ap* (*t*/*t*)	±*sc*/∓*sc* (+*g*/−*g*), (−*g*/+*g*)
Fine differentiation	±*sc*/±*sc* (+*g*/+*g*), (−*g*/−*g*)	*ap*/±*sc* (*t*/+*g*), (*t*/−*g*)	*ap*/*ap* (*t*/*t*)	
	favored	disfavored		

Fig. 7.19. *Systematic conformational analysis of the phosphodiester fragment of a polynucleotide single strand, in which the O-atoms in the chain are provided with additional bonds to monovalent phantom ligands* (*t*: *trans*, *g*: *gauche*)

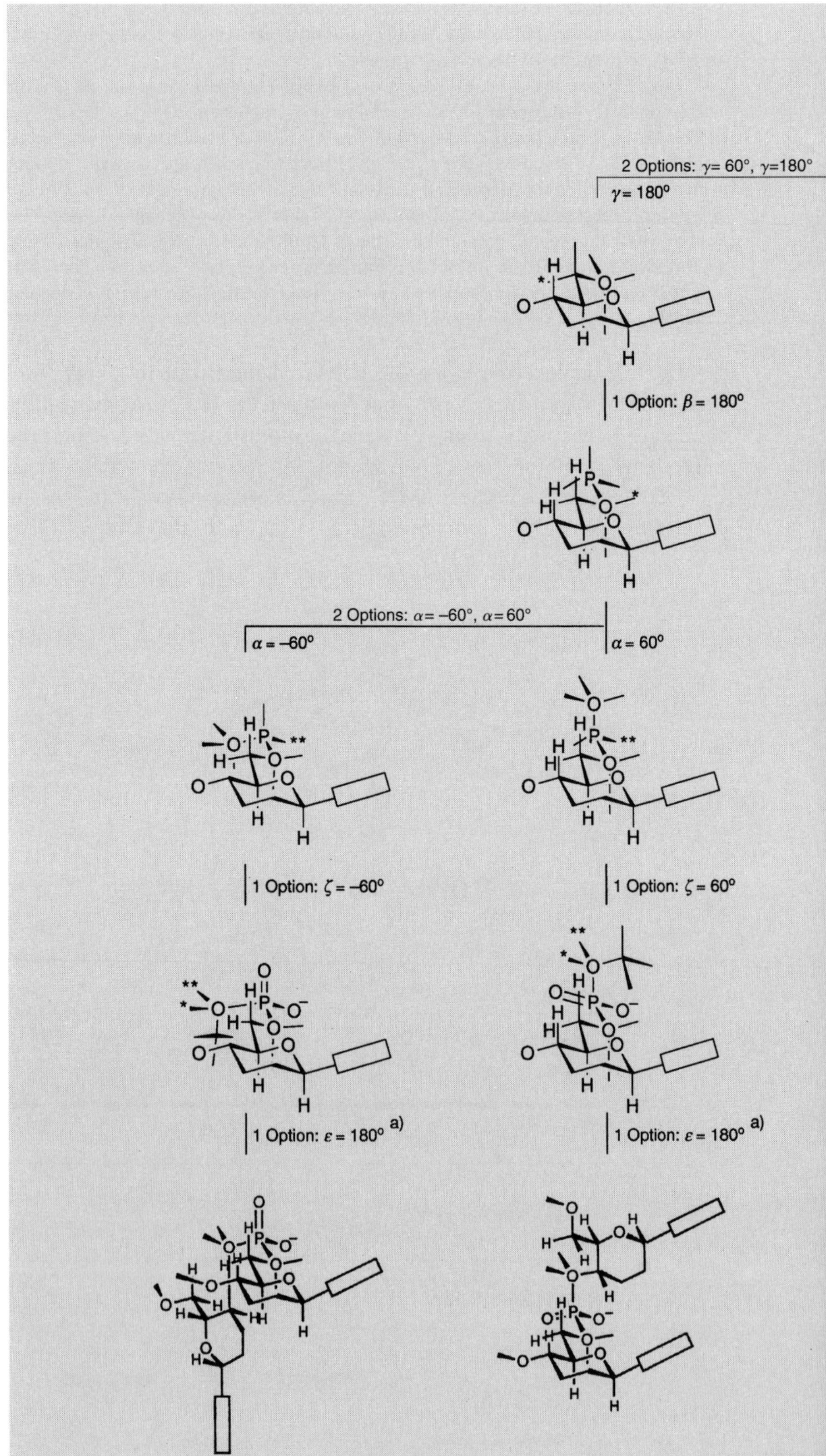

Fig. 7.20. *Decision-taking steps in finding the favored conformations of a homo-DNA oligonucleotide. a)* Option with the sterically least disadvantageous (but obligatory) 1,5-repulsion. *b)* Only 'allowed' if the subsequent torsion angle amounts to 180°.

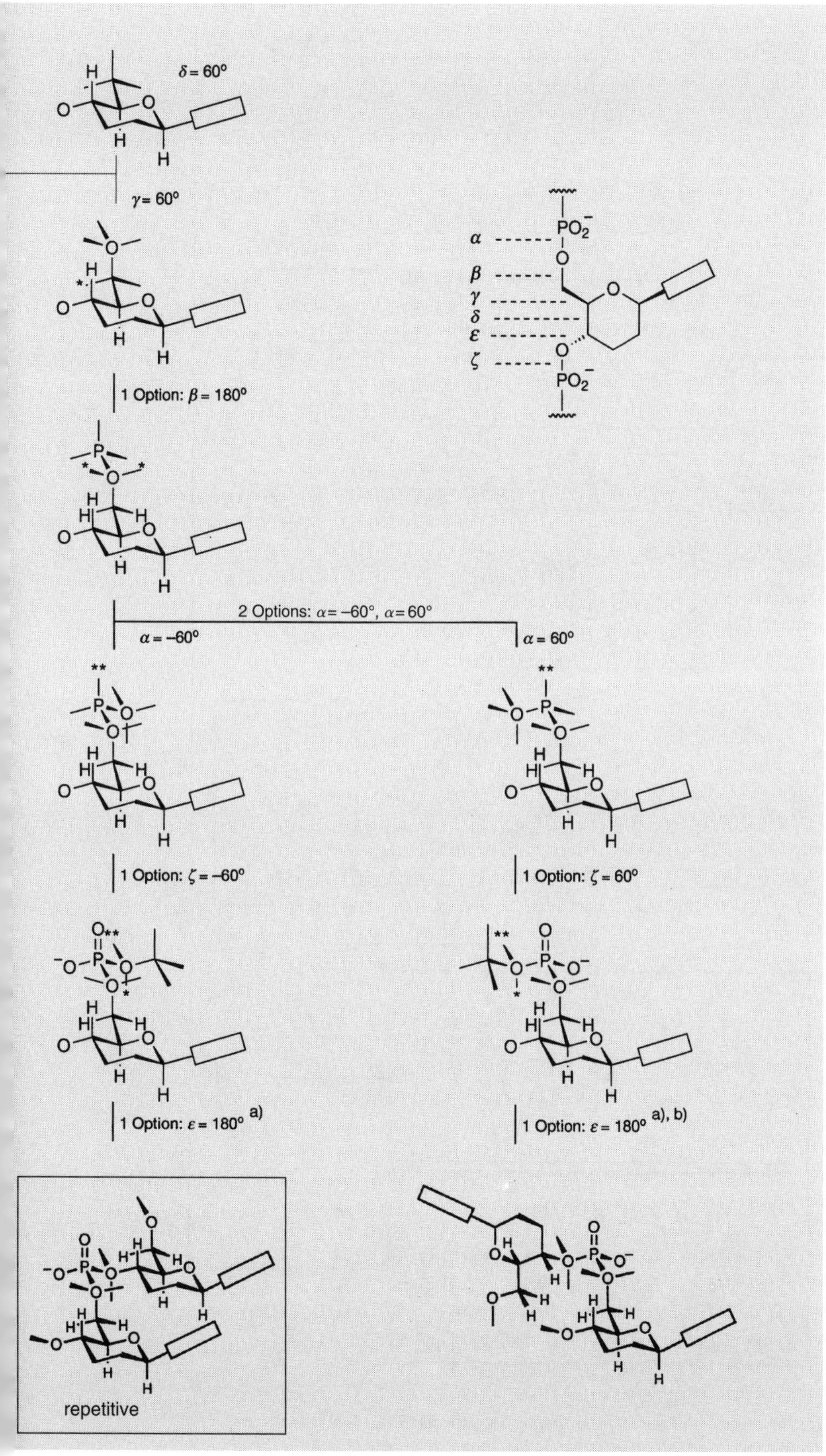

$\delta = 60°$

$\gamma = 60°$

α β γ δ ε ζ

PO_2^-

1 Option: $\beta = 180°$

2 Options: $\alpha = -60°$, $\alpha = 60°$

$\alpha = -60°$

$\alpha = 60°$

1 Option: $\zeta = -60°$

1 Option: $\zeta = 60°$

1 Option: $\varepsilon = 180°$ a)

1 Option: $\varepsilon = 180°$ a), b)

repetitive

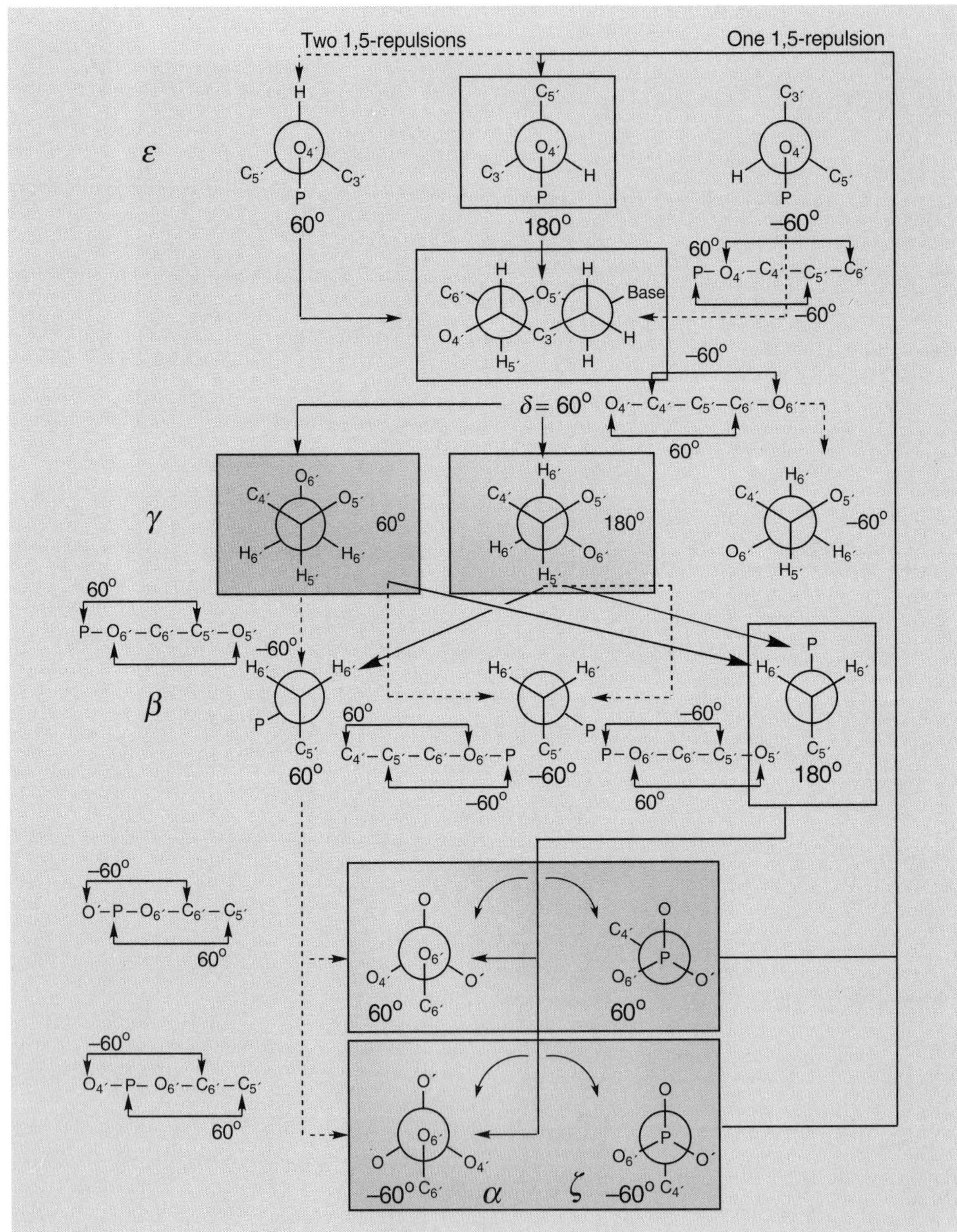

Fig. 7.21. Newman *projections as visual aids for the decision-taking steps in finding the four idealized conformations of a homo-DNA oligonucleotide with minimal* Pitzer *and* Newman *strain, taking account of the anomeric effects on the phosphodiester unit*

'Allowed' *Newman* projections are connected by heavy printed lines. Dashed lines connect *Newman* projections in which at least one must be excluded because of high *Newman* strain. In this latter case, the five-center unit, displaying *Newman* strain, is shown in conjunction with its two inadmissible torsion angles. Consideration of each of the adjacent torsion angles during systematic conformational analysis (*Sect. 7.2.2*) excludes those *Newman* projections with a gray background. The combinations with a gray background, of γ (+60° or 180°) and α, ζ (+60°, +60° or −60°, −60°), give those four conformational units which are 'allowed'.

However, of these four conformations, only one fulfills the last criterion which still must be taken into account: the criterion of *conformational repetition*. An oligonucleotide sequence is only capable of pairing, if the constitutional repeating unit also repeats itself conformationally. In the pairing conformation, the oligonucleotide backbone is linear (*Fig. 7.22*).

It is not possible to construct an unhindered oligomer strand from the three conformationally *non-repetitive* fragments in *Fig. 7.22*, as these fragments would hinder themselves intracatenately.

The idealized model of a segment of an oligonucleotide of homo-DNA type is shown in *Fig. 7.23*, in conjunction with that one of a segment of DNA type. The one has a torsion angle δ of 60° and a linear zig-zag chain as its backbone. The other has a

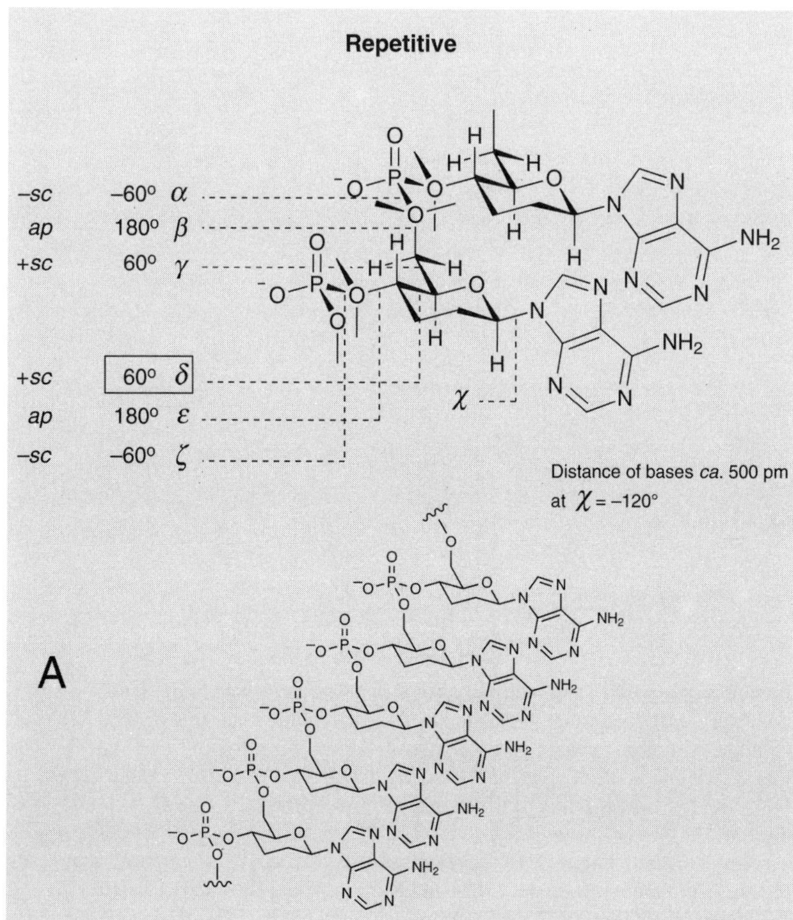

Fig. 7.22. *The four conformational units of a homo-DNA single strand selected from 486 formally possible conformations using systematic conformational analysis:* **A**: *repetitive;* **B**, **C**, *and* **D**: *non-repetitive*

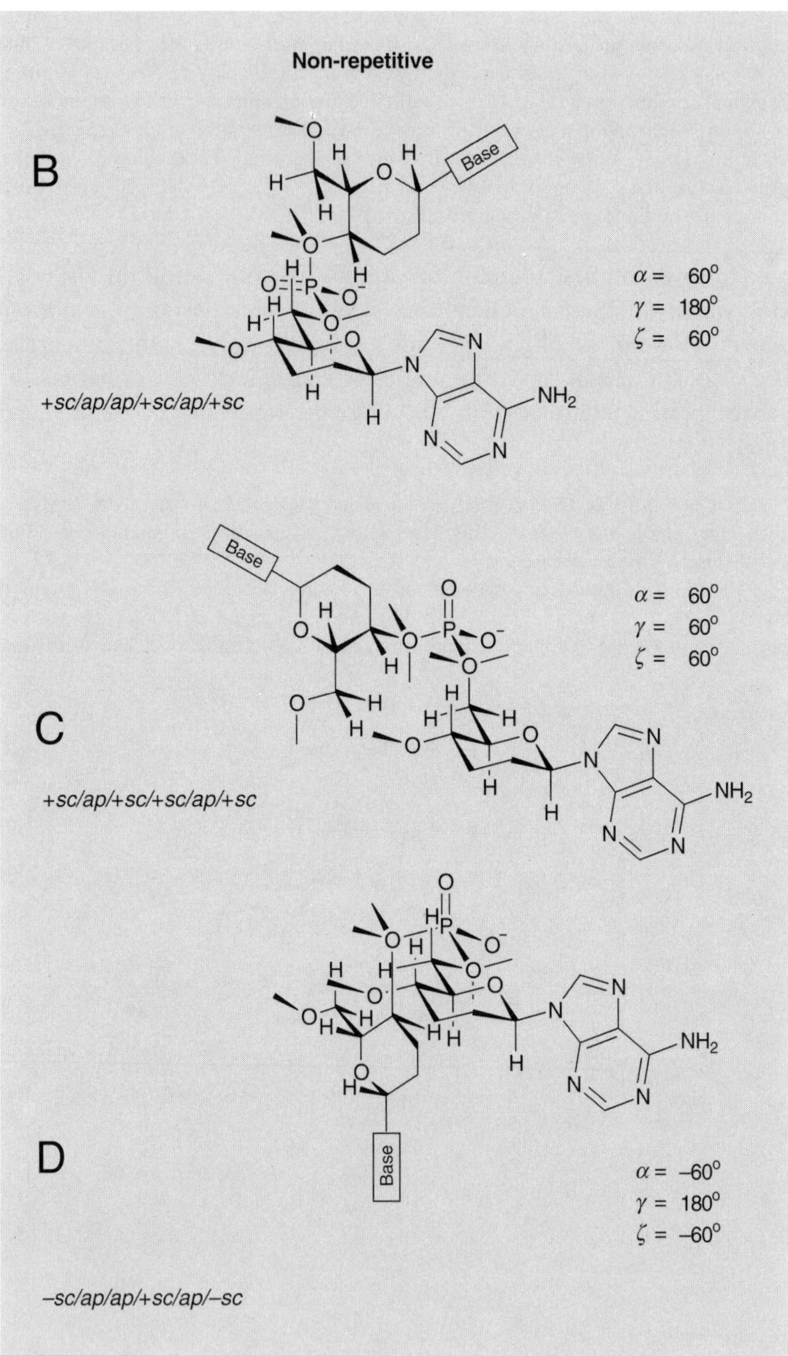

Fig. 7.22. (*cont.*)

torsion angle $\delta > 60°$, and, therefore, a helical chain as its backbone. The C(4′)–C(3′) bond acts as a hinge, when the torsion angle δ or ν_3 of the sugar ring (*Fig. 10.22*) alters. Oscillatory motion occurs particularly easily about this hinge in the mobile furanose ring, but is much more restricted in the rigid pyranose ring. This applies also to the DNA duplex, the flexibility of which is much greater than that one of the homo-DNA duplex.

The idealized homo-DNA single strand, constructed out of conformationally repetitive mononucleotide units, has an idealized, linear secondary structure. In reality, however, a homo-DNA single strand will only be quasi-linear, as any real devia-

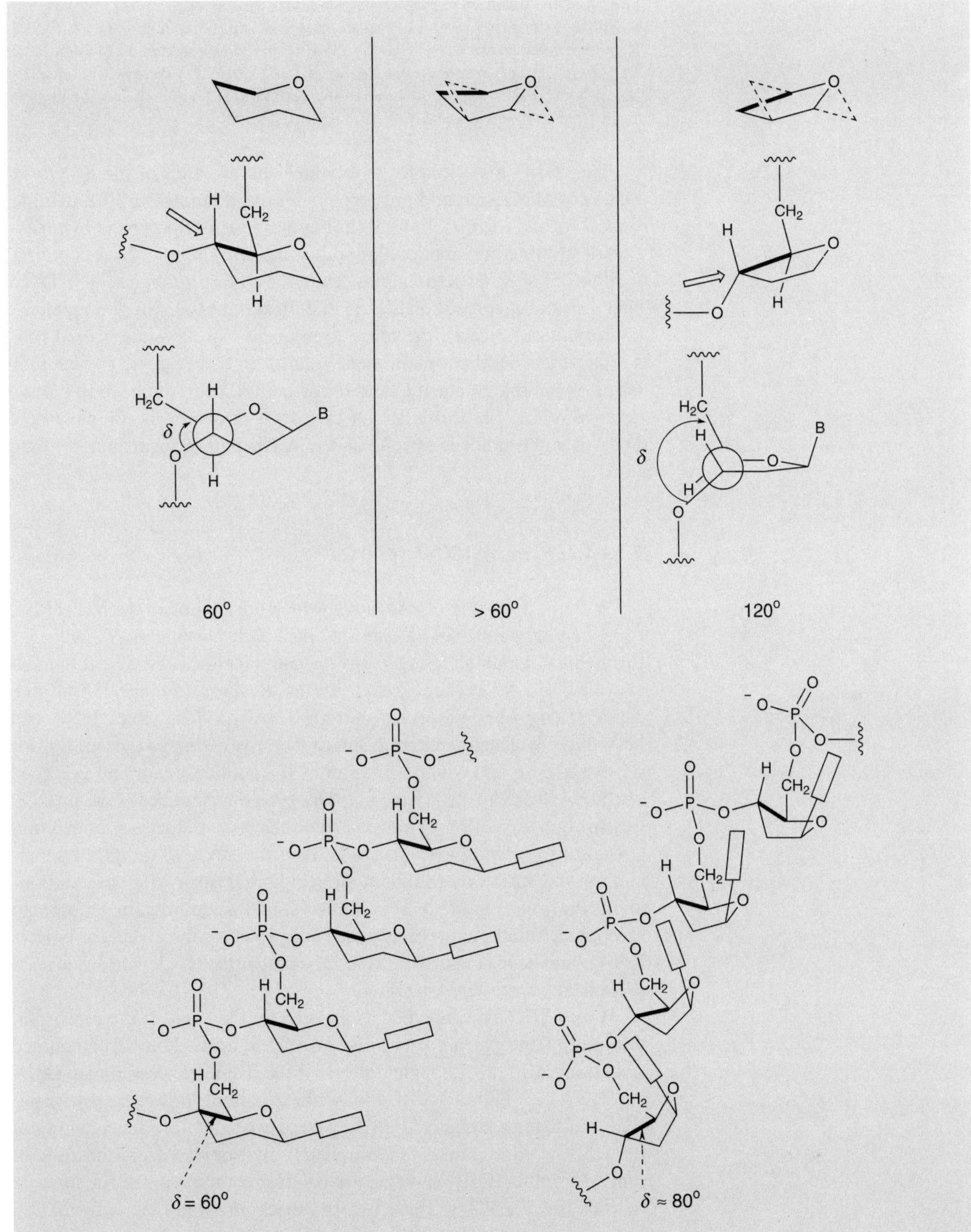

Fig. 7.23. *Influence of the endocyclic torsion angle δ on the backbone conformation of a DNA (or a RNA) single strand viewed from the model of idealized homo-DNA*

tion from the idealized values of its bond and torsion angles leads, by virtue of repetitive intensification, to a helical secondary structure (see text to *Fig. 7.10*). 1,5-Repulsion between the C(3′)−CH_2 group of the pyranose ring and an O-atom of the RO group of the phosphodiester ligand at C(4′) (*Fig. 7.21*) disrupts the idealized staggering and is an important cause of the deviation, at least of the torsion angles ζ and α from their *sc*-values of −60°.

The clarity and elegance with which the question of the helicity of DNA was addressed and answered is a superb example of the attitude, typical for the chemist, towards building up regular polymers by repetition of simple conformational modules and working out the exceptional qualities of such structures by examination of homologues. The DNA single strand is already helical in the idealized model of the repetitive conformation, because the sugar component has a five-membered ring, so that the exocyclic torsion angle δ must, in this case, be greater than 60°. It is capable of pairing, and the structural pre-organization plays a decisive role in the formation of the DNA double helix. Single strand DNA is right-handed, as long as the participating sugar is D-configurated.

7.3.1.1.2. Pyranosyl-RNA (p-RNA)

With the discovery of *ribozymes*, which brought them the *Nobel* Prize for Chemistry in 1989, *Sidney Altman* and *Thomas R. Cech* broke through a dogma prevailing in biology and biochemistry: no protein, no enzyme. As early as the mid-sixties, *Carl R. Woese*, *Francis Crick*, and *Leslie E. Orgel* had suspected that an RNA world, in which RNA had been the only genetic material, simultaneously acting as the catalyst for self-replication and for the creation of the peptide bond, had preceded our DNA/RNA/protein world. 2′-Deoxyribonucleotides come into being through the action of various ribonucleotide reductases on ribonucleotides: the step always proceeds from the RNA to the DNA series, and not the other way round, in all living organisms. After speculations about the *origins of life in an RNA world* had begun to gain widespread acceptance, the key question posed was whether the structure type of RNA could have constituted itself, or whether the RNA world was itself preceded by a pre-RNA world.

Homo-DNA was intended as a model for the study of the conformation and of the pairing capability of nucleic acids. Knowledge gained from it revealed the fundamental role of the five-membered ring in DNA and RNA. 2,3-Dideoxy-β-D-allopyranose cannot, however, be considered as a sugar which could already have existed before the dawning of the RNA world: unlike the completely hydroxylated aldohexoses of general formula $(CH_2O)_6$ which are product components of the *formose reaction* (see *Fig. 7.24*). Therefore, in Zürich, the study of conformation and pairing capabilities of pyranose oligonucleotides was extended into systems with suitable fully hydroxylated components: systems with allose, altrose, and glucose as the hexose components. Experimental find-

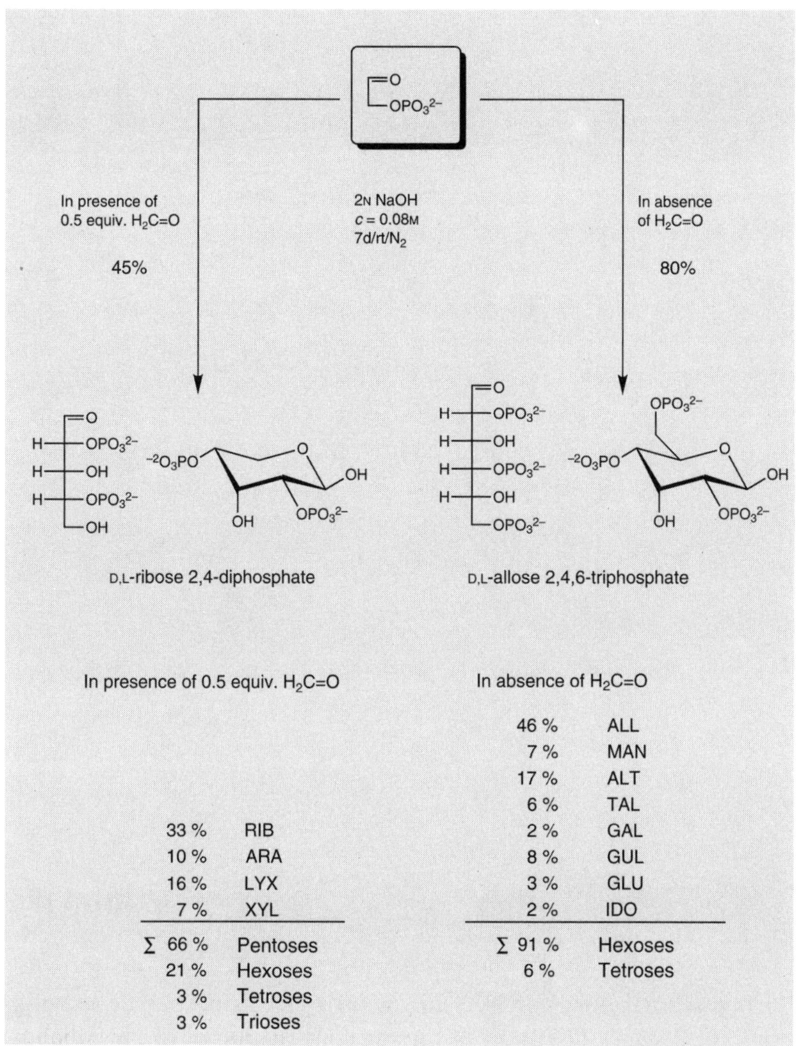

Fig. 7.24. *Conditions under which mainly* D,L-*ribose 2,4-diphosphate and* D,L-*allose 2,4,6-triphosphate arise out of glycolaldehyde phosphate*

ings gleaned from these systems were finally to lead to the synthesis and the examination of its pairing properties of pyranosyl-RNA (p-RNA), an oligonucleotide system isomeric with its natural counterpart RNA (*Fig. 7.25*).

Allopyranosyl- and altropyranosyl-oligonucleotides are of interest, because D,L-altrose and D,L-allose constitute the major part of the 2,4,6-triphosphorylated aldolization product of glycolaldehyde phosphate (*Fig. 7.24*). If the aldolization of glycolaldehyde phosphate is carried out in the presence of formaldehyde, however, pentoses make up two thirds of the 2,4-diphosphorylated reaction product, D,L-ribose alone being one third (*Fig. 7.24*).

The properties of RNA are subject to the influence of the pentofuranose sugar D-ribose. But what are the properties of the ribopyranosyl isomer of RNA, p-RNA, in which, for consistency, the sugar centers C(2′) and C(4′), rather than C(3′) and C(5′), are linked by phosphodiester groups?

Fig. 7.25. *Sections from the constitutionally isomeric ribofuranosyl (3′ → 5′)- and ribopyranosyl (2′ → 4′)-oligonucleotides*

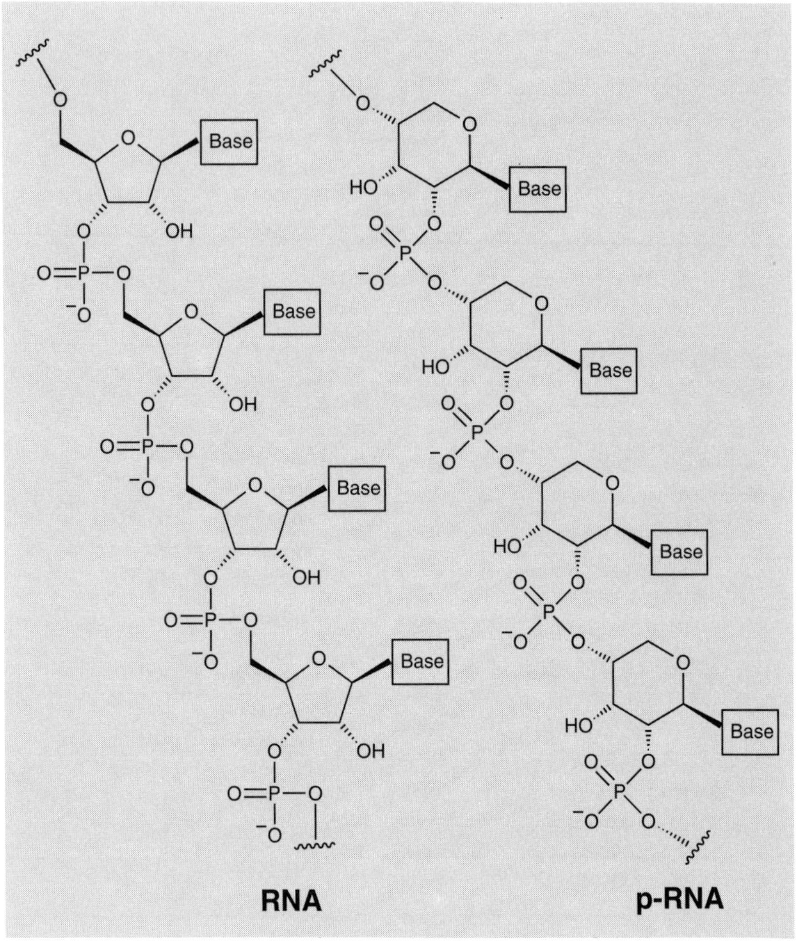

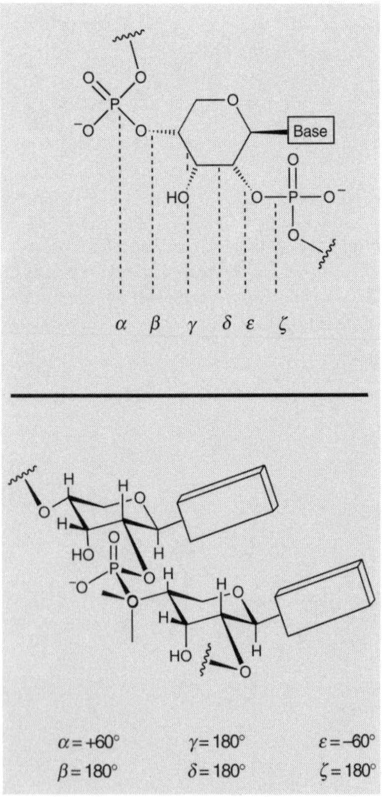

Fig. 7.26. *Specification* (above) *and numerical data* (below) *for the torsion angles in the repeating unit of the idealized, linear single strand of p-RNA*

In p-RNA, unlike in RNA, the sugar ring participates *via cis*-configuration between C(2′) and C(4′) in the construction of the sugar-phosphate backbone (*Fig. 7.25*).

Systematic conformational analysis, performed on p-RNA in a similar manner as accomplished on homo-DNA (*Sect. 7.3.1.1.1*), begins with a total of 162 ($= 2 \cdot 3^4$) single strands composed of idealized conformational units only. It reduces to nine single strands with minimum strain, of which, however, only one is capable of pairing according to the *Watson-Crick* pattern. Its conformational repeating unit shows the torsion angles given in *Fig. 7.26*, and it corresponds to the *+sc/ap*-phosphodiester-type conformation (*Fig. 7.27*).

7.3.1.2. Base Pairing as a Means of Molecular Recognition

A polynucleotide has several donor and acceptor groups for H-bonds in each of its nucleobases. This results in a combinatorial variety for self-organization. Two nucleotides, which join together, form a base pair, belonging to one among four constitutional types (*Watson-Crick*, *reverse Watson-Crick*, *Hoogsteen*, or *reverse Hoogsteen*; *vide infra*). A

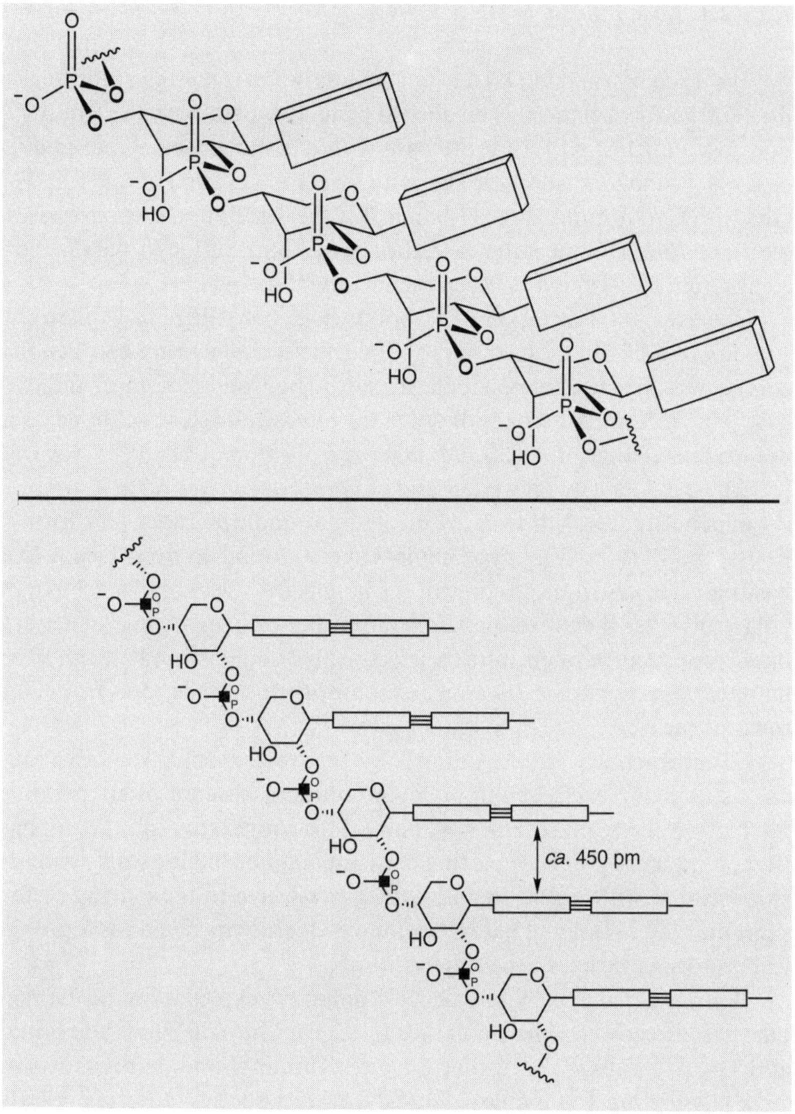

Fig. 7.27. *Schematic representation of the idealized, linear single strand of p-RNA* (above) *with emphasis on its pairing capability* (below)

polynucleotide single strand may combine with a second single strand to form a duplex, with a duplex to form a triplex, or with a triplex to form a quadruplex. Two neighboring single strands within this supermolecule are either parallel or antiparallel to one another, and form a double strand, which, depending on the conformational situation about the *N*-glycosidic bonds, belongs to one among four conformational types (*anti*/*anti*, *anti*/*syn*, *syn*/*anti*, or *syn*/*syn*).

Information on the strength of donor and acceptor groups is given in *Chapt. 15.2.6*.

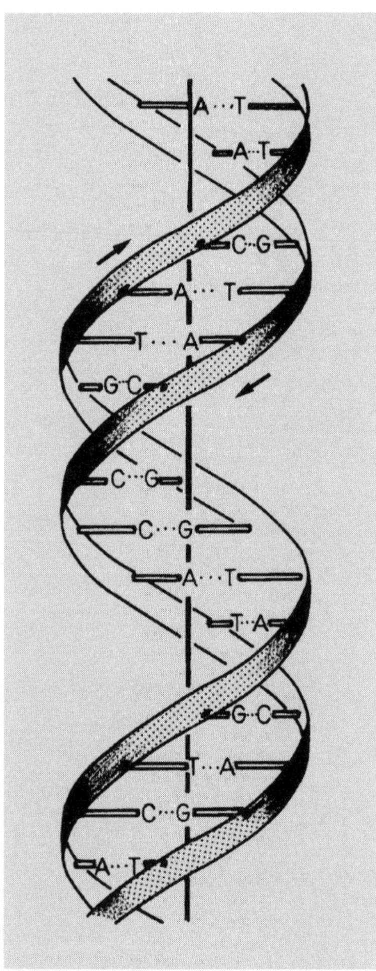

Fig. 7.28. *Section of the DNA duplex after* J. D. Watson *and* F. H. C. Crick

7.3.1.2.1. Base Pairing in DNA

The double strand of DNA (*Fig. 7.28*) shows base pairing according to the *Watson-Crick* pattern. This should be understood to mean a constitution-specific pairing of a purine base with a complementary pyrimidine base: particularly of Ade with Thy, with formation of two H-bonds, and of Gua with Cyt, forming three H-bonds (for the description of nucleobases, see *Table 10.15*). In the latter case, the largest contribution to stabilization comes from the H-bond between the two ring N-atoms.

However, the formation of H-bonds does not suffice to explain the stability of A/T- and, especially, of G/C-nucleoside pairing between the complementary nucleotide single strands. The isomorphism of the two base pairs means that the topology of the phosphodiester backbone does not need to change, if Ade (Gua) takes the position of Thy (Cyt) (or *vice versa*), or if Ade/Thy takes that one of Gua/Cyt (or *vice versa*).

Kinetic studies with isolated nucleobases indicate rapid pair formation of the *Watson-Crick* type: almost every mutual approach leads to a base pair. However, the lifetime of the base pairs is very brief (< 1 μs). In polar solvents, therefore, pair formation between two complementary bases, not incorporated into their respective single strands, cannot be made certain: hydration (in aqueous solution) pushes the equilibrium far towards the side of the individual bases.

Complementary single strands, by contrast, form thermodynamically stable duplexes even in aqueous solution, because of cooperative interactions between several base pairs. This cooperative interaction has the consequence that base pairing between oligonucleotide single strands is associated with a gain in *free enthalpy*, relative to the pairing of the same number of isolated bases. Optimal cooperativity demands topological pre-organization of the single strands.

Watson-Crick-type base pairing supplies the explanation for the regularities, already established earlier by *Erwin Chargaff*, that Ade (Gua) and Thy (Cyt) always occur in equivalent amounts, and that the ratio of Ade/Thy to Gua/Cyt is characteristic for every species, differing greatly in phylogenetically far-removed species (*Chargaff Rules*; *Table 7.3*).

The base pairs between the backbones of the single strands function structurally like the rungs of a rope ladder. Base pairing dictates a 1:1 relationship between the two strands. The single strands are of antiparallel ordering. Each base in the single strand of *sense*-sequence has its complementary base in that one with *antisense*-sequence. For example, a single strand with the *sense*-sequence (3′–5′)d(A-T-G-C-T-T-G) pairs (hybridizes) with the complementary single strand of *antisense*-sequence (5′–3′)d(T-A-C-G-A-A-C) (*Fig. 7.29*).

In the specification of the single strands:

- (3′–5′) or, more precisely, (3′ → 5′) means bridging of the nucleoside residues dA, dC, dG, and dT by the phosphodiester group.
- d (more precisely d²Rib*f*) is the description of the sugar, here: 2-deoxy-β-D-ribofuranose (for the nomenclature of nucleic acids, see *Chapt. 10.5.2*, especially the information going with *Figs. 10.17–10.19*).

Table 7.3. *Original Table by* E. Chargaff (*Structure and Function of Nucleic Acids as Cell Constituents, Federation Proc.* **1951,** *10,* 654)

Source	Adenine to Guanine	Thymine to Cytosine	Adenine to Thymine	Guanine to Cytosine	Purine to Pyrimidine
Ox	1.29	1.43	1.04	1.00	1.1
Human	1.56	1.75	1.00	1.00	1.0
Chicken	1.45	1.29	1.06	0.91	0.99
Salmon	1.43	1.43	1.02	1.02	1.02
Wheat	1.22	1.18 [a)]	1.00	0.97 [a)]	0.99
Yeast	1.67	1.92	1.03	1.20	1.0
Haemophilus influenzae	1.74	1.54	1.07	0.91	1.0
E. coli K-12	1.05	0.95	1.09	0.99	1.0
Avian Tubercule bacillus	0.4	0.4	1.09	1.08	1.1
Serratia marcescens	0.7	0.7	0.95	0.86	0.9
Schatz bacillus	0.7	0.6	1.12	0.89	1.0

[a)] Calculated using the sum of cytosine and methylcytosine. Considering cytosine alone, the resulting ratio of thymine to cytosine is 1.62, and of guanine to cytosine 1.33.

Each single strand, with the specific sequence of its bases (*structural syntax*) is a carrier of information, which, during the process of base pairing (according to the rules of *structural grammar*), may be selectively recognized by its complementary single strand. Information transfer in supramolecular base pairs is dealt with in *Chapt. 16*.

Content and style of the communication of *J. D. Watson* and *F. H. C. Crick* on the molecular structure of DNA are so unusual that it has been reproduced in *Fig. 7.30*.

Watson's bestseller '*The Double Helix*' (1968) told the story of the discovery of the structure of DNA to the general public from the point of view of one of its protagonists. *Crick*'s '*What Mad Pursuit*' (1988) described his view of how the structure and function of DNA had influenced the development of modern biology. In 1987 there even appeared a film: a BBC 'docudrama' entitled '*Life Story*', called '*Double Helix*' in its American edition.

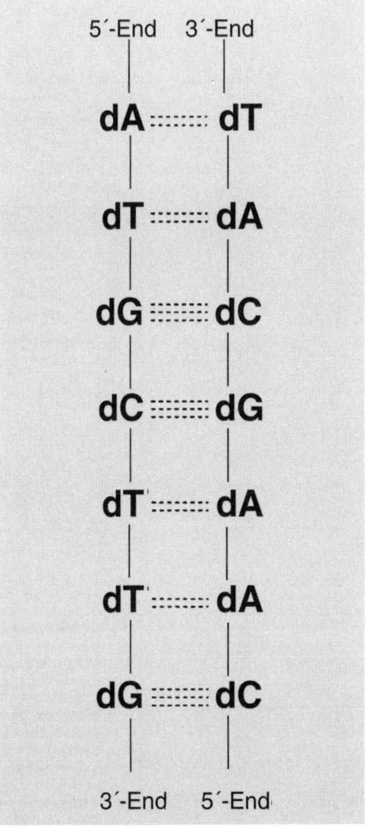

Fig. 7.29. Watson-Crick *base pairing between antiparallel arranged DNA single strands with two (three) H-bonds between dA and dT (dG and dC)*

No. 4356 April 25, 1953 NATURE 737

MOLECULAR STRUCTURE OF NUCLEIC ACIDS

A Structure for Deoxyribose Nucleic Acid

WE wish to suggest a structure for the salt of deoxyribose nucleic acid (D.N.A.). This structure has novel features which are of considerable biological interest.

A structure for nucleic acid has already been proposed by Pauling and Corey[1]. They kindly made their manuscript available to us in advance of publication. Their model consists of three intertwined chains, with the phosphates near the fibre axis, and the bases on the outside. In our opinion, this structure is unsatisfactory for two reasons: (1) We believe that the material which gives the X-ray diagrams is the salt, not the free acid. Without the acidic hydrogen atoms it is not clear what forces would hold the structure together, especially as the negatively charged phosphates near the axis will repel each other. (2) Some of the van der Waals distances appear to be too small.

Another three-chain structure has also been suggested by Fraser (in the press). In his model the phosphates are on the outside and the bases on the inside, linked together by hydrogen bonds. This structure as described is rather ill-defined, and for this reason we shall not comment on it.

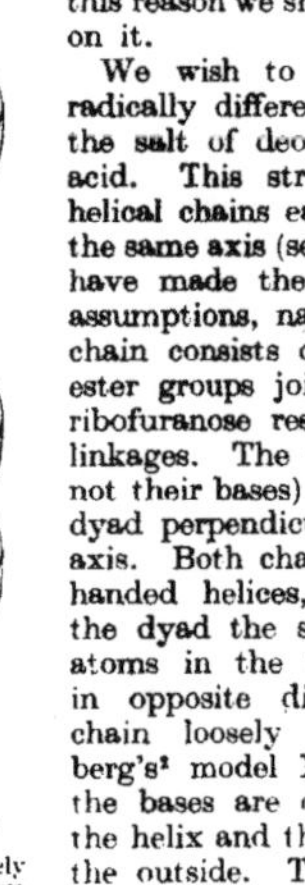

This figure is purely diagrammatic. The two ribbons symbolize the two phosphate—sugar chains, and the horizontal rods the pairs of bases holding the chains together. The vertical line marks the fibre axis

We wish to put forward a radically different structure for the salt of deoxyribose nucleic acid. This structure has two helical chains each coiled round the same axis (see diagram). We have made the usual chemical assumptions, namely, that each chain consists of phosphate diester groups joining β-D-deoxyribofuranose residues with 3′,5′ linkages. The two chains (but not their bases) are related by a dyad perpendicular to the fibre axis. Both chains follow right-handed helices, but owing to the dyad the sequences of the atoms in the two chains run in opposite directions. Each chain loosely resembles Furberg's[2] model No. 1; that is, the bases are on the inside of the helix and the phosphates on the outside. The configuration of the sugar and the atoms near it is close to Furberg's 'standard configuration', the sugar being roughly perpendicular to the attached base. There is a residue on each chain every 3·4 A. in the z-direction. We have assumed an angle of 36° between adjacent residues in the same chain, so that the structure repeats after 10 residues on each chain, that is, after 34 A. The distance of a phosphorus atom from the fibre axis is 10 A. As the phosphates are on the outside, cations have easy access to them.

The structure is an open one, and its water content is rather high. At lower water contents we would expect the bases to tilt so that the structure could become more compact.

The novel feature of the structure is the manner in which the two chains are held together by the purine and pyrimidine bases. The planes of the bases are perpendicular to the fibre axis. They are joined together in pairs, a single base from one chain being hydrogen-bonded to a single base from the other chain, so that the two lie side by side with identical z-co-ordinates. One of the pair must be a purine and the other a pyrimidine for bonding to occur. The hydrogen bonds are made as follows: purine position 1 to pyrimidine position 1; purine position 6 to pyrimidine position 6.

If it is assumed that the bases only occur in the structure in the most plausible tautomeric forms (that is, with the keto rather than the enol configurations) it is found that only specific pairs of bases can bond together. These pairs are: adenine (purine) with thymine (pyrimidine), and guanine (purine) with cytosine (pyrimidine).

In other words, if an adenine forms one member of a pair, on either chain, then on these assumptions the other member must be thymine; similarly for guanine and cytosine. The sequence of bases on a single chain does not appear to be restricted in any way. However, if only specific pairs of bases can be formed, it follows that if the sequence of bases on one chain is given, then the sequence on the other chain is automatically determined.

It has been found experimentally[3,4] that the ratio of the amounts of adenine to thymine, and the ratio of guanine to cytosine, are always very close to unity for deoxyribose nucleic acid.

It is probably impossible to build this structure with a ribose sugar in place of the deoxyribose, as the extra oxygen atom would make too close a van der Waals contact.

The previously published X-ray data[5,6] on deoxyribose nucleic acid are insufficient for a rigorous test of our structure. So far as we can tell, it is roughly compatible with the experimental data, but it must be regarded as unproved until it has been checked against more exact results. Some of these are given in the following communications. We were not aware of the details of the results presented there when we devised our structure, which rests mainly though not entirely on published experimental data and stereochemical arguments.

It has not escaped our notice that the specific pairing we have postulated immediately suggests a possible copying mechanism for the genetic material.

Full details of the structure, including the conditions assumed in building it, together with a set of co-ordinates for the atoms, will be published elsewhere.

We are much indebted to Dr. Jerry Donohue for constant advice and criticism, especially on interatomic distances. We have also been stimulated by a knowledge of the general nature of the unpublished experimental results and ideas of Dr. M. H. F. Wilkins, Dr. R. E. Franklin and their co-workers at King's College, London. One of us (J. D. W.) has been aided by a fellowship from the National Foundation for Infantile Paralysis.

J. D. WATSON
F. H. C. CRICK
Medical Research Council Unit for the Study of the Molecular Structure of Biological Systems,
Cavendish Laboratory, Cambridge.
April 2.

[1] Pauling, L., and Corey, R. B., *Nature*, **171**, 346 (1953); *Proc. U.S. Nat. Acad. Sci.*, **39**, 84 (1953).
[2] Furberg, S., *Acta Chem. Scand.*, **6**, 634 (1952).
[3] Chargaff, E., for references see Zamenhof, S., Brawerman, G., and Chargaff, E., *Biochim. et Biophys. Acta*, **9**, 402 (1952).
[4] Wyatt, G. R., *J. Gen. Physiol.*, **36**, 201 (1952).
[5] Astbury, W. T., Symp. Soc. Exp. Biol. 1, Nucleic Acid, 66 (Camb. Univ. Press, 1947).
[6] Wilkins, M. H. F., and Randall, J. T., *Biochim. et Biophys. Acta*, **10**, 192 (1953).

Fig. 7.30. *First original publication by* J. D. Watson *and* F. H. C. Crick *in* Nature **1953**

7.3.1.2.2. Base Pairing in Homo-DNA

Homo-DNA single strands preferentially form, like (but not with) DNA single strands, duplexes of antiparallel orientation. Whilst in the latter case the *Watson-Crick* rules give the order of thermodynamic stability for the nucleoside pairs

$$dG \cdot dC > dA \cdot dC,$$

for the former case, the particular order is

$$ddG \cdot ddC > ddA \cdot ddA \approx ddG \cdot ddG > ddA \cdot ddT$$

As d is used for the official *IUPAC* symbol d^2Rib*f*, dd stands for the official *IUPAC* symbol d^2d^3Glc*p*.

From the constitutional point of view, the most conspicuous difference in pairing behavior between homo-DNA and DNA oligonucleotides is the purine/purine self-pairing of the ddA and ddG nucleosides. This difference in pairing behavior in the two homologous series leads to the important insight that the *Watson-Crick* rules are not entirely a result of the molecular properties of the nucleobases, but that they are also subject to the constitution of the sugar-phosphate chain. This conclusion is confirmed by the finding that two single strands, one containing only purine bases, the other only pyrimidine bases, will not pair with one another, if the one has the homo-DNA backbone and the other the DNA backbone.

The pairing behavior in the homo-DNA series is so different from that one in the DNA series as to make the constitution of the resulting homo-DNA supermolecule open to question. For the case of the supramolecular coupling of the two nucleosides ddA and ddT, there are *a priori* the four constitutional pairing combinations already (*Sect. 7.3.1.2*) mentioned. The 16 variants with different combinations of *pairing type*, *base conformation* (*syn* or *anti*; for the conformational specification of the nucleic acids, see *Chapt. 10.5.2*), and *strand orientation* (parallel or antiparallel) are compiled on the next page.

For ddA/ddT pairing, the favored nucleoside pair found in the homo-DNA oligonucleotide duplex is of the *Watson-Crick* pattern, both bases being of *anti*-orientation and the two single strands of antiparallel order.

In the homo-DNA series, ddG/ddC pairing takes place in the *Watson-Crick* manner, like dG/dC pairing in the DNA series, with *anti*-orientation of the two bases, antiparallel ordering of the two strands, and formation of three H-bonds in either case. The ddA/ddA and ddG/ddG pairing in the homo-DNA series follows the reverse-*Hoogsteen* pattern, both with *anti*-orientation of the two bases and antiparallel ordering of the two single strands.

The observation that duplexes with purine/pyrimidine base pairs are thermodynamically more stable in the homo-DNA series than in the DNA series is determined primarily by entropy: the conformational spaces occupied by the single and double strands are more similar to one another in homo-DNA than in DNA.

Watson-Crick	*anti* *anti* antiparallel	*syn* *anti* parallel	*anti* *syn* parallel	*syn* *syn* antiparallel
reverse *Watson-Crick*	*anti* *anti* parallel	*syn* *anti* antiparallel	*anti* *syn* antiparallel	*syn* *syn* parallel
Hoogsteen	*anti* *anti* parallel	*syn* *anti* antiparallel	*anti* *syn* antiparallel	*syn* *syn* parallel
reverse *Hoogsteen*	*anti* *anti* antiparallel	*syn* *anti* parallel	*anti* *syn* parallel	*syn* *syn* antiparallel

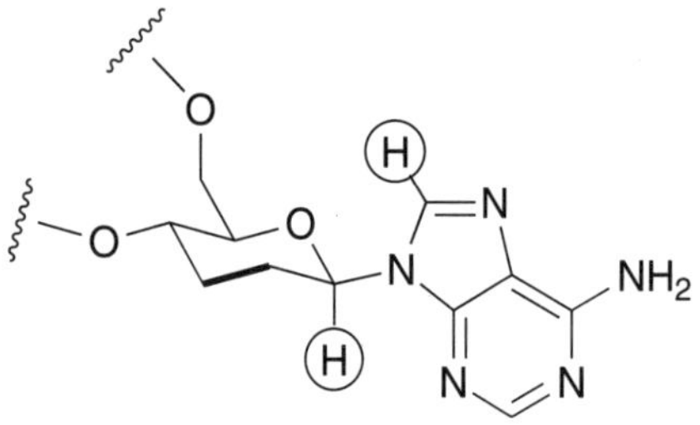

anti 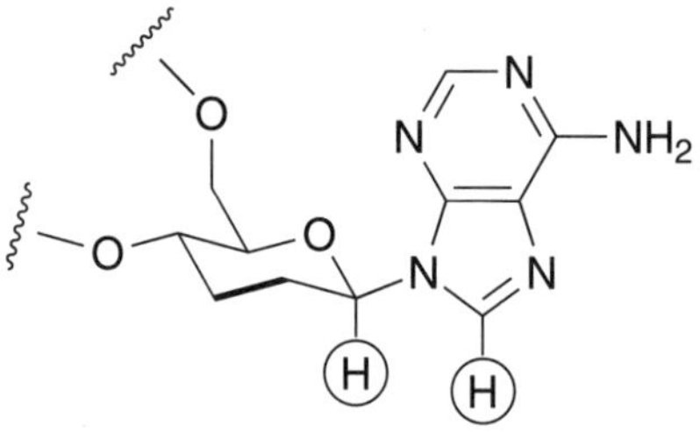 *syn*

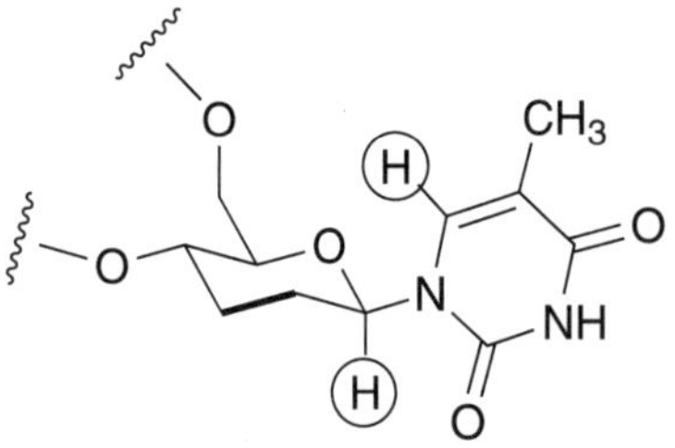

anti 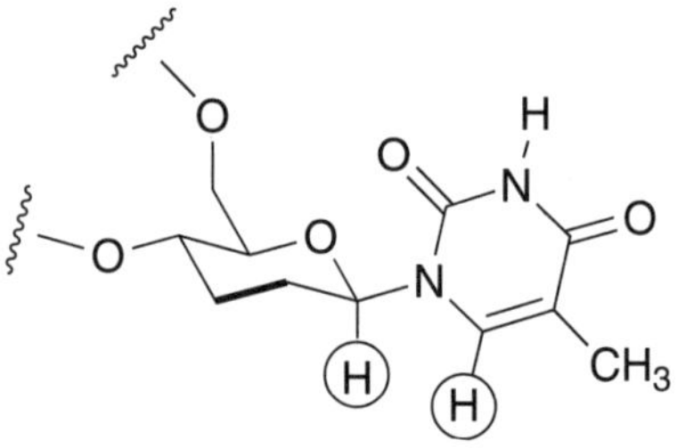 *syn*

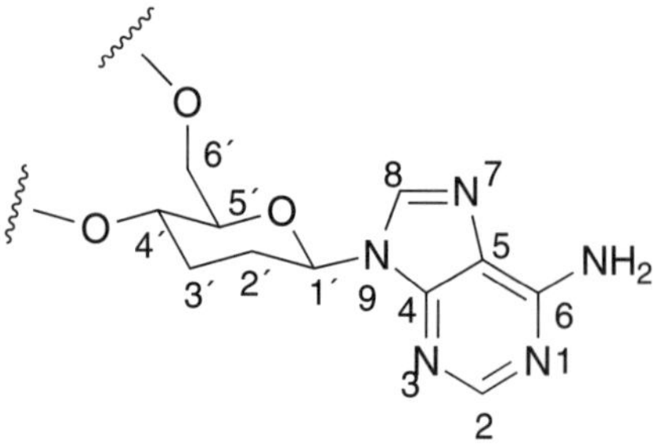

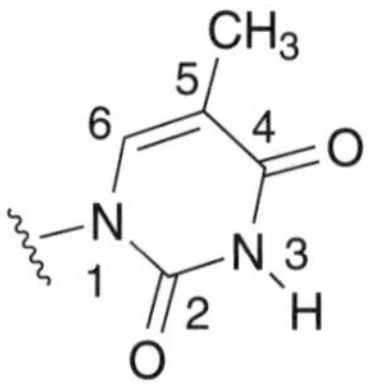

Adenine/Thymine Base Pairing

Watson-Crick — *anti* *anti* antiparallel

reverse *Watson-Crick* — *anti* *anti* parallel

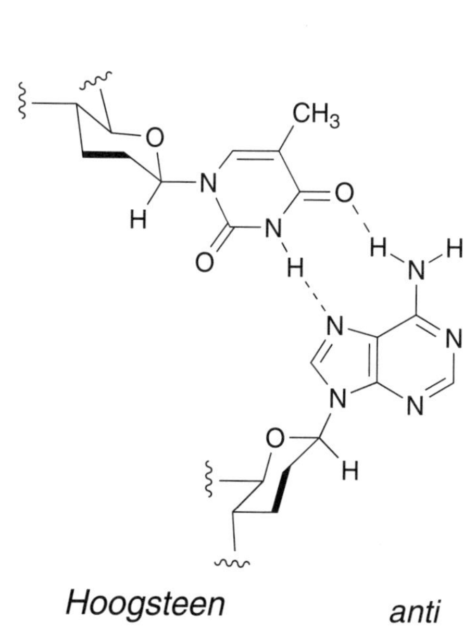

Hoogsteen — *anti* *anti* parallel

reverse *Hoogsteen* — *anti* *anti* antiparallel

Quasi-Isomorphism of the Reverse Adenine/Adenine and Reverse Guanine/Guanine *Hoogsteen* Base Pairings

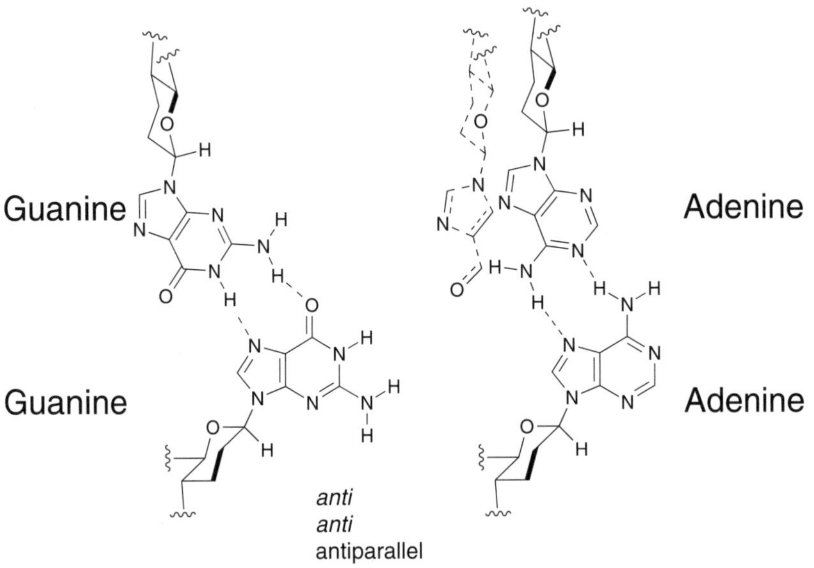

Guanine Adenine

Guanine Adenine

anti
anti
antiparallel

During the systematic conformational analysis of the homo-DNA single strand (*Sect. 7.3.1.1.1*), the topological selection criterion for pairing has already been taken into account. There, the $-sc/ap/+sc/+sc/ap/-sc$-conformation **A** (*Fig. 7.22*), the only remaining representative, conformationally repetitive and optimally capable of pairing among a set of 486 formally possible options, was identified. If, besides the favored $\pm sc/\pm sc$-phosphodiester fragment (*Fig. 7.19*), the $ap/\pm sc$-arrangement is also permitted, then the $ap/ap/ap/+sc/ap/-sc$-conformation (*Fig. 7.31*) is acquired.

A comparison of the homo-DNA single strands of idealized conformational types of *Fig.* 7.22, **A**, and *Fig. 7.31* with one another shows that both are linear, and, further, that the mutually parallel planes of the bases all, to differing degrees, incline towards the axis of the strand: *ca.* 60° in the first instance and *ca.* 45° in the second one. The distance between neighboring bases is only reduced by an insignificant amount by this and remains far from the relevant distance in the various conformers of DNA.

Schematic representations of models of the linear strands with inclined strand and base-pair axes (*Fig. 7.32*, upper row) show the geometrical reason for linear homo-DNA duplexes having the antiparallel strand orientation in preference over the parallel, and this all the more so, the steeper the incline.

Homo-DNA duplexes which are derived from the idealized conformational type (*Fig. 7.31*) through base pairing remain linear in the idealized duplex: the model for such a double strand represents a quasi-two-dimensionally chiral structure (*Fig. 7.32*, center). The model for a double strand assembled out of idealized, linear single strands of conforma-

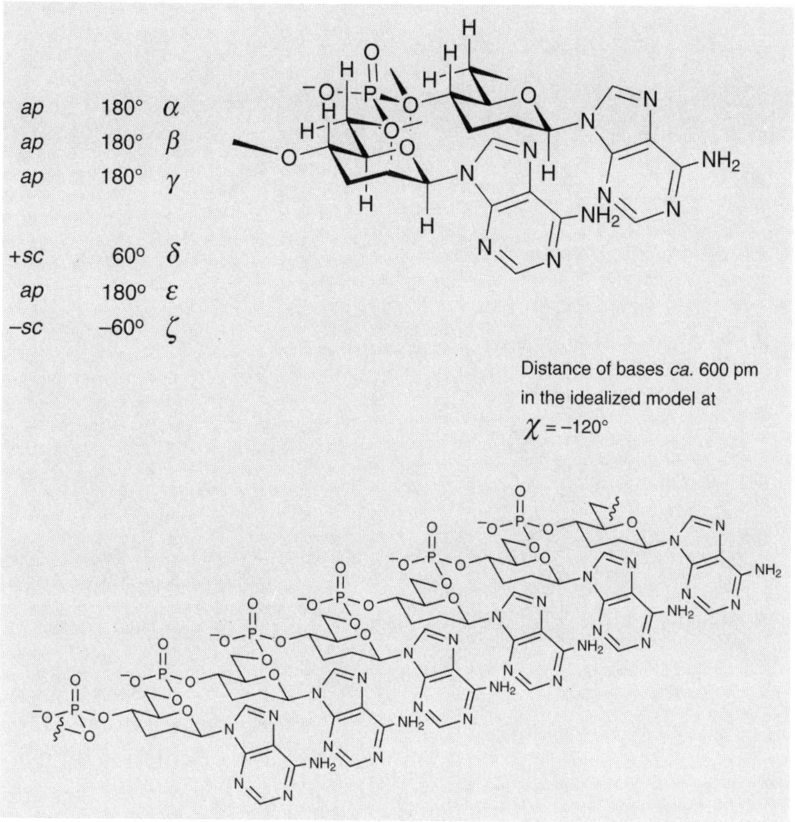

Fig. 7.31. *The idealized, linear homo-DNA single strand, made up out of repeating units of the* ap/ap/ap/+sc/ap/−sc-*conformation type and possessing the ability to form idealized, linear duplexes*

tional type **A** (*Fig. 7.22*) contains 'steps', and is three-dimensionally chiral in its pairing topology. So that a continual base pairing can exist, the antiparallel ordered single strands must become 'deidealized' (*Fig. 7.32*, below): then a helical duplex (*Sect. 7.2.2*) can be formed.

In both conformations of fragments of idealized, linear homo-DNA single strands (*Fig. 7.22*, **A**, and *Fig. 7.31*), the parallel oriented bases are too far away from one another (*ca.* 500–600 pm) to bring about any measure of base-pair stacking to which the stabilization in DNA (for a base distance of 350–400 pm) is attributed. *Fig. 7.33* illustrates ways in which the distance of parallel oriented nucleobases might be lessened, through tilting, undulation, or helicalization.

NMR-Spectroscopic findings do, in fact, support the idea of a dynamic conformational equilibrium in a homo-DNA duplex. The conformational flexibility of the duplex is entropically stabilizing and counters the enthalpically destabilizing effect of a decreased degree of base-pair stacking relative to the DNA duplex.

The sugar component of homo-DNA (type **A** and *Table 7.2*) is not viewed as a potentially prebiological natural product: it merely served as a model in *Eschenmoser*'s laboratory for the development of systematic conformational analysis of single- and double-strand nucleic acids. Completely hydroxylated hexopyranoses (type **B**, for example), in contrast, would be potentially prebiological sugars, but could be disregarded as nucleic-acid building blocks.

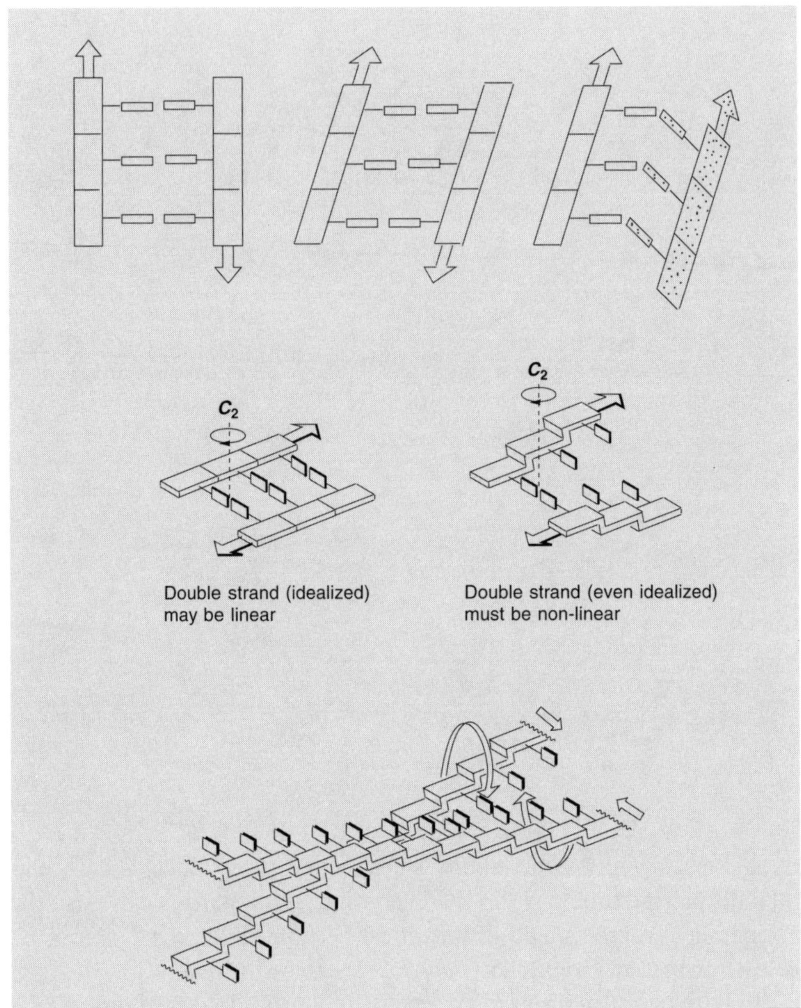

Fig. 7.32. *Antiparallel pairing of idealized, linear single strands*

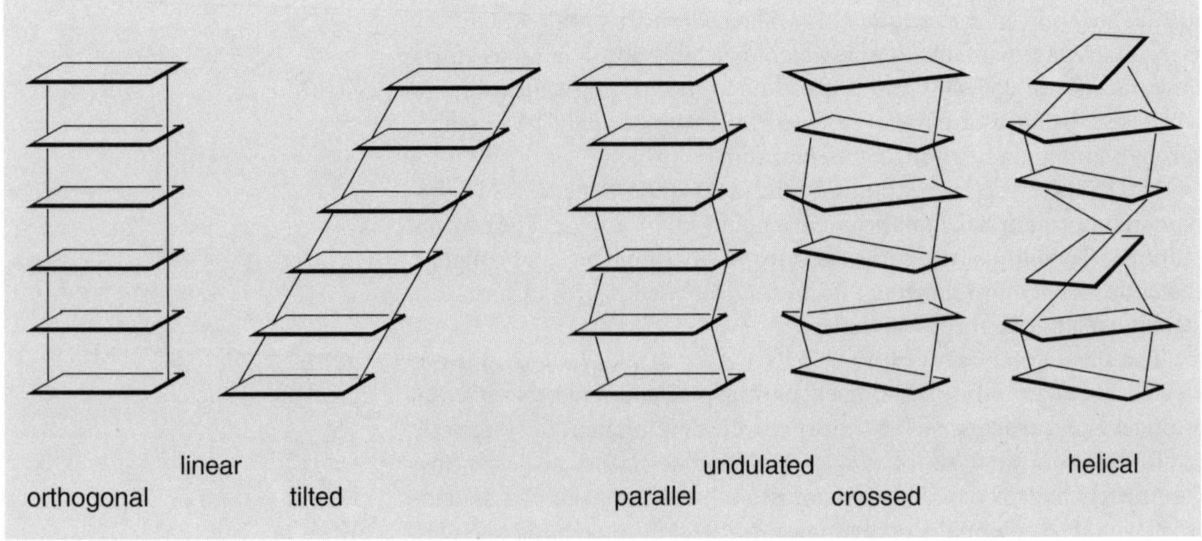

Fig. 7.33. *Shortening of the distances between base pairs in linear duplex models by tilting, undulation, or helicalization*

Results from the Zürich laboratory have shown that, of the eight diastereoisomeric aldohexopyranoses of the D-series (*Chapt. 5.2*), none, after incorporation into the sugar-phosphate backbone, forms duplexes comparable in thermodynamic stability with that one of DNA or RNA. The OH groups at C(2′) exert a destabilizing influence on the duplex formation of hexopyranosyl (4′→6′)-oligonucleotides as a consequence of steric hindrance: equatorially oriented OH groups more strongly than their axially oriented counterparts. It is, therefore, not surprising that 3′-deoxy-β-D-allopyranosyl oligonucleotides (type **C**) behave similarly to β-D-allopyranosyl oligonucleotides (type **B**) in reverse-*Hoogsteen* base pairing, and totally differently from 2′-deoxy-β-D-allopyranosyl oligonucleotides (type **D**). *Eschenmoser* has answered the question he himself proposed: '*Why pentose- and not hexose-nucleic acids?*', with the succinct conclusion that the hexopyranose sugars have 'too many atoms'. Similar considerations lead one to the conclusion that the DNA duplex is more stable than an RNA duplex. Indeed, bulges, loops, and intracatenate base pairing are encountered much more frequently in RNA than in DNA.

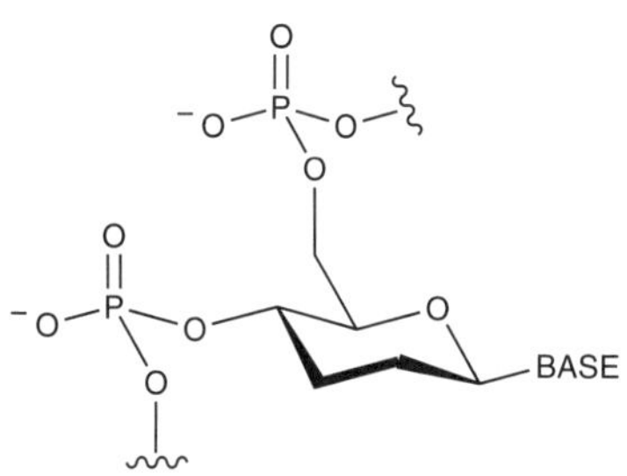

A 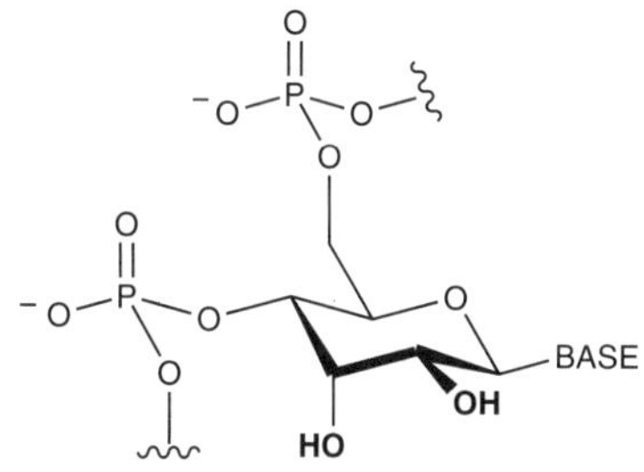 **B**

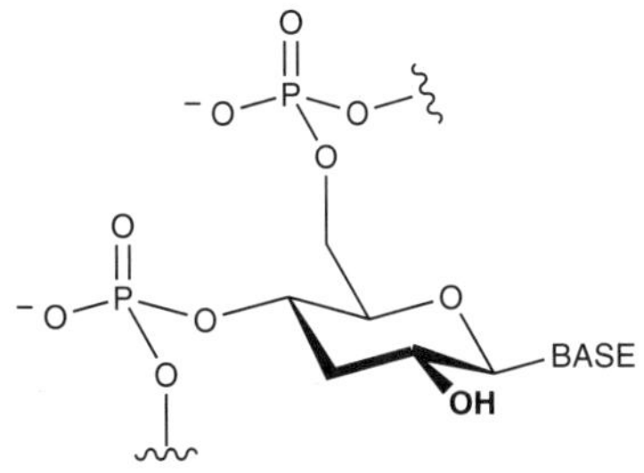

C 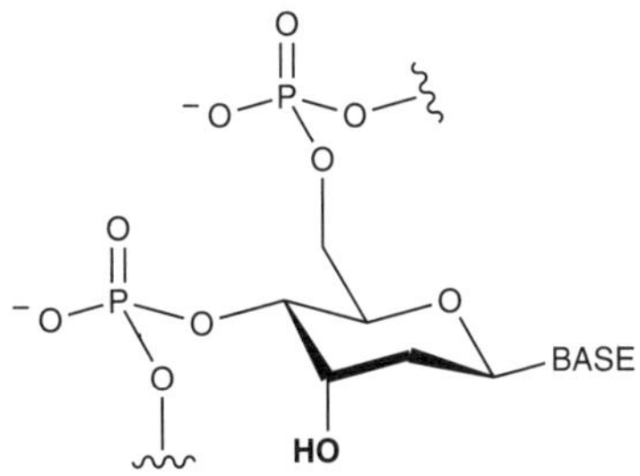 **D**

7.3.1.2.3. Base Pairing in p-RNA

According to model inspection (*Fig. 7.27*), pyranosyl-RNA (p-RNA) can form an idealized, linear duplex with antiparallel orientation of the two single strands by base pairing in the *Watson-Crick* manner. The observation that the purine/purine self-pairing (reverse-*Hoogsteen* base pairing) found in homo-DNA does not, however, take place in p-RNA leads to the conclusion that the backbone of p-RNA is less flexible. Every repeating unit of the backbone includes six flexible bonds in the case of DNA, there are only five flexible bonds in the case of homo-

DNA, and only just four ones in the case of p-RNA. In fact, *Watson-Crick*-type base pairing of p-RNA leads to purine/pyrimidine duplexes thermodynamically more stable than those ones of RNA, or DNA, and even of homo-DNA. It is, therefore, not surprising that p-RNA is the most selective of all the nucleic acids mentioned, when it comes to *Watson-Crick*-type duplex formation. The reason for this is the relatively high rigidity of the p-RNA backbone and the large inclination in the duplex of the strand and base-pair axes, a result due to the different constitution ((2′ → 4′) rather than (4′ → 6′) bridging). The steep inclination makes intercatenate base stacking possible (see *Fig. 7.27*). High selectivity in base pairing is the prerequisite for high-fidelity information transfer.

In search of the origin of RNA, p-RNA (as 'proto'-RNA) could prove to be a precious treasure, at least if it were established experimentally that p-RNA could transfer its own structural information and possessed sequence-determined catalytic properties. According to *Eschenmoser*, an intraduplex p-RNA → RNA isomerization with retention of base sequence in formation is mechanistically feasible.

7.3.1.2.4. From the Idealized Homo-DNA Model to the Backbone Conformation of the Single Strand in DNA Duplexes

The idealized chair conformation of cyclohexane served as a reference model for the systematic conformational analysis of steroids (*Chapt. 6.2*) as well as for pyranosides (*Chapt. 6.6.2* and *Sect. 7.3.2*) and can be extended to cover other natural-product classes in which six-membered rings occur as important structural elements. It is only appropriate to commence the systematic conformational analysis of polynucleotide duplexes using the homo-DNA single strand as a model (*Sect. 7.3.1.1.1*).

The idealized model of a homo-DNA single strand in its conformation prone to base pairing is linear and contains the *−sc/ap/+sc/+sc/ap/−sc*-fragment (*Fig. 7.22*, **A**) as its repeating unit. Widening (in a thought experiment) the torsion angle δ (clockwise) from 60° to higher values transforms the linear chain into a (right-handed) helix. This is obligatory when the pyranose ring is replaced by a furanose ring (*Fig. 7.23*). Comparison of the linear backbone conformation of the homo-DNA single strand with that one in DNA duplexes (using data from the X-ray crystal-structure analyses stored in the *Brookhaven Protein Database*) makes clear that it shows similarities to the corresponding 'deidealized' torsion angles in the DNA duplex. As DNA duplexes of conformational type A (see *Fig. 7.36*) deviate by +18° in average, and DNA duplexes of conformational type B by +62°, from the idealized value of 60° of a linear single strand, the idealized fragment of homo-DNA (**A** in *Fig. 7.34*) agrees very well with a corresponding fragment from the A-DNA duplex (**B**, δ = 78°); in any case, better than with fragments of B-DNA duplexes (**C**, δ = 122°; or **D**, δ = 140°).

Fig. 7.35 shows the idealized conformation **A** of the homo-DNA fragment with δ = 60°, as well as conformations in which the pyranose ring – in order to enable suitable manipulation of the endocyclic angle δ – has adopted the boat conformation with δ = 60° (**B**), δ = 120° (**C**), and δ = 180° (**D**). The fictitious conformations **B**, **C**, and **D**, altered merely in the six-membered sugar ring, are shown next to the real conformations with five-membered sugar rings, **B′** of A-DNA (with δ = 78°, C(3′)-*endo*-conformation of the furanose ring and *sc*-orientation of the vicinal backbone substituents), **C′** of B-DNA (with δ = 122°, C(1′)-*exo*-conformation of the furanose

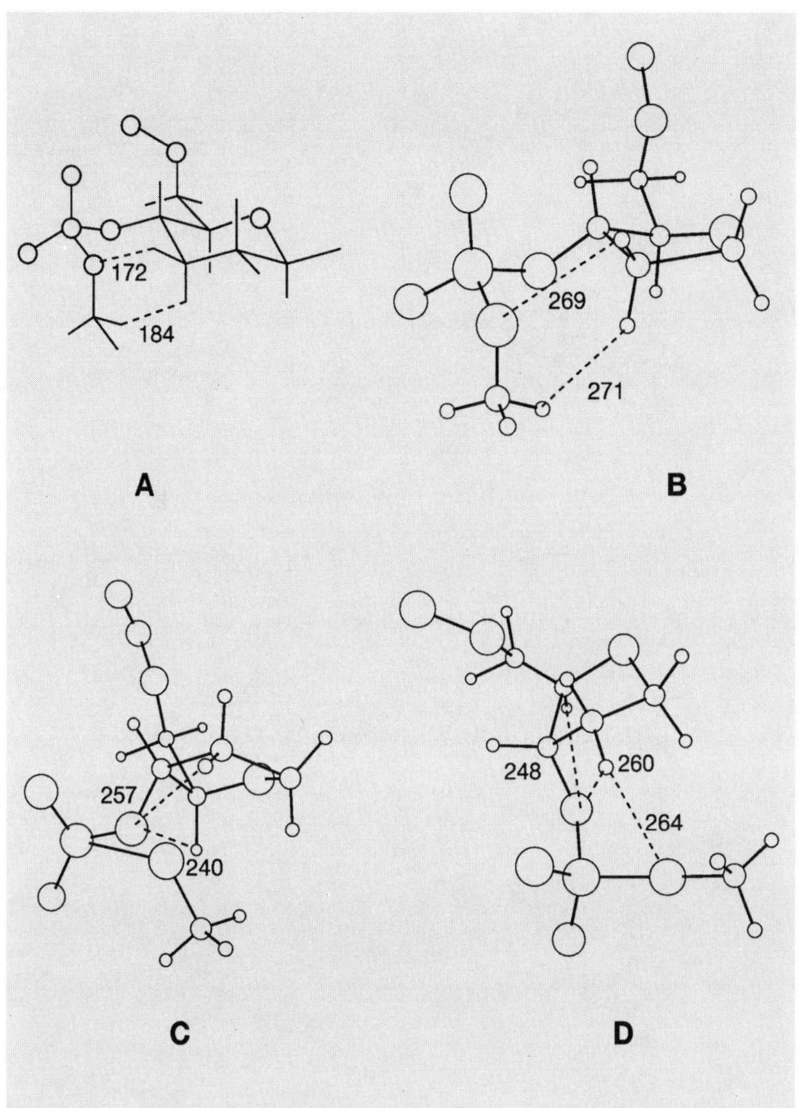

Fig. 7.34. *Comparison of the backbone unit of the homo-DNA model* (**A**) *with those ones of A-DNA* (**B**) *and B-DNA* (**C** and **D**) *duplexes*. The numbers give the H···H and H···O distances [pm].

ring and *ac*-orientation of the vicinal backbone substituents), and **D**′ of B-DNA (with $\delta = 140°$, C(2′)-*endo*-conformation of the furanose ring and *ac*-orientation of the vicinal backbone substituents).

The backbone conformation in the single strands of the A-DNA duplex comes quite close to that one of homo-DNA. Structural pre-organization of the single strand plays an important role in the formation of the DNA double helix.

It has been foreshadowed in the preceding exposition that the supramolecular nucleic-acid duplexes can occur in different conformations known as A-, B-, or Z-DNA duplexes. These conformationally more complex duplexes will now be briefly introduced (*Fig. 7.36*).

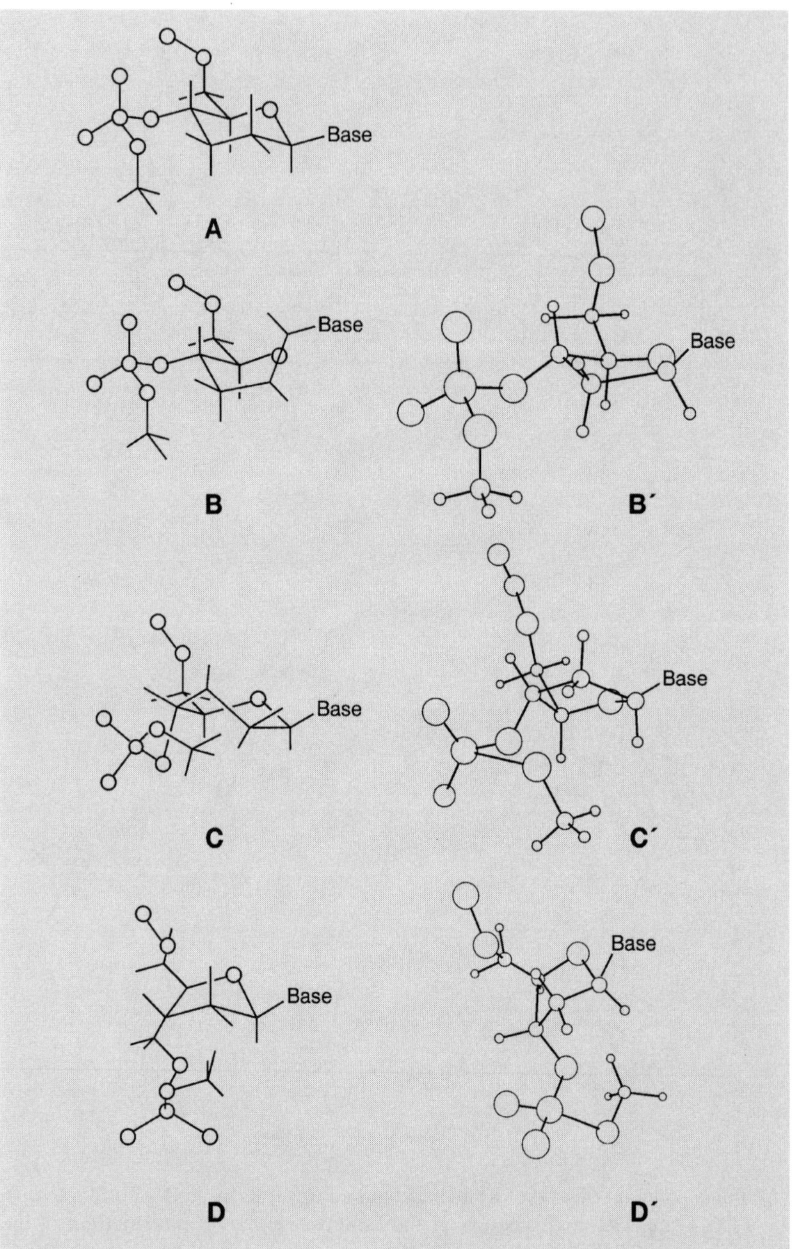

Fig. 7.35. *Backbone units of the homo-DNA model* (**A**) *with manipulated pyranose boat conformations* (**B–D**) *compared with backbone units from A-DNA* (**B′**) *and B-DNA* (**C′** and **D′**) *duplexes*

B-DNA Duplexes. The sugar-phoshate backbone of the two antiparallel arranged single strands is right-handed – as 2′-deoxy-β-ribofuranose is present as the D-enantiomer – in the *Watson-Crick* double helix. The nucleobases occur in the *anti*-conformation, the sugar ring in the C(2′)-*endo*-conformation. The duplex has a diameter of 2 nm. The complementary bases are similar in shape (*isomorphous*) and held together by H-bonds. The planes of the base pairs are approximately perpendicular to the helix axis, which, as a two-fold pseudorotational axis, penetrates the centers of the base-pair systems. Rotation of a base pair by 180° about the helix axis brings the sugar-phosphate fragments into superposition. Neighboring base pairs are 340 pm apart from one another, and the distance between two sequential phosphate groups amounts to *ca.* 670 pm. A complete helix turn incorporates ten (10.5 under physiological conditions) base pairs. Each base pair is twisted by 36° relative to those ones at either side of it.

The supermolecule shows two peripheral grooves, approximately equally deep but of different widths. The *major groove* is nearly 1-nm wide. Major and *minor grooves* are accessible, without hindrance, from the exterior. The highly hydrated B-DNA (95% relative humidity) contains two chains of H_2O molecules linked to each other in the major groove and one of them in the minor groove.

A-DNA Duplexes. These are typical of the conformation of the dehydrated DNA double helix (75% relative humidity), and also of DNA/RNA hybrids and RNA/RNA duplexes. (Because of steric hindrance between the C(2′)–OH group of the β-D-ribofuranose and the phosphate bridge leading to the adjacent sugar ring, the B-DNA duplex conformation is out of the question for participation by RNA single strands.) The sugar rings are now present in the C(3′)-*endo*-conformation, and hence adjacent phosphate residues in a single strand are pressed *ca.* 100 pm closer together (relative to the B-DNA duplex), so that a full turn encompasses between 11 and 12 base pairs, as opposed to 10.5 in B-DNA. Unlike in the B-DNA conformation, the base pairs in the A-DNA conformation are significantly shifted towards the major groove and inclined towards the helix axis, which does not transect the base pairs, but passes through a hollow space in the center of the supermolecule. The two grooves are thereby highly differentiated from each other, the major groove being very deep, but narrow, with restricted access. The minor groove is superficial, flat, and easily accessible from the exterior.

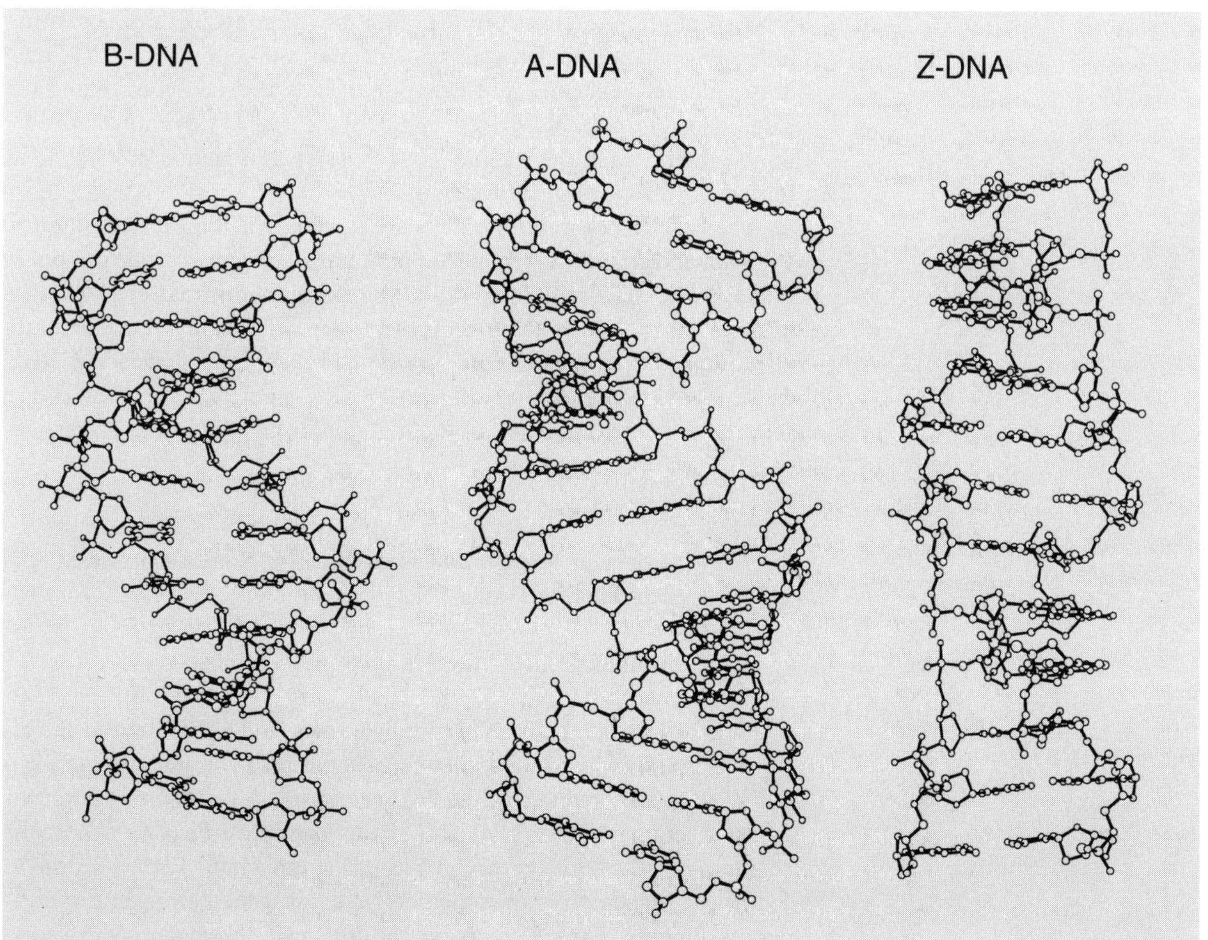

Fig. 7.36. *B-, A-, and Z-DNA duplexes*

Z-DNA Duplexes. Unlike A- and B-DNA duplexes, these are left-handed, and have acquired their name from the zig-zag arrangement of their sugar-phosphate backbone. They are also distinct from the A- and B-DNA duplexes in their nucleobase conformations: the nucleobases adopt alternate *syn-* and *anti*-orientations. Pyrimidine nucleosides exist in the usual *anti*/C(2′)-*endo*-conformation, the purine nucleosides, however, in the *syn*/C(3′)-*endo*-conformation. The base pairs are twisted by 180° in the Z-duplex, relative to the B-duplex. All this means that the Z-DNA repeating unit comprises a dinucleotide rather than a mononucleotide residue. The major groove can scarcely be made out; in reality it is part of the surface. The minor groove is deep, but extremely narrow.

The structural differences between the three prototype conformations of nucleic acids have consequences for their function. For the formation of supermolecules – between a DNA duplex and a protein, for example – the first crucial factor is which of the two grooves is easily accessible. In addition, it is important to know which atoms are available there for intermolecular interactions. The atoms and atom groups in the grooves are those ones located on the edges of the nucleobase pairs. Space filling and the ability of atom groups to play the part of acceptors (A) or donors (D) for H-bonds are steric or constitutional factors which determine the readable information content of both grooves (*Fig. 7.37*). DNA-Binding proteins, peptide analogs, oligonucleotides, or other molecules which enter into supramolecular interaction with DNA may drastically influence the conformation of individual components of the resulting supermolecule. Conformational changes of this kind distort the double helix, vary the profile of the major and minor grooves, and thereby intervene in the function of DNA.

The base pairs of a double helix are stacked. Stacking effects stabilize a duplex and counter the repulsion between the negatively charged phosphate groups. To maximize stacking effects, conformational changes have to be put up with. Since purine/purine stacking is particularly advantageous, it should come as no surprise that sequence-specific causes for conformational change do also occur. Stacking forces are referred to in the discussion of the function of DNA duplexes in replication (*Chapt. 16*).

7.3.1.3. Molecular Recognition as the Basis of Non-Stochastic Searching for Drugs

7.3.1.3.1. Paradigm Shift in the Search of Active Substances

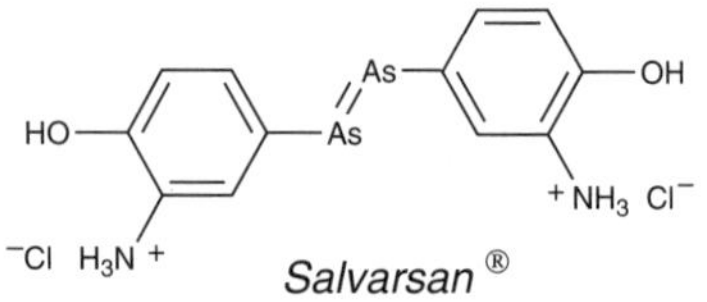

HO, As, As, OH, $^{+}NH_3$ Cl^-, ^{-}Cl H_3N^+

Salvarsan®

Chemotherapeutics were originally understood as synthesized chemical compounds which could kill microorganisms in blood or tissue without substantially damaging the host organism. A systematic search for selective 'magic bullets' of this kind was begun by *Paul Ehrlich* at the beginning of the 20th century in Frankfurt am Main. *Ehrlich* found, in *Salvarsan*, a chemotherapeutic effective against the pathogen responsible for syphilis. *Ehrlich* counts as the *father of chemotherapy*: because of his postulation of the relationship between the structure and the

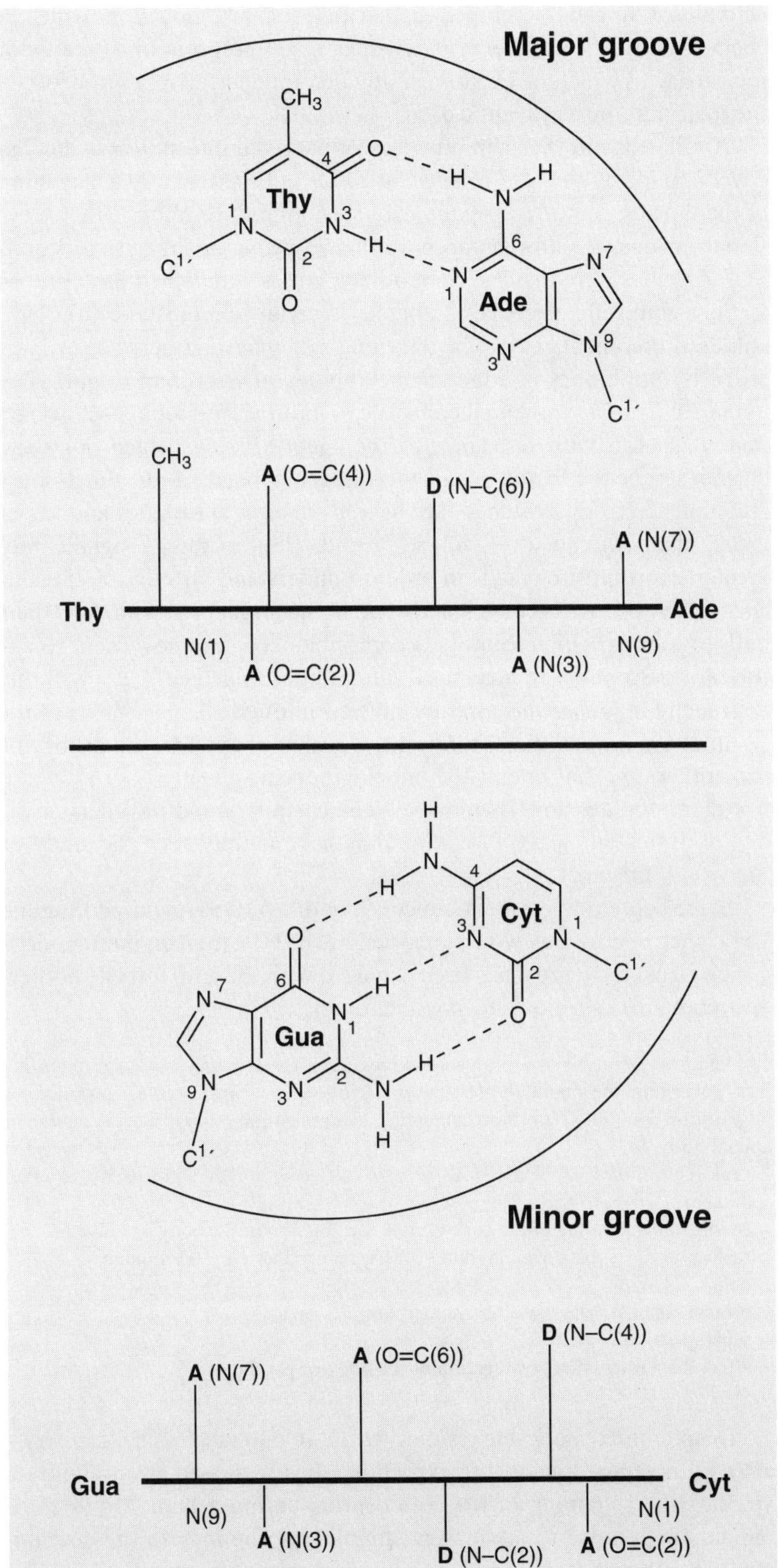

Fig. 7.37. *Schematic represention of acceptor and donor groups of H-bonds within the grooves of the DNA duplex*

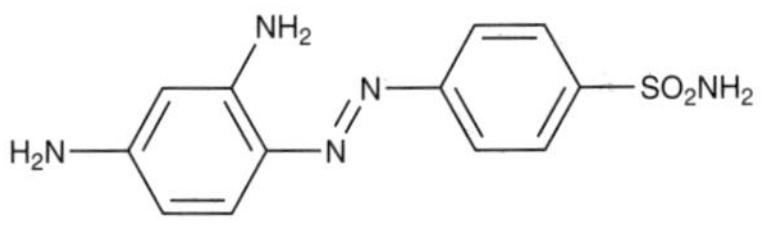

Prontosil®

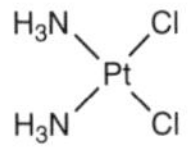

Cisplatin

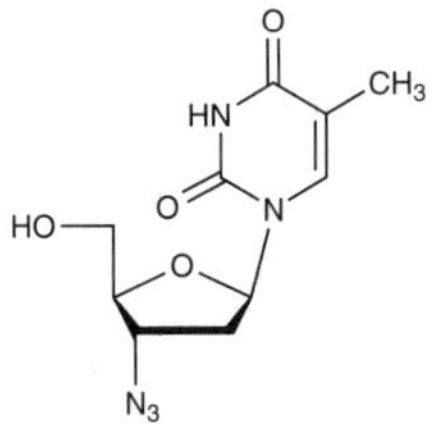

AZT

activity of a chemotherapeutic, because of his model in which a chemotherapeutic agent exerted its effect on a 'receptor', and because of the care with which he experimentally determined the relation between the toxic and the therapeutic dose.

A milestone in the chemotherapy of bacterial infection was the discovery, made by *Gerhard Domagk* at *Bayer* in Elberfeld at the beginning of the thirties, that *Prontosil* (an azo dye containing, among other functional groups, a sulfonamide group) is effective against streptococci. Sulfonamides were the first chemotherapeutics with which bacterial infections could effectively be combated. In certain circumstances they are still used therapeutically today, although they have largely been superseded by antibiotics (secondary metabolites of microbial origin). The applicability of a synthetic therapeutic or natural antibiotic ceases, when mutant species with resistance to the agent arise, at which point the chemotherapeutic battle against pathogenic bacteria or fungi must recommence from scratches. The search for new antibiotics and, especially, chemotherapeutics will never end. Besides this, a significantly greater effort must be made in order to understand how antibiotics and chemotherapeutics become inactivated in the target organism, how their path to their specific receptors becomes blocked, and how these receptors are susceptible to structural alteration. Only then will a suitable approach for *rational drug design* have been found.

In 1969, *Barnett Rosenberg* opened a new chapter in the history of chemotherapy: that one of the anticarcinogenic effectivity of *Cisplatin* (*cis*-diaminoplatinum(II) chloride). The chemistry and molecular biology of anticarcinogenic Pt compounds have mainly been the study of *Stephen J. Lippard.*

Some degree of therapeutic success against AIDS (Acquired Immune Deficiency Syndrome), which is caused by the HIV (human immunodeficiency virus) retrovirus, has been achieved with the antiviral chemotherapeutic AZT (3′-azido-3′-deoxythymidine).

Today's definition of a drug – as expressed in the German 'Arzneimittelgesetz' (law governing the manufacture and prescription of drugs) – is a substance, or preparation containing substances intended, by application on or inside the human or animal body, to:

- heal, relieve, prevent, or diagnose diseases, suffering, physical incapacity, or other ailment;
- enable understanding of the nature, state, or function of the body or of the psyche;
- replace active substances or fluids otherwise produced by the human or animal body;
- protect against, eliminate, or render harmless pathogens, parasites, or foreign substances;
- affect the nature, state, or function of the body or psyche.

Despite the remarkable results attributable to medicinal chemistry – after all, average human life expectancy has increased from about 40 years at the beginning of the 20th century to more than 70 years – it remains imperative to adopt new stimuli from biology to the development of new strategies, in order to achieve greater efficiency in designing new drugs. Today, on average, only one in 7,000 newly synthesized compounds is pharmacologically interesting. Average development costs of a marketable therapeutic are estimated as more than 100 million $. It has often been said that the medicinal chemist is like a riverboat gambler: against overwhelming odds, he believes that the next chemical he synthesizes and submits for biological evaluation will be the long-sought breakthrough drug.

The picture sketched by *Emil Fischer*, of searching for a key to fit an existing lock, must be reinterpreted, as regards the story so far, as more of a search for a possibly existent lock to fit a coincidentally found key. Rational drug design as envisaged by *Ehrlich* assumes a detailed understanding of primary-structure/tertiary-structure relationships and of supramolecular interactions between active agent and its complementary receptor, as well as – in general terms – a deep familiarity, on the part of the chemist, with supramolecular cell chemistry. To design strategies for targeted chemotherapy, it is essential to understand disease as a disturbance in the area of supramolecular cell chemistry.

There is no generally accepted definition of the term 'disease'. The same applies to the term 'health'. The preamble to the charter of the *World Health Organization* (WHO) offers: '*Health is the state of complete physical, psychological, and social well-being, and not merely the being free of sickness and affliction*'.

The genome (see *Chapt. 16*) of a cell contains in its genes the instructions for protein synthesis. Like the satisfactory methods for the preparation of organic chemicals, these instructions consist of sequences of orders, which are expressed in sentences. The sentences in the language of the cell contain words, which – as in 'human language' – possess no semantic meaning outside of the context of these sentences. The sentences of the genome are formulated in the syntax of the nucleic acids. The words used are nucleotide triplets. With the aid of supramolecular base pairing, the chemical information present in the gene is *copied*, *transcribed* from the DNA script into the RNA script (more seldom, *vice versa*), and *translated* from nucleic-acid syntax into that one of proteins. As stated in the *central dogma of molecular biology*, the cell uses the *antisense*-DNA strand as a transcription matrix for the preparation of a *sense*-RNA strand. Against this backdrop, rational drug design has more than one chance of preventing the synthesis of an unwanted protein.

By base pairing between an *in vitro* synthesized oligonucleotide and a DNA double strand, forming a triplex, or between the oligonucleotide and an RNA single strand – either *viral RNA* or *mRNA* – forming a duplex, it is possible to block replication and/or protein synthesis. Therefore, oligonucleotides which, through triplex or duplex formation, can alter the normal conduct of *primary* or *secondary semantides* – the nucleic acids of the genome or of the mRNA – can, in principle, combat the cause of a disease rather than its symptoms. If one knows the sequence of the nucleic-acid section to be blocked, then, thanks to *Watson-Crick* base pairing, one knows the constitution of the oligonucleotide one has to synthesize: a situation which meets the minimum requirements of rational drug design. Therapeutics of this type, functioning on an oligonucleotide basis, are known as *code blockers*.

Chapt. 16 gives information about the transfer of genetic information in supramolecular base pairs, the *central dogma of molecular biology*, and the *specification of semantophoretic molecules*.

Proteins which enter into supramolecular interactions with DNA or RNA as regulators influence the transmission or use of genetic information in the living cell. As an alternative to the synthesis of ligands which complex directly with nucleic acids, the synthetic chemist interested in chemotherapeutics will also envisage ligands which form supermolecules with these regulators and hence intervene indirectly in biological processes.

Proteins which function as enzymes, catalyzing metabolic pathways, may be activated (by *agonists*) or deactivated (by *antagonists*). Like the ligands for regulators mentioned above, ligands for enzymes cannot be collectively reduced to one common structural denominator.

The systematic search for new drugs begins with the search for a *lead structure*, the structure of an effective substance, from one of the usual sources of natural products. This lead structure is modified with the aim of increasing its therapeutic safety. The modified structure is then viewed by the synthetic chemist as the target structure. A particular synthetic pathway must be worked out for each structurally altered compound.

Recently, rather than synthesize one individual compound, interest has centered on the simultaneous synthesis of a whole ensemble of structurally related compounds (*diversomers*). This proceeds – usually in an automated process – either by the use of newly combined (recombinant) nucleic-acid molecules in biological synthesis or by the use of a combinatorial strategy in chemical synthesis. The mixture (*substance library*) thus obtained may contain hundreds, thousands, or even millions of individual compounds. Before these substance libraries are separated, and their individual components identified, the complex substance mixture is examined using sensitive probes (*screening*, using antibodies, receptors, or other complexing partners) to ascertain the biological activity of individual components. When a result is positive, then measures must be taken to synthesize, isolate, and identify the desired active compound at an appropriate scale. While the procedure outlined earlier has been named *rational drug design*, the simultaneous synthesis of multicomponent products might be called *'irrational' drug design*, then. As the latter expression almost indubitably leads to an undesired association, however, the term *stochastic search for new, biologically active molecules* ought to be used instead, with *non-stochastic search* substituting now for *rational drug design*. In this context, it may suffice to refer to the preparation of linearly constituted macromolecules by a strategy of encoded combinatorial synthesis. Encoded combinatorial synthesis instigates that each molecule synthesized carries with it a record of how it was prepared. The addition of each building block is recorded by a particular, reliably identifiable *reporter molecule*.

Aspects of Organic Chemistry, Volume 3: *Synthesis*, will describe in detail how strategies of combinatorial synthesis can lead to whole 'libraries' of structurally similar molecules. Here, we shall merely outline schematically the achievement of *W. Clark Still* at *Columbia University*, which makes use of a six-round peptide chain elongation, involving seven amino acids and 42 reporter molecules in a solid-phase synthesis (*Fig. 7.38*).

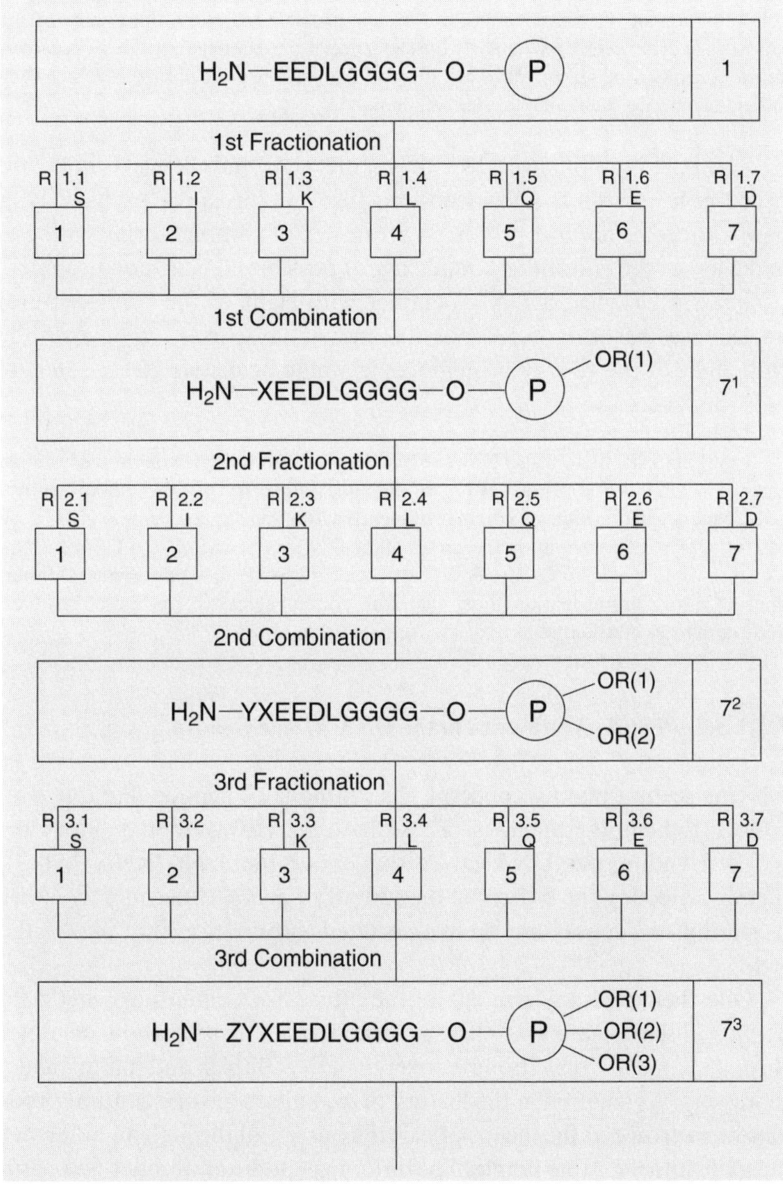

Fig. 7.38. *Schematic representation of the first three rounds of a combinatorial solid-phase synthesis after* W. C. Still

Each of the six rounds of synthesis begins with the fractionation of a particular amount of *Merrifield* polymer beads (P), charged, by means of an ester group, with a peptide fragment of constant sequence GGGGLDEE (see *Table 10.18*). It is distributed (because of the seven amino acids, S, I, K, L, Q, E, and D to be used) over seven reaction vessels. In each of these vessels, a reporter molecule R1–R7 (R1.1–R1.7 for the first round, R6.1–R6.7 for the sixth round) is attached also by means of an ester group to the *Merrifield* beads. Subsequently the chain elongation takes place: each vessel with its preordained amino acid. In this way, each of the total of $6 \cdot 7$ synthetic steps may be assigned its own reporter molecule. Afterwards, the *Merrifield* beads of all seven reaction vessels are recombined and freshly redistributed again between seven reaction vessels, for the next round of the synthesis. This alternation of fractionation and (re)combination leads to a combinatorial synthesis strategy in which, in this particular instance, $7^6 = 117{,}649$ different peptides are accessible. After selective removal of the reporter molecules from the selected *Merrifield* beads and analytical determination of the reporter molecules, it is possible to follow the course of the synthesis in every detail, particularly the sequence of the extended peptides, characteristic for each type of *Merrifield* beads. In other words, a 'peptide library' in which each individual species is already 'labeled' is obtained.

Thanks to combinatorial synthesis, the medicinal chemist is not compelled in his search for new lead structures, to 'prospect' exclusively in new sources of natural products. Strategies of combinatorial synthesis make use of the enormous complexity of possible constitutional isomers (*Chapt. 2.4*). Hence, a DNA octamer with eight of the (four) natural nucleobases has $4^8 = 65{,}536$ possible individual species, an octapeptide with eight of the twenty proteinogenic amino acids has $20^8 = 2.56 \cdot 10^{10}$ individual species.

Combinatorial factors also play a role in the groove-binding interaction of peptide analogs in the DNA minor groove or of oligonucleotides in the DNA major groove. The frequency with which a particular sequence within the human genome of 10^9 base pairs occurs is dependent upon the length of the DNA section (*Table 7.4*). For a DNA sequence of 17 base pairs, there exist, statistically, 8,589,934,592 different combinations. Oligo(d^2rib*f*)nucleotides must, therefore, contain at least 17 nucleotide residues, if they are to be considered as reagents for genome analysis.

7.3.1.3.2. Peptide Antibiotics in the DNA Minor Groove

This section mainly concerns the synthesis of peptide-like (*peptide-mimetic*) chemotherapeutics. These bind selectively to regulatory sequences within the DNA/DNA duplex or the DNA/RNA hybrid, thereby affecting the switching on and off of genes, consequently either triggering or suppressing their associated information flow inside the cell.

One chemotherapy strategy of the future, for combating genetically derived diseases, aims at the sequence-specific deactivation of genes through complex formation between the DNA duplex and 'tailor-made' repressors. As so often in the history of organic chemistry, natural products have provided the means of carving out a trail through an otherwise virtually impenetrable jungle. The di- or tripeptidic antibiotics netropsin (as the dihydrochloride) and distamycin A (as the hydrochloride), each of which forms a highly stable supermolecule with the B-DNA duplex (but not with other DNA-duplex conformational types nor with DNA single strands or with RNA), thereby inhibiting replication and transcription in the particular sequence area. They served, in *Peter B. Der-*

Table 7.4. *Statistical Frequency with which a Particular Base Sequence May Occur*

Number of Base Pairs	Number of Combinations
1	2
2	10
3	32
4	136
5	512
6	2,080
7	8,192
8	32,896
9	131,072
10	524,800
11	2,097,152
12	8,390,656
13	33,554,432
14	134,225,920
15	536,870,912
16	2,147,516,416
17	8,589,934,592
18	34,359,869,440
19	137,438,953,472
20	549,756,338,176
21	2,199,023,255,552
22	8,796,095,119,360
23	35,184,372,088,832

The statistical frequency with which the base sequence of a DNA section occurs is dependent on the number of base pairs in that section. Taking account of the fact that A and T, and also C and G, pair with each other, a DNA section with n base pairs has

for n odd $(4^n)/2$,
for n even $(4^n)/2 + (4^{n/2})/2$

possible combinations.

van's laboratory at the *California Institute of Technology* in Pasadena, as *lead compounds* for the synthesis of similarly constructed peptides (**III**–**V**) with altered properties regarding DNA complexation.

The *lead structure* for the design of a whole series of polypeptides with variable B-DNA recognition specificity is based on an X-ray crystal-structure analysis of the 1:1 complex of netropsin and the B-DNA duplex given in *Fig. 7.39*. The X-ray crystal-structure analysis shows that the crescent ligand has fitted well into the minor groove of the right-handed DNA double helix – more precisely, where sequential (dA,dT) segments occur. The H-atoms of the carboxamido groups participate in three-center H-bonds with N(3) of Ade and O at C(2) of Thy.

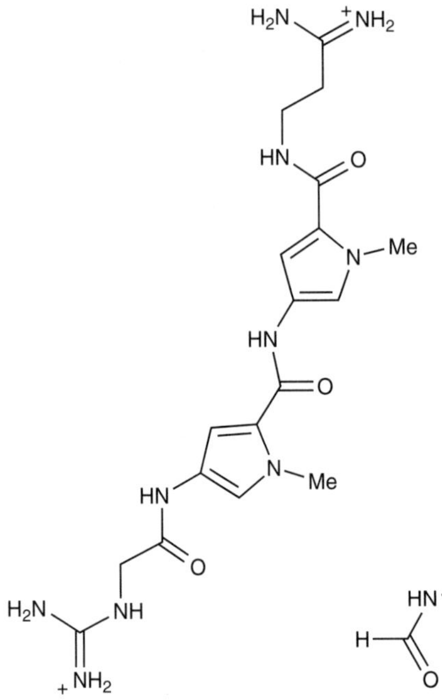

Netropsin (**I**)

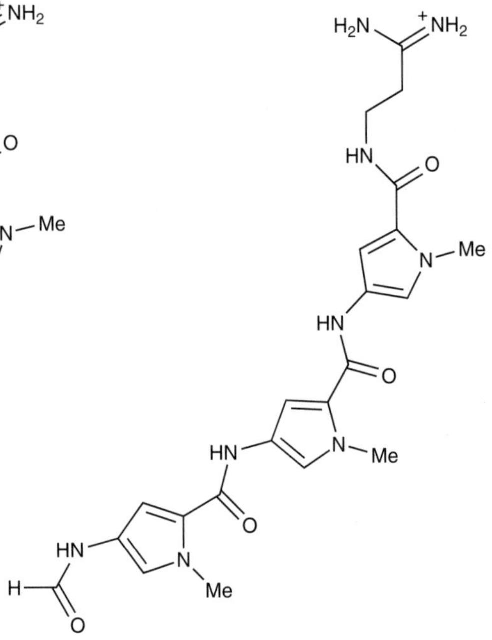

Distamycin A (**II**)

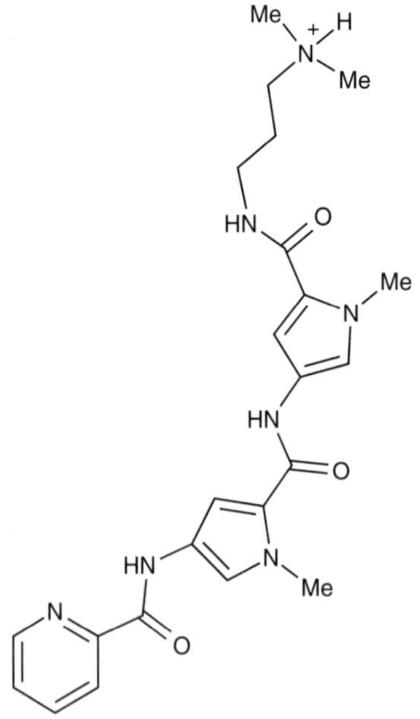

III

(after *Dervan*: 'Pyridine-2-carboxamidonetropsin', 2-PyN)

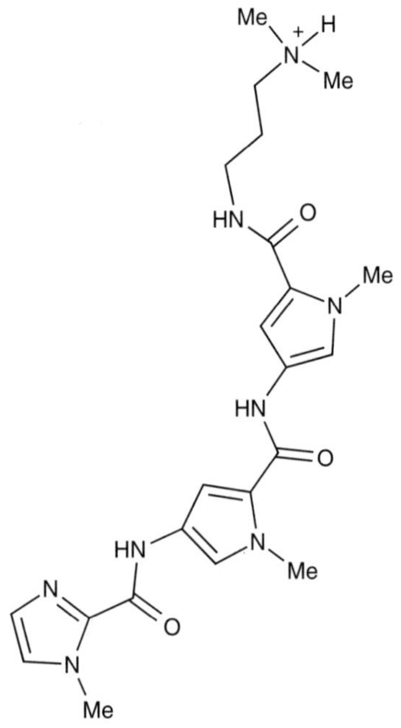

IV

(after *Dervan*: '*N*-Methylimidazole-2-carboxamidonetropsin', 2-ImN)

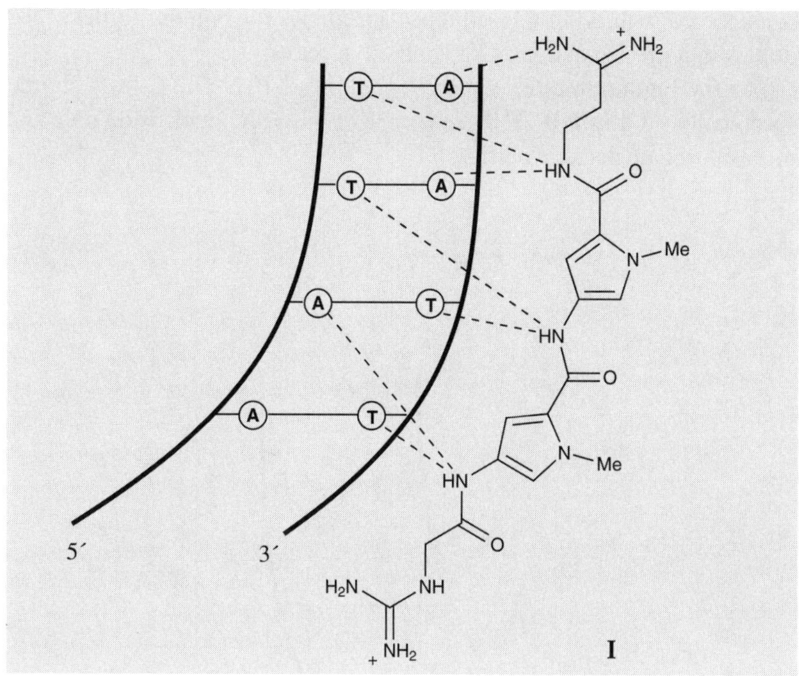

Fig. 7.39. *Schematic representation of the supermolecule made of B-DNA and netropsin* (**I**)

Netropsin (or distamycin A, too) merits consideration as a lead compound. As it only binds to four base pairs (*Fig. 7.39*), and as the complementary base-pair quartet occurs, statistically, once every 136 base pairs (*Table 7.4*), the resulting selectivity is too small for a therapeutic application of the two peptide antibiotics.

Increasing the number of repetitive *N*-methylpyrrole-2-carboxamido units expands, as expected, the binding region for (dA,dT)-rich DNA. As a rule, *n* amido groups of a ligand each enter into one bond with $(n + 1)$ DNA base pairs. Hence, the smallest recognition unit for an *N*-methylpyrrole-2-carboxamide consists of two base pairs. If two H-bond-acceptor groups may be located on adjacent nucleobases, but in different single strands, there exists a total of ten combinations: (dA,dA), (dA,dT), (dA,dC), (dA,dG), (dT,dT), (dT,dC), (dT,dG), (dC,dC), (dG,dG), and (dC,dG). Experimental studies have shown that, for *N*-methylpyrrole-2-carboxamido units, the tendency to form three-center H-bonds follows the descending order: (dA,dT) ≫ (dA,dA) > (dT,dT) > (dA,dC), (dT,dC), (dT,dG) ≫ (dA,dG), (dC,dG), (dC,dC), (dG,dG). The acceptor potential for three-center H-bonds in the supramolecular *N*-methylpyrrole-2-carboxamide/DNA combination follows the descending order: (3′–5′)(dT,dT) ≫ (dT,dA), (dA,dT) > (dG,dA), (dG,dT), (dC,dT) ≫ (dC,dA), (dC,dC), (dG,dC), (dC,dG). At a given length of the oligo(*N*-methylpyrrole-2-carboxamide) chain, it will certainly encounter difficulty in fitting into the DNA minor groove.

To design synthetic repressors, it is useful not only to consider how the binding region with (dA,dT)-rich DNA sequences may be enlarged, but also whether relatively small structural modifications to the ligands can succeed in altering the sequence-specificity for binding to the DNA duplex in a precise manner. The answer is yes.

As the binding models in *Fig. 7.40* show, 2-PyN and 2-ImN should specifically bind to a B-DNA sequence of a dG/dC nucleoside pair and three dA/dT nucleoside pairs.

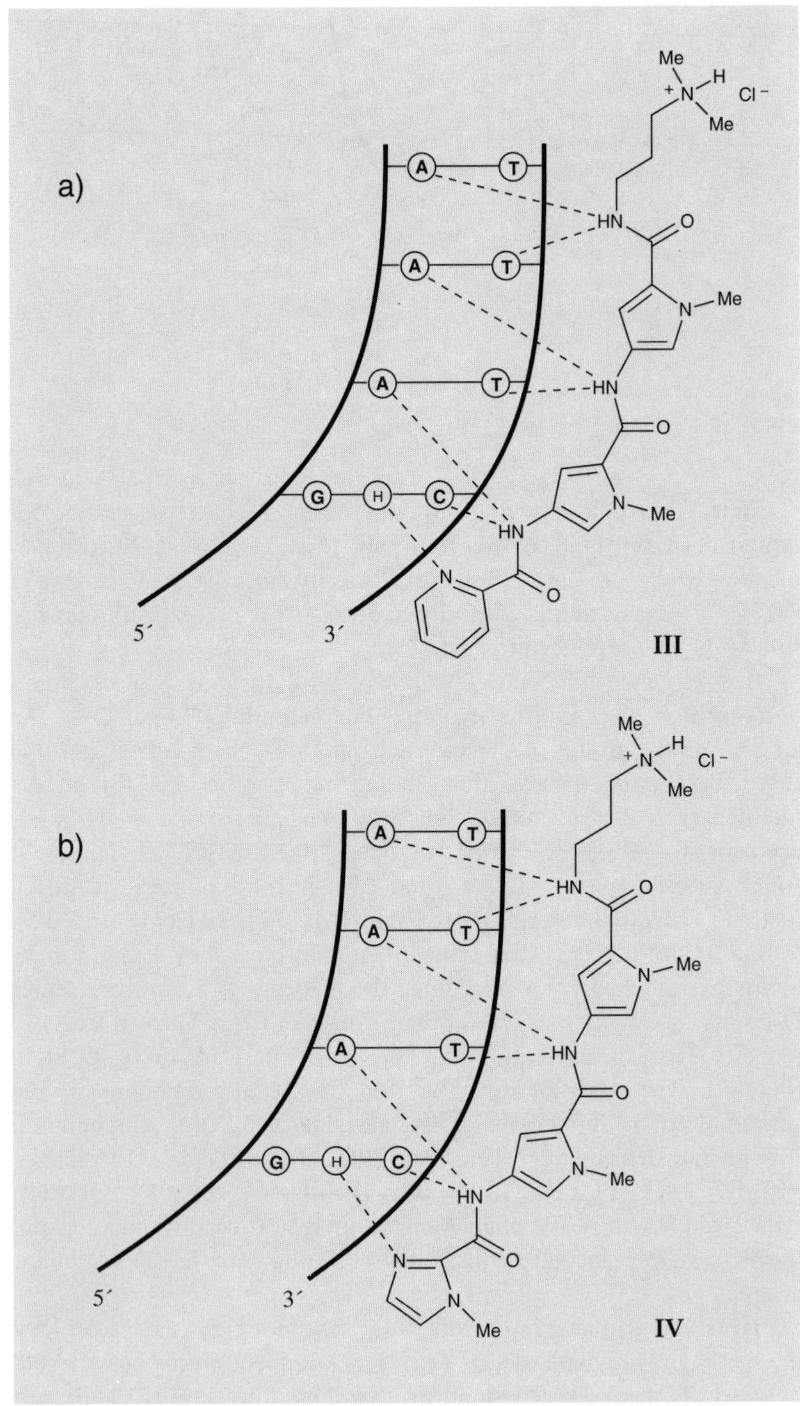

Fig. 7.40. *Schematic representation of the supermolecule made of B-DNA and* a) *'pyridine-2-carboxamidonetropsin'* (**III**) *and* b) *'*N*-methylimidazole-2-carboxamidonetropsin'* (**IV**)

In reality, it was found that 2-PyN interacts with three dA- or dT-rich duplex sections, all comprising five base pairs: (3′–5′)d(TTTTT), (3′–5′)d(AATAA), and (3′–5′)d(CTTTT). As well as these, a completely unexpected binding between 2-PyN and the sequence (3′–5′)d(TGTCA) was observed. This last sequence is recognized by 2-ImN with an even higher specificity: a result incomprehensible from the viewpoint of the 1:1 model described.

Just as the 1:1 model, and the impetus behind the synthesis of 2-PyN and 2-ImN, came from an X-ray crystal-structure analysis of the complex in which netropsin or distamycin A were participants, a 2D-NMR study, carried out at higher ligand concentration, led to a 2:1 model, in which two distamycin A molecules are bound to a central (3′–5′)d(AAATT) segment. The binding between 2-ImN and the (3′–5′)d(TGTCA) sequence may be explained in the context of the 2:1 model (*Fig. 7.41*).

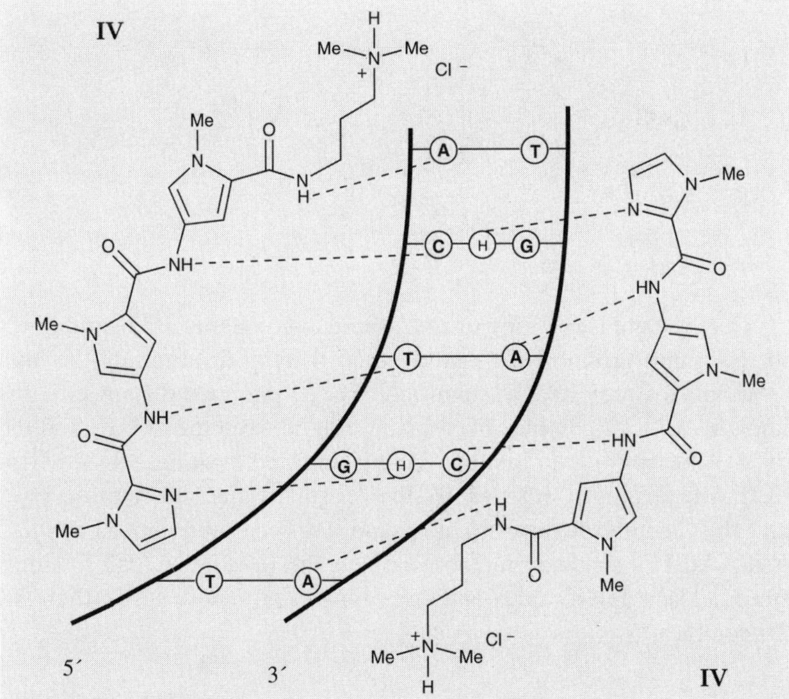

Fig. 7.41. *Schematic representation of the supermolecule made of B-DNA and two antiparallel arranged '*N*-methylimidazole-2-carboxamidonetropsin' ligands* (**IV**) (2:1 complex)

The supermolecule in which a pair of antiparallel oriented, constitutionally identical ligands have inserted themselves into the minor groove of a B-DNA segment differs structurally from the 1:1 supermolecule in two respects: the minor groove is approximately twice as wide in the former case as in the latter, and, instead of the three-center H-bonds encountered in the 1:1 complex, each of the two single strands binds exclusively to one ligand in the 2:1 complex.

Besides the 2:1 supermolecule, composed of two identical ligands and DNA, 1:1:1 supramolecular systems consisting of two distinct ligands and DNA (for example, of distamycin A, '2-imidazoledistamycin', and the duplex of *Fig. 7.42*) should also be taken into account.

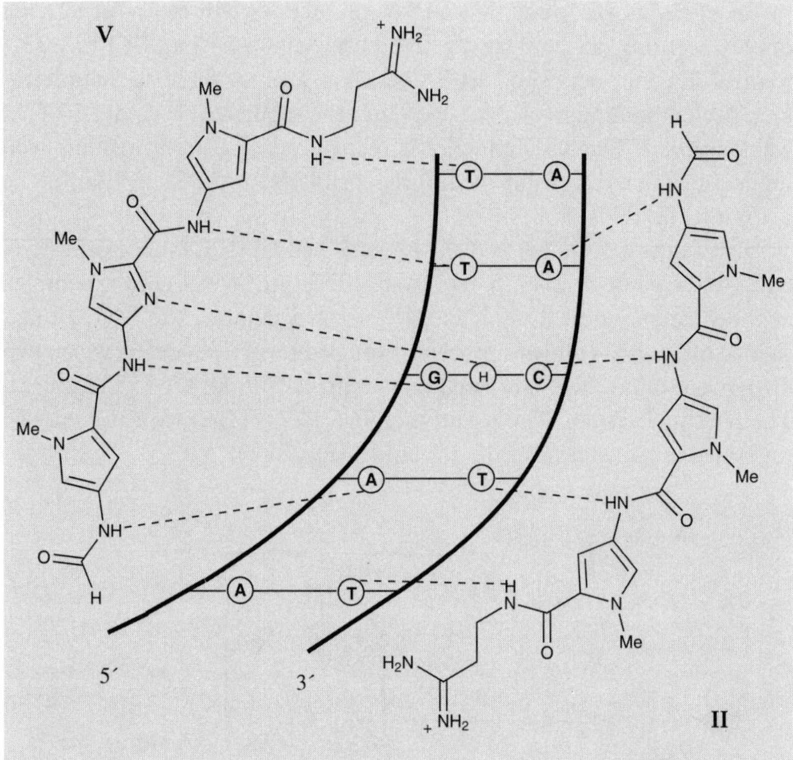

Fig. 7.42. *Schematic representation of the 1:1:1 supermolecule of distamycin A* (**II**), *'2-imidazoledistamycin'* (**V**), *and B-DNA*

Here the imidazole ring of the '2-imidazoledistamycin' ligand binds to the amino group of the central Gua, thereby determining the high specificity of molecular recognition. The '2-imidazoledistamycin'/distamycin A/DNA complex may distinguish between the dG/dC and the dC/dG nucleoside pair, as the binding sites and both ligands are symmetrically arranged relative to the central dG/dC nucleoside pair, and the '2-imidazoledistamycin' ligand has been taken up on the (3′–5′)d(AAGTT) single strand, but not on the (3′–5′)-d(AACTT) single strand. The 1:1:1 complex is thermodynamically more stable than the 2:1 complex.

7.3.1.3.3. Oligonucleotides in the DNA Major Groove: Triple Helices

Poly(U) and poly(A) combine in aqueous solution to form a double-helical supramolecular system. Depending on the temperature and the concentration of the counter-ion, however, a triple-helical 2:1 supramolecular system may be obtained instead. X-Ray diffraction on such strings shows that the structure of a *Watson-Crick* duplex is present. The third strand settles in the DNA major groove, by *Hoogsteen* pairing in parallel orientation to the 'homopurine strand' (*Fig. 7.43*).

Success in hindering those DNA-binding proteins participating in replication, transcription, and cell division in their supramolecular interaction with the DNA duplex by oligodeoxynucleotides could mean a

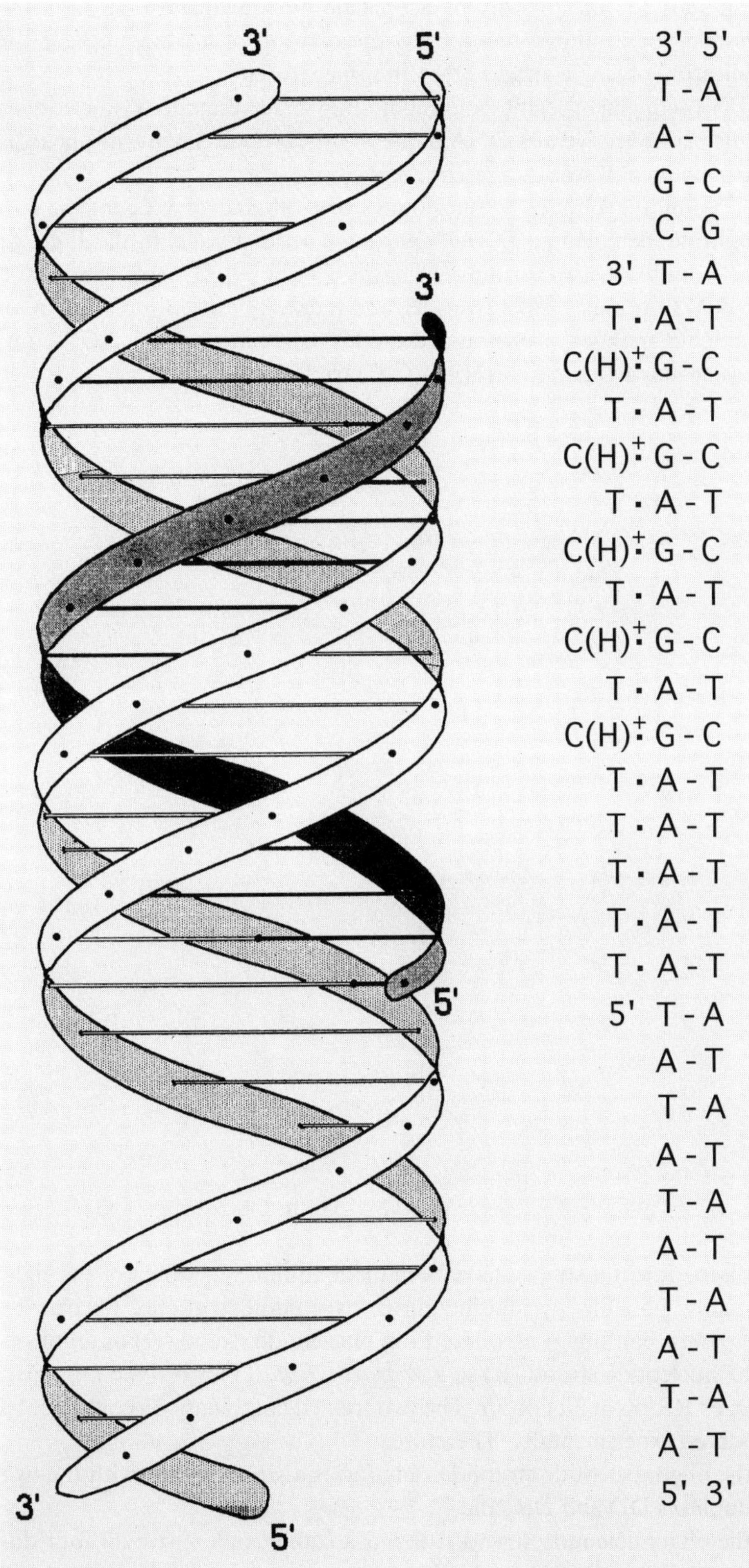

Fig. 7.43. *Triple-helical 1:1:1 supermolecule*

fresh start for the therapy of neoplastic or viral disease. It might be possible, for example, to suppress the division of a cancer cell or the replication of single-strand DNA in some viruses.

The sequence-specific binding of tailor-made oligodeoxynucleotides in the major groove of a DNA duplex, with formation of a triple helix, is the main area of research in *Dervan*'s laboratory.

Pyrimidine oligodeoxynucleotides are capable of recognizing homopurine sections in a DNA duplex, and settle parallel to the *Watson-Crick* double helix, forming *Hoogsteen* H-bonds. In this manner, a dT/d(A/T) triplex results from dT and a d(A/T) nucleoside pair. In an analogous manner, a protonated $dC(H)^+$ associates with a d(G/C) nucleoside pair to form a $dC(H)^+/d(G/C)$ triplex.

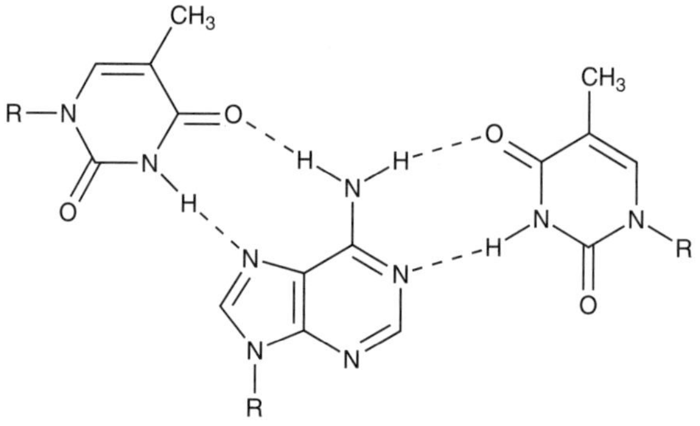

dT · d(A/T)

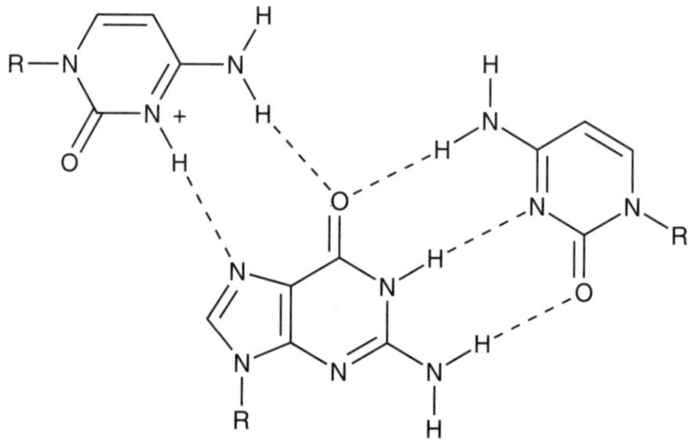

$dC(H)^+ \cdot d(G/C)$

	DDD	
DDR	[DRD]	RDD
RRD	RDR	[DRR]
	RRR	

Experimental studies have been made to find out which of the eight formally possible pyrimidine/purine/pyrimidine triplexes formed by *Hoogsteen* binding of an oligo(d^2rib*f*)nucleotide strand (D) or an oligo(rib*f*)nucleotide strand (R) to a *Watson-Crick* duplex of type DD, DR, RD, or RR do in fact occur. The two framed combinations could not be observed experimentally. Therefore:

- the oligonucleotide strand D only forms a stable triplex with the two duplexes DD and DR, and
- the oligonucleotide strand R forms a stable triplex with all four duplexes DD, DR, RD, and RR.

In the just used shorthand writing of triplexes, the first letter denotes the pyrimidine single strand bound in *Hoogsteen* manner. The second letter denotes the purine strand, and the third one the pyrimidine strand, in the *Watson-Crick* duplex.

In the six observed triplexes, the oligonucleotides D and R are arranged parallel to the purine single strand in the *Watson-Crick* duplexes DD, DR, RD, and RR. If the *Watson-Crick* duplex contains DNA in its purine single strand, then DNA and RNA single strands are bound sufficiently strongly. If the *Watson-Crick* duplex contains RNA in its purine single stand, then another RNA single strand can bind, but not a DNA single strand.

? · d(T/A)

Oligo(rib*f*)nucleotides exert a different effect on the sugar-phosphate backbone conformation, compared with that of the oligo(d^2rib*f*)nucleotide, because of the OH group at C(2′) (*Sect. 7.3.1.2.3*). Duplexes in the A-conformation are favored in the first case. The triple helix also holds the *Watson-Crick* duplex in the A-conformation, and so it is understandable that an oligo(rib*f*)nucleotide single strand binds to the duplex better than does an oligo(d^2rib*f*)nucleotide single strand. The nucleobase Thy (in DNA) has, unlike Ura (in RNA), a Me group at C(5), which exerts an additional stabilizing effect on the supermolecule (with duplex or triplex character) through hydrophobic stacking forces.

Purine oligonucleotides are also capable of recognizing homopurine sections in a DNA duplex, settling antiparallel to the purine strand in the major groove of the *Watson-Crick* double helix, forming reverse-*Hoogsteen* H-bonds.

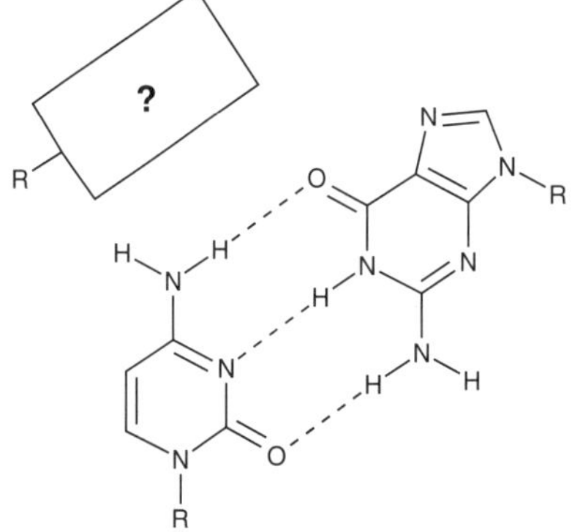

? · d(C/G)

The sequence-specific recognition of a DNA double helix by oligonucleotide-induced formation of a triple helix is, therefore, restricted to purine sections. Specific recognition of all of the base pairs of a DNA duplex requires the use of non-natural deoxyribonucleosides, which must be specially custom-made to fit.

If the synthesized oligonucleotide is equipped at the 3′- or the 5′-end with covalently bound 'chemical scissors', it is possible to selectively cleave the double strand. The double function – sequence-specific binding and sequence-specific cleavage – makes functionalized oligo(d^2rib*f*)-nucleotides possible tools for genome analysis.

The gene responsible for *Chorea Huntington* (*St. Vitus' dance*) was located in 1983 on human chromosome No. 4. Its structural elucidation was achieved ten years later in the way outlined.

7.3.1.3.4 *Antisense*-Oligonucleotides

One way to block replication and protein synthesis is to exploit the effect of oligo(d^2rib*f*)nucleotides on double-stranded DNA, forming triple helices. Another possibility might be to inhibit protein synthesis, using *antisense*-oligonucleotides to induce duplexes with single-stranded RNA, the nucleobases of which are of *sense*-ordering. *In vitro* studies, both using cell cultures and using a cell-free system, performed as early as 1978 by *Paul C. Zamecnik* and *Mary L. Stephenson*, ensured that such thoughts did not remain idle speculation for long. A 13-meric oligo-(d^2rib*f*)nucleotide of suitable sequence was found to be capable of inhibiting the growth of *Rous Sarcoma* virus, by specific hybridization with its complementary RNA, blocking the synthesis of a protein whose production was encoded in the RNA.

The way to a new class of antiviral drugs seemed open. Animal viruses may be divided into six different classes (*Table 7.5*). The various viruses differ from each other in the program which determines in what manner the genetic material of the viral nucleic acids of the genome (DNA or RNA, single or double strand, with *sense*- or *antisense*-sequence) transforms into that one of the viral mRNA, showing *sense*-ordering of its nucleobases. Blocking of mRNA by *antisense*-oligonucleotides ought to be possible in viruses of all classes.

Table 7.5. *Classification of Animal Viruses (s: sense, a: antisense)*

	DNA viruses		RNA viruses			
Nucleic-acid class	I	II	III	IV	V	VI
Genome	DNA Double strand	DNA Single strand (*s* or *a*)	RNA Double strand	RNA Single strand (*s*)	RNA Single strand (*a*)	RNA Single strand (*s*)
Intermediate copy	none	DNA Double strand	none	RNA Single strand (*a*)	none	1. DNA Single strand (*a*) 2. DNA Double strand
mRNA	Single strand (*s*)					

However, before a therapeutic application of *antisense*-oligonucleotides can be seriously considered, the following questions must have been answered affirmatively.

Is the proposed target sequence for the antisense-*oligonucleotide to be synthesized topologically accessible at all?* This question is not in any way trivial, as RNA usually exhibits a folded structure and, *a priori*, the exposed disposition of the target sequence, which is required for duplex formation to take place, is not known. Even if duplex formation in the desired sequence region does occur, the question still arises of whether it exists for long enough to inhibit translation of the sequence section to be blocked. The duplex lifetime plays an especially significant role, if the molecular mechanism suppressing translation is based on a steric blocking of those sections of mRNA with which RNA-binding proteins would otherwise interact. It seems to be of less significance for the alternative mechanism, in which cellular *RNase H* is able to separate the RNA portion from the RNA/DNA hybrid hydrolytically.

Does the synthesized antisense-*oligonucleotide fulfill the structural requirements which permit its* in vivo *application?* This question examines whether the potential *antisense*-therapeutic can penetrate through membranes to gain access to the cell interior, while avoiding reaction with cellular nucleases.

Are affinity and specificity with which the synthesized antisense-*oligonucleotide should bind to its complementary mRNA sequence section sufficient for targeted therapeutic application?* The affinity principally grows with increasing chain length of the *antisense*-oligonucleotide and with the number of G/C nucleoside pairs in the resulting duplex. It should be taken into account that the number of base pairings may be smaller than the number of chain units able to recognize the complementary single strand, and that the specificity of the *antisense*-oligonucleotide is dependent on whether – for the elimination of mispaired bases – the association/dissociation equilibrium is reached quickly enough, before the respective RNA hydrolysis begins.

Is the desired antisense-*oligonucleotide accessible at an acceptable cost, in large scale and using 'good manufacturing practice'? Antisense*-oligonucleotides of any desired base sequence may be produced by automated solid-phase synthesis. Because *antisense*-oligonucleotides are efficiently hydrolyzed, *in vivo*, by nucleases, attempts are often made to produce – also by solid-phase synthesis – structurally modified DNA analogs which are less reactive under physiological conditions than *antisense*-oligonucleotides. Essentially, structural alterations may be made on the sugar, the nucleobase, or on the phosphodiester bridge.

'Good manufacturing practice' (GMP) covers fundamental instructions for the manufacture and quality control of medicines, defined by the *World Health Organization* (*WHO*).

7.3.1.4. Structurally Modified Oligonucleotides

7.3.1.4.1. Oligonucleotides Modified in the Sugar Component

It is possible to conceive of many ways in which the sugar component of the sugar-phosphate backbone may be structurally changed. The greatest differences occur when various sugars of different constitutions are used. This is the case in the series RNA (β-D-ribofuranose), DNA (2-deoxy-β-D-ribofuranose), and homo-DNA (2,3-dideoxy-β-D-allopyranose). The difference is more subtle when the same sugar, but in different constitution or configuration, is used. An example of the former case is the series RNA (β-D-ribofuranose) and p-RNA (β-D-ribopyranose); an example of the latter case, when the nucleobase is not connected *via* a β-glycosidic bond, but *via* an α-glycosidic one, to a sugar of identical constitution.

In previous sections, RNA and DNA, DNA and homo-DNA, and RNA and p-RNA were compared with one another. We learned hereby that the additional OH group at C(2′) made RNA more reactive than

DNA, and we are not surprised to learn now that RNA has a greater recognition capacity for the formation of specific supramolecular duplexes. For homo-DNA, DNA, RNA, and p-RNA, we have seen the differences (*Sect. 7.3.1.2.3*) which are not confined to the conformation of the individual single strands, but which also cover the selectivity of the particular base pairing, as a result of molecular recognition.

Thus, each nucleic acid has its own specific molecular identity and its own base-pairing rules. This also applies to oligo(2-deoxy-α-D-ribofuranosyl)nucleotides. These assemble into double helices in which the single strands are oriented antiparallel. They also hybridize with oligo(2-deoxy-β-D-ribofuranosyl)nucleotides, but in such a manner that the two single strands are oriented parallel to one another. The resulting duplexes do not suppress translation, however.

7.3.1.4.2. Oligonucleotides Modified in the Base Component

In light of the emphasis placed on the structural requirements the sugar-phosphate backbone has to fulfill to make an oligonucleotide eligible as a pairing partner, with itself or with other oligo- or polynucleotides, interest is awakened in whether nucleobases other than the natural ones may be used.

2-Aminopurine may, without steric complications, take over the role of 6-aminopurine (= adenine) in the shaping of a DNA double helix with *Watson-Crick* base pairing. The fact that 2-aminopurine is not a 'letter' in the 'alphabet' of natural nucleobases may be biosynthetically contrived. It uncovers according to *Eschenmoser* the remarkable *coincidence*, in the natural nucleobases, *of specific suitability for the biologically fundamental function of base pairing on the one hand, and of immanent qualification for being self-constituted from elementary organic compounds on the other hand.*

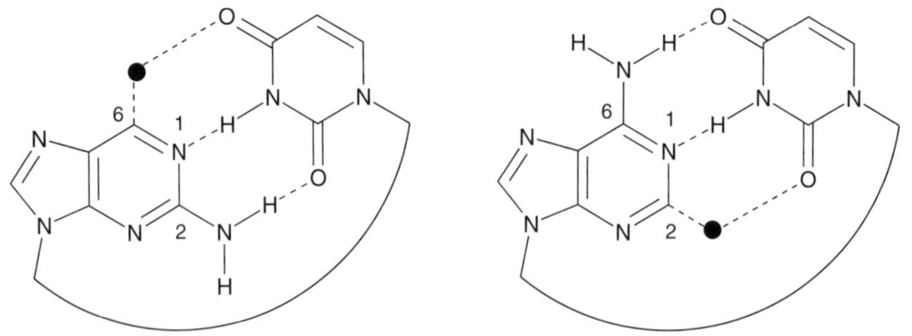

Nucleosides of 2,6-diaminopurine and uracil pair smoothly, forming three H-bonds (**A**). *Steven A. Benner* has succeeded in enzymatic incorporation of isoguanine and isocytosine (**B**) into DNA and RNA, and in their duplex formation by *Watson-Crick* base pairing. He accomplished a similar interaction between the nucleosides of 2,4-diaminopyrimidine and xanthine (**C**), or 1-methyl-1,4-dihydropyrazolo[4,3-*d*]pyrimidine-5,7-dione (**D**), respectively.

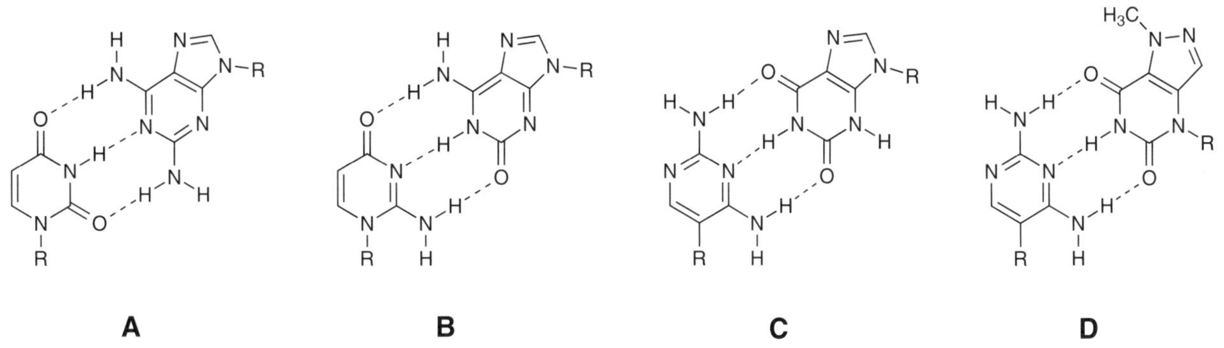

A **B** **C** **D**

7.3.1.4.3. Oligonucleotides Modified in the Phosphate Component

Considerable attention has been devoted to the preparation of thermodynamically stable and (particularly towards nucleases) unreactive *antisense*-oligonucleotide analogs (also possessing the ability to penetrate through cell membranes) in which the phosphate group had been structurally altered or replaced by another group entirely free of P.

Some examples, in which one or both O-atoms not directly involved in the bridge between two sugar residues have been replaced by other atoms or atom groups, are given below. New stereogenic centers are often introduced, with the consequence that, for n new stereogenic centers, 2^n diastereoisomers are to be expected. In the methyl phosphonates, and in the phosphotriesters, the bridge between two adjacent sugar rings is electrically uncharged, a fact which ought to facilitate the penetration of membranes (*Sect. 7.3.1.3.4*). The exchange of O for S leading to the phosphorothioates and phosphorothionates is a measure which should immediately be apparent to the chemist, who finds inspired by the periodic table of the elements (see *Table* on the inner front cover). The phosphorodithioates enable the introduction of new stereogenic centers to be avoided.

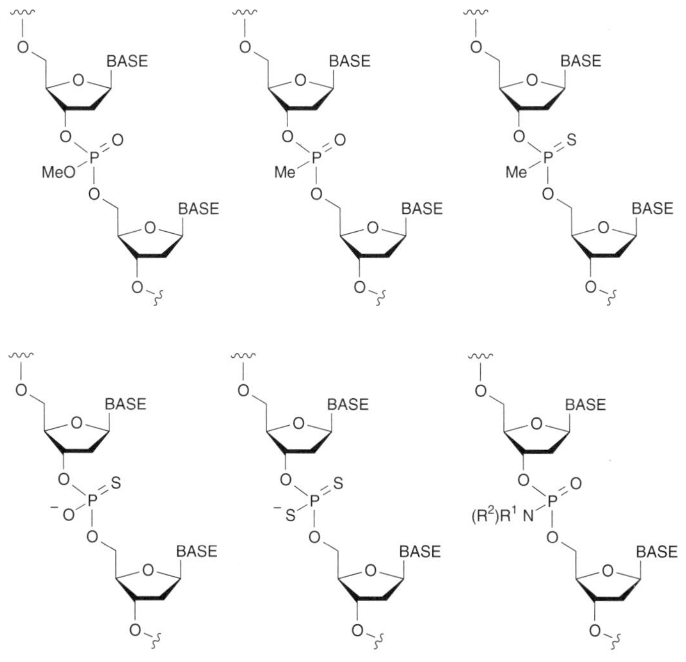

Constitutional modifications which essentially should not alter the conformation of the particular main chain are presented below.

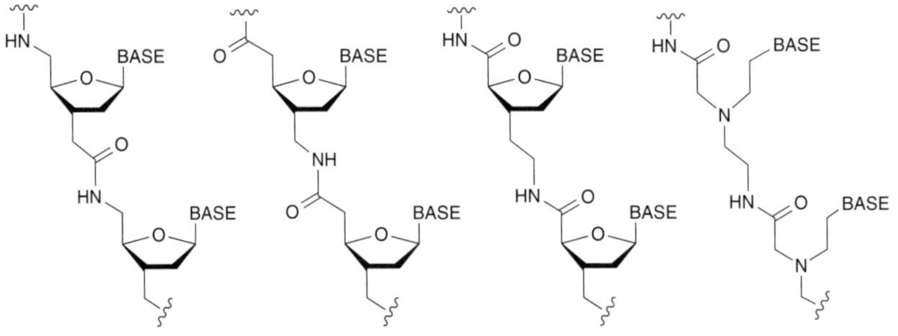

In search of analogs of DNA or RNA, polymers have been synthesized which no longer contain the sugar-phosphate backbone, and in which the nucleobases are covalently bound to polyamide chains.

7.3.2. Polysaccharides

Monosaccharides have already been dealt with in detail. Questions concerning their configuration were answered in *Chapt. 5*, concerning their conformation in *Chapt. 6.6.2*. *Chapt. 10.5.3* summarizes how carbohydrates may be described using formulae or names (with the corresponding descriptors).

7.3.2.1 The Diamond-Lattice Matrix as Starting Point for Conformational Analysis

If, for a disaccharide (or an oligosaccharide, generally), the participating monosaccharides and their acetal connectivity(ies) are known, and the knowledge that the participating pyranose rings will each adopt that chair conformation in which the greatest possible number of substituents have equatorial orientations applied, then *Pitzer* strain-free conformations worthy of consideration for systematic conformational analysis may be obtained by fitting into the diamond-lattice matrix (*Sect. 7.2.2*, *7.3.1.1*, and *Chapt. 10.3.3*). Account need only be taken of the requirements that:

- *Newman* strain is as small as (constitutionally) possible,
- the order of priority *ap* > *sc* applies for the C–C–O–C fragment, and, because of the anomeric effect, the order of priority *sc* > *ap* applies for the O–C–O–C fragment,
- cooperative formation of intracatenate H-bonds may invert stability orders postulated without consideration of H-bonds.

Wide-ranging studies by *Yoshito Kishi* have made it clear that the steric interaction of the X–C(4) bond with C(2′) is stronger than that with O(5′), and that this finding remains valid, even when – as in 2′-deoxy sugars – no 1,5-repulsion between C(4) and O(2′) can exist. This result has been established by NMR spectroscopy, from the extensive agreement between *O*-glycosides and their *C*-glycoside analogs, especially synthesized for comparison purposes. It agrees with the assertion (*vide supra*) that the energetically favored conformations of the atom sequences C(2′)–C(1′)–O(4)–C(4) and O(5′)–C(1′)–O(4)–C(4) should be *ap* and *sc*, respectively.

To gain an insight into the conformational behavior of the 1,4-linked polysaccharides *amylose* (from starch) and *cellulose*, the disaccharide parent compounds methyl α-maltoside (= methyl O^4-α-D-glucopyranosyl-α-D-glucopyranoside) and methyl α-cellobioside (= methyl O^4-β-D-glucopyranosyl-α-D-glucopyranoside) (X = O in each case), together with the analogous *C*-glycosides (X = CH_2 in each case), will be subjected to systematic conformational analysis.

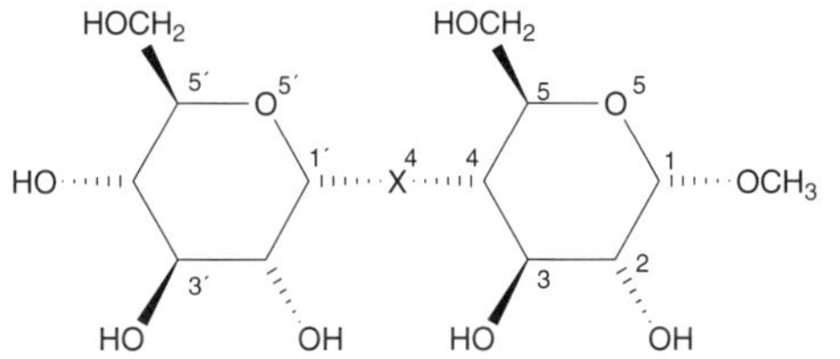

X = O: Methyl O^4-α-D-glucopyranosyl-α-D-glucopyranoside (= Methyl α-maltoside)

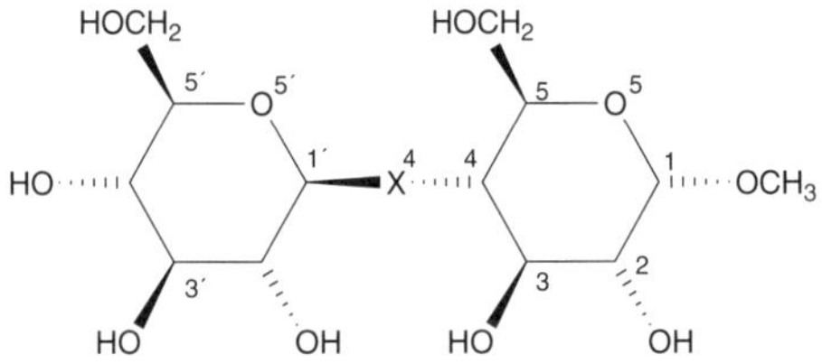

X = O: Methyl O^4-β-D-glucopyranosyl-α-D-glucopyranoside (= Methyl α-cellobioside)

7.3.2.2. Systematic Conformational Analysis of Methyl α-Maltoside and Its Analogous *C*-Glycoside

Of the three idealized conformational types without *Pitzer* strain, differing in their arrangement about the axial C(1′)–O(4) bond (*Fig. 7.44*), the one with *ap*-orientation of the C(2′)–C(1′)–O(4)–C(4) sequence (*Fig. 7.44*, **A**) should be assigned priority over the other two, in agreement with the selection criteria already specified (*Sect. 7.3.2.1*). This is easily justified, on sharper conformational resolution of the O–C–O atom sequence. Sharper conformational resolution means here – without wishing to anticipate an interpretation of the anomeric effect (*Chapt. 8.4.3*) – fitting the O–C–O sequence into the diamond-lattice matrix. It implies that each O-atom is assigned two phantom ligands (*Sect. 7.3.1.1.1*), and that the favored conformation is to be regarded as that one in which the bond between one of the phantom ligands and its associated O-atom is of *ap*-orientation relative to the bond between the connected C-atom and the next O-atom (*Fig. 7.44*, **A**, and *Table 7.6*). For a disaccharide with an axial C(1′)–O(4) bond, the *sc*-orientation of the O(5′)–C(1′)–O(4)–C(4) sequence is stabilized by the anomeric effect, both by O(4) and, in the reversed direction, by O(5′). Conformational type **A** (*Fig. 7.44*) is also distinguished by having the least possible *Newman* strain.

The other conformational type to fulfill the topological requirements for the anomeric effect to operate twice shows a torsion angle C(2′)–C(1′)–O(4)–C(4) of +60° (*Fig. 7.44*, **B**, and *Table 7.7*). However, it has such a high *Newman* strain, and 1,6-repulsions on top of that, that it may confidently be discounted.

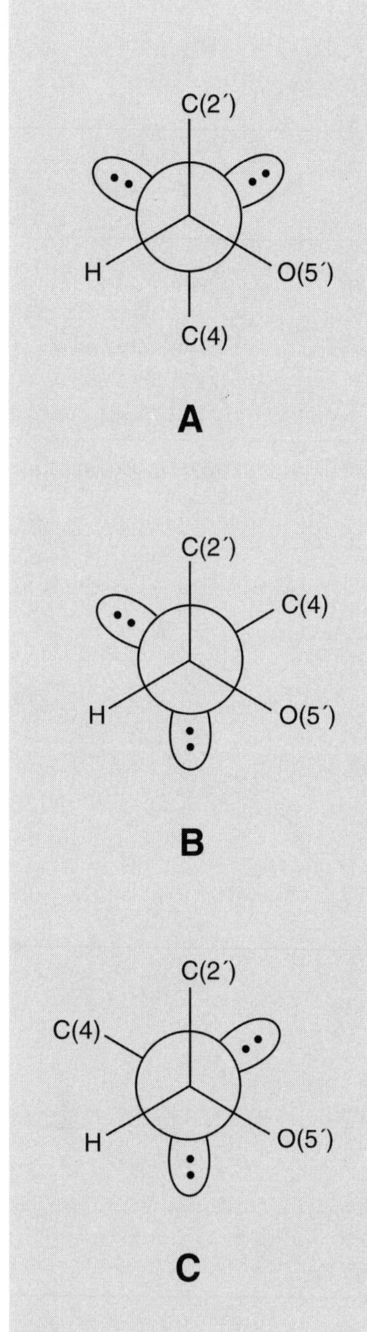

Fig. 7.44. Newman *projections of idealized, staggered conformations of 1,4-linked disaccharides with an axial C(1′)–O(4) bond and C(2′)–C(1′)–O(4)–C(4) torsion angles of 180°* (**A**), *+60°* (**B**), *and −60°* (**C**)

Table 7.6. *Characteristics of the Three Idealized, Staggered Conformations of Methyl α-Maltoside with C(2′)–C(1′)–O(4)–C(4) Torsion Angle of 180° (Fig. 7.44,* **A***) and Various Torsion Angles about the O(4)–C(4) Bond*

Diamond-lattice fitting	*Fig. 7.45*, **I**	*Fig. 7.45*, **II**	*Fig. 7.45*, **III**
C(1′)–O(4)–C(4)–C(5)	180°	−60°	+60°
C(1′)–O(4)–C(4)–C(3)	+60°	180°	−60°
1,5-Repulsions	O(3)····C(1′)	C(5)····O(5′) C(6)····C(1′)	C(3)····O(5′)
1,6-Repulsions	–	C(6)····O(5′)	–
1,7-Repulsions	–	–	O(3)····C(5′)
H-Bonds	–	–	–

The conformational type with a torsion angle C(2′)–C(1′)–O(4)–C(4) of −60° (*Fig. 7.44*, **C**, and *Table 7.8*), compared to the favored conformation with a corresponding torsion angle of 180° (*Fig. 7.44*, **A**), is not only disadvantaged in this respect, but also by the disappearance of one of the two stabilizing effects in the atom sequence O–C–O. As well as this, it displays considerable *Newman* strain.

The idealized conformations of methyl α-maltoside, each with a C(2′)–C(1′)–O(4)–C(4) torsion angle of 180°, are reproduced in the three staggered orientations **I**–**III** about the hinge of the O(4)–C(4) bond in *Fig. 7.45* (X = O) (see *Table 7.6*).

Table 7.7. *Characteristics of the Three Idealized, Staggered Conformations of Methyl α-Maltoside with C(2′)–C(1′)–O(4)–C(4) Torsion Angle of +60° (Fig. 7.44,* **B***) and Various Torsion Angles about the O(4)–C(4) Bond*

C(1′)–O(4)–C(4)–C(5)	180°	−60°	+60°
C(1′)–O(4)–C(4)–C(3)	+60°	180°	−60°
1,5-Repulsions	O(3) and C(5′) coincide	C(6) and C(3′) coincide	C(3)····C(2′) C(4)····C(3′) C(5)····O(5′) C(4)····C(5′)
1,6-Repulsions			C(3)····C(3′) C(5)····C(5′)
1,7-Repulsions			many
H-Bonds			–

Table 7.8. *Characteristics of the Three Idealized, Staggered Conformations of Methyl α-Maltoside with C(2')–C(1')–O(4)–C(4) Torsion Angle of –60° (Fig. 7.44,* **C***) and Various Torsion Angles about the O(4)–C(4) Bond*

C(1′)—O(4)—C(4)—C(5)	180°	–60°	+60°
C(1′)—O(4)—C(4)—C(3)	+60°	180°	–60°
1,5-Repulsions	O(3) and O(2′) coincide	C(4)····O(2′) C(6)····C(1′)	C(5)····C(2′) C(4)····O(2′)
1,6-Repulsions		–	C(5)····O(2′)
1,7-Repulsions		–	C(6)····C(3′)
H-Bonds		–	–

In conformation **II** there exist two 1,5-repulsions. As well as these, a particularly inconvenient 1,6-repulsion is present. Conformations **I** and **III** (*Fig. 7.45* and *Table 7.6*) both show only one 1,5-repulsion. The C(1′)–O(4) bond in conformation **III** participates in two *sc*-fragments (through the O(4)–C(4) bond with both C(3) *and* C(5)), but only in one *sc*-fragment in conformations **I** and **II** (through the O(4)–C(4) bond with C(3) *or* with C(5)). All in all, conformation **I** should be the least destabilized one. NMR Studies have shown that, in solution at room temperature, more than one conformation is present. In any case, conformation **I** ought to predominate as soon as the OH group at C(3), and hence the associated 1,5-repulsion, is removed. This supposition has been confirmed experimentally for X = CH_2.

While no X-ray crystal-structure analysis exists for methyl α-maltoside, however, methyl β-maltoside (= methyl O^4-α-D-glucopyranosyl-β-D-glucopyranoside) has been studied this way (*Fig. 7.46*).

In the latter case, the C(2′)–C(1′)–O(4)–C(4) torsion angle of –128.2° deviates markedly from the *ap*-orientation expected according to rule. The X-ray crystal-structure analyses of α-maltose (= O^4-α-D-glucopyranosyl-α-D-glucopyranose) and β-maltose (= O^4-α-D-glucopyranosyl-β-D-glucopyranose) display C(2′)–C(1′)–O(4)–C(4) torsion angles of –125.5° and –116.6°, respectively. The deviation from the *ap*-value cannot, therefore, be ascribed to a conformational alteration at the anomeric center C(1), but to the transition from vacuum (diamond-lattice fitting) or aqueous/methanolic solution (NMR spectroscopy) to the solid state (X-ray crystal-structure analysis). In the crystal structure of methyl β-maltoside monohydrate, three H-bonds exist between the two neighboring monosaccharide residues. One, between the OH groups at C(2′) and at C(3), represents a direct connection between the two pyranose rings. The other two, each with an H_2O molecule as an acceptor, bridge the two OH groups at C(6′) and at C(6). As a consequence of these H-bonds, a conformation deviating strongly from the idealized one is found.

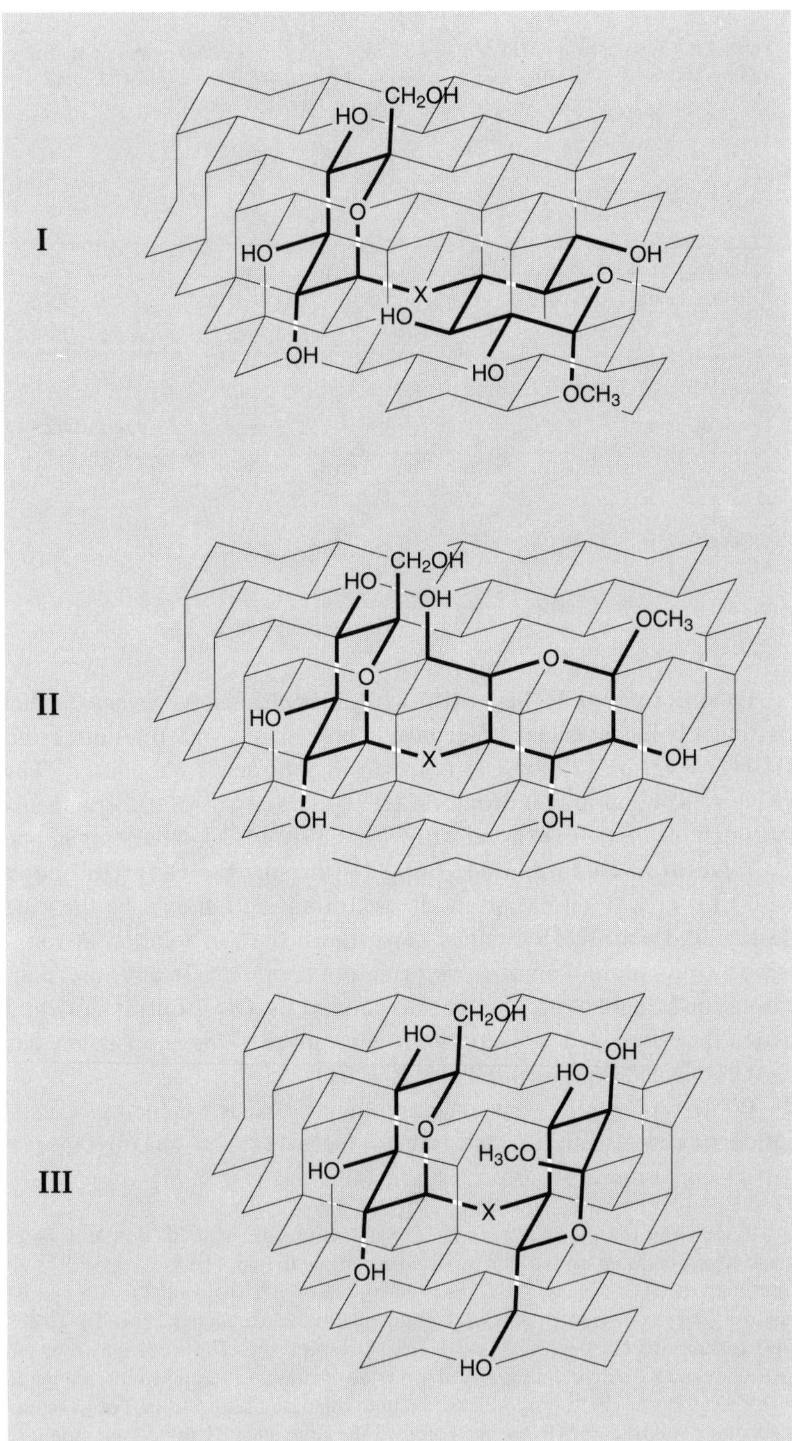

Fig. 7.45. *Diamond-lattice fitting of the conformations* **I–III** *of the disaccharide methyl α-maltoside* (X = O) *and its analogous* C-*glycoside* (X = CH_2)

The crystal structure of the amylose polymer (complexed with KOH and H_2O; *Fig. 7.47*), however, exhibits no intracatenate H-bonds of this kind, and its conformation corresponds better to that one which would be predicted on the grounds of systematic conformational analysis (C(2′)–C(1′)–O(4)–C(4) torsion angle *ca.* −144°, C(1′)–O(4)–C(4)–C(5) torsion angle *ca.* −150°).

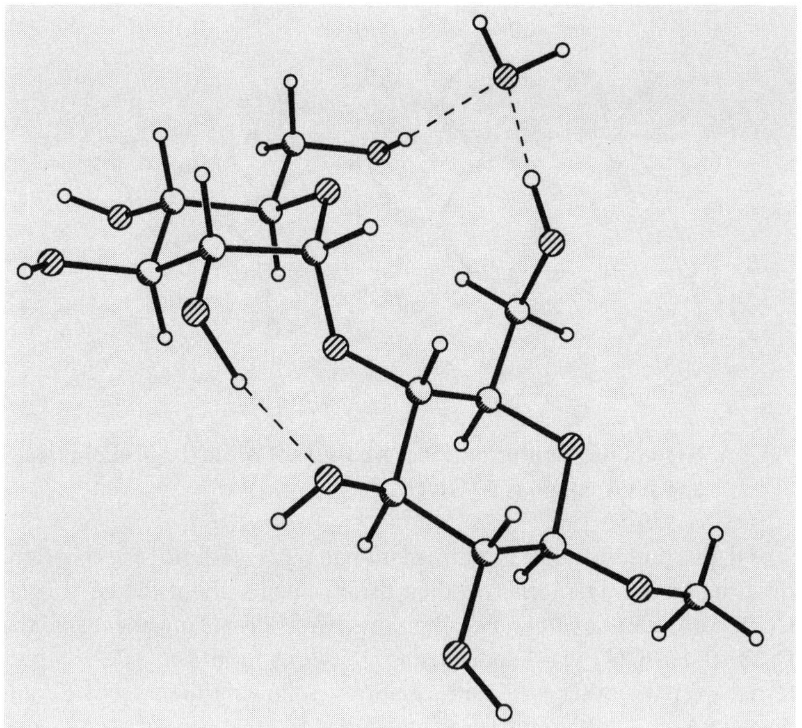

Fig. 7.46. *Crystal structure of the disaccharide methyl β-maltoside monohydrate*

Consequently, the ring O-atoms of two adjacent monosaccharide residues (O(5), O(5′), *etc.*) are each situated on the same side of the extended molecular strand: an arrangement which leads to a helix. The amylose complex forms a 6_1-helix, composed of asymmetric units, each consisting of three α-glucosyl monomers (*Fig. 7.48*), coiled around a two-fold screw axis.

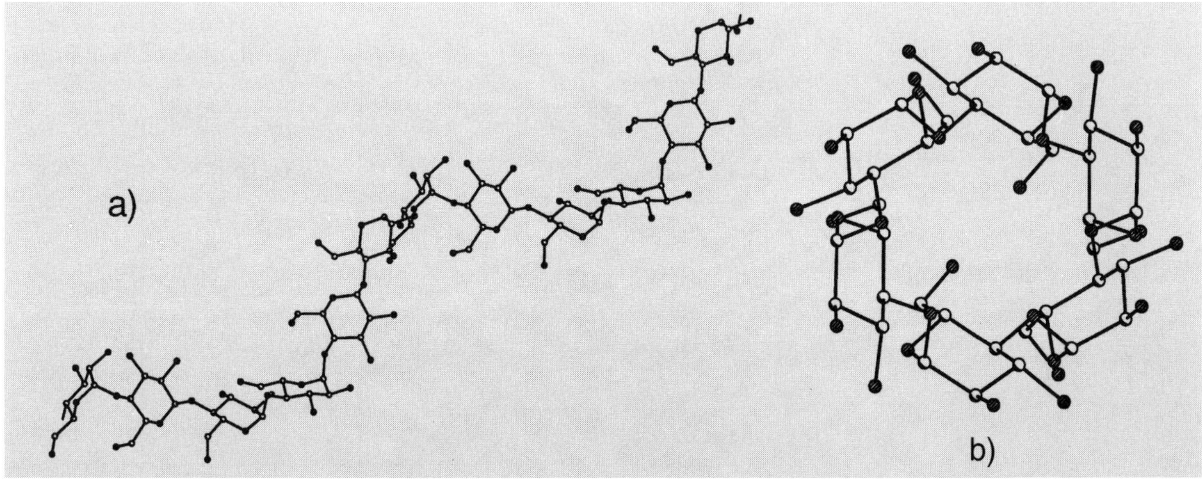

Fig. 7.47. *Section* (four asymmetric units) *from the crystal structure of amylose* (complexed with KOH and H_2O). No intracatenate H-bonds are present, intercatenate H-bonds with the solvent do not exert any crucial effect on the conformation and so are not indicated. *a*) View perpendicular to the long molecular axis. *b*) Projection along that molecular axis (two units are positioned exactly behind the ones illustrated; the two-fold symmetry is clearly to be seen).

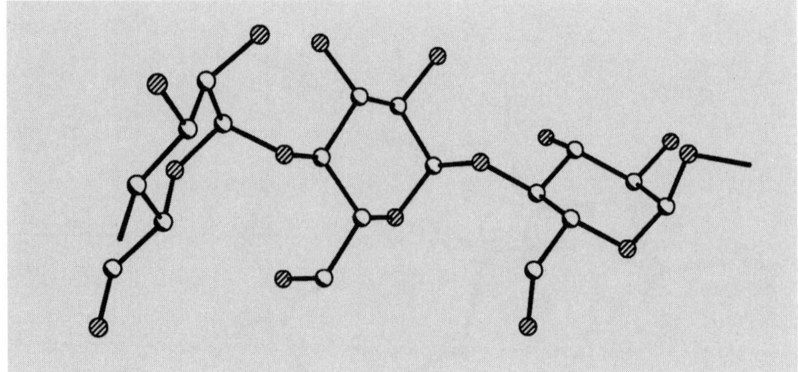

Fig. 7.48. *Section* (one asymmetric unit) *from the crystal structure of amylose* (complexed with KOH and H_2O)

7.3.2.3. Systematic Conformational Analysis of Methyl α-Cellobioside and Its Analogous *C*-Glycoside

Of the three idealized conformational types with no *Pitzer* strain, differing from one another in their arrangements about the equatorial C(1′)–O(4) bond (*Fig. 7.49*), the one with *ap*-arrangement of the C(2′)–C(1′)–O(4)–C(4) sequence (*Fig. 7.49*, **A**, and *Table 7.9*) is preferred over the other two conformations, following the selection rules given in *Sect. 7.3.2.1*. This is easily justified by relatively small *Newman* strain and the anomeric effect. For a disaccharide with an equatorial C(1′)–O(4) bond, however, stabilization of the *sc*-orientation of the O(5′)–C(1′)–O(4)–C(4) sequence may only arise through the anomeric effect of O(5′).

The second conformational type has a C(2′)–C(1′)–O(4)–C(4) torsion angle of −60° (*Fig. 7.49*, **B**, and *Table 7.10*). While favored by the anomeric effect, the conformation is, however, severely disadvantaged by the *sc*-arrangement of the C(2′)–C(1′)–O(4)–C(4) sequence, as well as by high *Newman* strain.

Table 7.9. *Characteristics of the Three Idealized, Staggered conformations of Methyl α-Cellobioside with C(2′)–C(1′)–O(4)–C(4) Torsion Angle of 180° (Fig. 7.49,* **A**) *and Various Torsion Angles about the O(4)–C(4) Bond*

Diamond-lattice fitting	*Fig. 7.50*, **I**	*Fig. 7.50*, **II**	*Fig. 7.50*, **III**
C(1′)—O(4)—C(4)—C(5)	−60°	+60°	180°
C(1′)—O(4)—C(4)—C(3)	180°	−60°	+60°
1,5-Repulsions	C(6)····C(1′)	C(5)····O(5′)	C(3)····O(5′) O(3)····C(1′)
1,6-Repulsions	–	–	O(3)····O(5′)
1,7-Repulsions	–	–	–
H-Bonds	–	–	O(3)—H····O(5′)

Table 7.10. *Characteristics of the Three Idealized, Staggered Conformations of Methyl α-Cellobioside with C(2')–C(1')–O(4)–C(4) Torsion Angle of –60° (Fig. 7.49,* **B***) and Various Torsion Angles about the O(4)–C(4) Bond*

C(1')—O(4)—C(4)—C(5)	–60°	+60°	180°
C(1')—O(4)—C(4)—C(3)	180°	–60°	+60°
1,5-Repulsions	C(5)····O(5') C(6)····C(1')	C(3)····O(5') C(5)····C(2')	C(3)····C(2') O(3)····C(1')
1,6-Repulsions	C(6)····O(5')	–	O(3)····C(2')
1,7-Repulsions	–	C(6)····O(2')	–
H-Bonds	–	–	–

The third conformational type, with a C(2')–C(1')–O(4)–C(4) torsion angle of +60° (*Fig. 7.49*, **C**, and *Table 7.11*), does not experience any stabilization by the anomeric effect, and, additionally, is disfavored because of the *sc*-orientation of the C(2')–C(1')–O(4)–C(4) sequence. In any case, it is endowed with significant *Newman* strain.

Table 7.11. *Characteristics of the Three Idealized, Staggered conformations of Methyl α-Cellobioside with (C(2')–C(1')–O(4)–C(4) Torsion Angle of +60° (Fig. 7.49,* **C***) and Various Torsion Angles about the O(4)–C(4) Bond*

C(1')—O(4)—C(4)—C(5)	–60°	+60°	180°
C(1')—O(4)—C(4)—C(3)	180°	–60°	+60°
1,5-Repulsions	C(6) and O(2') coincide	C(3)····C(2') C(4)····O(2')	C(4)····O(2') O(3)····C(1')
1,6-Repulsions		C(3)····O(2')	–
1,7-Repulsions		–	–
H-Bonds		–	–

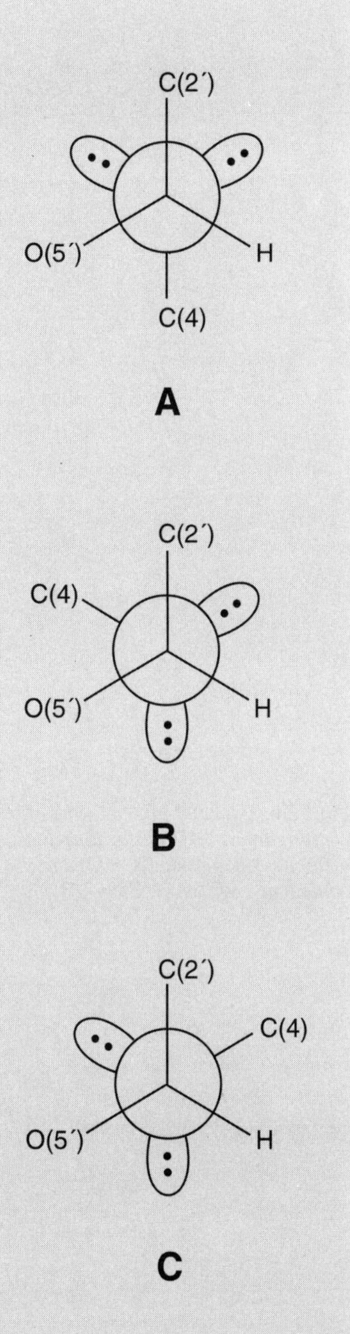

Fig. 7.49. Newman *projections of idealized, staggered conformations of 1,4-linked disaccharides with an equatorial C(1')–O(4) bond and C(2')–C(1')–O(4)–C(4) torsion angles of 180°* (**A**), *–60°* (**B**), *and +60°* (**C**)

Fig. 7.50 gives the idealized conformations **I–III** of methyl α-cellobioside with the *ap*-fixed C(2')–C(1')–O(4)–C(4) sequence and staggered orientations about the O(4)–C(4) bond in each case, together with their fittings into the diamond lattice (see *Table 7.9*).

Of the conformations **I**, **II**, and **III**, **III** may quickly be discarded. In it there can be seen two 1,5-repulsions, one 1,6-repulsion, and, furthermore, an *sc*-orientation between C(1') and C(3) about the O(4)–C(4) bond. For conformations **I** and **II**, one 1,5-repulsion is present in each

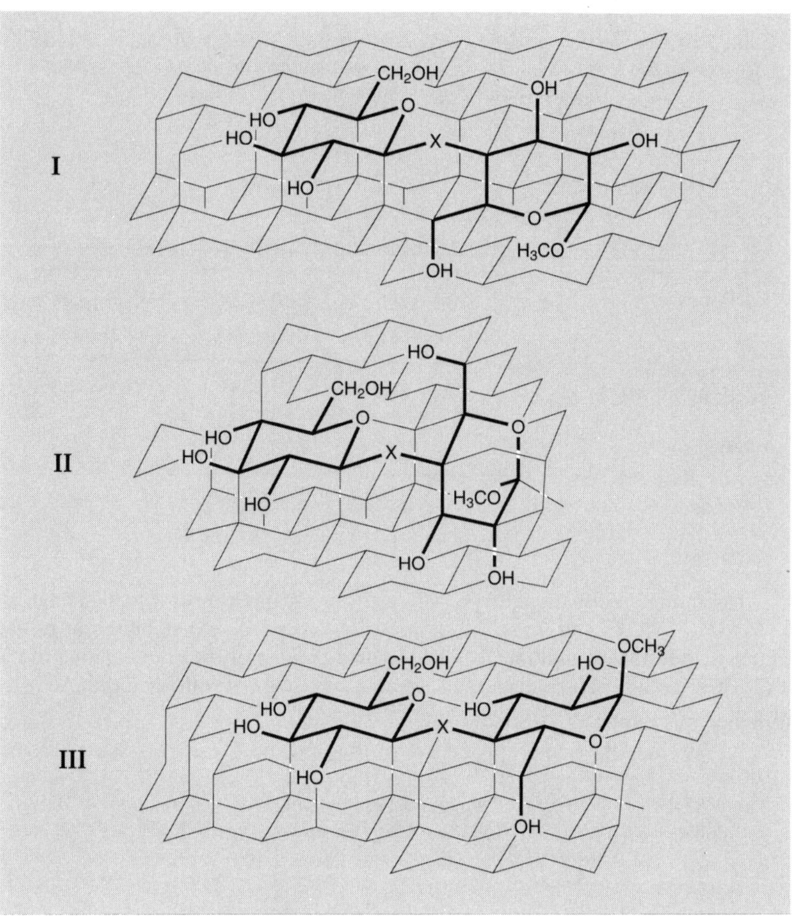

Fig. 7.50. *Diamond-lattice fitting of the conformations* **I–III** *of the disaccharide methyl α-cellobioside* (X = O) *and the analogous* C-*glycoside* (X = CH_2)

case. Conformation **I** has one *sc*-orientation about the O(4)–C(4) bond, conformation **II** has two *sc*-orientations. In solution, a mixture of conformations **I** and **II** is to be anticipated, with conformation **I** slightly predominating.

The single-crystal X-ray structure analysis of methyl α-cellobioside has not been reported. That one of the methanol solvate of methyl β-cellobioside (= methyl O^4-β-D-glucopyranosyl-β-D-glucopyranoside; *Fig. 7.51*) offers the following torsion angles: C(2′)–C(1′)–O(4)–C(4) = 152.0°, O(5′)–C(1′)–O(4)–C(4) = −91.1°, C(1′)–O(4)–C(4)–C(3) = 80.3°, and C(1′)–O(4)–C(4)–C(5) = −160.7°.

A glance at the crystal structure is sufficient to show that the conformation present here would be excluded by systematic analysis of those conformations which fit into the diamond lattice. The conformation is determined by, among other things, the three-center H-bond between the OH group at C(3) and O(5′) (strong) and between the OH groups at C(3) and C(6′) (weak). A search in the *Cambridge Structural Database* provides considerable documentary evidence that the majority of oligosaccharides comprised of equatorial/equatorial 1,4-linked monosaccharide residues show H-bonding between the OH group at C(3) and O(5′), and that this effects a similarly large deviation on the oligosaccharides concerned from conformation **III** outlined in *Fig. 7.50*. Looking more closely at the environment of the H-bond, and counting the seven-membered ring formed as a six-membered ring, ignoring the bridge H-atom, then a diamond-lattice fitting with *trans*-arrangement of the fused rings can be imagined. A further consequence of the H-bond is the alternating orientation of the

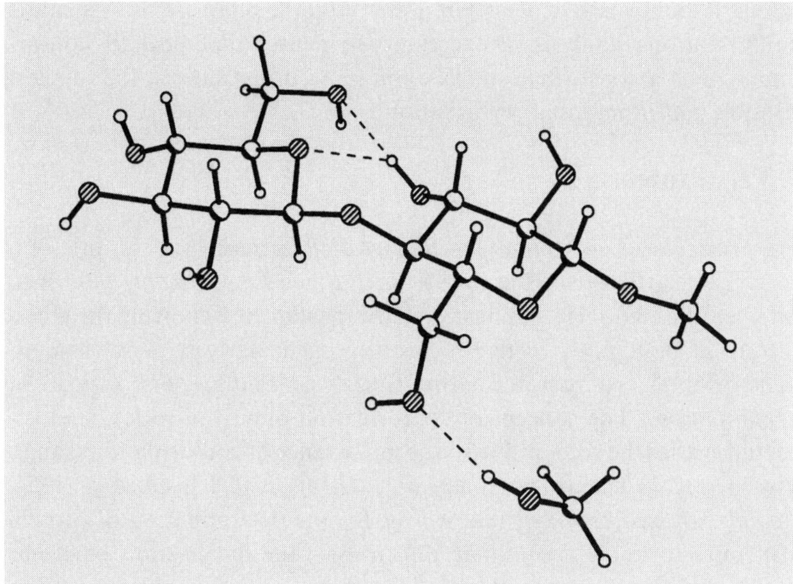

Fig. 7.51. *Crystal structure of the disaccharide methyl β-cellobioside as methanol solvate.* The H-bond between the OH groups of the solvent MeOH and that one at C(6) has no influence on the conformation.

pyranose rings (*Fig. 7.51*): the ring O-atoms of adjacent monosaccharide residues lie on opposite sides of the long molecular axis. In the case of the polymer *Ramie* cellulose, this causes – after rotation about the respective O(4)–C(4) bonds to widen the gap between O(3) and O(5′) and narrow that one between O(2′) and O(6) – the adoption of a quasi-linear arrangement of the sugar backbone (*Fig. 7.52*). Consequently, the crystal structure of *Ramie* cellulose contains another conformation-determining intracatenate H-bond: between the OH groups at C(6) and C(2′), C(6′) and C(2″), and so on.

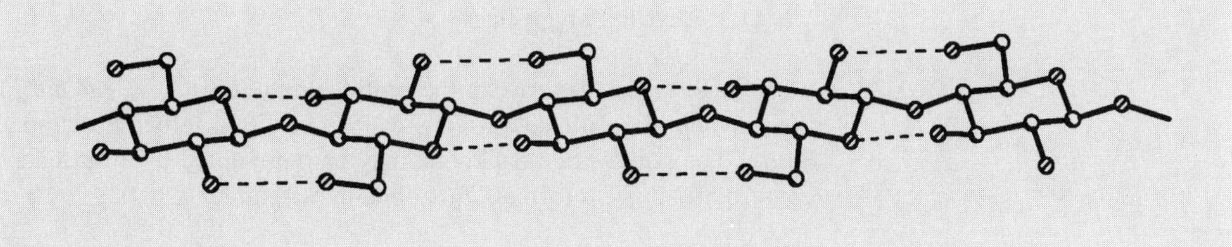

Fig. 7.52. *Section* (five asymmetric units) *of the crystal structure of* Ramie *cellulose*

7.3.2.4. The Variety of H-Bonds in Oligo- and Polysaccharides

The small selection of examples in the preceding sections has demonstrated that relationships within a crystal lattice can lead to conformations which differ greatly from those obtained from conformational analysis *in vacuo* or in aqueous solution. While the O-atoms in the rings and in the glycoside bonds are only capable of acting as acceptor groups in the formation of H-bonds, the OH groups, able to rotate, may function either as acceptor or donor groups in the formation of inter- or intracatenate H-bonds. The formation of intracatenate H-bonds is also the prime cause of reversal of order of stability (proposed without taking H-bonds into account) in non-crystalline phases. For systematic conformational analysis, it is necessary to identify that idealized conformation which, for formation of H-bonds, needs to undergo the smallest possible conformational perturbation.

7.3.3. Proteins

'*Molecular Architecture and Biological Reactions*' was the title of a remarkable article from the year 1946, in which *Linus Pauling* published his scientific credo. He emphasized the importance of knowing the architecture of biologically active molecules; particularly that of proteins. (The original text uses the terms *Gestalt* or *configuration* instead of *conformation*.) The concept of conformation only won widespread acceptance after the recognition of the importance of conformational analysis in steroid chemistry (*Chapt. 6.2*). In his *Nobel* Lecture of 1954, *Pauling* stressed how essential it was, for the determination of protein structure, to know the atomic dimensions, the localization of atoms which might participate in H-bonds, and the restriction of free rotation in the constitutional repeating unit of the amino-acid residue, as well as to test the resulting conclusions on physically relevant molecular models.

In the following sections, we will describe how, first with the aid of characteristic data from X-ray crystal-structure analysis, and, next, with the aid of systematic conformational analysis, the idealized conformation of a polypeptide backbone is worked out.

7.3.3.1. Pragmatic Conformational Analysis

7.3.3.1.1. Acyclic Polyglycines

Because of the planarity of the peptide unit (*Chapt. 8.1.1* and *8.4.1*), the polypeptide chain of proteins cannot be fitted into the diamond lattice. The polypeptide backbone may be represented by an extended conformation with a formal C=N bond in each amide group.

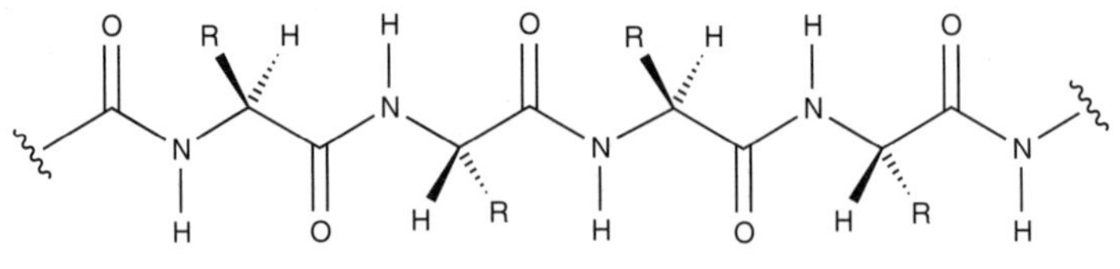

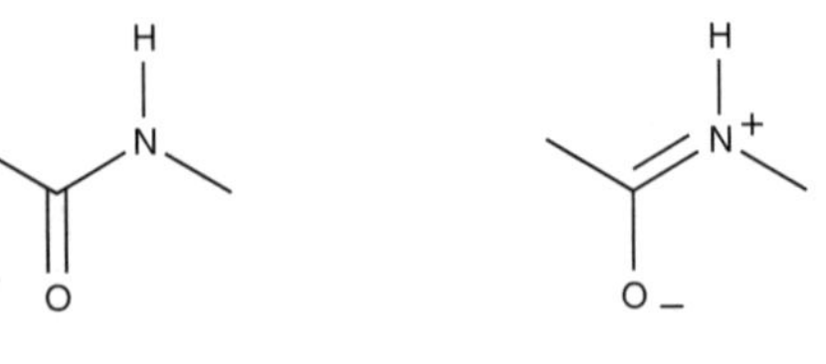

A repetitive constitutional pattern can be seen in the polypeptide backbone of proteins: after two centers of coordination number 3, there always follows a center of coordination number 4. In the case of polyglycine (R = H), the idealized, linear conformation with torsion angles $\phi = \psi = \omega = 180°$ can confidently be assumed. This can be justified in the following way: histograms of the C(=O)–O–C–C torsion angles in esters containing the $-CH_2$–O–C(=O)– group and the C(=O)–N–C–C torsion angles in amides containing the $-CH_2$–NH–C(=O)– group, using information from the *Cambridge Structural Database*, show that:

- the C(=O)–O–C–C torsion angle for the stated ester type occurs most frequently in the region of 180°,
- the C(=O)–N–C–C torsion angle for the stated amide type is scattered over a range between 60° and 180°, with a peak in the region of 80°.

If the ester case is viewed as normal, the amide case, however, being easily prone to perturbation by H-bonds, then the value 180° should apply for ϕ in idealized glycine fragments in polypeptides. The torsion angles ϕ (and ψ) in the crystal structure of Gly-Gly-Gly approaches this ideal figure quite closely (*Fig. 7.53*).

Fig. 7.53, b, shows a section from the crystal structure of Gly-Gly-Gly. It can be seen that two complementary single strands associate together in an antiparallel ordering, without substantial conformational changes, with formation of two pairs of H-bonds, so that the N-terminus of one single strand matches the C-terminus of the other one.

Tripeptides which comprise other amino acids besides Gly deviate more strongly from the idealized, linear conformation. This information can be extracted, for example, from the torsion angles of DL-Leu-Gly-Gly, relative to those ones of Gly-Gly-Gly (*Fig. 7.53*), and from the *Ramachandran* diagram for Gly residues in diverse polypeptides. This kind of diagram, in which ψ plotted over ϕ, is discussed in *Chapt. 10.5.4.2*.

7.3.3.1.2. Acyclic Polypeptides with Substituted Amino-Acid Residues

Upon introduction of side chains on C^{α}, the polypeptide backbone adopts an altered idealized conformation compared to that one of polyglycine. It has long been known that, in esters (containing the –CH(R)–O–C(=O)– group) and amides (containing the –CH(R)–NH–C(=O)– group), in the favored conformation both the C=O and C–H bonds, placed in 1,3-positions, are more or less parallel oriented. This suggestion is reminiscent of the allowed conformations of alkanes (*Sect. 7.2.2*), in which ligands bound to the 1- and 3-positions are arranged parallel, provided they are not both non-H-atoms (avoidance of *Newman* strain).

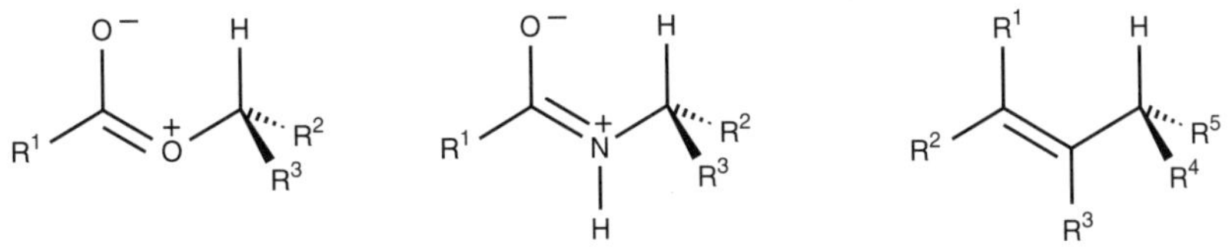

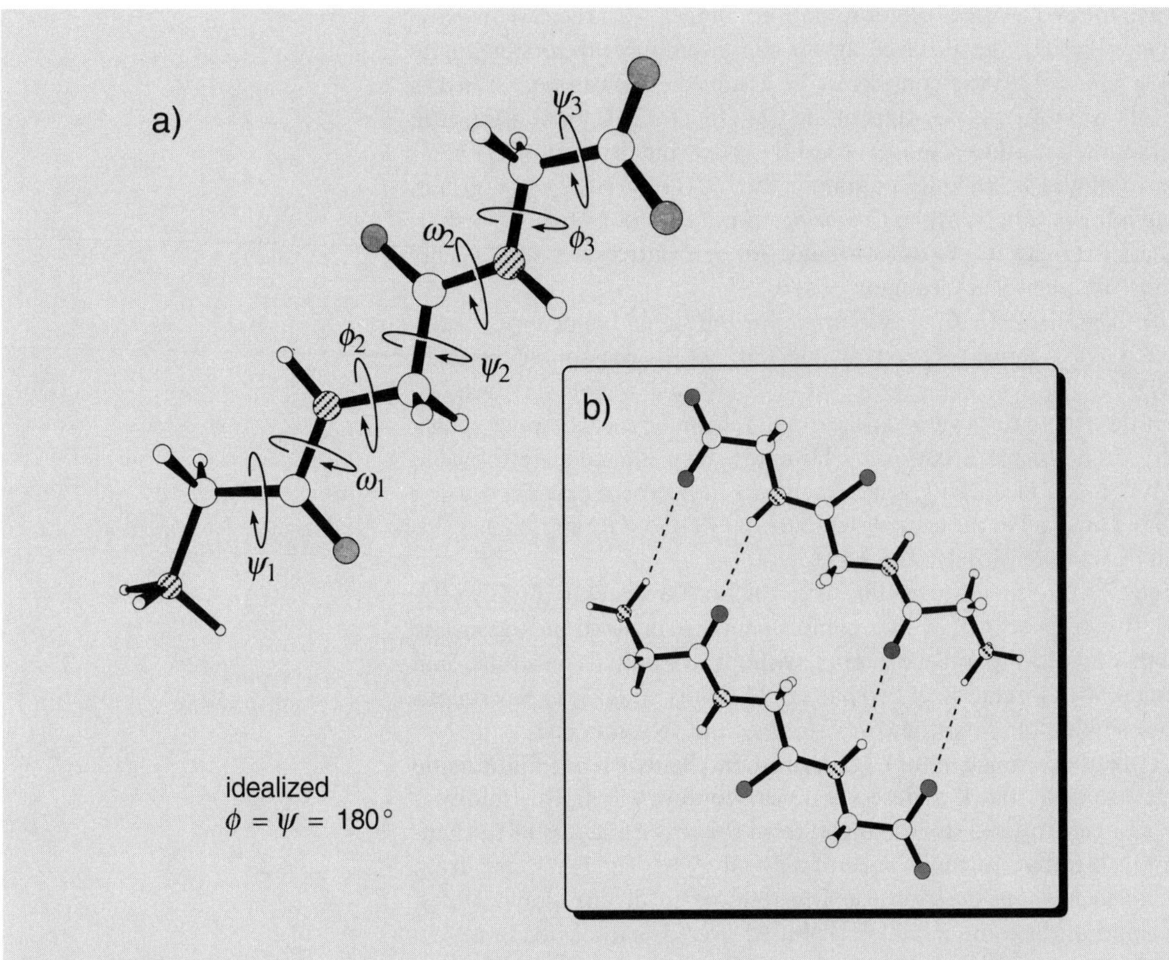

	Torsion angles [°]		
	Gly-Gly-Gly		DL-Leu-Gly-Gly
ψ_1	–150	(–162)	–130
ω_1	–176	(176)	172
ϕ_2	178	(–166)	–165
ψ_2	–172	(175)	–172
ω_2	–179	(–176)	–178
ϕ_3	173	(173)	–164
ψ_3	–173	(–169)	–177

Fig. 7.53. *Molecular structure of the Gly-Gly-Gly tripeptide* a) *idealized,* b) *in the crystal with two symmetry-independent molecules, and comparison of the torsion angles with those of* DL-*Leu-Gly-Gly each in the crystal*

From the fact that the O=C–O–C and C(=O)–O–C–H torsion angles in esters, and the O=C–N–C and C(=O)–N–C–H torsion angles in amides of relevant substitution fluctuate considerably around the value 0°, however, a very flat energy trough and facile deformability of the respective *sp*-conformation must be assumed.

The idealized conformation surmised for L-Ala-L-Ala-L-Ala is represented in *Fig. 7.54, a*. In this case as well, two complementary single strands associate together in an antiparallel ordering, without drastic conformational changes to occur (*Fig. 7.54, b*).

If the double strand section of the crystal structure of the L-Ala-L-Ala-L-Ala tripeptide is compared with that one of the Gly-Gly-Gly tripeptide, it can be concluded that the latter combination is extended, the former folded. The distance between C_1^α and C_3^α shrinks from 727 pm (718 pm) in Gly-Gly-Gly through 720 pm (699 pm) in L-Ala-L-Ala-L-Ala to 695 pm in β-poly(L-alanine).

7.3.3.2. Systematic Conformational Analysis

In their constitution, proteins are regular polymers, which – ignoring their side chains – are composed of repetitive amino-acid residues (with the backbone centers N_i, C_i^α, and C_i, *Chapt. 10.5.4.2*). The conformation of each main-chain section, comprising the five atoms C_{i-1} to N_{i+1} (*Fig. 7.55*), may essentially be described by the two sequential torsion angles ϕ and ψ. Distinction between those (ϕ, ψ) combinations which, according to the rules of conformational analysis, are 'allowed', and those which are 'forbidden' can be made in a graphic manner by means of a *Ramachandran* diagram (see *Fig. 10.27*). If particular torsion-angle pairs for sequential repeating units recur several times, then a conformationally regular polymer or polymer section is present. Conformationally regular polymers display typical elements of secondary structure. What biologists understand by *secondary structure* has been briefly referred to at the end of *Chapt. 10*.

7.3.3.2.1. Avoidance of *Newman* and *Pitzer* Strain

The two dominant requirements for
- the lowest possible *Pitzer* strain and
- the lowest possible *Newman* strain,

once more, arise as the key selection criteria for the determination of 'allowed' conformations in proteins.

Before conformational analysis can begin, it is necessary to clarify what is meant, in this context, by *Pitzer* strain. Previously, this could be done by simple reference to ethane, butane, and pentane. The non-H-atoms present were all characterized by the coordination number 4. For proteins, however, a case is encountered which has not yet been met: neither in alkanes, nor in cycloalkanes, nor in nucleic acids, nor in carbohydrates. For proteins, each repeating unit is found to contain centers (N_i and C_i) of coordination number 3, on either side of the

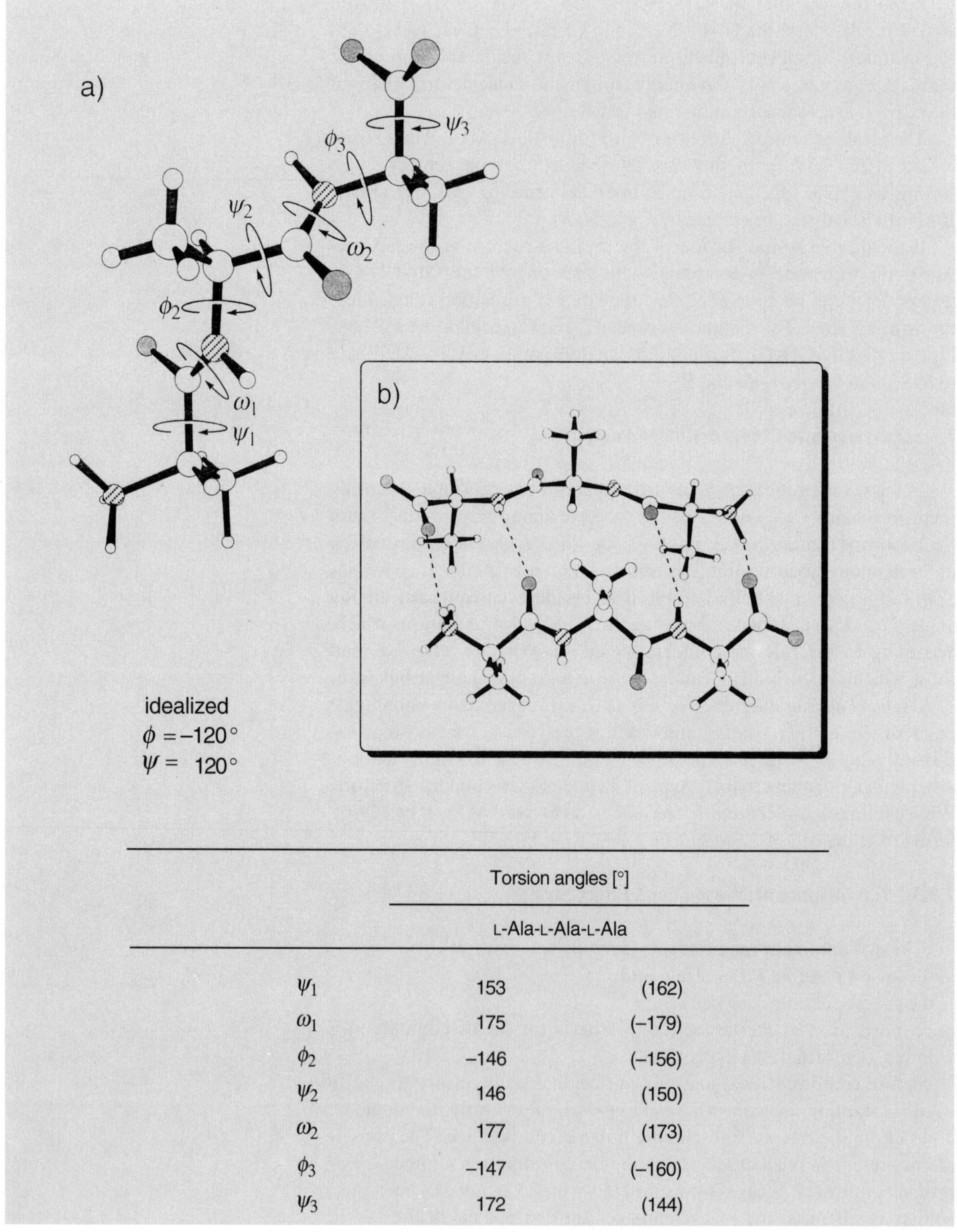

	Torsion angles [°]	
	L-Ala-L-Ala-L-Ala	
ψ_1	153	(162)
ω_1	175	(−179)
ϕ_2	−146	(−156)
ψ_2	146	(150)
ω_2	177	(173)
ϕ_3	−147	(−160)
ψ_3	172	(144)

Fig. 7.54. *Molecular structure of the* L-*Ala*-L-*Ala*-L-*Ala tripeptide* a) *idealized,* b) *in the crystal with two symmetry-independent molecules*

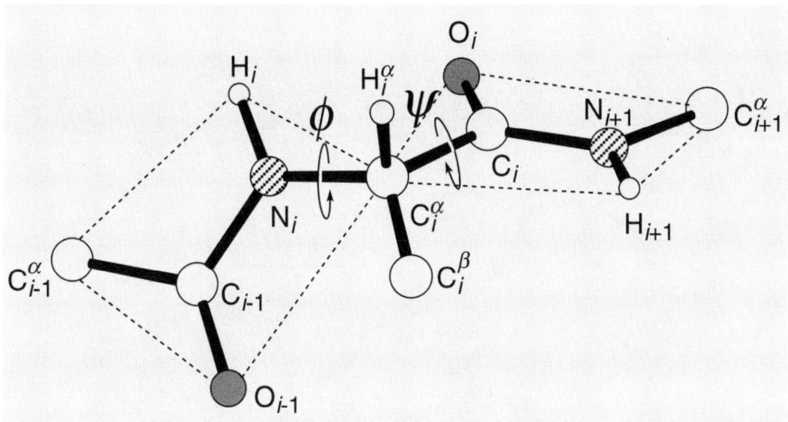

Fig. 7.55. *The determining structure element of polypeptides*

C_i^α-atom, itself of coordination number 4. Consequently, in either case, only one of the ligands – H_i or C_{i-1} at N_i, O_i or N_{i+1} at C_i – on each center of coordination number 3 may sit 'in the gap' between two ligands at C_i^α; the other is then compelled to eclipse the third ligand at C_i^α.

While butane, on adopting the torsion angles of the three staggered conformations ($\theta = \pm 60°, 180°$), is stabilized by the sum total and characterized by clear potential-energy minima, the energy minima relating to torsion about a bond between two centers of different coordination number are much less pronounced, because of opposing energy contributions. Furthermore, for butane it is known that the eclipsed arrangement of two C–C bonds is associated with higher *Pitzer* strain than the eclipsed arrangement of a C–C and a C–H bond.

Fig. 7.56, a, shows the six idealized *Newman* projections with a formal N_i=C_{i-1} bond, from which the torsion angle ϕ may easily be taken. Those conformations with $\phi = 0°$ and $\phi = 120°$ may be ignored for the conformational analysis of a polypeptide backbone: not only because of the eclipsed arrangement of the N_i–C_{i-1} and C_i^α–C_i or C_i^α–C_i^β bonds, but also especially, because of 1,5-repulsion between O_{i-1} and C_i or C_i^β.

In evaluating the local conformational regions about the torsion angle ψ with a formal C_i=N_{i+1} bond (*Fig. 7.56, b*) it should be borne in mind that two non-H-atoms are bound to the center of coordination number 3. Of the six idealized conformations, those remaining, with minimum *Pitzer* strain, are the ones in which either the C_i–N_{i+1} or the C_i–O_i bond is *sp*-oriented to the C_i^α–H_i^α bond: the conformations with $\psi = 120°$ and $\psi = -60°$.

The requirement that *Newman* and *Pitzer* strain should be minimized has reduced the number of combinations of ϕ and ψ per repeating unit from 36 to 8. Because the combinations $\phi = 60°$, $\psi = 120°$, and also $\phi = 60°$, $\psi = -60°$ lead to 1,5-repulsions between O_i and C_{i-1}, and between N_{i+1} and C_{i-1}, respectively, these may similarly be discarded (see the two (ϕ,ψ) combinations marked with an × in the *Ramachandran* diagram (see also *Chapt. 10.5.4.2*) in *Fig. 7.57*).

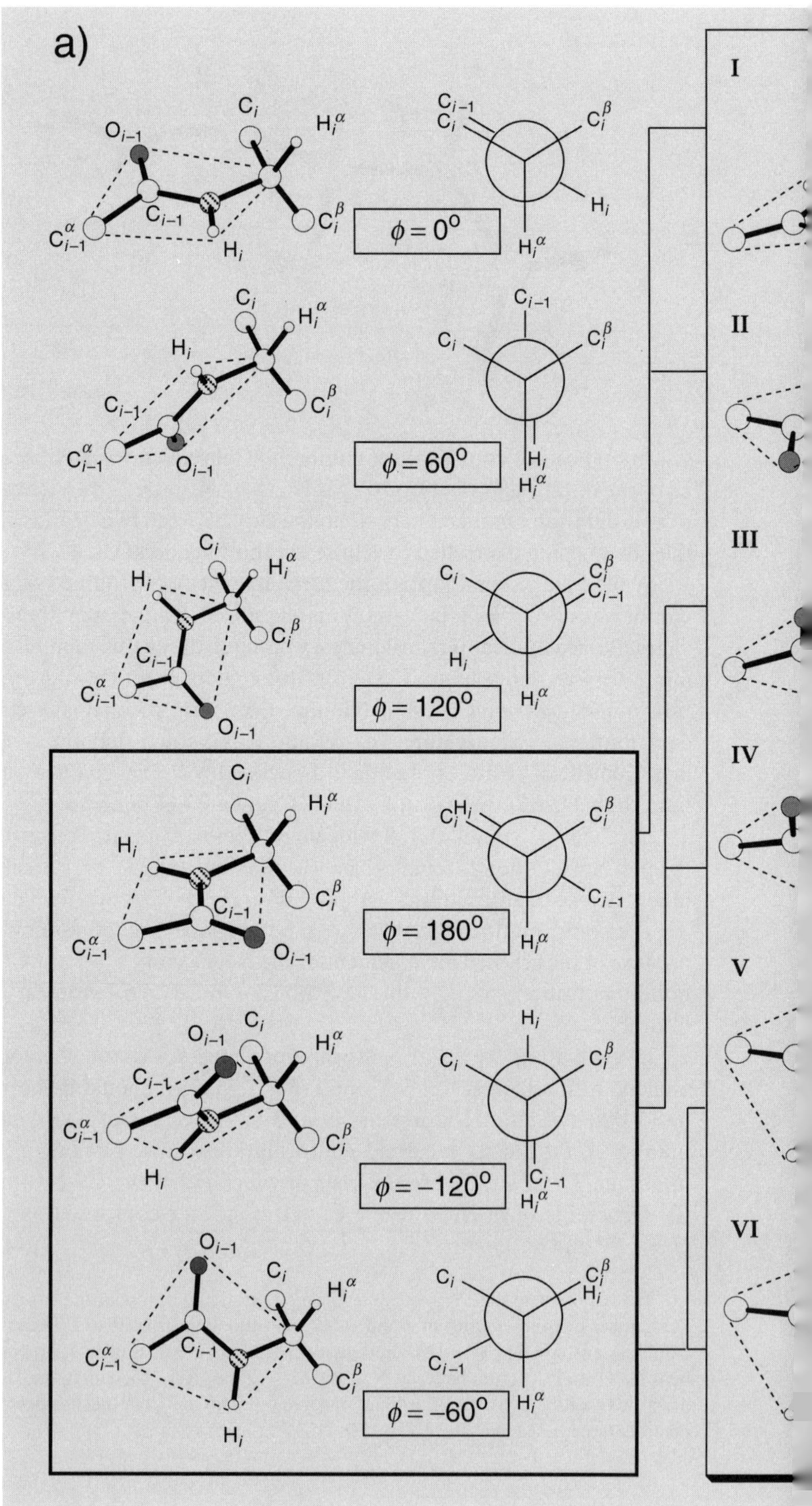
a)
O_{i-1}
C_i
H_i^α
C_{i-1}
C_{i-1}^α
H_i
C_i^β
$\phi = 0^o$
$\phi = 60^o$
$\phi = 120^o$
$\phi = 180^o$
$\phi = -120^o$
$\phi = -60^o$
I
II
III
IV
V
VI

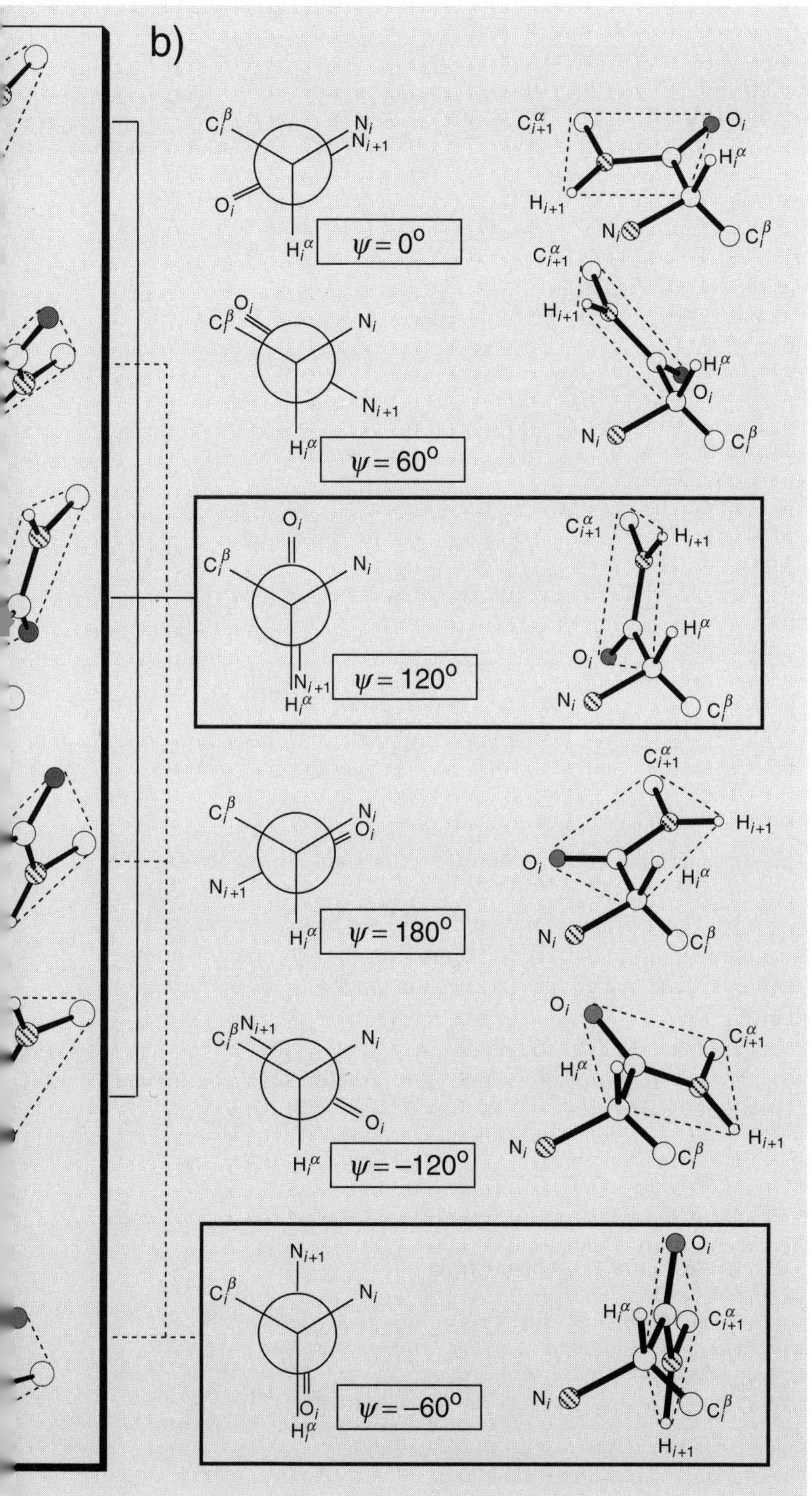

Fig. 7.56. *The six conformational sections* (**I–VI**), *remaining from the total of 36 combinations, after consideration of the requirement for minimum* Pitzer *and* Newman *strain of ϕ and ψ, each changing in 60° steps*

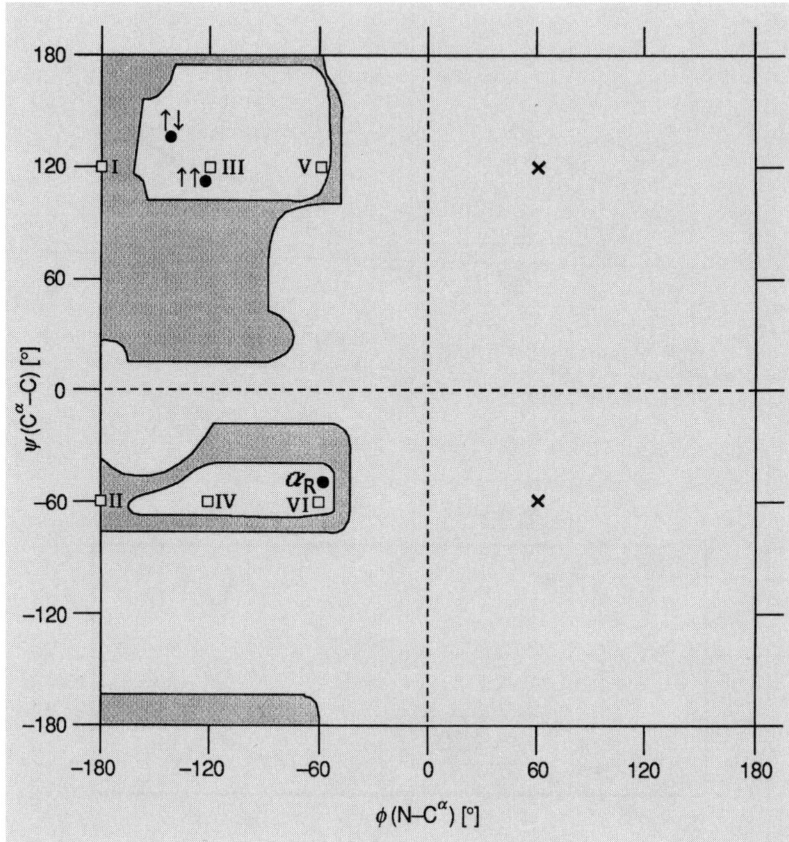

Fig. 7.57. Ramachandran *diagram with the conformational sections* **I–VI**

The six remaining conformational sections are represented as the main-chain sections **I–VI** in *Fig. 7.56*; as the five-center chain incorporating the backbone atoms of the amino-acid residues, as well as the two flanking atoms C_{i-1} and N_{i+1}. To further narrow the selection, it is appropriate to introduce two additional selection criteria, to be applied concomitantly:

- lowest possible 1,3-allylic strain and
- optimal pre-organization of the secondary structure for the formation of H-bonding patterns.

7.3.3.2.2. Avoidance of 1,3-Allylic Strain

For 1,1-disubstituted prop-2-enyl (= allyl) systems, one of the allylic bonds in the energetically favored conformation – avoiding *Newman* strain as much as possible – is *sp*-oriented to the double bond. *Francis Johnson* (1968) and *Reinhard W. Hoffmann* (1989) use the term '1,3-allylic strain', and have collected instances of 1,1-disubstituted allyl systems sufficiently varied in their structures to make apparent the rules applying for the preference of particular conformations. According to these rules,

- for 3-methylbut-1-ene, the order of stability is:

3-Methylbut-1-ene:	Reference compound for the conformational region with the torsion angle ψ	

global minimum > local minimum > maximum

with a slight preference for the conformation in the global minimum over that in the local minimum.

– for (Z)-4-methylpent-2-ene, the order of stability is:

(*Z*)-4-Methylpent-2-ene:	Reference compound for the conformational region with the torsion angle ϕ	

global minimum >> local minimum > maximum

with a strong preference for the conformation in the global minimum over that in the local minimum.

It can be seen from 2,3-dimethylbut-1-ene that a non-H-atom at C(2) does not alter the ascertained order of stabilities for 3-methylbut-1-ene.

2,3-Dimethylbut-1-ene

In either case, the conformation in the global minimum is characterized by the C(1)–H or C(1)–Me and C(3)–H bonds lying in a plane, or – to put it in another way – the C(3)–H bond being oriented in an eclipsed manner to the C(1)=C(2) bond. In the conformation in the local minimum (as in 3-methylbut-1-ene), a C(3)–Me bond is oriented in an eclipsed fashion to the C(1)=C(2) bond, provided *Newman* strain would not be hereby originated (as in (Z)-4-methylpent-2-ene). 3-Methylbut-1-ene is, of course, envisaged as the reference compound for the conformational region of the C_i=N_{i+1} bond, which includes the torsion angle ψ, while (*Z*)-4-methylpent-2-ene serves as the reference compound for the conformational region of the $N_i{=}C_{i-1}$ bond, containing the torsion angle ϕ. In consequence, it follows that the additional requirement for minimum 1,3-allylic strain is of much greater importance for the conformational region incorporating torsion angle ϕ than it is for the conformational region with torsion angle ψ.

Table 7.12 shows *prima facie* which of the main-chain sections **I**–**VI** of *Fig. 7.56* have 1,3-allylic strain in the local conformational regions for the torsion angles ϕ and ψ.

Table 7.12. *1,3-Allylic Strain in the Conformational Regions of ϕ and ψ of Conformational Sections* **I**–**VI**

Conformational section of *Fig. 7.56*	1,3-Allylic strain of the conformational region	
	for ϕ	for ψ
I	180° : +	120° : –
II	180° : +	–60° : +
III	–120° : –	120° : –
IV	–120° : –	–60° : +
V	–60° : +	120° : –
VI	–60° : +	–60° : +

Only conformational section **III** is completely free of any 1,3-allylic strain. All of the other conformational sections display 1,3-allylic strain, either in the sensitive conformational region of ϕ (**I** and **V**), in the less strain-prone conformational region of ψ (**IV**), or even in both regions (**II** and **VI**).

Regarding the complete set of causes of strain, the polypeptide strand of type **III** is energetically most favored. In comparison with polypeptide strands of types **I**, **II**, **V**, and **VI**, those ones of type **IV** are less strained. However, it is still necessary to consider the ability of each of the idealized polypeptide strands, with a backbone of regularly repeating conformational units and torsion-angle pairs from the main-chain sections **I**–**VI**, of forming as many inter- or intracatenate H-bonds between N–H donor and C=O acceptor groups as possible, without having to undergo major conformational alterations.

7.3.3.2.3. Optimal Pre-organization for the Formation of H-Bonds

The idealized main-chain sections **I**–**VI** discussed above, each with regularly recurring conformational repeating units (without allowing for H-bonds), are all helical, with the exception of **III**. *Table 7.13* summarizes the respective helix parameters (see *Fig. 7.10*), notes their handedness, and indicates in which instances the formation of stabilizing H-bonds may be expected to lead to a regular pattern of idealized secondary structure.

Table 7.13. *Helix Parameters for the Idealized Polypeptide Strands of Conformational Sections* **I–VI**

Polypeptide type (see *Fig. 7.56*)	Helix				H-Bonds	
	Handedness	n	h [pm]	p [pm]	intracatenate	intercatenate
I (180°, 120°)	right	2.7	344	929		
II (180°, −60°)	left	3.9	252	983		
III (−120°, 120°)	—	2.0	331	662		optimal
IV (−120°, −60°)	left	5.0	97	485	$N_i \cdots O_{i+4}$ 323 pm $H_i–N_i–O_{i+4}$ 60°	
V (−60°, 120°)	left	2.8	288	806		
VI (−60°, −60°)	right	4.0	119	476	$N_{i+4} \cdots O_i$ 253 pm $H_{i+4}–N_{i+4}–O_i$ 39° $N_{i+5} \cdots O_i$ 325 pm $H_{i+5}–N_{i+5}–O_i$ 28°	

Assuming that the formation of idealized, regular patterns of secondary structure by means of H-bonds occurs most frequently in those idealized main-chain sections which, in doing so, undergo the smallest possible degree of conformational alteration, **III** and **VI** would be reckoned to be the favored conformational sections.

From the *Ramachandran* diagram in *Fig. 7.57*, it may be seen that the conformation of the parallel aligned polypeptide single strands in a β-pleated sheet only deviates insignificantly from the idealized, linear conformation of the polypeptide single strand of type **III**. Similarly, the right-handed helix of idealized polypeptide backbone of type **VI** differs only slightly from the right-handed α-helix widely distributed among polypeptides.

7.3.3.3. Pleated Sheets

The main-chain section of type **III** (*Fig. 7.58*) is free of 1,3-allylic strain. It is, moreover, optimally suited for self-organization: with oligopeptide single strands of the same type, held together in supramolecular β-pleated sheets by intercatenate H-bonds.

The four centers C^{α}_{i-1}, C_{i-1}, N_i, and C^{α}_i of the polypeptide backbone, together with their bonds to H^{α}_{i-1}, O_{i-1}, H_i, and H^{α}_i, lie in a plane. This has the effect that each of two bonds among the mentioned ones, at centers next to each other are *ap*-oriented, at centers next but one to each other are parallel oriented.

The $C^{\alpha}_i–H^{\alpha}_i$ bond is part of two neighboring planes. It is flanked on one side by the $C_{i-1}–O_{i-1}$ bond, and on the other by the $N_{i+1}H_{i+1}$ bond. If the three-dimensional space is divided by the plane including the C=O groups into two half-spaces, the side chains at the C^{α}-atoms of the polypeptide backbone alternately change half-space.

Just as short single strands in the crystal lattice of oligopeptides are able to arrange themselves into supramolecular structure patterns by means of inter- or intracatenate H-bonds, proteins are often in situations

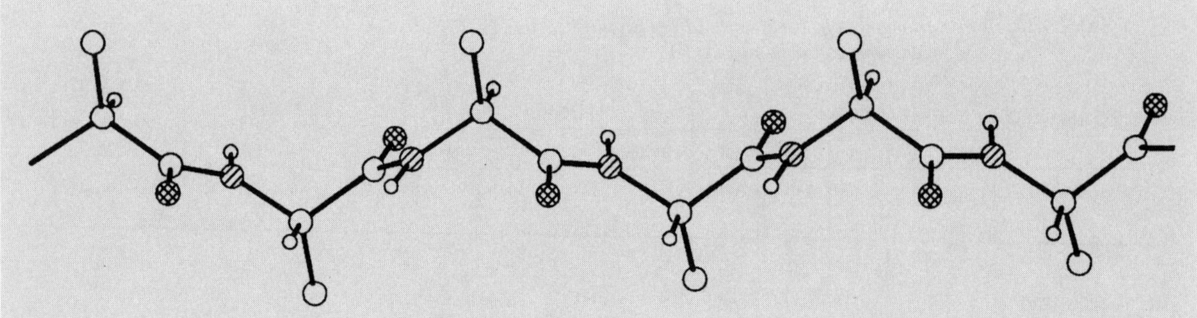

Fig. 7.58. *Idealized, linear polypeptide chain with regular recurrence of the torsion angles* $\phi = -120°$, $\psi = +120°$ (with conformational section **III** of *Fig. 7.56*)

where they may make so-called pleated sheets by formation of intercatenate H-bonds. Here, the strand sequences, self-organizing into the indicated pattern of secondary structure as a concomitant of the formation of intercatenate H-bonds may be oriented parallel (*Fig. 7.59*) or antiparallel (*Fig. 7.60*).

In antiparallel pleated sheets, the H-bonds are perpendicular to the polypeptide chain, in such a way that, after every small distance between two adjacent H-bonds, there follows a larger distance.

While the combination of an 'idealized pair of (ϕ,ψ) torsion angles' comes to $\phi = -120°$, $\psi = 120°$, the combination of a 'canonical pair of torsion angles' amounts to $\phi = -119°$, $\psi = 113°$ for the parallel and $\phi = -139°$, $\psi = 135°$ for the antiparallel pleated sheet. 'Canonical pair of torsion angles' means the particular pair obtained by statistical analysis of protein crystal structures in which the H-bonds are present.

7.3.3.4. Helices

Because polypeptide chains are endowed with a wealth of donor (N–H) and acceptor (C=O) groups for the formation of H-bonds, an intracatenate type of secondary-structure pattern is to be anticipated, as soon as the necessary prerequisites are met. These are:

- that the polypeptide chain is long enough for intracatenate H-bond formation,
- that the polypeptide chain is pre-organized in such a manner that intracatenate H-bonds between sequentially remote, but spatially close atoms may smoothly be formed without major alterations in conformation.

The polypeptide strand of type **VI** (see *Fig. 7.61* and cover illustration), as a right-handed helix, is optimally pre-organized for the formation of intracatenate H-bonds. Without allowing for H-bonds of this type, the N···O distance is 253 pm and the H–N···O angle 39°. An insignificant conformational alteration leads to the natural $3{,}6_{13}$-helix (*α-helix*; *Fig. 7.62*). For this fine adjustment, the N···O distance over the now permitted H-bond is lenghtened from 253 pm to 302 pm, the

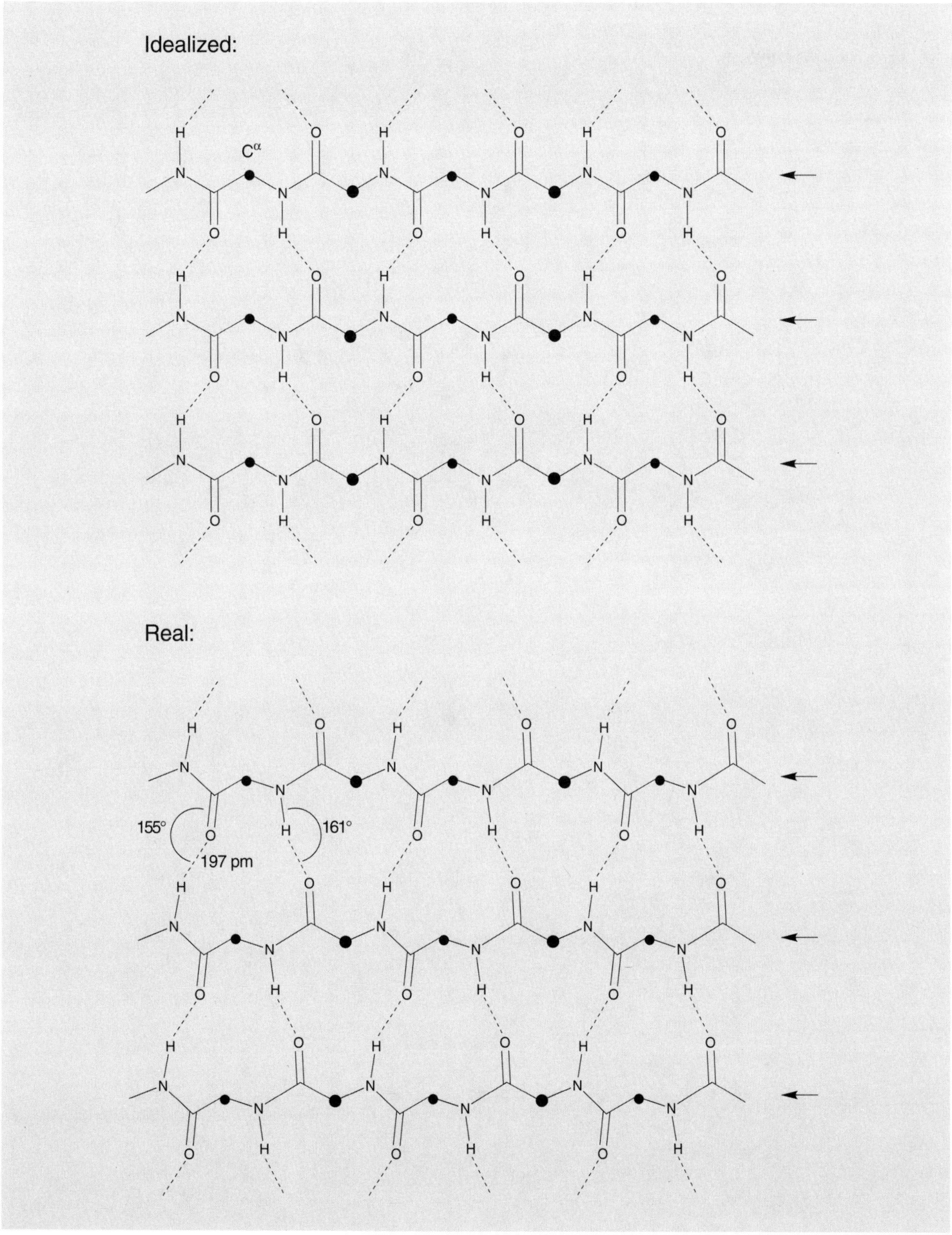

Idealized:

C^{α}

Real:

155°

161°

197 pm

Fig. 7.59. *Parallel β-pleated sheet*. The C^{α}-atoms represented as large (small) circles are placed above (below) the plane of the paper

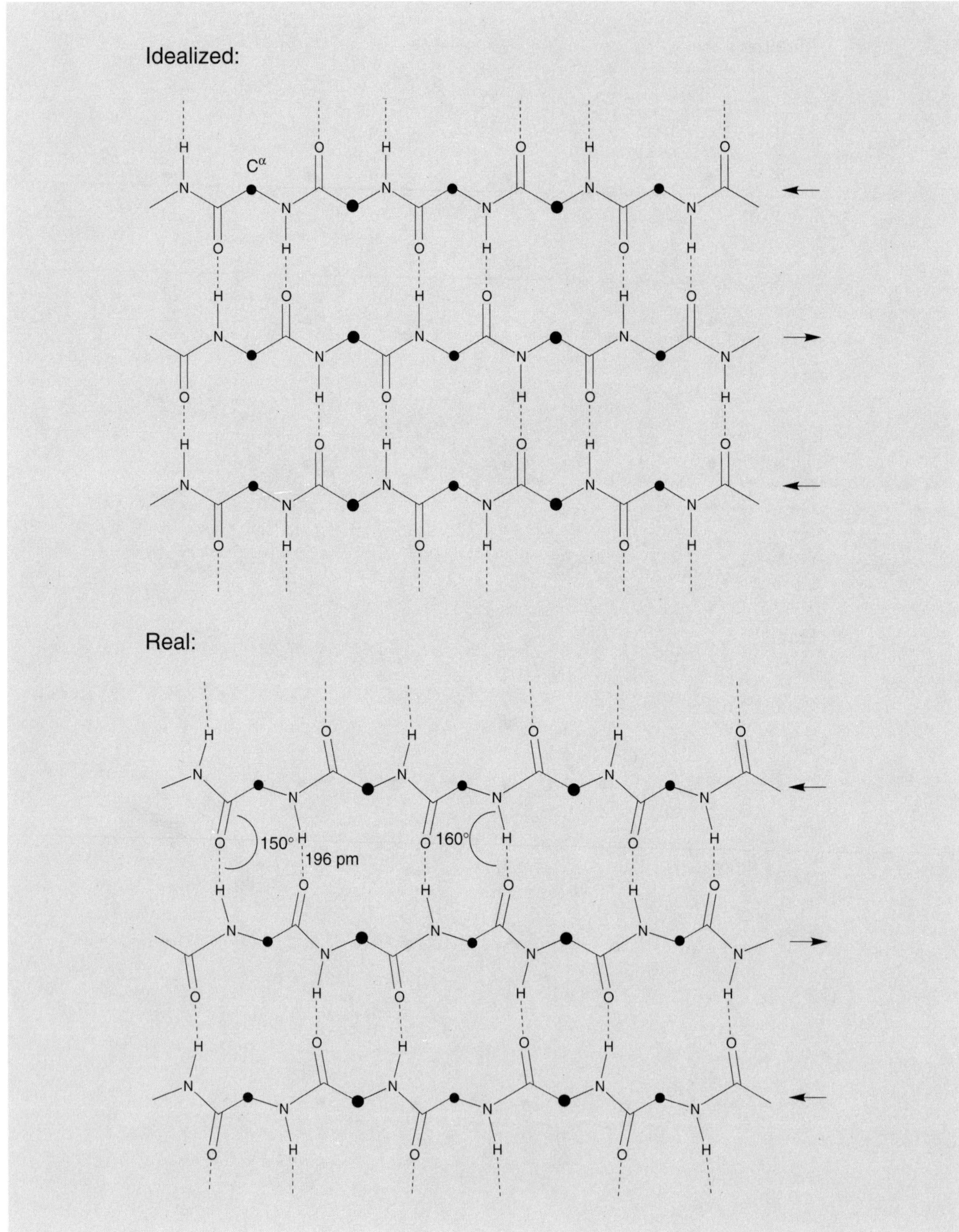

Fig. 7.60. *Antiparallel β-pleated sheet*. The C^{α}-atoms represented as large (small) circles are placed above (below) the plane of the paper

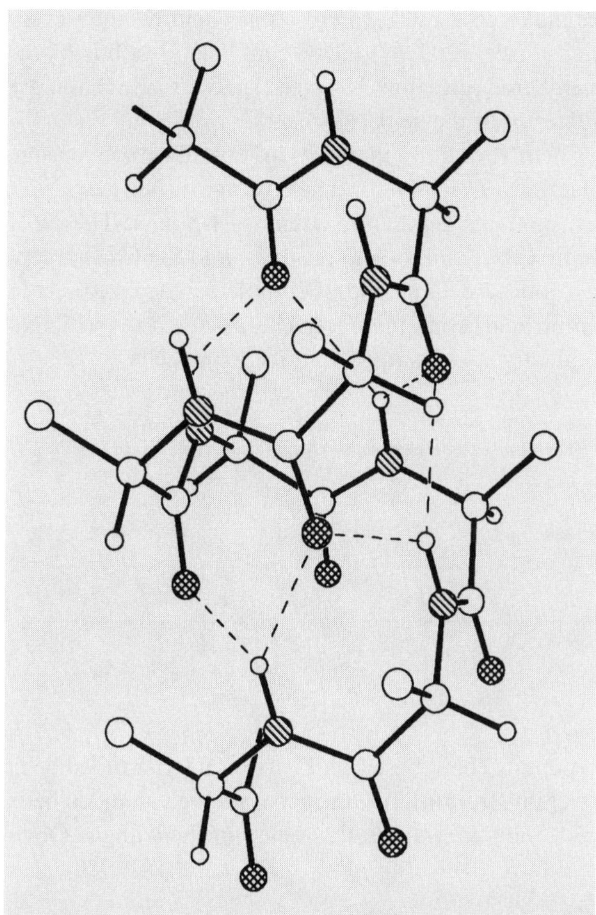

Fig. 7.61. *Idealized, right-handed 4_{13}-helix with regularly recurring torsion angles $\phi = \psi = -60°$* (with conformational section **VI** of *Fig. 7.56*) *without allowance for H-bonds*

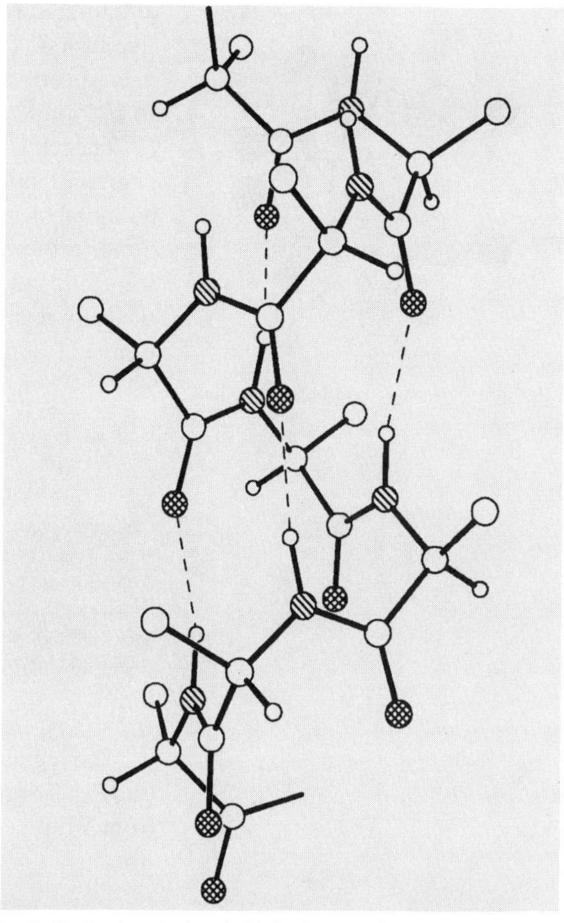

Fig. 7.62. *Real, right-handed $3{,}6_{13}$-helix with regularly recurring torsion angles $\phi = -57°$, $\psi = -47°$, with allowance for intracatenate H-bonds*

H–N···O angle diminishing from 39° to 9°. (An idealized polypeptide H-bond being characterized by an H–N···O angle of 0° and an N···O distance of 290 pm.)

The torsion angles of the idealized type-**VI** helix lie in the vicinity of the 'canonical pair of torsion angles': $\phi = -57°$, $\psi = -47°$ (see *Table 7.12*). The deviation of the 'canonical pair of torsion angles' from the 'idealized pair of torsion angles' ($\phi = -60°$, $\psi = -60°$) has the effect of reducing the number of constitutional repeating units per helix turn (n in *Fig. 7.10*): the helix becomes more slender.

Each H-bond in the α-helix is formed between the carbonyl O-atom of an nth amino-acid residue and the N–H group of the $(n + 4)$th amino-acid residue. They are oriented virtually parallel to the helix axis, in which every C=O group points towards the C-terminus. The helix, therefore, exhibits a dipole moment: positive at the N-terminus, negative at the C-terminus. Each amino-acid residue is shifted along the helix axis by $h = 150$ pm from its neighbors, on rotation of 100°, so that 3.6

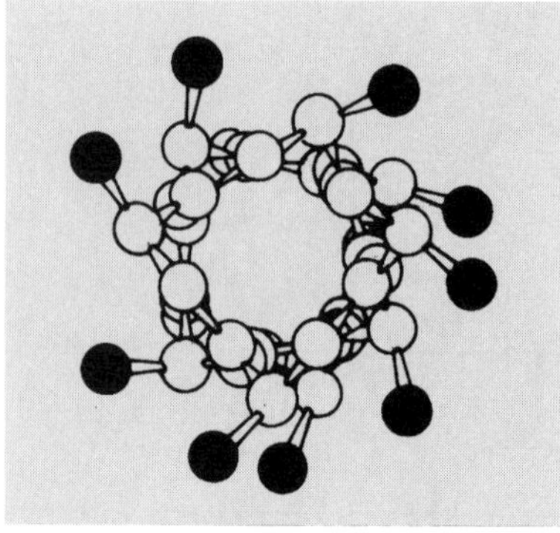

Fig. 7.63. *α-Helix* (viewed along the helix axis)

amino-acid residues make up a full turn (18 amino-acid residues every five turns). The pitch p (see *Fig. 7.10*) is 540 pm. The H-bonds of the α-helix form a 13-membered ring (thus $3{,}6_{13}$-helix) and the side chains are to be found on the exterior of the helix (*Fig. 7.63*).

The tendency for helix formation increases first of all with increasing degree of polymerization (*Table 7.14*). The average α-helix, with 12 amino-acid residues, contains eight intracatenate H-bonds. The N–H donor groups in the first four amino-acid residues, and the C=O acceptor groups in the last four, are all devoid of complementary partners in the helix region. Amino acids containing side chains suited to participation in H-bonds frequently act as 'stopgaps' outside the helix.

Table 7.14. *Number of H-Bonds Dependent on the Number of Amino-Acid Residues in the α-Helix*

Number of amino-acid residues	< 5	5	6	7	8	20	40
Number of H-bonds	0	1	2	3	4	16	36
% of the intramolecularly utilized H-bonds	0	20	33	43	50	80	90

The regular polypeptide chain based on **IV** forms a left-handed helix (see *Fig. 7.64* and cover illustration). In contrast to the right-handed helix from **VI**, it is free of 1,3-allylic strain in the region of the ϕ angle. Obvi-

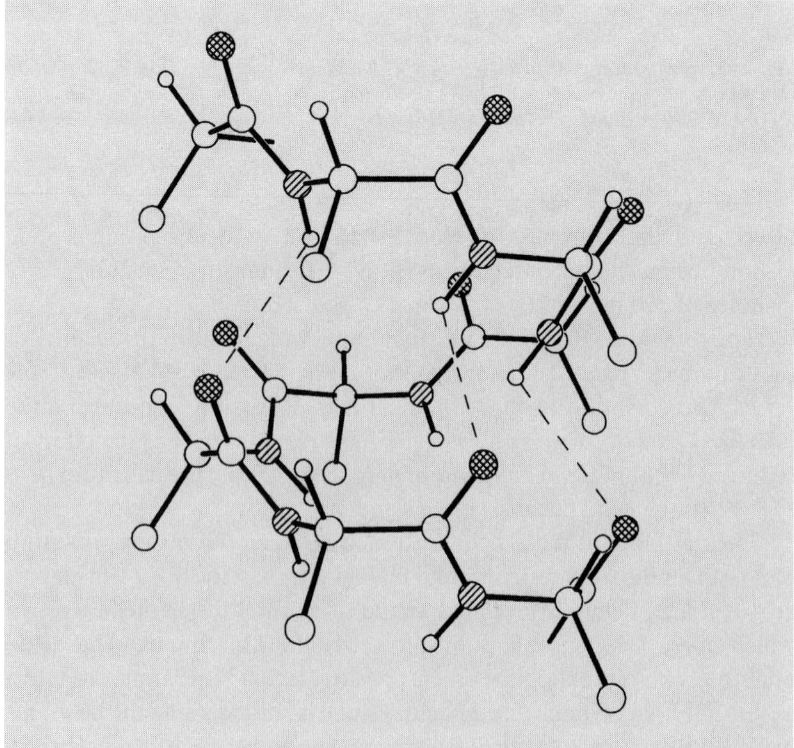

Fig. 7.64. *Idealized, left-handed 5_{17}-helix with regularly recurring torsion angles $\phi = -120°$, $\psi = -60°$* (with conformational section **IV** of *Fig. 7.56*)

ously, this advantage does not suffice for favoring of **IV** over **VI**, as the N···O distance of 323 pm and the H–N···O angle of 60° are too far away from the ideal values in an H-bond (in amides).

The *Brookhaven Protein Database* does give some (ϕ,ψ) combinations of −120°, −60° in amino-acid (except for Gly and Pro) residues of more than 2500 proteins. But no instance is to be found in which the pairs of torsion angles characteristic of **IV** occur more than three times in a row: consequently, it is not possible to speak of a helix. Formation of a helix clearly requires the cooperative formation of a whole series of H-bonds, and this requires an optimally pre-organized polypeptide backbone. On consulting the above mentioned database, it can be seen that polypeptide strands of type **V** (*Fig. 7.65, c*) do show a weak tendency for helix formation. In contrast, polypeptide strands of types **I** and **II** (*Fig. 7.65, a* and *b*) are not sufficiently preshaped for self-organization by formation of H-bonds of intercatenate or intracatenate character, and are not disposed to undergo transformation to conformational types of regular secondary structure. The histograms (*Fig. 7.66*) make clear that (ϕ,ψ) pairs of torsion angles characteristic for the pleated sheets and the α-helix vastly predominate over all others (*a*), especially upon inquir-

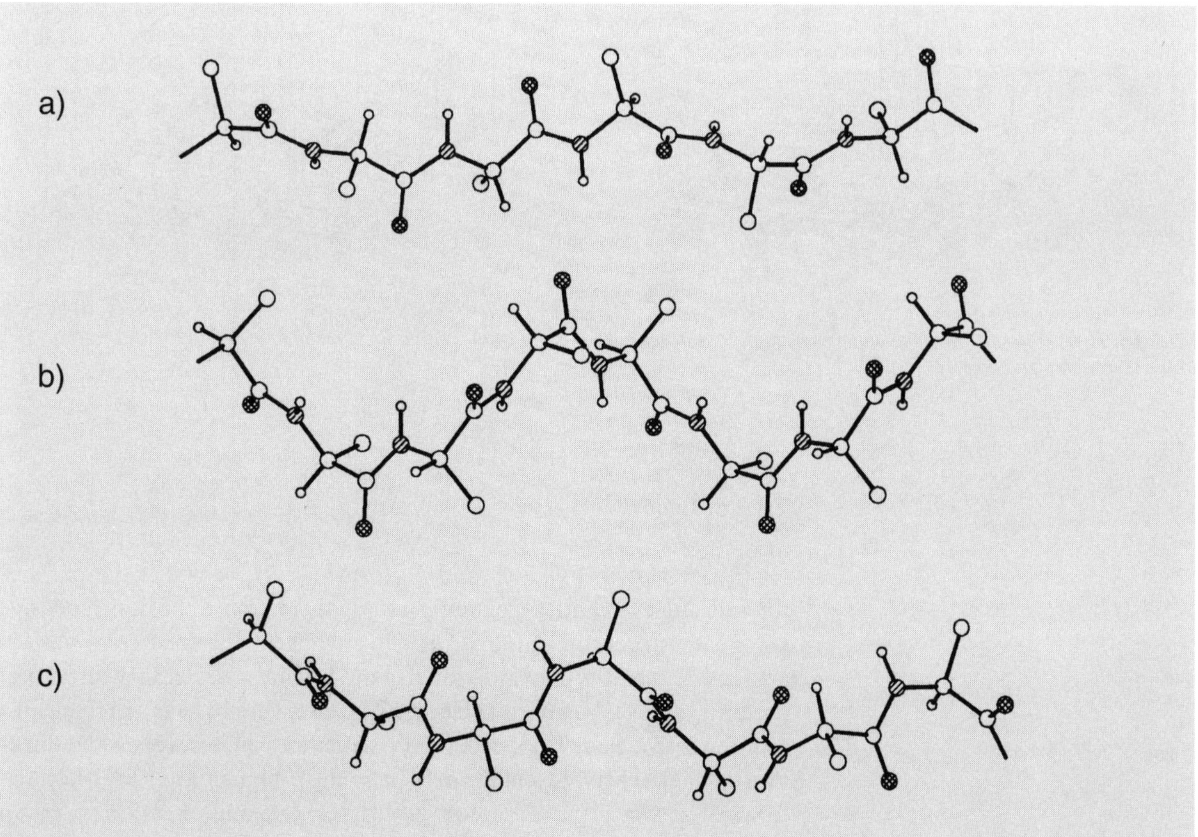

Fig. 7.65. *Idealized, helical polypeptide chains with regularly recurring torsion angles* a) $\phi = 180°$, $\psi = +120°$ (right-handed, with conformational section **I** of *Fig. 7.56*); b) $\phi = 180°$, $\psi = -60°$ (left-handed, with conformational section **II** of *Fig. 7.56*); c) $\phi = -60°$, $\psi = +120°$ (left-handed, with conformational section **V** of *Fig. 7.56*)

ing how often the same pairs of torsion angles follow at least four times in succession (*b*).

Experience has shown that Ala, Glu, Leu, and Met favor the formation of an α-helix, while Gly, Pro, Ser, and Tyr do not. α,α-Disubstituted glycine derivatives in which one of the two substituents is a Me group, or where both substituents are part of one and the same three-, five-, or six-membered ring, favor the formation of α-helical conformations as soon as they are incorporated into a polypeptide.

Besides the $3{,}6_{13}$-helix, there also exists the less widespread right-handed 3_{10}-helix. In this, a H-bond between the C=O group of an nth and the N–H group of the $(n+3)$th amino-acid residue completes a ten-membered ring.

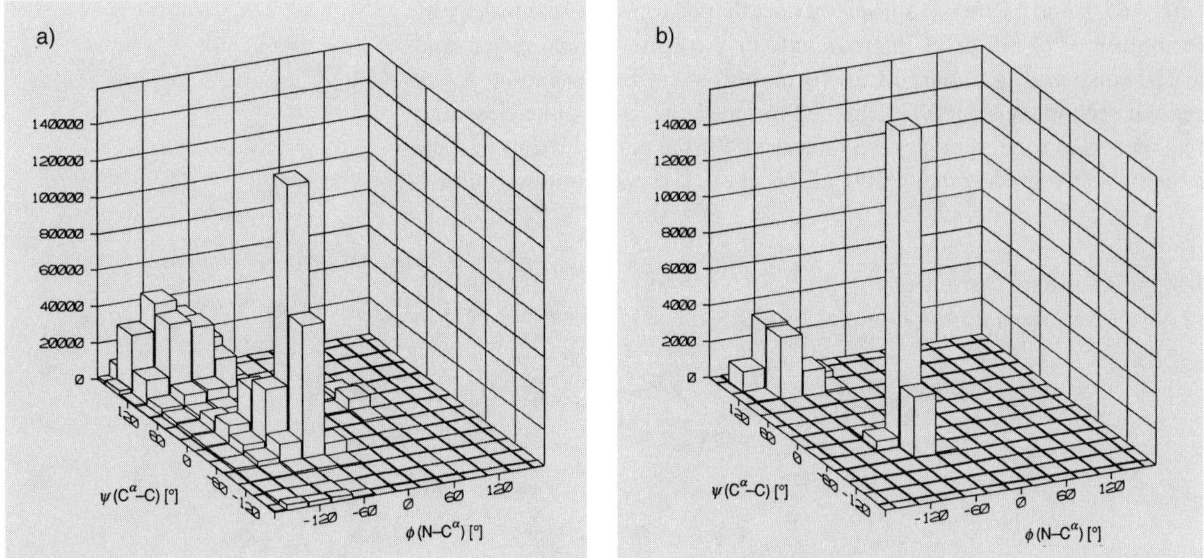

Fig. 7.66. *Histograms of* (ϕ,ψ) *combinations in protein structures* a) *for individual amino-acid residues* (except Gly and Pro), b) *for repetitive structures* (*cf. Fig. 7.57*)

7.3.3.5. Loops

In contrast to *α-helices* and *β-pleated sheets*, in which the pairs of torsion angles ϕ and ψ regularly recur, there also exist non-repetitive secondary-structural patterns, in which each of the successive amino-acid residues has its own individual ϕ and ψ values. Secondary-structural patterns of this type are called loops. The term should be taken to mean a non-repetitive element of secondary structure which reverses the direction of a polypeptide chain, and in such a manner that a H-bond is formed between the acceptor group $(C{=}O)_i$ and the donor group $(N{-}H)_{(i+n)}$. *Table 7.15* gives the relative distances between acceptor and donor groups for the so-called *α*-, *β*-, and *γ-loops*, together with the number of centers per H-bridged loop.

Table 7.15. *Characterization of Various Types of Loops*

Trivial notation	H-Bonds between amino-acid residues	Ring size
α-Loops	$5 \rightarrow 1$	13
β-Loops	$4 \rightarrow 1$	10
γ-Loops	$3 \rightarrow 1$	7

The *α-loop* resembles a section from an *α-helix*. The C=O group of an *n*th amino-acid residue is connected by an H-bond to the N–H group of the $(n+4)$th amino-acid residue, completing a 13-membered ring. *β-Loops* contain an H-bond between the C=O group of an *n*th amino-acid residue and the N–H group of the $(n+3)$th amino-acid residue in a ten-membered ring. *Fig. 7.67* shows two versions of a β-loop.

The β_I-type in *Fig. 7.67* can be adopted by any of the proteinogenic amino acids. Here the second amino-acid residue takes on an α-helical conformation, the third a 3_{10}-conformation.

In the β_{II}-type (*Fig. 7.67*), a 'forbidden' local conformational fragment – 1,5-repulsion between the O-atom of the second amino-acid residue and the C^{β}-atom of the third one – is only avoided when Gly plays the part of the third amino acid. Gly is indeed frequently found in this position in *β-loops*. Another option for avoiding this 1,5-repulsion is having the D-configuration present in the second or third amino acid. Finally, the repulsion is also avoided if the C_2^{β}- and/or C_3^{β}-centers are part of a ring: when the role of second and/or third amino acid is taken by Pro.

The rarely occurring *γ-loop* contains an H-bond connecting the C=O group of an *n*th amino-acid residue with the N–H group of the $(n+2)$th amino-acid residue.

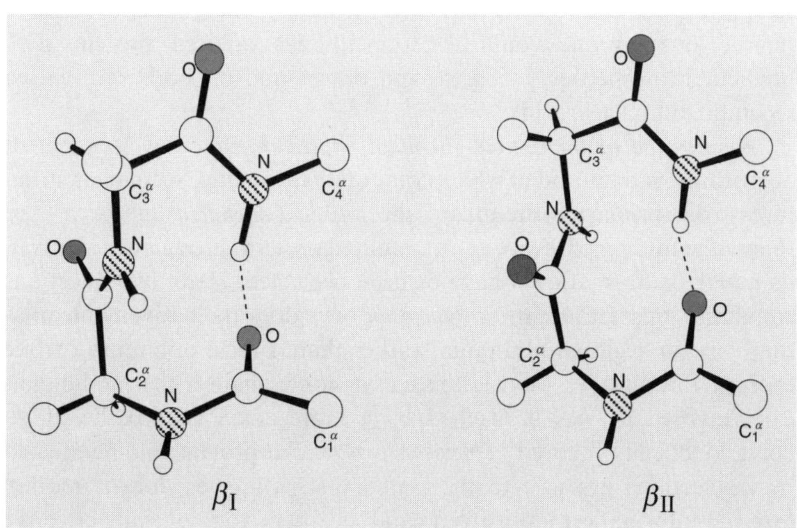

Fig. 7.67. *β-Loops*

Using NMR spectroscopy to measure NOE effects, it is possible, by determining specific distances between relevant protons, to identify both the repetitive elements of regular secondary-structure as well as irregular secondary-structure patterns (*Chapt. 13.4*).

7.3.3.6. Protein Folding

The polydimensional conformational space, which is available to a regular macromolecule with a linearly constituted backbone containing the ensemble of all torsion-angle combinations, can exceedingly be reduced by systematic conformational analysis relying only on 'allowed' combinations of torsion angles. This has been demonstrated on the examples of pyranosyl-oligonucleotides (*Sect. 7.3.1.1*) and oligopeptides (*Sect. 7.3.3.2*).

The α-helix and the two types of β-pleated sheets, with their characteristic patterns of H-bonds, were proposed as early as 1951 by *L. Pauling* and *R. B. Corey*, using insight gained from X-ray crystal-structure analysis to design a blueprint for the construction of physically relevant molecular models.

Discrete conformations are assigned to oligopeptides, even at the stage of consideration of primary structure by systematic conformational analysis. They should be of great value as starting conformations for the study of the early stages of protein folding in which a substantial amount of secondary-structure formation already occurs. (See *Chapt. 10.5.4.2* for the meaning of the terms primary and secondary structure.)

In previous efforts to predict the secondary structure of any given polypeptide of known primary structure, using statistical analysis of already solved protein structures, appropriate algorithms for computer-assisted pattern recognition have only had a success rate of about 50%. Apparently, neural network techniques have very recently been able to increase the rate to almost 70%. Even this result, however, remains far from the objective of direct derivation of biologically active tertiary structure from amino-acid sequence. In reality, the goal is even more remote: one day one would like to synthesize 'tailored' proteins with predictable properties, using protein design and methods of creating recombinant nucleic acids.

Besides the *mathematical-chemical approach* – the development of algorithms, with the aid of which computers can deduce secondary structures from primary structures – the *physical-chemical approach* – to follow the folding pathway of an unfolded protein through transients to its native conformation – has not been neglected. Here, two questions command greatest attention. *Question one*: does the native conformation exist in a global minimum, rather than a local minimum of free energy? Put another way: is the native conformation thermodynamically favored or was it reached by a more convenient pathway, so being kinetically favored? *Question two*: is the protein folding steered by instructions intrinsic to the primary structure, or does it require external information input from catalytically or topologically effective folding auxiliaries? Because general answers, valid for all instances, cannot be expected, some particularly well documented cases will be emphasized.

Ribonuclease A (*RNase A*) is a protein with only 124 amino-acid residues, among them eight cysteins. The protein may be denatured using excess 2-mercaptoethanol in 8M urea solution. Four disulfide bridges are reductively cleaved.

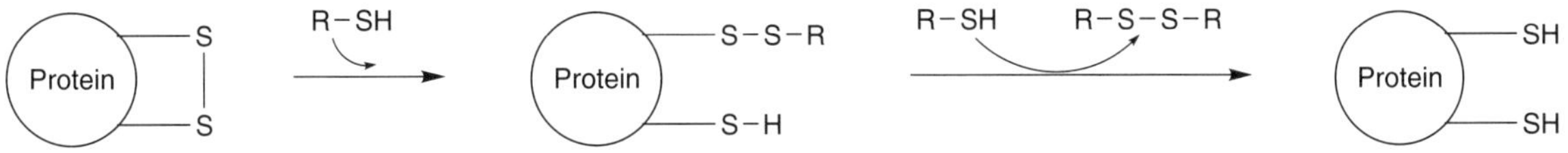

The resulting mixture of random conformations shows practically no more enzymatic activity. In the absence of 2-mercaptoethanol and urea, the purified protein reverts into completely active RNase A within a few hours. The four S–S bridges come back into being during this renaturation. The renaturation requires considerable time, as eight Cys residues have 105 formal options open to them in forming four disulfide bridges. In contrast, *in vivo* synthesis of RNase A requires only a few minutes. The native conformation is probably situated in the global minimum, and the pathway to it smoothed by a disulfide isomerase. *Protein Disulfide Isomerases* (*PDI*) accelerate – in other cases as well – posttranslational modifications of binding and cleaving of disulfide bridges, functioning as cellular catalysts. They are widely distributed in the lumen of the *endoplasmic reticulum*, an intracellular membrane system comprising about half of the total cellular membrane content. For this trail-blazing studies into RNase A, *Christian B. Anfinsen* received the *Nobel* prize for chemistry in 1972.

At room temperature, *in vitro* protein folding, the slow isomerization of prolyl–peptidyl bonds is a commonly encountered phenomenon. For example, one such case is to be found in the refolding of *RNase T1* from *Aspergillus oryzae*. *Gunter Fischer* found out in 1984 that refolding of this kind was accelerated by *Peptidyl-Prolyl Isomerases* (*PPI*). The presence of PPI has been demonstrated in all organisms and tissues, so that their catalytic *in vivo* action may be assumed.

Pro is the only one of the proteinogenic amino acids in which the amide N-atom (after incorporation into a peptide) is completely substituted. This has consequences for the structure and dynamics of peptides with Pro residues. The Pro residue occupies an extreme position in the *Ramachandran* diagram, and the barrier for isomerization about the prolyl–peptidyl bond is somewhat lower than in peptides without Pro. While the *ap*-conformation practically never (0.1%) occurs in peptides in which Pro does not play a part, it is encountered in Pro-containing peptides in equilibrium (between 10 and 30%). In the native conformations of structurally well studied, biologically active proteins, the *antiperiplanar* prolyl–peptidyl bond is present to an extent of at least about 7%.

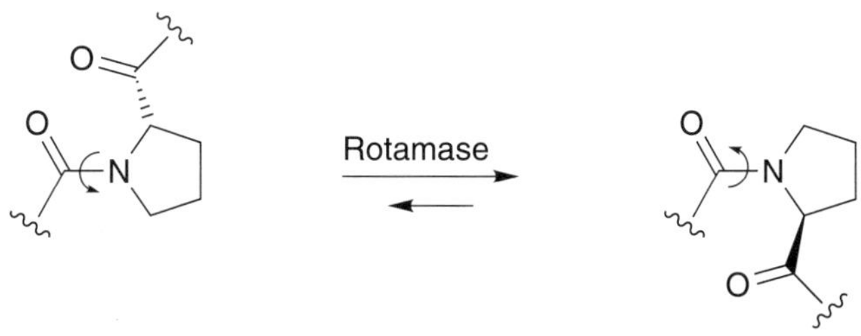

A 'twisted' amide bond is assumed as transition structure for *sp*⇌*ap* isomerization about the prolyl–peptidyl bond. This also applies for enzymatic isomerization by PPI.

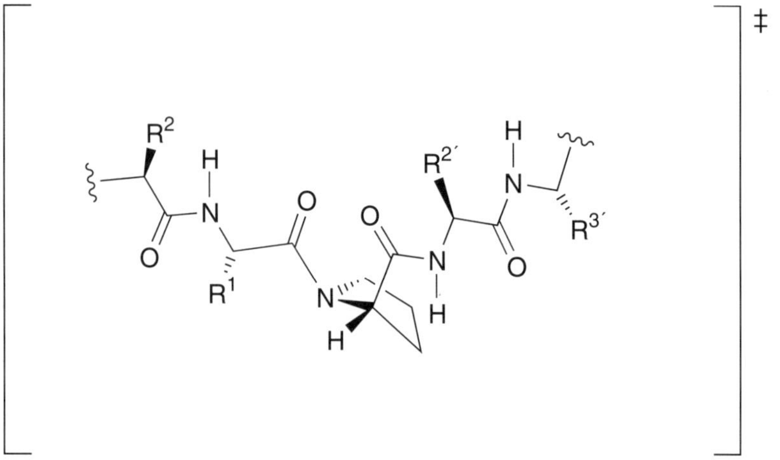

Finally, comparison of the wild type of RNase T1 with one of its variants in which no *ap*-oriented prolyl–peptidyl group was present demonstrated the rate-limiting character of *in vitro* isomerization and the catalysibility of this rearrangement by PPI.

After many single observations, the impression grows that protein folding in a reduced conformational space effectively proceeds after local elements of secondary structure have come into being in a sequence-specific manner. These local regions of secondary structure act as part of a scaffold, on which the further folding process can proceed more easily. The rate-limiting steps of the entire folding process occur shortly before reaching the native conformation, with its hydrophobic nucleus and involve processes concerned with docking one substructural section onto another. Towards the end of the pathway from the unfolded protein to the native conformation, an ensemble of conformations, in equilibrium with the native conformation, is encountered. It is advisable, therefore, to regard proteins as fluctuating systems of a whole ensemble of conformers.

Is *in vivo* self-organization of proteins so effective that only the native, biologically active conformation arises? This question cannot be answered unambiguously. However, it is assumed that enzymes in the endoplasmic reticulum relatively quickly can degrade incorrectly folded proteins, which are spatially disposed to the action of *proteases* (*hydrolases* able to cleave peptide bonds) in a favorable way.

In order that proteins synthesized in a cell interior may pass through membranes, and that particularly long protein chains do not associate incorrectly – judged from the perspective of the sought-after native conformation – through intra- or intercatenate association, there exists a group of function-related proteins which plays an auxiliary role. Why this group of folding supporters, or better 'incorrect-folding inhibitors' has been called *molecular chaperones* follows from an explanation of the meaning of the word recently put forward:

'The word chaperone is normally used to describe a particular and largely outdated form of social behavior by human beings. The *Oxford English dictionary* (2nd edition) describes chaperone, or more correctly chaperon, as a person, usually a married or elderly woman, who, for the sake of propriety, accompanies a young unmarried lady in public as guide and protector. Thus, the traditional role of the human chaperone, if described in biochemical terms, is to prevent improper interactions between potentially complementary surfaces.'

These chaperones selectively form supermolecules with freshly synthesized proteins, subject to their folding state, in order to suppress detrimental aggregation or assist desired conformational alterations. The action of two chaperones, already known under the name of *heat-shock proteins* (*Hsp*), on the protein *rhodanase (thiosulfate-sulfur transferase)* provides an example of this. *Hsp 70* initially binds to unfolded rhodanase, *Hsp 60* afterwards binds to the partially folded rhodanase. The first chaperone endeavors to inhibit deviation from the 'prescribed' folding pathway, the latter chaperone is oriented towards finding the 'correct' folding pathway. Meanwhile, is has been possible to verify that folding auxiliaries of this type intervene not only in easily studied *in vitro* folding, but also in more arduous *in vivo*-folding experiments, in which these auxiliaries are present in substoichiometric concentrations only.

In short: for the three important types of slow processes in protein folding – disulfide rearrangement, isomerization of prolyl–peptidyl bonds, and continual association of regions of secondary structure into domain patterns – there are available for the cell three types of folding supporters or 'incorrect-folding inhibitors': PDI, PPI, and chaperones.

7.4. Helicates

Duplex formation in nucleic acids is the most prominent example of molecular recognition. Chain-shaped molecules constitution-specifically interact as complementary partners and spontaneously organize into supermolecules. Self-organization of molecules into supermolecules requires a minimum size (quite a few donor and acceptor groups) and topology (conformationally preorganized or easily preformable molecular partners) of the participating molecules, as well as a gain in free energy upon self-organization. Recently, an intensive search for chemical systems with a tendency to spontaneous self-organization has commenced.

A technology successfully geared towards the transition from the structure of complementary molecules, *via* that one of supramolecular complexes, to the structure, or even the habit, of crystals can today only be described as wishful thinking at best. However, it should not be overlooked that more or less systematic efforts to design *crystal* structures on the basis of *molecular* structure and then to make them reality (*crystal engineering*; *Chapt. 12.2.2*) are increasing in number and intensity.

There are cases in which H-bonds assure the coherence of the resulting supramolecular complexes. Predictability of supramolecular structures of this type is founded upon the spatially preferred orientations of H-bonds and the formation of the supermolecule under thermodynamically determined conditions. *George M. Whitesides*' programmatic studies of the melamine/cyanuric acid complex (see *Chapt. 15.2.4*) belong in this category. However, there also exist cases in which, for example,

(*R*,*R*)- or (*S*,*S*)-cyclohexane-1,2-diamine and (*S*,*S*)-cyclohexane-1,2-diol assemble in the crystalline state into unexpected triple helices, as reported by *Stephen Hanessian*.

Spontaneous self-organization into double- or triple-helical metal complexes – helicates – has been observed in the laboratories (Strasbourg and Paris) of *Jean-Marie Lehn*. Oligo(2,2′-bipyridyl) single strands of varying chain lengths, after addition of copper(I) or nickel(II) ions, assemble constitution-specifically into the corresponding double or triple helicate, with constitutionally identical ligands in every case (*Fig. 7.68*). The presence, in each case, of an O-atom in the bridge between two bipyridyl residues should benefit the helical pre-organization of the ligands (*Sect. 7.2.2*), and hence complex formation.

Helicates with constitutionally identical ligands (*homo-helicates*) take precedence over those with constitutionally non-identical ligands (*hetero-helicates*). Favoring of discrimination in recognition between self and non-self may be traced, essentially, to the high enthalpy of complex formation, in which all of the bipyridyl residues reach a maximum of coordination. Entropy is strongly reduced in helicate formation, as the number of molecules diminishes greatly. The smaller variety (2/2, 2/2, 3/3, 3/3) on forming homo-helicates, compared with the greater variety (2/2, 2/3, 3/2, 3/3) on forming homo- and hetero-helicates, while offering no entropic advantages, does afford enthalpic gains.

7.5. Conclusion

Paul J. Flory's observation in the *Nobel* Lecture of 1974, concerning the *conformation of linear macromolecules* – '*The rules of chemical valency, even in their simplest form, permit … anticipation of the existence of macromolecular structure*' – gives no hint of the prejudices of prominent representatives of organic chemistry, against which *Hermann Staudinger* had to argue for years before his views won general acceptance. To this day, macromolecules, even biomacromolecules, are the Cinderellas in the texts of organic chemistry. Organic chemists' predilection for molecular systems which can be precisely studied experimentally, and whose behavior can be exactly described biased the development of an autonomous discipline of 'macromolecular chemistry', leaving the study of complex interrelationships in living systems to biologists, or even biologically interested physicists. If one considers that scientific disciplines are, in the widest sense of the word, political institutions, that demarcate areas of academic territory, allocate the privileges and responsibilities of expertise, defining which scientific skills and knowledge are appropriate to which scientists, and which problems are scientifically relevant, then it is scarcely to be wondered that the explanation of complex interrelationships in living systems has come to be made with less and less input from organic chemists.

The ideal pathway from micro- to macromolecular compounds of linearly constituted backbone is offered by conformational analysis. For

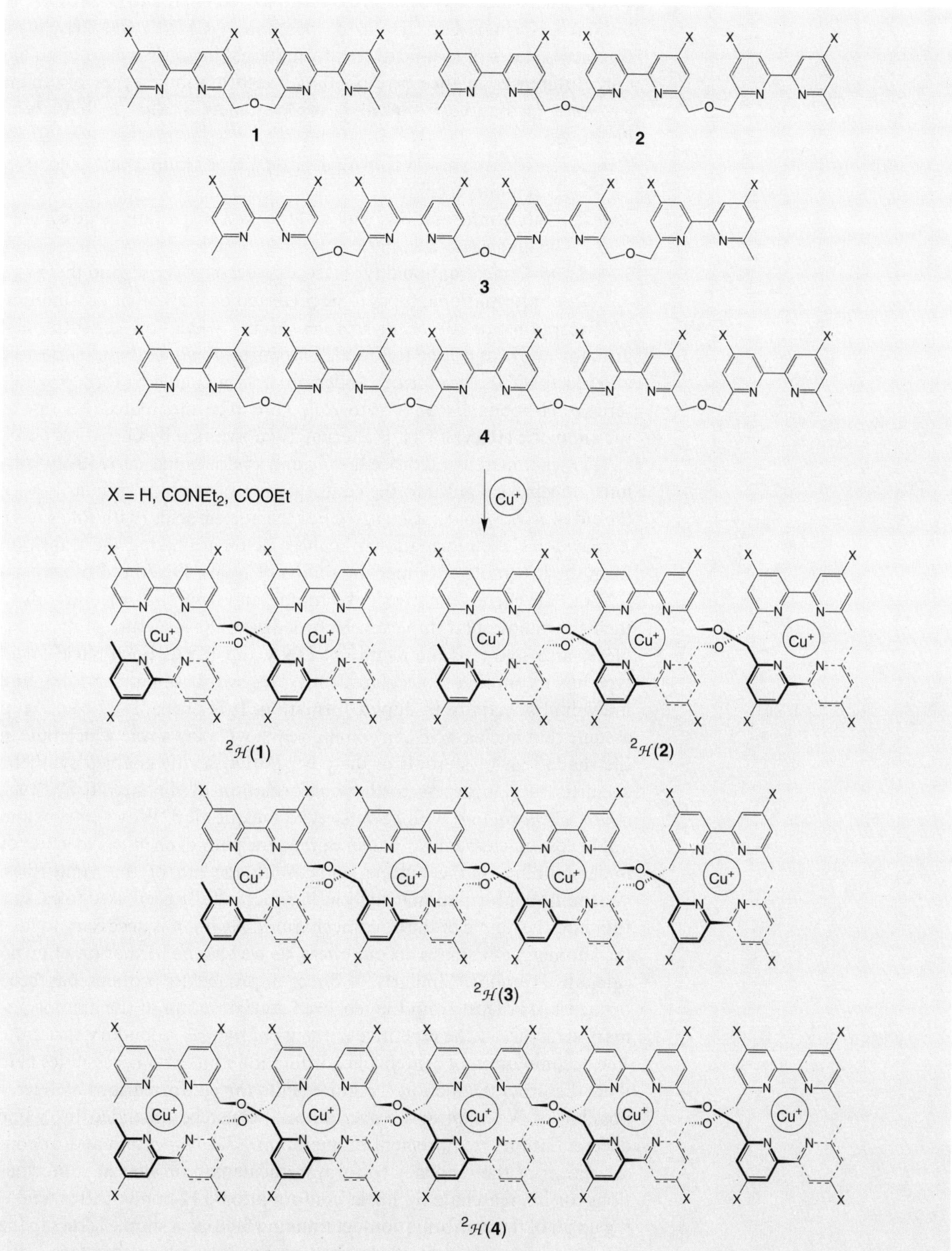

Fig. 7.68. *Homo-helicates as favored products of molecular self-organization* ($^2\mathcal{H}$: double helicate)

compounds which may be described with the aid of ethano ($-CH_2CH_2-$), butano ($-CH_2CH_2CH_2CH_2-$), or pentano ($-CH_2CH_2CH_2CH_2CH_2-$) fragments, rules of systematic conformational analysis, based upon energy differences in the corresponding conformational types of ethane (→ *Rule 1*), butane (→ *Rule 3*), or pentane (→ *Rule 2*) have been developed.

Rule 1: Only conformations with as little *Pitzer* strain as possible to be taken into account.

Rule 2: Only conformations with as little *Newman* strain as possible to be taken into account.

Rule 3: *ap*-Conformational types to be preferred over *sc*, and these two conformational types to be preferred over all other possibilities.

Exceptions to all of these rules occur, when, on constitutional grounds, they cannot be followed. Exceptions to *Rules 2* and *3* must be allowed for if the conformer under discussion is able to form a whole series of H-bonds. Exceptions to *Rule 3* are to be anticipated (because of the anomeric effect) if CH_2 is alternately substituted by O.

Turner's increment procedure, using cyclohexane derivatives, permits a deeper insight into the conformation of steroids (primarily with the aid of *Rules 1* and *3*, *Chapt. 6*). For the nucleic acids (with *Rules 1–3*), *Eschenmoser*'s conformational analysis helps in making the transition from the low-molecular repeating units of homo-DNA and pyranosyl-RNA to the corresponding single strands, and from the single strands to their supramolecular duplexes. Knowledge of the structure of these duplexes also sharpens the picture of DNA and RNA. Each nucleic-acid type has its specific molecular identity, its own base-pairing rules, and individual selectivity in duplex formation. It is probably not unfair to assume that nucleic acids, in coming years, will play a role which puts in the shade that of steroids in the past. *Derek Barton* enabled synthetic chemists to gain access to the 'conformation of the steroid nucleus', *Albert Eschenmoser*, with his answer to the question 'Why pentose- and not hexose-nucleic acids?' has smoothed the path of chemists to a deeper understanding of the nucleic acids. With the aid of the same rules, systematic conformational analysis has successfully been used to carve a trail into the jungle of the oligosaccharides. Again, it is necessary to take the anomeric effect into account here, as well as the formation of intracatenate H-bonds. Similarly, a direct approach to proteins has been opened up, at least from the idealized conformation in the area of primary structure to the repetitive elements of regular secondary structure. The accumulation of centers of coordination number 3 necessitates here that, the gross distinction – requirements for conformations as free as possible of *Newman* and *Pitzer* strain – must be amended by a fine differentiation – requirement of minimum 1,3-allylic strain and/or consideration of the tendency to form intracatenate (in helical conformations) or intercatenate (in linear conformations) H-bonds. After weighing up all of the conformation-determining factors, a simple access to the two repetitive elements of idealized protein secondary structure – the β-pleated sheet and the α-helix – is found.

Supramolecular chemistry owes *J.-M. Lehn* not only the *Cartesian* clarity of his coining of terms, but also – wholly in the spirit of *Pierre Eugène Marcelin Berthelot* – non-natural objects, synthesized especially for the experimental study of self-organization.

Apart from the obvious intention to better acquaint the medicinally interested chemist with biomacromolecules and their supramolecular interactions with other biomacromolecules or with synthetic ligands or receptors, the primary intention of the authors in *Chapt. 7* has been to clarify the manner and the extent to which the rules (the *grammar*) of systematic conformational analysis enable the determination of the formal number of torsion-angle combinations worth considering, together with associated chemical information (*structural syntax*).

Further Reading

To Sect. 7.1.

H. Staudinger, *Nobel Lecture: Macromolecular Chemistry* in *Nobel Lectures Chemistry 1942–1962*, p. 397, Elsevier Publ. Comp., Amsterdam, 1964;
– Arbeitserinnerungen, Hüthig Verlag, Heidelberg, 1961.

J.-M. Lehn, *Nobel Lecture: Supramolecular Chemistry – Space and Perspectives – Molecules, Supermolecules, and Molecular Devices, Angew. Chem. Int. Ed.* **1988,** *27*, 91.
– *Supramolecular Chemistry, Science* **1993,** *260*, 1762.

To Sect. 7.2.

K. Ziegler, *Nobel Lecture: Consequences and development of an invention* in *Nobel Lectures Chemistry 1963–1970*, p. 6, Elsevier Publ. Comp., Amsterdam, 1972.

G. Natta, *Nobel Lecture: From the stereospecific polymerization to the asymmetric autocatalytic synthesis of macromolecules* in *Nobel Lectures Chemistry 1963–1970*, p. 27, Elsevier Publ. Comp., Amsterdam, 1972.

'IUPAC-Basic Definitions of Terms Relating to Polymers', Pure Appl. Chem. **1974,** *40*, 477.

'IUPAC-Nomenclature of Regular Single-Strand Organic Polymers', Pure Appl. Chem. **1976,** *48*, 373.

'IUPAC-Stereochemical Definitions and Notations Relating to Polymers', Pure Appl. Chem. **1981,** *53*, 733.

A. D. Jenkins, K. L. Loening, *Nomenclature* in *Comprehensive Polymer Science*, Vol. 1 (Ed. G. Allen, J. Bevington, C. Booth, C. Price), p. 13, Pergamon Press, Oxford, 1989.

M. Farina, *The Stereochemistry of Linear Macromolecules* in *Topics in Stereochem.* **1987,** *17*, 1.

H.-G. Elias, *Makromoleküle*, 4. Aufl., Hüthig & Wepf, Basel, 1981.

E. Müller (Ed.), *Methoden der Organischen Chemie (Houben-Weyl)*, 4. Aufl., Bd. XIV/1 (1961) und XIV/2 (1993): *Makromolekulare Stoffe*, Thieme Verlag, Stuttgart.

7.2.2.

P. J. Flory, *Nobel Lecture: Spatial Configuration of Macromolecular Chains* in *Nobel Lectures Chemistry 1971–1980*, p. 156, World Scientific, Singapore, 1993.

A. Abe, R. L. Jernigan, P. J. Flory, *Conformational Energies of n-Alkanes and the Random Configuration of Higher Homologs Including Polymethylene, J. Am. Chem. Soc.* **1966,** *88,* 631.

M. Saunders, *Stochastic Exploration of Molecular Mechanics Energy Surfaces. Hunting for the Global Minimum, J. Am. Chem. Soc.* **1987,** *109,* 3150.

P. W. Smith, W. C. Still, *The Effect of Substitution and Stereochemistry on Ion Binding in the Polyether Ionophor Monensin, J. Am. Chem. Soc.* **1988,** *110,* 7917.

R. W. Hoffmann, *Flexible Molecules with Defined Shape – Conformational Design, Angew. Chem. Int. Ed.* **1992,** *31,* 1124.

C. R. Noe, C. Miculka, J. W. Bats, *Helicity of oligomeric formaldehyde, Angew. Chem. Int. Ed.* **1994,** *33,* 1476.

To Sect. 7.3.

L. Pauling, *Molecular Architecture and Biological Reactions, Chem. Eng. News* **1946,** *24,* 1375.

7.3.1.

G. M. Blackburn, M. J. Gait (Ed.), *Nucleic Acids in Chemistry and Biology,* Oxford University Press, Oxford, 1990.

7.3.1.1.1.

A. Eschenmoser, *Kon-Tiki-Experimente zur Frage nach dem Ursprung von Biomolekülen,* Verhandlungen der Gesellschaft Deutscher Naturforscher und Ärzte, 116. Versammlung, Berlin, 1990, S. 135, Wissenschaftl. Verlagsgesellschaft, Stuttgart, 1991;
– *Warum Pentose- und nicht Hexose-Nucleinsäuren? Nachr. Chem. Tech. Lab.* **1991,** *39,* 795.

A. Eschenmoser, E. Loewenthal, *Chemistry of Potentially Prebiological Natural Products, Chem. Soc. Rev.* **1992,** *21,* 1.

A. Eschenmoser, M. Dobler, *Warum Pentose- und nicht Hexose-Nucleinsäuren? Teil I: Einleitung und Problemstellung, Konformationsanalyse für Oligonucleotid-Ketten aus 2′,3′-Dideoxyglucopyranosyl-Bausteinen ('Homo-DNS') sowie Betrachtungen zur Konformation von A- und B-DNS, Helv. Chim. Acta* **1992,** *75,* 218.

M. Böhringer, H.-J. Roth, J. Hunziker, M. Göbel, R. Krishnan, A. Giger, B. Schweizer, J. Schreiber, C. Leumann, A. Eschenmoser, *Warum Pentose- und nicht Hexose-Nucleinsäuren? Teil II: Oligonucleotide aus 2′,3′-Dideoxy-β-D-glucopyranosyl-Bausteinen ('Homo-DNS'): Herstellung, Helv. Chim. Acta* **1992,** *75,* 1416.

7.3.1.1.2.

S. Altman, *Nobel Lecture: Enzymatic Cleavage of RNA by RNA, Angew. Chem. Int. Ed.* **1990,** *29,* 749.

T. R. Cech, *Nobel Lecture: Self-splicing and Enzymatic Activity of an Intervening Sequence RNA from Tetrahymena, Angew. Chem. Int. Ed.* **1990,** *29,* 759.

C. R. Woese, *The Origins of the Genetic Code,* Harper and Row, New York, 1967.

F. H. C. Crick, *The Origin of the Genetic Code, J. Mol. Biol.* **1968,** *38,* 367.

L. E. Orgel, *Evolution of the Genetic Apparatus, J. Mol. Biol.* **1968,** *38,* 381.

F. H. Westheimer, *Polyribonucleic acids as enzymes, Nature* **1986,** *319,* 534.

W. Gilbert, *The RNA world, Nature* **1986,** *319,* 618.

P. Reichard, *From RNA to DNA, Why So Many Ribonucleotide Reductases? Science* **1993,** *260,* 1773.

G. F. Joyce, A. W. Schwartz, S. L. Miller, L. E. Orgel, *The case for an ancestral genetic system involving simple analogues of the nucleotides, Proc. Natl. Acad. Sci. USA* **1987,** *84,* 4398.

J. A. Doudna, S. Couture, J. W. Szostak, *A Multisubunit Ribozyme That is a Catalyst of and Template for Complementary Strand RNA Synthesis, Science* **1991,** *251,* 1605.

L. E. Orgel, *Molecular replication, Nature* **1992,** *358,* 203.
S. Pitsch, S. Wendeborn, B. Jaun, A. Eschenmoser, *Why Pentose- and Not Hexose-Nucleic Acids? Part VII: Pyranosyl-RNA ('p-RNA'), Helv. Chim. Acta* **1993,** *76,* 2161.
R. F. Gesteland, J. F. Atkins (Ed.), *The RNA World – The Nature of Modern RNA Suggests a Prebiotic RNA World,* Cold Spring Harbor Laboratory Press, Plainview, 1993.

7.3.1.2.1.

J. D. Watson, F. H. C. Crick, *Molecular Structure of Nucleic Acids. A Structure for Deoxyribose Nucleic Acid, Nature* **1953,** *171,* 737.
F. Crick, *DNA: Test of Structure? Science* **1970,** *167,* 1694;
– *An Error in Model Building, Nature* **1967,** *213,* 798.
M. Eigen, *Selforganization of Matter and the Evolution of Biological Macromolecules, Naturwiss.* **1971,** *58,* 465.
G. Schwarzenbach, *Der Chelateffekt, Helv. Chim. Acta* **1952,** *35,* 2344.
J. D. Watson, *The Double Helix,* Weidenfeld and Nicolson, London, 1968.
F. Crick, *What Mad Pursuit,* Basic Books, New York, 1988.
A. Sayre, *Rosalind Franklin and DNA,* Norton & Comp., New York, 1975.
M. Jackson, *Life Story,* BBC Film, 1987.

7.3.1.2.2.

J. Munziker, H.-J. Roth, M. Böhringer, A. Giber, U. Diederichsen, M. Göbel, R. Krishnan, B. Jaun, C. Leumann, A. Eschenmoser, *Warum Pentose- und nicht Hexose-Nucleinsäuren? Teil III: Oligo(2',3'-dideoxy-β-D-glucopyranosyl)nucleotide ('Homo-DNS'): Paarungseigenschaften, Helv. Chim. Acta* **1993,** *76,* 259.
G. Otting, M. Billeter, K. Wüthrich, H.-J. Roth, C. Leumann, A. Eschenmoser, *Warum Pentose- und nicht Hexose-Nucleinsäuren? Teil IV: Homo-DNS: ^{1}H-, ^{13}C-, ^{31}P- und ^{15}N-NMR-spektroskopische Untersuchung von ddGlc(A-A-A-A-A-T-T-T-T-T) in wässriger Lösung, Helv. Chim. Acta* **1993,** *76,* 2701.
A. Eschenmoser, *Towards a chemical etiology of the structure of nucleic acids, Chem. Biol.* **1994,** *15. April*: Introductory issue.

7.3.1.2.4.

N. R. Cozzarelli, J. C. Wang, *DNA Topology and its Biological Effects,* Cold Spring Harbor Laboratory Press, Cold Spring Harbor, 1990.

7.3.1.3.1.

P. Ehrlich, *Nobel Lecture 1908: Partial cell functions* in *Nobel Lectures Physiology or Medicine 1901–1921,* Elsevier Publ. Comp., Amsterdam, 1967.
E. Bäumler, *Auf der Suche nach der Zauberkugel,* Econ-Verlag, Düsseldorf, 1963.
G. Domagk, *Nobel Lecture 1947: Further progress in chemotherapy of bacterial infections* in *Nobel Lectures Physiology or Medicine 1922–1941,* Elsevier Publ. Comp., Amsterdam, 1965.
Science 1994, *15. April*: *Frontiers in Biotechnology: Resistance to Antibiotics.*
I. Bertini, H. B. Gray, S. J. Lippard, J. S. Valentine, *Bioinorganic Chemistry,* University Science Books, Mill Valley, CA, 1994;
– *Chemistry and molecular biology of platinum anticancer drugs, Pure Appl. Chem.* **1987,** *59,* 731.
H. Mitsuya, S. Broder, *Strategies for antiviral therapy in AIDS, Nature* **1987,** *325,* 773.
J. Thesing, *Trendumkehr in der Arzneimittelforschung, Naturwissensch.* **1977,** *64,* 601.
J. Drews, *Naturwissenschaftliche Paradigmen in der Medizin – Die Rolle der Chemie, Swiss. Chem.* **1991,** *13,* 33.
D.M. Spencer, T.J. Wandless, S.L. Schreiber, G.R. Crabtree, *Controlling Signal Transduction with Synthetic Ligands, Science* **1993,** *262,* 1019.

J. Travis, *Making Molecular Matches in the Cell, Science* **1993,** *262,* 989.

M. L. Riordan, J. C. Martin, *Oligonucleotide-based therapeutics, Nature* **1991,** *350,* 442.

S. Brenner, R. A. Lerner, *Encoded combinatorial chemistry, Proc. Natl. Acad. Sci. USA* **1992,** *89,* 5381.

D. Pei, H. D. Ulrich, P. G. Schultz, *A Combinatorial Approach Toward DNA Recognition, Science* **1991,** *253,* 1408.

M. H. J. Ohlmeyer, R. Swanson, L. W. Dillard, J. C. Reader, G. Asouline, R. Kobayashi, M. H. Wigler, W. C. Still, *Complex Synthetic Chemical Libraries Indexed with Molecular Tags, Proc. Natl. Acad. Sci. USA* **1993,** *90,* 10922.

A. Borchardt, W. C. Still, *Synthetic Receptor Binding Elucidated with an Encoded Combinatorial Library, J. Am. Chem. Soc.* **1994,** *116,* 373.

R. M. Baum, *Combinatorial Approaches Provide Fresh Leads for Medicinal Chemistry, Chem. Eng. News* **1994,** *7. Februar,* S. 20.

7.3.1.3.2.

M. L. Kopka, C. Yoon, D. Goodsell, P. Pjura, R. E. Dickerson, *The molecular origin of DNA-drug specificity in netropsin and distamycin, Proc. Natl. Acad. Sci. USA* **1985,** *82,* 1376.

M. Coll, C. A. Frederick, A. H.-J. Wang, A. Rich, *A bifurcated hydrogen-bonded conformation in the d(A-T) base pairs of the DNA dodecamer d(CGCAAATTTGCG) and its complex with distamycin, Proc. Natl. Acad. Sci. USA* **1987,** *84,* 8385.

P. B. Dervan, *Sequence Specific Recognition of Double Helical DNA. A Synthetic Approach, Nucleic Acids and Molecular Biology,* Vol. 2 (Ed. F. Eckstein, D. M. J. Lilley), Springer-Verlag, Berlin, 1988.

J. G. Pelton, D. E. Wemmer, *Structural characterization of a 2:1 distamycin A·d(CGCAAATTGGC) complex by two-dimensional NMR, Proc. Natl. Acad. Sci. USA* **1989,** *86,* 5723.

W. S. Wade, M. Mrksich, P. B. Dervan, *Design of Peptides That Bind in the Minor Groove of DNA at 5′-(A,T)G(A,T)C(A,T)-3′-Sequence by a Dimeric Side-by-Side Motif, J. Am. Chem. Soc.* **1992,** *114,* 8783.

M. Mrksich, P. B. Dervan, *Antiparallel Side-by-Side Heterodimer for Sequence-Specific Recognition in the Minor Groove of DNA by a Distamycin/1-Methylimidazole-2-carboxamide-netropsin Pair, J. Am. Chem. Soc.* **1993,** *115,* 2572.

B. H. Geierstanger, T. J. Dwyer, Y. Bathini, J. W. Lown, D. E. Wemmer, *NMR Characterization of a Heterocomplex Formed by Distamycin and Its Analog 2-ImD with d(CGCAAGTTGGC):d(GCCAACTTGCG): Preference for the 1:1:1 2-ImD:Dst:DNA Complex over the 2:1 2-ImD:DNA and the 2:1 Dst:DNA Complexes, J. Am. Chem. Soc.* **1993,** *115,* 4474.

J. W. Lown, *Targeting the DNA Minor Groove for Control of Biological Function: Progress, Challenges, and Prospects, Chemtracts Org. Chem.* **1993,** *6,* 205.

7.3.1.3.3.

G. Felsenfeld, D. R. Davies, A. Rich, *Formation of a Three-Stranded Polynucleotide Molecule, J. Am. Chem. Soc.* **1957,** *79,* 2023.

S. Arnott, P. J. Bond, *Structures for poly(U)-poly(A)-poly(U) triple stranded polynucleotides, Nature New Biol.* **1973,** *244,* 99.

O. Kennard, W. N. Hunter, *Single-Crystal X-ray Diffraction Studies of Oligonucleotides and Oligonucleotide-Drug Complexes, Angew. Chem. Int. Ed.* **1991,** *30,* 1254.

H. E. Moser, P. B. Dervan, *Sequence-Specific Cleavage of Double Helical DNA by Triple Helix Formation, Science* **1987,** *238,* 645.

L. J. Maher, III, B. Wold, P. B. Dervan, *Inhibition of DNA Binding Proteins by Oligonucleotide-Directed Triple-Helix Formation, Science* **1989,** *245,* 725.

L. C. Griffin, P. B. Dervan, *Recognition of Thymine-Adenine Base Pairs by Guanine in a Pyrimidine Triple Helix Motif, Science* **1989,** *245,* 967.

P. B. Dervan, *Oligonucleotide Recognition of Double-helical DNA by Triple-helix Formation* in *Oligodeoxynucleotides – Antisense Inhibitors of Gene Expression* (Ed. J. S. Cohen), MacMillan Press, Houndmills, 1989.

J. F. Gusella, N. S. Wexler, P. M. Conneally, S. L. Naylor, M. A. Anderson, R. E. Tanzi, P. C. Watkins, K. Ottina, M. R. Wallace, A. Y. Sakaguchi, A. B. Young, I. Shoulson, E. Bonilla, J. B. Martin, *A polymorphic DNA marker genetically linked to Huntington's disease, Nature* **1983,** *306,* 234.

The Huntington's Disease Collaborative Research Group, *A Novel Gene Containing a Trinucleotide Repeat That is Expanded and Unstable on Huntington's Disease Chromosomes, Cell* **1993,** *72,* 971.

7.3.1.3.4.

P. Zamecnik, M. L. Stephenson, *Inhibition of Rous sarcoma virus replication and cell transformation by a specific oligodeoxynucleotide, Proc. Natl. Acad. Sci. USA* **1978,** *75,* 280.

M. L. Stephenson, P. Zamecnik, *Inhibition of Rous sarcoma viral RNA 'translation' by a specific oligodeoxynucleotide, Proc. Natl. Acad. Sci. USA* **1978,** *75,* 285.

C. Cheong, G. Varani, I. Tinoco, Jr., *Solution structure of an unusually stable RNA hairpin 5'GGAC(UUCG)GUCC, Nature* **1990,** *346,* 680.

J. S. Cohen, *Strategies and Realities* in *Oligodeoxynucleotides – Antisense Inhibitors of Gene Expression,* MacMillan Press, Houndmills, 1989.

H. M. Weintraub, *Antisense RNA and DNA, Scientific American* **1990,** *Januar,* 34.

E. Uhlmann, A. Peyman, *Antisense Oligonucleotides: A New Therapeutic Principle, Chem. Rev.* **1990,** *90,* 543.

J. Engels, *Antisense Oligonucleotide; Krankheit – Fehler in der Informationsübertragung, Nachr. Chem. Tech. Lab.* **1991,** *39,* 1250.

A. S. Moffat, *Making Sense of Antisense, Science* **1991,** *253,* 510.

D. Herschlag, *Implications of Ribozyme Kinetics for Targeting the Cleavage of Specific RNA Molecules In Vivo: More Isn't Always Better, Proc. Natl. Acad. Sci. USA* **1991,** *88,* 6921.

T. M. Woolf, D. A. Melton, C. G. B. Jennings, *Specificity of antisense oligonucleotides in vivo, Proc. Natl. Acad. Sci. USA* **1992,** *89,* 7305.

J. F. Milligan, M. D. Matteucci, J. C. Martin, *Current Concepts in Antisense Drug Design, J. Med. Chem.* **1993,** *36,* 1923.

C. A. Stein, Y.-C. Cheng, *Antisense Oligonucleotides as Therapeutic Agents – Is the Bullet Really Magical? Science* **1993,** *261,* 1004.

S. A. Narang (Ed.), *Synthesis and Applications of DNA and RNA,* Academic Press, Orlando, 1987.

7.3.1.4.

P. Strazewski, C. Tamm, *Replication Experiments with Nucleotide Base Analogues, Angew. Chem. Int. Ed.* **1990,** *29,* 36.

7.3.1.4.2.

C. Switzer, S. E. Moroney, S. A. Benner, *Enzymatic Incorporation of a New Base Pair into DNA and RNA, J. Am. Chem. Soc.* **1989,** *111,* 8322.

J. A. Piccirilli, T. Krauch, S. E. Moroney, S. A. Benner, *Enzymatic incorporation of a new base pair into DNA and RNA extends the genetic alphabet, Nature* **1990,** *343,* 33.

L. E. Orgel, *Adding to the genetic alphabet, Nature* **1990,** *343,* 18.

R. Dagani, *Genetic alphabet for RNA gains two letters, Chem. Eng. News* **1990,** *22. January,* p. 25.

7.3.1.4.3.

P. E. Nielsen, M. Egholm, R. H. Berg, O. Buchardt, *Sequence-Selective Recognition of DNA by Strand Displacement with a Thymine-Substituted Polyamide, Science* **1991,** *254,* 1497.

M. Egholm, O. Buchardt, L. Christensen, C. Behrens, S. M. Freier, D. A. Driver, R. H. Berg, S. K. Kim, B. Norden, P. E. Nielsen, *PNA hybridizes to complementary oligonucleotides obeying the Watson-Crick hydrogen-bonding rules, Nature* **1993,** *365,* 566.

D. J. Patel, *Marriage of convenience, Nature* **1993,** *365,* 490.

F. Flam, *Can DNA Mimics Improve on the Real Thing? Science* **1993,** *262,* 1647.

A. De Mesmaecker, A. Waldner, J. Lebreton, P. Hoffmann, V. Fritsch, R. M. Wolf, S. M. Freier, *Amide bridging, a new type of modification of oligonucleotide backbones, Angew. Chem. Int. Ed.* **1994,** *33,* 226.

7.3.2.

P. G. Goekjian, Tse-Chong Wu, Y. Kishi, *Conformational Similarity of Glycosides and Corresponding C-Glycosides, J. Org. Chem.* **1991,** *56,* 6412.

P. G. Goekjian, Tse-Chong Wu, Han-Young Kang, Y. Kishi, *Preferred Conformation of Carbon Analogues of Isomaltose and Gentobiose, J. Org. Chem.* **1991,** *56,* 6422.

Y. Wang, P. G. Goekjian, D. M. Ryckman, W. H. Miller, S. A. Babirad, Y. Kishi, *Conformational Analysis of 1,4-Linked Carbon Disaccharides, J. Org. Chem.* **1992,** *57,* 482.

T. Haneda, P. G. Goekjian, S. H. Kim, Y. Kishi, *Synthesis and Conformational Analysis of Carbon Trisaccharides, J. Org. Chem.* **1992,** *57,* 490.

G. A. Jeffrey, *Crystallographic Studies of Carbohydrates, Acta Cryst., Sect. B* **1990,** *46,* 89;
– *Experimental and Theoretical Bases for Accurate Modeling – An Experimentalist Looks at Modeling* in *Computer Modeling of Carbohydrate Molecules* (Ed. A. D. French, J. W. Brady), American Chemical Society, Washington, DC, 1990.

A. D. French, *The Crystal Structure of Native Ramie Cellulose, Carbohydrate Res.* **1978,** *61,* 67.

F. H. Allen, O. Kennard, R. Taylor, *Systematic Analysis of Structural Data as a Research Technique in Organic Chemistry, Acc. Chem. Res.* **1983,** *16,* 146.

7.3.3.

R. E. Dickerson, I. Geis, *Proteins: Structure, Function, and Evolution,* 2nd Edn., The Benjamin/Cummings Publ. Comp., Menlo Park, CA, 1983.

G. E. Schulz, R. H. Schirmer, *Principles of Protein Structure,* Springer-Verlag, New York, 1979.

7.3.3.1.

W. B. Schweizer, J. D. Dunitz, *Structural Characteristics of the Carboxylic Ester Group, Helv. Chim. Acta* **1982,** *65,* 1547.

P. Chakrabarti, J. D. Dunitz, *Structural Characteristics of the Carboxylic Amide Group, Helv. Chim. Acta* **1982,** *65,* 1555.

T. Srikrishnan, N. Winiewicz, R. Parthasarathy, *Crystal and molecular structure of glycylglycylglycine, Int. J. Peptide Protein Res.* **1982,** *19,* 103.

K. N. Goswami, V. S. Yadava, V. M. Padmanabhan, DL-*Leucylglycylglycine (LGG), Acta Cryst., Sect. B* **1977,** *33,* 1280.

J. K. Fawcett, N. Camerman, A. Camerman, *The Structure of the Tripeptide* L-*Alanyl*-L-*alanyl*-L-*alanine, Acta Cryst., Sect. B* **1975,** *31,* 658.

S. Arnott, S. D. Dover, A. Elliott, *Structure of β-Poly-*L-*alanine: Refined Atomic Co-ordinates for an Anti-parallel Beta-pleated Sheet, J. Mol. Biol.* **1967,** *30,* 201.

7.3.3.2.

L. Schäfer, C. Van Alsenoy, J. N. Scarsdale, *Molecular structures and conformational analysis of the dipeptide N-acetyl-N′-methylglycylamide and the significance of local geometries for peptide structures, J. Chem. Phys.* **1982,** *76,* 1439.

J. N. Scarsdale, C. Van Alsenoy, Y. J. Klimkowski, L. Schäfer, F. A. Momany, *Optimized Molecular Structures and Conformational Analysis of N′-Acetyl-N-*

methylalaninamide and Comparison with Peptide Crystal Data and Empirical Calculations, J. Am. Chem. Soc. **1983,** *105,* 3438.

H.-J. Böhm, S. Brode, *Ab Initio Calculations on Low-Energy Conformers of N-Acetyl-N′-methylalaninamide and N-Acetyl-N′-methylglycinamide, J. Am. Chem. Soc.* **1991,** *113,* 7129.

L. Schäfer, S.Q. Newton, M. Cao, A. Peeters, C. Van Alsenoy, K. Wolinski, F.A. Momany, *Evaluation of the Dipeptide Approximation in Peptide Modeling by Ab Initio Geometry Optimizations of Oligopeptides, J. Am. Chem Soc.* **1993,** *115,* 272.

K. B. Wiberg, E. Martin, *Barriers to Rotation Adjacent to Double Bonds, J. Am. Chem. Soc.* **1985,** *107,* 5035.

7.3.3.2.2.

F. Johnson, *Allylic Strain in Six-membered Rings, Chem. Rev.* **1968,** *68,* 375.

R.W. Hoffmann, *Allylic 1,3-Strain as a Controlling Factor in Stereoselective Transformations, Chem. Rev.* **1989,** *89,* 1841.

J.L. Broeker, R.W. Hoffmann, K.N. Houk, *Conformational Analysis of Chiral Alkenes and Oxonium Ions: Ab Initio Molecular Orbital Calculations and an Improved MM2 Force Field, J. Am. Chem. Soc.* **1991,** *113,* 5006.

7.3.3.2.3.

M. Levitt, J. Greer, *Automatic Identification of Secondary Structure in Globular Proteins, J. Mol. Biol.* **1977,** *114,* 181.

W. Kabsch, C. Sander, *Dictionary of Protein Secondary Structure: Pattern Recognition of Hydrogen Bond and Geometrical Features, Biopolymers* **1983,** *22,* 2577.

7.3.3.4.

L. Pauling, *Nobel Lecture Chemistry 1954: Modern Structural Chemistry* in *Nobel Lectures Chemistry 1942–1962,* Elsevier Publ. Comp., Amsterdam, 1964.

L. Pauling, R.B. Corey, *The Configuration of Polypeptide Chains in Proteins* in *Progress in the Chemistry of Organic Natural Products* (Ed. L. Zechmeister), **1954,** *11,* 180.

F. Crick, *The Impact of Linus Pauling on Molecular Biology: A Reminiscence* in *The Chemical Bond – Structure and Dynamics* (Ed. A. Zewail), Academic Press, San Diego, 1992.

D. Obrecht, C. Spiegler, P. Schönholzer, K. Müller, H. Heimgartner, F. Stierli, *A New Approach to Enantiomerically Pure Cyclic and Open-Chain (R)- and (S)-α,α-Disubstituted α-Amino Acids, Helv. Chim. Acta* **1992,** *75,* 1666.

7.3.3.5.

I.L. Karle, *X-Ray Analysis: Conformation of Peptides in the Crystalline State* in *The Peptides,* Vol. 4, Academic Press, New York, 1981.

J.S. Richardson, D.C. Richardson, *Principles and Patterns of Protein Conformation* in *Prediction of Protein Structure and The Principles of Protein Conformation* (Ed. G.D. Fasman), Plenum Press, New York, 1989.

7.3.3.6.

B. Rost, R. Schneider, C. Sander, *Progress in protein structure prediction? Trends in Biol. Science* **1993,** *18. April,* 120.

D. Freedman, *AI Helps Researchers Find Meaning in Molecules, Science* **1993,** *261,* 844.

C.B. Anfinsen, *Nobel Lecture: Studies on the Principles that Govern the Folding of Protein Chains, Nobel Lectures Chemistry 1971–1980,* p. 55, World Scientific, Singapore, 1993.

R. Jaenicke, *Protein Folding: Local Structures, Domains, Subunits, and Assemblies, Biochem.* **1991,** *30,* 3147.

S.W. Englander, *In Pursuit of Protein Folding, Science* **1993,** *262,* 848.

G. Fischer, F.X. Schmid, *The Mechanism of Protein Folding. Implications of In Vitro Refolding Models for De Novo Protein Folding and Translocation in the Cell, Biochem.* **1990,** *29,* 2205.

N. J. Bulleid, *Protein Disulfide-Isomerase: Role in Biosynthesis of Secretory Proteins, Adv. Protein Chem.* **1993,** *44,* 125.

J. F. Brandts, H. R. Halvorson, M. Brennan, *Consideration of the Possibility That the Slow Step in Protein Denaturation is Due to Cis-Trans Isomerism of Proline Residues, Biochem.* **1975,** *14,* 4953.

G. Fischer, H. Bang, C. Mech, *Nachweis einer Enzymkatalyse für die cis-trans-Isomerisierung der Peptidbindung in prolinhaltigen Peptiden, Biomed. Biochim. Acta* **1994,** *10,* 1101.

T. Kiefhaber, H.-P. Grunert, U. Hahn, F. X. Schmid, *Replacement of a Cis Proline Simplifies the Mechanism of Ribonuclease T1 Folding, Biochem.* **1990,** *29,* 6475.

M. K. Rosen, S. L. Schreiber, *Natural products as probes in the study of cellular functions. Investigation of immunophilins, Angew. Chem. Int. Ed.* **1992,** *31,* 384.

R. L. Stein, *Mechanism of Enzymatic and Nonenzymatic Proline Cis-Trans Isomerization, Adv. Protein Chem.* **1993,** *44,* 1.

F. X. Schmid, L. M. Mayr, M. Mücke, E. R. Schönbrunner, *Prolyl Isomerases: Role in Protein Folding, Adv. Protein Chem.* **1993,** *44,* 25.

R. J. Ellis, *Proteins as molecular chaperones, Nature* **1987**, *328*, 378.

R. J. Ellis, S. M. van der Vies, *Molecular Chaperones, Ann. Rev. Biochem.* **1991**, *60*, 321.

J. Martin, F. U. Hartl, *Protein Folding in the cell: molecular chaperones pave the way, Current Biol. Structure* **1993**, *1*, 161.

M.-J. Gething, J. Sambrook, *Protein Folding in the cell, Nature* **1992**, *355*, 33.

E. A. Craig, *Chaperones: Helpers Along the Pathways to Protein Folding, Science* **1993**, *260,* 1902.

D. A. Agard, *To Fold or Not to Fold, Science* **1993**, *260*, 1903.

R. Pain, *Further up the kinetic pathway, Nature* **1992**, *356*, 664.

R. J. Ellis, *Chaperoning nascent proteins, Nature* **1994**, *370*, 96.

To Sect. 7.4.

G. M. Whitesides, J. P. Mathias, C. T. Seto, *Molecular Self Assembly and Nanochemistry: A Chemical Strategy for the Synthesis of Nanostructures, Science* **1991**, *254*, 1312.

J. A. Zerkowski, J. C. MacDonald, C. T. Seto, D. A. Wierda, G. M. Whitesides, *Design of Organic Structures in the Solid State: Molecular Types Based on the Network of Hydrogen Bonds Present in the Cyanuric Acid Melamine Complex, J. Am. Chem. Soc.* **1994**, *116*, 2382.

S. Hanessian, A. Gomtsyan, M. Simard, S. Roelens, *Molecular Recognition and Self-Assembly by "Weak" Hydrogen Bonding: Unprecedented Supramolecular Helicate Structures from Diamine/Diol Motifs, J. Am. Chem. Soc.* **1994**, *116*, 4495.

J.-M. Lehn, A. Rigault, J. Siegel, J. Harrowfield, B. Chevrier, D. Moras, *Spontaneous assembly of double-stranded helicates from oligobipyridine ligands and copper(I) cations: Structure of an inorganic double helix, Proc. Natl. Acad. Sci. USA* **1987**, *84*, 2565.

J.-M. Lehn, A. Rigault, *Helicates. Four- and five-membered double-helix complexes of copper(I) and poly(pyridine) ligands, Angew. Chem. Int. Ed.* **1988,** *27,* 1095.

U. Koert, M. M. Harding, J.-M. Lehn, *DNA deoxyribonucleohelicates: self-assembly of oligonucleosidic double-helical metal complexes, Nature* **1990**, *346*, 339.

R. Krämer, J.-M. Lehn, A. Marquis-Rigault, *Self-recognition in helicate self-assembly: Spontaneous formation of helical metal complexes from mixtures of ligands and metal ions, Proc. Natl. Acad. Sci. USA* **1993**, *90*, 5394.

J.-M. Lehn, *Self-assembly of double helical, triple helical, and deoxyribonucleo-helicate architectures, Chem. Biol.* **1994**, *15. April*: Introductory issue.

E. C. Constable, *Sodium springs a surprise, Nature* **1994**, *367*, 415.

8 The Qualitative MO Model

8.1. Limits of the Structure Model of the Classical Organic Chemistry

Even on introducing the structure model of the classical organic chemistry (*Chapt. 2.2*), it was briefly mentioned that questions would be encountered which would be unanswerable using this model as a basis. In such a case, it is necessary to abandon the tried and trusted model and to move on to another one. In science, progress has the meaning of a progression of steps, each leading, by 'trial and elimination of error' (*Sir Karl (Raimund) Popper*) to a new level of understanding. If the advance is very great, then – to make this fact clear – the term 'scientific revolution' (*Thomas S. Kuhn*) may be applied. Even a scientific revolution still has significant conservative features, as the new way of looking at a given problem must be at least as capable of formulation of every deduction made, as before, using the old model. If the new level of awareness is essentially an enhanced version of its predecessor, then criteria for the evaluation for both, and hence also for the progress made, will be available. *Imre Lakatos* advanced from *Popper*'s 'naive falsification' to a 'sophisticated falsification'. According to this, one theory should only be replaced by another in a case of 'progressive refinement of the problem'. Before it can be accepted, the new theory should differ from the old one in possessing an enhanced empirical content. This process of subjecting theories to a rationale of critical assessment may also be applied to research programs. A research program has merited its support, if it has led to a progressive advancement of the problem.

We have previously encountered five problems which cannot be explained using the structure model of the classical organic chemistry. They are recapitulated below. Solutions may be found in *Sect. 8.4*.

8.1.1. The Planar Arrangement of Atoms of the Amide Group

It may be inferred from the microwave spectrum of formamide that all six atoms of this molecule lie in a plane.

X-ray crystal-structure analyses of crystalline peptides show that the structural element described as a peptide unit (*Chapt. 10.5.4.2*) is periplanar. The planarity of the peptide unit and the relatively high rotational barrier about the (O=C)–N bond, both fundamental for the structure and function of proteins, cannot be explained from the perspective of the structure model of the classical organic chemistry.

8.1.2. Anomeric Effects

The similarities identified by the conformational analyst on comparison of carbocyclic and heterocyclic compounds with one another were shown in *Chapt. 6.6*. They are particularly easy to be seen in six-membered, saturated systems: idealized staggering and the decreasing order of preference in four-atom modules for the sequence *ap* > *sc* (*Chapt. 10.3.2*) form a common denominator for assessment by conformational analysis. The observation that the proportion of those anomers with an axial orientation of the OH group at the anomeric center is greater than would be expected under application of the rules derived using cyclohexane derivatives (*Chapt. 6.6.2*) leads one to suspect that, for instances in which CH_2 groups are alternately replaced by O-atoms, an exceptional effect, the so-called *anomeric effect*, must be operating. This anomeric effect, however, is not limited to the OH group as a substituent at C(1) nor to the C(1)–O bond (*Table 8.1*).

Table 8.1. *Determination of the Equilibrium of Anomers by the Substituents at the Anomeric Center*

Compound	Substituent at C(1)	% Axial orientation
D-Glucopyranose[a])	OH	36
Methyl D-glucopyranoside[b])	OMe	67
Penta-*O*-acetyl-D-glucopyranoside[c])	OAc	86
Tetra-*O*-acetyl-D-glucopyranosyl chloride[d])	Cl	94

[a]) 25°C in H_2O. [b]) 25°C in MeOH. [c]) 25°C in AcOH/Ac_2O 1:1. [d]) 30°C in MeCN.

The magnitude of the anomeric effect decreases in the sequence halogen > PhCOO > AcO > AcS > MeO > RS > HO > H_2N > MeO_2C > pyridinium, even reversing for the pyridinium group. In this last case, the term *reverse anomeric effect* is used.

The *endo-anomeric effect* favors the *sc*-conformation of C(5)–O(5)–C(1)–O(1) over the *ap*-conformation. A similar effect – now called the *exo-anomeric effect* – favors the *sc*- over the *ap*-conformation for the structural fragment O(5)–C(1)–O(1)–R. Since the *ap*-conformation is energetically more favorable than the *sc*-conformation in the C(2)–C(1)–O(1)–R moiety, the conformations **A1**, from the **A** series with axial orientation of the RO group, and **E1**, from the **E** series with equatorial orientation of the RO group, are favored (*Table 8.2*).

Table 8.2. *Favored Conformations* (framed) *Caused by the* exo-*Anomeric Effect.* **A3** is 'forbidden' because of *Newman* strain, **E2** is disfavored because of *sc*-orientation of R relative to C(2), **A2** and **E3** are, furthermore, disfavored by the missing efficacy of the *exo*-anomeric effect.

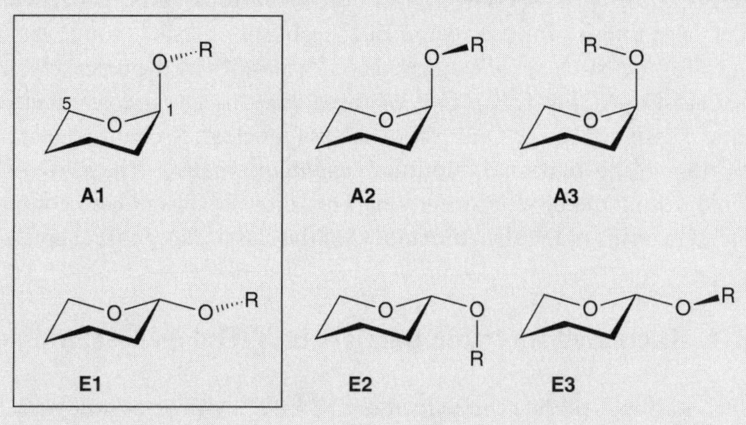

	A1	E1	A2	A3	E2	E3
Conformational type						
C(5)–O(5)–C(1)–O(1)	–*sc*	*ap*	–*sc*	–*sc*	*ap*	*ap*
O(5)–C(1)–O(1)–R	–*sc*	+*sc*	*ap*	+*sc*	–*sc*	*ap*
C(2)–C(1)–O(1)–R	*ap*	*ap*	+*sc*	–*sc*	+*sc*	–*sc*

The *exo*-anomeric effect plays a role of considerable significance for oligosaccharides, and was taken into account, for example, in the conformational analysis of methyl α-maltoside and methyl α-cellobioside (*Chapt. 7.3.2.2* and *7.3.2.3*).

Anomeric effects also extend their reach into acyclic compounds in which one or more CH_2 groups are replaced alternately by O-atoms. In contrast to the order of increasing energy for the conformations of pentane (*cf. Fig. 7.7*):

$$ap/ap > ap/\pm sc > \pm sc/\pm sc > \pm sc/\mp sc$$

the conformations of dimethoxymethane follow the order

$$\pm sc/\pm sc > ap/\pm sc > ap/ap > \pm sc/\mp sc$$

This deviation may be traced back to the *general anomeric effect*. It follows directly, on comparison of the linear conformation of poly(methylene) with the helical conformation of poly(oxymethylene) (*Chapt. 7.2.2*).

endo-Anomeric, *exo-anomeric*, and *general anomeric effects* are also subsumed under the term *gauche effect*. This term emphasizes the preference of the *sc*-conformation over the *ap*-conformation for particular sequences. None of these effects is explicable by the structure model of the classical organic chemistry.

8.1.3. The Favored Conformation of the Repeating Unit of a Polynucleotide

Homo-DNA offers an easy approach to the conformation of nucleic acids (*Chapt. 7.3.1.1.1*). On application of various rules, a particular order of stabilities for the region of the phosphodiester group was assumed for the entire set of staggered conformations of the repeating unit of homo-DNA (*Fig. 7.19*). One of these rules makes use of phantom ligands, assigned to the O-atoms of the polynucleotide chain, and states that the conformational module *(phantom ligand)−O−P−O(−R)* should adopt the *ap*-conformation. This directive cannot be comprehended in terms of the structure model of the classical organic chemistry.

8.1.4. Increased Inversion Barriers in *N*-Halogeno-aziridines

Compounds of the general formula N(ABC), with an asymmetrically substituted N-atom of coordination number 3, withstood, for a considerable time, all attempts to separate them, as racemic mixtures, into their enantiomers: although ammonia and amines have pyramidal structures. To explain this, it was supposed that nitrogen compounds of this type can racemize through inversion (*Chapt. 3.3.1*), and, consequently, that the inversion barrier must lie far below 100 kJ/mol (*Chapt. 6.3.3*). Subsequently, however, it has, in fact, proved possible to separate the two diastereoisomers of 7-chloro-7-azabicyclo[4.1.0]heptane **3** and **4** (see *Chapt. 3.3.1*) from one another. The impact of the Cl-atom cannot be explained on the basis of the structure model of the classical organic chemistry.

8.1.5. High Symmetry and High Thermodynamic Stability of Benzene

In the structure model of the classical organic chemistry, a crucial role is played by the criterion of experimental verification of the required number of isomers. For disubstituted benzene derivatives, only three constitutional isomers are ever found, rather than four (for identical substituents) or even five (for different substituents). The larger numbers should be expected, assuming the correctness of the *Kekulé* structure. It may be concluded that the benzene skeleton belongs to the symmetry point group D_{6h}, and not to the symmetry point group D_{3h}. The obvious assumption, also made by *Kekulé*, that two *Kekulé* structures existed in rapid equilibrium with one another, does not explain why benzene displays a completely unexpectedly high thermodynamic stability, not at all interpretable by the structure model of the classical organic chemistry. Thus, there exists a 'benzene problem' (see *Chapt. 14*).

8.2. Change in the World-View of Physics

Aristoteles supposed matter infinitely divisible. *Leukipp* represented the opposite viewpoint: in the division of matter, one would ultimately strike atoms. In the following 2,000 years, sides were continually taken for one or the other philosophical view. At the beginning of the 19th century, *John Dalton*, in his *law of multiple proportions* (*Chapt. 2.1*), found a powerful argument for the atomic theory. The observation that chemical elements in chemical compounds always occur in particular proportions could be explained simply by elements only being divisible as far as atoms, and compounds as far as molecules, and that such a molecule contained a particular number of atoms of the corresponding elements. At the beginning of this century, the last remaining doubts as to the validity of atomic theory were removed by *Albert Einstein*'s interpretation of *Brown*ian molecular motion, *Joseph J. Thomson*'s discovery of electrons, *Einstein*'s interpretation of the photoelectric effect by the quantum theory of light, and *Sir Ernest Rutherford*'s explanation for radioactive α-emission. Experiments, which could be repeated whenever desired, had refuted *Aristoteles*' view. The philosopher *K. R. Popper* emphatically demonstrated the epistemological importance of falsification as opposed to verification. He distinguished between theories according to their distinct quality. By a good theory, he meant one which, in principle (by experimentation, for example), could be refuted. The hypotheses of *Aristoteles* and *Leukipp*, at the time, were good theories in this respect. Experimental observations agreed with the latter, but not, however, with the former one.

The success of *Sir Isaac Newton*'s *theory of gravity* in the field of mechanics led the *Marquis de Laplace*, at the beginning of the 19th century, to assume that the universe was wholly deterministic: a consequence of scientific laws. *de Laplace* supposed that, with the help of a set of scientific laws, every event in the universe might be predicted, as long as the state of the universe at one particular point in time was completely known. The fact that the constellation of the solar system at any desired point in time could be calculated, with the aid of *Newton*'s laws, from the then current positions of sun and planets complied with the doctrine of scientific determinism.

That this doctrine would have to be given up, at least in microscopic systems, became clear when it was shown that electrons, up until then regarded as particles (corpuscles), behaved in diffraction experiments in a manner which, according to *James Clerk Maxwell*'s *theory of electromagnetism*, could only be expected of waves. Conversely, *Einstein*, in his interpretation of the photoelectric effect, demonstrated that electromagnetic radiation also possessed some of the properties of corpuscles. This corpuscle, called *light quantum* or *photon*, extended the existing series of elementary particles, which until then consisted of electrons, protons, and neutrons. The sum of the photons, endowed with a particular energy, corresponds to the total energy of the radiation. According to the theory of *Max Planck*, each quantum possesses a discrete energy, and according to *Einstein*, this energy is higher, the higher the frequency of the observed radiation.

Quantum theory is able to correctly predict the experimentally observed distribution of light from a blackbody. Its effect on the putative determinism, though, was first made apparent by *Heisenberg*'s uncertainty principle. If one wants to know the future position and momentum of a particle, it is necessary to know its present position and momentum. This can be measured by, for example, letting light strike the particle. Some light will be scattered, and the position of the particle can hereby be worked out. However, the position can only be determined to lie within the range of the wavelength of the used light. So, light of shorter wavelength is needed. The shorter the wavelength, however, the higher the energy of the photon and the stronger the jolt given to the electron when the photon is scattered by it. Hence, the more accurately it is attempted to measure the position of the particle, the more inaccurately can its momentum be determined. According to *Werner Heisenberg*, the product of the inaccuracy of the position and of the inaccuracy of momentum of the particle can never be lower than a particular minimum value: the *Planck* constant. This restriction is independent of the manner in which the measurements are undertaken. *Heisenberg*'s uncertainty principle embodies a fundamental and essential characteristic of our universe. Even 60 years after its description, this characteristic has not been given its due in the collective conscious. Writers and essayists, but also philosophers and psychologists, and the greater part of mankind with them, still largely allow this aspect of the world to pass them by.

The uncertainty principle definitively put an end to the idea of cosmological determinism. It is self-evident that it will never be possible to exactly predict future events, if it is never possible to exactly describe the state of the universe at a given moment.

In physics, it was necessary to replace *Newton*ian mechanics with the quantum or wave mechanics developed by *Werner Heisenberg*, *Erwin Schrödinger*, and *Paul Dirac*. In quantum mechanics, rather than work with separate determinations of position and momentum, quantum states are used, in which position and momentum are considered in combination. The different approach brings a different result in its wake. Instead of a single, discrete result, a series of discrete results (*eigenvalues*), with differing statistical probabilities, is obtained. Consequently, it is not possible to predict the specific result of a single measurement. Instead, result A is expected with a particular probability, result B with another particular probability, as long as the same measurements are carried out on a large number of identical systems. Quantum mechanics brings an unavoidable degree of unpredictability and uncertainty into science.

Although light consists of waves, it behaves, in accordance with *Einstein*'s photon model, as though it were composed of particles: it may only be absorbed or emitted quantumwise. *Heisenberg*'s uncertainty principle has the consequence that particles behave in certain respects like waves: they have no definite position, but are 'smeared' over space with certain distribution probabilities. Quantum mechanics underpins a completely new mathematical approach in which the description of the real world no longer rests on the distinction between waves and particles.

After *Thomson* had tracked down the negatively charged electron at the end of the 19th century, and *Rutherford* the positively charged atomic nucleus some years later, it was possible to propose the *Rutherford* atomic model. It contains an *atomic nucleus of nuclear charge number Z* (number of protons), making up virtually the entire atomic mass in a tiny proportion of the atomic volume. The nuclear charge number *Z* is identical to the number of electrons in the *electron shells*, so that the atom, overall, is neutral. So that the atom should be stable and, usually, possess a significant lifetime, a picture has developed, in analogy to the *Newton*ian mechanical planet system, of an atomic nucleus around which the electron (in the H-atom) or electrons (in other atoms) move(s) at (a) relatively large distance(s), like the planets around the sun. However, because the electron represents the fundamental particle of negative charge, and because an accelerating charge produces a radiation wave (orbiting around a nucleus would constitute acceleration towards it), the atom would be continually loosing energy: the consequence of *Maxwell*ian electrodynamics. This would be at the cost of the total energy of the circling electrons, resulting in their falling into the nucleus and coming to rest. To bring the stability of the atom into harmony with the *Rutherford* atomic model, *Niels Bohr* – contrary to experience from the macroscopic world but in agreement with *Planck*'s ideas of *flux discreteness* – presumed, in the *Bohr-Rutherford* atomic model, that electrons were restricted to moving in certain, discrete orbits (each assigned its so-called *Bohr* radius) around the atomic nucleus. *Bohr*'s conceptual *coup de force* thus compelled the atom to be stable, invalidating at a stroke the energy loss deriving from *Maxwell*ian electrodynamics, and with it any bearing that the established theories of light and optics prevailing in the macroscopic world might have on the microscopic world of atoms.

Richard P. Feynman was of the view that the fact that *Newton*ian mechanics and *Maxwell*ian electrodynamics of the macroscopic world had to be replaced by quantum mechanics and quantum electrodynamics in the microscopic world is characteristic of the uncomfortable aspects of nature, which go against the grain of common sense.

Heisenberg's uncertainty principle offers some plausibility for the stability of the atom. If the electrons were located in the atomic nucleus, then their position would be known. According to the uncertainty principle, they would then have to be endowed with a high kinetic energy. With this energy, however, they would break free from the atomic nucleus. In reality, a compromise is instigated, and the electrons trundle around the atomic nucleus, without it being possible to determine where they are located and how fast they are moving.

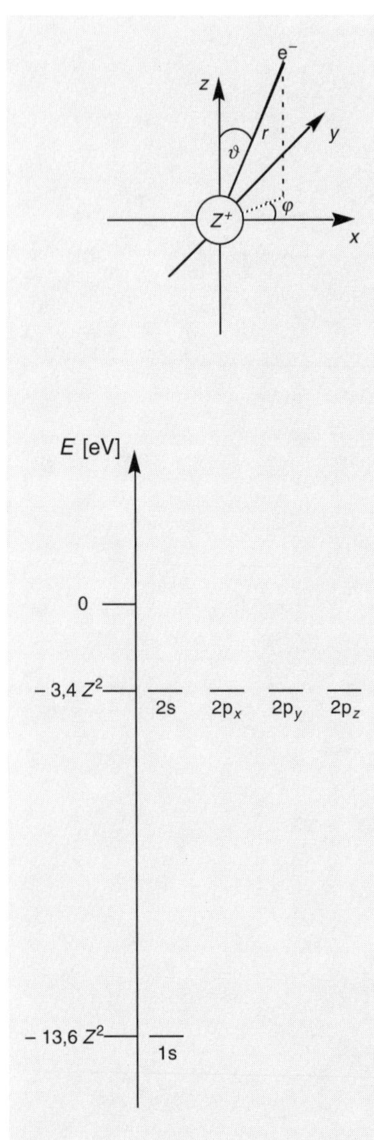

Fig. 8.1. *The five lowest-energy atomic orbitals obtained by solution of the* Schrödinger *equation for an electron in the* Coulomb *potential of an atomic nucleus*

8.3. Atomic and Molecular Orbitals (AOs and MOs)

Quantum mechanics states that the motion in space of a free particle of low mass – such as an electron – cannot be described using a trajectory from *Newton*ian law, but instead by a function $\psi(\boldsymbol{r},t)$. This function may be approached by solving the *Schrödinger* equation after having made various approximations. The square $|\psi|^2 = \psi^\star \cdot \psi$ ($\psi^\star$ is the complex conjugate function of ψ) corresponds to the probability density $\mathrm{d}W/\mathrm{d}V$, thereby giving, for every position, the probability of encountering the particle there. It is appropriate to work with mutually orthogonal, normalized ψ-functions. Mathematically, this means:

$$\int \psi_i^\star \psi_j \mathrm{d}V \begin{cases} = 0, \text{ for } i \neq j \\ = 1, \text{ for } i = j \end{cases}$$

and states that the volume integral becomes zero if the functions are not identical. For identical functions, the volume integral is normalized to 1, because the probability of finding the electron anywhere in space is 1.

8.3.1. One-Electron Atoms

The solution of the *Schrödinger* equation for an electron moving in the *Coulomb* potential of a Z-fold positively charged nucleus gives an infinitely large number of possible ψ-functions, each with a *discrete* energy (*energy eigenvalues*). These functions of state (*eigenfunctions*) are also called *orbitals*. The number of orbitals with the same energy (*degenerate orbitals*) grows with increasing energy. The five orbitals lowest in energy are designated 1s, 2s, $2p_x$, $2p_y$, and $2p_z$ (*Fig. 8.1*).

The position of the electron relative to the nucleus may be unambiguously defined by giving its *Cartesian* coordinates x, y, z or its polar coordinates r, ϑ, φ.

8.3.1.1. The Ground State of the H-Atom

The (real) function ψ_{1s}, for the lowest-energy eigenvalue, describes the ground state of the H-atom with the nuclear charge number $Z = 1$. With an energy of -13.6 eV relative to the energy of the fully dissociated H-atom (arbitrarily set at zero), it is dependent only on the distance r, not on the angles ϑ and φ (*Fig. 8.1*). The probability $\mathrm{d}W$ of finding the electron in any given volume element $\mathrm{d}V$, of distance r from the nucleus, is

$$\mathrm{d}W = \psi_{1s} \cdot \psi_{1s} \mathrm{d}V$$

The probability of finding the electron in the entire set of volume elements at distance r, *i.e.*, in a spherical shell with inner radius r and outer radius $r + \mathrm{d}r$, is

$$\mathrm{d}W = \psi_{1s} \cdot \psi_{1s} \cdot 4\pi r^2 \mathrm{d}r$$

The functions ψ_{1s}, ψ_{1s}^2, and $\psi_{1s}^2 \cdot 4\pi r^2$ are represented in *Fig. 8.2.* They reveal an apparent paradox. The probability density dW/dV is at its maximum at $r = 0$, hence at the nucleus. On examining the distance of the electron from the nucleus, however, it can be seen that the probability of encountering the electron at distance $r = 0$ is zero. The most probable distance of the electron from the nucleus in the H-atom ground state results at 53 pm (the *Bohr* radius a_0). In the *Bohr* model of the H-atom, though, the electron always has this separation from the atomic nucleus, while the quantum-mechanical model gives the result that the electron has a 20% probability of being found in a spherical shell of thickness of ± 10 pm on either side of the *Bohr* orbit.

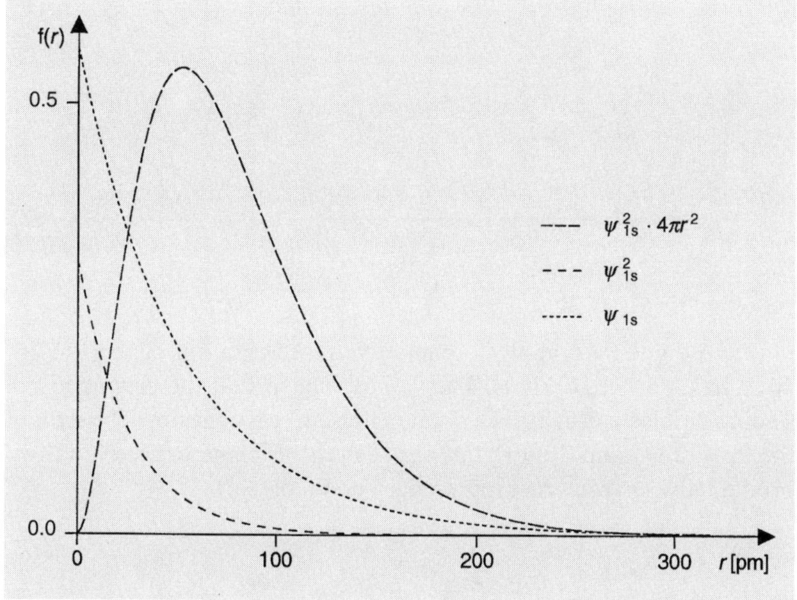

Fig. 8.2. *Representation of the functions ψ_{1s}, ψ_{1s}^2, and $\psi_{1s}^2 \cdot 4\pi r^2$ for the H-atom*

For the graphic representation of s-orbitals most frequently used in chemistry, a sphere of finite radius is chosen so as to give a 90% probability of finding the electron within that sphere.

8.3.1.2. Higher Energy States of the H-Atom

Many excited states, in which the electron is described by different eigenfunctions (such as ψ_{2s}), exist besides the electronic ground state. The functions ψ_{2s}, ψ_{2s}^2, and $\psi_{2s}^2 \cdot 4\pi r^2$ are represented in *Fig. 8.3.* The key difference between the two orbitals 1s and 2s is that the function ψ_{2s} has a value of zero at $r = 106$ pm, because of a change in sign, so that, at this distance from the nucleus, the probability density of finding the electron is equal to zero. The relative sign of the eigenfunction is of importance for the qualitative MO model, so it must always be given: either directly, by '+' or '−', or indirectly by *hatching* or *non-hatching*. Because it is the square of the eigenfunction that is decisive for the electron density, it makes no difference which particular region is denoted with '+' and which with '−'. The horizons where the function value equals zero, because of a sign change in the eigenfunction, are called *nodal planes*.

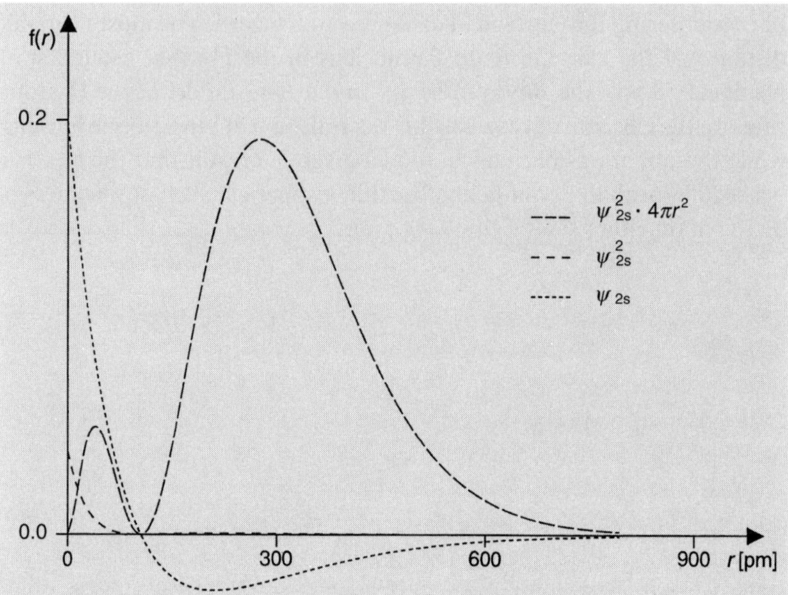

Fig. 8.3. *Representation of the functions* ψ_{2s}, ψ^2_{2s}, *and* $\psi^2_{2s} \cdot 4\pi r^2$ *for the H-atom*

Fig. 8.4 gives a graphical, qualitative representation of the 1s-, 2s-, $2p_x$-, $2p_y$-, and $2p_z$-orbitals. They are drawn so that the probability of finding the electron at a given point on the curve is constant. The signs of the $2p_x$-, $2p_y$-, and $2p_z$-orbitals show that the plane between the two orbital halves in each case represents a nodal plane.

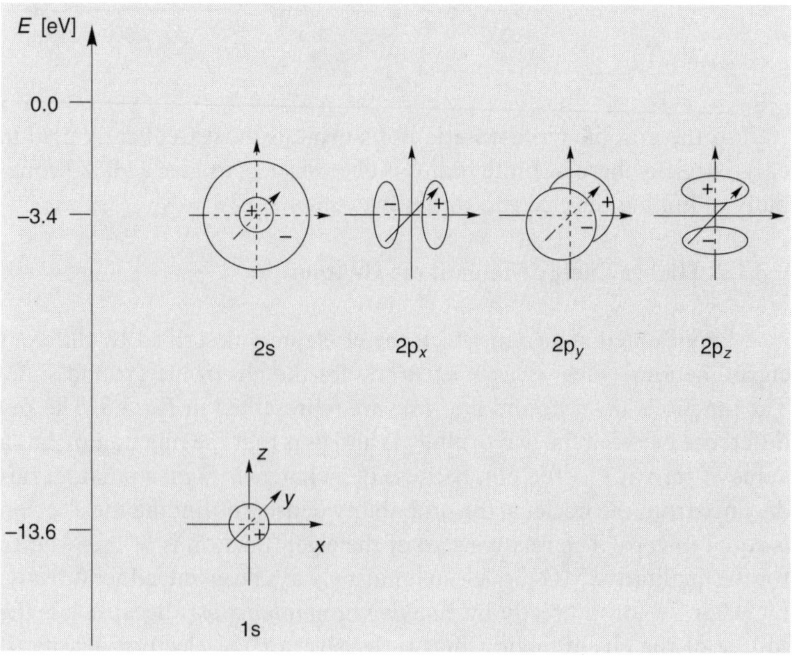

Fig. 8.4. *Qualitative representation of the five lowest-energy H-atom orbitals*

8.3.1.3. Atoms with Higher Nuclear Charge Numbers

Up to now, the significance of orbitals has been demonstrated using the example of the H-atom with the nuclear charge number $Z = 1$. The knowledge gained from this also applies for other one-electron atoms, with the essential difference that
- the energy of a particular orbital decreases proportionally to Z^2
- the orbital, while retaining its form, shrinks by a factor of $1/Z$.

The shrinking of orbitals upon increasing nuclear charge number is easily inferable. The electron will increasingly be attracted towards the nucleus, and so will reside closer, on average, to the nucleus at any given time.

8.3.1.4. Occupied and Unoccupied Orbitals and Their Energies

A one-electron atom can only exist in particular states, which may each be described by a state function and its associated energy. The electron is said to be located in a particular orbital ψ. The energy of the system is, therefore, identified with the orbital energy of ψ. There also exist other, unoccupied, orbitals. These, similarly, are assigned orbital energies, corresponding to the energy of the one-electron atom, in that state, were that orbital occupied. This description, separating orbitals from electrons, is only exact for one-electron systems, but, because of its simplicity, it is also used for many-electron systems – then only as an approximation, however.

8.3.2. Many-Electron Atoms

While the *Coulomb* force, acting on an electron, is always, in one-electron systems, directed towards the nucleus, an individual electron in a many-electron system is subject to a vectorial combination of forces: the force directed towards the nucleus (assumed to be at rest) and the repulsive forces between the electrons, which move relative to one another. The resulting total force operating on the individual electron is, therefore, no longer centrosymmetrical. Many-center problems of this kind cannot be solved even under the assumptions of *Newton*ian mechanics. It is, therefore, not surprising that the *Schrödinger* equation for many-electron atoms, even for the He-atom, is, intrinsically, no longer analytically solvable.

It has, therefore, been necessary to introduce such approximations which are mathematically manageable, and which afford usable solutions through linear combinations. One such is represented by the 'orbital approximation'.

8.3.2.1. The Orbital Approximation

In the simplest case, electron/electron interactions are ignored. Physically, this means that each individual electron – as in the case of the one-electron atom – is considered only to 'feel' the attraction to the nucleus, and no longer the repulsion of the other electrons. In other words, one single problem with one eigenfunction exists for each one of the n electrons, and this eigenfunction may be applied as in a one-electron system to afford the solution for that electron.

Hence, the eigenfunction $\psi(\boldsymbol{r}_1, \boldsymbol{r}_2 \ldots \boldsymbol{r}_n)$, simultaneously dependent on the coordinates of all n electrons, may be approximated by the product of n functions, each dependent solely on the coordinates of one individual electron:

$$\psi(\boldsymbol{r}_1, \boldsymbol{r}_2 \ldots \boldsymbol{r}_n) \xrightarrow[\text{approximation}]{\text{orbital}} \varphi_1(\boldsymbol{r}_1) \cdot \varphi_2(\boldsymbol{r}_2) \ldots \varphi_n(\boldsymbol{r}_n)$$

These functions φ_1–φ_n, because they resemble the functions of the H-atom, are described as hydrogen-like one-electron functions. Although the product does not fulfill the *Pauli* principle (*Sect. 8.3.2.2*), we will refrain from introducing the so-called *Slater* determinants, as these are not necessary for our discussion.

The advantage of this orbital approximation lies in its being founded on the already known one-electron functions. Electrons can once more be filled into one-electron orbitals and the total energy of the system calculated by addition of the energies of the orbitals occupied. This procedure, however, unlike that one for an one-electron atom, only represents a crude approximation, because of the neglect of electron/electron interactions. The model of the so-called *effective nuclear charge* offers the simplest way of allowing, at least partially, for these interactions in the context of the orbital approximation. It makes the assumption that a reduced, *effective* nuclear charge, rather than the full charge, acts on individual electrons in the many-electron atom. The reason for this lies in the time-averaged deshielding of the actual nuclear charge by the remaining electrons. The mutual influence on electron movement, the so-called electron correlation, is deliberately ignored.

Like in the one-electron atom, energy and 'size' of individual atomic orbitals are dependent on nuclear charge – not the real, but the effective one. Effective nuclear charge is determined by

- *the number and 'nature' of electrons.* The deshielding effect of an electron in the 1s-orbital is stronger for an electron in a 2p-orbital than for an electron in the 2s-orbital. Consequently, the effective nuclear charge acting on an electron in the 2s-orbital is greater than that on an electron in a 2p-orbital of the same atom, so that the energetic degeneration of the 2s- and 2p-orbitals is lost on transition to the many-electron atoms. This explains the splitting of the various energy levels in many-electron atoms: the basis for the 'Aufbau' principle of the periodic system of the elements (see inner front cover).
- *The actual nuclear charge.* The greater the actual nuclear charge, the greater the effective nuclear charge – despite the increasing number of deshielding electrons – as an additional electron is only capable of deshielding less than one complete additional nuclear charge. Hence, for example, the effective nuclear charge in the He-atom amounts to *ca.* 1.7. This means that the atomic orbitals become 'smaller' and lower in energy the more one proceeds in the series from lithium to fluorine.

8.3.2.2. Occupation Rules and Electronic Structure

There exist particular limitations and rules for the allowed combinations of the various possible one-electron functions:

- The *Pauli principle* dictates that electrons in the same atom or molecule may only be described by one and the same one-electron function if they differ in their *spin*. As an electron can only have two different spin directions (↑ or ↓, characterized by the spin quantum numbers $s = +1/2$ or $-1/2$), it follows that one orbital can only be 'occupied' (*populated*) by a maximum of two electrons, which must then exhibit opposite spins.
- The most stable electron population for a given set of orbitals is that one in which the electrons, in pairs (with opposite spins), occupy the orbitals in ascending order of energy (hence, starting at the lowest-lying orbital).
- For occupation of two degenerate orbitals by two electrons, the more stable ordering is that one in which the two electrons exhibit the same spin, and so each of the two orbitals is occupied by one electron (*Hund*'s rule).

There are several possible ways in which a given set of orbitals may be occupied by a given number of electrons (*Fig. 8.5*). The different populations are each characterized, according to the orbital approximation, as a product of one-electron functions, and described in terms of electronic structure (*electronic configuration*, as the physicist would say). The ground state of the Li-atom, for example, is described by the electronic structure $1s^2\,2s$.

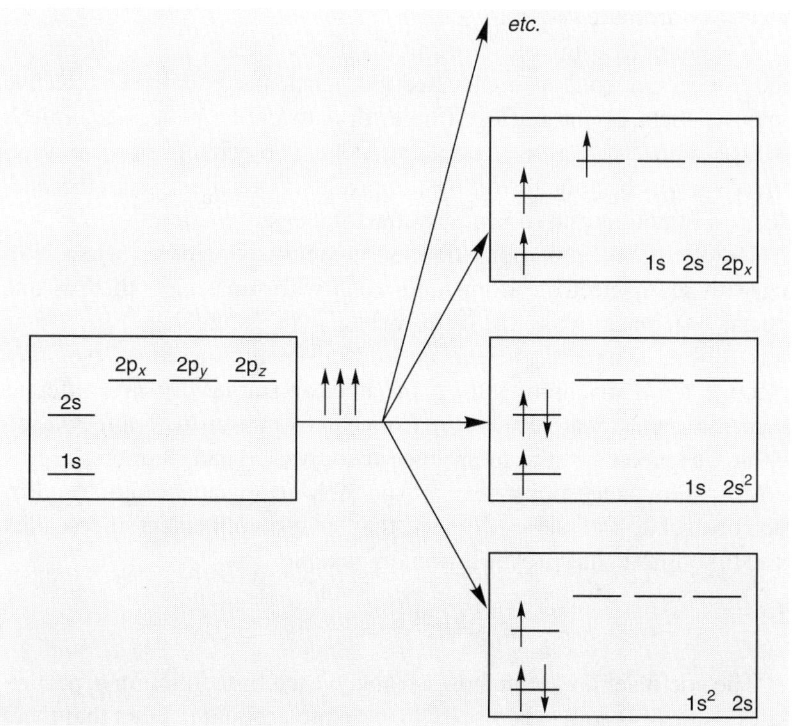

Fig. 8.5. *Possible patterns of occupation for three electrons in a given set of orbitals*

8.3.3. Molecules

8.3.3.1. The LCAO-MO Approach

A molecule consisting of *m* atomic nuclei and *n* electrons is assigned an eigenfunction

$$\psi\,(\boldsymbol{R}_1, \boldsymbol{R}_2 \ldots \boldsymbol{R}_m, \boldsymbol{r}_1, \boldsymbol{r}_2 \ldots \boldsymbol{r}_n)$$

dependent on the coordinates $\boldsymbol{R}_k$ of the nuclei and the coordinates $\boldsymbol{r}_l$ of the electrons. Because of their very much smaller mass and inertia, the electrons can adapt to reorientation of the nuclei practically without any time-lapse, *i.e.*, from the point of view of the nuclei, the movement of the electrons is 'infinitely' rapid. From the electrons' perspective, in contrast, the movement of the nuclei is 'infinitely' slow, so that, for the behavior of the electrons, nuclear motion is unimportant: only the immediate nuclear framework is of significance. Because of this, nuclear and electronic motions to a very good approximation may be separated (*Born-Oppenheimer* approximation):

$$\psi(\boldsymbol{R}_1 \ldots \boldsymbol{R}_m, \boldsymbol{r}_1 \ldots \boldsymbol{r}_n) = \chi(\boldsymbol{R}_1 \ldots \boldsymbol{R}_m) \cdot \psi'(\boldsymbol{r}_1 \ldots \boldsymbol{r}_n)$$

The second factor of this product describes the behavior of electrons in a fixed nuclear framework.

For a molecular framework of nuclei, there are – like in atoms – various 'allowed' electronic structures, each with its own discrete energy. The energies of all possible nuclear frameworks form the energy hypersurface – also called the potential-energy surface – of the molecule in its electronic ground state or one of its excited states in the energy/atomic-nucleus coordinate system.

It is possible, using one-electron functions ϕ_i, each giving the probability of an electron's being located in a particular part of the effective potential field of the nuclear framework, to define *molecular orbitals* (MOs), similarly to atomic orbitals (AOs). The orbital approximation offers various options to exploit the products of suitably selected one-electron functions and so produces the total eigenfunction.

Despite the orbital approximation's extensive analogies, when dealing with many-electron atoms and when with molecules, there is one crucial difference: while the set of orbitals (1s, 2s, $2p_x$, *etc.*) for atoms only differs in extent and energy from atom type to atom type, the effective fields originating from the nuclear framework are different from molecule to molecule and from conformation to conformation within one molecule, and so produce an individual and characteristic set of MOs ϕ_i for every molecule in any possible arrangement of nuclei. For the construction of these MO sets, the linear combination of so-called base functions φ_j has proved especially reliable:

$$\phi_i = \sum_j c_{ij}\, \varphi_j$$

The coefficients c_{ij} state how 'strongly' each base function φ_j participates in the MO ϕ_i. It is necessary to take into account the fact that there are several different sets of base functions. For chemists, used to conceive molecules being constructed out of atoms, the procedure in which the desired one-electron functions of a molecule are built up out of the known one-electron functions of its participating atoms, following the modular practice (use of building blocks), has won general acceptance, being known as LCAO-MO (*L*inear *C*ombination of *A*tomic *O*rbitals to *M*olecular *O*rbitals). The number of MOs calculated must always correspond to the number of AOs used in their construction.

8.3.3.2. Orbital Symmetry

A symmetry operation transposes an orbital ψ into a new orbital ψ'. For non-degenerate orbitals is

$$\psi' = c \cdot \psi$$

where c represents a constant, characteristic for the symmetry operation in question, which may have the value +1 or −1. If the orbital does not alter its sign on execution of the symmetry operation ($c = +1$), it is *symmetrical* with respect to this symmetry operation. If, however, it alters its sign, then it is *antisymmetrical* ($c = -1$). This is demonstrated in *Table 8.3*, using the example of the $2p_x$-AO: this orbital is symmetrical with respect to rotation about the x axis and reflection across the xy or xz plane, and antisymmetrical with respect to reflection across the yz plane, inversion and rotation about the y or z axis.

In the drawing up of correlation diagrams (*vide infra*), the symmetry of MOs plays a decisive role. Each MO has been constructed in such a way that it belongs to the same symmetry point group as does the molecule (or its model) which it serves to describe. Its behavior relative to the respective symmetry operations can then be characterized by one of the irreducible representations of the point group concerned (see *Chapt. 11.4*). Use is made of this classification in *Sect. 8.3.3.4.3*, *Chapt. 14.4.2*, and *Chapt. 14.5*, though the fact is not expressly mentioned there.

8.3.3.3. Correlation Diagrams

8.3.3.3.1. Bonding, Antibonding, and Nonbonding Molecular Orbitals

The selection of a suitable method for numerical calculation of MOs is dependent on the particular problem. In many cases, it suffices to use results from the qualitative MO model, in which MOs are worked out through linear combination of the AOs of participating atoms. The following rules apply for the linear combination of AOs to MOs:

- The AOs to be combined to an MO must be of the same irreducible representation of that symmetry point group to which the molecule (or molecular fragment) concerned belongs.
- For combination, it is best to use AOs which, energetically, are not very far removed from each other. Besides this, only those AOs are used for which there exists a region of space in which both exhibit a

Table 8.3. *Symmetry Behavior of the $2p_x$-orbital*

Symmetry operation		Result	c
Identity operation	E	Symmetrical	+1
Reflection	$\sigma(xy)$	Symmetrical	+1
Reflection	$\sigma(xz)$	Symmetrical	+1
Reflection	$\sigma(yz)$	Antisymmetrical	−1
Rotation	$C_2(y)$	Antisymmetrical	−1
Rotation	$C_2(z)$	Antisymmetrical	−1
Rotation	$C_\infty(x)$	Symmetrical	+1
Inversion	i	Antisymmetrical	−1

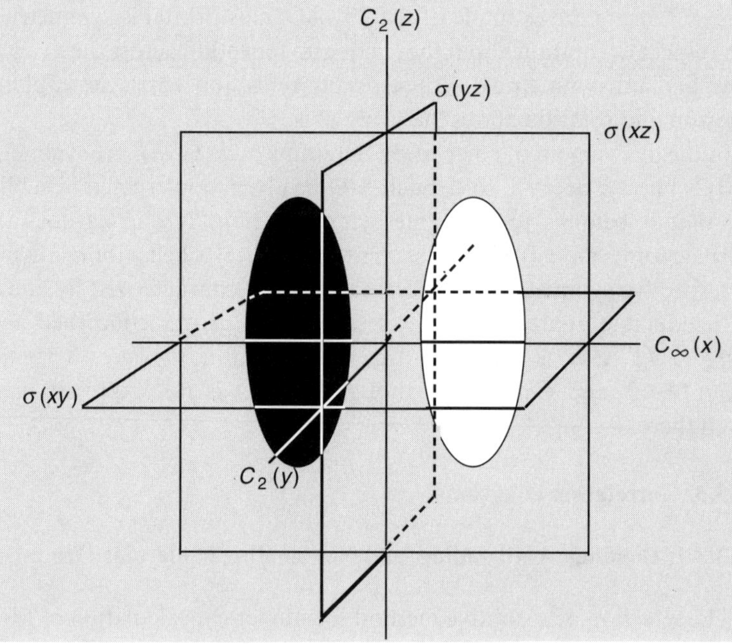

considerable electron density. The extent of this spatial interpenetration (*overlap*) of two AOs φ and φ' is given by the so-called *overlap integral*:

$$\int \varphi\varphi'^{*}\mathrm{d}V$$

It is a measure of the lowering in energy of those linear combinations in which both of the AOs in the area of overlap are of the same sign, and also a measure of the raising of the energy of those linear combinations in which they are of opposite sign. The first type of MOs is known as *bonding*, the latter as *antibonding* MOs.

The stabilizing or destabilizing energy may be calculated using a different integral, and is often simply called the *resonance integral*. The resonance integral does not have any direct relationship to the resonance theory mentioned in *Chapt. 14*.

In linear combination of AOs which fulfill the mentioned prerequisites, the choice of the coefficients is limited by the requirement for orthogonality in the resulting MOs. The MOs of the H_2 molecule may illustrate this point.

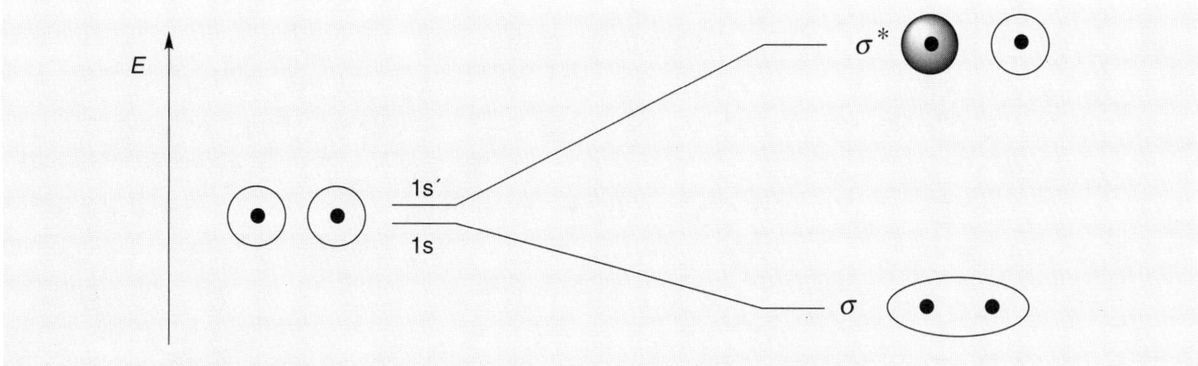

Fig. 8.6. *Positive and negative linear combinations of two 1s-orbitals*

The two 1s-orbitals of the H-atoms participating in the molecule will serve as AOs. They are represented in the correlation diagram of *Fig. 8.6* (left). The two orbitals fulfill all of the conditions mentioned above. They are of the same symmetry and energy, and interpenetrate significantly in the H_2 molecule. We are going to form a positive and a negative linear combination (right):

$$\sigma = (1s + 1s')/\sqrt{2} \text{ and } \sigma^* = (1s - 1s')/\sqrt{2}.$$

The two normalized linear combinations are orthogonal to one another, as the volume integral

$$\int((1s + 1s')/\sqrt{2})\cdot((1s - 1s')/\sqrt{2})dV$$

becomes zero. As the AOs are normalized, consideration of the coefficients of the AOs in the two MOs is sufficient to check this conclusion. For the σ-MO, these are $1/\sqrt{2}$ for both the 1s- and the 1s′-AO, in the σ^*-MO, however, they are $1/\sqrt{2}$ for the 1s-orbital and $-1/\sqrt{2}$ for the 1s′-AO. By addition of the products of the AO coefficients of the two MOs, for the above integral, we obtain

$$2\cdot(1/\sqrt{2})\cdot(1/\sqrt{2}) + 2\cdot(1/\sqrt{2})\cdot(-1/\sqrt{2}) = 0.$$

The positive linear combination σ is a bonding MO. The increased electron density between the two nuclei leads to a decrease in the energy of this orbital. The negative linear combination σ^*, in contrast, has a nodal plane between the two nuclei, leading to an increase in energy for this antibonding orbital.

Similar rules to those applying for occupation of AOs are valid for these two MOs: these lead to four different possible electronic structures (*Fig. 8.7*). The complete symbol for denoting an electronic structure also gives the total spin $S = \Sigma s$, by a superscript to the left. It is not given directly, however, but in form of the spin multiplicity $2S + 1$.

Total Spin	Spin Multiplicity	Notation
0	1	*Singlet*
1	3	*Triplet*

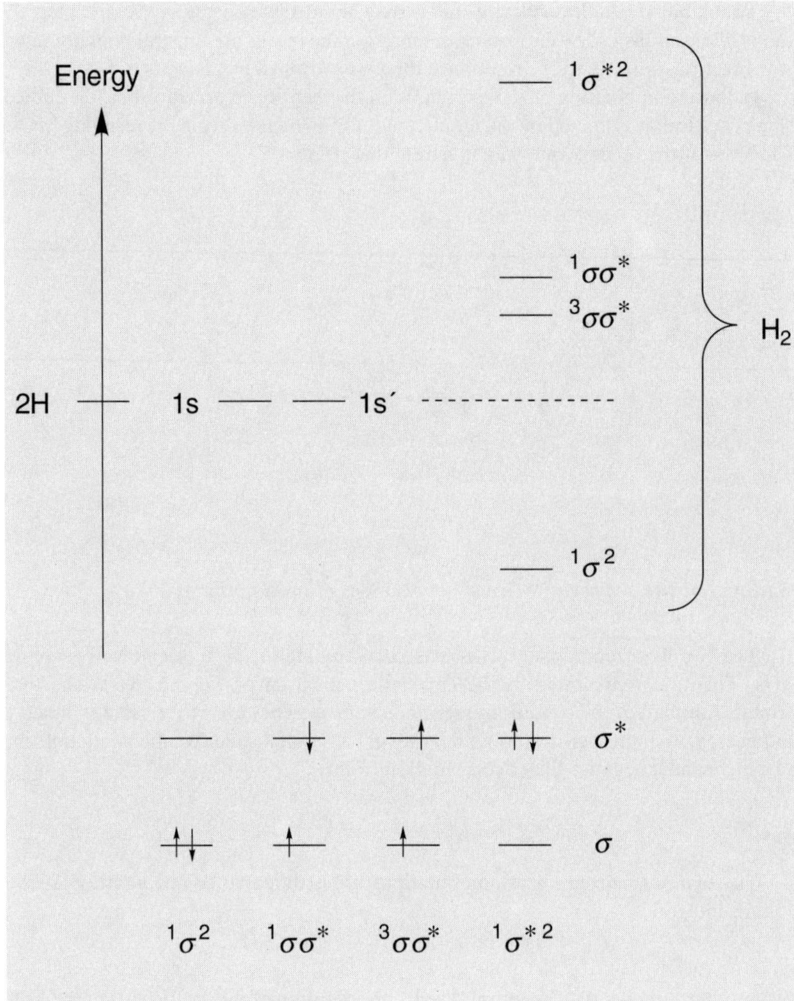

Fig. 8.7. *Possible patterns of occupation for the two electrons after overlapping of two 1s-orbitals, together with the relative energies of the possible electronic structures of the H_2 molecule*

The destabilization incurred by overlapping of two orbitals of opposite sign is always greater in magnitude than the stabilization due to two corresponding overlapping orbitals of the same sign. Therefore, comparison of the energies of the various electronic structures of the H_2 molecule with the energy of two isolated H-atoms permits the following conclusions to be drawn:

- The electronic structure $^1\sigma^2$ (the ground state) of the H_2 molecule is more stable than two isolated H-atoms.
- The structures $^1\sigma\sigma^*$ and $^3\sigma\sigma^*$ are less stable than two isolated H-atoms: this means that, on transition of an electron from the σ-MO to the σ^*-MO, dissociation of the H_2 molecule is to be anticipated.

Qualitatively, the same correlation diagram results for a He_2 molecule. The ground state of the He_2 molecule would be described by the electronic structure $^1(\sigma^2\sigma^{*2})$, which would be less stable than the two separate He-atoms with the electronic structures $1s^2$ and $1s'^2$; this explains in the context of the MO model, why a H_2 molecule does exist, whereas a He_2 molecule does not.

The decrease or increase in energy of MOs is dependent on the degree of overlap of the AO participants: the greater the degree of overlap, the stronger the change in energy. The extent of the overlap is determined by the size of the AOs and their mutual orientations. In the case of the H_2 molecule, the overlap of the two 1s-AOs increases with decreasing internuclear separation, so that the energy of the σ-MO initially decreases, while that one of the σ^*-MO increases continually. The equilibrium distance between the two atoms in the ground state of the H_2 molecule is characterized by this tendency favoring the smallest possible internuclear separation being compensated for by repulsion between the two atomic nuclei (*Fig. 8.8*).

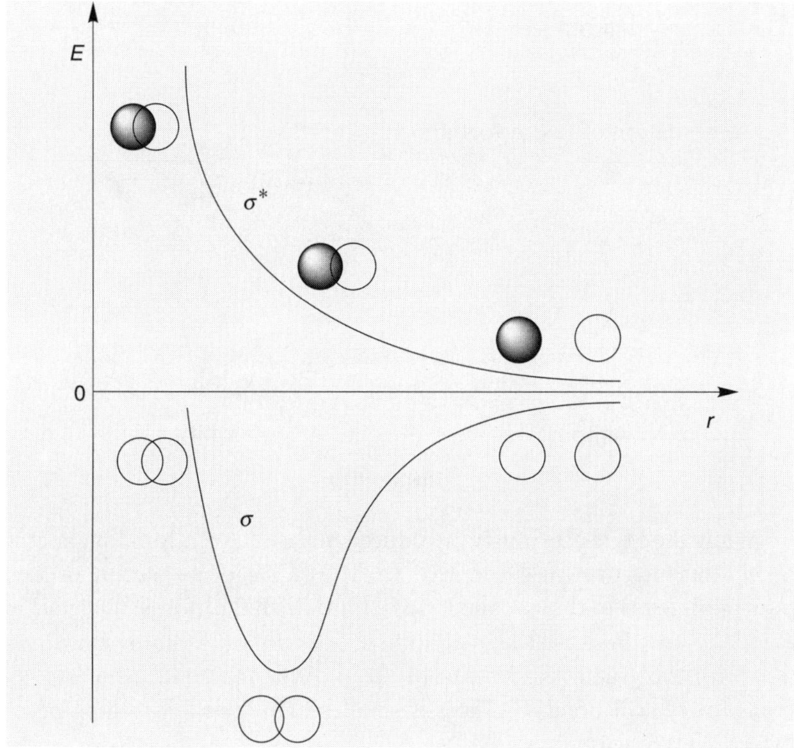

Fig. 8.8. *Energy change of the σ- and σ*-MOs depending on the internuclear distance*

With this result, which gives the total energy of the H_2 molecule and the equilibrium distance between the two protons upon occupation of the bonding MO by an electron pair, the *covalent bond* in the H_2 molecule is specified as an electron-pair bond, and the *valence bond concept*, intimately bound up with the *molecular concept* of the chemist, placed on a physical foundation.

Formation of a chemical bond is also possible with one electron in the σ-MO. For such a H_2^+ ion, the MO model predicts that the equilibrium internuclear distance must be greater than that one in the H_2 molecule, as the same internuclear repulsion is countered, in the former case, by a bond with only one electron, and in the latter case, by one with two electrons. It is indeed possible, under certain conditions, to produce a H_2^+ ion, whose internuclear distance, at 106 pm, lies significantly above the value of 74 pm for the H_2 molecule.

It is possible to execute other AO combinations, in a manner similar to that one used in the combination of two s-AOs to form a σ-MO and a σ^*-MO.

Combination of an s- and a p_x-orbital:

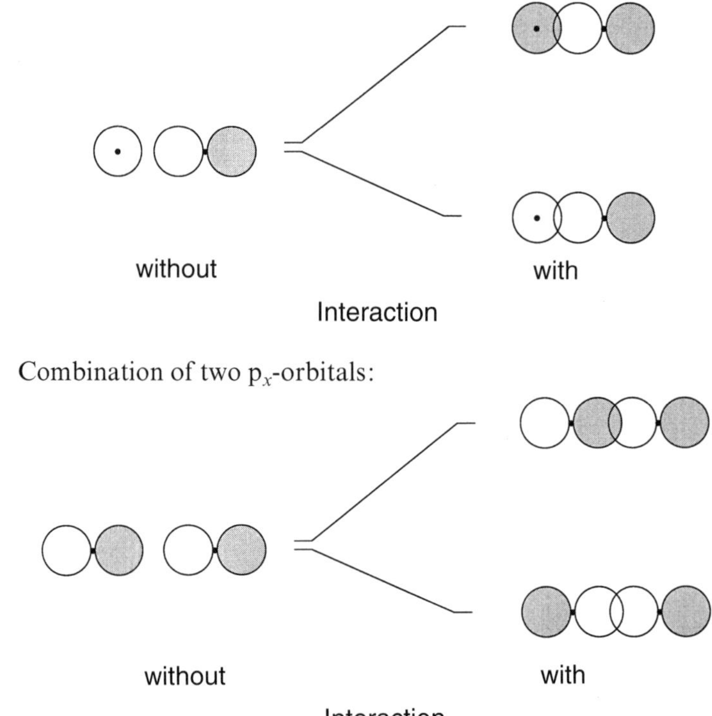

Combination of two p_x-orbitals:

While the MOs previously introduced possessed rotational symmetry about the internuclear bond axis (σ-MOs), the combination of two p_z-orbitals leads to the so-called π-MOs, in which the atomic nuclei lie in a nodal plane. Because the overlap here does not occur along the direction with the greatest extension of the p_z-AOs, the splitting of energy levels for typical bond distances is smaller than for the overlapping of two p_x-AOs to form two σ-MOs.

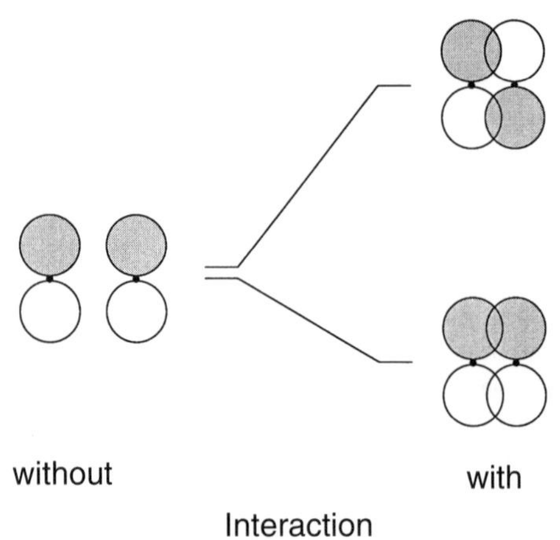

However, the four following combinations of AOs each show different symmetry behavior relative to rotation about the bond axis.

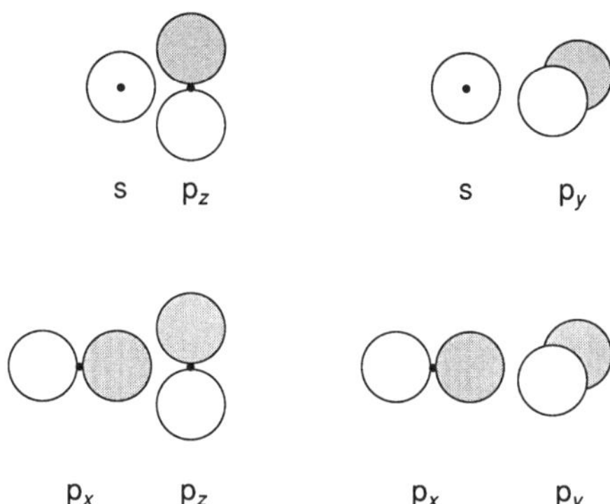

It is inevitable in such cases that, in the area of overlap, identical volumes of compatible and opposite sign of both AOs will be present, so that, overall, no stabilization energy will be engendered. This is demonstrated on the example of the overlapping of an s-AO with a p_z-AO on the right.

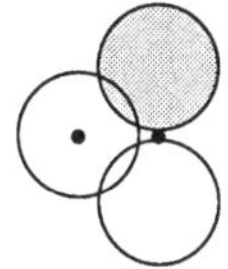

The combination to an MO of p_z on one atom and p_y on another is – once more because of different symmetry behavior – 'forbidden'. We shall examine the symmetry operations $C_2(x)$, $\sigma(xy)$, and $\sigma(xz)$:

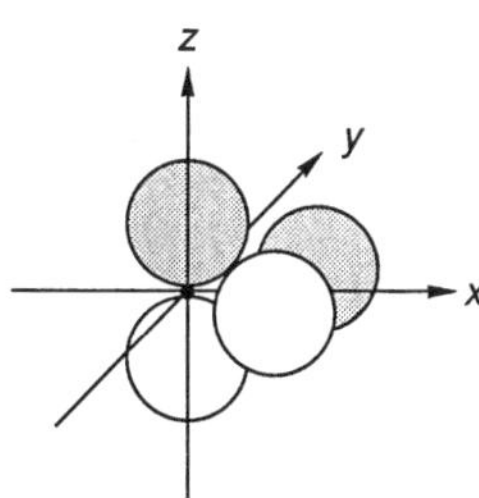

While both AOs are antisymmetrical relative to $C_2(x)$, p_y behaves symmetrically upon operation of $\sigma(xy)$, and antisymmetrically on execution of $\sigma(xz)$. The opposite applies for p_z.

For the linear combination of two 1s-AOs to the σ- and σ^*-MOs of the H_2 molecule, the coefficients were identical in magnitude. For the combination of orbitals of different type, or belonging to atoms of different elements, the coefficients are no longer of equal degree. In general, the AO with the lower energy has a larger coefficient in the bonding MO than that one with higher energy. Because of the orthogonality of MOs, the reverse is true for the antibonding MO. Exact numerical values of the coefficients of AOs may only be obtained by quantitative calculation (see *Aspects of Organic Chemistry*, Volume 4: *Methods of Structure Determination*). We will nevertheless give the results of such calculations for some examples, especially for those ones introduced at the beginning of this chapter.

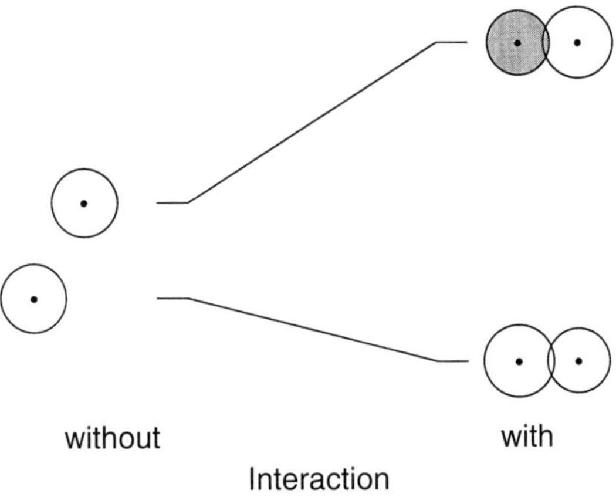

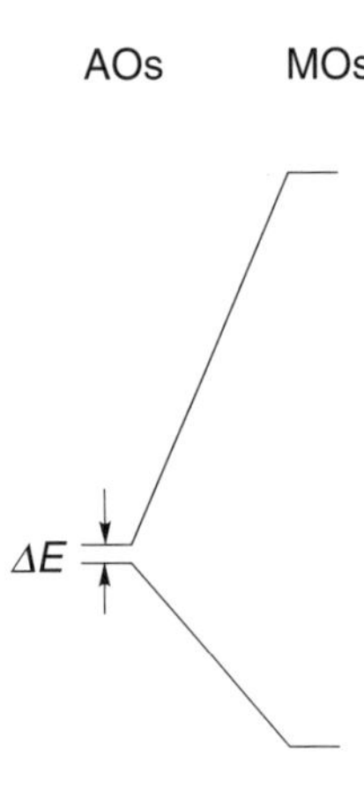

The degree to which two MOs are stabilized or destabilized, relative to the contributing AOs, is dependent on the energy difference between these two AOs: the greater the energy difference, the smaller the stabilization or destabilization effect.

We have now learned the most important rules necessary for the qualitative construction of the MOs of a given framework of nuclei by means of a correlation diagram. These rules form the core of the qualitative MO model.

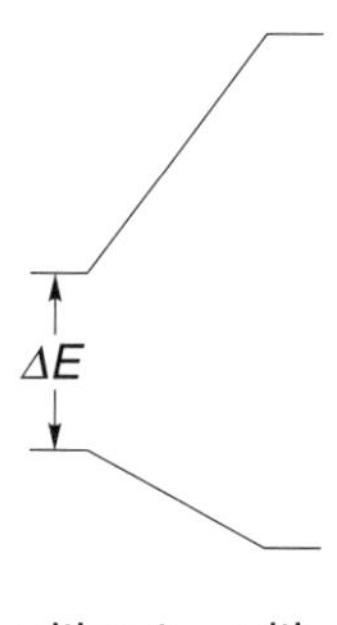

In the following section, correlation diagrams will be constructed for linear and non-linear methylene. Here, in addition to *bonding* and *antibonding* MOs, *nonbonding* MOs will be encountered.

8.3.3.3.2. Correlation Diagram for Linear Methylene

There are two ways in which to split the proposed framework of nuclei of assumed linear methylene into two fragments, **A** and **B**, with the appropriate fragment orbitals (a fragment orbital is an orbital only extending over a part of a molecule).

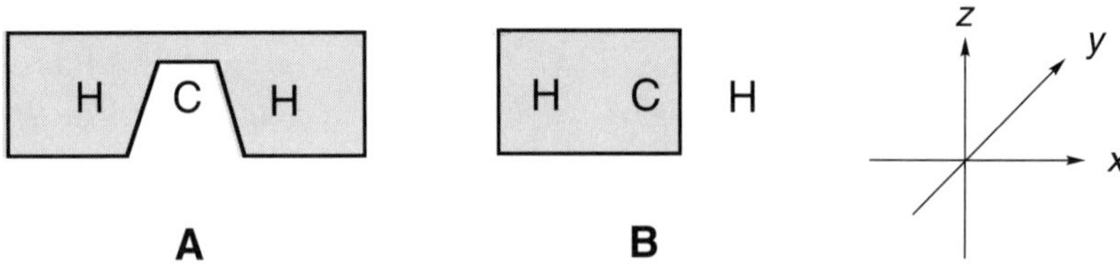

In principle, these two ways are equivalent. However, option **A**, shown at the left hand side above, is the more convenient one for the construction of a correlation diagram, for reasons of symmetry. As shown in *Fig. 8.9*, linear methylene is expediently broken down into the fragments of a *C-atom* and an *extended* H_2 *molecule*, and the associated fragment orbitals are arranged according to their energies.

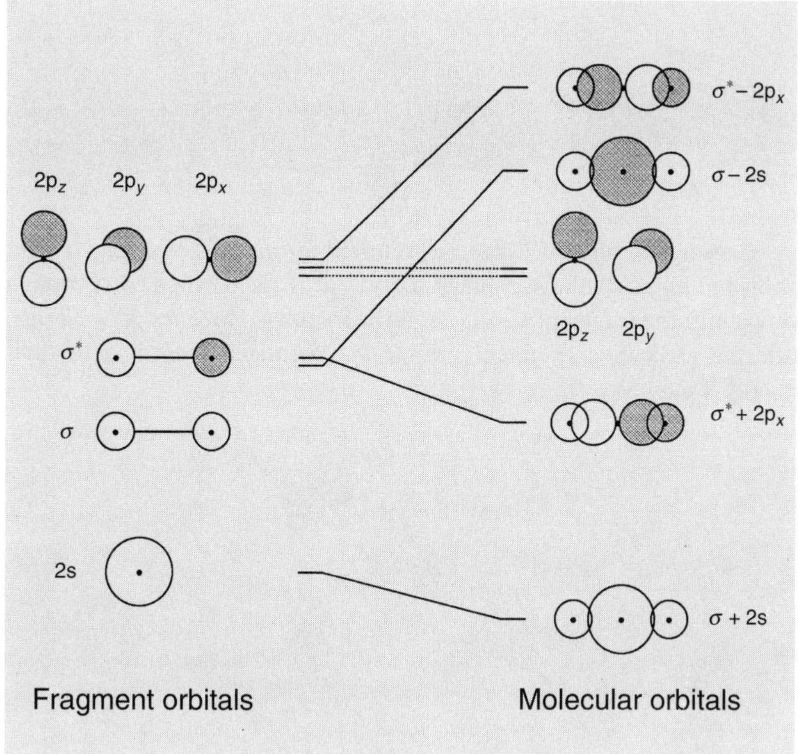

Fig. 8.9. *Orbital-correlation diagram for linear methylene*

To simplify the correlation diagram which needs to be constructed, the analysis will not be carried out in the symmetry point group $D_{\infty h}$, but in the subgroup D_{2h}: this alters nothing in the final result.

The classification of the fragment orbitals relative to the symmetry operations of the symmetry point group D_{2h} (three reflections, three C_2 rotations, and an inversion) immediately shows that

- the 2s-orbital of the fragment *C-atom* and the σ-orbital of the fragment *extended H_2 molecule* both exhibit the same symmetry behavior relative to all symmetry operations, and consequently undergo splitting of energies on overlapping, forming a bonding combination $\sigma + 2s$ and an antibonding combination $\sigma - 2s$;
- the $2p_x$-orbital of the fragment *C-atom* and the σ^*-orbital of the fragment *extended H_2 molecule* also exhibit identical behavior relative to all symmetry operations, and consequently undergo splitting of energies on overlapping, forming a bonding combination $\sigma^* + 2p_x$ and an antibonding combination $\sigma^* - 2p_x$;
- the fragment orbitals $2p_y$ and $2p_z$ find no partner with identical symmetry behavior in the fragment *extended H_2 molecule*, and hence remain unaltered (*nonbonding MOs*).

In accordance with the occupation rules, the ground state of assumed linear methylene will be given by the electronic structure $(\sigma + 2s)^2\,(\sigma^* + 2p_x)^2\,2p_y\,2p_z$, with the two electrons in the $2p_y$- and $2p_z$-orbitals having parallel spin (*triplet state*).

8.3.3.3.3. Correlation Diagram for Non-Linear Methylene

In this case there are also two possible ways (**A** and **B**), in which to split the proposed framework of nuclei of assumed non-linear methylene into two segments with appropriate fragment orbitals.

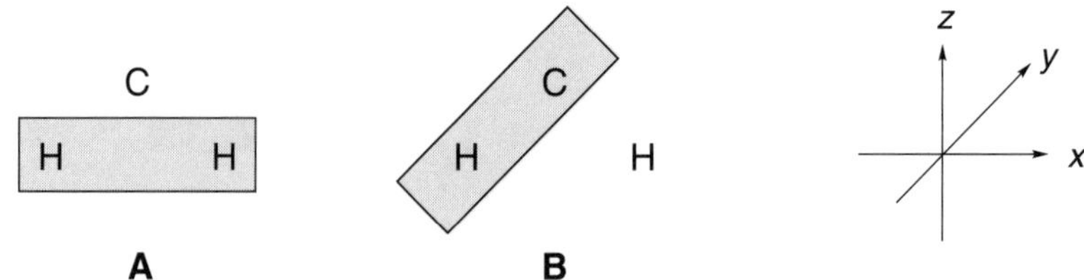

Once again, option **A** is to be preferred for reasons of symmetry. As shown in *Fig. 8.10*, the non-linear methylene is again expediently broken down into the fragment *C-atom* and the fragment *extended H_2 molecule*, and the associated fragment orbitals are arranged according to their energy levels.

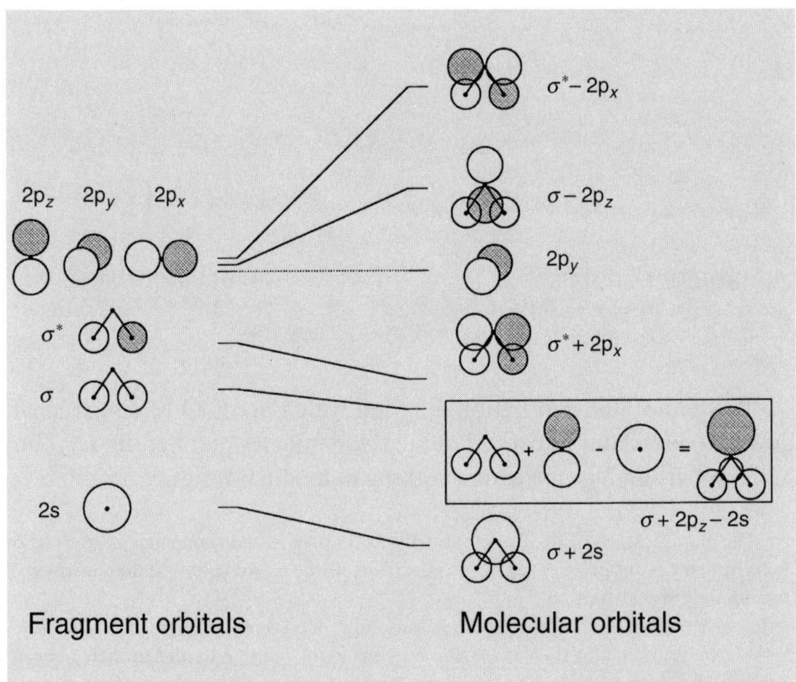

Fig. 8.10. *Orbital-correlation diagram for non-linear methylene*

Classification of the fragment orbitals relative to the symmetry operations of the symmetry point group C_{2v} (two reflections and the C_2 rotation) immediately shows that

- the $2p_y$-orbital has no partner in the fragment *extended H_2 molecule*, and so remains unchanged as a nonbonding orbital;
- the $2p_x$-orbital enters into interaction exclusively with the σ^*-orbital, and these two orbitals split, into the bonding combination $\sigma^* + 2p_x$ and the antibonding combination $\sigma^* - 2p_x$;
- the three remaining fragment orbitals σ, 2s, and $2p_z$ all display identical symmetry behavior.

Going by what has been said previously, it would be possible to arrive at the proposition of simply overlapping the σ- and the 2s-orbitals into the bonding combination $\sigma + 2s$ and the antibonding combination $\sigma - 2s$ and, equally, the σ- and the $2p_z$-orbitals into the bonding combination $\sigma + 2p_z$ and the antibonding combination $\sigma - 2p_z$. However, by doing this, a total of four final orbitals would have been created out of three initial ones, and this is not allowed.

Multiple correlations of this kind occur frequently when using the qualitative MO model, in which case, generally, only interactions between energetically similar orbitals are taken into account. In this instance, this means that the interaction between 2s and $2p_z$ is ignored. This leaves overlapping between

- the energetically highest $2p_z$-orbital with the energetically close-lying σ-orbital, in an antibonding sense, forming the orbital $\sigma - 2p_z$,
- the energetically lowest-lying 2s-orbital with the energetically close-lying σ-orbital, in a bonding sense, forming the orbital $\sigma + 2s$,
- the σ-orbital and the higher-lying $2p_z$-orbital, in a bonding sense, and, at the same time, the lower-lying 2s-orbital, in an antibonding sense, forming the orbital $\sigma + 2p_z - 2s$, as shown framed in *Fig. 8.10*.

The ground state of assumed non-linear methylene will then be given by the electronic structure $(\sigma + 2s)^2\,(\sigma + 2p_z - 2s)^2\,(\sigma^* + 2p_x)^2$ (*singlet state*).

8.3.3.3.4. *Walsh* Diagrams

Fig. 8.11 shows the MOs of linear methylene (left) and of non-linear methylene (right), together with their associated correlation lines. These correlation lines are obtained by connecting orbitals of identical symmetry behavior (relative to those symmetry operations common to both linear and non-linear methylene) to one another. It should be borne in mind that only correlation lines between orbital pairs of different symmetry behavior are permitted to cross. A diagram of this kind, showing the change in the nodal and energetic characteristics of the MOs on changing the arrangement of nuclei, is called a *Walsh* correlation diagram. In particular, it permits conclusions to be drawn about the relationship between molecular and electronic structure, giving, for example, an answer to the question of whether a molecule of three atoms is linear or non-linear. (*Nodal characteristics* refer to the localization of nodal planes in an MO.)

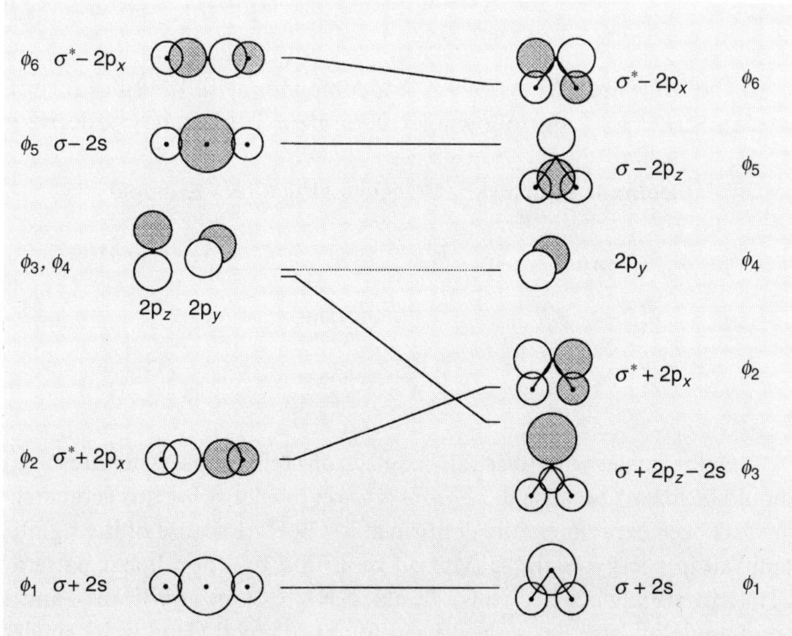

Fig. 8.11. Walsh *correlation diagram for methylene*

In the construction of the MOs of linear and non-linear methylene we could just as easily have put a Li-, Be-, B-, N-, O-, or F-atom in the place of the fragment *C-atom*. The qualitative appearance of the correlation diagram and the resulting orbitals would only have been affected slightly by this. Hence, the *Walsh* correlation diagram represented in *Fig. 8.12* describes the transition, quite generally, from a linear to a non-linear XH_2 molecule.

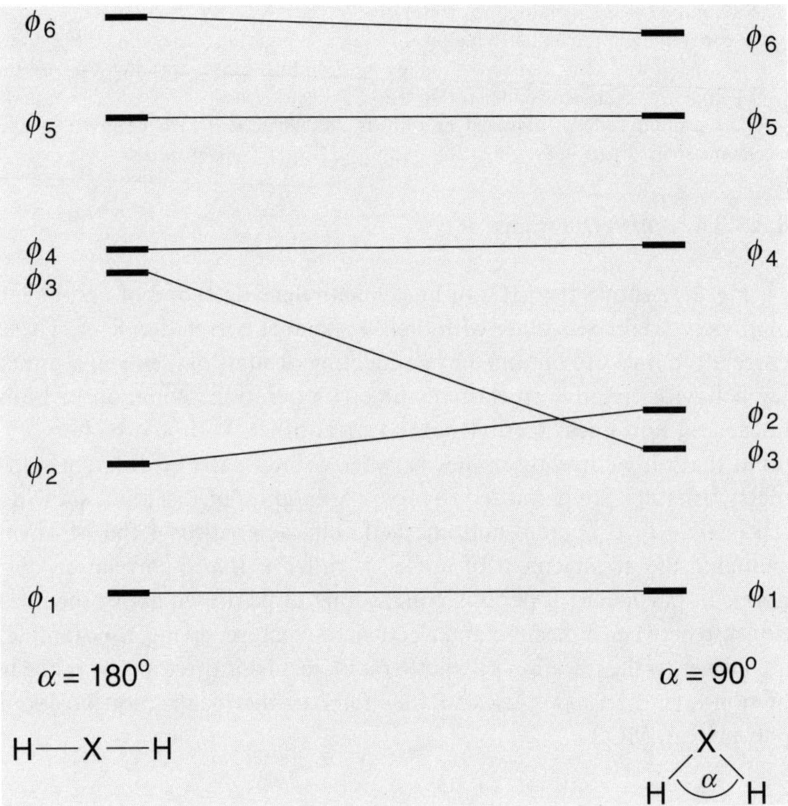

Fig. 8.12. Walsh *correlation diagram for* XH_2 (X represents an atom of the second period)

If the MOs are successively filled, complying with the rules formulated in *Sect. 8.3.2.2*, the following structural predictions may be derived:

Electronic structure	Molecular structure	Example
$\phi_1^2\ \phi_2^2$	linear	H—Be—H
$\phi_1^2\ \phi_2^2\ \phi_3^2$	non-linear	H–C̄–H
$\phi_1^2\ \phi_2^2\ \phi_3^2\ \phi_4^2$	non-linear	H–O–H

XH_2 molecules with four valence electrons (electronic structure $\phi_1^2\,\phi_2^2$) should be linear, because the ϕ_2-MO is lower in energy for this geometry: this has been experimentally confirmed for BeH_2. Because of the significant fall in energy of the ϕ_3-MO on transition to a non-linear pattern, CH_2 (with six valence electrons) should, in contrast, be non-linear – and a bond angle of 105° has indeed been found for methylene in its singlet state (*singlet methylene*; electronic structure $\phi_1^2\ \phi_2^2\ \phi_3^2$). The observation that methylene in the triplet state (*triplet methylene*; electronic structure $\phi_1^2\ \phi_2^2\ \phi_3\ \phi_4$), in which the ϕ_3-MO is only singly occupied, exhibits a bond angle of 136° is also consistent with the *Walsh* correlation diagram. The transition from six to eight electrons (from CH_2 to H_2O) should not cause any changes in structure, as the ϕ_4-MO is nonbonding, both in the linear and in the non-linear case. This conclusion is also borne out, as the bond angles in singlet methylene and in water are practically identical. Therefore, whether an XH_2 molecule is linear or non-linear, in accordance with the *Walsh* correlation diagram, solely dependent on the number and distribution of its valence electrons.

Change in electronic structure			**Change in molecular structure**
for different numbers of electrons			
$\phi_1^2\ \phi_2^2$	→	$\phi_1^2\ \phi_2^2\ \phi_3^2$	α becomes smaller
$\phi_1^2\ \phi_2^2\ \phi_3^2$	→	$\phi_1^2\ \phi_2^2\ \phi_3^2\ \phi_4^2$	α remains constant
for equal numbers of electrons			
$\phi_1^2\ \phi_2^2$	→	$\phi_1^2\ \phi_2\ \phi_3$	α becomes smaller
$\phi_1^2\ \phi_2^2\ \phi_3^2$	→	$\phi_1^2\ \phi_2^2\ \phi_3\ \phi_4$	α becomes larger

If it were desired to predict also the relative energies of singlet and triplet methylene, as well as their molecular structures, then it would be necessary to take account of two opposing effects:

- the energy difference between the two MOs ϕ_3 and ϕ_4 on transition from $^1CH_2 \rightarrow {}^3CH_2$,
- the smaller degree of electrostatic repulsion of the two unpaired electrons in triplet methylene, compared with the electron pair in singlet methylene.

Considerations of this kind expect too much of the qualitative MO model. Experimental findings show that methylene possesses a triplet ground state, *ca.* 40 kJ/mol lower in energy than the lowest singlet state.

Besides *bonding* and *antibonding* orbitals, correlation diagrams may also include *nonbonding* MOs (*Sect. 8.3.3.3.1*). As a rule, heteroatoms such as O, N, or Cl also have electron pairs in nonbonding MOs in their compounds. These electron pairs are specified *lone pairs*, to stress this fact. The geometry of molecules of, for example, type H–X–X–H is determined by the total number of bonding electrons and lone pairs. In such cases, *Walsh* diagrams state that *ethyne* should be linear, *diimine* non-linear but planar, and *hydrogen peroxide* non-planar.

Having constructed the MOs of methylene from AOs, and also the associated *Walsh* correlation diagram for linear ($D_{\infty h}$) and non-linear (C_{2v}) methylene, it is appropriate to summarize the *rules of the qualitative MO model* regarding the drawing up and interpretation of correlation diagrams.

Ethyne

Diimine

Hydrogen peroxide

- The MOs of interest are accessible through linear combination of their underlying AOs (or suitable fragment orbitals).
- Two (or more) orbitals of atoms or molecular fragments will overlap with one another, as long as they show the same symmetry behavior. The resulting MOs are either symmetrical or antisymmetrical relative to the symmetry operation appropriate to the molecule concerned.
- Two (or more) orbitals of atoms or molecular fragments will overlap increasingly – leading to lowering of energy of the bonding MOs – the better matched they are geometrically.
- Two (or more) orbitals of atoms or molecular fragments will enter all the more strongly (weakly) into interaction with one another, the less (more) they differ from each other energetically.
- The MOs – as usual – will be occupied from the bottom orbital upwards, in compliance with the rules of *Hund* and *Pauli* (*Sect. 8.3.2.2*).
- Only valence electrons are taken into account. (The total number of electrons of an atom or a molecule consists of the number of *core electrons* and the number of *valence electrons*: the C-atom (ethane), for example, has 2 (4) core electrons and 4 (14) valence electrons.) Electron/electron interactions are ignored.
- Qualitatively, MO correlation diagrams give the same picture for a whole class of molecules (of type H–X–H or H–X–X–H, for example). The individual members of this class differ in the number of their valence electrons, which must be accommodated in the orbital system considered.
- The total electronic energy of a molecule is obtained by taking account of the respective orbital energy contribution of each valence electron.

It may be inferred from the last-stated rule that it is possible to extend the qualitative discussion of electronic structure, and its related molecular properties, onto the quantitative level. What is gained in precision for an individual compound is lost on any generalization to a whole compound class.

8.3.3.4. The HMO Model and Its Simplifications

Erich Hückel's MO model (HMO model) makes use of an approximation founded in quantum mechanics, which supplies qualitative (or, at best, semiquantitative) answers to questions concerning electronic structure (especially of conjugated, unsaturated molecules).

The HMO model (see *Erich Hückel*'s monograph, as well as that one of *Edgar Heilbronner* and *Hans Bock*) has played a role of a significance which can hardly be overstated in the development of organic structure chemistry in the 1950s and 1960s.

An extensive selection of simple structures is to be found in the set of tables *Heilbronner/Straub, Hückel Molecular Orbitals* (Springer-Verlag, Berlin, 1966), listing in a user-friendly fashion the results of the calculations of one-electron eigenfunctions and one-electron eigenvalues using *Hückel*'s procedure.

8.3.3.4.1. The σ/π Separation

For the HMO model, the LCAO-MO approach (*Sect. 8.3.3.1*) is applied, with some simplifications, to π-electron systems. The first simplification is the conceptual *separation of σ- and π-systems*. In a molecule, assumed planar and defined as lying across the xy plane, the s-, p_x-, and p_y-AOs are symmetrical relative to the reflection in the molecular plane;

the p_z-AOs, however, antisymmetrical. Hence, in the construction of MOs from these AOs using the LCAO-MO procedure, no mixing ensues between the s-, p_x-, and p_y-AOs on the one hand and the p_z-AOs on the other. It is possible to draw up two separate correlation diagrams:

- one for the symmetrical AOs, from which the symmetrical (relative to the molecular plane) σ-MOs are constructed, and
- one for the antisymmetrical AOs, from which the antisymmetrical (relative to the molecular plane) π-MOs are constructed.

Because of the orthogonality of the σ- and the π-orbitals, it is possible to handle the σ-molecular skeleton separately from the π-electron system.

Following this σ/π separation, those bonds represented by σ-MOs are indicated merely by lines (in the same way as in the structure model of the classical organic chemistry), only the π-MOs are constructed explicitly out of the p_z-AOs. This restriction to π-MOs is justified in that

- a series of chemical questions which can be asked within the framework of the MO model can be adequately answered only with the assistance of the *highest occupied molecular orbital* (HOMO) and/or the *lowest unoccupied molecular orbital* (LUMO),
- in systems, free from heteroatoms, with π-electrons, HOMO and LUMO, as a rule, are π-MOs.

It has indeed proved useful, in the description of the electronic structure of organic compounds, to start out from a fundament made out of σ-bonds and to set up delocalized π-electron systems in the appropriate locations (and possibly lone electron pairs). Consequently, the π-electrons of conjugated, unsaturated systems may be filled into MOs,

- which either – in unison with the formula symbols of the classical organic chemistry – are strictly *localized* in the region of the individual C=C bonds, or
- very differently from the formula symbols of the classical organic chemistry, are *delocalized* over the entire conjugated, unsaturated structure region.

The HMO theory is able to explain the discrepancy between the calculated energies of systems with strictly localized C=C bonds (of the fictitious *Kekulé* benzene, for example, see *Chapt. 14.3.2*) and the experimentally determined enthalpies of formation of conjugated, unsaturated compounds (such as real benzene) in a simplified manner, purely in terms of the delocalization of π-electrons.

Hückel's method has proved itself to be very useful for handling conjugated π-systems. π-MOs constructed using the HMO model are dependent on the topology of the molecules, but not on the number of π-electrons. Ignoring electron/electron repulsion, the total energy E_π may be obtained directly from the energies of the π-MOs ε_i:

$$E_\pi = \sum_i \mathrm{b}_i \cdot \varepsilon_i$$

The occupation number b_i gives the number of electrons (0, 1, or 2) occupying the ith orbital. The energies are expressed with the assistance of two empirical parameters α and β, where

- α represents the energy liberated when an electron occupies the $2p_z$-AO of a C-atom (*Coulomb integral*),
- β represents the energy liberated when an electron delocalizes itself over further (adjacent) AOs (*resonance integral*),
- α and β are negative.

From the HMO energy-level diagram for ethene (*Fig. 8.13*), it can be seen that, from the two $2p_z$-AOs, each with the relative energy α, there are formed two π-MOs, with energies $\varepsilon_1 = \alpha + \beta$ (bonding) and $\varepsilon_2 = \alpha - \beta$ (antibonding). Ignoring the overlap integral, the destabilization of the antibonding MO is equivalent in magnitude to the stabilization of the bonding MO. The total ground-state π-electron energy is consequently $E_\pi = 2(\alpha + \beta) = 2\alpha + 2\beta$, and the energy gain upon formation of the π-bond $\Delta E_\pi = 2\beta$.

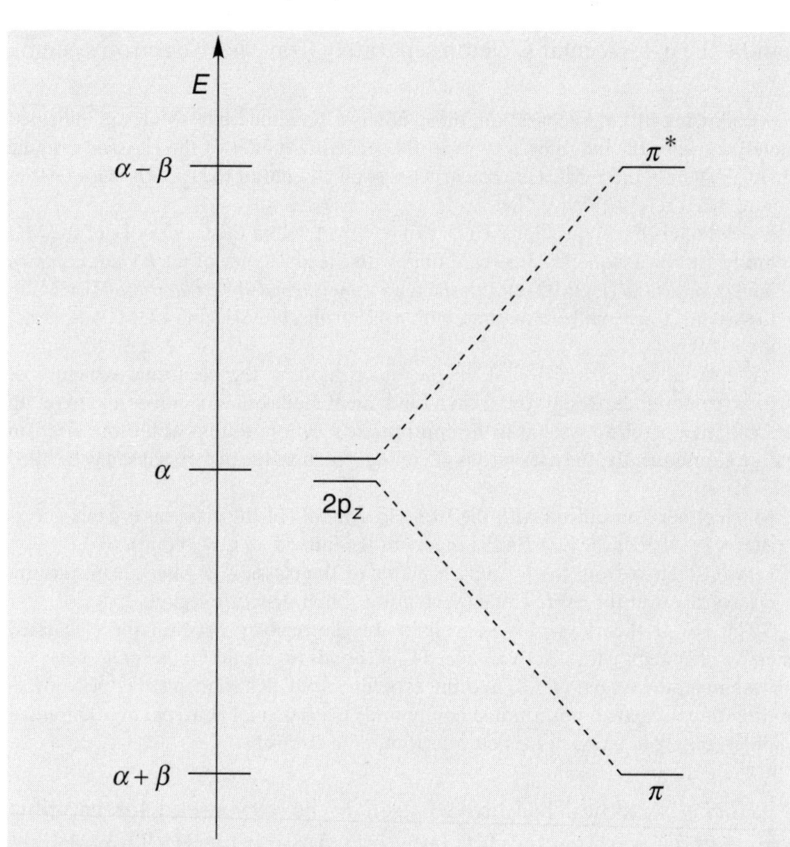

Fig. 8.13. *HMO Energy-level diagram for ethene*

The depictions of π and π^* (in the margin) correspond to an HMO approach under the more restricted condition that the AOs are not allowed, simultaneously, to adopt finite values in the same volume element (*zero differential overlap approximation*).

Conjugated, unsaturated compounds have repeatedly been stressed. The concept of *conjugation* has a constitutional meaning, and it was originally related to an alternating sequence of single and (at least two) double bonds. The prototype of conjugated, unsaturated compounds is buta-1,3-diene (for the class of acyclic, unsaturated organic compounds) or benzene (*Chapt. 14.3.2*; for the class of carbocyclic, unsaturated compounds). Nowadays, the concept of conjugation includes systems in which centers possessing lone electron pairs are adjacent to one another (*Sect. 8.3.3.3.4*) or to a grouping with multiple bond(s).

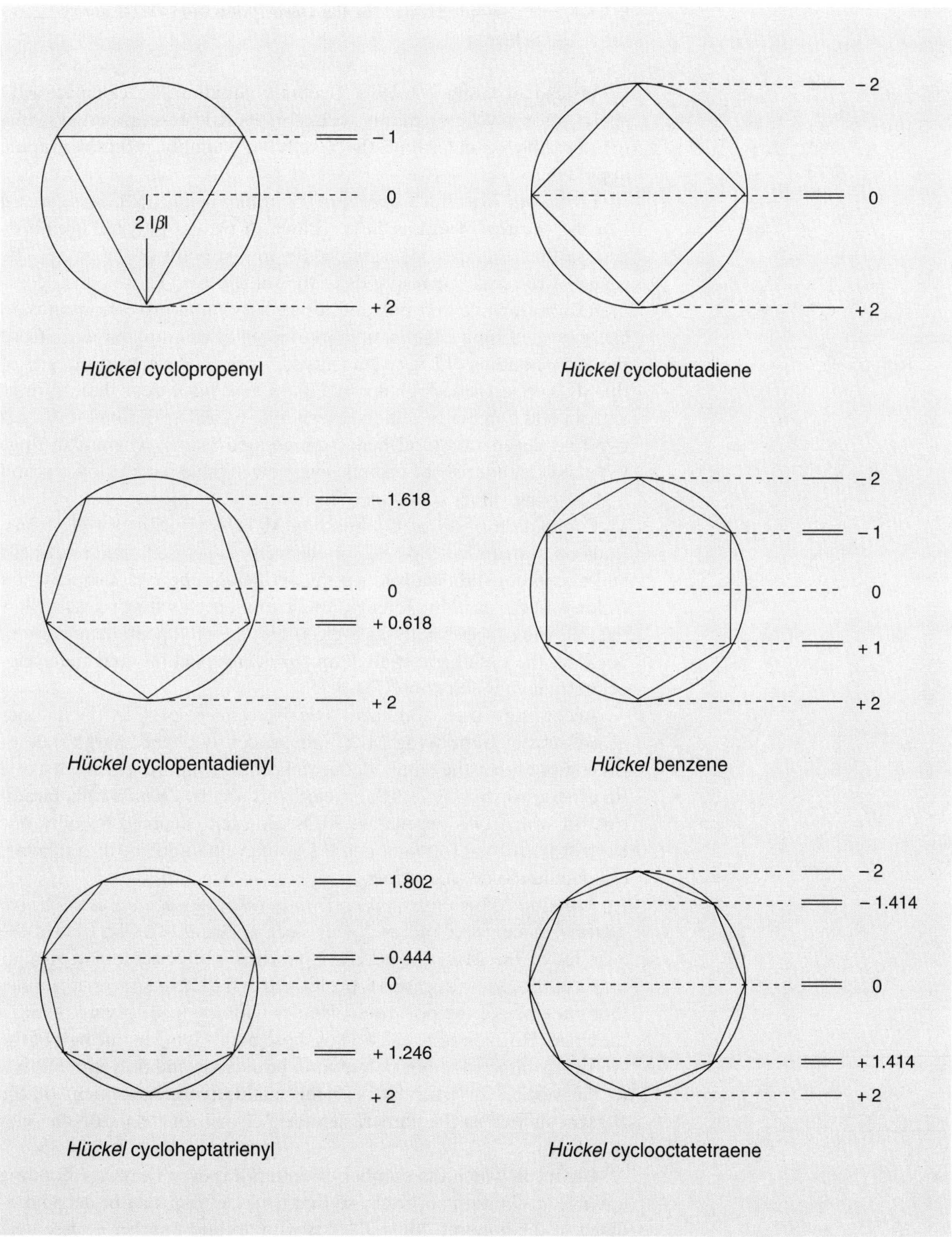

Fig. 8.14. Frost-Musulin-*type mnemonic device for the description of MO-energy schemes for cyclic, conjugated, unsaturated systems of the* Hückel *type* (|β| is the energy unit used)

8.3.3.4.2. Mnemonic Device for the Description of HMO-Energy Schemes

Instead of using *Hückel*'s algebraic equations to calculate MO energies for acyclic or monocyclic, conjugated, unsaturated systems, it is possible to determine the results very simply, using a graphic approach.

Frost *and* Musulin*'s approach to cyclic, conjugated, unsaturated systems.* A circle of radius $2|\beta|$ is drawn. A regular polygon is embedded, so that one apex is located at the lowest point of the circle. The center of the circle corresponds to the value $\varepsilon = \alpha$. Those points common to both circle and polygon, upon horizontal projection onto an energy scale (using either simple trigonometry or graph paper), afford the corresponding *Hückel* eigenvalues, *i.e.*, the one-electron energies of the MOs concerned. A glance at *Fig. 8.14* makes it clear that, in rings with an odd number of centers, every eigenvalue bar the lowest-lying is two-fold degenerate (and hence represented twice), whereas in rings with an even number of centers, every eigenvalue bar the lowest- and highest-lying values is likewise degenerate.

Occupation of all of the bonding MOs, each with two electrons, leads to systems with $4n + 2$ π-electrons in 'closed' electron shells and maximum stabilization. To this set belong the cyclopropenylium cation and the cyclobutadienylium dication, each with two, as well as the following species with six π-electrons: the cyclopentadienyl anion, benzene, the cycloheptatrienylium (*tropylium*) cation, and the cyclooctatetraenylium dication (*Fig. 8.15*).

Accordingly then, stabilization is determined solely by the number of π-electrons in bonding MOs, independently of the charge state of the compound or the number of centers in the ring. Systems with 4 or 8 (in general $4n$; $n = 1, 2, 3 \ldots$) π-electrons are, by *Hund*'s rule, biradicals (in which two degenerate MOs are each occupied by only one electron), such as, for example, (square) cyclobutadiene and (planar) cyclooctatetraene (*Fig. 8.16*).

Frost *and* Musulin*'s approach to acyclic, conjugated, unsaturated systems.* A circle of radius $2|\beta|$ is once again drawn. To obtain the energies of the MOs for acyclic π-systems with n centers, a regular polygon with $2n + 2$ centers is positioned in the circle, in such a manner that one apex of the polygon coincides with the lowest-lying point in the circle. Horizontal projection of those points lying in one half of the (vertically bisected) circle common to both circle and polygon, but not to the vertical bisector, affords the qualitative arrangement of the π-MOs, as well as the numerical values of their one-electron energies (*Fig. 8.17*).

Chains in which the number of centers n is even show $n/2$ bonding and $n/2$ antibonding orbitals, so that n π-electrons may be accommodated in the bonding MOs. Chains with an odd number n of centers have $(n - 1)/2$ bonding, $(n - 1)/2$ antibonding MOs, and one nonbonding MO (*Fig. 8.18*).

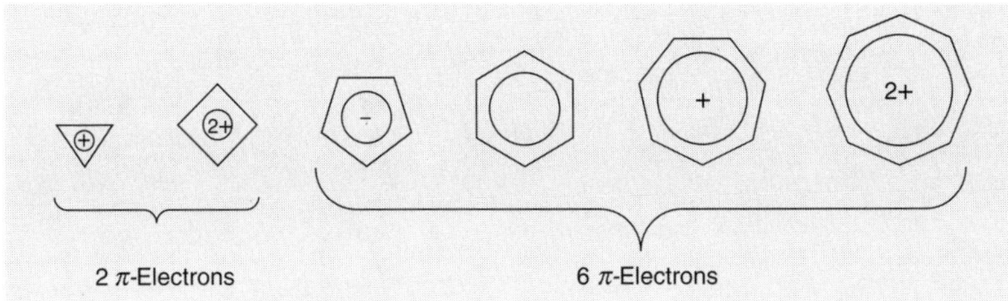

Fig. 8.15. *Cyclic, conjugated molecules and ions with* $4n + 2$ *π-electrons*

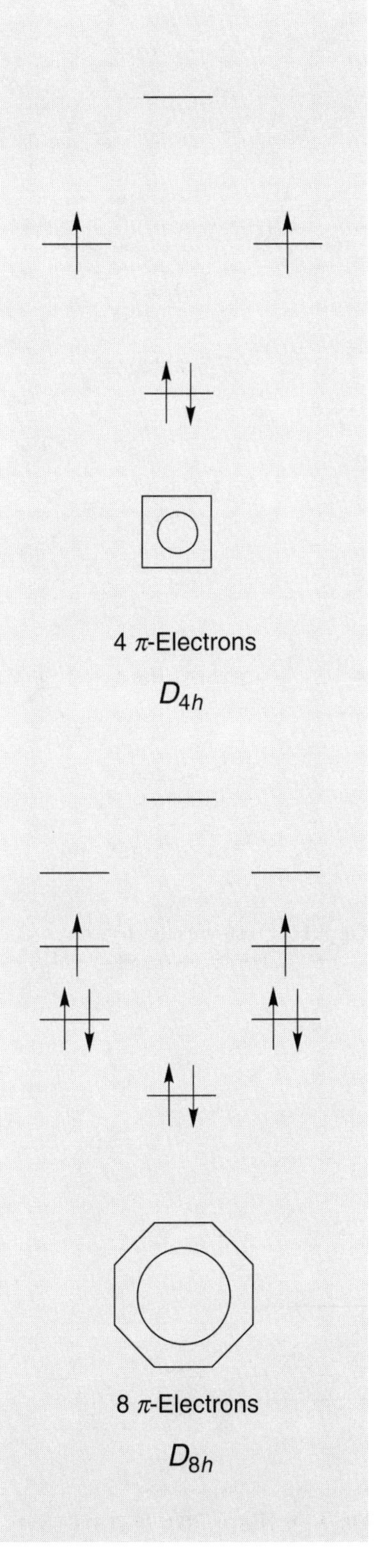

Fig. 8.16. *Cyclic, conjugated molecules with* $4n$ *π-electrons*

Heilbronner*'s approach to cyclic, conjugated, unsaturated systems of the* Möbius *type.* A strip twisted through 180° and then closed into a ring has *Möbius* topology. *Edgar Heilbronner* has demonstrated the potential existence of cyclic molecules with conjugated π-electron systems and *Möbius* topology. If the number of centers of the ring is even, then the MOs which arise are exclusively degenerate. If the number of ring members is odd, however, all of the MOs are degenerate, with the single exception of the highest one. Rather than use *Heilbronner*'s algebraic equation to calculate MO energies for cyclic, conjugated, unsaturated systems with *Möbius* topology, the results may be determined simply using the graphic technique of *Howard E. Zimmerman*.

A regular polygon is positioned in the circle of radius $2|\beta|$ in such a manner that the apexes of the polygon correspond to points on the circle and that one vertex of the polygon, in the lower half of the circle, is horizontal. Horizontal projection of the points common to circle and polygon onto the energy scale gives the one-electron energies of the bonding, nonbonding, and antibonding MOs. *Fig. 8.19* illustrates that, in a *Möbius* system, all of the bonding orbitals are doubly occupied when the number of π-electrons amounts to $4n$.

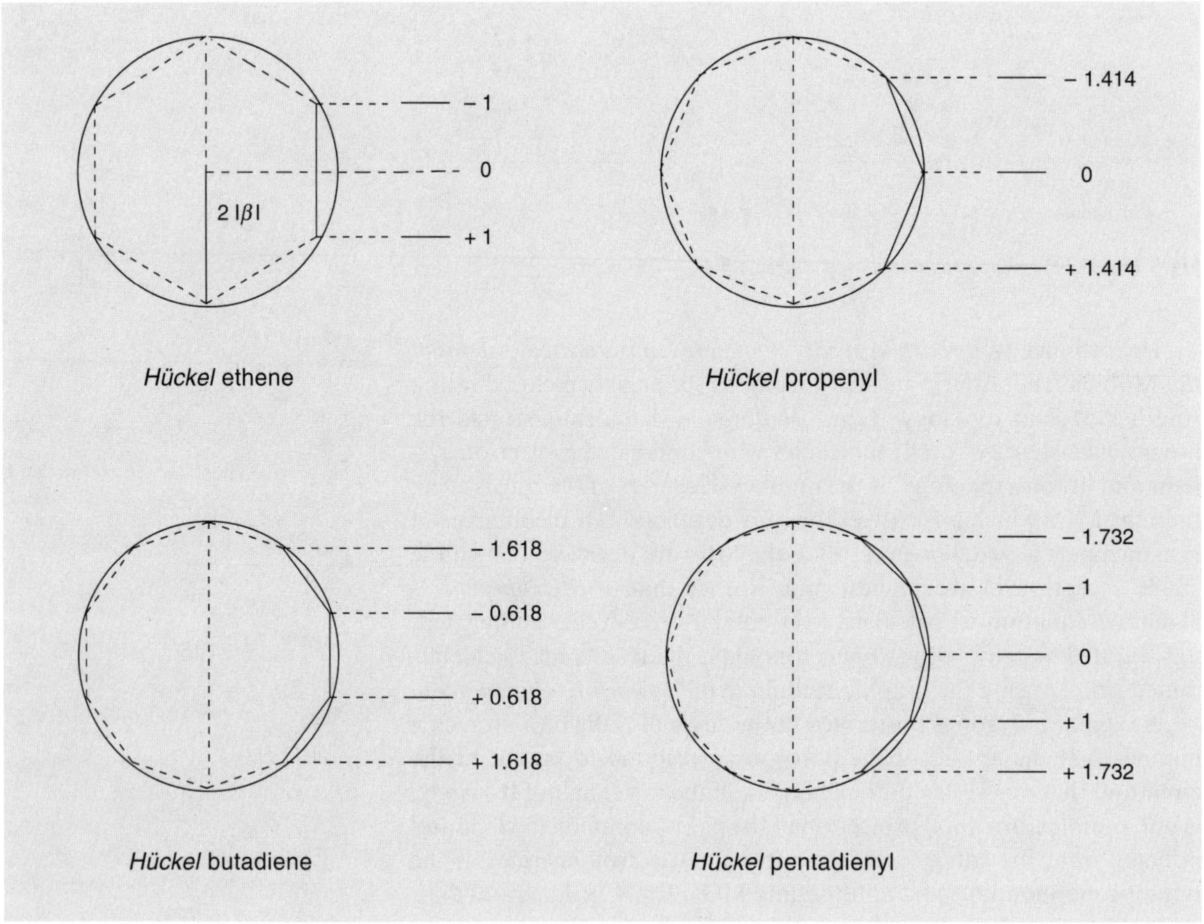

- 1
0
+ 1
2 |β|
Hückel ethene
- 1.414
0
+ 1.414
Hückel propenyl
- 1.618
- 0.618
+ 0.618
+ 1.618
Hückel butadiene
- 1.732
- 1
0
+ 1
+ 1.732
Hückel pentadienyl

Fig. 8.17. Frost-Musulin-*type mnemonic device for the description of MO-energy schemes for acyclic, conjugated, unsaturated systems* ($|\beta|$ is the energy unit used)

E [β]
-1
0
1
n
2
3
4
5

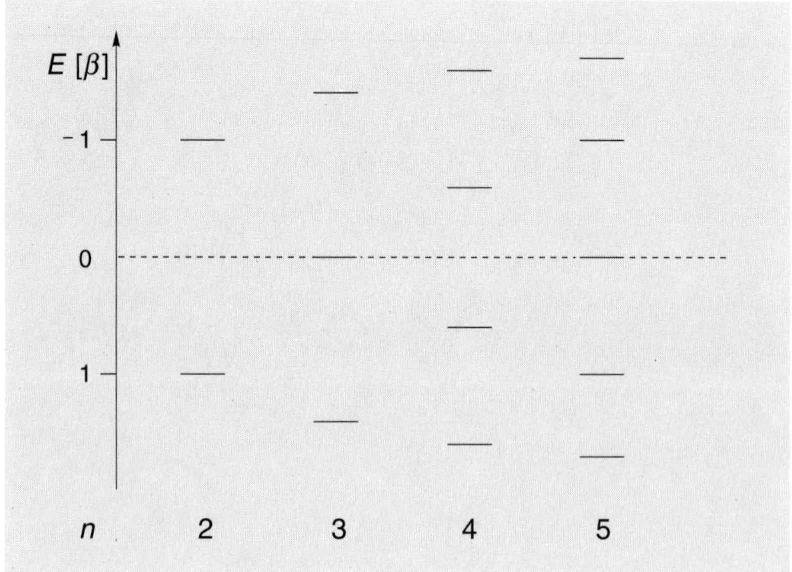

Fig. 8.18. *MO Splitting for acyclic, conjugated, unsaturated systems into bonding, antibonding, and nonbonding types, depending on the number of centers* n

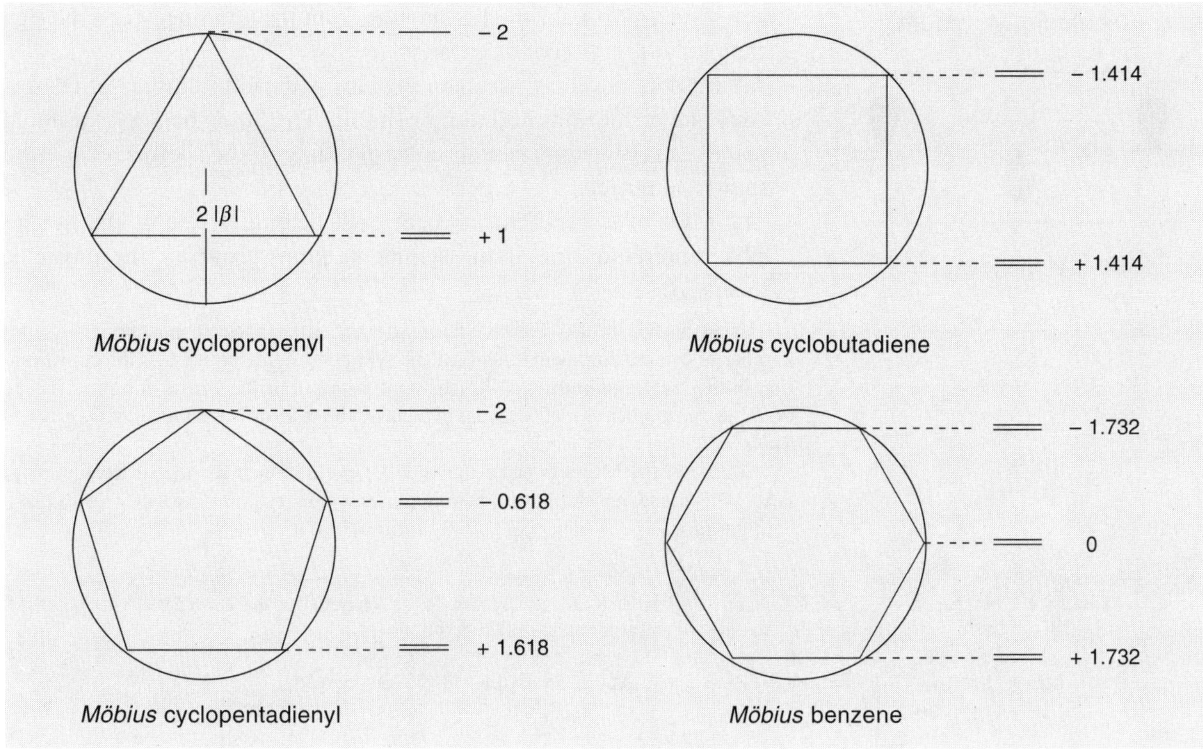

Fig. 8.19. Heilbronner-Zimmerman-*type mnemonic device for the description of MO-energy schemes for cyclic, conjugated, unsaturated systems of the* Möbius *type* ($|\beta|$ is the energy unit used)

8.3.3.4.3. Simple Rules for the Description of Nodal Properties of the Orbitals of an HMO Set

For the description of nodal properties, it is necessary to take into account whether the molecular system is cyclic or acyclic, symmetrical or asymmetrical, has an odd or an even number of centers, and may be denoted as alternating (a center arbitrarily marked with an asterisk is always followed by an unmarked center and *vice versa*) or non-alternating. The following rules apply to conjugated, unsaturated, periplanar molecules (or their models):

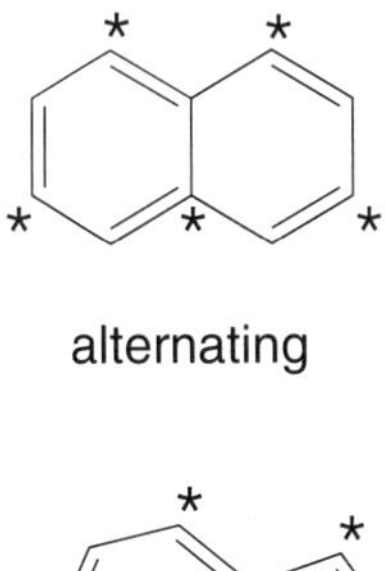

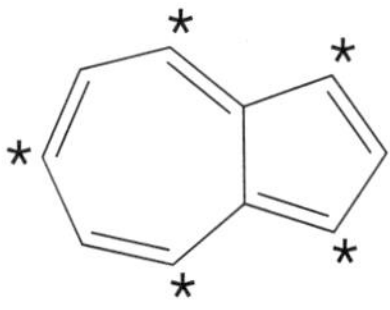

- A π-electron system generated out of n p_z-AOs by linear combination has available n π-MOs.
- In the MO with the lowest one-electron energy, all of the AOs are of the same sign.
- The number of nodal planes increases with increasing one-electron energy, being one integer lower than the running number of the associated MO for acyclic, conjugated, unsaturated π-electron systems. For cyclic, conjugated, unsaturated π-electron systems with n centers, the MO of lowest one-electron energy (ϕ_1) has no nodal plane. For π-electron systems with an even number of centers, the MO of highest one-electron energy ($\phi_{(n+2)/2}$) has $n/2$ nodal planes. The pairs of degenerate MOs have J nodal planes ($J = 1, 2, 3 \ldots (n-2)/2$).
- In alternating π-electron systems, each bonding MO has an associated

Pentadienyl system

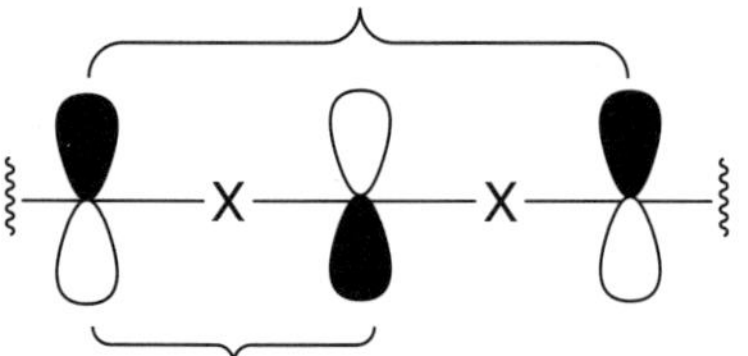

Propenyl system

antibonding MO, in which – starting from the bonding MO – the sign changes at every second center.

- Alternating, acyclic π-electron systems with odd numbers of centers have one or more nonbonding orbitals. The latter possesses a nodal plane at every second center, corresponding to the coefficient pattern shown on the left.
- The MOs of a π-electron system each belong to one of the irreducible representations of the symmetry point group of the molecule (*Sect. 8.3.3.2*).

Using acyclic buta-1,3-diene, with an alternating π-electron system, an even number of centers and membership of the symmetry point group C_{2v} (the conformation in the local minimum) or C_{2h} (the conformation in the global minimum) as an example, the application of the rules explained above affords the sets of MOs given in *Table 8.4*.

Similarly, the MO sets given in *Table 8.5* are obtained for the stereoisomers of hexa-1,3,5-triene, which likewise belong to the symmetry point group C_{2v} (with (3*Z*)-configuration) or C_{2h} (with (3*E*)-configuration).

Table 8.4. *HMO Sets for* C_{2h}- *and* C_{2v}-*Buta-1,3-diene on Application of the Rules Explained in the Text* (Σ AO-IA: sum of the bonding (b) and antibonding (a) interactions (IA))

MO	Irred. represent.	Buta-1,3-diene (C_{2h})	Σ AO-IA	Buta-1,3-diene (C_{2v})	Irred. represent.	MO
ϕ_4	B_g		0 b 3 a		A_2	ϕ_4
ϕ_3	A_u		1 b 2 a		B_1	ϕ_3
ϕ_2	B_g		2 b 1 a		A_2	ϕ_2
ϕ_1	A_u		3 b 0 a		B_1	ϕ_1

Table 8.5. *HMO Sets for (E)- and (Z)-Hexa-1,3,5-triene on Application of the Rules Explained in the Text* (Σ AO-IA: sum of the bonding (b) and antibonding (a) interactions (IA))

MO	Irred. represent.	(3*E*)-Hexa-1,3,5-triene (C_{2h})	Σ AO-IA	(3*Z*)-Hexa-1,3,5-triene (C_{2v})	Irred. represent.	MO
ϕ_6	B_g	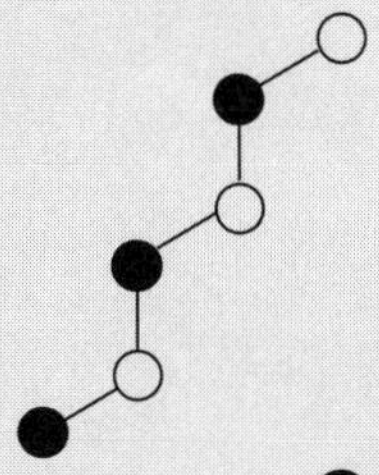	0 b 5 a	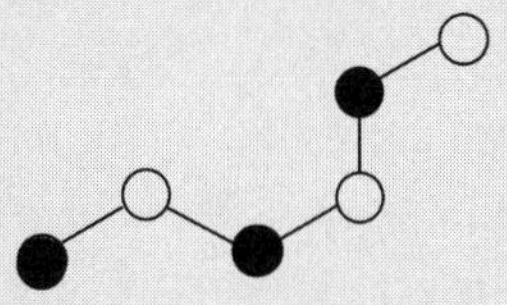	A_2	ϕ_6
ϕ_5	A_u	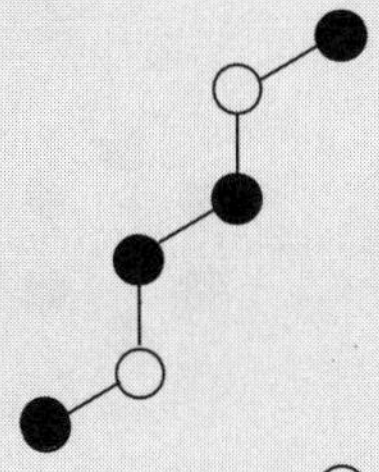	1 b 4 a	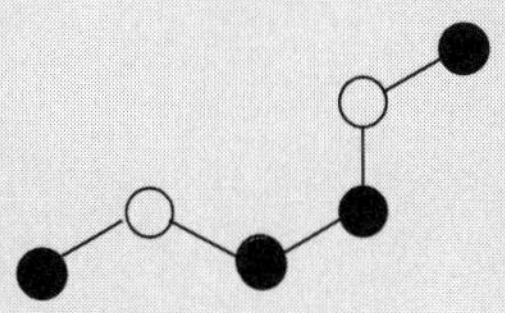	B_1	ϕ_5
ϕ_4	B_g	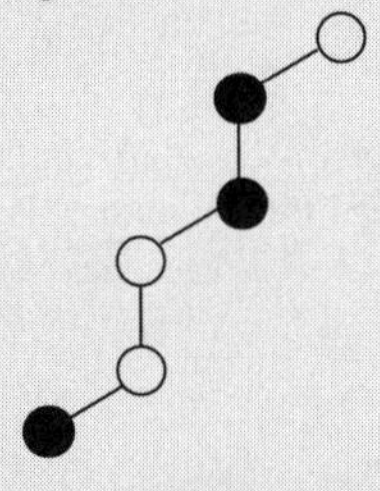	2 b 3 a	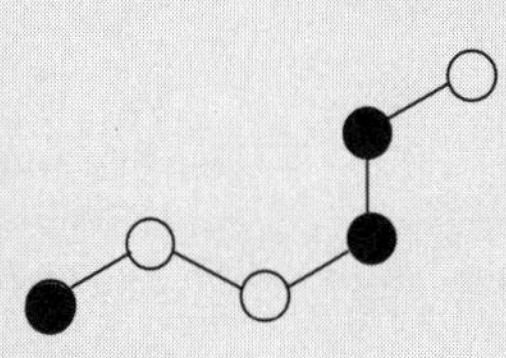	A_2	ϕ_4
ϕ_3	A_u	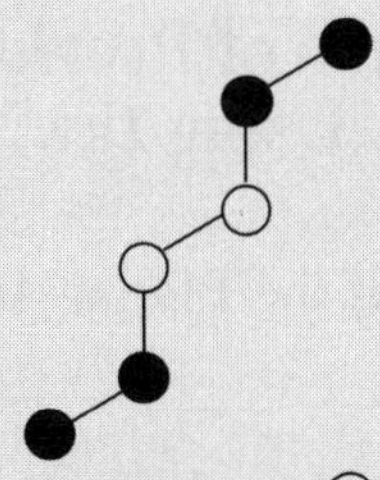	3 b 2 a	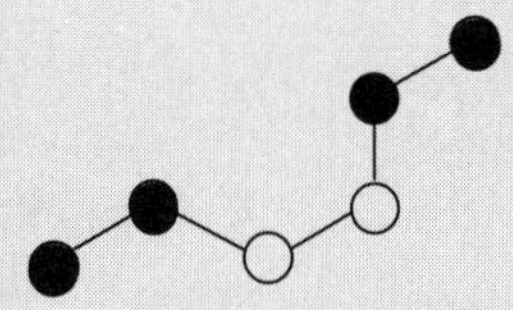	B_1	ϕ_3
ϕ_2	B_g	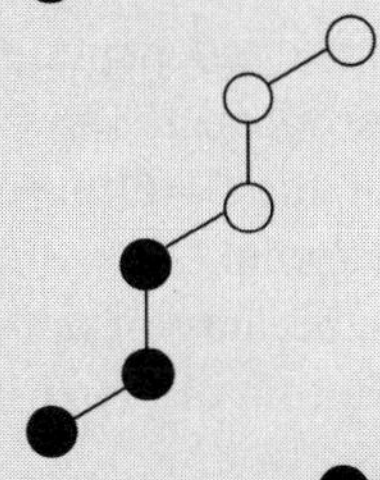	4 b 1 a	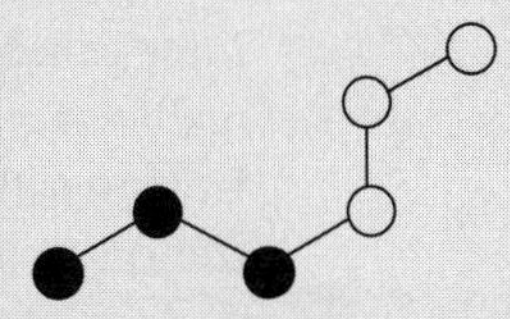	A_2	ϕ_2
ϕ_1	A_u	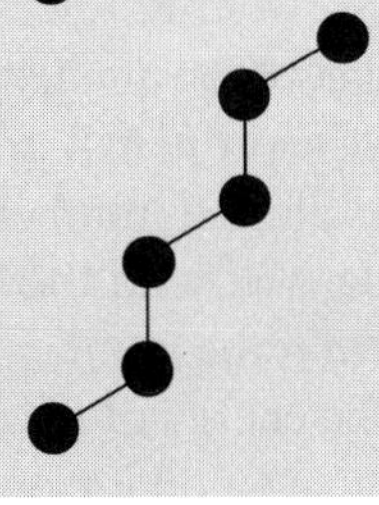	5 b 0 a	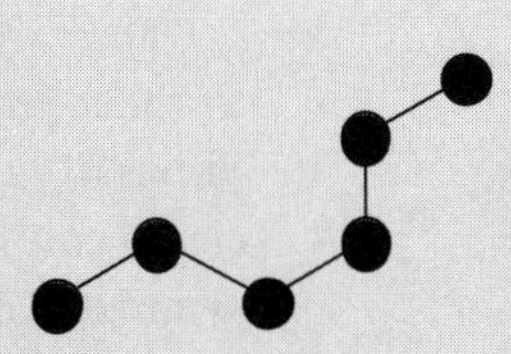	B_1	ϕ_1

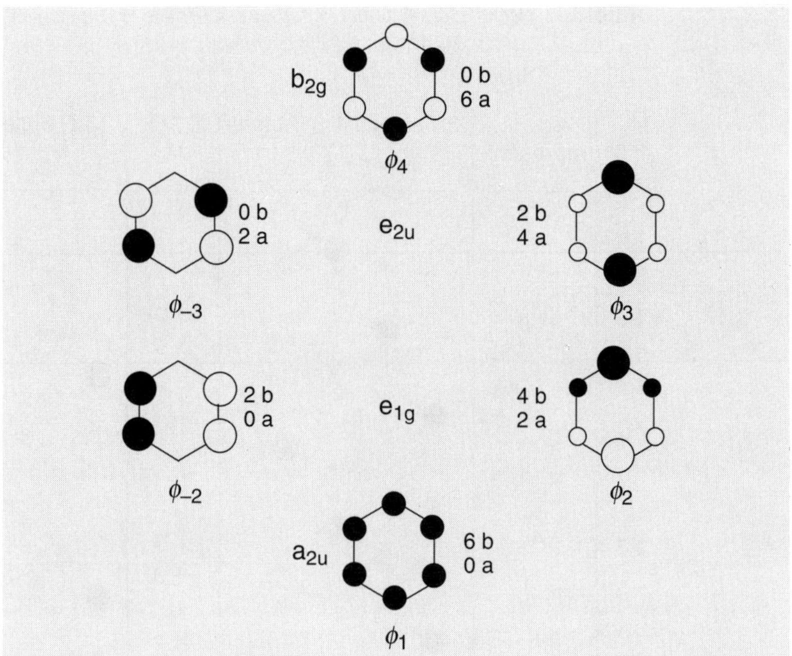

Fig. 8.20. *HMO Set of benzene with indication of the sum of bonding* (b) *and antibonding* (a) *interactions*

For benzene, the MO set of *Fig. 8.20* applies. It can be seen that the MO ϕ_1 contains no nodal plane except the *xy* plane, the next highest MOs ϕ_{-2} and ϕ_2 each contain one, the following ϕ_{-3} and ϕ_3 each have two, and, finally, the MO ϕ_4 contains three nodal planes.

8.4. Interpretation of Previously Inexplicable Phenomena Using the MO Model

8.4.1. Stabilization of the Planar Amide Group Through Conjugation

The conjugated, unsaturated peptide group contains a non-linear three-center system (heteropropenyl system) with four electrons (formally, two π-electrons in the C=O group and a so-called *lone pair*). Coplanarity of the amide group (*Sect. 8.1.1*) enables maximum delocalization of these electrons, because of optimum overlapping of the three p_z-AOs, at the cost of adoption of planarity by the N-atom and its three proximal atoms.

The π-MOs ϕ_1, ϕ_2, and ϕ_3 of the amide bond (*Fig. 8.21*, left) are related to the π-MOs of the allyl(= propenyl) group. However, because of the different effective nuclear charge numbers of C, N, and O, the coefficients can no longer be derived with the help of the simple, qualitative symmetry rules. The rule that the number of nodal planes (excluding the *xy* plane) increases with the energy of the corresponding MO is fulfilled: ϕ_1 has none, ϕ_2 one, and ϕ_3 two nodal planes.

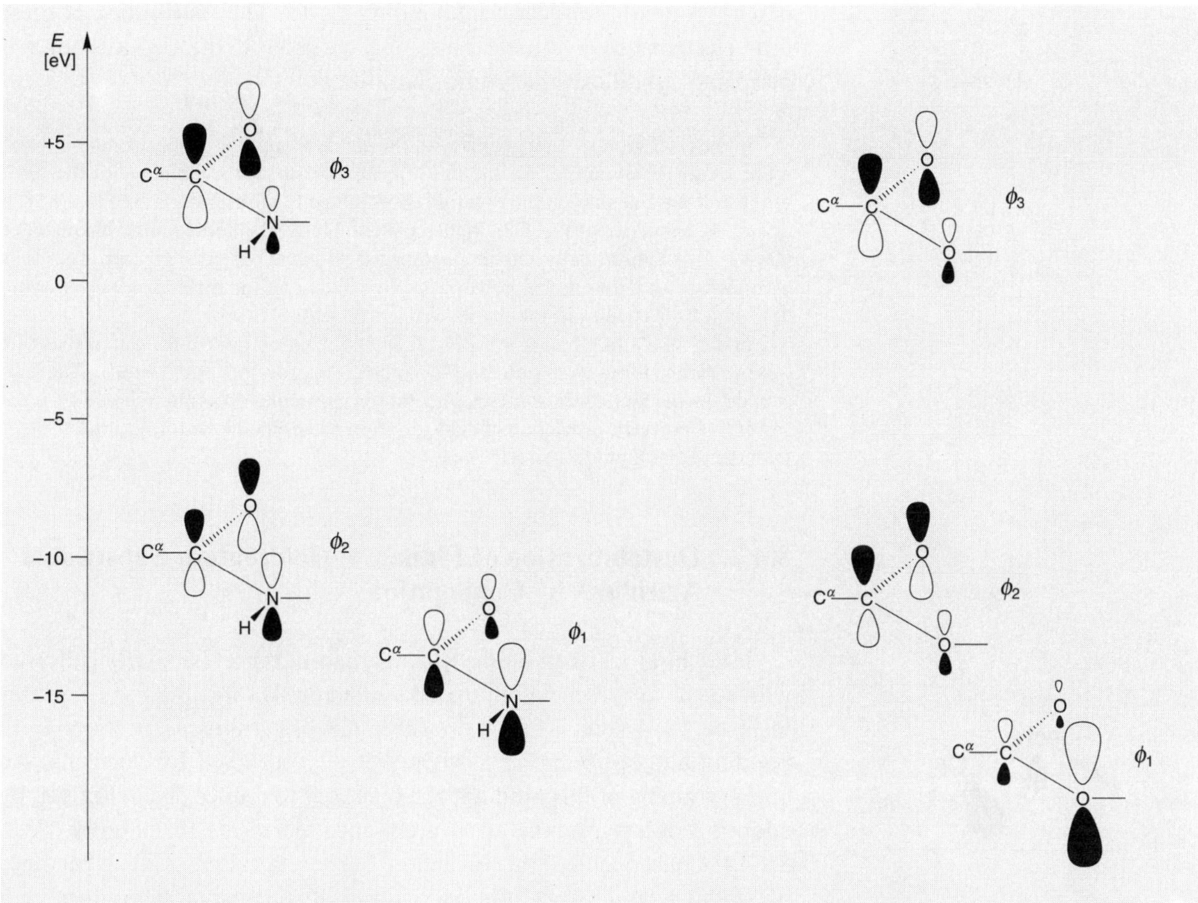

Fig. 8.21. *Graphical representation of the three propenyl-like π-MOs of the amide* (left) *and ester* (right) *groups, together with the associated one-electron energies*

Quantitative calculation, allowing for the effective nuclear charge number, gives the following coefficients for the three MOs for the amide group, from the N-, C-, and O-p_z-AOs: ϕ_1: 0.87 for N, 0.44 for C, 0.22 for O; ϕ_2: 0.42 for N, −0.47 for C, −0.78 for O; ϕ_3: 0.24 for N, −0.77 for C, 0.59 for O. It is possible to demonstrate the orthogonality of the three MOs by summation of the products of the AO coefficients. Hence, for the two MOs ϕ_1 and ϕ_2: $0.87 \cdot (0.42) + 0.44 \cdot (-0.47) + 0.22 \cdot (-0.78) = 0$. Equivalent calculations for ϕ_1 and ϕ_3, or for ϕ_2 and ϕ_3, similarly give a total of 0. The MO coefficients of the ester group (*vide infra*) are ϕ_1: 0.97 for O (ester), 0.23 for C, 0.06 for O; ϕ_2: 0.19 for O (ester), −0.58 for C, −0.79 for O; ϕ_3: 0.15 for O (ester), −0.78 for C, 0.61 for O.

The conjugated amide system contains four electrons. Hence, the ϕ_1 and ϕ_2 MOs are each occupied by two electrons. The depiction of the peptide bond with C−N and C=O bonds in terms of the structure model of the classical organic chemistry is refined under MO considerations. The AOs of C and O transform into the two occupied MOs, both with the same relative sign in either case. The signs of the coefficients of the AOs of C and N are identical in the lowest occupied MO, opposed in the highest occupied MO. The less pronounced double-bond character between C and N found by the MO approach corresponds to the existence of a C–N bond in the classical bonding model. The distribution of these four electrons over three atoms, only possible if the peptide bond is periplanar, stabilizes that conformation.

In the ester group, isoelectronic with the amide group, this delocalization is much more weakly pronounced. To the first approximation, the π-electrons of the C=O group and the lone pair on the two-fold coordinated O-atom are localized in the MOs ϕ_2 and ϕ_1, respectively (*Fig. 8.21*, right). Consequently, the double-bond character of (O=)C–O is substantially reduced compared with the (O=)C–N bond. This is in agreement with the experimentally observed decrease of the rotational barrier about this bond from *ca.* 80 kJ/mol in amides to *ca.* 35 kJ/mol in esters.

In the discussion of protein folding (*Chapt. 7.3.3.6*), a twisted transition structure was mentioned for *sp* ⇌ *ap* isomerization (about the prolyl–peptidyl bond). This finds support in the increased isomerization rate in non-polar solvents, relative to polar solvents. Geometric restriction of delocalization makes the twisted structure less polar than the planar amide structure.

8.4.2. Destabilization of Planar *N*-Heteroatom-Substituted Aziridines by Conjugation

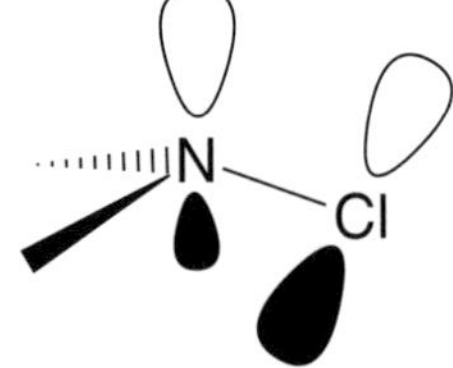

If the N-atom of an amine has an adjacent center, similarly endowed with a *lone pair*, then a conjugated system exists (*Sect. 8.3.3.4.1*). In this instance, the system is of the *two center/four electron type*, in which both bonding and antibonding π-MOs are fully occupied by electrons. An atom grouping of this kind has the tendency to reduce delocalization by adoption of a pyramidal structure, hence increasing its stability (*Sect. 8.1.4*). Consequently, the inversion barrier is raised, and it becomes possible to demonstrate the coexistence of stereoisomeric amines with three constitutionally distinct ligands, or even to separate them from one another. This was verified in *Eschenmoser*'s laboratory. A key role is, therefore, ascribed to this *conjugative destabilization* of the planar transition structure by directly adjacent heteroatoms, each with at least one lone pair, in answering an old problem.

8.4.3. The Causes of the Anomeric Effects

Lone pairs (*Sect. 8.3.3.3.4*) may participate in electron donor/electron acceptor interaction, representing a directed electron flow through the adjacent bond into the next but one. Since a delocalization of this kind, from a phantom ligand playing the part of the electron donor, over three chemical bonds to an electron acceptor, has been shown empirically to be optimal when the phantom ligand at the lone pair is situated in an *ap*-orientation relative to the acceptor, the four-atom fragment concerned will, if molecular flexibility permits, adopt this conformation.

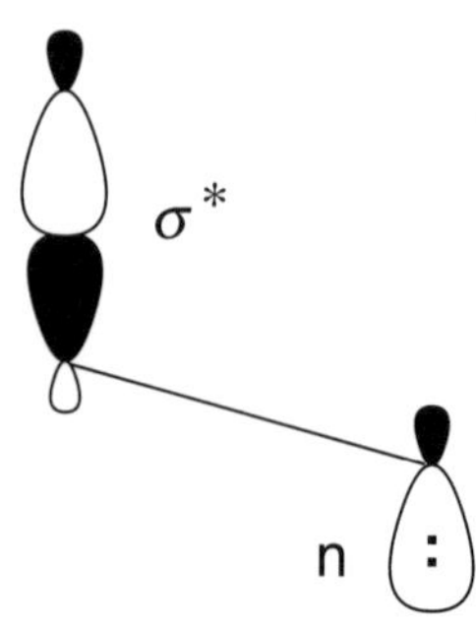

For conformational fragments with alternating O-atoms, diverse anomeric effects demand an order of energies which deviates from that one where CH_2 takes the place of O (*Sect. 8.1.2*). They may be explained by a stable conformation, in which an n→σ^* delocalization comes into being, caused by the *ap*-orientation of an MO with a lone pair on a heteroatom, relative to a low-lying antibonding MO (σ^*-MO) between the directly adjacent atom and the overnext one.

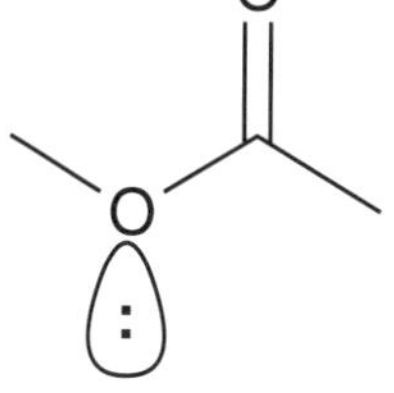

A word about esters: because of $n \rightarrow \pi^*$ delocalization, the ester group (like the amide group) is planar. The $n \rightarrow \sigma^*$ delocalization leads to the *sp*-conformation being energetically favored over the *ap*-conformation.

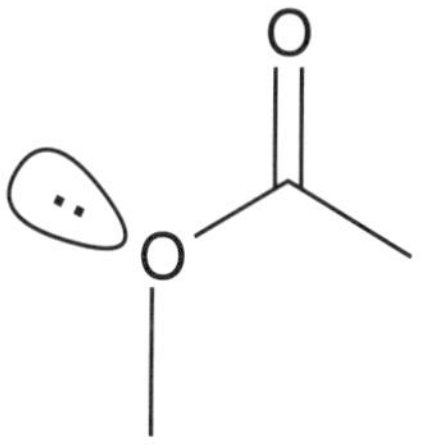

O

O

sp

8.4.4. The Preferred Conformation of the Repeating Unit of a Homo-DNA Single Strand

During the conformational analysis of the homo-DNA single strand (*Chapt. 7.3.1.1.1*), the order of stabilities for the phosphodiester group was given (in *Fig. 7.19*) without any explanation. This omission will now be rectified.

O

O

ap

The elaborate description (making use of phantom ligands) is made simpler by inclusion of the lone pairs on the various O-atoms. *Fig. 8.22* contains the conclusions of *Fig. 7.19*, together with deductions which demonstrate how, besides *steric* evaluation criteria, a *stereoelectronic* criterion ($n \rightarrow \sigma^*$ stabilization) has been applied.

The stereoelectronic criterion involves, generally speaking, the formulation of a conformation in which the maximum possible stabilization is ensured by the greatest possible degree of overlap of suitable orbitals. In this specific case, it especially involves taking account of the *general anomeric effect*. The conformations which are stereoelectronically favored are those in which $n \rightarrow \sigma^*$ delocalization comes into being, *i.e.*, those in which a lone pair on the O-atom of one RO group has an *ap*-orientation relative to the bond between the P-atom and the O-atom of the other RO group. The fact that the σ^*-MO of the bond between P and an O-atom of coordination number 2 (the O-atom of the RO group) is considered, rather than the σ^*-MO of the bond between P and an O-atom of coordination number 1 (a negatively charged O-atom) is due to the greater effective nuclear charge of the former O-atom, which engenders a lower-lying σ^*-MO of the bond concerned.

8.4.5. Annulenes

The term '*annulene*' was introduced by *Franz Sondheimer*. To the compound class of annulenes belong monocyclic, conjugated, unsaturated hydrocarbons of the general molecular formula $(CH)_{2m}$. Answers to questions of the electronic structure of annulenes are of particular interest. It has proved appropriate to arrange annulenes in two subgroups: annulenes with $4n + 2$ and those with $4n$ π-electrons. Moreover, monocyclic, unsaturated systems which are both planar and conjugated are divided into *aromatic* and *antiaromatic* compounds. According to *Erich Hückel* aromatic compounds are those ones which are each stabilized in comparison with the respective acyclic reference compound with the same number of C-atoms and π-electrons, because of a higher degree of HMO π-electron energy. Antiaromatic compounds are those ones

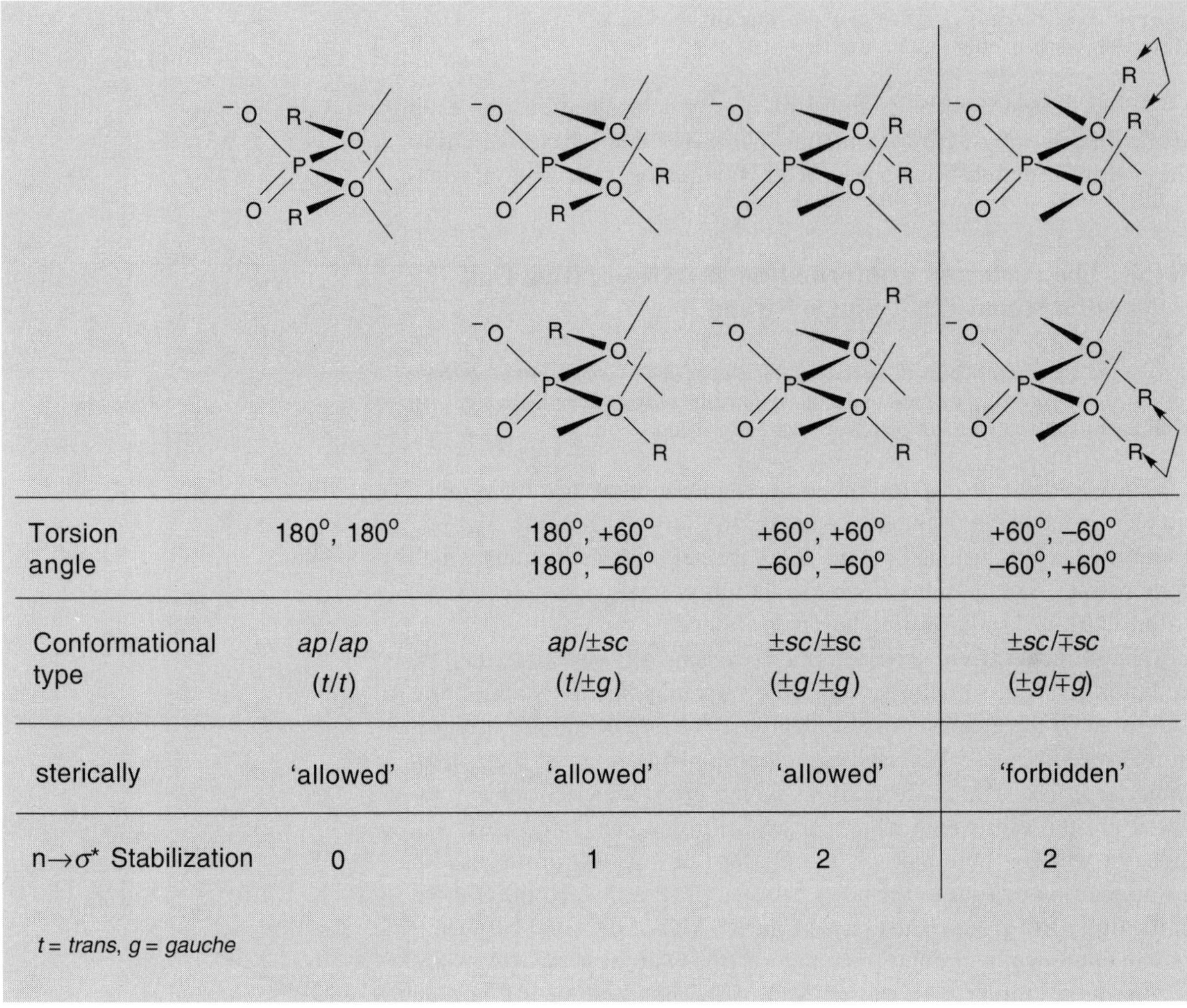

Torsion angle	180°, 180°	180°, +60° 180°, −60°	+60°, +60° −60°, −60°	+60°, −60° −60°, +60°
Conformational type	*ap*/*ap* (*t*/*t*)	*ap*/±*sc* (*t*/±*g*)	±*sc*/±*sc* (±*g*/±*g*)	±*sc*/∓*sc* (±*g*/∓*g*)
sterically	'allowed'	'allowed'	'allowed'	'forbidden'
n→σ* Stabilization	0	1	2	2

t = *trans*, *g* = *gauche*

Fig. 8.22. *Systematic conformational analysis of the phosphodiester fragment of a polynucleotide single strand* (for the O-atoms in the chain, not only the bonds to R but also the bonds each to a phantom ligand are given)

which, as proposed by *Ronald Breslow*, are destabilized through having a lower degree of HMO π-electron energy, each compared with the appropriate acyclic reference compound with the same number of C-atoms and π-electrons.

8.4.6. The 'Benzene Problem'

In the planar arrangement of its ring atoms (of coordination number 3), benzene is free of *Baeyer* strain. In the HMO model it is energetically distinguished by having a total of six electrons available for inclusion in three bonding π-MOs. This means that here is a case in which the *Butlerov* requirement '*one compound – one formula*' cannot be fulfilled. A detailed discussion of the benzene problem and the potential of organic chemistry for the generation of theories takes place in *Chapt. 14*.

8.4.7. Delocalization of σ-Electrons

Cases which were inexplicable in terms of the structure model of the classical organic chemistry (*Sect. 8.1.1–8.1.5*) provided the incentive to cease procrastination with consideration of the nature of the chemical bond. The qualitative MO model helped the process along. For annulenes with $4n + 2\ \pi$-electrons (*Sect. 8.4.5*), the matter is one of π-electron delocalization. For the coplanar amide bond (*Sect. 8.4.1*), $n \rightarrow \pi^*$ delocalization enters the picture. For the various anomeric effects (for *endo*- and *exo*-anomeric effects in carbohydrates and, generally, in acetals: *Sect. 8.4.3*; for the general anomeric effect in the repeating units of polynucleotides: *Sect. 8.4.4*), all of which lead to energetic favoring of the *sc*-pattern over the *ap*-arrangement, an $n \rightarrow \sigma^*$ delocalization is the cause. For alkanes – as we have repeatedly seen – the reverse order applies: the *ap*-conformation is preferred over the *sc*. Examples here are supplied by butane, methylcyclohexane, pentane, and hexane. The descending order of stabilities, in every one of these cases, is:

$$H/H > H/CH_3 > CH_3/CH_3$$

(*Newman* strain; see *Chapt. 7.2.2*). This finding has been tacitly put down to steric causes. Steric effects seem plausible enough: however, in themselves, they do not tell the whole story. In addition to these, there exists a $\sigma \rightarrow \sigma^*$ delocalization, whose contribution to stabilization is at its greatest, for alkanes, in the *ap*-conformation. One CC/CC interaction and two HC/CH interactions are found in this conformation. In the *sc*-conformation, however, one HC/CH interaction and two CC/CH interactions are present. A CC/CH interaction contributes less to stabilization than does the average of the contributions of CC/CC and HC/CH interactions.

8.5. Conclusion

Chemists have noticed the diversity of organic chemical structures. They have established clarity by means of numerous structural classes. By organizing the incalculably vast number of organic compounds into suitable structural classes, a body of knowledge was assembled, which permitted appreciation of how far this differentiability on grounds of structural characteristics is able to reflect differences in properties. In this way, with chemists' intuition, and without inquiring into the nature of chemical bonds, there was developed the structure theory of the classical organic chemistry. This intensified the perception that the discrete nature of chemical compounds could be traced back to the existence of specific molecules, in which atoms of particular elements, in constant ratios and characteristic spatial arrangements, were to be encountered.

Occasionally, problem cases were hit upon – like, for example, the peptide bond, the anomeric effect, or benzene – whereupon it became evident that it was no longer possible to evade producing a physically founded answer to the question of the nature of the chemical bond. *Electron theory of the chemical bond*, taking account not only of electrons participating in bonds, but also of lone pairs in an inventory of the entire set of valence electrons present in a molecule, attempts to modernize the predictions of the structure model of the classical organic chemistry. The same can be said of *resonance theory* (*Chapt. 14.1*), which attempts to rectify the empirical observation that, now and again, a single formula symbol from the structure archive of the classical organic chemistry is not by itself suited to adequately interpret the bonding properties of a compound under discussion. However, the trend by which the discipline of structural chemistry developed into a branch of physics was no longer to be retarded at this.

The *qualitative MO model* offers the chemist the physical basis for a deeper understanding of the relationship between the structural details of a chemical compound and its properties. The correlation between molecular and electronic structure may be worked out with little effort, thus demonstrating that the geometry of, for example, a three (type H–X–H)- or four (type H–X–X–H)-atom molecule is solely dependent on the number and distribution of valence electrons in the appropriate MOs. The results provided by *Walsh* correlation diagrams, using this approach, have found expression in *Walsh*'s rules. In an aesthetically very pleasing manner, the *HMO model*, with its various simplifications, reveals structural patterns which permit subtle distinctions to be made within and between the various classes of compounds.

Like the structure model of the classical organic chemistry, the qualitative MO model, where bonding, overlap, symmetry, and nodal properties play a fundamental role, makes it possible to treat complex structures as assemblages of simple repeating units, using correlation diagrams as a structural orientating aid.

The fact that, besides *steric* effects, *electronic* and even *stereoelectronic* effects have now entered the picture is not only beneficial for understanding problem cases of the classical organic chemistry, but will also be of great use in the discussion of chemical reactions (*Aspects of Organic Chemistry*, Volume 2: *Reactivity*).

At the close of this essay on the structure of organic compounds, the philosopher mentioned in *Chapt. 1.1* will be prepared now to testify the essence of structure theory: the chemist is concerned with molecules. It is true that he or she does not know what they look like *in reality*, but, by translating ideas into models, it is possible, with the assistance of the latter, to draw conclusions about molecular properties. Molecular properties may be tested experimentally, and experimental testing shows how useful the underlying model really is. Thus, utility is the sole criterion for judging which model should be applied in any given case (see *Chapt. 14.5*).

Further Reading

To Sect. 8.1.

K. R. Popper, *The rationality of scientific revolutions* in *The Herbert Spencer Lectures 1973* (Ed. R. Harrè), p. 72, Oxford University Press, Oxford, 1975.

H. Bondi, *What is progress in science?* in *The Herbert Spencer Lectures 1973* (Ed. R. Harrè), p. 1, Oxford University Press, Oxford, 1975.

T. S. Kuhn, *The Structure of Scientific Revolutions,* 2nd Edn., The University of Chicago Press, Chicago, 1970.

I. Lakatos, *Die Methodologie wissenschaftlicher Forschungsprogramme* (Ed. J. Worrall, G. Currie), Vieweg-Verlag, Braunschweig, 1982.

I. B. Cohen, *Revolution in Science,* The Belknap Press of Harvard University Press, Cambridge, Mass., 1985.

8.1.1.

L. Pauling, R. B. Corey, *The Configuration of Polypeptide Chains in Proteins, Fortschr. Chem. Organ. Naturstoffe* **1954,** *11,* 180.

8.1.2.

J. T. Edward, *Stability of glycosides to acid hydrolysis, Chem. Ind. (London)* **1955,** 1102.

R. U. Lemieux, *Rearrangement and Isomerizations in Carbohydrate Chemistry* in *Molecular Rearrangements,* Vol. 2 (Ed. P. de Mayo), p. 709, Interscience Publ., New York, 1964.

8.1.4.

J. Meisenheimer, W. Theilacker, *Stereochemie des Stickstoffs* in *Stereochemie* (Ed. K. Freudenberg), Franz Deuticke, Leipzig, 1933.

J. Meisenheimer, L.-H. Chou, *Versuche zur Spaltung substituierter cyclischer Äthylenimine in optisch aktive Isomere, Liebigs Ann. Chem.* **1939,** *539,* 70.

J. F. Kincaid, F. C. Henriques, Jr., *The Stability toward Racemization of Optically Active Compounds with Especial Reference to Trivalent Nitrogen Compounds, J. Am. Chem. Soc.* **1940,** *62,* 1474.

V. Prelog, P. Wieland, *Über die Spaltung der* Tröger*'schen Base in optische Antipoden, Helv. Chim. Acta* **1944,** *27,* 1127.

To Sect. 8.2.

S. W. Hawking, *A Brief History of Time,* Bantam Press, London, 1988.

R. P. Feynman, *QED – The Strange Theory of Light and Matter,* Princeton University Press, Princeton, 1985.

R. P. Feynman, R. B. Leighton, M. Sands, *The Feynman Lectures on Physics,* Vol. 1–3, Addison-Wesley Publ. Comp., Reading, Mass., 1963.

To Sect. 8.3.

W. Kutzelnigg, *Einführung in die Theoretische Chemie, Bd. 2: Die chemische Bindung,* Verlag Chemie, Weinheim, 1978.

M. Klessinger, *Elektronenstruktur organischer Moleküle,* Verlag Chemie, Weinheim, 1982.

8.3.3.3.4.

B. M. Gimarc, *Applications of Qualitative Molecular Orbital Theory, Acc. Chem. Res.* **1974,** *7,* 384.

E. Amitai-Halevi, *Orbital Symmetry and Reaction Mechanism,* Chapt. 4, Springer-Verlag, Berlin, 1992.

8.3.3.4.

E. Hückel, *Grundzüge der Theorie ungesättigter und aromatischer Verbindungen,* Verlag Chemie, Berlin, 1938.

E. Heilbronner, H. Bock, *The HMO-Model and its Application,* Wiley-Interscience, London, 1976.

A. S. Streitwieser, *Molecular Orbital Theory for Organic Chemists,* Wiley, New York, 1961.

8.3.3.4.2.

A. A. Frost, B. Musulin, *A Mnemonic Device for Molecular Orbital Energies, J. Chem. Phys.* **1953,** *21,* 572.

E. Heilbronner, *Hückel Molecular Orbitals of Möbius-Type Conformations of Annulenes, Tetrahedron Lett.* **1964,** 1923.

H. E. Zimmerman, *Quantum Mechanics for Organic Chemists,* Academic Press, New York, 1975.

To Sect. 8.4.

8.4.1.

L. Pauling, *Recent Work on the Configuration and Electronic Structure of Molecules with some Applications to Natural Products, Fortschr. Chem. Organ. Naturstoffe* **1939,** *3,* 203.

8.4.2.

A. Eschenmoser, D. Felix, *Slow inversion at pyramidal nitrogen. Isolation of diastereomeric 7-chloro-7-azabicyclo[4.1.0]heptanes at room temperature, Angew. Chem. Int. Ed.* **1968,** *7,* 224.

J.-M. Lehn, J. Wagner, *Hindered Nitrogen Inversion in N-Halogenoaziridines and in N-Halogenoazetidines, Chem. Commun.* **1968,** 148.

A. Rauk, L. C. Allen, K. Mislow, *Pyramidal Inversion, Angew. Chem. Int. Ed.* **1970,** *9,* 400.

J. B. Lambert, *Pyramidal Atomic Inversion, Topics in Stereochem.* **1971,** *6,* 19.

8.4.3.

C. Romers, C. Altona, H. R. Buys, E. Havinga, *Geometry and Conformational Properties of Some Five- and Six-Membered Heterocyclic Compounds Containing Oxygen or Sulfur, Topics in Stereochem.* **1969,** *4,* 39.

S. David, O. Eisenstein, W. J. Hehre, L. Salem, R. Hoffmann, *Superjacent Orbital Control. An Interpretation of the Anomeric Effect, J. Am. Chem. Soc.* **1973,** *95,* 3806.

O. Eisenstein, N. T. Anh, Y. Jean, A. Devaquet, J. Cantacuzène, L. Salem, *Lone Pairs in Organic Molecules: Energetic and Orientational Non-Equivalence, Tetrahedron* **1974,** *30,* 1717.

W. A. Szarek, D. Horton (Ed.), *Anomeric Effect,* American Chemical Society, Washington, DC, 1979.

A. J. Kirby, *The Anomeric Effect and Related Stereoelectronic Effects at Oxygen,* Springer-Verlag, Berlin, 1983.

8.4.4.

A. Eschenmoser, M. Dobler, *Warum Pentose- und nicht Hexose-Nucleinsäuren? Teil I: Einleitung und Problemstellung, Konformationsanalyse für Oligonucleotid-Ketten aus 2',3'-Dideoxyglucopyranosyl-Bausteinen ('Homo-DNS') sowie Betrachtungen zur Konformation von A- und B-DNS, Helv. Chim. Acta* **1992,** *75,* 218.

8.4.7.

R. Hoffmann, *Interaction of Orbitals through Space and through Bonds, Acc. Chem. Res.* **1971,** *4,* 1.

To Sect. 8.5.

A. D. Walsh, *The Electronic Orbitals, Shapes, and Spectra of Polyatomic Molecules,* Part I: *AH_2 Molecules, J. Chem. Soc.* **1953,** 2260; Part II: *Non-hydride AB_2 and BAC Molecules, J. Chem. Soc.* **1953,** 2266; Part III: *HAB and HAAH Molecules, J. Chem. Soc.* **1953,** 2288; Part IV: *Tetratomic Hydride Molecules, AH_3, J. Chem. Soc.* **1953,** 2296; Part V: *Tetratomic, Non-hydride Molecules, AB_3, J. Chem. Soc.* **1953,** 2301; Part VI: *H_2AB Molecules, J. Chem. Soc.* **1953,** 2306; Part. VIII: *Pentatomic Molecules: CH_3I, J. Chem. Soc.* **1953,** 2321; Part IX: *Hexatomic Molecules: Ethylene, J. Chem. Soc.* **1953,** 2325.

9 Documentation and Retrieval of Chemical Knowledge

9.1. *Beilstein Handbook of Organic Chemistry*

The *Beilstein Handbook of Organic Chemistry* has already been cited in *Chapt. 1.2* as a possible information source regarding the question of how the field of organic chemistry and its compounds may be delineated. To give an impression of the scope of the work, it should be pointed out that, at the beginning of 1996, *Beilstein* had attained the imposing figure of 495 individual subvolumes, comprising a total of *ca.* 410,000 pages.

9.1.1. The Printed Handbook

It seems appropriate to mention briefly how the *Beilstein Handbook of Organic Chemistry* is constructed. The Handbook, in production for over one hundred years, is divided into different series, each of which relates to a particular time period (*Table 9.1*).

The *V. Supplementary Series* is in production at the moment. This series marked the transition to the English language of the *Beilstein-In-*

Table 9.1. *Chronological Classification of the* Beilstein *Handbook* (4th Edition)

Series	Abbreviation	Fully reviewed literature	Color of title section on spine
Main work	H	until 1909	green
I. Supplementary series	E I	1910 – 1919	red
II. Supplementary series	E II	1920 – 1929	white
III. Supplementary series	E III	1930 – 1949	blue
III./IV. Supplementary series*)	E III/IV *)	1930 – 1959	blue/black
IV. Supplementary series	E IV	1950 – 1959	black
V. Supplementary series	E V	1960 – 1979	red

*) Vol. 17–27 of Supplementary Series III and IV, covering the heterocyclic compounds are combined in a joint issue.

stitut für Literatur der Organischen Chemie in Frankfurt am Main. Prior to 1984, the work was known by its German title of *Beilsteins Handbuch der Organischen Chemie.*

Each series is itself divided into 27 volumes (often these are in turn divided into subvolumes), in which individual classes of compounds and the arrangement of individual compounds within these classes are ordered according to the *Beilstein System*. The *Beilstein System* should be understood as a series of rules, with the aid of which any and every carbon compound may be assigned a unique place in the entirety of carbon compounds. This whole is divided into 4720 *System Numbers*, which may be used as an aid to orientation in *Beilstein*.

The *Beilstein System* is based on the three main classes of organic compounds: the acyclic or open-chain carbon compounds, the isocyclic or carbocyclic compounds, and the heterocyclic compounds. In compounds of the last mentioned class, the element carbon has been replaced by other elements in one or more ring atoms (*Table 9.2*).

Table 9.2. *The Main Division of the* Beilstein *Handbook*

***Beilstein* Classification** Compound class	Volumes	System No.
Acyclic compounds	1 – 4	1 – 449
Isocyclic compounds	5 – 16	450 – 2358
Heterocyclic compounds	17 – 27	2359 – 4720

Sorting within these main classes is based on the concept of functional classes. This concept determines the subcategory of organic compounds, as characterized by the nature and number of their functional groups (*Table 9.3*). Since the time of *Justus von Liebig*, functional groups have been understood to mean atoms or atom groups occupying positions which, in the basic skeleton (of the parent compound), are taken by H-atoms. The functional groups are responsible, to a considerable degree, for the physical, chemical, or biological properties exhibited by compounds of the particular functional class.

All carbon compounds appearing in the scientific literature are included in *Beilstein*, provided that they are reliably characterized, have been isolated in a sufficiently pure form, possess a definite structural formula, and their constitution is known. An exception is made in the case of macromolecular carbon compounds, as well as for CO, CO_2, carbonic acid and its salts, and for carbides. Organometallic compounds are only dealt with in a limited way in *Beilstein*, as these are more exhaustively handled by the *Gmelin-Institut für Anorganische Chemie und Grenzgebiete*, a part of the *Max-Planck-Gesellschaft zur Förderung der Wissenschaften*, located in Frankfurt am Main. *Table 9.4* illustrates which organometallic compounds had been registered and reviewed by *Gmelin* up until 1991.

Table 9.3. *Classification of Organic Compounds by Main and Functional Classes According to the* Beilstein System

Class of the registry compounds	Sign of the functional group		Beilstein volume No. for main section: A Acycles	B Isocycles	C Heterocycles – Nature and number of ring heteroatoms: 1O*)	2O*) 3O*)	1N	2N	3N 4N	1N, 1O*), 1N, 2O*), ----, 2N, 1O*), 2N, 2O*), ----, further heteroatoms**)
1 Compounds without functional groups		—	1	5	17	19	20	23	26	27
2 Hydroxy compounds		—OH		6						
3 Oxo compounds	=O	=O		7			21	24		
		=O + —OH		8	18			25		
4 Carboxylic acids	$\lessdot$O / OH	$\lessdot$O / OH	2	9			22			
		$\lessdot$O/OH + —OH; $\lessdot$O/OH + =O; $\lessdot$O/OH + =O + —OH	3	10						
5 Sulfinic acids		$—SO_2H$	4	11						
6 Sulfonic acids		$—SO_3H$								
7 Seleninic and selenonic acids tellurinic acids		$—SeO_2H$, $—SeO_3H$ $—TeO_2H$'								
8 Amines	$—NH_2$	$—NH_2$		12						
		$[—NH_2]n$; $—NH_2$ + —OH		13						
		$—NH_2$ =O; $—NH_2$ + $\lessdot$O $—NH_2$ + ...		14						
9 Hydroxylamines and dihydroxyamines		—NH-OH; —N(OH)OH		15						
10 Hydrazines		$—NH—NH_2$								
11 Azo compounds		—N=NH		16						
12 Diazonium compounds		$[—N\equiv N]^{\oplus}$								
13 Compounds with groups of three or more N-atoms		$—NH—NH—NH_2$; $N(NH_2)_2$; $—N=N—NH_2$ usw.								
14 Compounds containing carbon directly bonded to P, As, Sb, and Bi	e.g.	$—PH_2$; PH—OH; $—P(OH)_2$; $—PH_4$; - - - ; $PO(OH)_2$								
15 Compounds containing carbon directly bonded to Si, Ge, and Sn	e.g.	$—SiH_3$; $—SiH_2(OH)$; - - -								
16 Compounds containing carbon directly bonded to elements of the 3rd–1st A-groups of the periodic table	e.g.	$—BH_2$; —BH(OH) ; - - - ; $—Mg^{\oplus}$								
17 Compounds containing carbon directly bonded to elements of the 1st–8th B-groups of the periodic table	e.g.	—HgH; $—Hg^{\oplus}$; - - -								

*) instead of O also S, Se, Te **) *e.g.* B, Si, P; but not S, Se, Te

Beilstein undertakes a process of critical evaluation of all published data and discoveries relating to each individual compound. Because of this meticulous examination, through which incompatible results from different publications can be recognized, and obvious mistakes and contradictions eliminated, *Beilstein* achieves a degree of reliability superior to that of any other, purely bibliographical documentation and reference work, such as, for example, *Chemical Abstracts*.

The *Centennial Index*, a general register of *Beilstein* Series H–E IV, providing a comprehensive guide to organic compounds registered in *Beilstein* from literature of the period 1779–1959, appeared in 1991 and 1992.

9.1.2. The SANDRA Program

Searching for information in the *Beilstein* Handbook is made exceptionally easy by using the computer program SANDRA, which may be run on *IBM*-PCs or on *IBM*-compatible systems.

Table 9.4. *List of Organometallic Compounds Given in the* Gmelin Handbook of Inorganic and Organometallic Chemistry; *Status: 1992.* (The numbers given denote the atomic number and the *Gmelin System Number.*)

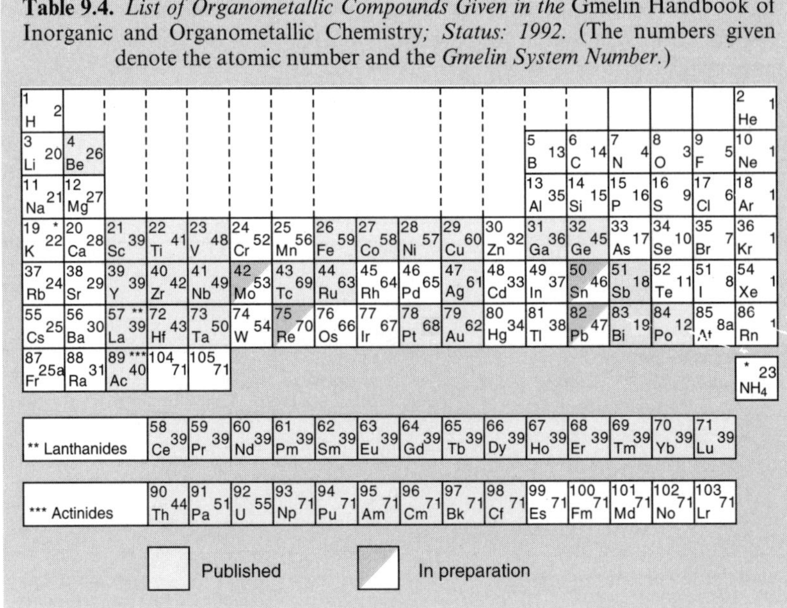

1 H 2																	2 He 1
3 Li 20	4 Be 26											5 B 13	6 C 14	7 N 4	8 O 3	9 F 5	10 Ne 1
11 Na 21	12 Mg 27											13 Al 35	14 Si 15	15 P 16	16 S 9	17 Cl 6	18 Ar 1
19 * K 22	20 Ca 28	21 Sc 39	22 Ti 41	23 V 48	24 Cr 52	25 Mn 56	26 Fe 59	27 Co 58	28 Ni 57	29 Cu 60	30 Zn 32	31 Ga 36	32 Ge 45	33 As 17	34 Se 10	35 Br 7	36 Kr 1
37 Rb 24	38 Sr 29	39 Y 39	40 Zr 42	41 Nb 49	42 Mo 53	43 Tc 69	44 Ru 63	45 Rh 64	46 Pd 65	47 Ag 61	48 Cd 33	49 In 37	50 Sn 46	51 Sb 18	52 Te 11	51 I 8	54 Xe 1
55 Cs 25	56 Ba 30	57 ** La 39	72 Hf 43	73 Ta 50	74 W 54	75 Re 70	76 Os 66	77 Ir 67	78 Pt 68	79 Au 62	80 Hg 34	81 Tl 38	82 Pb 47	83 Bi 19	84 Po 12	85 At 8a	86 Rn 1
87 Fr 25a	88 Ra 31	89 *** Ac 40	104 71	105 71													* NH_4 23

** Lanthanides	58 Ce 39	59 Pr 39	60 Nd 39	61 Pm 39	62 Sm 39	63 Eu 39	64 Gd 39	65 Tb 39	66 Dy 39	67 Ho 39	68 Er 39	69 Tm 39	70 Yb 39	71 Lu 39
*** Actinides	90 Th 44	91 Pa 51	92 U 55	93 Np 71	94 Pu 71	95 Am 71	96 Cm 71	97 Bk 71	98 Cf 71	99 Es 71	100 Fm 71	101 Md 71	102 No 71	103 Lr 71

Published | In preparation

The constitutional formula of an organic compound is entered graphically, with the aid of a 'mouse', to obtain a statement of
- the relevant Handbook series,
- the corresponding volume and subvolume,
- the sum formula,
- the concordance reference to the relevant page in the main work,
- the appropriate system number or system-number range,
- additional characteristics for classification: degree of saturation and number of C-atoms of the basic reference compound.

From Version 3.0, SANDRA additionally permits '*structure browsing*' in the *Beilstein*-ONLINE database.

9.1.3. The *Beilstein*-ONLINE Database

Since 1988, the published volumes of the *Beilstein* Handbook have been made available, in stages, in the complementary form of a database. The database will contain all information in critically evaluated form, as is to be found in the Handbook.

The database, like the Handbook, is compound-oriented. Each documentation unit (Dokumentationseinheit: DE) in the database corresponds to a title compound in the Handbook and/or the combined unchecked information (see below) available about this compound in the literature of the *V.* and/or *VI. Supplementary Series.* A DE consists of:
- the *Beilstein* Registry Number (BRN),
- the structural formula (in the form of a connectivity matrix, including configuration, if known),
- the preparation and reactions,
- physical data,
- the bibliography: all Handbook and primary literature citations. Each preparation or reaction and each reported property in the DE is given in conjunction with at least one primary literature citation.

Every DE can contain up to 400 fields. 80 fields have numerical content, 250 contain keywords, the rest consist of text fields.

Because of its data-structure, searches can be made in the *Beilstein* Database in a number of different ways. Any combination of queries relating to
- structure or substructure searching,
- numerical range searching (for physical data),
- text-related searching (such as bibliographic searching)

is possible.

At present, factual data (these are chemical properties and physical characteristics) of compounds from the Main Work and the *I.* to the *IV. Supplementary Series* are available in electronic form.

As well as these, information in unchecked form (not yet critically assessed) relating to the time periods of the *V.* and *VI. Supplementary Series*, from 1960 onwards, is offered. At present, the database covers 3.8 million compounds with their associated factual data. By the end of 1994, the database has been brought up to a fully up-to-date state. It now covers more than 6 million compounds.

Access to the *Beilstein* Database may be obtained either online (through *Scientific Technical Network* (*STN*) International, DIALOG, and ORBIT) or on an in-house computer system.

To afford greater topicality, primary data from *ca.* 80 key journals are offered quarterly on *CD-ROM* with a retrieval system which can be used on a personal computer. This service, entitled *Current Facts*, covers the twelve months prior to the release of each *CD-ROM* and has been available since 1990.

Current Facts offers information about structures (up to 300,000) and factual data. Included in the factual data offered is information about chemical reactions, preparations, numerical data, and – new for *Beilstein* – biological information. Factual data searching proceeds in the same way as on the ONLINE database and may be directed, both in structure and in substructure searching, towards particular stereoisomers and tautomers. The response usually takes a few seconds and is presented in a user-friendly fashion.

Current Facts output contains *autonames*: *IUPAC* names for the relevant compounds, generated with the aid of the program AUTONOM (*AUTO*matic *NOM*enclature). AUTONOM may also be used separately on *IBM*-compatible PCs. Any organic structure may be entered by the user, regardless of whether the related compound is known or not. In most cases (*ca.* 80%), AUTONOM responds with a name approved by the *International Union of Pure and Applied Chemistry* (*IUPAC*). Otherwise it declines to name the compound (*ca.* 20%: too difficult). Computer-assisted translation of the information contained in a structural formula into that one of a *IUPAC* name is unquestionably the problem solver of the future.

9.2. The *Chemical Abstracts Service* (*CAS*) of the *American Chemical Society* (*ACS*)

9.2.1. The Printed Media

The strength of *Beilstein* is its critical assessment. The strength of *Chemical Abstracts Service* is its swift provision of information from international chemical literature (since 1907). A reference publication (*Abstract Issue*) appears every week (*Fig. 9.1*), with publications
- in odd-numbered issues, from biochemistry and organic chemistry,
- in even-numbered issues, from macromolecular chemistry, applied chemistry, chemical engineering, and physical, inorganic, and analytical chemistry being updated fortnightly.

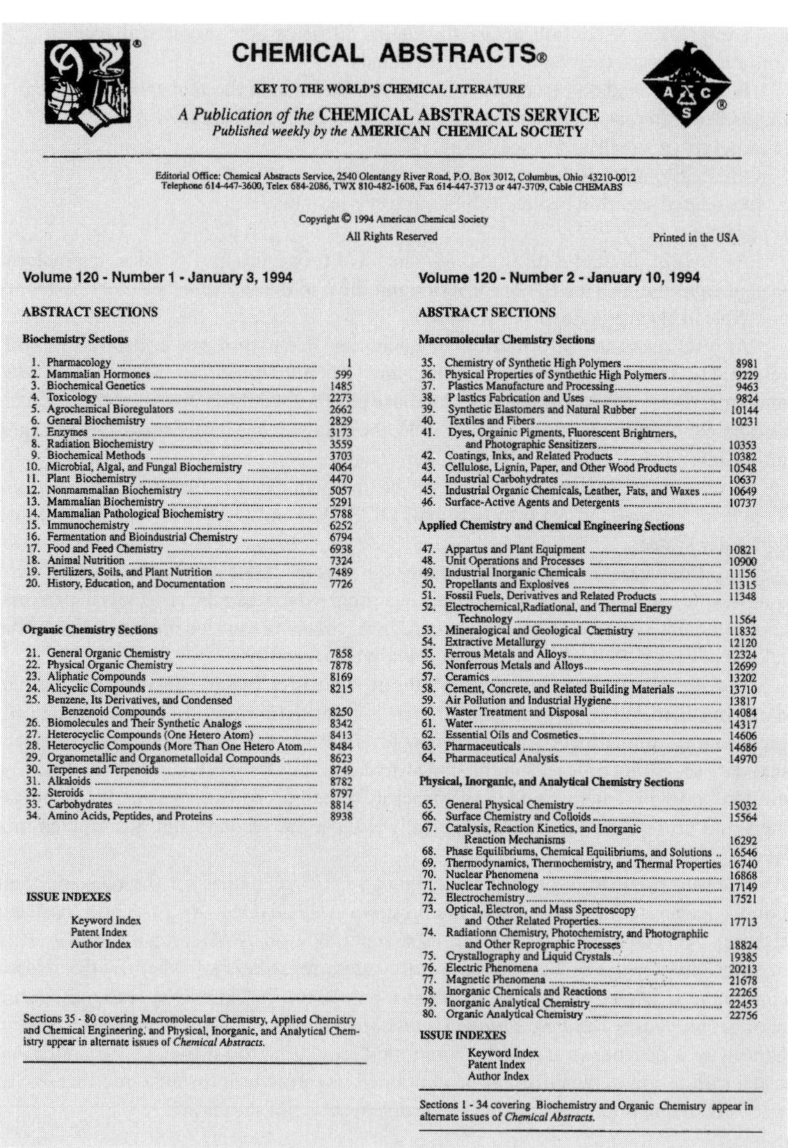

CHEMICAL ABSTRACTS®

KEY TO THE WORLD'S CHEMICAL LITERATURE

A Publication of the CHEMICAL ABSTRACTS SERVICE
Published weekly by the AMERICAN CHEMICAL SOCIETY

Editorial Office: Chemical Abstracts Service, 2540 Olentangy River Road, P.O. Box 3012, Columbus, Ohio 43210-0012
Telephone 614-447-3600, Telex 684-2086, TWX 810-482-1608, Fax 614-447-3713 or 447-3709, Cable CHEMABS

Copyright © 1994 American Chemical Society
All Rights Reserved

Printed in the USA

Volume 120 - Number 1 - January 3, 1994

ABSTRACT SECTIONS

Biochemistry Sections

1.	Pharmacology	1
2.	Mammalian Hormones	599
3.	Biochemical Genetics	1485
4.	Toxicology	2273
5.	Agrochemical Bioregulators	2662
6.	General Biochemistry	2829
7.	Enzymes	3173
8.	Radiation Biochemistry	3559
9.	Biochemical Methods	3703
10.	Microbial, Algal, and Fungal Biochemistry	4064
11.	Plant Biochemistry	4470
12.	Nonmammalian Biochemistry	5057
13.	Mammalian Biochemistry	5291
14.	Mammalian Pathological Biochemistry	5788
15.	Immunochemistry	6252
16.	Fermentation and Bioindustrial Chemistry	6794
17.	Food and Feed Chemistry	6938
18.	Animal Nutrition	7324
19.	Fertilizers, Soils, and Plant Nutrition	7489
20.	History, Education, and Documentation	7726

Organic Chemistry Sections

21.	General Organic Chemistry	7858
22.	Physical Organic Chemistry	7878
23.	Aliphatic Compounds	8169
24.	Alicyclic Compounds	8215
25.	Benzene, Its Derivatives, and Condensed Benzenoid Compounds	8250
26.	Biomolecules and Their Synthetic Analogs	8342
27.	Heterocyclic Compounds (One Hetero Atom)	8413
28.	Heterocyclic Compounds (More Than One Hetero Atom	8484
29.	Organometallic and Organometalloidal Compounds	8623
30.	Terpenes and Terpenoids	8749
31.	Alkaloids	8782
32.	Steroids	8797
33.	Carbohydrates	8814
34.	Amino Acids, Peptides, and Proteins	8938

ISSUE INDEXES

Keyword Index
Patent Index
Author Index

Sections 35 - 80 covering Macromolecular Chemistry, Applied Chemistry and Chemical Engineering, and Physical, Inorganic, and Analytical Chemistry appear in alternate issues of *Chemical Abstracts*.

Volume 120 - Number 2 - January 10, 1994

ABSTRACT SECTIONS

Macromolecular Chemistry Sections

35.	Chemistry of Synthetic High Polymers	8981
36.	Physical Properties of Synthethic High Polymers	9229
37.	Plastics Manufacture and Processing	9463
38.	P lastics Fabrication and Uses	9824
39.	Synthetic Elastomers and Natural Rubber	10144
40.	Textiles and Fibers	10231
41.	Dyes, Orgainic Pigments, Fluorescent Brightmers, and Photographic Sensitizers	10353
42.	Coatings, Inks, and Related Products	10382
43.	Cellulose, Lignin, Paper, and Other Wood Products	10548
44.	Industrial Carbohydrates	10637
45.	Industrial Organic Chemicals, Leather, Fats, and Waxes	10649
46.	Surface-Active Agents and Detergents	10737

Applied Chemistry and Chemical Engineering Sections

47.	Appartus and Plant Equipment	10821
48.	Unit Operations and Processes	10900
49.	Industrial Inorganic Chemicals	11156
50.	Propellants and Explosives	11315
51.	Fossil Fuels, Derivatives and Related Products	11348
52.	Electrochemical,Radiational, and Thermal Energy Technology	11564
53.	Mineralogical and Geological Chemistry	11832
54.	Extractive Metallurgy	12120
55.	Ferrous Metals and Alloys	12324
56.	Nonferrous Metals and Alloys	12699
57.	Ceramics	13202
58.	Cement, Concrete, and Related Building Materials	13710
59.	Air Pollution and Industrial Hygiene	13817
60.	Waster Treatment and Disposal	14084
61.	Water	14317
62.	Essential Oils and Cosmetics	14606
63.	Pharmaceuticals	14686
64.	Pharmaceutical Analysis	14970

Physical, Inorganic, and Analytical Chemistry Sections

65.	General Physical Chemistry	15032
66.	Surface Chemistry and Colloids	15564
67.	Catalysis, Reaction Kinetics, and Inorganic Reaction Mechanisms	16292
68.	Phase Equilibriums, Chemical Equilibriums, and Solutions	16546
69.	Thermodynamics, Thermochemistry, and Thermal Properties	16740
70.	Nuclear Phenomena	16868
71.	Nuclear Technology	17149
72.	Electrochemistry	17521
73.	Optical, Electron, and Mass Spectroscopy and Other Related Properties	17713
74.	Radiationn Chemistry, Photochemistry, and Photographiic and Other Reprographic Processes	18824
75.	Crystallography and Liquid Crystals	19385
76.	Electric Phenomena	20213
77.	Magnetic Phenomena	21678
78.	Inorganic Chemicals and Reactions	22265
79.	Inorganic Analytical Chemistry	22453
80.	Organic Analytical Chemistry	22756

ISSUE INDEXES

Keyword Index
Patent Index
Author Index

Sections 1 - 34 covering Biochemistry and Organic Chemistry appear in alternate issues of *Chemical Abstracts*.

Fig. 9.1. *The contents' pages of two sequential issues of* Chemical Abstracts *with its 80 sections*

The *Abstracts* are divided into 80 different *sections*. Detailed bibliographic data accompany a usually brief outline of the content. The latter is intentionally non-judgmental and should serve merely to help the reader decide if he or she wishes to read the associated original publication.

It is possible to effectively and reliably retrieve information stored in *Chemical Abstracts*, with the aid of a structured index system (*Fig. 9.2*):

- on publications by a particular author in the *Author Index*,
- on individual chemical compounds in the *Chemical Substance Index*, the *Formula Index*, and the *Index of Ring Systems*,
- on patents and their relationship with patent applications in other countries in the *Patent Index* (prior to 1981: *Numerical Patent Index* and *Patent Concordance*),
- for literature searches on specific themes in the *General Subject Index* or the *Keyword Index*.

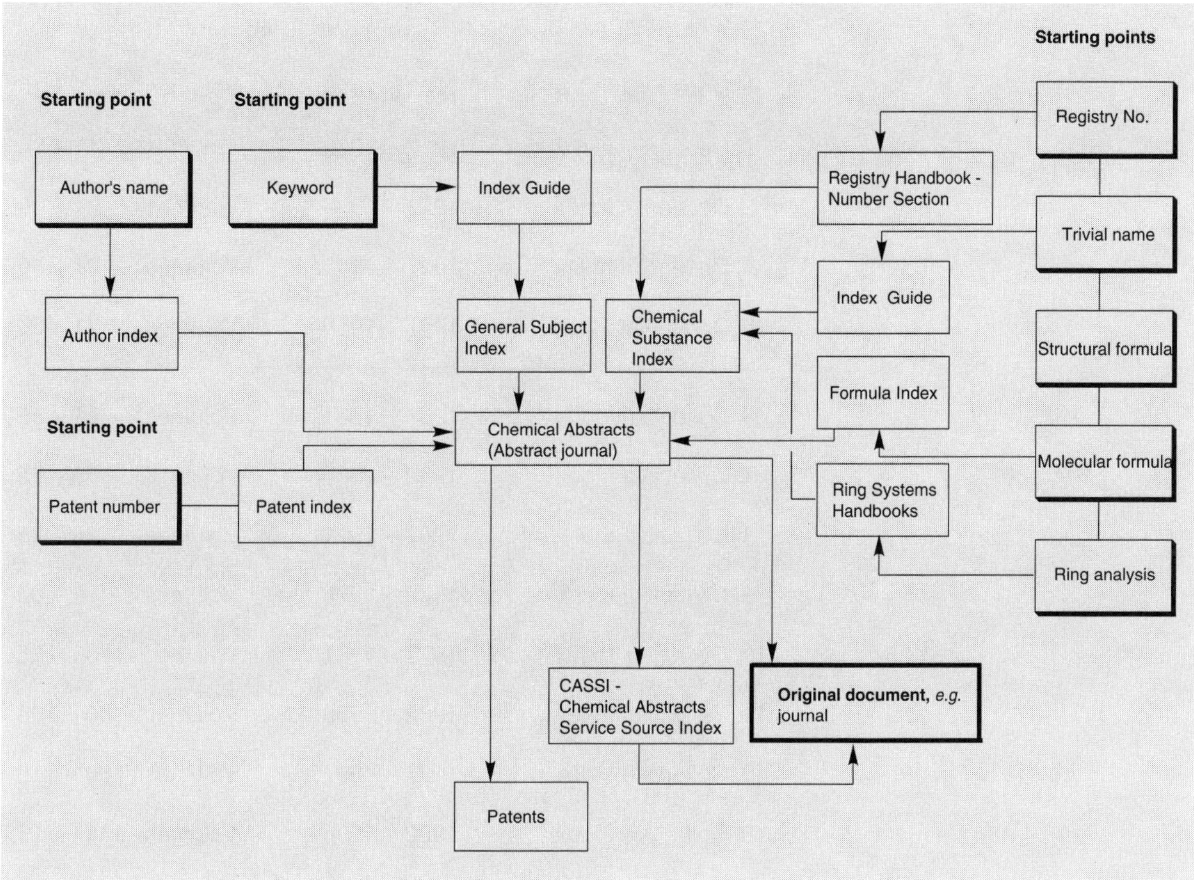

Fig. 9.2. *Guide for information retrieval from* Chemical Abstracts

The indexes are to be found

- in each weekly issue (*Issue Index*),
- in the six-monthly *Volume Indexes*,
- in the five- or ten-yearly *Collective Indexes* and *Decennial Indexes* (*Table 9.5*).

It is necessary to be aware that

- the *Keyword Index* of a single issue is limited to keywords contained in the title and the abstract,
- the *Volume Indexes* contain terms from the original documentation which have been selected using a *controlled vocabulary*.

The user of the printed media is only able to achieve complete information retrieval once the *Volume Index* has been published. Each *Collective Index* comes with its own *Index Guide* which:

- enables cross-references to be made between trivial names or names in earlier usage and the currently (exclusively) used names of chemical compounds,
- indicates the keywords used by *Chemical Abstracts* for synonymous expressions,
- explains the main guidelines for refereeing policy which have been followed in a particular period.

The *Index Guide* is regularly updated, extended, and republished.

It should be noted that the *accession numbers* have only been consecutively listed, one per abstract, since 1967. Before then, a number was used to denote each column in an issue and a lower-case letter indicated the approximate line in this column. Between 1967 and the end of 1990, 9.5 million publications were excerpted. Since 1965, every compound reported in the literature has been assigned its own *registry number*: at the end of 1993 almost 12.8 million *registry numbers* had been assigned.

Table 9.5. *Decennial and Collective Indexes of* Chemical Abstracts

1. Decennial Index	1907 – 1916	Volumes	1 – 10
2. Decennial Index	1917 – 1926	Volumes	11 – 20
3. Decennial Index	1927 – 1936	Volumes	21 – 30
4. Decennial Index	1937 – 1946	Volumes	31 – 40
5. Decennial Index	1947 – 1956	Volumes	41 – 50
6. Collective Index	1957 – 1961	Volumes	51 – 55
7. Collective Index	1962 – 1966	Volumes	56 – 65
8. Collective Index	1967 – 1971	Volumes	66 – 75
9. Collective Index	1972 – 1976	Volumes	76 – 85
10. Collective Index	1977 – 1981	Volumes	86 – 95
11. Collective Index	1982 – 1986	Volumes	96 – 105
12. Collective Index	1987 – 1991	Volumes	106 – 115
13. Collective Index	1992 – 1996	Volumes	116 – 125

9.2.2. The Electronic Media

The abstracts which have appeared in *Chemical Abstracts* since 1967 have been collated into a database, in which online-searches may be performed extremely quickly, *via* computer.

Bibliographical data, the other entries contained in the indexes of *Chemical Abstracts'* printed material, and also the texts of the abstracts have been stored in bibliographic databases at *Scientific Technical Network (STN) International/CAS* ONLINE, or at other commercial information providers. These may be accessed by the information user with various licenses for use, knowledge of search strategies, a suitable terminal, and the necessary funds for payment of bills for the computer telephone link: a process called *online retrieval.*

Interactive dialog

- does not render knowledge of the use of *CAS*-printed media dispensable: *Chemical Abstracts'* collected material dating from before 1967 is not completely available in a database,
- opens up additional search options to those ones which may be made in the printed media: almost all terms which have been stored may be used as search objectives, and may be combined in a search logic substantially more complex than is possible 'by hand' in the printed media (*multidimensional search profile*).

As already mentioned, since 1965 *CAS* has followed a policy of registering all chemical compounds which appear in the chemical literature; recently those compounds described exclusively in the literature from before 1965 have also been registered. Individual registry numbers (*CA-RN*) are assigned even to each stereoisomer. Every registered compound is stored in a *Compound Index Database* which, similarly, is available for online retrieval. If the registry number, or the English systematic name or synonym of a chemical compound is known, then the structural formula, molecular formula, and further synonyms may be accessed, as well as the most recent publications for which the compound has been indexed. The structures of compounds are stored in such a way that a search may also be made by drawing a structural formula on a graphics terminal. This database is enlarged by *ca.* 10,000 new registry numbers weekly.

9.3. The *Science Citation Index*

Almost every scientific publication contains a list of references (bibliography) stating the exact sources of literature cited in that publication. The *Science Citation Index*, published by the *Institute for Scientific Information* (*ISI*) in Philadelphia, makes it possible to find every publication (from 1955 onwards) in whose list of references a particular, earlier paper has been cited. It is reasonable to assume that the content of the cited and the citing literature are related. A network can thus be obtained, from which the influence of an earlier publication on later works may be deduced. The *Science Citation Index* covers the fields of the biosciences, chemistry, the geosciences, physics, physiology, engineering, and applied sciences. At present, over 600,000 publications from about 3,200 scientific journals from these fields of study are assessed annually, the literature citations from the list of references of each publication being recorded. In the *Science Citation Index* there is to be found, under the name of the first author (only initials of first names are given), for each of his or her publications, a list of those later publications in which that work has been cited. The bimonthly issues of the *Science Citation Index* are condensed into an annual volume at the end of each year.

The complete work consists of four mutually complementary indexes, which each address the same material from different viewpoints. The alphabetically ordered *Source Index* contains the complete bibliographical information for each article: author, title (in English), the journal in which the article appeared (with volume number, issue number, page number, and year of publication), and the number of literature citations contained in the article. The address of the author is given as well (anonymous articles are listed alphabetically by journal at the beginning of the *Source Index*). A two-letter language code precedes the title, given in English.

In the *Permuterm Subject Index*, there follows a guide to each source article by means of a few words from the (English) title of the article. These keywords are simply followed by the name of the author. The complete bibliographic details relating to the article of interest can then be taken from the *Source Index*.

The original *Citation Index* lists cited authors (in alphabetical order) with their publications (arranged chronologically), and gives the journal in which the publication by the citing author appeared.

The *Corporate Index* provides a list of authors by the name of their institutions, with a brief pointer to their publications. The institutions are also listed by geographical location.

SCI-SEARCH is a database equivalent to the printed *Science Citation Index*, which, like *CAS* ONLINE, can be used for computer-assisted online retrieval. At present it contains information from 1974 onwards.

Further Reading

To Sect. 9.1.

Kennen Sie Beilstein? Springer-Verlag, Berlin.

The Short Cut to Locating a Compound in Beilstein Handbook of Organic Chemistry, Springer-Verlag, Berlin.

This is Gmelin, Springer-Verlag, Berlin.

SANDRA, Structure and Reference Analyser for Users of the Beilstein Handbook of Organic Chemistry, Springer-Verlag, Berlin.

W. Liebscher, *Die Beschreibung von Substanzen – die Ambivalenz der Nomenklatur,* Mitteilungsblatt der GDCh-Fachgruppe 'Chemie – Information – Computer', Frankfurt am Main, 1992, Nr. 21, 4.

A. Lawson, *Systematic Nomenclature from F.K. Beilstein to AUTONOM,* Mitteilungsblatt der GDCh-Fachgruppe 'Chemie – Information – Computer', Frankfurt am Main, 1992, Nr. 21, 27.

E. Zass, *Chemische Weltliteratur auf CD-ROM, Nachr. Chem. Tech. Lab.* **1991,** *39,* 1152;

– *On Searching the Literature – Using the Computer (and Your Head) to Retrieve Structures, References, Reactions, and Data Online* in *A Guide for the Perplexed Organic Experimentalist* (Ed. H.J.E. Loewenthal), John Wiley & Sons, Chichester, 1990.

To Sect. 9.2.

Mit den Chemical Abstracts führen alle Wege zum Ziel, VCH Verlagsgesellschaft, Weinheim.

H. Schulz, *Von CA bis CAS ONLINE,* VCH Verlagsgesellschaft, Weinheim, 1985.

To Sect. 9.3.

How to Search the Science Citation Index, ISI Institute for Scientific Information, Offenbach am Main.

10 The Use of Formulae and Names for the Description of Molecules in the Context of the Structure Model of the Classical Organic Chemistry

'Chemical thinking' begins with the acceptance of discrete molecules and their atomic composition. In order to make 'chemical thoughts' as clear and unequivocal as possible, a *chemical iconography* (*chemical-formula language*) has been instituted and developed. The symbols of this chemical iconography have been an important means of coding the information content of a molecular structure since the dawn of chemical science. For communication of chemical information to proceed reliably, it is important that the sender and the receiver are using the same terms covering the same meanings. The *International Union of Pure and Applied Chemistry* (*IUPAC*) ensures that the symbols of chemical iconography are clearly specified and, after a certain test phase, are bindingly laid down.

The semiotic specification of molecules using formulae unites all chemists despite the various obstacles of their natural languages, but nevertheless acts as a barrier to comprehension, not only for the lay public, but also, frequently, for scientists of other disciplines who are interested in chemical compounds and their properties. The language of chemistry connects physics, *via* chemistry, with biology: all of the areas of human knowledge concerned with phenomena which can be interpreted at the molecular level. Journalists, judges, environmentalists, every involved citizen should possess at least a measure of knowledge of the chemical 'code of symbols'. The involved citizen, wishing to understand and give opinions on chemical problems and their solutions, cannot but avoid picking up some knowledge how to use the language of chemical formulae. At the same time, chemists will have to make greater efforts than in the past to use chemical language as clearly and unambiguously as possible.

10.1. Constitutional Formulae

To describe which particular atoms, with their specific valencies, are connected together in molecules, it is usual to depict the relevant element symbols and connect them together with symbols for single, double, or triple bonds, expressing by the use of suitable signs whether chains, rings, or combinations of both are present (*Tables 10.1–10.3*).

Table 10.1. *Atoms, Their Symbols, Valencies, and Occurrence in Structural Units*

Element	Symbol	Structural unit	
Carbon	**C**	tetravalent	—C— (four bonds)
Nitrogen	**N**	trivalent	—N<
Oxygen	**O**	bivalent	—O—
Hydrogen	**H**	monovalent	H—

Table 10.2. *Bonds, Their Symbols, and Occurrence in Structural Units*

Bond	Symbol	Structural unit
Single bond	—	—C—C— (with vertical bonds on each C)
Double bond	=	>C=C<
Triple bond	≡	—C≡C—

In this manner, formula representations, showing clearly the *connectivity* (*constitution*) of the atoms comprising a molecule, are obtained. Constitutional formulae of this type rapidly prove themselves to be sources of clear, graphic structural information, the use of which chemists are no longer able to do without. They are particularly appropriate for illustrating the various constitutional isomers, for example, of molecular formula C_3H_6O (*Table 10.4*).

It has become a normal practice to substitute extended formulae by slightly abbreviated ones (*Table 10.5*) in which particular atom groups (CH_3, CH_2) are brought together as one. Alternatively, full pictograms, containing the essential structural elements in heavily simplified form, may be used (*Table 10.6*). Here, practically all that is written out are the heteroatoms and the H-atoms bound to them. The skeleton of C-atoms, with its H-atoms, is reduced to the representation of the C–C bonds. A further means of representing constitutional isomers makes use of *graphs*. For the first five members of the homologous series of alkanes,

Table 10.3. *Chains or Rings, Their Symbols, and Occurrence in Structural Units*

Constitutional type	Symbol	Structural unit	Constitutional type	Symbol	Structural unit
Chain	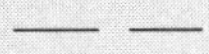 — —	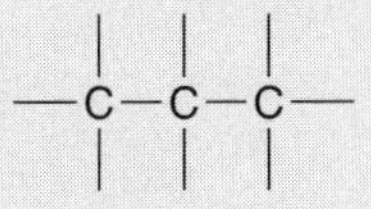 —C—C—C—	Ring	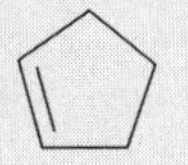	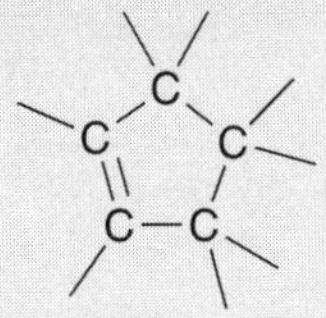
	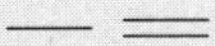 — =	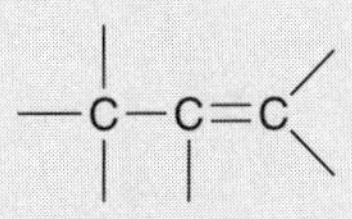 —C—C=C		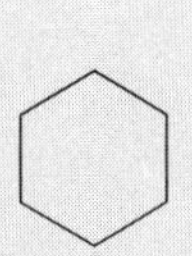	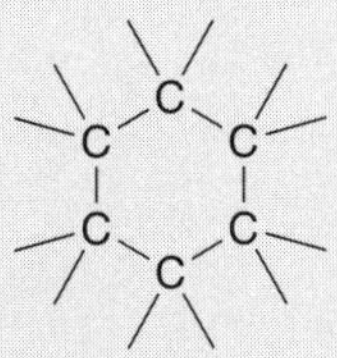
	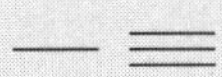 — ≡	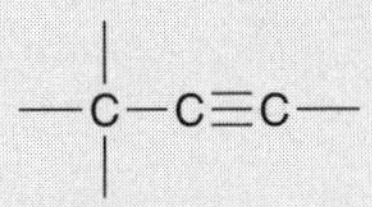 —C—C≡C—		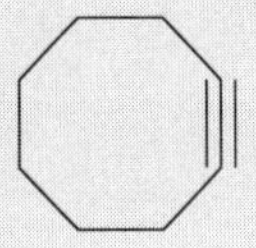	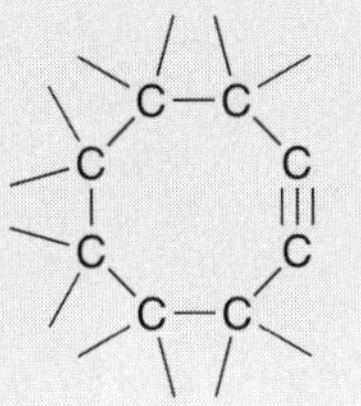
	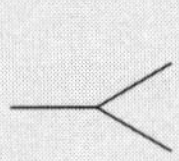 —<	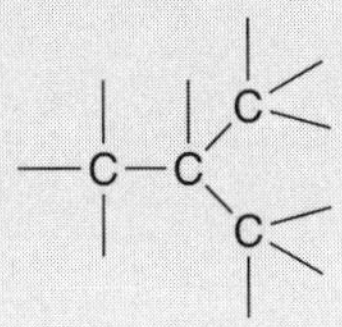 —C—C(C)C			

the number of trees (*Table 10.7*) agrees with the number of constitutional isomers.

The use of *graph theory*, however, gives the correct number of possible constitutional isomers only when the valencies of C and H are taken into account. For the molecular formula C_8H_{18}, this means that, instead

Table 10.4. *Extended Formula Representations for the Constitutional Isomers of* C_3H_6O

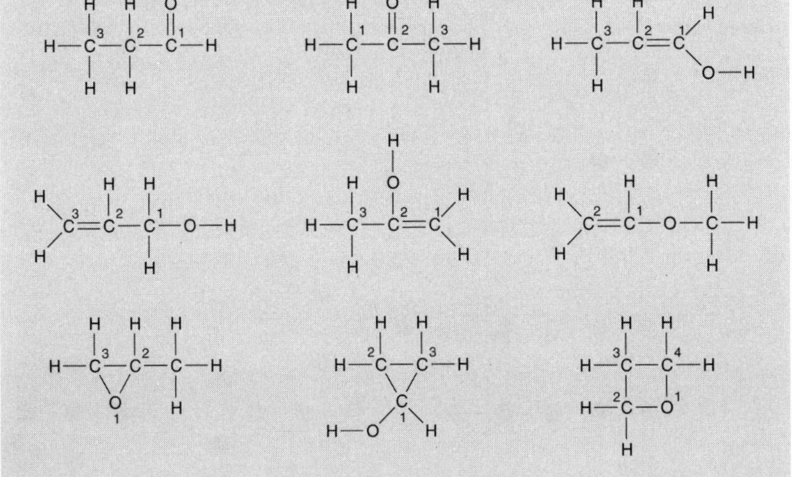

Table 10.5. *Slightly Abbreviated Formula Representations for the Constitutional Isomers of C_3H_6O*

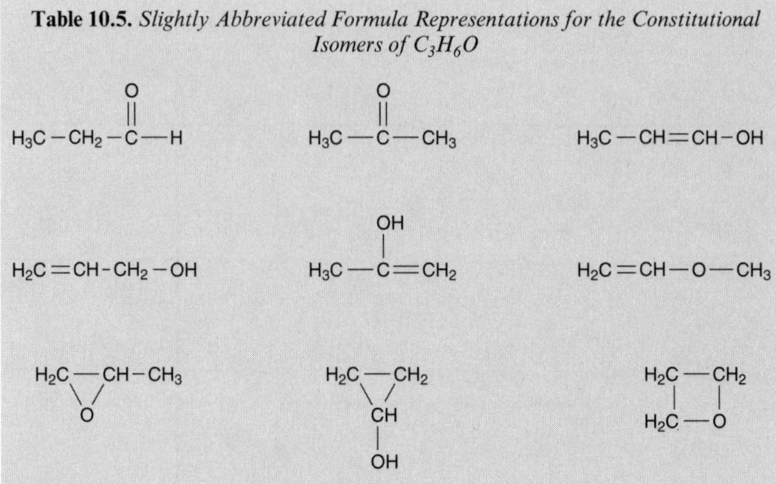

Table 10.6. *Heavily Abbreviated Formula Representations for the Constitutional Isomers of C_3H_6O*

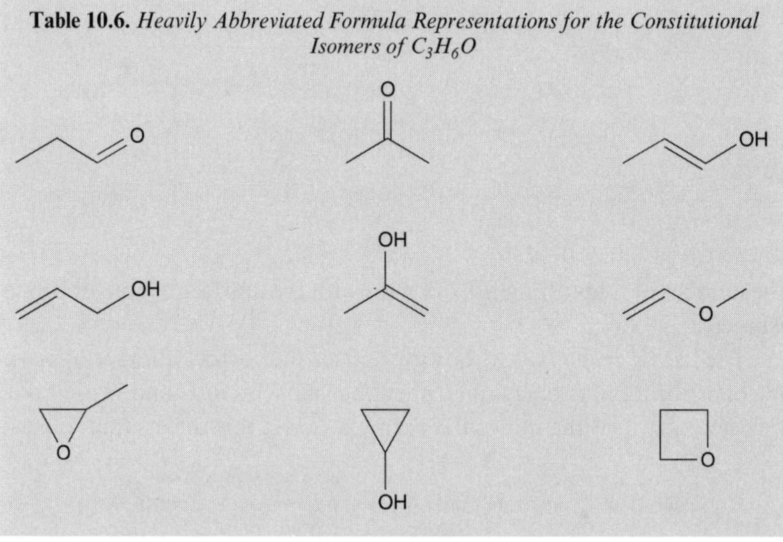

Table 10.7. *The Graph Theory Description of the First Five Members of the Homologous Series of Alkanes.* The number of *trees* corresponds to the number of constitutional isomers.

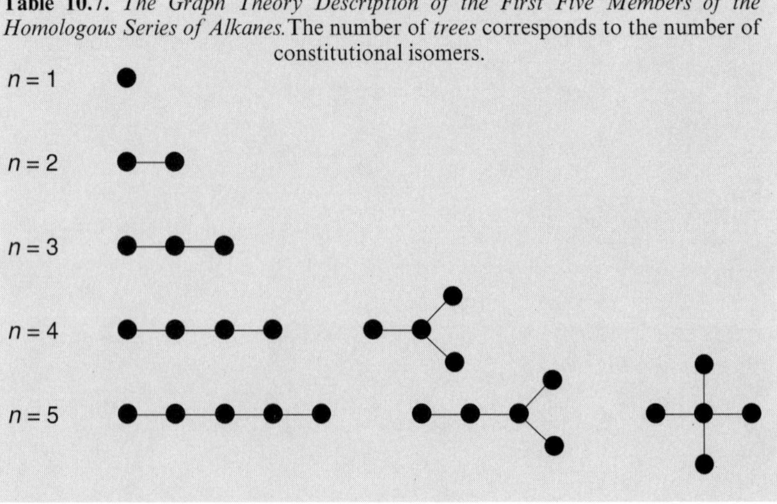

of the 23 constitutional isomers which would otherwise be predicted, only the 18 structures consisting solely of points of the orders 1 or 4 are possible (*Table 10.8*). Since computers with large memory capacities have been available, there has been increased activity directed at finding an algorithm which can give the correct number of constitutional isomers for each individual molecular formula (such as $C_{25}H_{52}$) belonging to a general series (C_nH_{2n+2}) and list their individual appearances.

Table 10.8. *For Compounds of Molecular Formula C_8H_{18}, the Number of* Trees *Corresponds to the Number of Constitutional Isomers, if only Points of the Order 1 or 4 (Valency of Hydrogen and Carbon) are Considered*

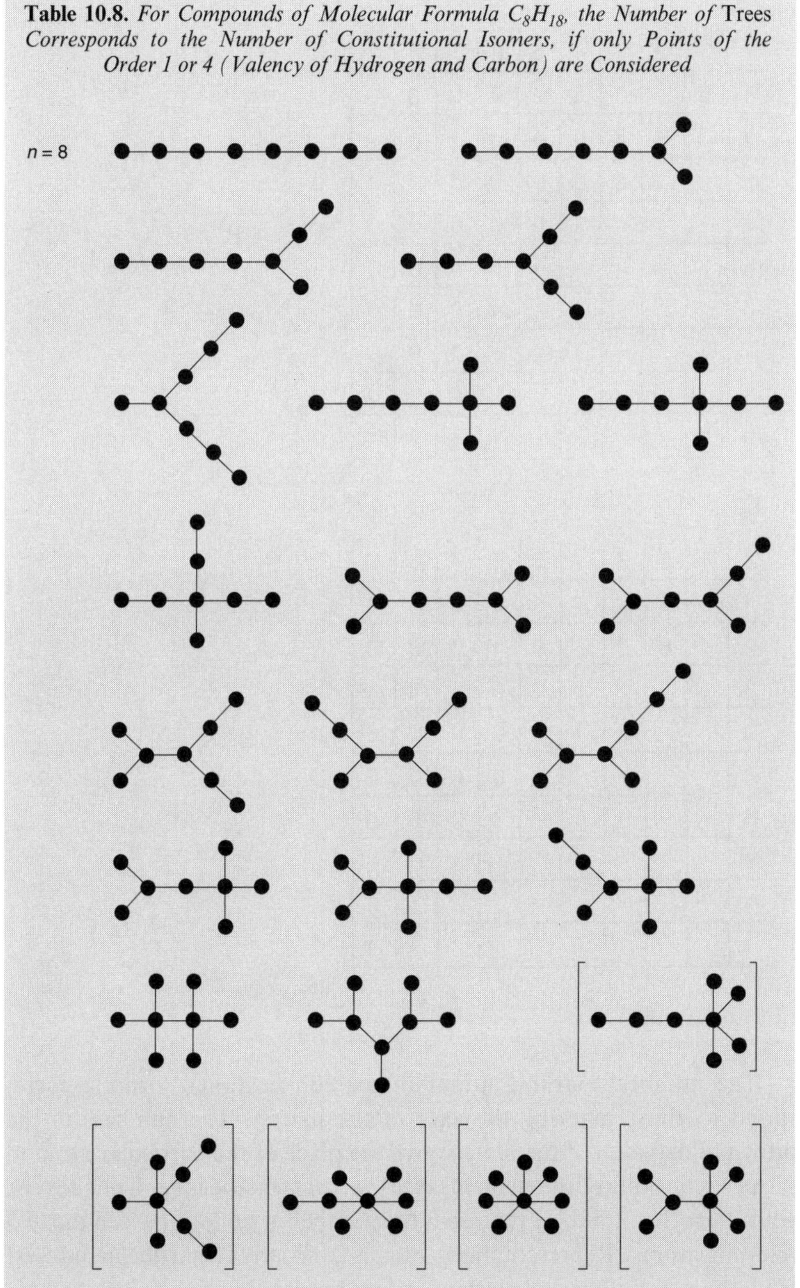

In connection with these different styles of depiction of formulae, one alternative means of representation of constitutional isomers should be mentioned. Using this method, different constitutions are not represented as pictographs, but as *matrices*. This can most easily be exemplified using the two constitutional isomers with the molecular formula C_3H_6 (*Table 10.9*).

Table 10.9. *Description of the Constitutions of Cyclopropane and Propene, Using a Matrix in Each Case*

	1	2	3	4	5	6	7	8	9
1	6	1	1	1	1	0	0	0	0
2		6	1	0	0	1	1	0	0
3			6	0	0	0	0	1	1
4				1	0	0	0	0	0
5					1	0	0	0	0
6						1	0	0	0
7							1	0	0
8								1	0
9									1

H8 H9
C3
H4 H7
C1—C2
H5 H6

	1	2	3	4	5	6	7	8	9
1	6	2	0	1	1	0	0	0	0
2		6	1	0	0	1	0	0	0
3			6	0	0	0	1	1	1
4				1	0	0	0	0	0
5					1	0	0	0	0
6						1	0	0	0
7							1	0	0
8								1	0
9									1

H6 H7
H4
C1=C2—C3—H8
H5
H9

The numbers marking atoms in the constitutional formula correspond to those marking the axes of the matrix. The numbers in the principal diagonal of the matrix give the place of the particular atom in the periodic table of the elements (see the *Table* on the inner front cover), while the other numbers represent the number of bonds between the two relevant atoms. Representations of this kind are of particular interest from the point of view of electronic data processing in chemistry.

Ludwig Wittgenstein stated that a picture can replace a description, and that the two are synonymous. *René Magritte* affirmed that a picture can symbolize an object, but that the two are not identical. *Magritte*'s painting showing a pipe with the statement '*Ceci n'est pas une pipe*' (*Fig. 10.1*) draws attention to this trivial fact whilst illustrating a pattern of behavior typical of a chemist: formula representations used by chemists are only substitutes for the actual molecules.

Fig. 10.1. *The Belgian painter* René Magritte, *by one of his paintings which incorporate words, expresses, what the chemist, too, became aware of: that* pictures *or* models *are instrumental in describing the world.*

Hence, just as the illustration in the margin (*Fig. 10.1*) merely symbolizes that object with the name 'pipe', the illustration below (*Fig. 10.2*) is not cyclohexane but merely a symbol for the molecule with the name 'cyclohexane'.

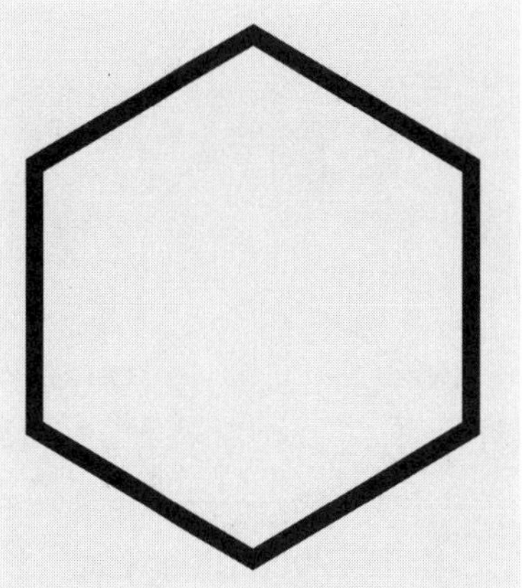

Fig. 10.2. *'This is not cyclohexane'.*

10.2. Configurational Formulae

The tetrahedron model devised for methane and its derivatives leaves the plane of the printed page and describes three-dimensional structures. Normally, in chemical formulae symbols

- *ordinary lines* are used to represent bonds within the plane of the page,
- *filled-in, wedge-shaped symbols* are used to represent bonds protruding out of the page towards the reader,
- *dashed, wedge-shaped symbols* are used to represent bonds which also leave the plane of the page, but away from the reader.

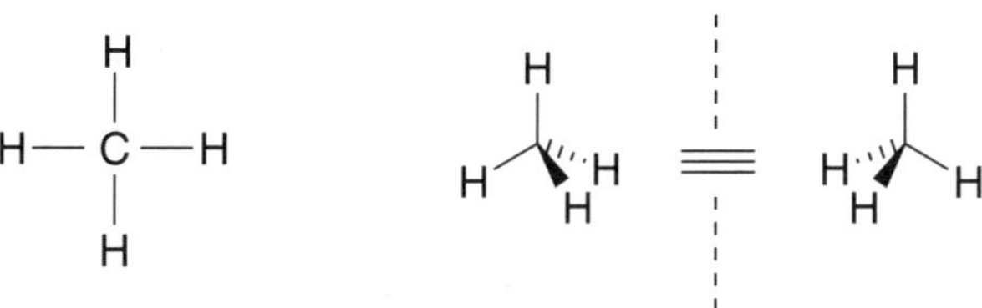

Stereoformulae of this type are described as *configurational formulae*.

10.2.1. Configurational Isomers with Stereogenic Centers

Configurational isomerism comes into consideration when at least one stereogenic center is present. Glyceraldehyde was originally selected as the prototype for the description of this kind of isomerism. Its two enantiomers are represented here in the form of *Fischer* projections.

CHO / HO—H / CH_2OH (*S*)   OHC / H—OH / HOH_2C (*R*)

E. Fischer introduced the projections named after him in the course of his work on sugars. He prescribed the following conventions:

- The longest carbon chain (main chain) is arranged vertically, so that the C-atom with the lowest locant is uppermost.
- For each C-atom in the main chain, the vertical (intracatenate) bonds point away from the observer (into the half-space below the plane of the paper).
- For each C-atom in the main chain, the horizontal (extracatenate) bonds point towards the observer (into the half-space in front of the plane of the paper).

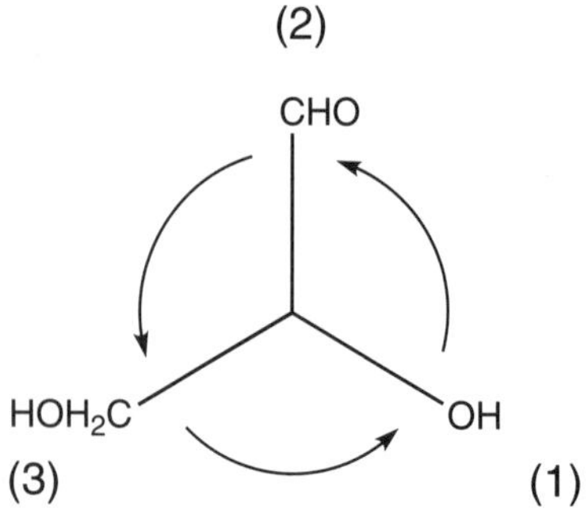

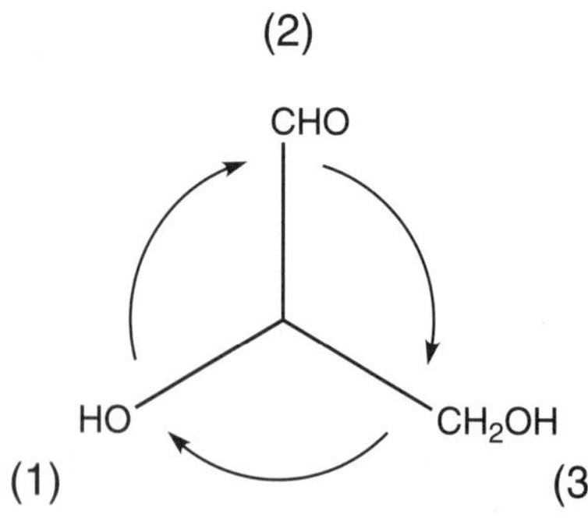

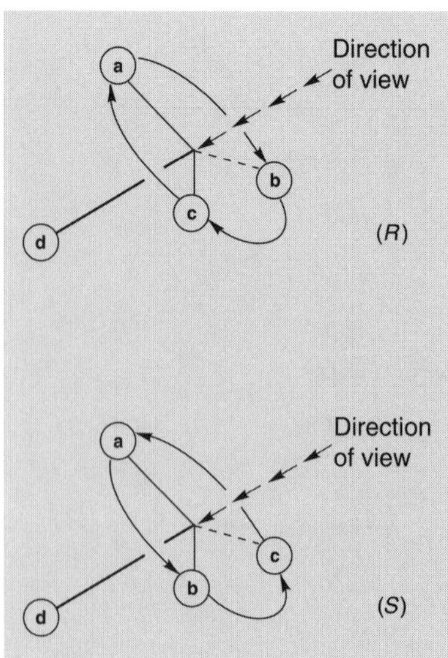

Fig. 10.3. *Arrangement of the ligands a, b, c, and d by priority*

If a particular constitution has two enantiomers, then a descriptor, permitting an unambiguous distinction between the two, is required. Here the notation of *Robert Sidney Cahn*, *Sir Christopher (Kelk) Ingold*, and *Vladimir Prelog* (*CIP* notation) has proven its worth. Basic to the *CIP* notation (or the (*R*/*S*)-convention) are *sequence rules*, which arrange the four substituents at a stereogenic center ('asymmetry center') into a sequence of falling priority. These rules state that:

- atoms of higher atomic number have priority over those ones of lower atomic number,
- atoms of higher mass number have priority over those ones of lower mass number,
- (*R*) has priority over (*S*), (*R*,*R*) or (*S*,*S*) over (*R*,*S*) or (*S*,*R*).

Of these three sequence rules, the second and third are only taken into account, if no clear priority has already been established by application of the preceding rule.

Once the priority (here a > b > c > d) at a stereogenic center has been established, the stereogenic center is then assigned a descriptor, (*R*) or (*S*), with the aid of the *chirality rule*. The central atom is considered to be at the hub of a 'steering wheel' and the bonds to the three ligands of highest priority (a, b, c) to be the spokes of the 'wheel' (*Fig. 10.3*). The fourth ligand d, of lowest priority, is then to be found at the foot of the 'steering column'. The observer views it all from the driver's perspective, through the center of the spokes towards the foot of the 'steering column'.

If the 'steering wheel' is then turned, so that the original point 'a' passes through 'b' and then through 'c', in the direction of falling priority, then the direction of rotation serves as a means of specifying the absolute configuration. An anticlockwise rotation results in the configuration receiving the (*S*) (derived from the Latin *sinister* = left) descriptor, whilst a clockwise rotation results in the (*R*) (from the Latin *rectus* = right) descriptor.

10.2.1.1. Acyclic Molecules with One Stereogenic Center

The following example (*Fig. 10.4*) has the constitution of 3-chloro-2-(hydroxymethyl)-4-methylpentan-1-ol. To specify the sense of chirality at a stereogenic center, the ligands must be brought into a definite order of priorities. For complex structures, a special, hierarchical graph (*digraph*), representing the type and connectivity of atoms in the form of a tree-diagram, is used for this purpose. To transpose information relating to the atoms adjacent to the stereogenic center (to the proximate atoms) from the *Fischer* projection, the original *Fischer* projection, with a vertically oriented main chain (**A**), must be converted into a modified *Fischer* projection with a horizontal main chain (**B**). A 90° rotation of the *Fischer* projection in the plane of the page inverts the absolute configuration! For structurally complex molecules, it is advantageous to use a simplified representation (**D**), rather than the complete formulation (**C**), leaving out the irrelevant H-atoms and replacing the C-atoms by their locants. The order of rank of the atoms is then determined by consideration of their topological distance from the stereogenic center, and then by the sequence rules. Atoms of the *n*th sphere take precedence over

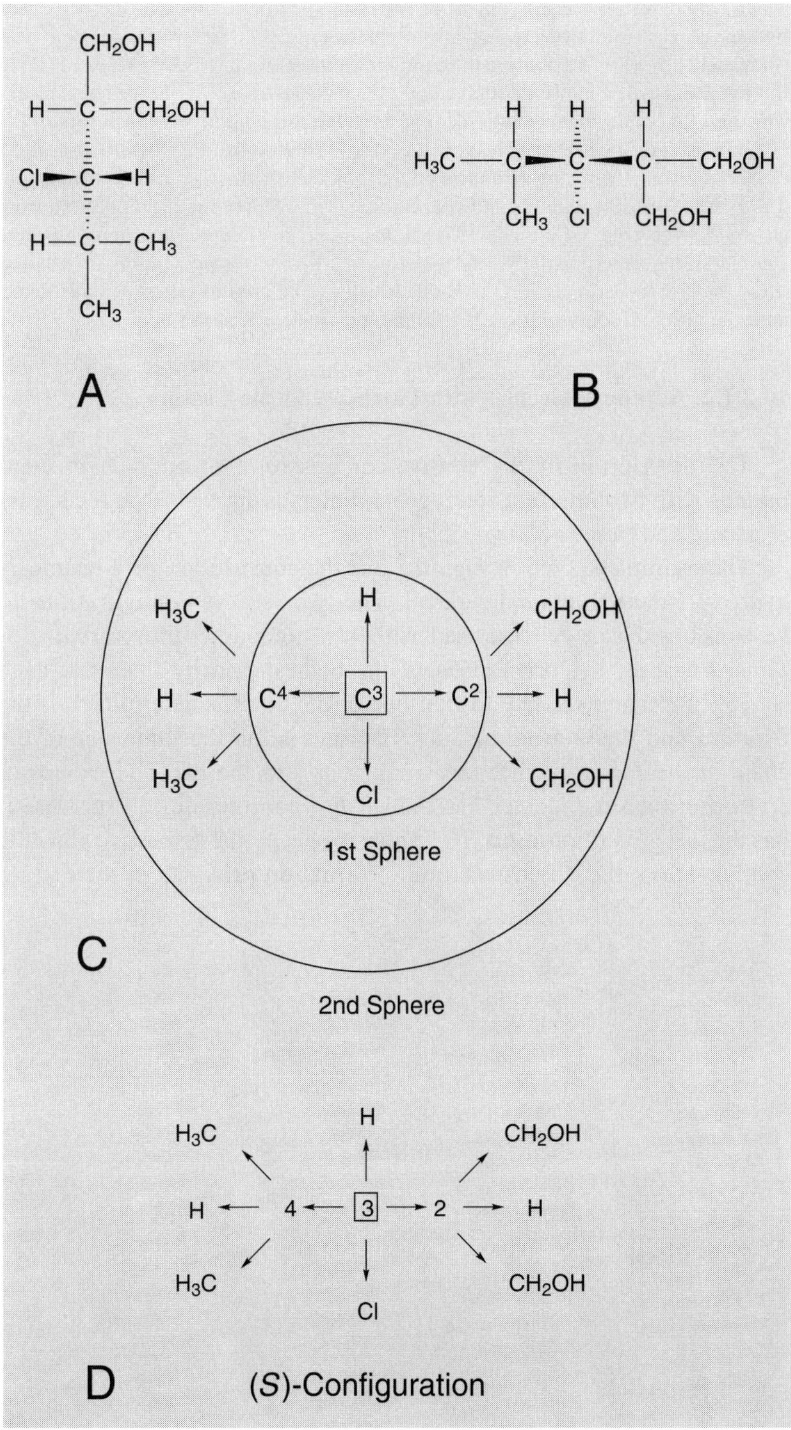

Fig. 10.4. *Determination of the (*S*)-configuration for the enantiomer with the constitution 3-chloro-2-(hydroxymethyl)-4-methylpentan-1-ol. A*) *Fischer* projection; *B*) modified *Fischer* projection; *C*) complete hierarchical digraph; *D*) simplified digraph.

atoms of the $(n + 1)$th sphere. To establish the sense of chirality, it is necessary to follow the branches of the hierarchical digraph until a sphere is reached, in which an unambiguous differentiation of the ligands is possible. In the present instance, (S)-configuration is assigned.

When the *Fischer* projection was first introduced, no experimental method existed which would permit the assignment of (R)- or (S)-configurations to the two enantiomers of glyceraldehyde (or of any other case). *Emil Fischer*'s assignment was completely arbitrary, attributing the configuration that we have just defined as (R) to the glyceraldehyde enantiomer that rotates the plane of linearly polarized light clockwise, and that one which we have defined as (S) to the enantiomer which rotates the plane of linearly polarized light anticlockwise. While aware that he stood a 50:50 chance of hitting the wrong arrangement, he took the risk anyway, as his primary aim was to make possible a simple and systematic specification of the sugars derived from glyceraldehyde (*Fig. 5.4*). Since 1951, it has been possible to determine absolute configurations experimentally, and we now know that *Fischer*, purely by chance, chose the correct arrangement. Methods which can be used to experimentally determine the absolute configuration of a compound are described in *Chapt. 12*.

10.2.1.2. Acyclic Molecules with Two Stereogenic Centers

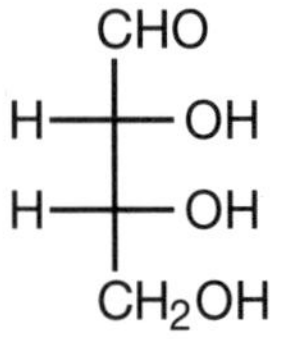

CHO
H—OH
H—OH
CH_2OH
Erythrose

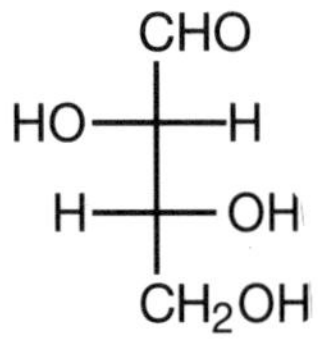

CHO
HO—H
H—OH
CH_2OH
Threose

The description of the relative configuration of open-chain compounds with two adjacent stereogenic centers is derived from the sugars *erythrose* and *threose* (*Chapt. 5.2*).

The example shown in *Fig. 10.5* has the constitution of 2-bromo-3-hydroxy-2-methylbutanedioic acid. The two relative configurations to be considered can be described with the stereodescriptors *erythro* or *threo*. For this, the orientation of the highest-priority ligand at each stereogenic center relative to that one of the other is determined. If the Br-atom and the O-atom of the OH group lie on the same side of the chain in the *Fischer* projection, analogously to the two OH groups of erythrose, then the isomer has the *erythro*-configuration; otherwise it has the *threo*-configuration. To be able to specify the respective absolute configuration, the stereostructural information provided in the *Fischer*

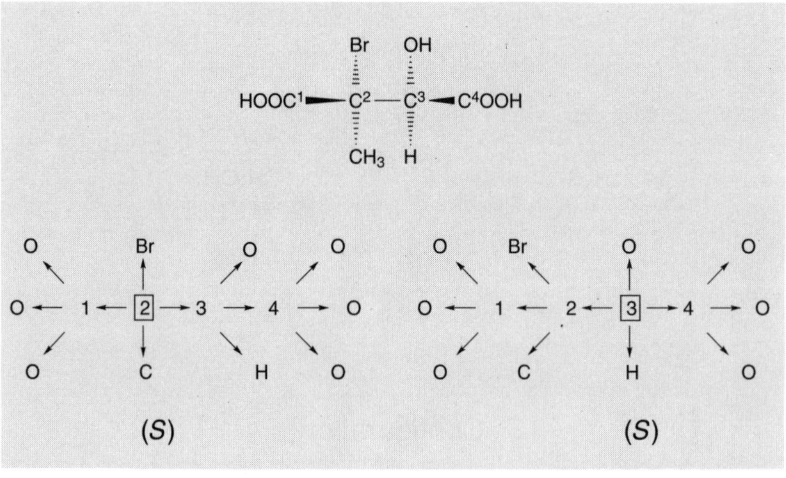

Br OH
$HOOC^1$—C^2—C^3—C^4OOH
CH_3 H

O ← 1 ← [2] → 3 → 4 → O (Br, C; O, O; O, H; O, O) (S)
O ← 1 ← 2 ← [3] → 4 → O (O, O; Br, C; O, H; O, O) (S)

Fig. 10.5. *Determination of the (2S,3S)-configuration for the stereoisomer with the constitution of 2-bromo-3-hydroxy-2-methylbutanedioic acid* (*above:* modified *Fischer* projection; *below:* simplified digraph)

projection (or the modified *Fischer* projection) must be transposed into the corresponding (simplified) digraph. The C-atoms of the C=O groups are given an additional O-atom each, to allow for the double bond; the provision of such phantom atoms (of atomic number zero) allows the number of ligands to be increased to four for all the C-atoms. In this instance, the (2*S*,3*S*)-configuration is identified.

To sum up, the complete information about the relative and absolute configurations of all the possible stereoisomers is presented in *Fig. 10.6*.

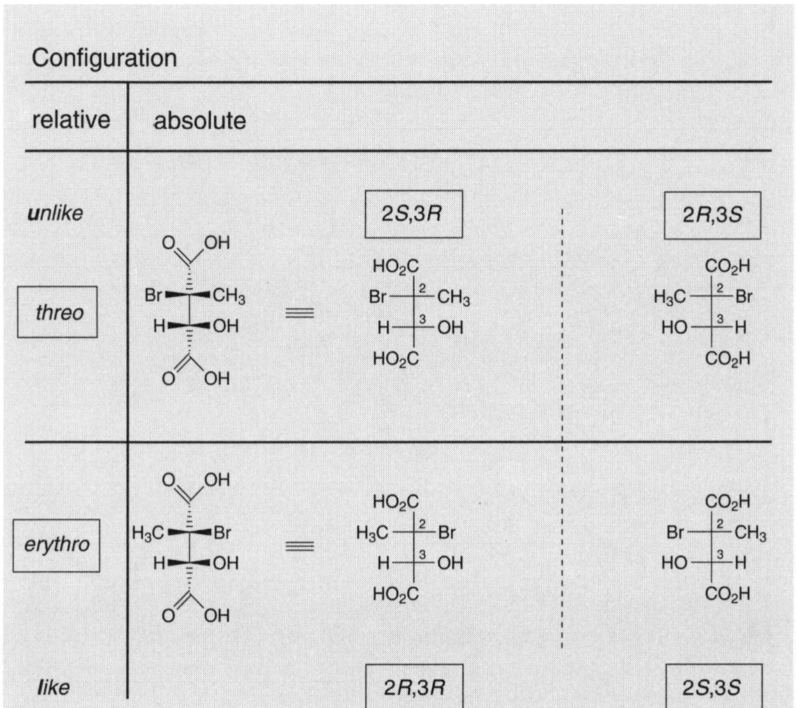

Configuration			
relative	absolute		
unlike		2*S*,3*R*	2*R*,3*S*
threo	Br, CH_3 / H, OH (wedge-bond) ≡	HO_2C / Br–2–CH_3 / H–3–OH / HO_2C	CO_2H / H_3C–2–Br / HO–3–H / CO_2H
erythro	H_3C, Br / H, OH (wedge-bond) ≡	HO_2C / H_3C–2–Br / H–3–OH / HO_2C	CO_2H / Br–2–CH_3 / HO–3–H / CO_2H
like		2*R*,3*R*	2*S*,3*S*

Fig. 10.6. *The four stereoisomers of 2-bromo-3-hydroxy-2-methylbutanedioic acid in the* Fischer *projection.* The transposition from 'wedge-bond' style to *Fischer* projection is shown for two diastereoisomers.

As may be inferred from *Fig. 10.6*, to specify the relative configuration the general notation *l* (*like*) and *u* (*unlike*) may be used instead of the special notation (*erythro* and *threo*). For that purpose, the descriptor pairs (*R*,*R*) or (*S*,*S*) are subsumed under *l*, and (*R*,*S*) or (*S*,*R*) under *u*. If the absolute configuration of the individual components of an enantiomeric pair ((2*S*,3*R*) and (2*R*,3*S*), say) is unknown, then the stereodescriptors (R^*) and (S^*) may be used, the specification of the first-named enantiomer arbitrarily beginning with (R^*).

10.2.1.3. Acyclic Molecules with Three or More Stereogenic Centers

The relative configuration of molecules with three or four stereogenic centers in an unbranched chain may be denoted with the aid of the carbohydrate stereodescriptors (*Fig. 5.4*): *ribo*, *arabino*, *xylo*, and *lyxo*, or *allo*, *altro*, *gluco*, *manno*, *gulo*, *ido*, *galacto*, and *talo*.

In the 'ladder-patterns' given on the next page, the horizontal lines represent the orientation of the reference ligands relative to the C-chain,

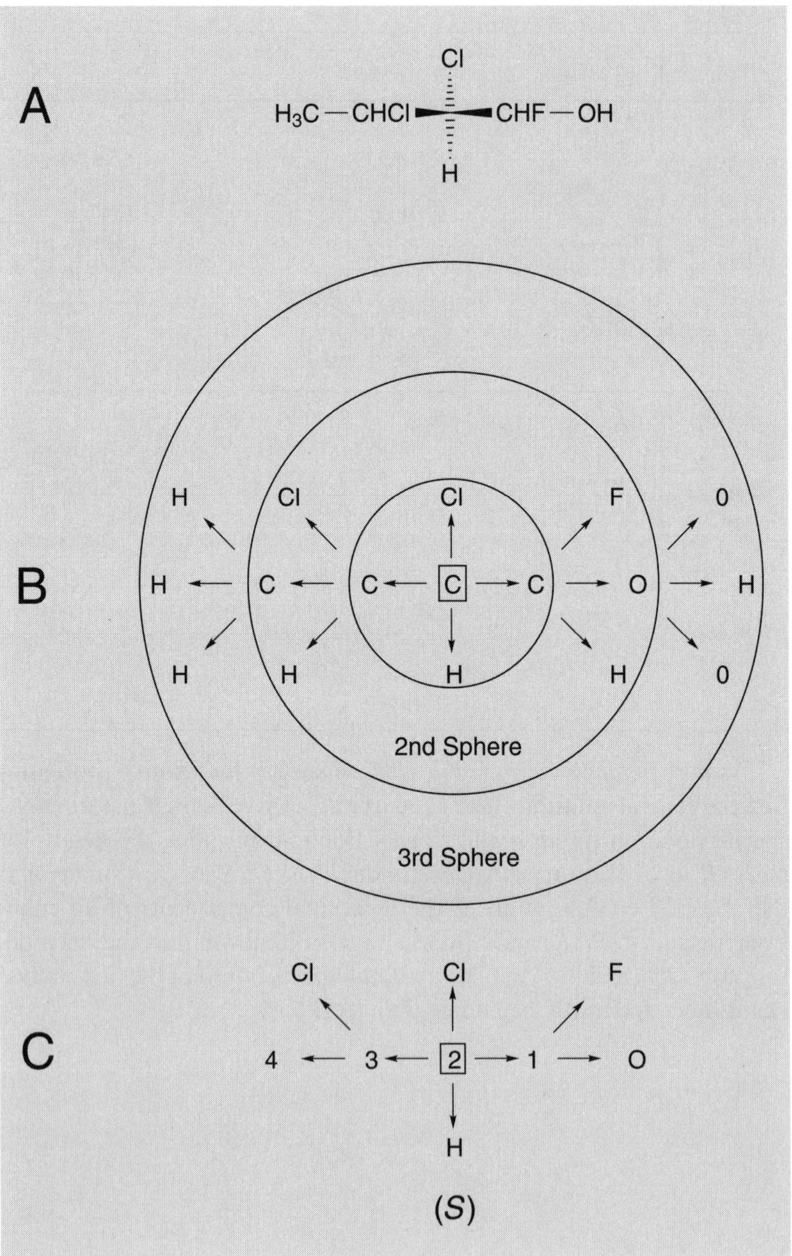

Fig. 10.7. *Determination of the (S)-configuration for the enantiomer with the constitution of 2,3-dichloro-1-fluorobutan-1-ol. A*) Modified *Fischer* projection; *B*) complete hierarchical digraph; *C*) simplified digraph.

shown as a *Fischer* projection. The atom with the lowest locant is to be found uppermost in the vertical chain.

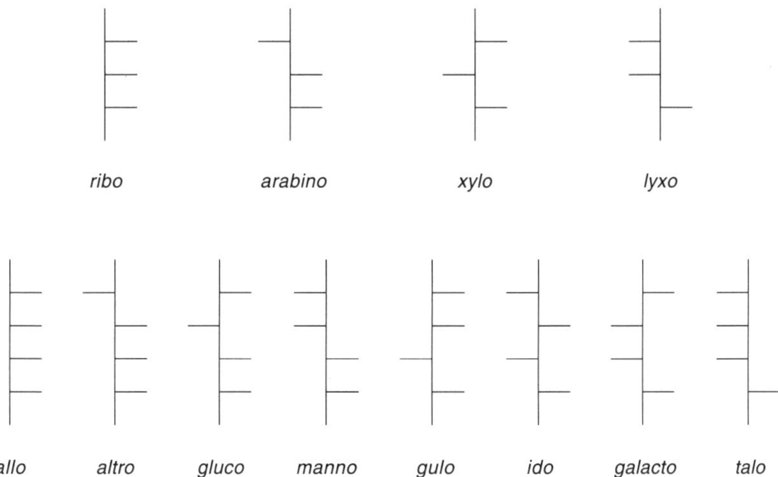

In the example of *Fig. 10.7*, one of the eight possible configurations sharing the constitution of 2,3-dichloro-1-fluorobutan-1-ol is pictured in the form of a simplified digraph. In this case, it is already possible to establish the order of ligand priority at the selected stereogenic center in the second sphere. According to the sequence rules, Cl has priority over F, and so the (*S*)-configuration results.

In contrast, the absolute configuration of stereoisomers inherent in the constitution of 4,6-dichloro-3-(hydroxymethyl)-5-methylhexan-2-ol (*Fig. 10.8*) cannot be determined until the third sphere. Because Cl > O, the (*S*)-configuration also follows here.

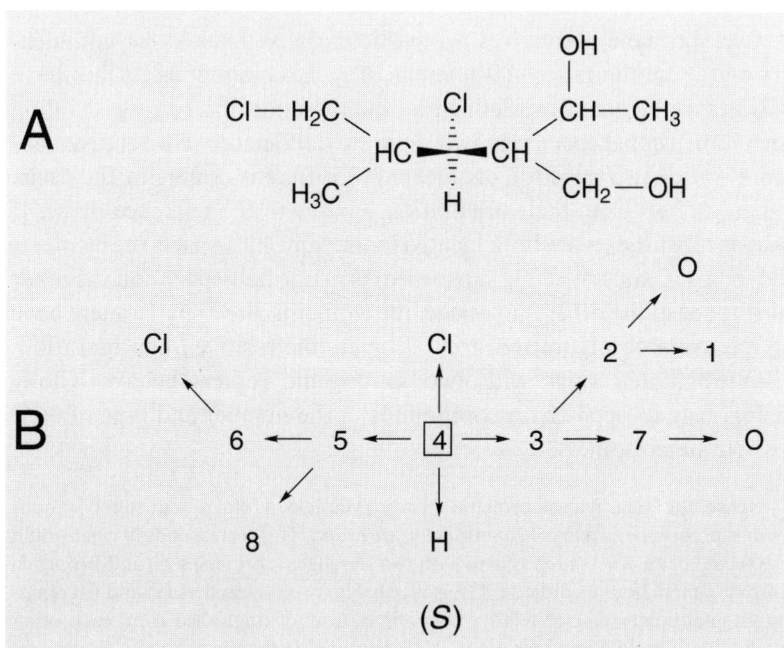

Fig. 10.8. *Determination of the (S)-configuration for the enantiomer with the constitution of 4,6-dichloro-3-(hydroxymethyl)-5-methylhexan-2-ol. A*) Modified *Fischer* projection; *B*) simplified digraph.

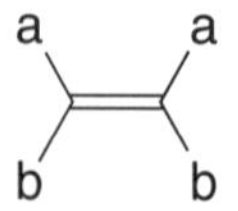

a a
b b
cis

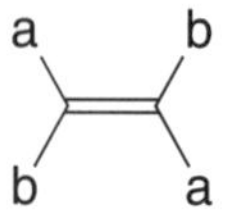

a b
b a
trans

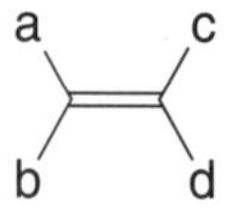

a c
b d
(*Z*)

a > b; c > d

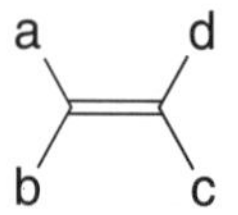

a d
b c
(*E*)

10.2.2. Configurational Isomers with Stereogenic Double Bonds

Older literature generally makes use of *cis*/*trans*-notation for the specification of stereoisomers with stereogenic C=C bonds. Difficulties are encountered, however, in cases such as four-fold substitution, as the order of priority of the ligands is indefinite. The (*E*/*Z*)-convention was introduced to regard all those cases in which the ligands were individually specified and not undefined, such as a, b or R^1, R^2. Here, the ligands at each of the central atoms are assigned a falling order of priorities (a > b; c > d) by application of the *sequence rules* (*Sect. 10.2.1*), with the plane of the two-membered ring serving as the reference plane. Assignment of the configuration is then made by using the ligand of higher priority at each C-atom as the reference ligand. The relative orientation of the two reference ligands then gives either the (*Z*)-configuration (*Z* from the German '*z*usammen' = together), when the two higher-priority ligands are located on the same side of the reference plane, or the (*E*)-configuration (*E* from the German '*e*ntgegen' = opposite), when they are on opposite sides.

A further *sequence rule* must be noted to allow for cases in which the relative configuration of stereogenic C=C bonds has to be taken into account in order to assign the correct (*R*)- or (*S*)-descriptor to a stereogenic center, according to the *chirality rule* given in *Sect. 10.2.1*. A (*Z*)-configurated C=C bond has priority over an (*E*)-configurated C=C bond. This sequence rule itself takes precedence over that one dealing with the order of priorities of substituents which contain themselves stereogenic centers.

10.2.3. Configurational Isomers of Cyclic Compounds

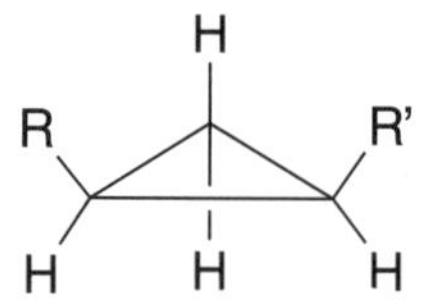

H
R R'
H H H
cis

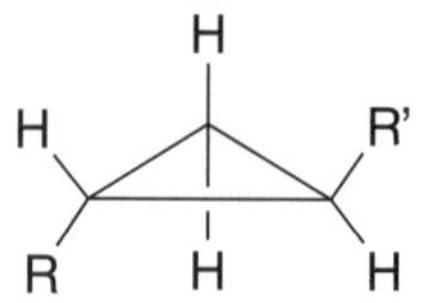

H
H R'
R H H
trans

Cyclopropane derivatives are particularly well suited for commentary on the configurational isomerism of cyclic compounds, as they have a definite reference plane defined by the three-membered ring, dividing three-dimensional space into two separate half-spaces. The relative configuration of two substituents located at different centers in the cyclic system follows from their orientation relative to the reference plane. If the two substituents are both located in the same half-space, the isomer is said to be *cis*, and when one substituent is in one half-space and the other substituent in the other half-space, the isomer is *trans*. *cis*-Isomers have the relative *u*-configuration, *trans*-isomers the relative *l*-configuration.

Disubstituted rings with two stereogenic centers behave wholly analogously to open-chain compounds in the number and type of their possible stereoisomers.

Hence, for 2-methylcyclopropane-1-carboxylic acid, a total of four stereoisomers, two diastereoisomeric pairs of enantiomers, are found. This corresponds to what would be expected of an open-chain system with two stereogenic centers, such as 2-bromo-3-hydroxy-2-methylbutanedioic acid (*Fig. 10.6*). The two *cis*-isomers (a) and (b) represent an enantiomer pair of relative *u*-configuration, distinguished from each other only by their absolute configurations. The same applies for the two *trans*-isomers (c) and (d) with the relative *l*-configuration. The absolute configurations may be taken from the following illustration.

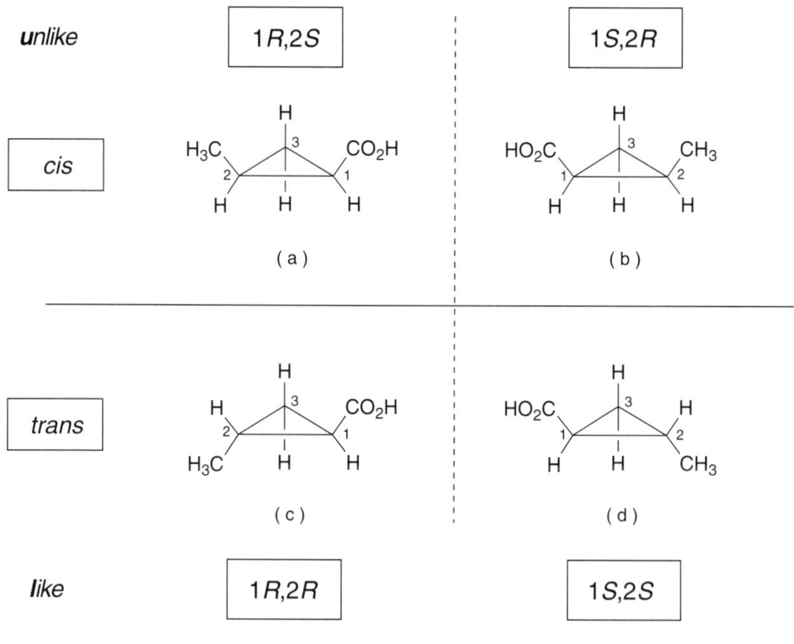

Constitution: 2-Methylcyclopropane-1-carboxylic acid
relative Configuration: *u* (*cis*)/ *l* (*trans*)
absolute Configuration: (1*R*,2*S*) or (1*S*,2*R*) / (1*R*,2*R*) or (1*S*,2*S*)

A further *sequence rule* must be noted to allow for cases in which the relative configurations of cyclic systems has to be taken into account in order to assign the correct ((*R*) or (*S*)) descriptor for absolute configuration to a stereogenic center, according to the *chirality rule* given in *Sect. 10.2.1.* A cyclic substituent of relative *cis*-configuration has priority over a cyclic substituent of relative *trans*-configuration. This sequence rule itself takes precedence over that one dealing with the order of priorities of substituents which contain themselves stereogenic centers.

For rings with more than two substituents at which stereoisomerism needs to be taken into account, it is necessary to select a *reference* ligand (at C(1)) to permit specification of the configuration. This reference ligand

Configuration

H3C, H (C-3); H, Cl (C-2); CO2H, H (C-1)

relative	2*t*-Chloro-3*c*-methylcyclo-propane-1*r*-carboxylic acid
absolute	(1*R*,2*S*,3*R*)-2-Chloro-3-methylcyclopropane-1-carboxylic acid

is assigned the descriptor *r* in the name of the relevant compound. The other ligands on the ring are indicated as *c* (for *cis*) or *t* (for *trans*), depending on their orientation relative to the reference ligand.

Up to now, we have looked at the *cis/trans*-convention as applied to three-membered rings, although it is also of value for the description of the relative configuration of larger-ring molecules. Rings with more than three members, however, are usually not planar. Statistical planarity is nonetheless assumed, though, as it permits great simplification in the nomenclature and projection of cyclic systems.

For the specification of the absolute configuration of cyclic structures, the procedure is different from that one for acyclic compounds. The rings present are 'opened' on both sides of the stereogenic center and duplicate atoms added to the branches of the graph thus obtained, permitting the cyclic system to be reduced to an open-chain partial structure. In the case of the enantiomer with the (cyclobutyl)(cyclopropyl)methanol constitution given in *Fig. 10.9*, the (*R*)-configuration can be identified in this way.

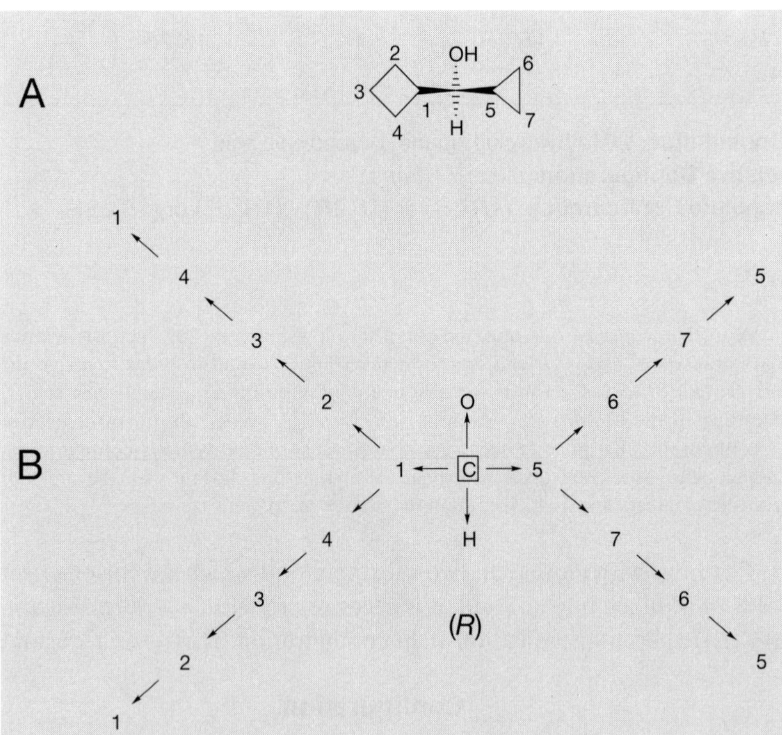

Fig. 10.9. *Determination of the (R)-configuration for the enantiomer with the (cyclobutyl)(cyclopropyl)methanol constitution. A*) Modified *Fischer* projection; *B*) simplified digraph.

If several stereogenic centers are present in a ring, then an individual graph must be prepared for each of them. In this way, the (2*S*,3*S*)-configuration may be determined for the stereoisomer with the 2-methyl-3-vinylcyclopentan-1-one constitution and the relative *trans*-configuration given in *Fig. 10.10*.

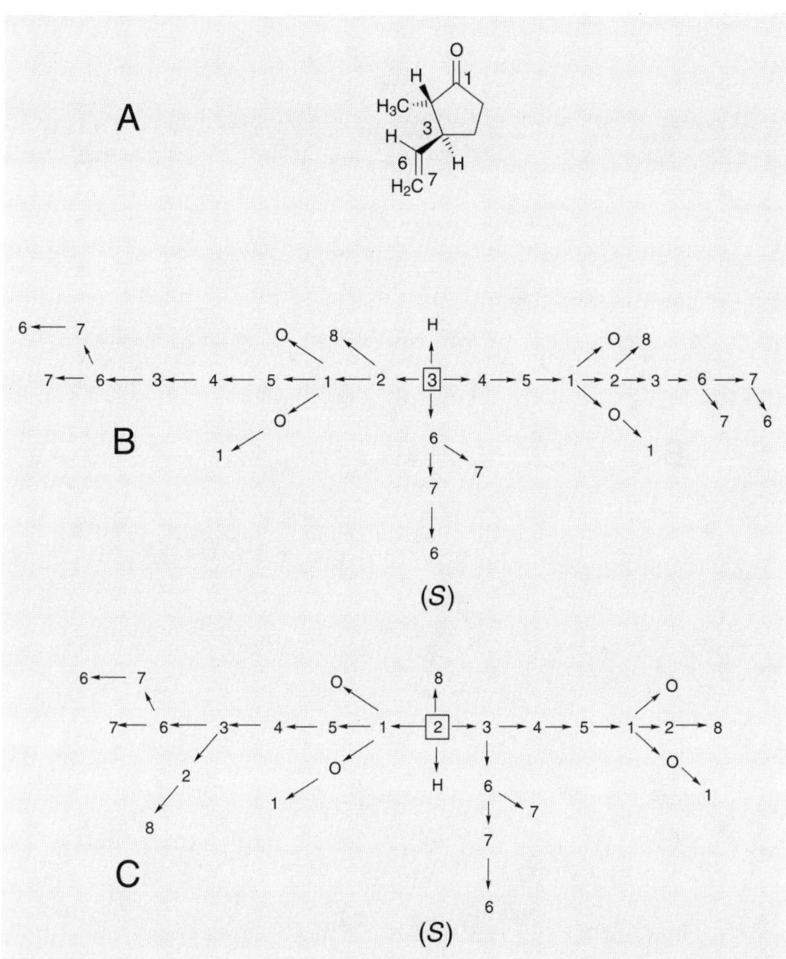

Fig. 10.10. *Determination of the (2S,3S)-configuration for the stereoisomer with the 2-methyl-3-vinylcyclopentan-1-one constitution. A*) Configurational formula; *B*) and *C*) simplified digraphs for the determination of the sense of chirality at C(3) and C(2), respectively.

Were an isopropenyl group, rather than the vinyl group, present at C(3), then the sense of chirality at C(3) would be changed from (*S*) to (*R*) as an 'artifact of the sequence rules' (*Fig. 10.11*).

10.2.4. Configurational Isomers of Allenes, Alkylidenecycloalkanes and Spiro Compounds

That enantiomers may occur in the cases of the appropriately substituted title compounds has been mentioned before in *Chapt. 2.6.3.2.*

CIP Notation is also used for the description of absolute configuration of stereoisomers of the allene type. To establish the priority of the ligands, the molecule is represented in the so-called *Newman* projection. This projection represents the molecule, viewed from the perspective of an imaginary axis running through the three atoms comprising the allene system. It makes no difference which end of the axis the projection is made from, as either viewpoint will give the same absolute configuration. An additional sequence rule of the *CIP* system for stereoisomers of the

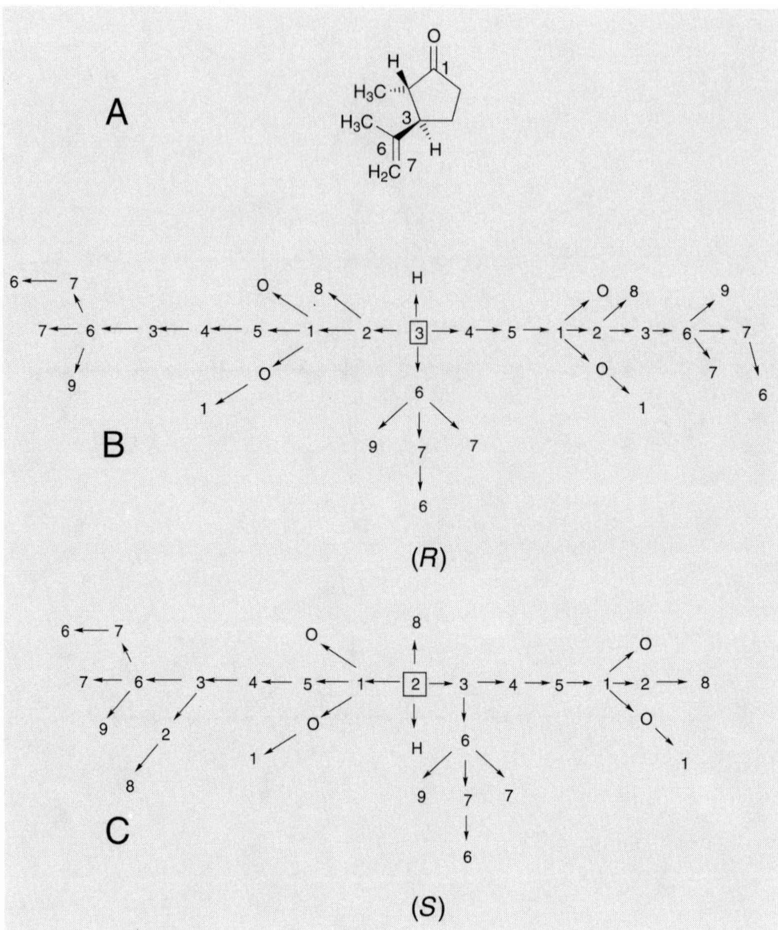

Fig. 10.11. *Determination of the (2*S*,3*R*)-configuration for the stereoisomer with the 3-isopropenyl-2-methylcyclopentan-1-one constitution. A*) Configurational formula; *B*) and *C*) simplified digraphs for the determination of the sense of chirality at C(3) and C(2), respectively.

allene type gives higher priority to the ligand pair closer to the observer, relative to the ligand pair further away. To establish the full priority sequence (1 > 2 > 3 > 4), the ligands at the two terminal C-atoms are ordered in a manner such that the ligands closer to the observer are assigned priorities 1 and 2, and those further away are given priorities 3 and 4. The direction of the circle described in moving through the falling priorities (1 > 2 > 3) gives the absolute configuration: (*R*) for clockwise, (*S*) for anticlockwise rotation. *Fig. 10.12* shows the procedure, using as an example the two enantiomers with the penta-2,3-diene constitution.

In the case of the two enantiomers of the constitution of (4-methylcyclohexylidene)acetic acid, one of the two-membered rings of the allene system has been replaced by a six-membered ring (*Fig. 10.13*). Unambiguous specification of the sense of chirality is achieved in a manner similar to that one above.

The sense of chirality of suitably substituted spiro compounds may be specified using a similar technique. *Fig. 10.14* demonstrates the method and unambiguous results for the two enantiomers with the spiro[3.3]heptane-2,6-dicarboxylic acid constitution.

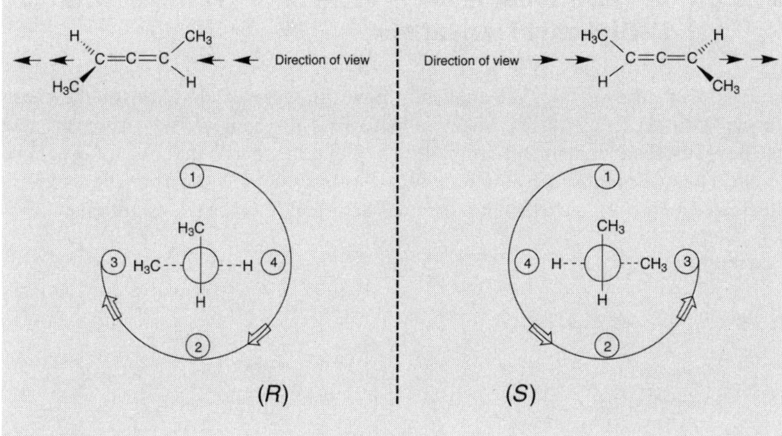

Fig. 10.12. *Determination of the sense of chirality of the two enantiomers of penta-2,3-diene*

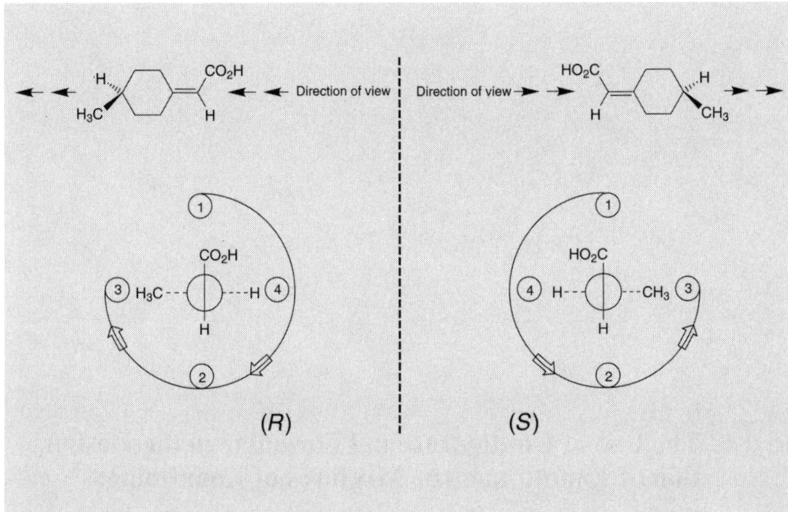

Fig. 10.13. *Determination of the sense of chirality of the two enantiomers of (4-methylcyclohexylidene)acetic acid*

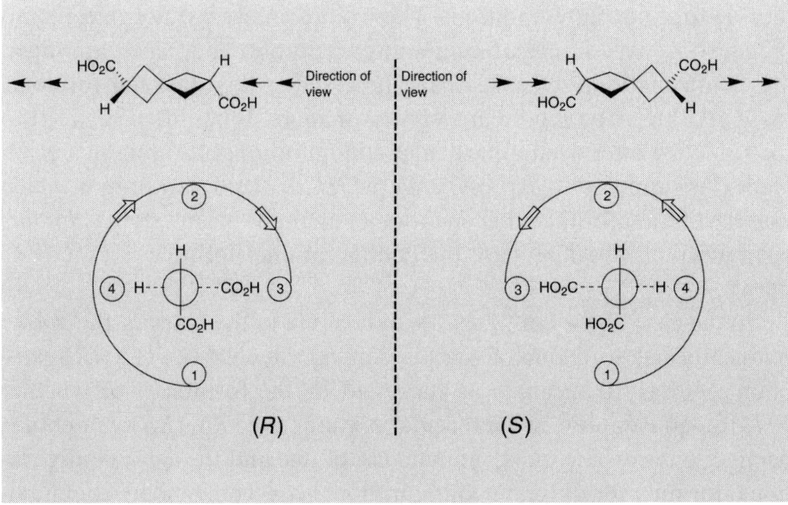

Fig. 10.14. *Determination of the sense of chirality of the two enantiomers of spiro[3.3]heptane-2,6-dicarboxylic acid*

10.2.5. Configurational Isomers of 2,6,2′,6′-Tetrasubstituted 1,1′-Biphenyl Derivatives

Fig. 10.15 demonstrates schematically how the sense of chirality for the enantiomers with the constitution of 6,6′-dimethyl-1,1′-biphenyl-2,2′-dicarboxylic acid may be established: assignment proceeds in a similar manner to that one for the allene system. The two enantiomers of 6,6′-dinitro-1,1′-biphenyl-2,2′-dicarboxylic acid were illustrated in *Fig. 2.8*: the left-hand formula corresponds to the (*R*)-enantiomer.

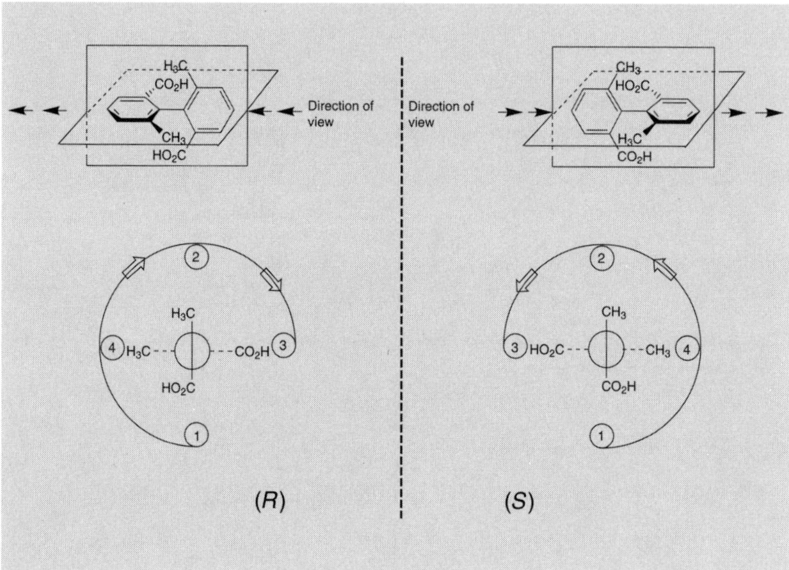

Fig. 10.15. *Determination of the sense of chirality of the two enantiomers of 6,6′-dimethyl-1,1′-biphenyl-2,2′-dicarboxylic acid*

10.2.6. The Use of Configurational Formulae in the Designation of Enantiomers or Mixtures of Enantiomers

The abstract structure of a chemical compound may be expressed with the aid of agreed symbols. The chemist illustrates these structural ideas in formula representations. These representations, as a rule, unambiguously convey information about constitution and, when appropriate, conformation. Over the years, however, configurational formulae (*Sect. 10.2*) have been used in a very lax manner. It has often been left to the reader to infer what a particular configurational formula in a given context is supposed to express: only the *relative* configuration, in which case the one and/or the other enantiomer might be meant, or the *absolute* configuration, in which case the configurational formula is exactly as drawn.

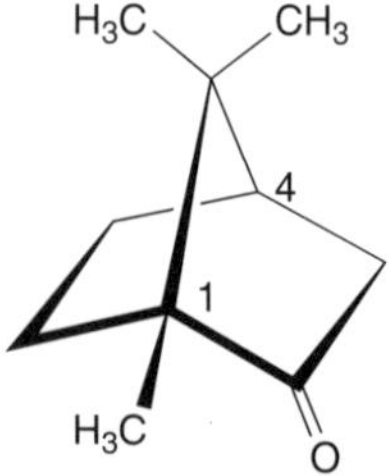

Formula representation

In the case of the configurational formula in the margin, the reader frequently has to deduce from the context whether (+)-(1*R*,4*R*)-camphor genuinely is meant – as suggested by the formula – or whether (−)-(1*S*,4*S*)-camphor or the racemic mixture (±)-(1*RS*,4*RS*)-camphor is being discussed. The indiscriminate use of one and the same configurational formula for different configurational concepts hinders communication between chemists and acts as an unnecessary barrier in conveying information concerning stereostructural problems to interested non-chemists.

Therefore, for further explanation involving the information content of a configurational formula, we shall pay close attention to the absolute configuration. In the event of it being necessary to discuss the enantiomer of a compound whose configurational formula is already pictured, we will refrain from using the corresponding illustration. Instead, we will assign a number (bold-faced Arabic number; **1** in this case) or letter (bold-faced) to the formula, and use this, together with the stereodescriptor *ent*. Mixtures, in which one or the other enantiomer predominates, will be described by using both the number without a prefix *and* the same number with the *ent*-descriptor. In this manner, it is easy to distinguish between cases in which the unequal enantiomer ratio is unknown (**1**/*ent*-**1** ≠ 1), cases in which it is only known which of the two enantiomers predominates (**1** > or < *ent*-**1**), and cases in which the ratio has been determined exactly (such as **1**/*ent*-**1** = 96:4).

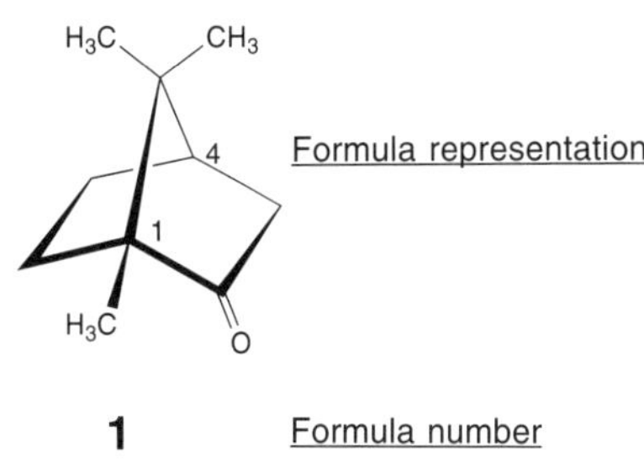

In the common special case of both enantiomers being present in (practically) equal proportions, the stereodescriptor *rac* will be put before the formula number (which only stands for one of the two enantiomers) to emphasize the special instance that the enantiomer mixture is the *racemic mixture*.

In short: the formula number **1** corresponds to the formula representation depicted and denotes (+)-(1*R*,4*R*)-camphor, *ent*-**1** corresponds to the (not illustrated) mirror image of the formula depicted and denotes (−)-(1*S*,4*S*)-camphor, and *rac*-**1** relates to both the depicted formula and its mirror image, and denotes the racemic mixture ((±)-(1*RS*,4*RS*)-camphor).

10.2.7. Bird's-Eye View Description of Configuration

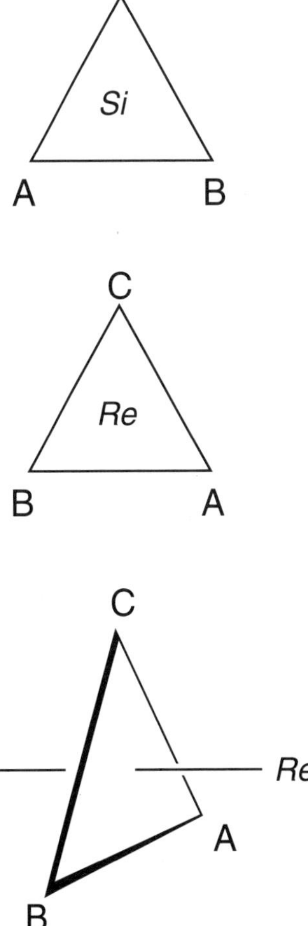

Molecules which may be represented by the simplex of two-dimensional space – by an equilateral triangle with three distinctly labeled apexes – are two-dimensionally chiral. The plane defined by such a triangle divides the three-dimensional space into two half-spaces. An observer, having the bird's-eye view of the triangle, is located in one of these half-spaces. If the path from A *via* B to C, as viewed by the observer, is clockwise, then the observer's perspective is specified as being *Re*. The half-space in which the observer is located in this case is also expediently assigned the *Re*-descriptor. In contrast, if the observer has a view of the triangle in which the path from A to B to C is anticlockwise, then the observer's perspective, as well as the half-space in which the observer is situated, is assigned the descriptor *Si*. The order of rank is similar to that one outlined in *Sect. 10.2.1*: A > B > C.

The example of acetaldehyde shows the conditions clear and plain. The plane of the molecule divides the three-dimensional space into two half-spaces. The perspectives from each of these half-spaces are enantiotopic; the half-spaces themselves are

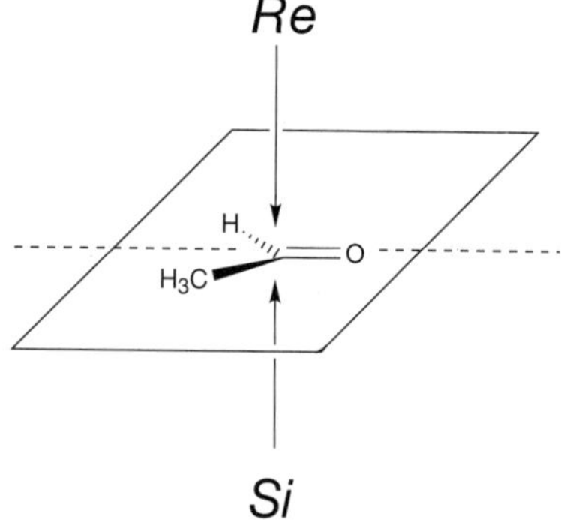

enantiomorphic. The view from the *Re*-half-space is of the *Re*-side, from the *Si*-half-space the *Si*-side is perceived.

Depending on whether the two half-spaces from which a molecule is viewed are identical, enantiomorphic, or diastereomorphic, the two faces are described as *equivalent* (*homotopic*), *enantiotopic*, or *diastereotopic* (*Chapt. 4.2*).

If a C=C unit is present, the specification of each of the two two-dimensionally stereogenic centers may be made using the descriptors *Re* and *Si*. Hence, because of the relationship of the descriptors to each other, not only the absolute, but also the relative configuration and topicity are established. The relative topicity (*Re,Re*) or (*Si,Si*) is put together by the descriptor *lk* (from *like*), the relative topicity (*Re,Si*) or (*Si,Re*) by the descriptor *ul* (from *unlike*).

In the example in the margin, the two views of (*Z*)-but-2-ene are homotopic, those ones of (*E*)-but-2-ene enantiotopic. The relative topicity is *ul* in the former, and *lk* in the latter case.

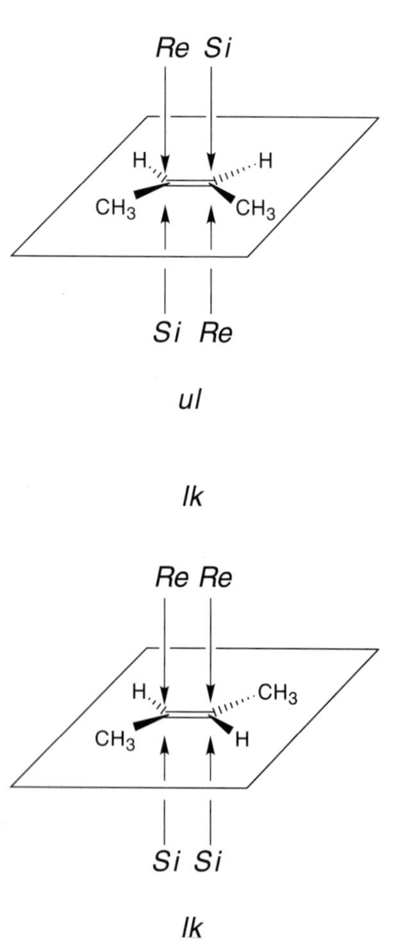

10.3. Conformational Formulae

10.3.1. Graphic Representations

It can be shown, using ethane as an example, that an infinite number of different conformations are passed through in rotation about the C–C bond. This structural diversity extends between the staggered conformation on one side and the eclipsed conformation on the other.

As well as the perspective representations of the two relevant ethane conformations, occasionally referred to as the 'saw-horse' representa-

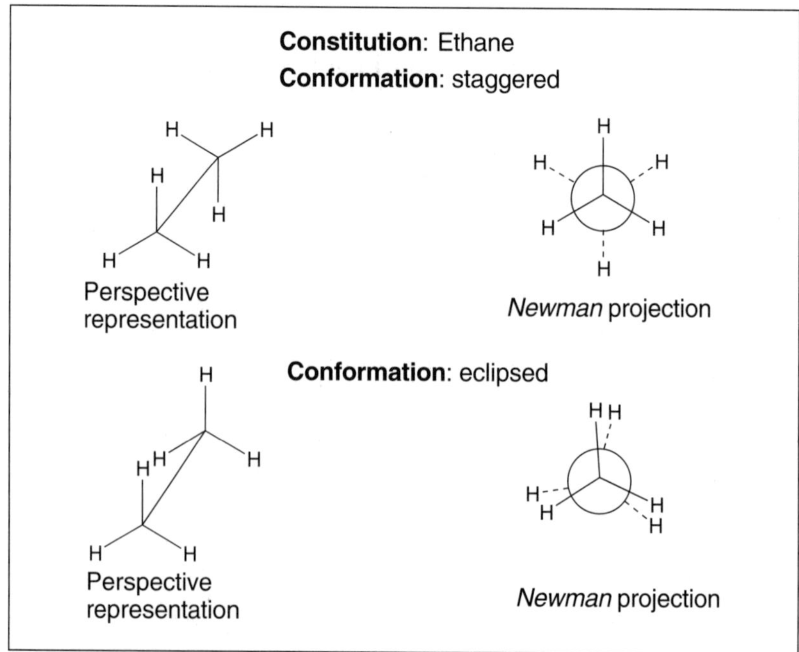

tions for reasons which are self-explanatory, the so-called *Newman* projections are often used.

For the *meso*- and the (2*S*,3*S*)-configurations of 2,3-dibromobutane, the staggered conformations are given in their perspective representations and as *Newman* projections. The eclipsed conformations are given in their perspective representations, as *Newman* projections and also as *Fischer* projections.

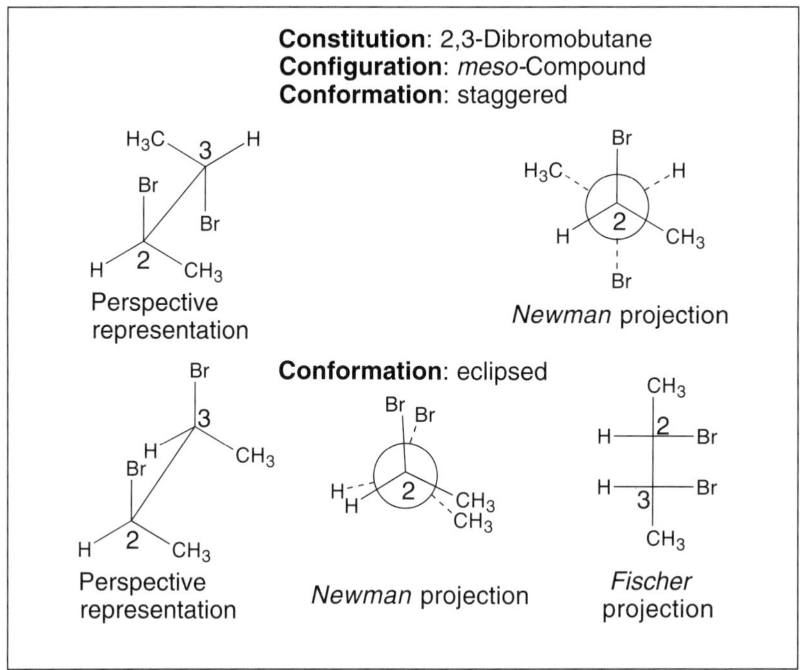

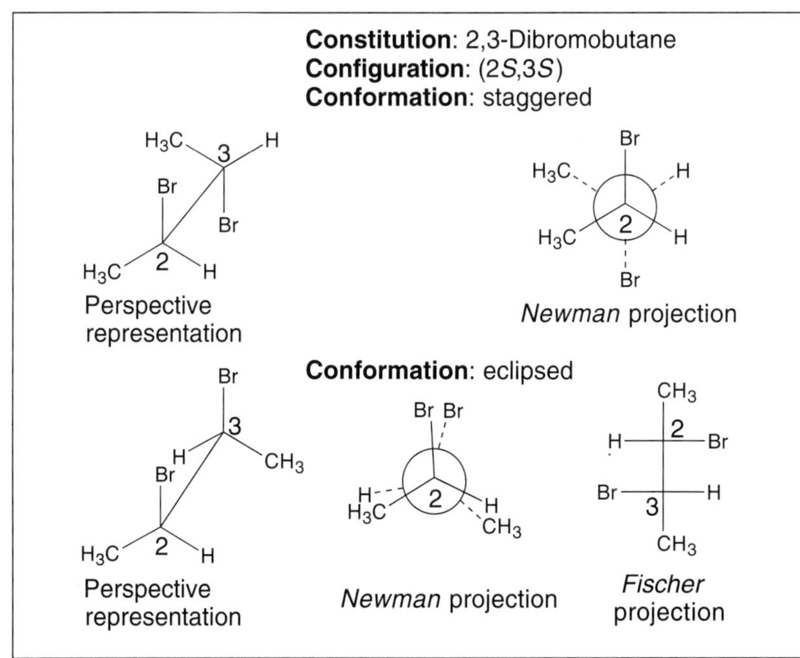

In the *Newman* projection, an observer faces the C(2)–C(3) bond from C(2) towards C(3). The plane of projection is perpendicular to the C(2)–C(3) bond. Bonds from C(2) to its ligands appear as radii of a circle in the projection plane. The bonds from C(3) to its ligands lie behind the plane of projection. To avoid these bonds being hidden behind those in front, a *syn-periplanar* (see *Table 10.10*) conformation is usually depicted in preference to the more accurate *synplanar* conformation.

10.3.2. The Torsion-Angle Notation of *Klyne* and *Prelog*

Torsion angles are *the* distinctive characteristics of a conformation. Their unambiguous description is a basic prerequisite for every communicable conformational analysis. Unlike ethane, butane exhibits an *energy profile* (*Fig. 10.16*) – a diagram which shows the dependence of the potential energy of a molecular system on a particular structure parameter (the torsion angle in this case) – with maxima and minima of unequal magnitudes. It is, therefore, not sufficient in this instance merely to distinguish between a staggered and an eclipsed conformational type.

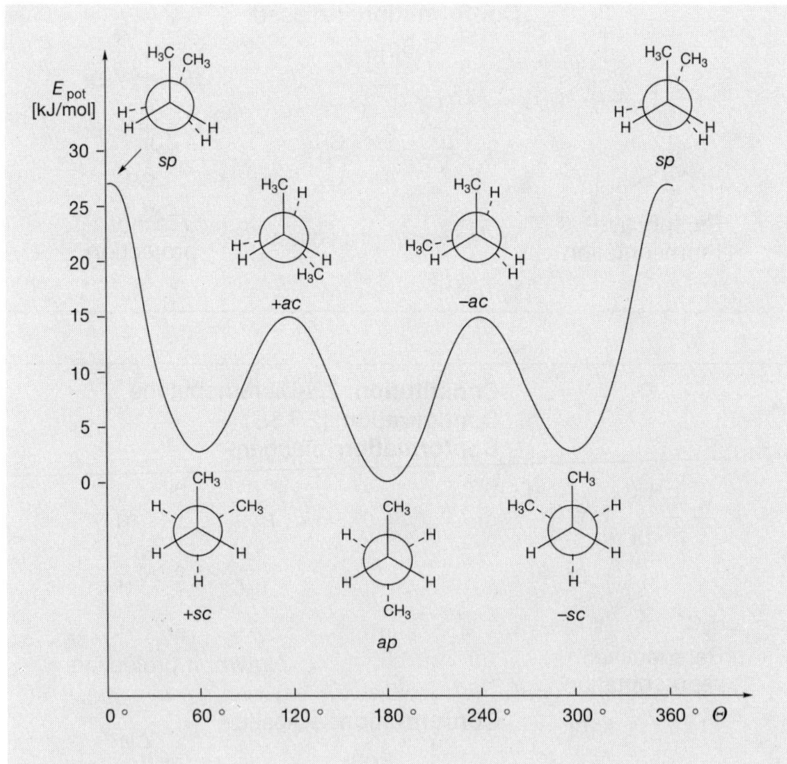

Fig. 10.16. *The energy profile of butane and the conformations corresponding to the maxima and minima*

As in other areas of stereochemistry, *Prelog*'s wisdom (*'We feel that, particularly for educational reasons, economy in coining new terms should be the order of the day'*) has also been of great use for avoiding misunderstandings in conformational description. A convention set up by *William Klyne* and *Vladimir Prelog* is based on a general system with four connected atoms A–X–Y–B. The defining parameter for the description of a conformation is the torsion angle Θ between A and B along the X–Y bond.

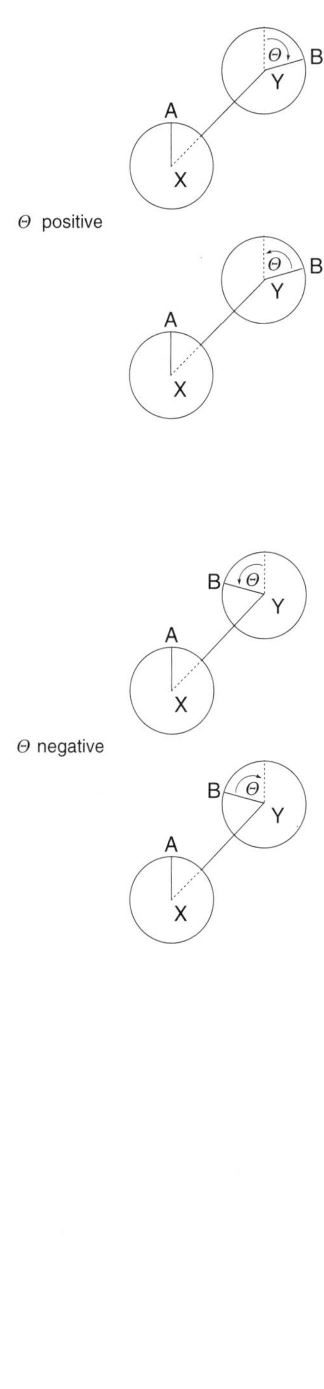

Both A and B are the substituents with highest priority according to the *CIP* notation, located at X and Y, respectively. The situation frequently arises that it is opportune to describe the positions of A and B relative to the bond X–Y, without knowing the exact torsion angle. In such an instance, the conformation may be described by the combination of three terms, based on the appropriate division of a circle into six segments (*Table 10.10*).

Table 10.10. *Torsion-Angle Notation of* Klyne *and* Prelog

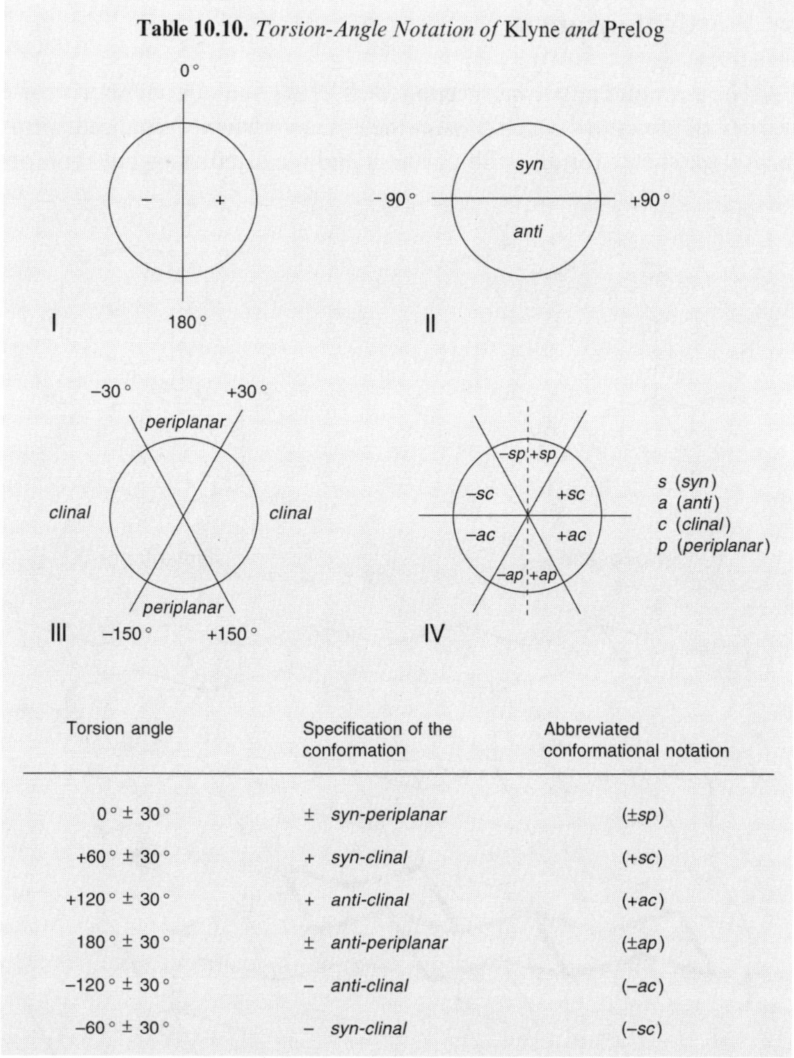

Torsion angle	Specification of the conformation	Abbreviated conformational notation
0 ° ± 30 °	± *syn-periplanar*	(±*sp*)
+60 ° ± 30 °	+ *syn-clinal*	(+*sc*)
+120 ° ± 30 °	+ *anti-clinal*	(+*ac*)
180 ° ± 30 °	± *anti-periplanar*	(±*ap*)
–120 ° ± 30 °	– *anti-clinal*	(–*ac*)
–60 ° ± 30 °	– *syn-clinal*	(–*sc*)

Hence, a '+' is used, if B is clockwise, at an angle between 0° and 180°, shifted from A; a '–' is used if B is anticlockwise shifted from A (I, *Table 10.10*). The *syn*- and *anti*-descriptors express whether Θ is less or more than 90° (II, *Table 10.10*). *Periplanar* (III, *Table 10.10*) stands for the regions between +30° and –30°, and +150° and –150°, while *clinal* denotes the regions between +30° and +150°, and –30° and –150°.

The extreme points of the butane energy profile (*Fig. 10.16*) have been assigned their respective conformations, given as *Newman* projections, and their relevant conformational descriptors according to the *Klyne-Prelog* convention. If the *ap*-conformation in the global minimum is given the relative energy value of 0 kJ/mol, then the +*sc*- and −*sc*-conformations in local minima have a value of 3.6 kJ/mol. The transitional conformations of rotation (with *ac*- or *sp*-orientation) have values of 15.1 and 26.4 kJ/mol, respectively.

10.3.3. The Diamond Lattice as a Mnemonic Device for the Graphic Representation of Idealized, Staggered Conformations

The diamond lattice has been used as a representational matrix for a variety of structures which possess the characteristics of idealized, tetrahedral geometry for all bond partners and idealized, staggered confor-

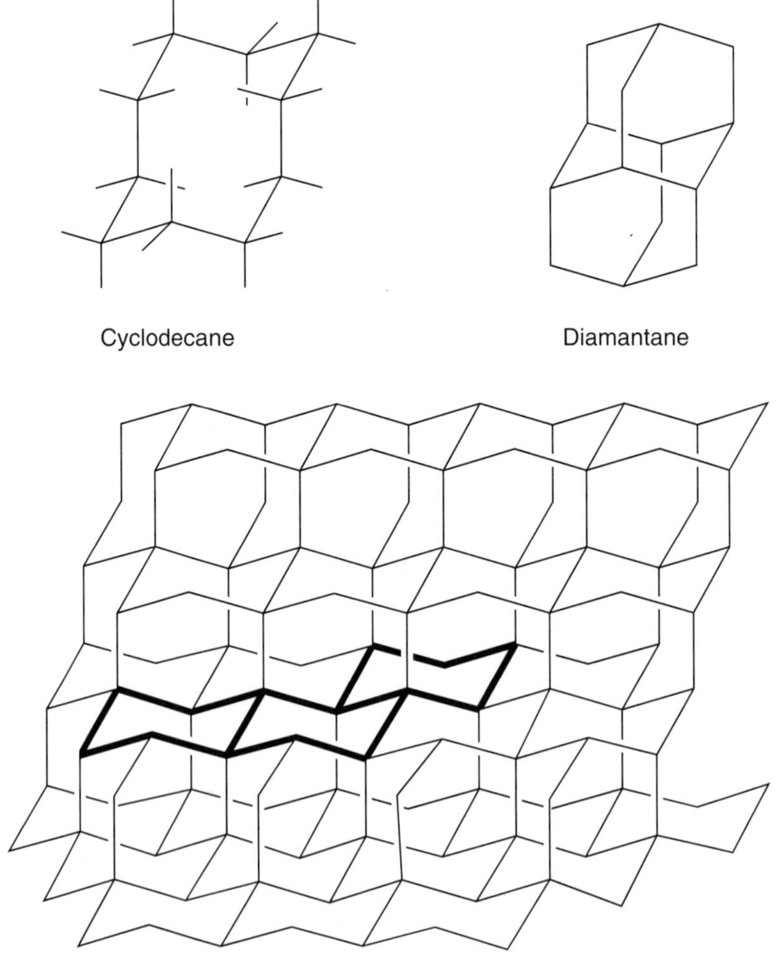

Section from the diamond lattice with *trans/anti/trans*-perhydrophenanthrene (*Chapt. 6.3.5.3*)

mations for all (equally long, to the first approximation) C–C bonds. It serves as a grid sorting out conformations free of *Pitzer* strain, excluding eclipsed conformations, but making no distinction between the subgroups of *ap*- and *sc*-staggering.

The diamond lattice can be used for the idealized representation of the chair conformation of cyclohexane, the two stereoisomeric decalins, the majority of the stereoisomeric perhydroanthracenes and perhydrophenanthrenes, also adamantane, diamantane, and cyclodecane, as well as the alkanes, poly(methylene), poly(propylene), and poly(oxymethylene), and even nucleic acids with pyranose sugars and pyranose carbohydrates. *V. Prelog* has so often successfully used the diamond lattice, which seems to have been utilized for the first time by *J.C. Speakman*, as a graphic tool for the representation of staggered conformations that 'a chemist close to him (*A. E.*)' has described the diamond lattice as '*the poor man's computer*'. Be that as it may, a computer program capable of identifying conformations of medium-to-large-ring cycloalkanes which would fit in a diamond-lattice structure was developed as early as 25 years ago. In the intervening period, computer programs devoted to *hunting for the global-minimum structure* have been further developed. Comparisons with spectroscopic findings have proven the usefulness of programs of this type. They have also shown, however, that, in non-crystalline states (and sometimes even in crystalline phase), more than one conformation must be taken into account.

10.3.4. Bird's-Eye View Description of Conformation

In recent years, a means of describing conformations which is connected to the torsion-angle concept (*Sect. 10.3.2*) has become increasingly popular. The conformation of a cyclic compound is drawn in bird's-eye view and described by giving the torsion angle of each bond (actually the angle described by the two bonds on either side of that bond) in the projected two-dimensional image of the conformation. A (+)-sign is used to denote that the bond further away is displaced in a clockwise direction relative to the nearer one, a (−)-sign to denote an anticlockwise relationship. The technique is best illustrated with the conformations relevant to cyclohexane, given in *Table 10.11*.

10.4. Names and Descriptors

Now that the means of describing the structure of molecules using constitutional, configurational, and conformational formulae have been demonstrated, it should be emphasized that, as has already been shown in examples, *systematic names* only give the constitution of a chemical compound. Details of (relative or absolute) configuration are conveyed by *stereodescriptors*. Trivial names, as a rule, contain structural information which goes beyond mere constitution. This fact becomes particularly apparent in the cases of sugars and steroids, compound classes previously regarded as fields of study for specialists.

Table 10.11. *Molecular Models of Cyclohexane Conformations with Their Respective Symmetry Elements and Torsion Angles*

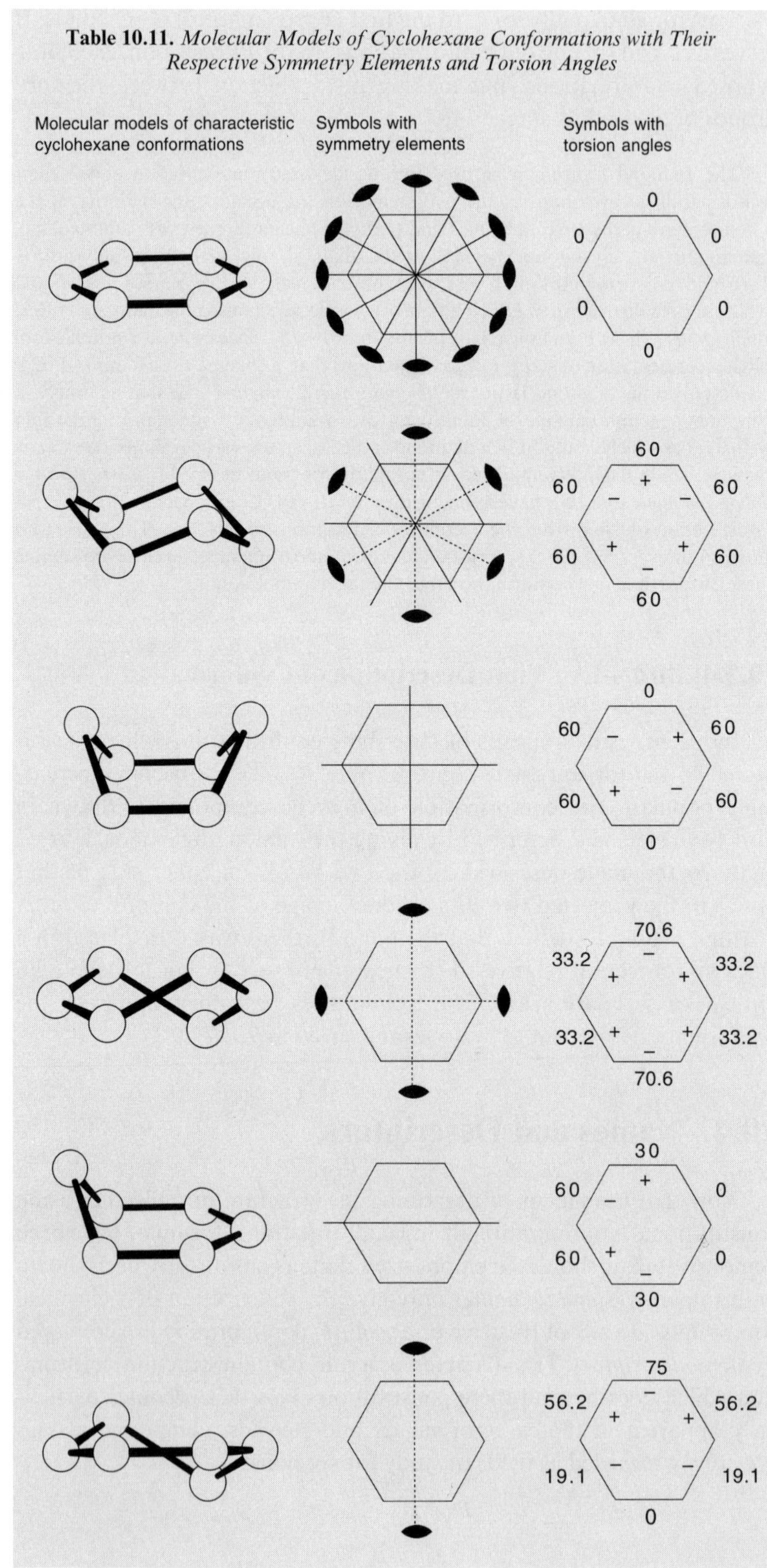

Name of the constitution systematic (upper row) trivial (lower row)	Stereodescriptors for the configuration relative (upper row) absolute (lower row)	*Chapter* or *Section*
2,3-Dihydroxybutanedioic acid (= Tartaric acid)	*meso*-2,3-Dihydroxybutanedioic acid; (2*R*,3*R*)-2,3-Dihydroxybutanedioic acid[a])	*2.6.2.1*
2-Bromo-3-hydroxy-2-methyl-butanedioic acid	*threo*-2-Bromo-3-hydroxy-2-methyl-butanedioic acid; (2*R*,3*S*)-2-Bromo-3-hydroxy-2-methylbutanedioic acid	*10.2.1.2*
1,2-Dimethylcyclopropane	*cis*-1,2-Dimethylcyclopropane; (1*R*,2*R*)-1,2-Dimethylcyclopropane[a])	*2.6.2.1*
2-Methylcyclopropane-1-carboxylic acid	*cis*-2-Methylcyclopropane-1-carboxylic acid; (1*R*,2*S*)-2-Methylcyclopropane-1-carboxylic acid	*10.2.3*
2-Chloro-3-methylcyclopropane-1-carboxylic acid	2*t*-Chloro-3*c*-methylcyclopropane-1*r*-carboxylic acid; (1*R*,2*S*,3*R*)-2-Chloro-3-methylcyclopropane-1-carboxylic acid	*10.2.3*

[a]) The absolute configuration of one of the two enantiomers is given here, as opposed to the relative configuration in the upper row.

10.5. Special Compound Classes

10.5.1. Steroids

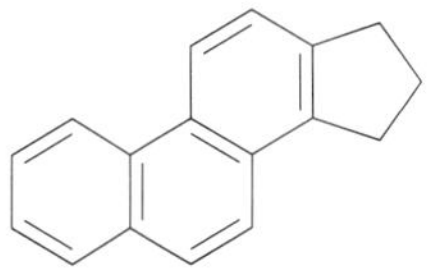

The class of steroids includes all compounds with a modified '1,2-cyclopentenophenanthrene' skeleton (original name; = 16,17-dihydro-15*H*-cyclopenta[*a*]phenanthrene in systematic nomenclature). Modifications may occur formally through the lack of one or more bonds, through enlargement and/or contraction of one or more rings. The unsystematic numbering of the individual C-atoms is the result of an original assignment of an incorrect ring skeleton to the steroids. After revision of the constitution, the original numbering was retained in order to avoid unnecessary confusion in the preparation and reading of publications.

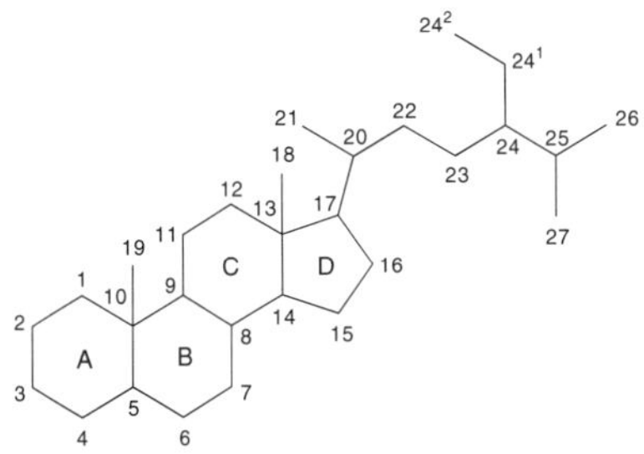

The atoms C(10) and C(13) usually bear Me groups, and a side chain is frequently found at C(17). Substituents located on the same side of the molecule as the angular Me groups are assigned the β-configuration, substituents on the opposite side the α-configuration (*Chapt. 6.2*).

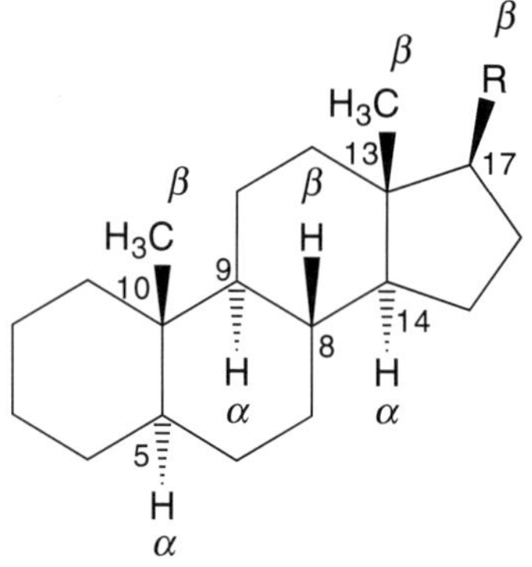

As, on the one hand, a system of fused cyclohexane rings is thermodynamically stable when it adopts the greatest possible number of chair conformations, and as, on the other hand, two *trans*-fused cyclohexane rings cannot invert, there exists a definite relationship between the configurational (α or β) and the conformational (*a* or *e*) orientation of a ligand for each position in the steroid system (*Table 10.12*).

Table 10.12. *Relationship Between the Configurational and Conformational Orientation of Ligands in Steroids*

	trans-A/B		*cis*-A/B			*trans*-A/B	*cis*-A/B
Position	α	β	α	β	Position	α	β
1	(a)	(e)	(e)	(a)	11	(e)	(a)
2	(e)	(a)	(a)	(e)	12	(a)	(e)
3	(a)	(e)	(e)	(a)	13	—	(a)
4	(e)	(a)	(a)	(e)	14	(a)	—
5	(a)	—	—	A (a); B (e)	15	(e*)	(a*)
6	(e)	(a)	(e)	(a)	16	—	—
7	(a)	(e)	(a)	(e)	17	(a*)	(e*)
8	—	(a)	—	(a)			
9	(a)	—	(a)	—			
10	—	(a)	—	A (e); B (a)			

Symbols assigned a * mean that the relationship stated is with regard to ring C.
(a) denotes axial, (e) denotes equatorial.

Table 10.13 gives the names and formula representations of the basic skeletons, which differ from each other in their substituents at C(10), C(13), or C(17).

Names and formula representations of some members of biologically important steroid families are listed in *Table 10.14*.

Names and symbols (α and β) are used to describe absolute configuration, so that the specification of stereogenic centers with the usual descriptors ((*R*) or (*S*)) is redundant in the steroid skeleton.

The rules of conformational analysis were developed for steroids. It is not by chance that *Chapt. 6* is entitled '*Conformational Analysis (Demonstrated on Steroids)*'.

Table 10.13. *Basic Skeletons of Steroids*

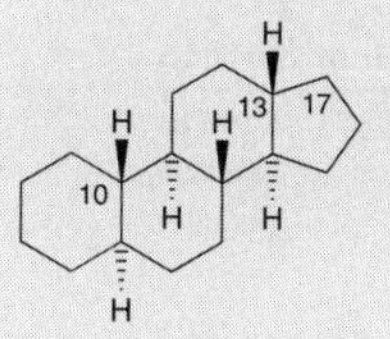

5α-Gonane

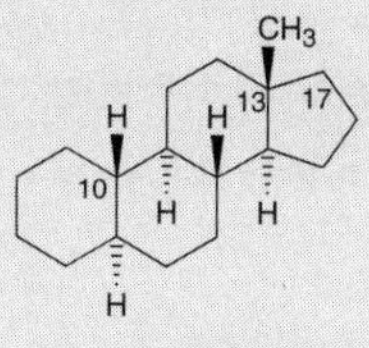

5α-Estrane

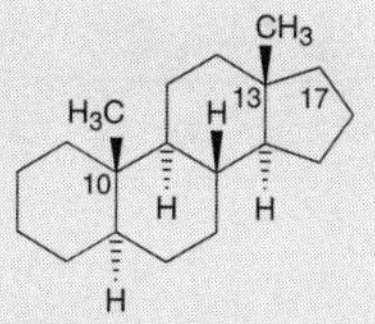

5α-Androstane

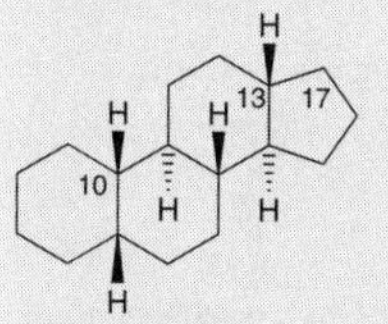

5β-Gonane

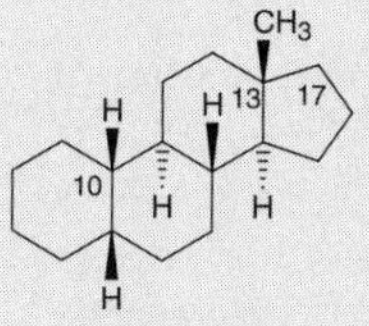

5β-Estrane

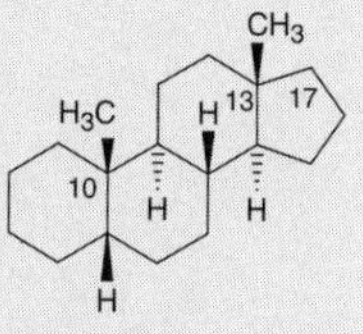

5β-Androstane

Side chain R	Sense of chirality	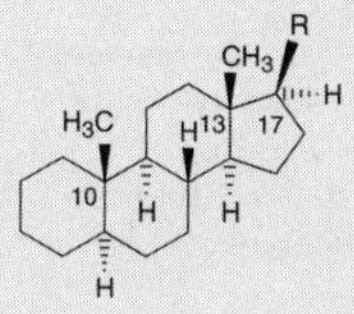	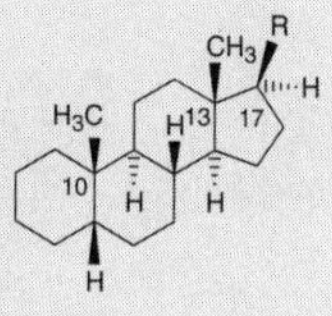
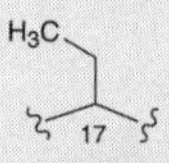	—	5α-Pregnane	5β-Pregnane
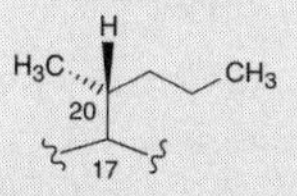	(20*R*)	5α-Cholane	5β-Cholane
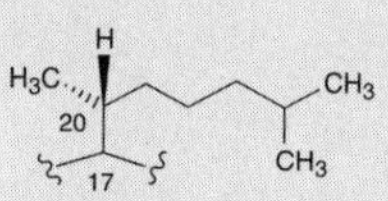	(20*R*)	5α-Cholestane	5β-Cholestane
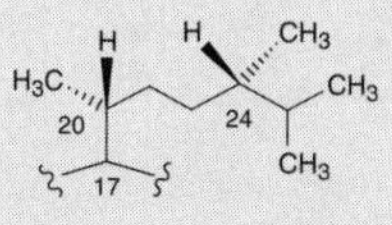	(20*R*,24*S*)	5α-Ergostane	5β-Ergostane
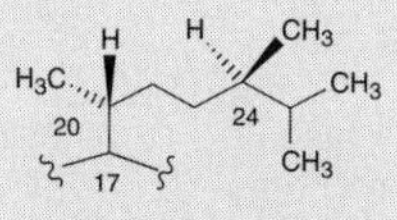	(20*R*,24*R*)	5α-Campestane	5β-Campestane
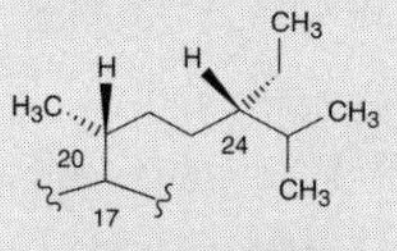	(20*R*,24*S*)	5α-Poriferastane	5β-Poriferastane
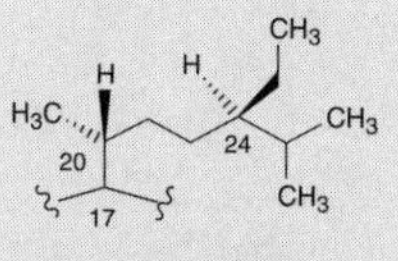	(20*R*,24*R*)	5α-Stigmastane	5β-Stigmastane
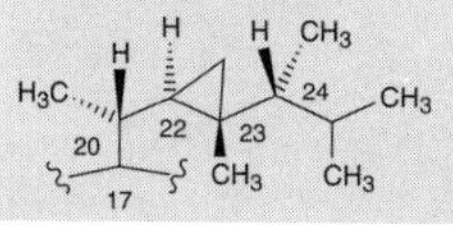	(20*S*,22*R*, 23*R*,24*R*)	5α-Gorgostane	5β-Gorgostane

Table 10.14. *Some Biologically Important Steroids*

Structure type	Example		Type of function
Estrane	18*a*-Homo-17α-ethynyl-17β-hydroxy-estra-4-en-3-one (= (–)-Norgestrel)	1	Sex hormone (Gestagen) Oral contraceptive
	1,3,5(10)-Estratriene-3,17β-diol (= (+)-Estradiol-17β)	2	Sex hormone (Estrogen)
	18*a*-Homo-17α-ethynyl-17β-hydroxy-estra-4,15-dien-3-one (= (–)-Gestoden)	3	Sex hormone (Gestagen) Oral contraceptive
	11β-[4-(Dimethylamino)phenyl]-17β-hydroxy-17α-(1-propynyl)-estra-4,9-dien-3-one (= (+)-Mifepristone; RU 486)	4	Sex antihormone (Progesterone antagonist)
Androstane	3α-Hydroxy-5α-androstan-17-one (= (+)-Androsterone)	5	Sex hormone (Androgen)
	17β-Hydroxy-4-androsten-3-one (= (+)-Testosterone)	6	Sex hormone (Androgen)
Pregnane	4-Pregnene-3,20-dione (= (+)-Progesterone)	7	Sex hormone (Gestagen)
	11β,17α,21-Trihydroxy-4-pregnene-3,20-dione (= (+)-Cortisol)	8	Glucocorticosteroid
	17α-Acetoxy-6-chloro-1α,2α-methylene-4,6-pregnadiene-3,20-dione (= (+)-Cyproteron acetate)	9	Sex antihormone (Androgen antagonist)
	(18*R*,20*S*)-18,11-Acetal-20,18-hemiketal of 11β,21-Dihydroxy-3,20-dioxo-4-pregnen-18-al (= (+)-Aldosterone)	10	Mineral corticosteroid
Cholane	3α,7α,12α-Trihydroxy-5β-cholane-24-carboxylic acid (= (+)-Cholic acid)	11	Bile acid
Cholestane	5-Cholesten-3β-ol (= (–)-Cholesterol)	12	Sterol from animal cells
9,10-Secocholestane	(1*S*,3*R*,5*Z*,7*E*)-9,10-Secocholesta-5,7,10(19)-triene-1,3,25-triol (= (+)-Calcitriol; 1α,25-Dihydroxycholecalciferol)	13	Calcitropic hormone
Ergostane	5,7,22-Ergostatrien-3β-ol (= (–)-Ergosterol)	14	Sterol from yeast cells
Stigmastane	5-Stigmasten-3β-ol (= (–)-β-Sitosterol)	15	Sterol from plant cells
Spirostane	(25*S*)-5-Spirosten-3β-ol (= (–)-Diosgenin)	16	Sapogenin

Table 10.14. *Some Biologically Important Steroids (cont.)*

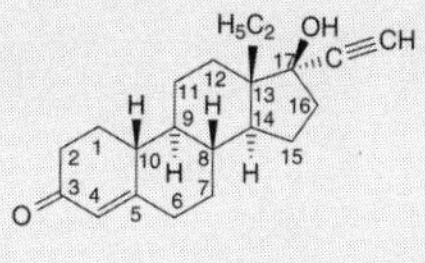

1

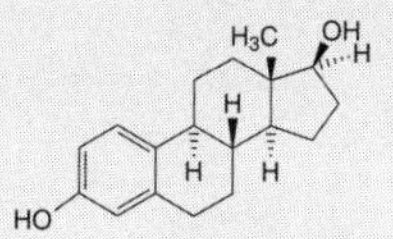

2

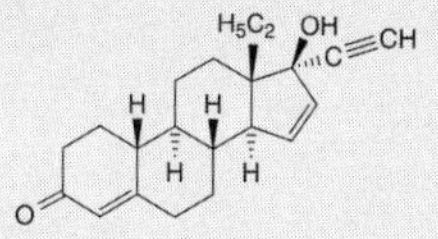

3

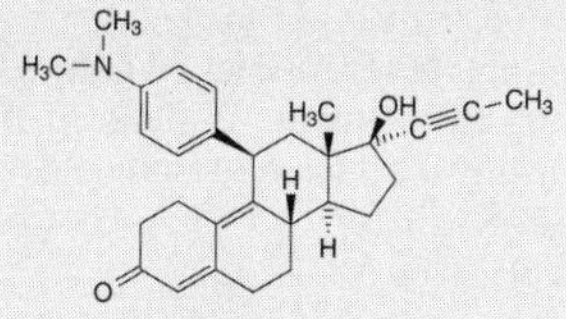

4

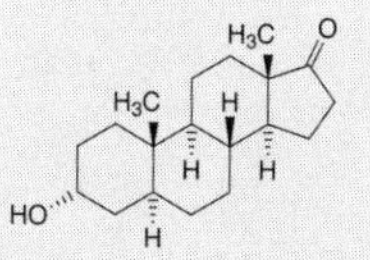

5

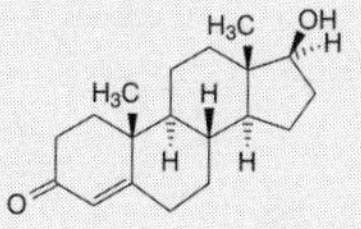

6

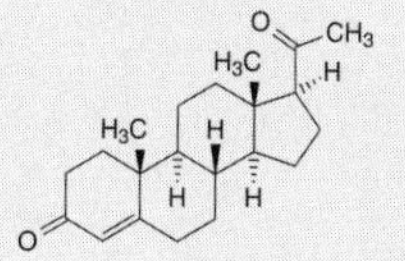

7

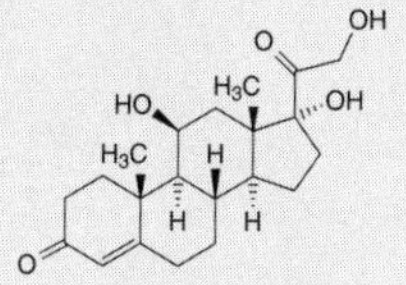

8

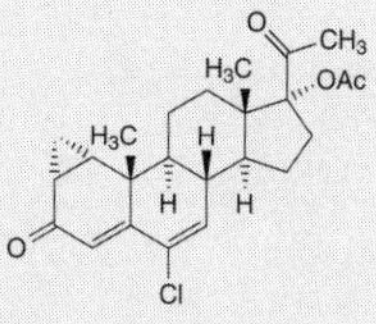

9

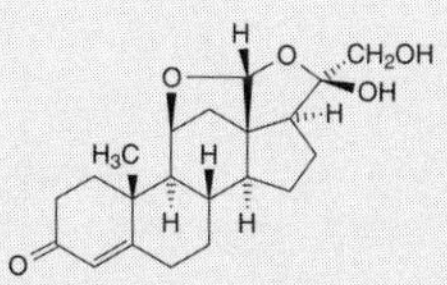

10

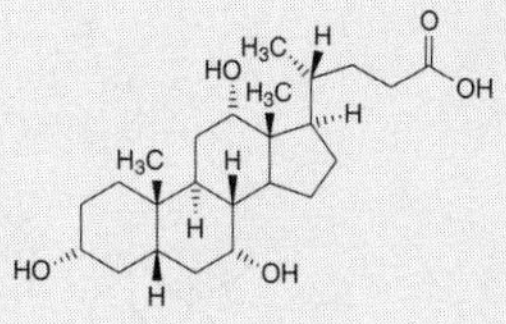

11

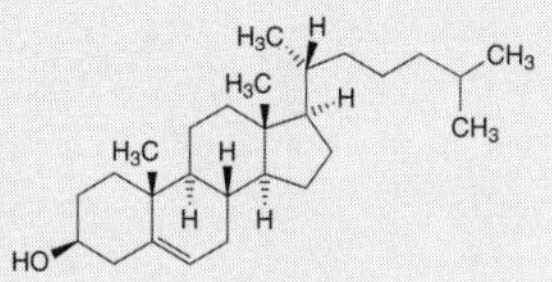

12

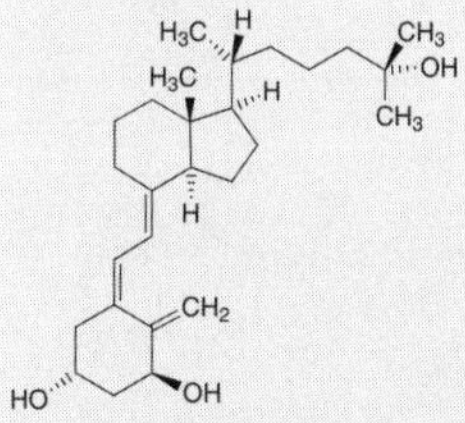

13

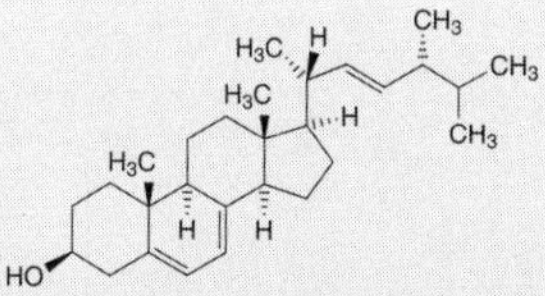

14

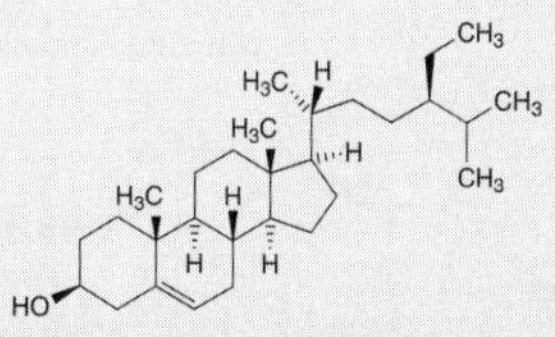

15

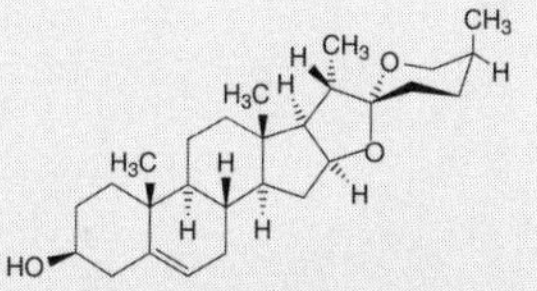

16

10.5.2. Nucleic Acids

Polymers with a linearly constituted sugar-phosphate backbone are regarded as belonging to the class of nucleic acids. In the naturally occurring nucleic acids, DNA (deoxyribonucleic acid) and RNA (ribonucleic acid), the sugars are 2-deoxy-β-D-ribofuranose and β-D-ribofuranose. The *constitutional repeating unit* is an O(5)-phosphorylated sugar in each case, bearing an *N*-glycosidically bound nucleobase at C(1) (*Fig. 10.17*).

For the synthesis of non-natural oligonucleotides with the hexopyranosyl (4′→6′)-constitution ('*Homo-DNA*'; *Fig. 10.18*), the sugar used is 2,3-dideoxy-β-D-*erythro*-hexopyranose.

For the non-natural oligonucleotides with the pentopyranosyl (2′→4′)-constitution (*p-RNA*; *Fig. 10.19*), the sugar is β-D-ribopyranose.

DNA contains the pyrimidine bases Thy and Cyt, and the purine bases Ade and Gua. In RNA, Ura takes the place of Thy. The phospho-

Example of a (3′–5′)-penta-β-D-ribofuranosyl-nucleotide single strand: (3′–5′)Rib*f*(C-U-G-A-G).
Example of a duplex of complementary (3′–5′)-penta-2′-deoxy-β-D-ribofuranosyl-nucleotide single strands: (3′–5′)d²Rib*f*(C-T-G-A-G)·(5′–3′)-d²Rib*f*(G-A-C-T-C).

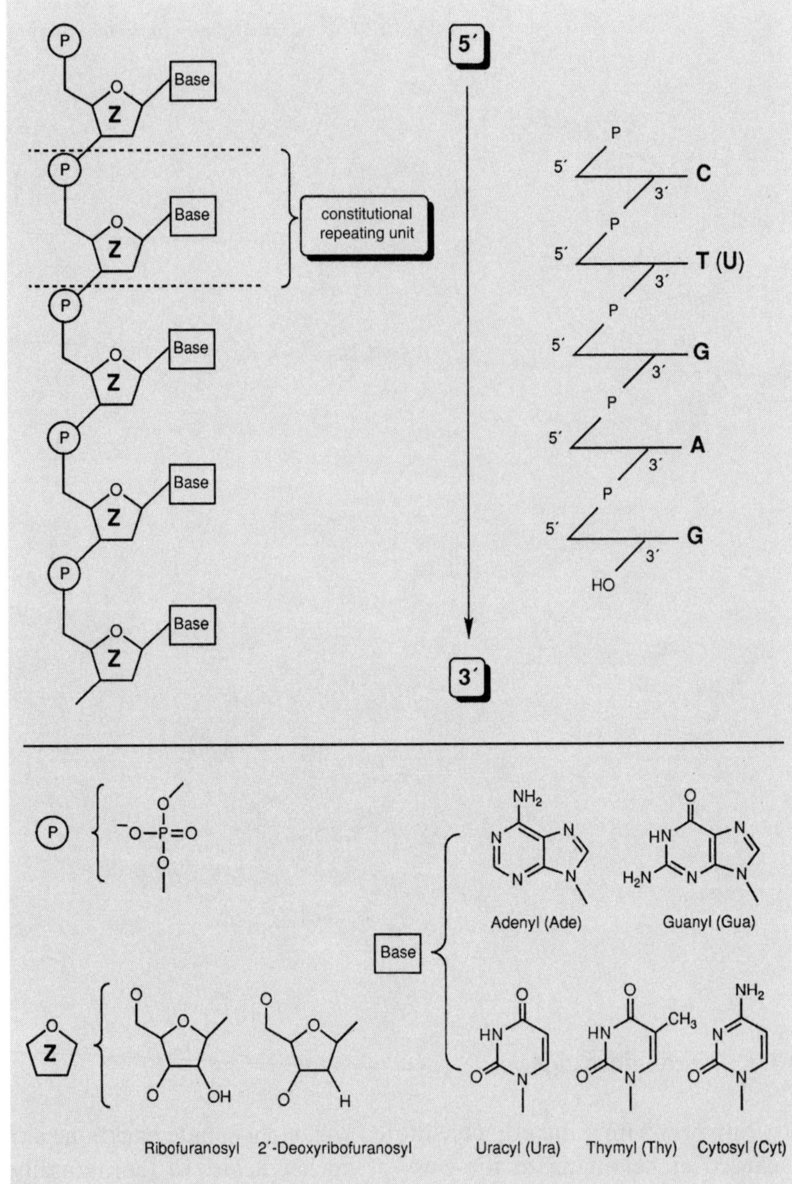

Fig. 10.17. *Schematic representation of a constitutional segment of a single strand of DNA or RNA*

ric-acid residue, whether singly or doubly esterified, is denoted by a *P* in the three-letter notation. In the one-letter notation, however, a more differentiating convention exists. A monosubstituted (terminal) phosphoric-acid residue is indicated by a p, while an (internal) phosphodiester residue, connecting the centers C(3′) and C(5′), may be expressed by a hyphen or an arrow indicating the connectivity (3′→5′), or by a comma (3′,5′), depending on whether or not the sequence is known.

The three-letter and one-letter notations for nucleobases, nucleosides, and nucleotides are given in *Table 10.15*.

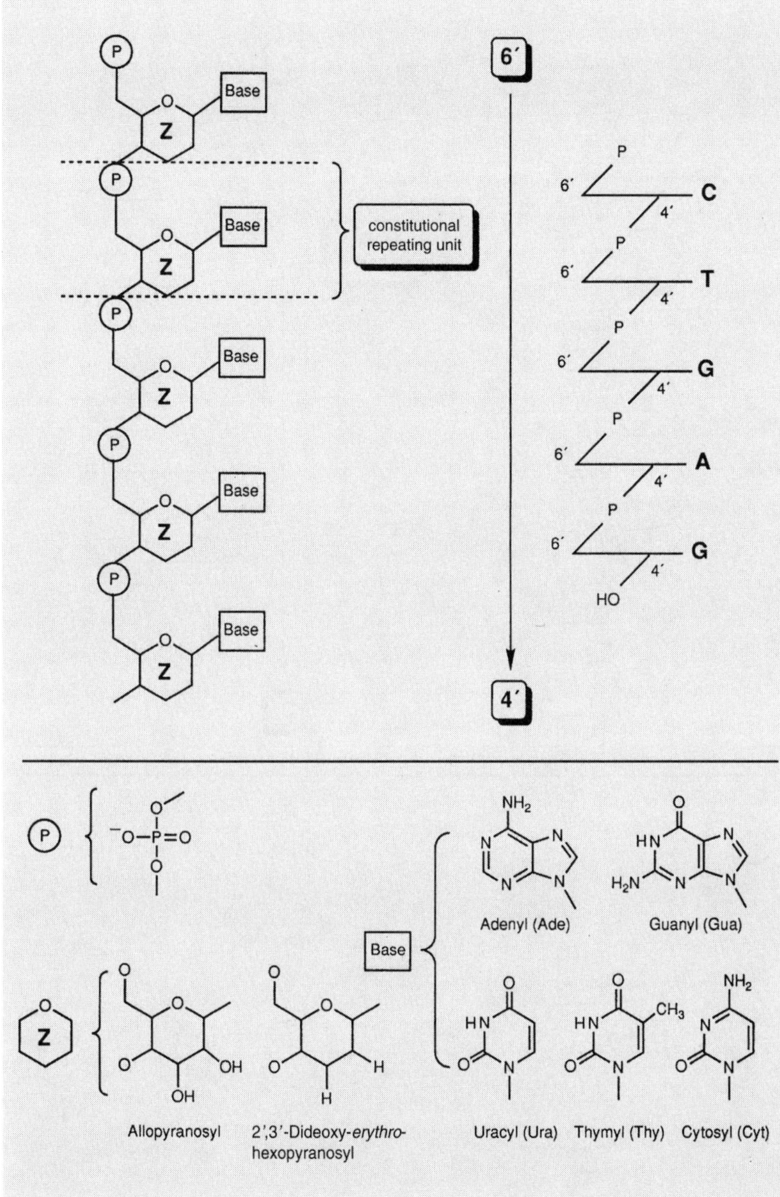

Example of a (4′–6′)-penta-2′,3′-dideoxy-β-D-*erythro*-hexopyranosyl-nucleotide single strand: (4′–6′)d^2d^3Glc*p*(C-T-G-A-G).
Example of a duplex of complementary (4′–6′)-penta-2′,3′-dideoxy-β-D-*erythro*-hexopyranosyl-nucleotide single strands: (4′–6′)d^2d^3Glc*p*-(C-T-G-A-G)·(6′–4′)d^2d^3Glc*p*-(G-A-C-T-C).

Fig. 10.18. *Schematic representation of a constitutional segment of a (4′→6′)-pyranosyl-oligonucleotide*

It has proven valuable in many cases to have available symbols which are wholly or partially unspecified as regards the nucleoside subclasses. For an unspecified purine nucleoside (pyrimidine nucleoside), the symbol R (Y) is used. If it is not known whether a purine or a pyrimidine nucleoside is present, or if one wants to keep the option open, then the symbol N is used. The numbering of the individual atoms is given in the examples of *Figs. 10.17–10.19* (see text in each of the related boxes), *Fig. 7.37*, *Fig. 10.23*, and *Sect. 10.5.3.2*.

The *configurational repeating unit*, in both DNA and homo-DNA, shows that two neighboring ring atoms of the sugar residue play a role in the construction of the sugar-phosphate backbone, in such a way that the backbone grows from each of the furanose or pyranose sugar rings in

Example of a (2′–4′)-penta-β-D-ribopyranosyl-nucleotide single strand: (2′–4′)Rib*p*(C-T-G-A-G).
Example of a duplex of complementary (2′–4′)-penta-β-D-ribopyranosyl-nucleotide single strands: (2′–4′)-Rib*p*(C-T-G-A-G)·(4′–2′)Rib*p*-(G-A-C-T-C).

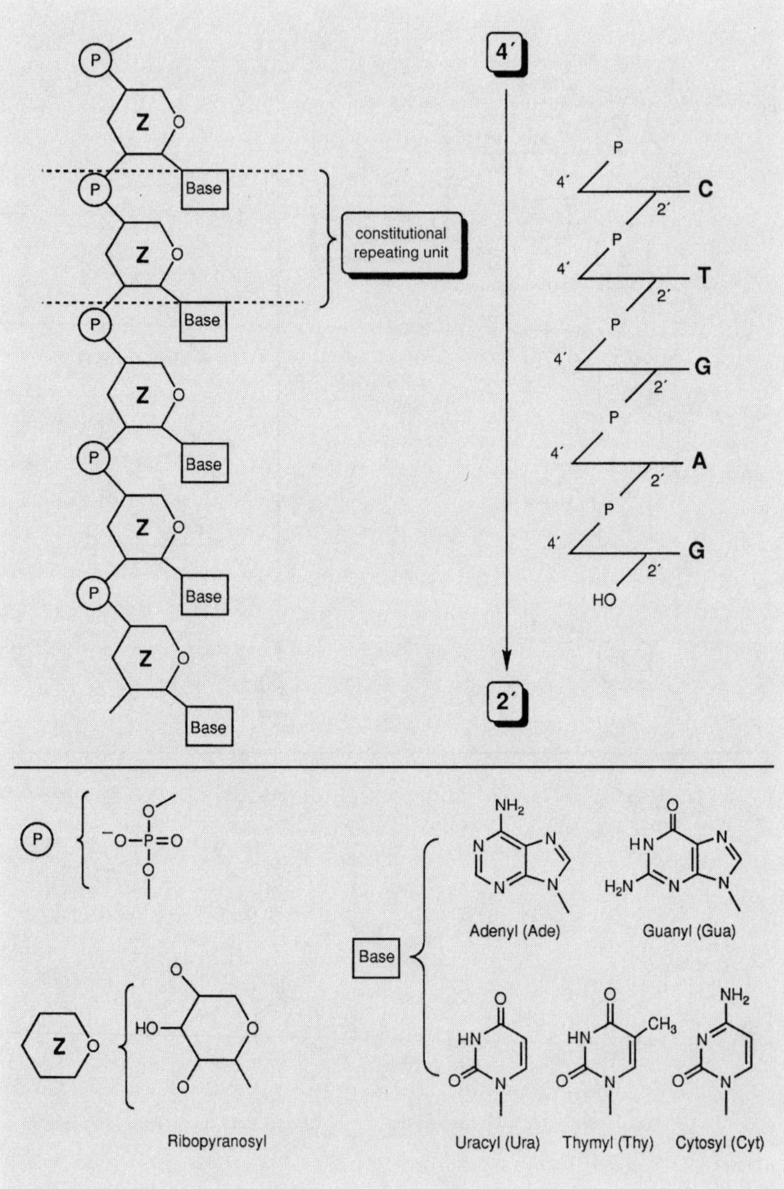

Fig. 10.19. *Schematic representation of a constitutional segment of a (2′ → 4′)-pyranosyl-oligonucleotide*

trans-orientation. The nucleobase is connected, with β-orientation, at C(1′) (*Fig. 10.20*).

The conformations of the pyranose and furanose rings are discussed in *Sect. 10.5.3.1* and *10.5.3.2*.

The backbone of a polynucleotide chain contains the *conformational repeating unit*, which covers an array of six bonds: P–O(5′), O(5′)–C(5′), C(5′)–C(4′), C(4′)–C(3′), C(3′)–O(3′), and O(3′)–P. The torsion angles about these bonds are called α, β, γ, δ, ε, and ξ (*Fig. 10.21*).

Table 10.15. *Specification of Nucleobases, Nucleosides, and Nucleotides According to the Three- or One-Letter Notation*

Base	Symbol	Ribonucleoside	Symbol		Ribonucleotide	Symbol	
Uracil	Ura	Uridine	Urd	U	Uridine-5´-phosphate	Urd-5´-*P*	pU
Cytosine	Cyt	Cytidine	Cyd	C	Cytidine-5´-phosphate	Cyd-5´-*P*	pC
Adenine	Ade	Adenosine	Ado	A	Adenosine-5´-phosphate	Ado-5´-*P*	pA
Guanine	Gua	Guanosine	Guo	G	Guanosine-5´-phosphate	Guo- 5´-*P*	pG
Base	**Symbol**	**2´-Deoxyribo-nucleoside**	**Symbol**		**2´-Deoxyribo-nucleotide**	**Symbol**	
Thymine	Thy	Thymidine	dThd	dT	Thymidine-5´-phosphate	dThd-5´-*P*	dpT
Cytosine	Cyt	Deoxycytidine	dCyd	dC	Deoxycytidine-5´-phosphate	dCyd-5´-*P*	dpC
Adenine	Ade	Deoxyadenosine	dAdo	dA	Deoxyadenosine-5´-phosphate	dAdo-5´-*P*	dpA
Guanine	Gua	Deoxyguanosine	dGuo	dG	Deoxyguanosine-5´-phosphate	dGuo-5´-*P*	dpG

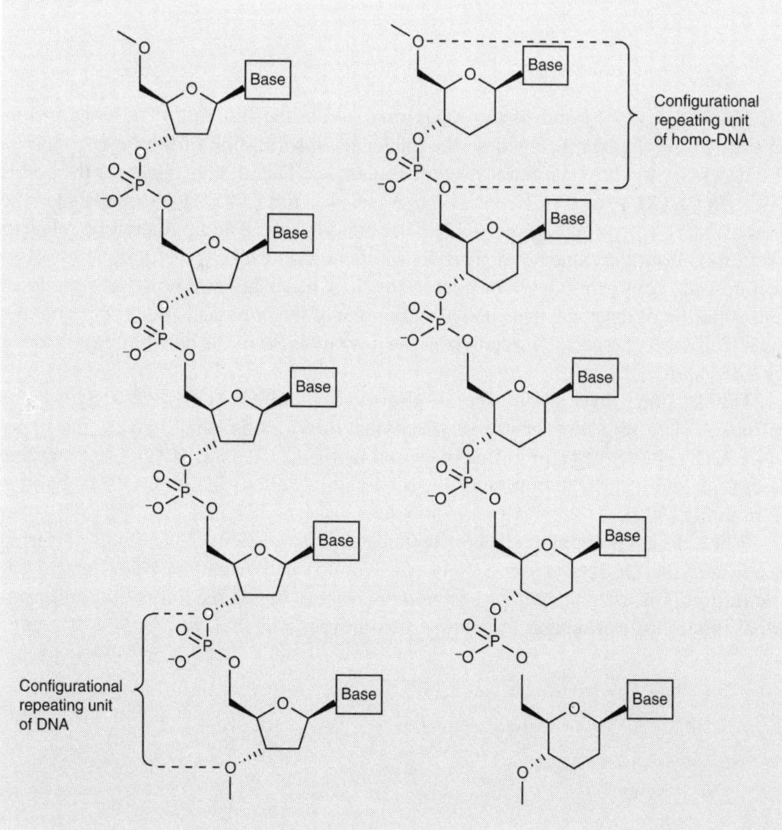

Fig. 10.20. *Schematic representation of a configurational segment of a (3′ → 5′)-furanosyl- and a (4′ → 6′)-pyranosyl-oligonucleotide*

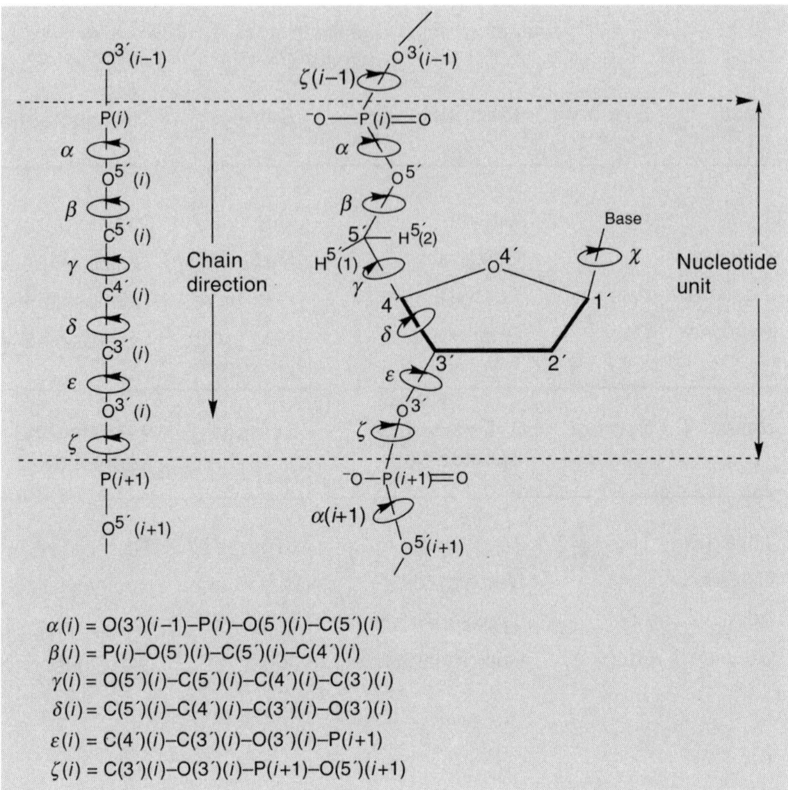

Fig. 10.21. *Schematic representation of a conformational segment of a (3′→5′)-furanosyl-oligonucleotide*

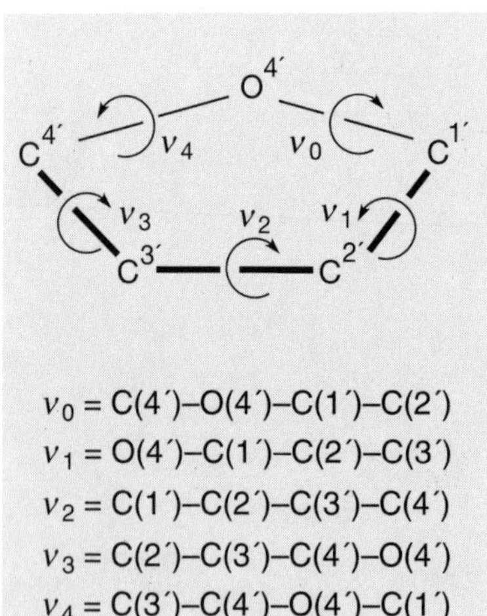

Fig. 10.22. *Torsion angles in the sugar ring*

The C(4′)–C(3′) bond of the sugar ring fulfills the function of a hinge in this repeating unit. To completely describe the local conformation of the sugar ring, it is also necessary to know the endocyclic torsion angles. The torsion angles for the bonds O(4′)–C(1′), C(1′)–C(2′), C(2′)–C(3′), C(3′)–C(4′), and C(4′)–O(4′) are assigned the symbols ν_0, ν_1, ν_2, ν_3, and ν_4 (*Fig. 10.22*). The torsion angles δ and ν_3, therefore, relate to a rotation about the same bond. This does not constitute an overdefinition, however, as, for some conformational considerations, it is necessary to know both the local conformation of the sugar ring, and also that one of the polynucleotide backbone: the sugar ring deserves special thought, because it is both part of the backbone and part of the side chain.

The torsion angle about the *N*-glycosidic bond (N–C(1′)) is assigned the symbol χ. The sequence of atoms specifying this torsion angle covers the array O(4′)–C(1′)–N(9)–C(4) for purine bases and O(4′)–C(1′)–N(1)–C(2) for pyrimidine bases (*vide supra*). The definition of the torsion angle about the *N*-glycosidic bond is given in *Fig. 10.23*.

When the torsion angle $\chi = 0°$, the O(4′)–C(1′) and N(9)–C(4) bonds in purine bases, and the O(4′)–C(1′) and N(1)–C(2) bonds in pyrimidine bases, are in *sp*-orientation. The assignments *syn* and *anti* are used to define the following conformational ranges for purine and pyrimidine derivatives; *syn*: $0° \pm 90°$; *anti*: $180° \pm 90°$.

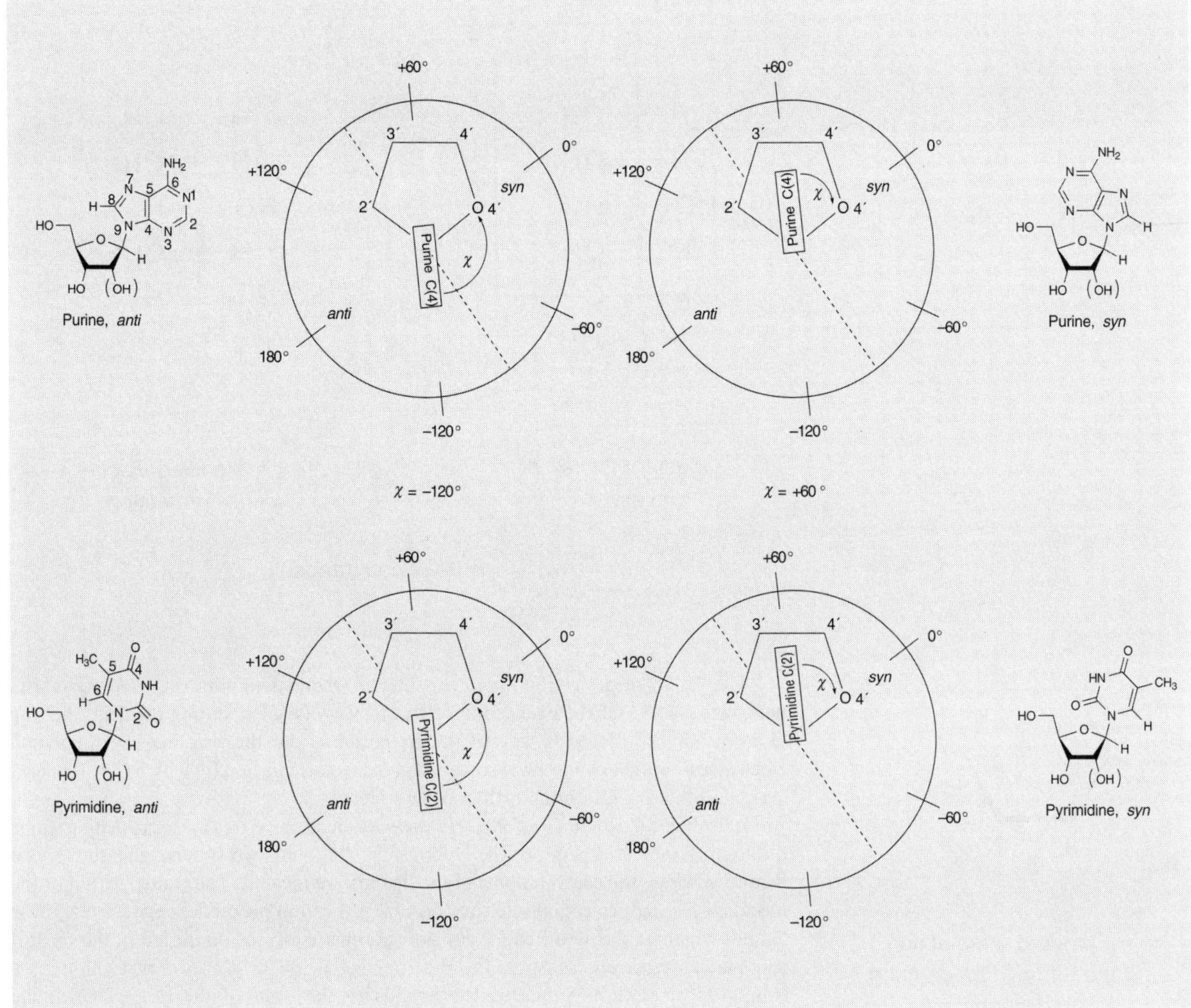

Fig. 10.23. *Specification of the torsion angle about the* N-*glycosidic bond*

10.5.3. Carbohydrates

10.5.3.1. Pyranoses

There are various different semiotic means of describing the tetrahydropyran-derived semiacetals found in monosaccharides: the *Fischer* projection (*Sect. 10.2.1*), the *Haworth* projection, and the symbol of the chair conformation. To convert *Fischer* projections into representations of chair conformations, the following procedure, demonstrated on α-D-glucopyranose, is used.

Firstly, it is necessary to ensure that all of the ring atoms lie in the same (vertical) plane: for this purpose, the original *Fischer* projection is modified by rearranging that tetrahedron with C(5) in the center.

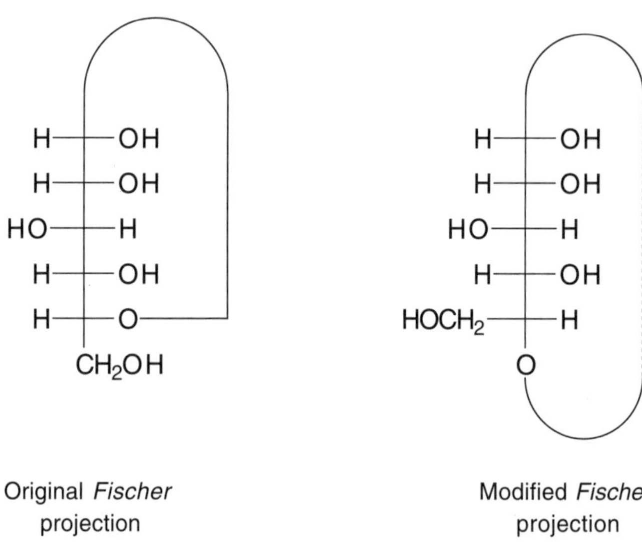

Original *Fischer* projection

Modified *Fischer* projection

α-D-Glucopyranose

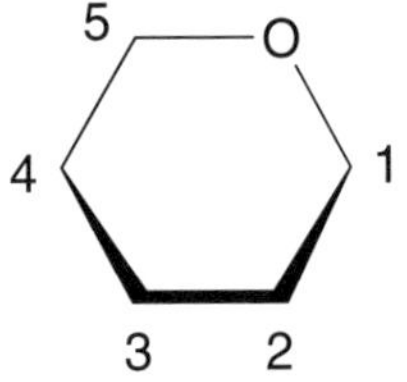

Standard orientation of the tetrahydropyran ring with conventional ring numbering

Haworth projection of α-D-glucopyranose

The *Haworth* projection assumes a planar tetrahydropyran ring, represented in perspective style. Of the twelve different possible ways of orienting the ring (the ring O-atom may be placed in any of the six corners, and the ring may be numbered clockwise or anticlockwise), the so-called standard orientation (in which C(1) is placed 'east' and the ring O-atom 'north-east') is selected.

It is helpful to think of the *Haworth* projection as really representing three different planes: the plane of the heterocyclic ring and two others; one above, one below, in which the central atoms of the ligands are located. The transposition of the modified *Fischer* projection into the *Haworth* projection proceedes very simply: those ligands which, in the modified *Fischer* projection, are located on the left of the vertical ring plane are merely transposed to the upper plane of the *Haworth* projection in its standard orientation. Similarly, the ligands to the right of the ring plane in the modified *Fischer* projection are relocated to the lower plane of the *Haworth* projection.

To avoid the unnecessary and inaccurate assumption of a planar tetrahydropyran ring, the symbol of the chair conformation is used instead of the *Haworth* projection. This merely involves moving C(1) and C(4) out of the reference plane to obtain a chair conformation corresponding to the *Haworth* projection. Two isomeric chair conformations are possible; these are unambiguously specified by the descriptors ${}^{1}C_{4}$ and ${}^{4}C_{1}$. *C* stands for chair conformation (other possibilities would be *B* for *b*oat, *S* for *s*kew, and *H* for *h*alf-chair), the numbers denote which of the two ring atoms, C(1) and C(4), lie above and below the reference plane (the reference plane is that one containing two parallel vertexes of the ring, selected such that the C-atom of lowest possible locant is situated outside of the reference plane). The two inverse chair conformers of α-D-glucopyranose and their mirror-image isomers are now determined by means of their formula representations and the stereodescriptors which may be assigned to their names.

Characterization of the configuration at the anomeric center proceeds as follows. If the OH group at the anomeric center and that one at the reference center (the stereogenic center with the highest locant) are situated on the same side in the *Fischer* projection, then the α-anomer is present; otherwise the β-anomer is present.

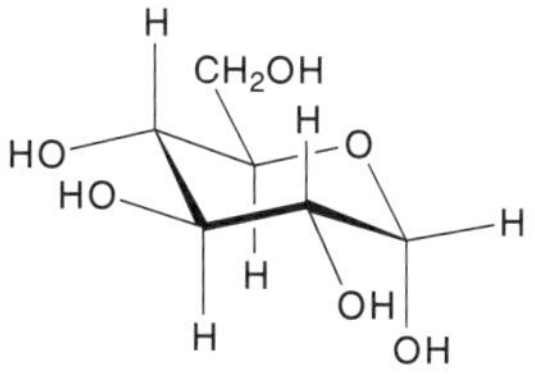

α-D-Glucopyranose - $^4C_1$

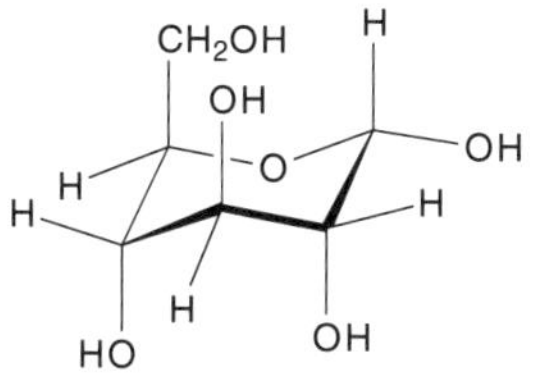

α-D-Glucopyranose - $^1C_4$

Mirror plane

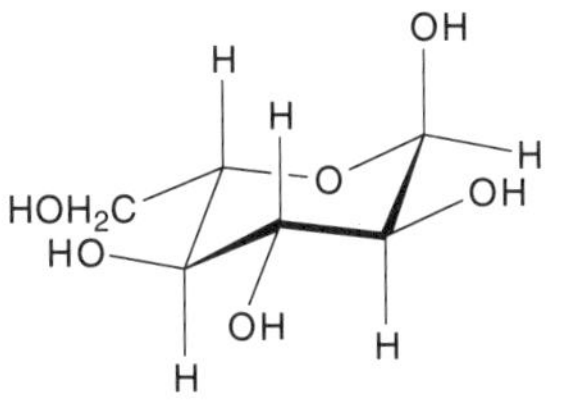

α-L-Glucopyranose - $^1C_4$

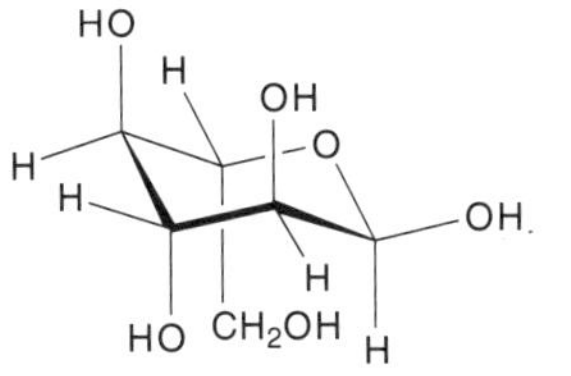

α-L-Glucopyranose - $^4C_1$

10.5.3.2. Furanoses

For five-membered monosaccharides, distinction is made between conformations *E* (for *e*nvelope) and *T* (for *t*wist). In the *E*-conformations, the reference plane is defined by four ring atoms. For the aldofuranoses, the following conformations are possible: ^1E, E_1, ^2E, E_2, ^3E, E_3, ^4E, E_4, ^OE, and E_O. The reference plane of the *T*-conformation, defined by three adjacent ring atoms, is so selected that one of the two extraplanar atoms lies above the plane and the other below. For the aldofuranoses, the following conformations are possible: $^OT_1$, ^1T_O, $^1T_2$, $^2T_1$, $^2T_3$, $^3T_2$, $^3T_4$, $^4T_3$, ^4T_O, and $^OT_4$.

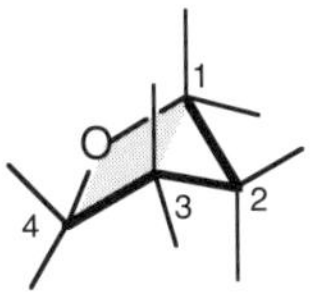

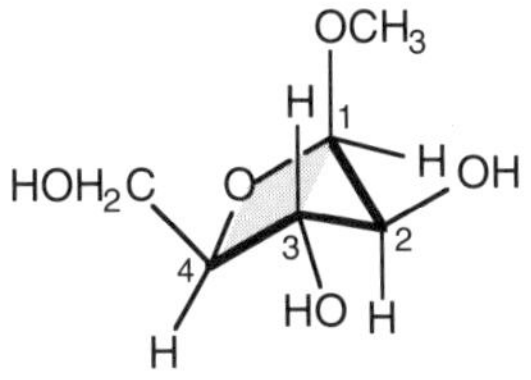

E_2

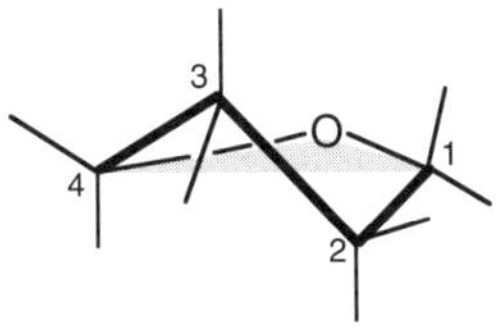

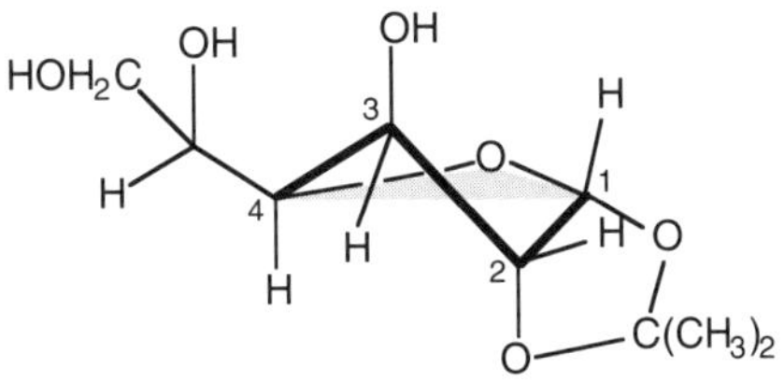

$^3T_2$

^1E means: C(1) is situated outside the plane in which the other four ring centers are to be found, and on that side, viewed from which the ring is numbered clockwise. E_1 means: C(1) is situated outside the plane in which the other four ring centers are to be found, and on that side, viewed from which the ring is numbered anticlockwise.

The furanose ring plays a special role in the chemistry of the nucleic acids (*Sect. 10.5.2*). Conformational alterations in the region of the sugar influence the overall conformation of DNA and RNA molecules.

The transition from one furanose conformation to another has a definite effect on the conformation of the sugar-phosphate chain, and hence on the biological function of the nucleic acids. Using the carbocyclic five-membered ring as a frame of reference for tetrahydrofuran, it can be concluded by analogy that the furanose ring system in nucleosides must be flexible, and that a pseudorotational cycle must exist. As the barrier to internal rotation is lower in the C–O bond than in the C–C bond, tetrahydrofuran conformations with an eclipsed arrangement around one of the two C–O bonds and a significant degree of protrusion in the C-atom opposite (C(2′)-*endo* or ^2E, and C(3′)-*endo* or ^3E) will be energetically favored (*Fig. 10.24*).

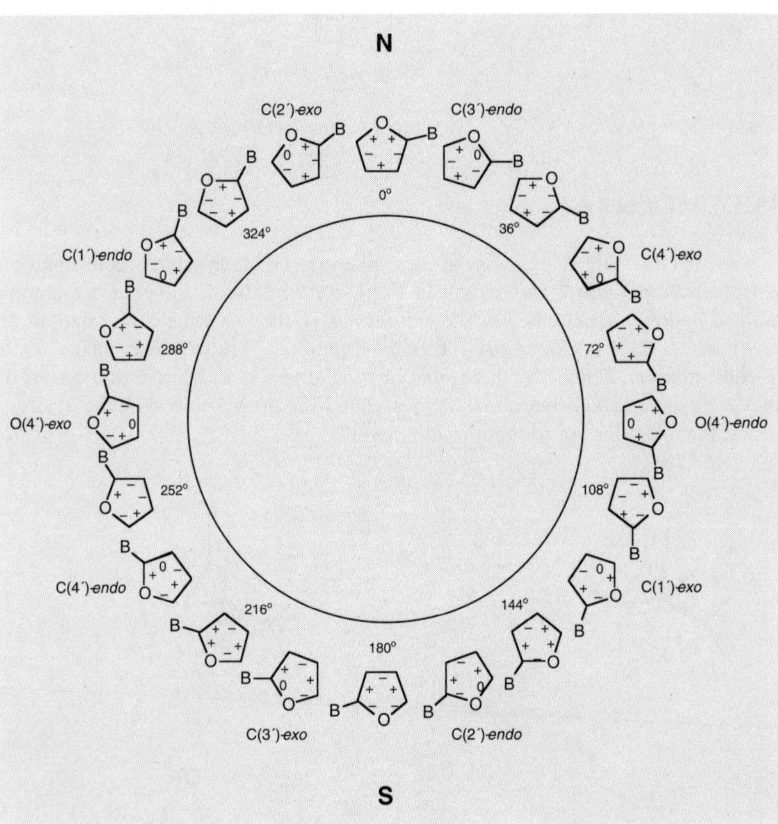

Fig. 10.24. *Pseudorotational cycle of the furanose sugar ring in nucleosides*

The earlier descriptors C(2′)-*endo* and C(3′)-*endo*, or C(3′)-*endo*/C(2′)-*exo* have more recently been replaced by the descriptors ^2E and ^3E, or $^3T_2$. The descriptors *endo* or *exo*, accordingly, mean the same thing as the number positioned above and before, or below and after, the letters *E* and *T*.

10.5.3.3. Polysaccharides

The term *polysaccharide* (*oligosaccharide*) refers to a macromolecule in which more than (up to) ten monosaccharide residues are connected to one another, each *via* an acetal bond, known as a *glycosidic bond.*

A glycosidic bond (formally, at least) is created by a condensation in which the OH group at the anomeric center (C(1′)) of one monosaccharide residue and one of the OH groups at the centers C(2) to C(6) of another monosaccharide residue participate. Even in the condensation of as few as two (or three) identical D-hexopyranoses, 11 different combinations, representing five different constitutions (or 176 different combinations), are possible.

Stereoformulae for the representation of oligo- and polysaccharides will be displayed using the example of the disaccharide β-maltose. In addition to the conformational representation and *Haworth* projection which we have already encountered, the so-called *Mills* projection is given.

Conformational formula

Haworth projection

Mills projection

The trivial name β-maltose corresponds to the systematic name O^4-α-D-glucopyranosyl-β-D-glucopyranose, abbreviated as α-D-Glcp-(1→4)-β-D-Glc. For the abbreviation, the trivial name of each monosaccharide component is shortened to three letters (*Table 10.16*).

Table 10.16. *Three-Letter Notations for Monosaccharides*

Pentoses:					
Ribose	Rib	Arabinose	Ara	Xylose	Xyl
Lyxose	Lyx				
Hexoses:					
Allose	All	Altrose	Alt	Glucose	Glc
Mannose	Man	Gulose	Gul	Idose	Ido
Galactose	Gal	Talose	Tal	Fructose	Fru
Fucose	Fuc	Muramic acid	Mur	Rhamnose	Rha
Nonoses:					
Neuraminic acid	Neu				

Examples of di-, tri-, and tetrasaccharides are listed in *Table 10.17*.

Table 10.17. *Trivial Names and Abbreviated Structure Descriptions of Di-, Tri-, and Tetrasaccharides*

Trivial name	Abbreviated structure description
Cellobiose	β-D-Glc*p*-(1⟶ 4)-D-Glc
Lactose	β-D-Gal*p*-(1⟶ 4)-D-Glc
Maltose	α-D-Glc*p*-(1⟶ 4)-D-Glc
Sucrose	α-D-Glc*p*-(1⟷ 2)-D-Fru*f*
Cellotriose	β-D-Glc*p*-(1⟶ 4)-β-D-Glc*p*-(1⟶ 4)-D-Glc
Maltotriose	α-D-Glc*p*-(1⟶ 4)-α-D-Glc*p*-(1⟶ 4)-D-Glc
Raffinose	α-D-Gal*p*-(1⟶ 6)-α-D-Glc*p*-(1⟷ 2)-β-D-Fru*f*
Cellotetrose	β-D-Glc*p*-(1⟶ 4)-β-D-Glc*p*-(1⟶ 4)-β-D-Glc*p*- -(1⟶ 4)-D-Glc
Maltotetrose	α-D-Glc*p*-(1⟶ 4)-α-D-Glc*p*-(1⟶ 4)-α-D-Glc*p*- -(1⟶ 4)-D-Glc

The configuration of the sugar (D or L), its ring size (pyranose = *p* or furanose = *f*), and the configuration at the anomeric center (α or β) are additionally specified. A horizontal arrow indicates the specific connectivity between the saccharide monomers. If the oligosaccharide branches, then a vertical arrow denotes the site of the bonds.

α-D-Gal*p*-(1→ 4)-β-D-Glc*p*NAc-(1→ 4)-D-Glc*p*
6
↑
1
β-D-Fru*f*

→ 4)-[α-D-Glc*p*]$_6$-(1 ⌒

The numbering of the monosaccharide residues in an oligo- or polysaccharide proceeds from right to left. This has the advantage that the numbering does not alter, if the chain is shortened or lengthened, which usually occurs at the left-hand end.

No polysaccharides consisting of more than six different monosaccharide residues (mostly pentose and hexose monosaccharides or their derivatives) are known at present. The monosaccharide residues tend to be regular repeating units. Because of the structural complexity of these molecules, however, the constitution of these repeating units has, in many cases, not yet been firmly established to the last detail. In addition, it is a fact that different specimens of different origins of seemingly the same polysaccharide can show slight differences in the degree of substitution of derivatives, in the chain length, and in composition. This phenomenon, which hardly has any noticeable effect on the overall structure, is described as *microheterogeneity*. The particular individuals of such a set could be called *diversomers*, as is the case for the product of an automatic combinatorial synthesis.

10.5.4. Proteins

Proteins are polymers with linear, unbranched main chains, in which amino-acid residues are linked together repetitively, by means of peptide bonds. Compounds with these bonds are also known as *peptides*. The terms protein and polypeptide may be used synonymously.

10.5.4.1. α-L-Amino Acids

All *proteinogenic amino acids* contain a central C-atom (C^{α}), to which the ligands CO_2H, NH_2 (NH in the case of proline), H, and the so-called side chain R (R = H in the case of glycine) are bound. Each individual amino acid owes its individuality to its side chain.

Trivial names are generally used for the twenty (see, however, *Chapt. 16.3.1*) proteinogenic amino acids. *Table 10.18* lists these trivial names together with the systematic names, and gives their *three-letter* and *one-letter notations*, as well as the *nucleic-acid triplets* which function as *codons* representing a specific amino acid in protein biosynthesis.

The numbering of the individual C-atoms of the amino-acid main chain (together with the side-chain branches) is shown for the examples of Ile, Pro, Tyr, and His.

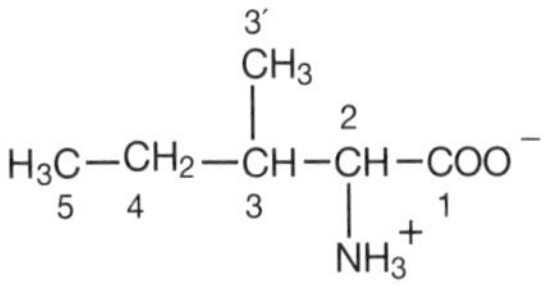

Ile

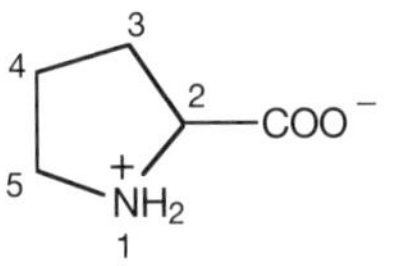

Pro

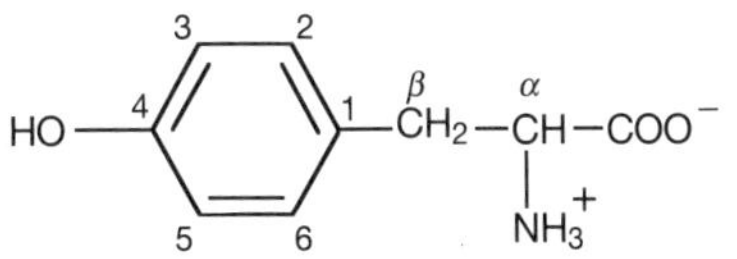

Tyr

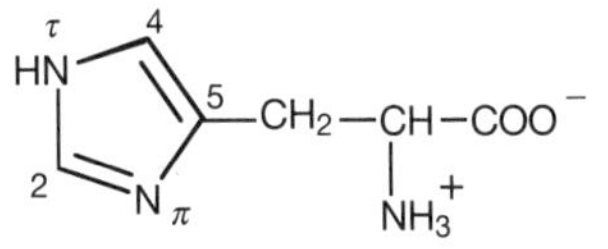

His

With the exception of Gly, the building blocks of polypeptides and proteins are chiral. Furthermore, they are not present as racemic mixtures. The absolute configuration at the C^{α}-atom is specified (as in sugars) with the letter D or L. D or L may be quickly and unambiguously determined from the *Fischer* projection, relating L-serine to L-glyceraldehyde (*Sect. 10.2.1*).

As indicated in *Table 10.18*, proteinogenic amino acids have the L-configuration.

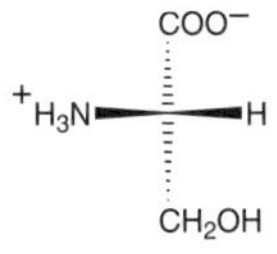

L-Serine

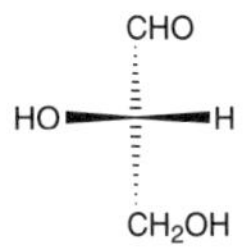

L-Glyceraldehyde

Amino acids with two stereogenic centers (Ile and Thr) are handled in a special way: if the relative configuration is altered, then the prefix *allo* is used.

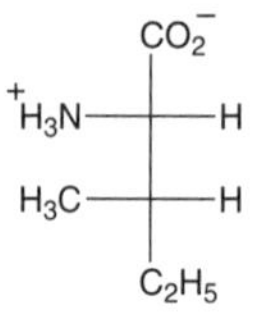

L-Isoleucine 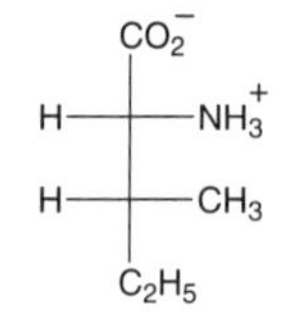 D-Isoleucine 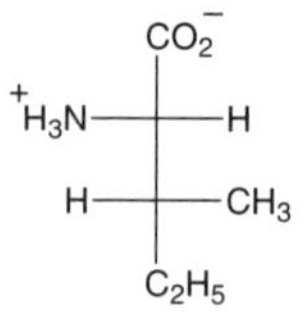 L-Alloisoleucine 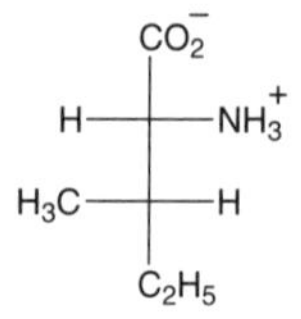 D-Alloisoleucine

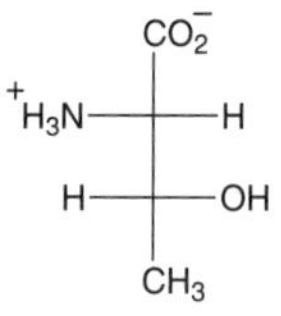

L-Threonine 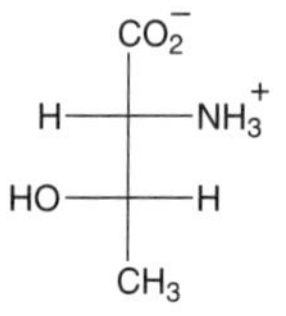 D-Threonine 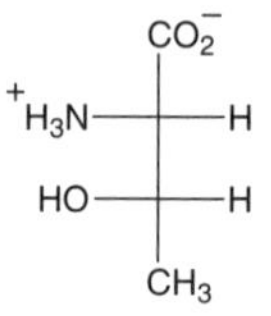 L-Allothreonine 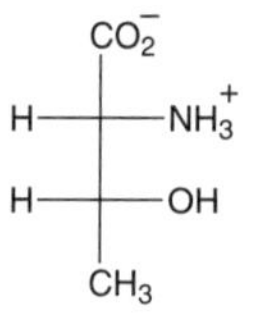 D-Allothreonine

Table 10.18. *Three-Letter and One-Letter Notations, and the Nucleic-Acid Triplet Codons for the Specified Amino Acids*

Trivial name	Systematic name	Symbol		Codons
Alanine	L-2-Aminopropanoic acid	Ala	A	GCA GCC GCG GCU
Arginine	L-2-Amino-5-guanidinopentanoic acid	Arg	R	AGA AGG CGA CGC CGG CGU
Asparagine	L-2-Amino-3-carbamoylpropanoic acid	Asn	N	AAC AAU
Aspartic acid	L-2-Aminobutanedioic acid	Asp	D	GAC GAU
Cysteine	L-2-Amino-3-mercaptopropanoic acid	Cys	C	UGC UGU
Glutamine	L-2-Amino-4-carbamoylbutanoic acid	Gln	Q	CAA CAG
Glutamic acid	L-2-Aminopentanedioic acid	Glu	E	GAA GAG
Glycine	Aminoethanoic acid	Gly	G	GGA GGC GGG GGU
Histidine	L-2-Amino-3-(1*H*-imidazol-4-yl)propanoic acid	His	H	CAC CAU
Isoleucine	L-2-Amino-3-methylpentanoic acid	Ile	I	AUA AUC AUU
Leucine	L-2-Amino-4-methylpenedioic acid	Leu	L	UUA UUG CUA CUC CUG CUU
Lysine	L-2,6-Diaminohexanoic acid	Lys	K	AAA AAG
Methionine	L-2-Amino-4-(methylthio)butanoic acid	Met	M	AUG
Phenylalanine	L-2-Amino-3-phenylpropanoic acid	Phe	F	UUC UUU
Proline	L-Pyrrolidine-2-carboxylic acid	Pro	P	CCA CCC CCG CCU
Serine	L-2-Amino-3-hydroxypropanoic acid	Ser	S	AGC AGU UCA UCC UCG UCU
Threonine	L-2-Amino-3-hydroxybutanoic acid	Thr	T	ACA ACC ACG ACU
Tryptophan	L-2-Amino-3-(1*H*-indol-3-yl)propanoic acid	Trp	W	UGG
Tyrosine	L-2-Amino-3-(4-hydroxyphenyl)propanoic acid	Tyr	Y	UAC UAU
Valine	L-2-Amino-3-methylbutanoic acid	Val	V	GUA GUC GUG GUU

Table 10.18. *Three-Letter and One-Letter Notations, and the Nucleic-Acid Triplet Codons for the Specified Amino Acids* (*cont.*)

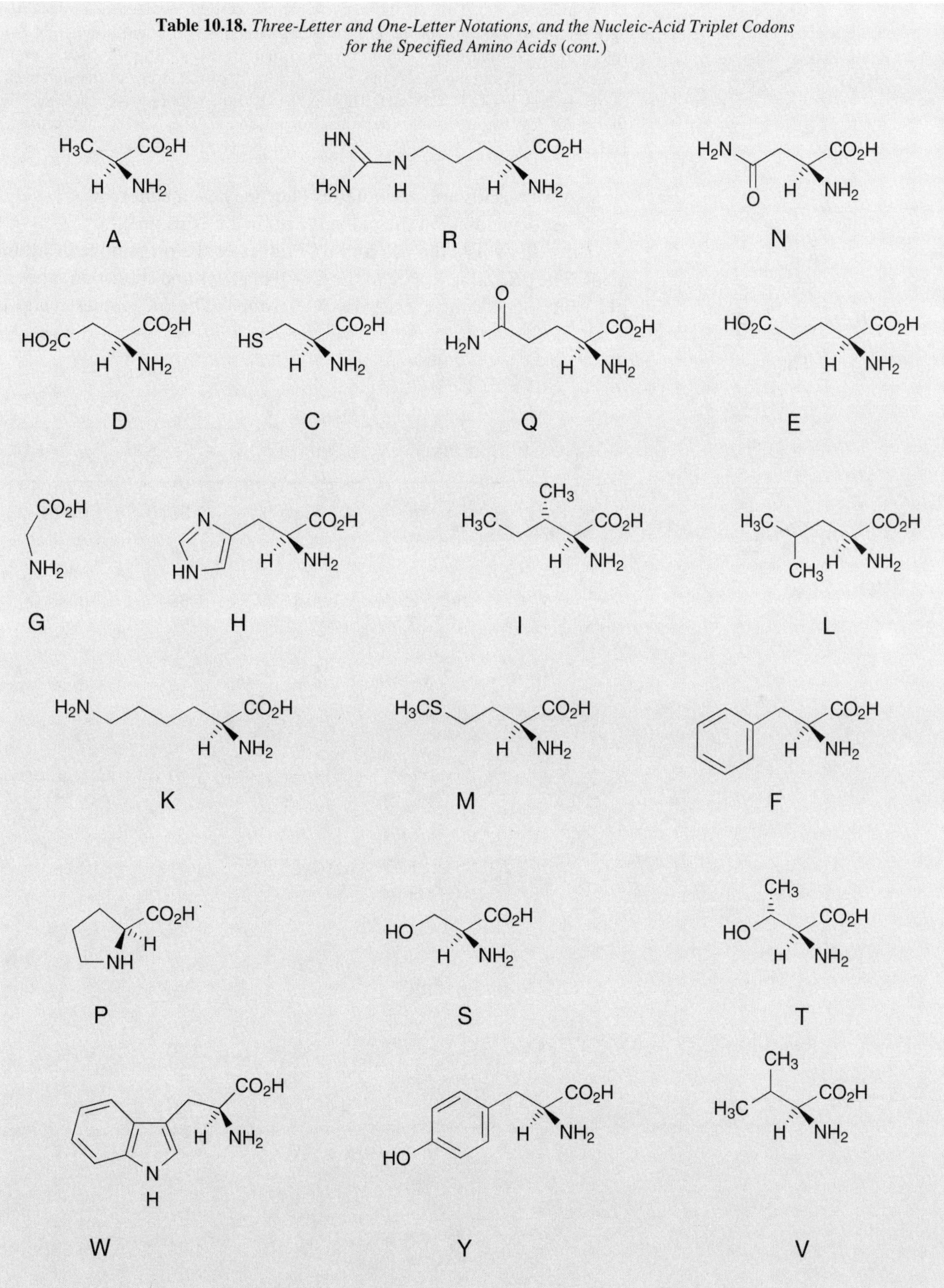

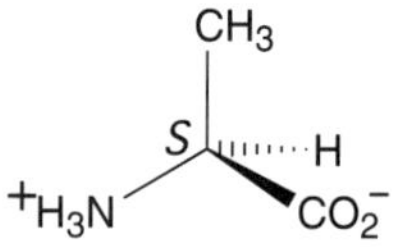

Ala

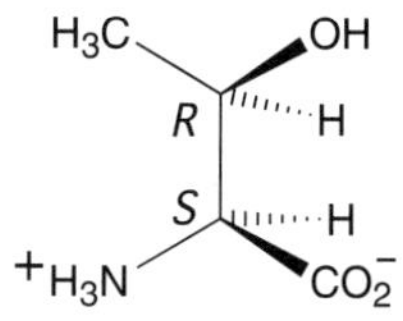

Thr

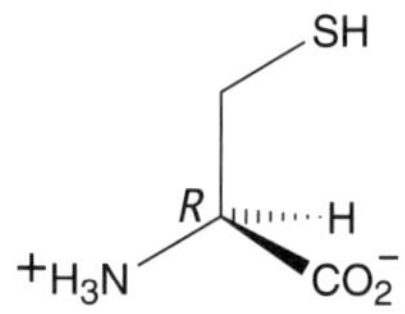

Cys

If an amino acid or its derivative contains several stereogenic centers, it is appropriate to make use of the *CIP* notation (*Sect. 10.2.1*). For transposition from D/L- into (*R*/*S*)-notation, it should be noted that, in the majority of cases, L-amino acids become (*S*)-amino acids: the order of priority of the atoms at the stereogenic center is $NH_2 > CO_2H > R > H$. However, L-cysteine becomes (*R*)-cysteine: here the order of priorities of the adjacent atoms at the stereogenic center is $NH_2 > CH_2SH > CO_2H > H$ ('artifact of the sequence rules'; *Sect. 10.2.3*).

α-Amino acids are amphoteric. They may each behave as an acid or as a base, depending on the pH value of their environment.

Table 10.19 lists the pK_1 and pK_2 values of the proteinogenic amino acids. The pK_1 values relate to the COOH groups and show that, above a pH value of 3.5, these exist as CO_2^- groups. The pK_2 values relate to

Table 10.19. *p*K *Values of the Proteinogenic Amino Acids*

Structure	Symbol	pK_1 C^{α}-COOH	pK_2 C^{α}-NH_3^+
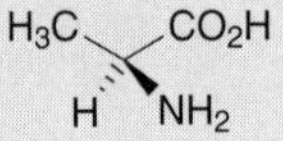 H_3C, CO_2H, H, NH_2	Ala	2.35	9.87
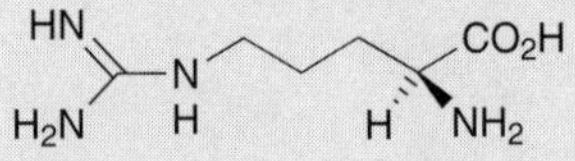 HN, H_2N, N, H, CO_2H, H, NH_2	Arg	1.82	8.99
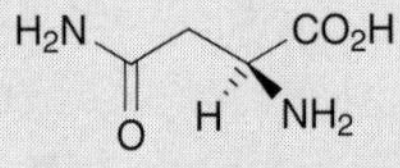 H_2N, O, CO_2H, H, NH_2	Asn	2.14	8.84
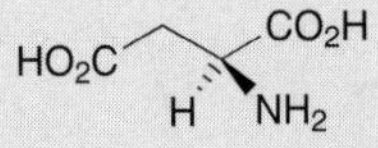 HO_2C, CO_2H, H, NH_2	Asp	1.99	9.90
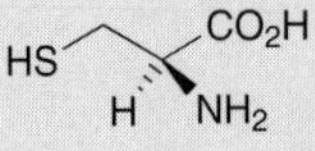 HS, CO_2H, H, NH_2	Cys	1.92	10.78
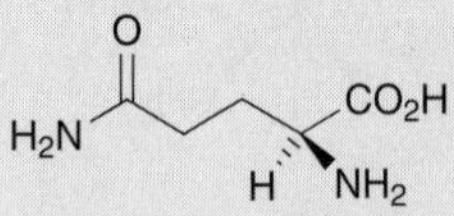 O, H_2N, CO_2H, H, NH_2	Gln	2.17	9.13
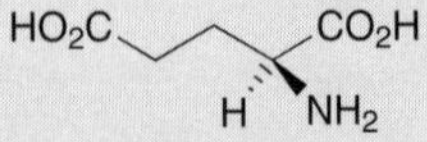 HO_2C, CO_2H, H, NH_2	Glu	2.10	9.47
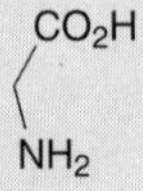 CO_2H, NH_2	Gly	2.35	9.78
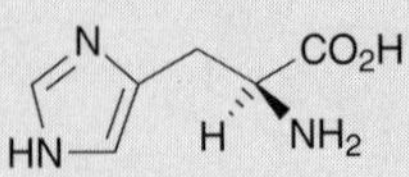 N, HN, CO_2H, H, NH_2	His	1.80	9.33

Table 10.19. p*K* *Values of the Proteinogenic Amino Acids* (*cont.*)

Structure	Symbol	pK_1 C^{α}-COOH	pK_2 C^{α}-NH_3^+
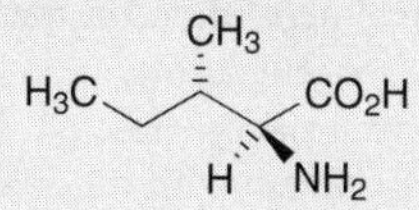	Ile	2.32	9.76
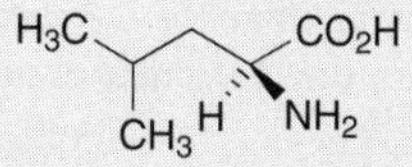	Leu	2.33	9.74
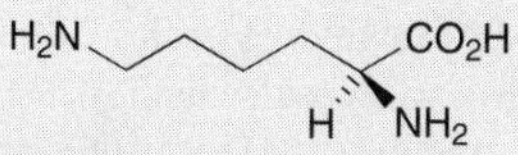	Lys	2.16	9.18
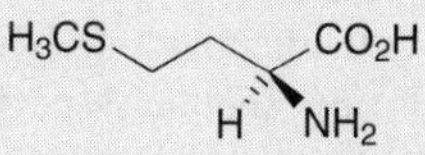	Met	2.13	9.28
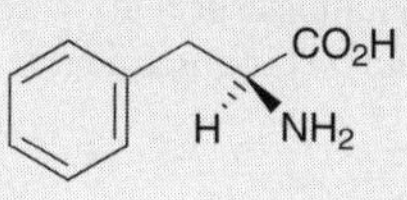	Phe	2.16	9.18
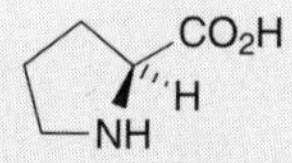	Pro	2.95	10.65
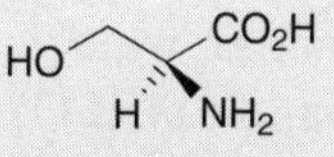	Ser	2.19	9.21
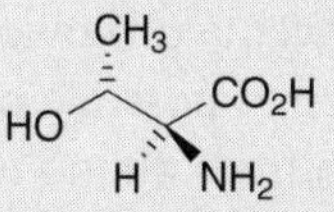	Thr	2.09	9.10
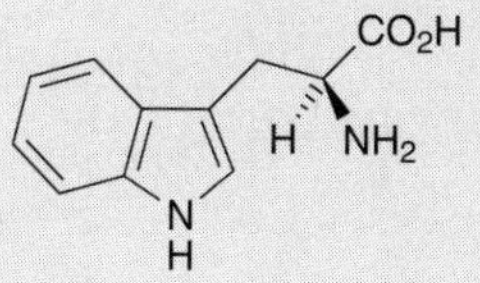	Trp	2.43	9.44
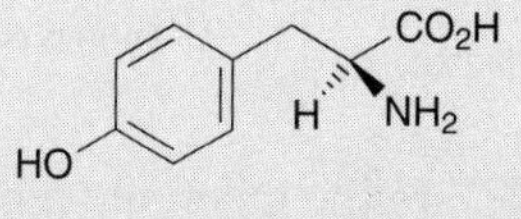	Tyr	2.20	9.11
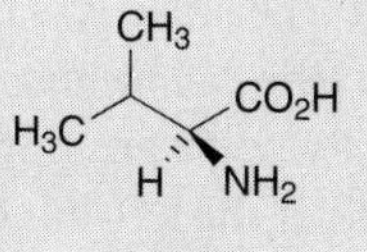	Val	2.29	9.74

the NH_2 groups and show that, below a pH value of 8.0, these exist as NH_3^+ groups. Although amino acids occur as zwitterions under normal conditions, they are, as a rule, depicted in formulae as the undissociated compounds.

Amino acids may be divided into different groups, depending on the nature of their side chains. Gly has neither a side chain nor a stereogenic center, and is thus the simplest amino acid. Ala, Val, Leu, and Ile each have an alkyl (methyl, isopropyl, isobutyl, or *sec*-butyl) side chain, which tends to avoid contact with water as much as possible (*hydrophobic property*). From this arises the impression that unpolar side chains are attracted to each other. Opinion is, however, that they are excluded by clusters of associated H_2O molecules. Discussion about the molecular basis of hydrophobicity is currently rather confused.

Phe, Trp, and Tyr possess a benzyl, (indol-3-yl)methyl or a 4-hydroxybenzyl side chain, respectively, and so also belong to α-amino acids with *hydrophobic side chains*, as do Pro and the two sulfur-containing representatives, Met and Cys, with CH_3S and SH group, respectively. Ser and Thr each have a hydroxyalkyl side chain, while Asn and Gln have a carbamoylalkyl side chain: these four constitute the family of α-amino acids with polar, but uncharged side chains. To the family of α-amino acids with positively charged side chains at neutral pH belong Lys (with aminobutyl), Arg (with guanidinopropyl), and His (with imidazolylmethyl side chain). Asp (with carboxymethyl) and Glu (with carboxyethyl side chain) belong to the family of α-amino acids with negatively charged side chains. Polar side chains, especially when charged, have the tendency to become solvated by H_2O, and are consequently known as *hydrophilic side chains*.

10.5.4.2. The Local Peptide Unit

α-Amino acids – their L-enantiomers, as a rule – are the monomeric building blocks for the regular head-to-tail polymers known as polypeptides or proteins. Peptides are, therefore, amides, which can arise, for example, through condensation (with elimination of H_2O) between the NH_2 group of one amino acid and the COOH group of another one. The iteration of the condensation process for the twenty proteinogenic amino acids leads to an unimaginably huge number of possible polypeptides. Hence, a protein consisting of four (one hundred) amino acids already has $20^4 = 160{,}000$ ($20^{100} \approx 10^{130}$) possible sequences.

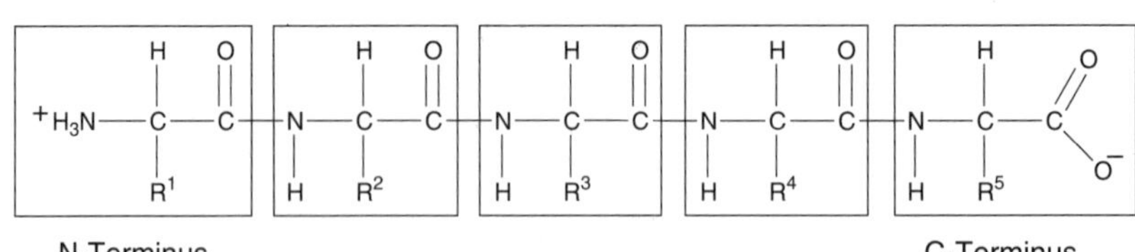

The *amino-acid residues* already mentioned are the indefinite repeating units of proteins: indefinite repeating units, because R is unspecified. Each protein has a free NH_2 group at one end (the *N-terminus*) and a free COOH group at the other (the *C-terminus*). Proteins are normally written in a way that the N-terminus is to be found at the left-hand end of the horizontally arranged chain.

To denote peptides simply, the three-letter notation (*Table 10.18*) of the constituent amino acids is used. Peptides with a bond between C(1) and N(2) (such as *N*-(L-α-glutamyl)glycine) require no special explanation.

Peptides with a bond other than between C(1) and N(2) (such as *N*-(L-γ-glutamyl)glycine) require special terminology. (The acyl residues of glutamic acid are assigned the letters α or γ, depending on whether the C(1)O_2H or the C(5)O_2H group is involved.)

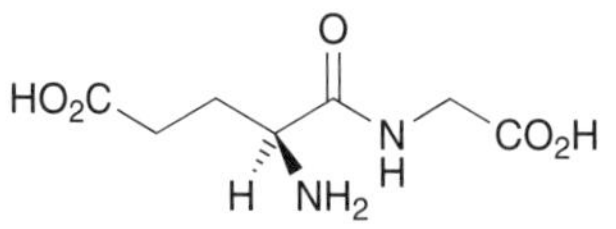

N-(L-α-Glutamyl)glycine Glu-Gly

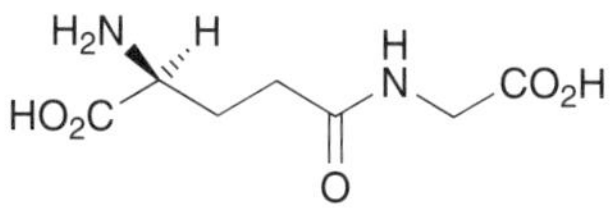

N-(L-γ-Glutamyl)glycine

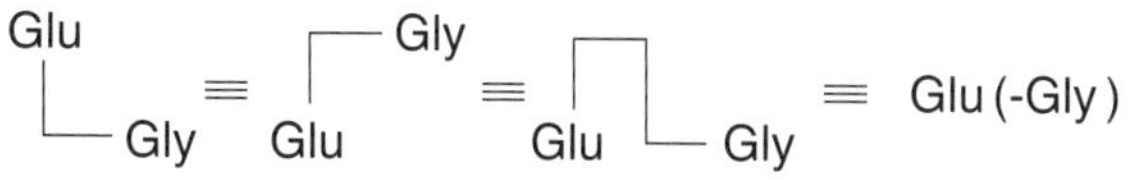

Glu(-Gly)

If, in a particular context, it is necessary to refer to a peptide with a reversed sequence to one already described (with name or formula representation, and the synonymous formula number (**1**, say)), then the formula number, in conjunction with the descriptor *retro*, may be used. Similarly, the enantiomer of a peptide already referred to may be denoted with the prefix *ent* (see *Sect. 10.2.6*).

Ala-Lys-Glu-Tyr-Leu
lupaciubin (**1**)

retro-**1**

ent-**1**

For cyclic peptides (cyclopeptides) it is necessary to pay attention to the correct formulation of the sequence. This is demonstrated using three representations of the peptide antibiotic *gramicidin S*.

cyclo(-Val-Orn-Leu-D-Phe-Pro-Val-Orn-Leu-D-Phe-Pro-)

Val-Orn-Leu-D-Phe-Pro-Val-Orn-Leu-D-Phe-Pro

→Val→Orn→Leu→D-Phe→Pro
Pro←D-Phe←Leu←Orn←Val←

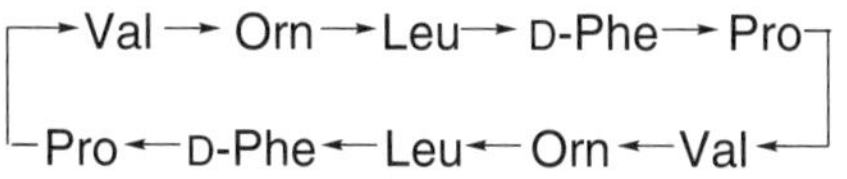

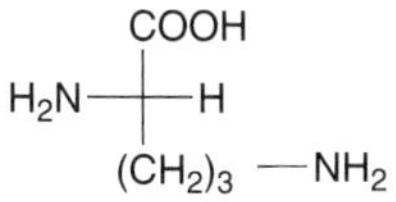

Orn = (+)-L-Ornithine

The amide bond in peptides is also known as the *peptide bond.* To ascribe importance to the peptide bond, the term *peptide unit* (–CHR–CO–NH–) is used in preference to that one of amino-acid residue (–NH–CHR–CO–). The peptide unit is a structural element which is encountered in all polypeptides, and which gives a certain rigidity to the polypeptide backbone. The atoms C^{α}_{i-1}, C_{i-1}, O_{i-1}, H_i, N_i, and C^{α}_i all lie almost in one plane (*Fig. 10.25*).

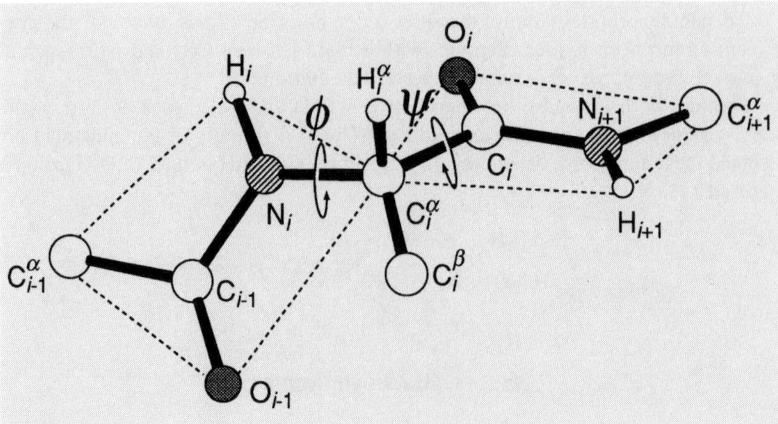

Fig. 10.25. *The determining partial structure in polypeptides*

Despite the regular recurrence of six neighboring atoms arranged in a planar manner, the backbone of polypeptides is not rigid. The C^α center plays the role of an element of mobility: all the bonds connected to it are single bonds and function as rotation axes. Rotations about the C^α–N and C^α–C(=O) bonds may be described by the two torsion angles: ϕ for C(=O)–N–C^α–C(=O) and ψ for N–C^α–C(=O)–N for each conformational repeating unit. If, finally, it is desired also to clarify whether the *sp*- or the *ap*-amide conformation is being discussed, then the torsion angle ω for C^α–C(=O)–N–C^α, with values of either *ca.* 0° or 180°, may also be given. To compare the topologies of several polypeptides with one another, the so-called (ϕ,ψ) diagrams (*Ramachandran* diagrams) are frequently used. *Fig. 10.26* shows the (ϕ,ψ) diagram for the amino-acid residue of Gly in oligopeptides.

It may be seen that the conformational space available to the unsubstituted Gly fragment is very large. This is the case at one end of the series of the proteinogenic amino acids. At the other end – the amino-acid residue here is fixed in the cramped torsion-angle range $\phi = -60° \pm 20°$ – lies the *N*-substituted amino acid Pro. All the other proteinogenic amino acids lie between the two extremes of Gly and Pro. *Fig. 10.27* gives the *Ramachandran* diagram showing the considerably restricted, relative to Gly, conformational spaces for the residues of all of the proteinogenic amino acids exept Gly and Pro.

The literature is divided concerning the numerical values of ϕ and ψ. According to the *IUPAC* recommendations of 1974, the idealized linear polypeptide backbone is characterized by the torsion angles $\phi = \psi = \omega = 180°$. The center of *Ramachandran* diagrams is accordingly defined at $\phi = \psi = 0°$.

The rotational barrier about the C(=O)–N bond is higher than would be expected for a C–N bond in general, amounting to *ca.* 80 kJ/mol. Of the two planar conformations with a formal C=N bond, the *sp*-oriented amide is thermodynamically favored over its *ap*-counterpart by *ca.* 8.5 kJ/mol.

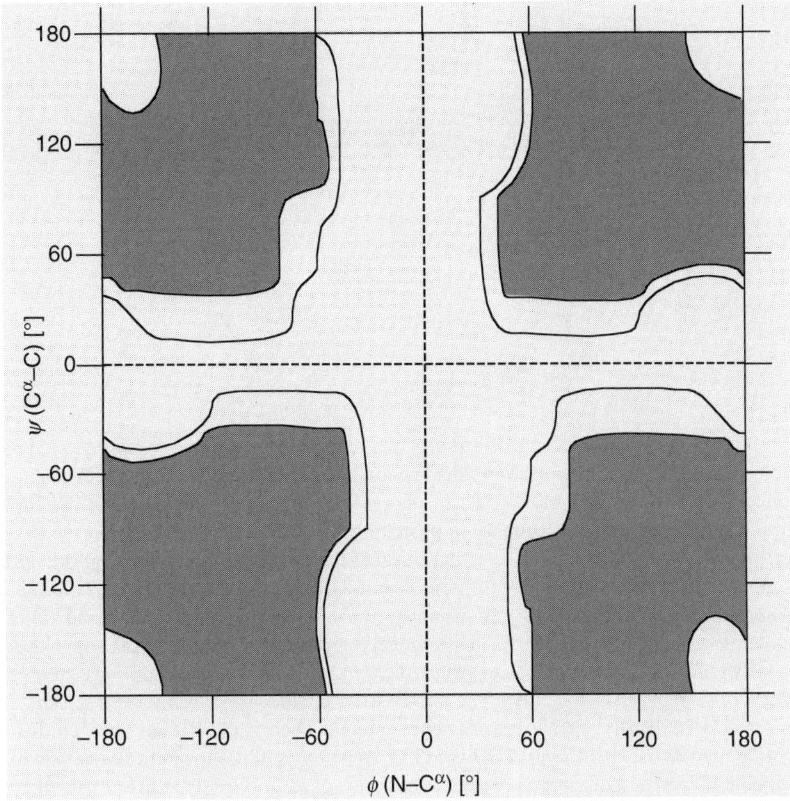

Fig. 10.26. Ramachandran *diagram for the amino-acid residue of Gly in oligopeptides indicating the distinct likelihood of certain combinations of* ϕ *and* ψ*:* ϕ/ψ with no adverse non-bonding interactions are shaded dark-gray, those ones with minor steric crowding are shaded light-gray.

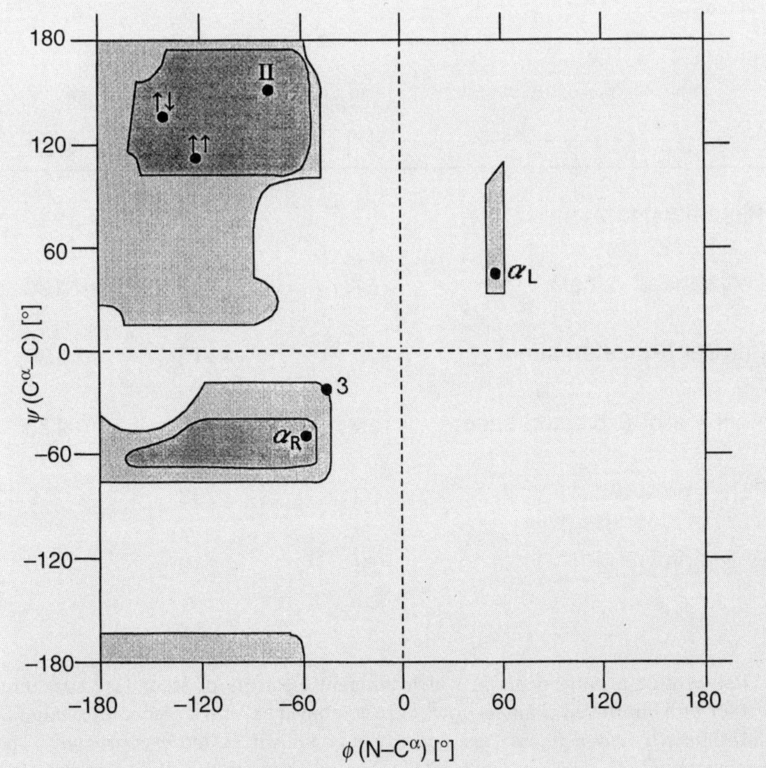

Fig. 10.27. Ramachandran *diagram for the proteinogenic amino acids other than Gly and Pro.* α_R: right-handed α-helix; α_L: left-handed α-helix; 3: 3_{10}-helix; II: left-handed poly(L-proline) II-helix; ↑↑: parallel and ↑↓: antiparallel β-pleated sheet.

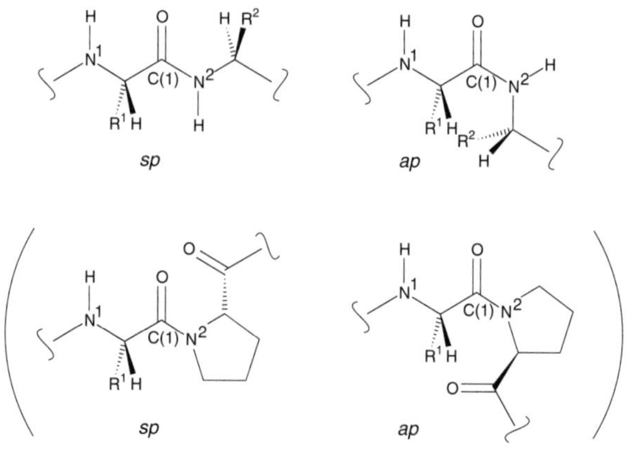

Biologists use the concepts of primary and secondary structure for regular macromolecules with nucleotide or amino-acid residues as repeating units. For chemists, the concept of *primary structure* consists of a series of clearly distinct aspects. For proteins, for example, the sequence of proteinogenic amino-acid residues represents a *constitutional* component, the L-configuration of the chirality centers a *configurational* component, and the planarity of the peptide bond (*Chapt. 8.1.1* and *8.4.1*) a *conformational* component. Biologists and chemists understand essentially the same thing under the concept of *secondary structure* : local regions of a given conformation within linearly constituted macromolecules, in which, in addition to non-repetitive elements (like loops), repetitive elements of secondary structure (such as *α-helices* and *β-pleated sheets*) are frequently and successively encountered. Their study is a matter of conformational analysis, and *Chapt. 7.3.3* contains an analysis of conformational details of proteins, as well as a discussion of diverse elements of secondary structure. *Table 10.20* gives values of characteristic torsion angles for typical elements of regular secondary structure.

Table 10.20. *Torsion Angles for Typical Structure Elements of Proteins*

	ϕ [°]	ψ [°]	ω [°]
Right-handed α-helix	–57	–47	180
Left-handed α-helix	+57	+47	180
Parallel β-pleated sheet	–119	+113	180
Antiparallel β-pleated sheet	–139	+135	–178
Poly(L-proline) I	–83	+158	0
Poly(L-proline) II	–78	+149	180

Polypeptide arrangements in which different elements of secondary structure, together with unordered chain sections, have combined to form larger conformational regions (*motifs* and *domains*) are consequently known as *tertiary structure*. The (molecular) *tertiary structure* represents the spatial orientation of the complete set of atoms in an individual polypeptide chain, including side chains, or of polypeptide chains which are covalently bound to one another. Finally, a (supramolecular) *quaternary structure* must be taken into account for those proteins which consist of subunits bound together in a non-covalent fashion. For structure-activity relationships, it is necessary to know the tertiary and quaternary structure of a protein molecule or protein supermolecule, respectively.

Further Reading

General

R. S. Cahn, C. Ingold, V. Prelog, *Specification of Molecular Chirality, Angew. Chem. Int. Ed.* **1966,** *5,* 385.

V. Prelog, G. Helmchen, *Basic Principles of the CIP-System and Proposals for a Revision, Angew. Chem. Int. Ed.* **1982,** *21,* 567.

V. Prelog, M. Kovacevi'c, M. Egli, *Lipophilic Tartaric Acid Esters as Enantioselective Ionophores, Angew. Chem. Int. Ed.* **1989,** *28,* 1147.

J. Rigaudy, S.P. Klesney, *IUPAC Nomenclature of Organic Chemistry*, Pergamon Press, Oxford, 1979.

M.V. Kısakürek (Ed.), *Organic Chemistry: Its Language and Its State of the Art,* Verlag Helvetica Chimica Acta, Basel, 1993.

G. Giese, *Beilstein's Index,* Springer-Verlag, Berlin, 1986.

Beilstein Handbook of Organic Chemistry: Prefix List, Springer-Verlag, Berlin.

Beilstein's Handbuch der Organischen Chemie: Stereochemische Bezeichnungsweisen, Springer-Verlag, Berlin.

To Sect. 10.1.

J. Lederberg, G. L. Sutherland, B. G. Buchanan, E. A. Feigenbaum, A. V. Robertson, A. M. Duffield, C. Djerassi, *Applications of Artificial Intelligence for Chemical Inference. The Number of Possible Organic Compounds, J. Am. Chem. Soc.* **1969,** *91,* 2973.

N. Trinajstic, Z. Jericevic, J. V. Knop, W. R. Müller, K. Szymanski, *Computer Generation of Isomeric Structures, Pure Appl. Chem.* **1983,** *55,* 379.

F. Harary, *Graph Theory,* Addison-Wesley Publ. Comp., Reading, 1972.

A.T. Balaban, *Chemical Applications of Graph Theory,* Academic Press, London, 1976.

L. Wittgenstein, *Philosophische Untersuchungen* in Werkausgabe Bd. 1 and *Eine Philosophische Betrachtung* in Werkausgabe Bd. 5, Suhrkamp Verlag, Frankfurt am Main, 1989.

P.-L. Luisi, R.M. Thomas, *The Pictographic Molecular Paradigm, Naturwissensch.* **1990,** *77,* 67.

H. Torczyner, *Magritte: The True Art of Painting,* Abradale Press, New York, 1985.

To Sect. 10.2.

R.S. Cahn, C. Ingold, V. Prelog, *Specification of Molecular Chirality, Angew.Chem. Int. Ed.* **1966,** *5,* 385.

V. Prelog, G. Helmchen, *Basic Principles of the CIP-System and Proposals for a Revision, Angew. Chem. Int. Ed.* **1982,** *21,* 567.

To Sect. 10.3.

10.3.2.

W. Klyne, V. Prelog, *Description of Steric Relationships Across Single Bonds, Experientia* **1960,** *16,* 521.

R. Bucourt, *The Torsion Angle Concept in Conformational Analysis, Topics in Stereochemistry* **1974,** *8,* 159.

10.3.4.

J. C. Speakman, *The Dissociation Constants and Stereochemistry of Some Stereoisomeric Dibasic Acids, J. Chem. Soc.* **1941,** 490.
J. Dunitz, V. Prelog, *Röntgenographisch bestimmte Konformationen und Reaktivität mittlerer Ringe, Angew. Chem.* **1960,** *72,* 896.
V. Prelog, *Specification of the Stereospecificity of Some Oxido-Reductases by Diamond Lattice Sections, Pure Appl. Chem.* **1964,** *9,* 119.
V. Prelog, M. Kovacević, M. Egli, *Lipophilic Tartaric Acid Esters as Enantioselective Ionophores, Angew. Chem. Int. Ed.* **1989,** *28,* 1147.
M. Saunders, *Medium and Large Rings Superimposable with the Diamond Lattice, Tetrahedron* **1967,** *23,* 2105.

To Sect. 10.5.

10.5.1.

L. Fieser, M. Fieser, *Steroids,* Reinhold, New York, 1959.
'Nomenclature of Steroids', Pure Appl. Chem. **1972,** *31,* 285; **1989,** *61,* 1783.
N. S. Bhaca, D. H. Williams, *Applications of NMR-Spectroscopy in Organic Chemistry,* Holden-Day, Inc., San Francisco, 1964.
P. Crabbè, *Optical Rotatory Dispersion and Circular Dichroism in Organic Chemistry,* Holden-Day, Inc., San Francisco, 1965.
Atlas of Steroid Structure, Vol. 1, 1975 (Ed. W. L. Duax, D. A. Norton), Vol. 2, 1984 (Ed. J. F. Griffin, W. L. Duax, C. M. Weeks), IFI/Plenum, New York.
H. Budzikiewicz, C. Djerassi, D. H. Williams, *Mass Spectrometry of Organic Compounds,* Holden-Day, Inc., San Francisco, 1967.
E. Heftmann (Ed.), *Modern Methods of Steroid Analysis,* Academic Press, New York, 1973.

10.5.2.

W. Saenger, *Principles of Nucleic Acid Structure,* Springer, New York, 1984.
'IUPAC-IUB Rules on Abbreviations and Symbols for Nucleic Acids, Polynucleotides and their Constituents', Pure Appl. Chem. **1974,** *40,* 277.
'IUPAC-IUB Rules on Abbreviations and Symbols for the Description of Conformations of Polynucleotide Chains', Pure Appl. Chem. **1983,** *55,* 1273.
S. Diekmann, *Definitions and nomenclature of nucleic acid structure parameters, The EMBO Journal* **1989,** *8,* 1.

10.5.3.

J. F. Kennedy (Ed.), *Carbohydrate Chemistry,* Clarendon Press, Oxford, 1988.
J. F. Stoddart, *Stereochemistry of Carbohydrates,* Wiley-Interscience, New York, 1971.
W. Pigman, D. Horton (Ed.), *The Carbohydrates,* 2nd Edn., Four Volumes, Academic Press, New York, 1972.
'Tentative Rules for Carbohydrate Nomenclature, Part I, 1969', Biochemistry **1971,** *10,* 3983, 4995.
'Nomenclature of Unsaturated Monosaccharides (Provisional)', Pure Appl. Chem. **1982,** *54,* 207.
'Nomenclature of Branched-Chain Monosaccharides (Provisional)', Pure Appl. Chem. **1982,** *54,* 211.
'Conformational Nomenclature for Five- and Six-Membered Ring Forms of Monosaccharides and Their Derivatives (Provisional)', Pure Appl. Chem. **1981,** *53,* 1901.
'Polysaccharide Nomenclature (Provisional)', Pure Appl. Chem. **1982,** *54,* 1523.
'Abbreviated Terminology of Oligosaccharide Chains (Provisional)', Pure Appl. Chem. **1982,** *54,* 1517.

'Symbols for Specifying the Conformation of Polysaccharide Chains (Provisional)', Pure Appl. Chem. **1983,** *55,* 1269.

S.H. DeWitt, J.S. Kiely, C.J. Stankovic, M.C. Schroeder, D.M. Reynolds Cody, M.R. Pavia, *'Diversomers': An approach to nonpeptide, nonoligomeric chemical diversity, Proc. Natl. Acad. Sci. USA* **1993,** *90,* 6909.

10.5.4.

R.E. Dickerson, I. Geis, *The Structure and Action of Proteins*, Benjamin/Cummings, Menlo Park, CA, 1969.

G.E. Schulz, R.H. Schirmer, *Principles of Protein Structure,* Springer, New York, 1979.

C. Levinthal, *Molecular Model-building by Computer, Scientific American* **1966,** *214,* 42.

'IUPAC-IUB Nomenclature and Symbolism for Amino Acids and Peptides', Pure Appl. Chem. **1984,** *56,* 596.

'IUPAC-IUB Abbreviations and Symbols for Description of Conformation of Polypeptide Chains', Pure Appl. Chem. **1974,** *40,* 293.

10.5.4.1.

Aminosäure-Bausteine des Lebens, Collection of transparencies of the Fonds der Chemischen Industrie, Frankfurt am Main, 1993.

10.5.4.2.

R.B. Corey, L. Pauling, *Fundamental Dimensions of Polypeptide Chains, Proc. Roy. Soc.* **1953,** *B141,* 10.

G.N. Ramachandran, V. Sasisekharan, *Conformation of Polypeptides and Proteins, Advances in Protein Chemistry* **1968,** *23,* 284.

11 Symmetry Point Groups and Space Groups

Symmetry is a common quality. Its presence is often associated with harmony and aesthetics. One needs only consider the rotational symmetry in flowers or the mirror symmetry in animals and humans (the right and left hands, for example). In chemistry, many phenomena may be reduced to a common denominator by consideration of symmetry. One reason for the popularity of symmetry-based analyses is that it requires no complicated mathematical techniques to use a symmetry argument. In previous chapters, symmetry arguments have found application in various contexts. The fundamental principles and rules upon which symmetry analysis is based are explained in this chapter.

11.1. Phenomenological Description of Molecular Symmetry

11.1.1. Symmetry Elements and Symmetry Operations

A *symmetry operation* is a geometric operation, the application of which, to a particular object, produces a spatial orientation of that object indistinguishable from the original orientation. For example (see next page): a rotation of 180° about an axis bisecting the bond angle H–O–H in the H_2O molecule leads to an exchange of positions of the two H-atoms H_a and H_b. As, in reality, these two atoms may not be distinguished from one another, the orientation resulting from this rotation is equivalent to the original. A further 180° rotation leads to an orientation identical with the original.

Upon rotation about an axis, all points on the object under observation change their positions, with the exception of those ones lying on the axis themselves. The set of points which do not change their positions during the course of a symmetry operation is known as the *symmetry element* associated with this operation. The symmetry-related properties of an object (a molecule, for example, or, better, a molecular model) may be described with the aid of a set of symmetry operations. Symmetry operations to be considered are: *rotation*, *reflection*, *inversion*, *rotary reflection*, and *rotary inversion*. These operations are briefly introduced below. It will be seen that, of these, a mere two symmetry operations

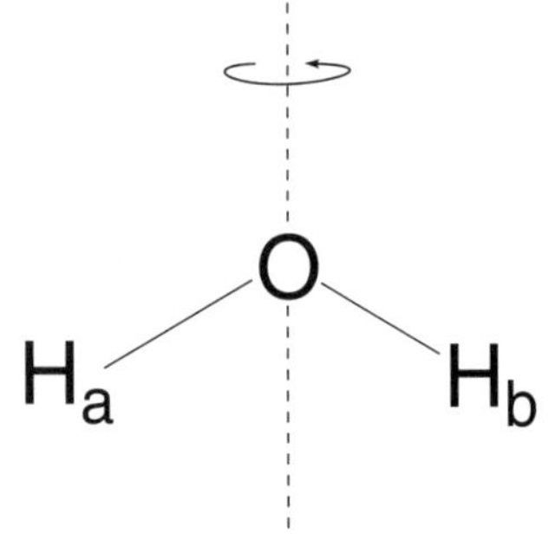

(*rotation* and *rotary reflection*) are sufficient to describe all of the symmetry properties of a molecule. Common to all symmetry operations mentioned is the fact that the position of the geometric center of the molecule does not change: all symmetry elements intersect this center.

11.1.1.1. The Identity Operation *E*

The *identity operation E* leaves the molecule unchanged. This symmetry operation is necessary to be able to mathematically express, say, the result of a 180° rotation carried out twice, and to describe, with the aid of group theory, that the result of a series of symmetry operations is itself also a symmetry operation.

11.1.1.2. The Rotation C_n

This symmetry operation describes a rotation of the angle $\theta = 360°/n$ about an n-fold rotation axis. In a rotation of 180° about a two-fold rotation axis, coinciding with the z axis of a *Cartesian coordinate system*, the coordinates of all points change as follows:

$$x, y, z \rightarrow -x, -y, z.$$

The rotation C_1, in which the angle of rotation is 360°, is identical with the identity operation E. If $n > 2$, then the sense of rotation must be taken into account, as it is then no longer trivial whether the rotation proceeds clockwise or anticlockwise. Examples of rotation axes may be found in the molecules cyclopropane, ethene, and ethyne (*Fig. 11.1*). The cyclopropane molecule (top) possesses a C_3 axis perpendicular to the molecular plane and three C_2 axes, each intersecting with one C-atom and bisecting one C−C bond. In the molecular model of ethene (center), three mutually perpendicular C_2 axes may be identified. Linear ethyne (bottom) displays a C_∞ axis and an infinite number of C_2 axes, all perpendicular to the former. All linear molecules possess a C_∞ axis, as they may undergo infinitely small rotations without changing orientation.

11.1.1.3. The Reflection σ

This symmetry operation occurs only in achiral molecules. The corresponding symmetry element (the mirror plane) bisects the molecule. All of the examples given until now in this chapter possess mirror symmetry. A mirror plane perpendicular to the z axis causes the following change in coordinates:

$$x, y, z \rightarrow x, y, -z.$$

If the mirror plane is perpendicular to the highest-order axis of rotation (usually assumed to be vertical), then it is assigned the descrip-

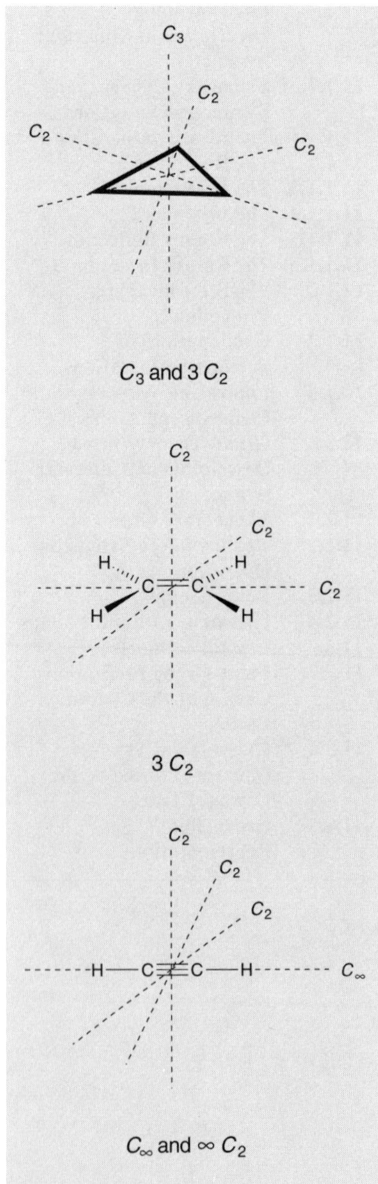

Fig. 11.1. *Examples of rotation axes*

tor σ_h (horizontal). If, on the other hand, the mirror plane includes the rotation axis, it is then known as σ_v (vertical); linear molecules possess an infinite number of mirror planes of this type.

11.1.1.4. The Inversion *i*

Inversion through an inversion center (the only point to remain unchanged by this operation) likewise is only possible in achiral molecules and leads to the following coordinate change:

$$x, y, z \rightarrow -x, -y, -z.$$

Of the examples previously described in this chapter, ethene and ethyne both possess an inversion center; however, cyclopropane and H_2O do not.

11.1.1.5. The Rotary Reflection S_n

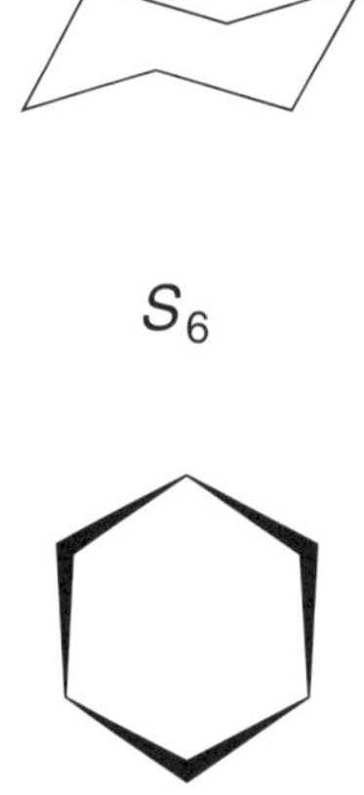

This symmetry operation consists of two suboperations combined: a rotation of $360°/n$ and a reflection across a plane perpendicular to the rotation axis. The two symmetry elements, C_n and σ_h, do not have to be present individually. Cyclohexane in the chair conformation contains an S_6 axis, CH_4 three S_4 axes; the S_6 or S_4 axes are harder to recognize at first glance than the C_3 or C_2 axes resulting from them.

11.1.1.6. The Rotary Inversion $S_{\bar{n}}$

This, similarly, is a combination of two suboperations: a rotation of $360°/n$ and an inversion. The two symmetry elements, C_n and i, need not be individually present in this case either. The inclusion of a reflection or an inversion dictates that rotary reflection or rotary inversion does not occur in chiral molecules.

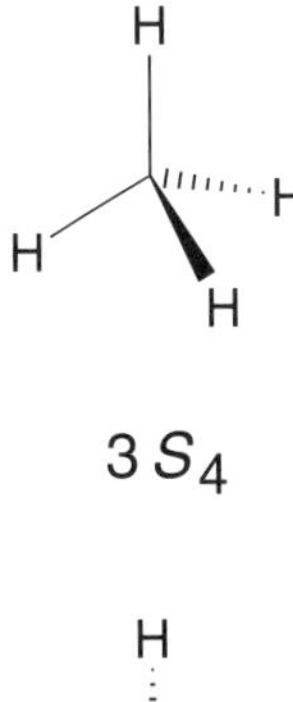

11.1.2. The Stereographic Projection

The stereographic projection has proven itself as an aid for the illustration, in two-dimensional form, of the entire set of symmetry elements present in a molecule. The molecule is placed inside an imaginary hollow sphere, the geometric center of the molecule coinciding with that one of the sphere. All symmetry elements, therefore, intersect the center of the sphere. The points or lines of intersection of the symmetry elements with the surface of the sphere are projected onto the equator plane, which is located perpendicularly to the highest-order axis of rotation (*Fig. 11.2*). In this process:

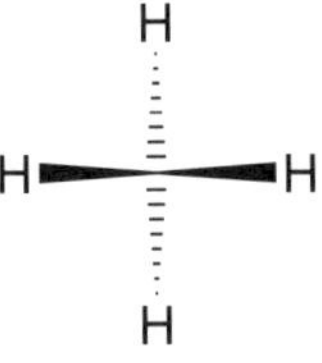

– A mirror plane σ_h must intersect the surface of the sphere at the equator. In such a case, the equator is emphasized in heavy print.
– A mirror plane σ_v becomes a line in the projection, a skewed mirror plane becomes an ellipse.
– An n-fold axis perpendicular to the equator plane intersects the sphere at its north and south pole, and is represented as a small n-sided regular polygon at the center of the equator plane. This polygon is drawn solid to denote a rotation axis, open to denote a rotary-reflection axis. A C_2-axis is characterized using the symbol ⬮.
– An n-fold axis in the equator plane intersects the sphere at two diagonally opposite points, which are marked with the corresponding n-sided polygon and connected to each other with a dashed line. This line is drawn solid, if the axis is part of a σ_v plane.

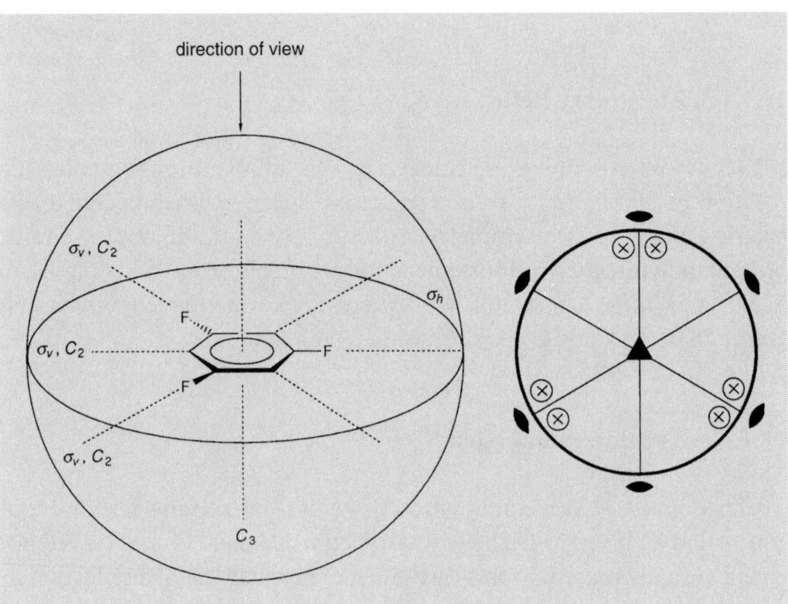

Fig. 11.2. *The stereographic projection of the symmetry elements of 1,3,5-trifluorobenzene* (right) *and its construction* (left)

A random point, which does not lie on a symmetry element, is now selected and the symmetry operations corresponding to the various symmetry elements carried out on it until the starting point is reached once more. So that information relating to the third dimension, lost in the process of constructing the projection, may at least be approximately retrieved, all points situated in the 'northern' hemisphere, relative to the observer, are marked with a cross (×), and all points in the 'southern' hemisphere with a circle (○). The pattern of crosses and circles which arises is characteristic of the combination of symmetry elements present.

Fig. 11.3 shows the stereographic projections resulting from the presence of axes of rotary reflection of order S_1, S_2, S_3, S_4, S_5, or S_6. The following rules may be discerned:

- S_1 generates the same point pattern as a reflection σ_h.
- S_2 generates the same pattern as an inversion i. This fact means that *any* axis through the inversion center constitutes an S_2 axis, and affords the opportunity of representing the inversion center, which does not intersect the surface of the sphere at any point, and which, therefore, would otherwise not appear in the stereographic projection, as an S_2 axis perpendicular to the equator plane.
- S_6 generates the same pattern as a combination of C_3 and an inversion i.
- S_3 or S_5 generates the same pattern as a combination of C_3 or C_5 with a reflection σ_h. This observation may be generalized: a rotary-reflection axis S_n, n = odd number, is nothing else than a combination of C_n and σ_h, and, therefore, does not need to be considered separately.

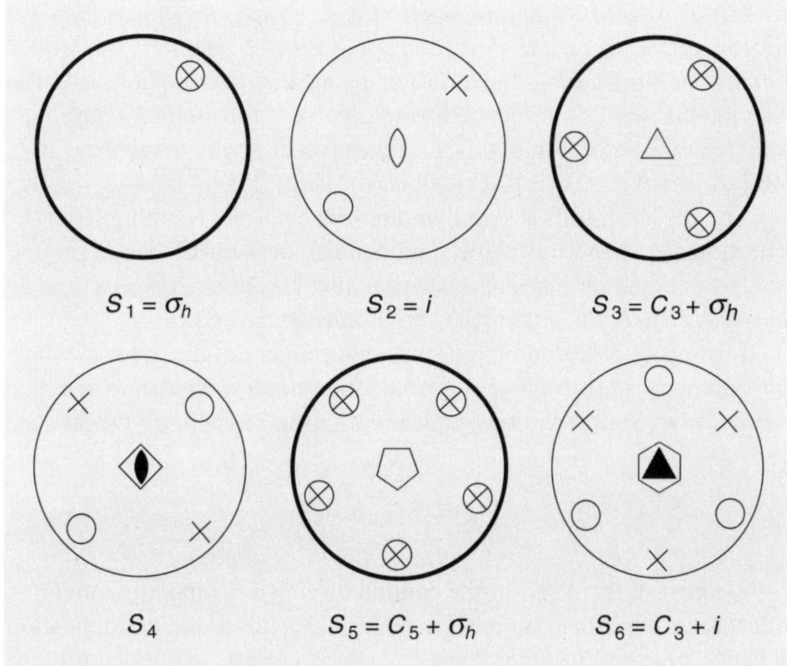

Fig. 11.3. *The stereographic projections resulting from the presence of rotary-reflection axes* S_n *(n = 1–6)*

Furthermore, it may be seen from *Fig. 11.3* that the point patterns shown may be generated just as well by rotary inversions. The following relationship applies:

$$S_{\bar{1}} \triangleq S_2$$
$$S_{\bar{2}} \triangleq S_1$$
$$S_{\bar{3}} \triangleq S_6$$
$$S_{\bar{4}} \triangleq S_4$$
$$S_{\bar{5}} \triangleq S_{10}$$
$$S_{\bar{6}} \triangleq S_3$$

We have, therefore, ascertained that:
- rotary inversion is equivalent to rotary reflection,
- σ is equivalent to S_1,
- i is equivalent to S_2.

The consequences of this are far-reaching: all of the symmetry operations discussed up to now may be described in terms of the rotation C_n (*proper rotation*) and the rotary reflection S_n (*improper rotation*).

11.1.3. Combinations of Symmetry Operations

We have already seen that the repeated application of particular symmetry operations includes other symmetry operations automatically. For example, C_4 and S_4, after two-fold application, correspond to a C_2 operation. S_3 implies C_3 and σ_h, and C_6 corresponds to C_2 or C_3 (after three- or two-fold application). Beyond these simple cases, the combination of symmetry operations often requires the existence of further, equivalent symmetry operations. If a symmetry operation X transposes a point with coordinates x_1, y_1, z_1 into x_2, y_2, z_2, and then another symmetry operation Y transposes the new point x_2, y_2, z_2 into x_3, y_3, z_3, then there must exist a third symmetry operation Z, which transposes x_1, y_1, z_1 directly into x_3, y_3, z_3. The sequential application of symmetry operations is called 'multiplication', the 'product' being read from right to left. The example just discussed may be formulated very concisely: $Y \circ X = Z$. A concrete case is the combination of an inversion with a subsequent C_2 rotation about the z axis $[C_2(z) \circ i]$: $C_2(z) \circ i\,[x, y, z] = C_2(z)\,[-x, -y, -z] = x, y, -z$. This result is synonymous with that one resulting from the reflection across a mirror plane σ_h perpendicular to the C_2 axis: $\sigma_h\,[x, y, z] = x, y, -z$. A mirror plane of this type is always generated if, as well as an inversion center, a C_n axis with even-numbered n exists.

If the result of the combination is independent of the order in which the symmetry operations are applied, then the multiplication is called *commutative*. Multiplication is indeed commutative in the above case, as:

$$C_2 \circ i = i \circ C_2 = \sigma_h$$
$$C_2 \circ \sigma_h = \sigma_h \circ C_2 = i$$
$$\sigma_h \circ i = i \circ \sigma_h = C_2$$

In contrast, the result of the combination of a C_3 operation and a σ_v reflection is dependent on the sequence of execution; the multiplication is, hence, not commutative. The combination of a C_3 axis and a σ_v plane generates two further vertical mirror planes σ_v' and σ_v''. Commencing from a point (×), selected at random, a C_3 operation, followed by a reflection across σ_v, leads to a position directly accessible through reflection of the initial point across σ_v'' (*Fig. 11.4*, top). Reversal of the sequence leads to another point directly accessible through reflection of the initial point across σ_v' (*Fig. 11.4*, bottom). Such a finding is valid for the majority of combinations. Only the following symmetry operations always commute:

- two sequential rotations about the same axis,
- two C_2 operations about perpendicular axes,
- two reflections in perpendicular mirror planes,
- C_n and σ_h,
- an inversion and any rotation or reflection.

Two mirror planes, σ_1 and σ_2, intersecting at any angle φ, transpose an object, *via* double reflection, into a position directly accessible from the starting position through a rotation C_n (with $n = 180°/\varphi$). The relevant rotation axis coincides with the line of intersection of the two mirror

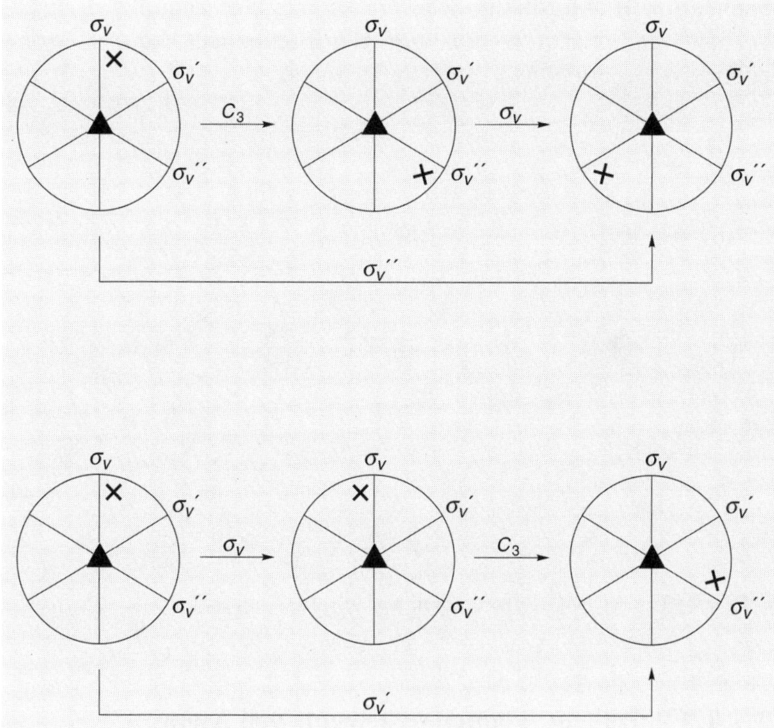

Fig. 11.4. *Combination of a C_3 operation and a σ_V reflection*

planes (*Fig. 11.5*). The sequence of the reflections determines whether the rotation is clockwise or anticlockwise. Hence, the multiplication is only commutative, if the mirror planes are perpendicular to each other ($\varphi = 90°$): then $n = 2$ and the direction of rotation in a C_2 operation is irrelevant.

Conversely, a C_n axis combined with a σ_v mirror plane gives rise to a second mirror plane, making an angle of $\varphi = 180°/n$ with the first one. By the same principle, considering the consequences of the combination of the C_n axis and this second mirror plane, altogether $n - 1$ new mirror planes are generated. For example, the addition of a vertical mirror plane to a C_4 axis doubles the number of points (×) in the stereographic projection. The symmetry of the point pattern demands the presence of three further vertical mirror planes (*Fig. 11.6*).

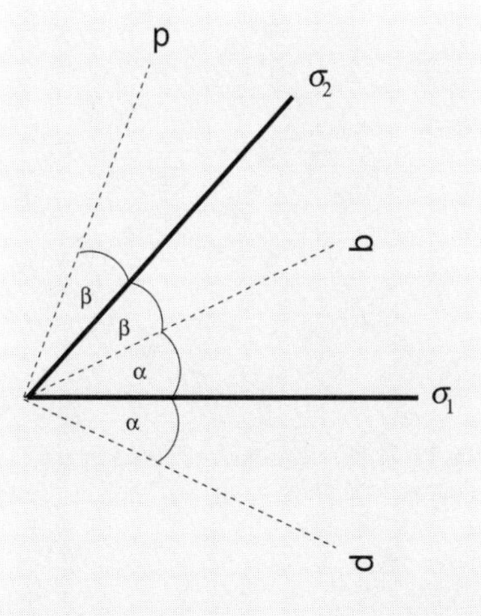

Fig. 11.5. *The result of a two-fold reflection across two mirror planes, situated at an angle $\varphi = \alpha + \beta$ to one another*

As rotations do not alter the absolute configuration of a chiral molecule, the result of two sequential rotations must also be a rotation; the three rotation axes are all intersecting at one point. Analysis of the combination of two C_2 axes, making an angle of φ with one another, leads to a result showing analogies to the already discussed case of two mirror planes: as here there arises a C_n axis, perpendicular to the two C_2 axes, and with an order of $n = 180°/\varphi$. This means that for $\varphi = 90°$ a C_2 axis, for $\varphi = 60°$ a C_3 axis, for $\varphi = 45°$ a C_4 axis, and for $\varphi = 30°$ a C_6 axis is generated. Conversely, a combination of a C_n axis and a perpendicular C_2 axis leads to a total of $n - 1$ new C_2 axes, making an angle of $\varphi = 180°/n$ with each other. This result may be generalized further:

combination of two rotation axes or two rotary-reflection axes engenders new rotation axes; combination of a rotation axis and a rotary-reflection axis generates new rotary-reflection axes.

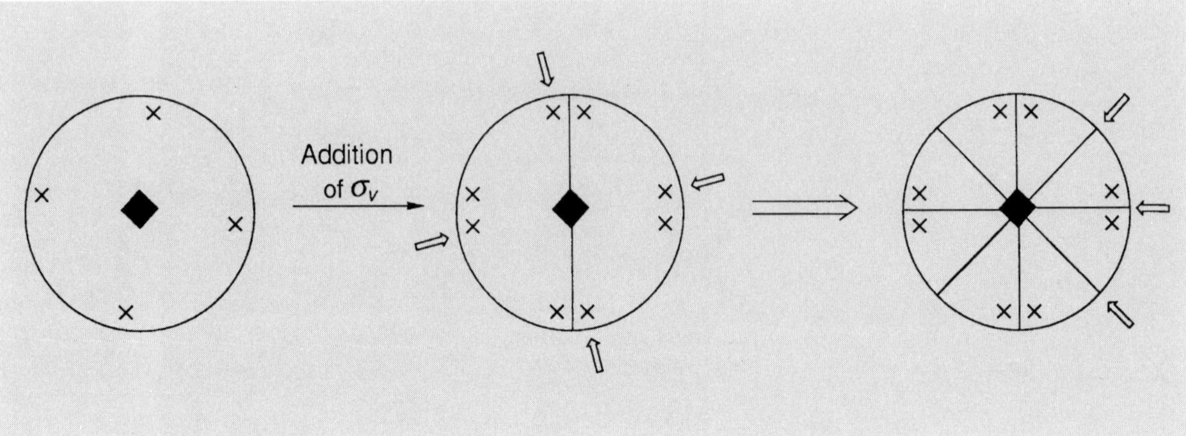

Fig. 11.6. *Addition of a vertical mirror plane to a* C_4 *axis.* The newly introduced points and mirror planes are marked with arrows.

11.1.4. Equivalent Symmetry Elements and Atoms

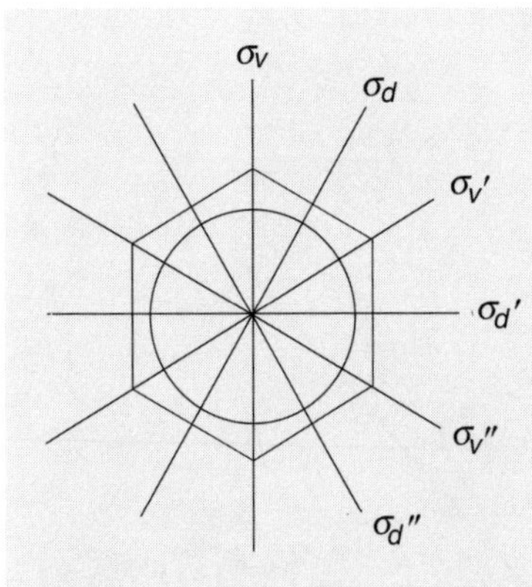

Fig. 11.7. *Vertical mirror planes in the benzene molecule*

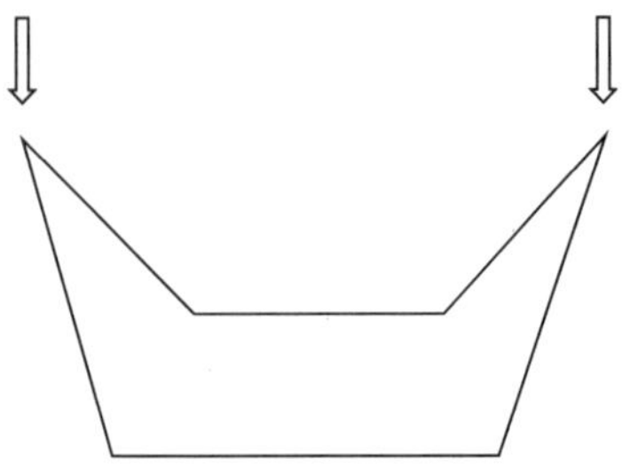

Two symmetry elements A and B, present in a molecule, are called equivalent if A can be transposed into B by application of a symmetry operation of the molecule. If A is equivalent to C, as well as to B, then B and C are also equivalent. Hence, the three C_2 axes in the cyclopropane molecule (*Fig. 11.1*) are equivalent, as they may be transposed into one another by the C_3 axis. In the benzene molecule, on the other hand, there exist two sets of vertical mirror planes: σ_v, σ_v', and σ_v'', and σ_d, σ_d', and σ_d'' (*d* stands for diagonal; *Fig. 11.7*). Two mirror planes within a set are equivalent to one another, but not the other possible pairs, as the former (but not the latter) may be transposed into one another by a C_6 operation.

Accordingly, equivalent atoms in a molecule are those atoms of the same atomic number (in case of isotopes of the same atomic mass) which may be exchanged with one another by means of symmetry operations. For example, all of the H-atoms in methane, cyclopropane, or benzene, and all the C-atoms in the chair conformation of cyclohexane are equivalent. In contrast, two different sets of C-atoms are to be found in the boat conformation of cyclohexane: the C-atoms at bow and stern cannot be transposed into those ones at the bottom by any symmetry operation of the boat model.

11.2. Group-Theory-Based Description of Molecular Symmetry

11.2.1. Elementary Group Theory

We have, up to now, described molecular symmetry phenomenologically. This certainly afforded important insights, but some questions remained unanswered, such as the number of possible combinations of symmetry operations. Group theory provides answers to such questions.

A *group*, in the mathematical sense, is a set of (group) elements, whose combinations are defined and which satisfy the following conditions:

1) The combination ('multiplication') of two elements of a group must result in an element of this group ($A \circ B = C$). If all of the multiplications are commutative ($A \circ B = B \circ A$), then the group is called *Abelian* (after the Norwegian mathematician *Niels Henrik Abel*). If not all multiplications are commutative, then the sequence of the multiplications is important.

2) One element of a group (the identity E) must commute with all elements, without having any effect: $E \circ X = X \circ E = X$.

3) The Associativity Law must apply to multiplication: $A \circ (B \circ C) = (A \circ B) \circ C$.

4) For each element there must exist an inverse element within the group: R is inverse to S, if $R \circ S = S \circ R = E$. Hence, S, naturally, is also inverse to R. This relationship between the elements R and S is expressed as $R = S^{-1}$ or $S = R^{-1}$.

Groups may be finite or infinite. The number of elements within a finite group is described as the order of the group. A finite group is completely and unambiguously defined, if all h elements and all h^2 combinations are known. Examples of groups include the integers with respect to addition as the combination ($E = 0$) and the real numbers (except for the number zero) with respect to arithmetic multiplication as the combination ($E = 1$). If the elements of a group are completely contained within a larger group, then the smaller one is called a subgroup. In finite groups, the order of a subgroup is a divisor of the order of the corresponding (super-)group.

11.2.2. Application to Symmetry Operations

A complete set of symmetry *operations* (not symmetry *elements*) of a molecule comprises a group. As one point – the geometric center of the molecule – remains unaltered in all symmetry operations described so far, these groups are known as *symmetry point groups*. The four prerequisites for the existence of a mathematical group are fulfilled, without exception:

1) The combination of two symmetry operations (their sequential execution) is equivalent to a symmetry operation of the molecule.

2) The identity operation can always be carried out.

3) The Associativity Law applies.

4) Each symmetry operation has an inverse operation. For any rotation, a rotation in the opposite sense about the same angle constitutes its inverse. Reflections and inversions (like the rotation C_2) are inverse to themselves.

Before we examine which point groups have to be considered at all, we must first establish some clarity about how many symmetry operations follow from a particular symmetry element. In general, it is far easier to recognize symmetry elements than to recognize symmetry operations.

- A mirror plane or an inversion center generates only one symmetry operation, as a double reflection or inversion leads back to the original situation.
- Each C_n axis generates n rotations (C_n, C_n^2, ...C_n^{n-1}, $C_n^n = E$), as only the n-fold application of a C_n operation leads back to the original orientation.
- Each S_n axis engenders n or $2n$ symmetry operations depending on whether n is even or odd.

  For even n, the result is: S_n, S_n^2, ...S_n^{n-1}, $S_n^n = E$.

  For odd n, the result is: S_n, S_n^2, ...$S_n^n = \sigma_h$, S_n^{n+1}, ...$S_n^{2n} = E$.

11.2.3. Point Groups

Designation of symmetry point groups usually proceeds with the aid of the convention of *Artur Schönflies*; crystallographers, however, favor the notation of *Carl Hermann* and *Charles Mauguin* (see *Sect. 11.3.1*). We shall discuss the possible symmetry groups in ascending order.

The simplest group is C_1, containing only one element (the identity operation E); this is the point group of asymmetrical molecules. If the molecule possesses a mirror plane or an inversion center as its only non-trivial symmetry element, then it is of group C_s or C_i (group order 2). If a rotation axis C_n or a rotary-reflection axis S_n is present, then we are dealing with the point group of the same name, of order n. S_n Groups need only be taken into account for even-numbered n, since an S_n axis with odd-numbered n does not represent an intrinsic symmetry element, but may be interpreted as a combination of C_n and σ_h (point group C_{nh}). The combination of C_n and σ_v is correspondingly named as C_{nv}. Both point groups are of order $2n$.

If a molecule, in addition to its principal rotation axis C_n, possesses C_2 axes perpendicular to the principal axis and no other symmetry elements, then this combination of order $2n$ is named as D_n. If a mirror plane perpendicular to the principal rotation axis is added (or the combination $C_n + \sigma_h + \sigma_v$ is present), then the number of group elements doubles once more and comes to $4n$ (point group D_{nh}). In this constellation, the C_2 axes coincide with the σ_v planes. Should this not be the case, so that the vertical mirror planes bisect the angles between adjacent C_2 axes, then the point group D_{nd}, of the same order $4n$, is the result. A characteristic of this point group is the presence of an S_{2n} axis of rotary reflection as a consequence of the combination of the symmetry elements indicated; this is worth mentioning because the point group D_{nd} can be recognized by this symmetry element.

All linear molecules possess a C_∞ axis and an infinite number of σ_v planes. If a horizontal mirror plane is lacking, then the molecule is of point group $C_{\infty v}$. If σ_h does exist (and hence also an inversion center and an infinite number of C_2 axes perpendicular to the C_∞ axis), then the relevant point group is called $D_{\infty h}$.

A case not yet considered is that one in which several C_n axes, where $n > 2$, are present. These are best analyzed by analogy with the five regular polyhedra (*Platonic solids*), in which all faces, vertices, and edges are equivalent; as these possess the property that their faces, each consisting of some regular n-sided polygon (with $n = 3$, 4, or 5), are perpendicular to the corresponding C_n axes. The sum of the number of faces and the number of vertices exceeds the number of edges by two in every case (*Fig. 11.8*).

The *tetrahedron* possesses the following symmetry elements: four C_3 axes, each traversing one vertex and the center of the face opposite, three mutually perpendicular S_4 axes traversing the centers of opposite edges, and six mirror planes, each including one edge of the tetrahedron. The resulting point group T_d contains 24 symmetry operations.

The *octahedron* possesses the following symmetry elements: three C_4 axes (and three S_4 axes parallel to them) going through opposite vertices, six C_2 axes going through the centers of opposite edges, three σ_h planes, each going through four vertices, four S_6 axes going through the centers of opposite faces, six vertical mirror planes, bisecting opposite edges, and an inversion center. These result in a total of 48 symmetry operations (point group O_h).

The *cube* is very closely related to the *octahedron*. It possesses the same symmetry elements, and hence belongs to the same point group. The symmetry elements relating to the vertices and the faces are different, however. While, for the cube, the C_4 axes transect the faces and the C_3 axes the vertices, the reverse applies for the octahedron.

Dodecahedron (*D*) and *icosahedron* (*I*) exist in a similar relationship to each other as *octahedron* and *cube*; they possess the same symmetry elements, but differently ordered: six S_{10} axes going through opposite faces (*D*) or vertices (*I*), ten S_6 axes through opposite vertices (*D*) or faces (*I*), 15 C_2 axes through opposite edges, 15 mirror planes, each containing two C_2 and two C_5 axes (as a consequence of the S_{10} axes), and an inversion center. A total of 120 symmetry operations (point group I_h) is the result.

In each of the groups T_d, O_h, and I_h there exists a subgroup, containing all of the C_n, but none of the S_n operations (including reflection and inversion), by which the group order is halved. These are accordingly called *T*, *O*, and *I*. If a σ_h plane (or an inversion center) is added to the symmetry elements of point group *T*, the result is the point group T_h. Like T_d, T_h constitutes a subgroup of O_h and contains 24 symmetry operations. There exists, therefore, a total of seven point groups for highly symmetrical molecules. It is very easy, by the way, to derive the number of symmetry operations contained within a point group from its stereographic projection: it is the same as the number of equivalent positions generated from a randomly selected starting point.

Membership of a particular point group may be ascertained for a molecule of known structure, with the aid of a 'symmetry separation procedure' (*Fig. 11.9*). The first step establishes whether a linear molecule is under scrutiny. If the molecule is non-linear, then the second

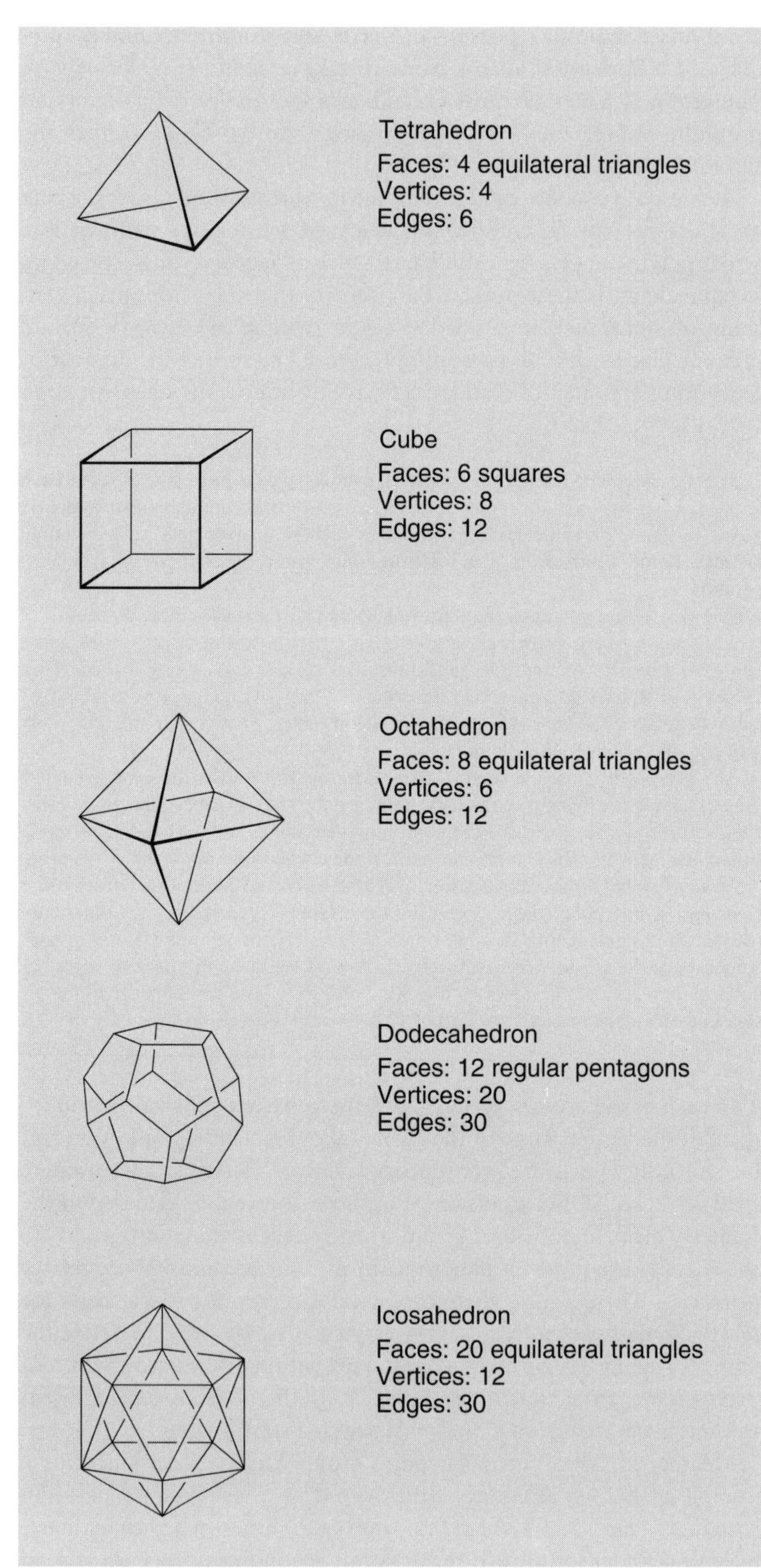

Fig. 11.8. *The five* Platonic solids

step of the 'separation procedure' tests whether more than one C_n axis with $n > 2$ is present and, hence, whether a point group for highly symmetrical molecules is appropriate. If this is not the case, then the symmetry elements present are examined. The 'separation procedure' thus leads unequivocally, with few questions, to the correct point group. This procedure was considered in detail in *Chapt. 3*. If it is applied to the examples discussed so far, then the following point groups result: C_{2v} for H_2O and the boat conformation of cyclohexane, D_{3h} for cyclopropane and 1,3,5-trifluorobenzene, D_{2h} for ethene, $D_{\infty h}$ for ethyne, D_{3d} for the chair conformation of cyclohexane, T_d for methane, and D_{6h} for benzene.

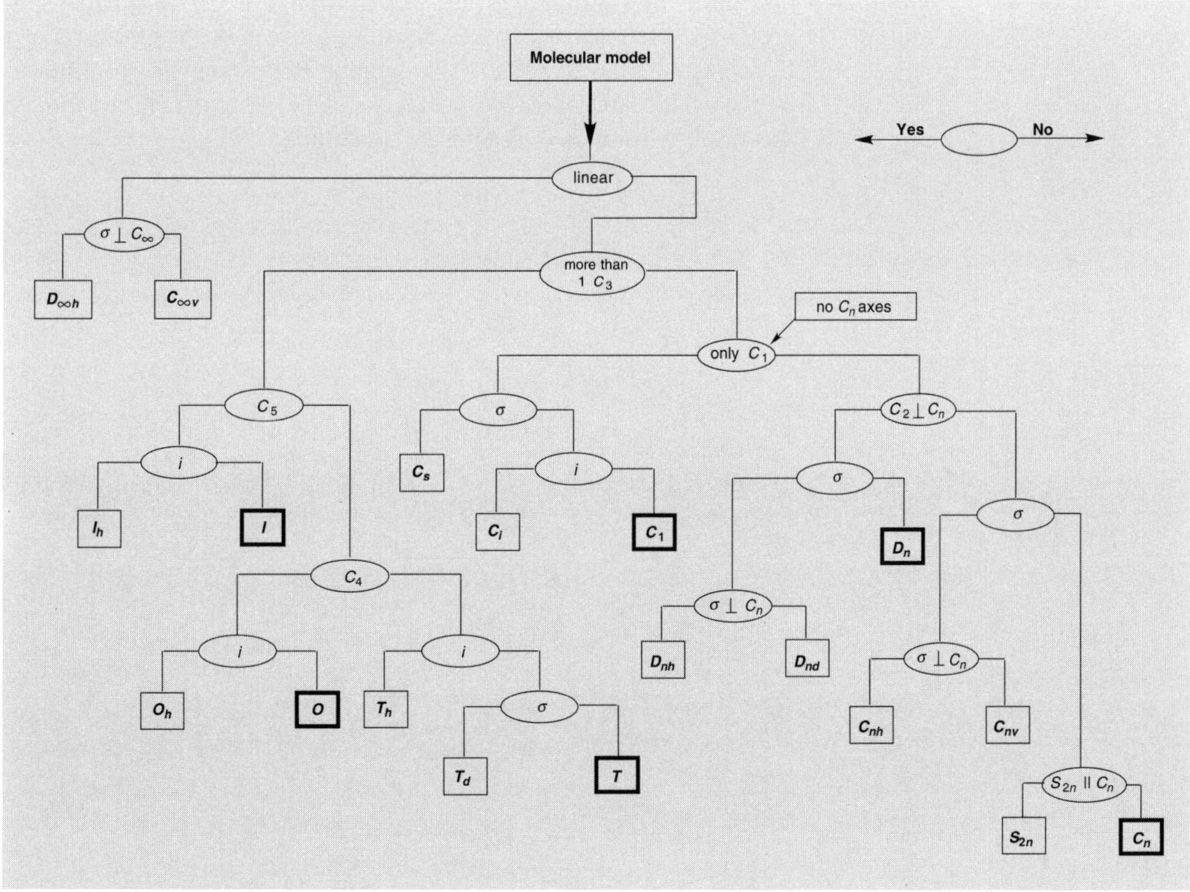

Fig. 11.9. *'Symmetry-separation procedure' for the determination of symmetry point groups.* Point groups of chiral molecules are emphasized using heavy print.

As an exercise in symmetry-property relationships, we shall look into the question of when a molecule possesses a dipole moment. As an experimentally determinable quantity, the dipole moment is invariant with respect to symmetry operations. It must, therefore, coincide with all of the symmetry elements present in the molecule. For the dipole moment not to be equal to zero, neither an inversion center nor a rotary-reflection axis may be present; also, not more than one rotation axis is permitted. Close consideration of *Fig. 11.9* reveals that, consequently, only the following point groups are allowed: C_1, C_s, C_n, C_{nv}, and $C_{\infty v}$. Hence, the measurement of the dipole moment of a molecule can lead to conclusions about its molecular symmetry.

11.2.4. Hierarchy of Point Groups

The question of which of two molecules is the 'more symmetrical' can only be answered if one point group is a subgroup of another. Otherwise, a comparison is inappropriate, as the decision as to whether a planar molecule with a C_6 axis is more or less symmetrical than one with a C_5 axis involves purely aesthetic considerations. By studying the symmetry operations of a group, it is possible to find all of its imaginable subgroups, and to represent the relationships between them in the form of a hierarchy diagram (*Fig. 11.10*). A diagram of this kind gives an indication of what degree of reduction of molecular symmetry may be expected as a result of rotation, vibration, or substitution.

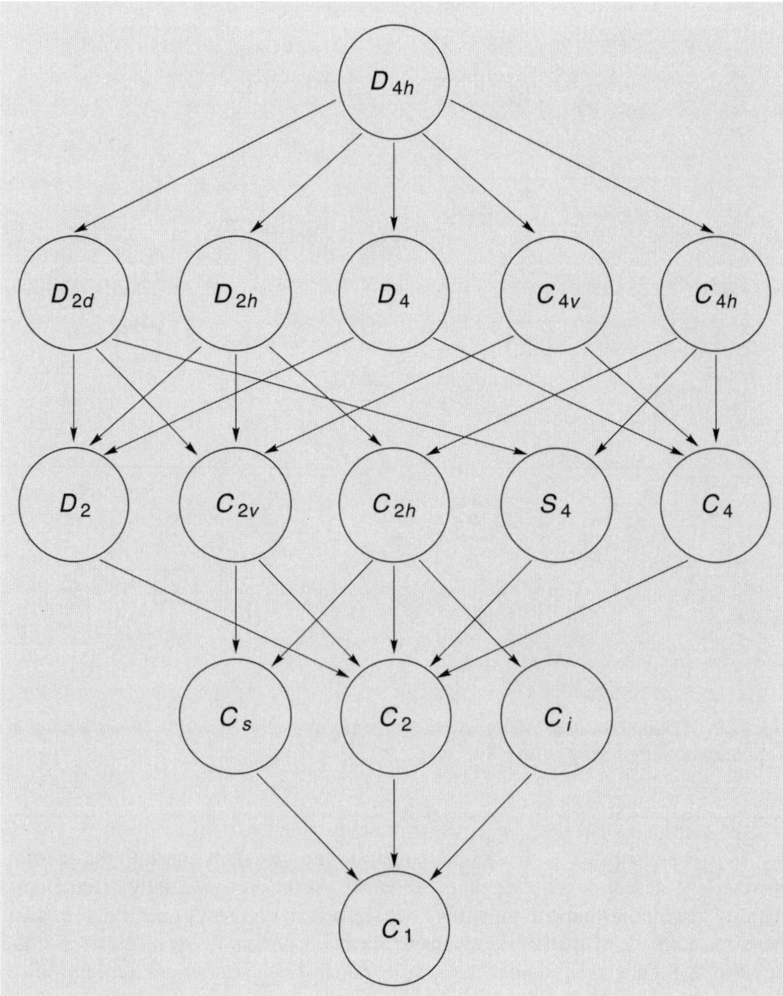

Fig. 11.10. *Hierarchy diagram for the point group* D_{4h}. The arrows point to possible subgroups. Each reduction of symmetry halves the order of the group.

11.3. Crystal Symmetry

11.3.1. Point-Group Limitations Caused by the Crystal Lattice

A crystal is characterized by a three-dimensional, periodic structure, described by three non-coplanar translation vectors $\boldsymbol{a}$, $\boldsymbol{b}$, and $\boldsymbol{c}$. These vectors define the *crystallographic unit cell*: the smallest unit, the translation of which yields the whole crystal structure. Depending on the symmetry of the unit cell, it is possible to distinguish between six crystal systems: *triclinic*, *monoclinic*, *orthorhombic*, *tetragonal*, *trigonal/hexagonal*, and *cubic*. If one corner of a unit cell is selected as the origin of coordinates, then the positions of the atoms at a distance $|\boldsymbol{r}|$ from the origin may be expressed using the relative coordinates x, y, and z:

$$\boldsymbol{r} = x\boldsymbol{a} + y\boldsymbol{b} + z\boldsymbol{c}\,(0 \leq x, y, z < 1)$$

For the description of finite objects (such as molecules or, better, molecular models), there exists an infinite number of point groups, because the order of an axis of rotation or rotary reflection is not subject to any limitations in this case. Entirely different in a crystal. There, the infinite three-dimensional periodicity is only compatible with the following symmetry elements: one-, two-, three-, four-, and six-fold rotation axes and rotary-reflection axes. This state of affairs can be better understood by considering that a surface can be completely covered by identical parallelograms (C_2), equilateral triangles (C_3), squares (C_4), or regular hexagons (C_6), leaving no gaps in between, but not with regular pentagons or heptagons. The number of point groups is hereby reduced to 32; these are compiled in *Table 11.1*, with their respective crystal systems assigned. The monoclinic crystal system is particularly well suited for C_2,

Table 11.1. Schönflies *and* Hermann-Mauguin *Symbols for the 32 Crystallographic Point Groups* (distributed across the various crystal systems). The relevant general point groups are listed on the left. The assignment of the cubic point groups is not directly derivable.

	triclinic	monoclinic	ortho-rhombic	tetragonal	trigonal	hexagonal	cubic
C_s		$C_s \equiv m$					
C_i	$C_i \equiv \bar{1}$						
C_n	$C_1 \equiv 1$	$C_2 \equiv 2$		$C_4 \equiv 4$	$C_3 \equiv 3$	$C_6 \equiv 6$	$T \equiv 23$
S_n				$S_4 \equiv \bar{4}$	$S_6 \equiv \bar{3}$		
C_{nv}			$C_{2v} \equiv mm2$	$C_{4v} \equiv 4mm$	$C_{3v} \equiv 3m$	$C_{6v} \equiv 6mm$	
C_{nh}		$C_{2h} \equiv 2/m$		$C_{4h} \equiv 4/m$		$C_{3h} \equiv \bar{6}$	$T_h \equiv m3$
						$C_{6h} \equiv 6/m$	
D_n			$D_2 \equiv 222$	$D_4 \equiv 422$	$D_3 \equiv 32$	$D_6 \equiv 622$	$O \equiv 432$
D_{nh}			$D_{2h} \equiv mmm$	$D_{4h} \equiv 4/mmm$		$D_{3h} \equiv \bar{6}m\,2$	$T_d \equiv \bar{4}3m$
						$D_{6h} \equiv 6/mmm$	$O_h \equiv m3m$
D_{nd}				$D_{2d} \equiv \bar{4}2m$	$D_{3d} \equiv \bar{3}m$		

the trigonal for C_3, the tetragonal for C_4, and the hexagonal for C_6 axes. The general point group D_{nd} is only represented by D_{2d} and D_{3d} in *Table 11.1*, as D_{4d} or D_{6d} contain an S_8 or S_{12} axis, and so are incompatible with the crystal lattice.

For each point group, the *Schönflies* notation used so far, together with the *Hermann-Mauguin* notation used by crystallographers, is given in *Table 11.1*. The latter notation is based on the following rules:

- An n-fold rotation axis is designated n (instead of C_n).
- Instead of rotary-reflection axes, the equivalent rotary-inversion axes are used, and are designated $\bar{n}$.
- Mirror planes are not called σ, but m. A σ_h plane is indicated with n/m, a σ_v plane with nm.
- The notation is as concise as possible; only the minimum number of symmetry elements necessary for unambiguous characterization is given. Hence, the *Schönflies* D_{2h} point group becomes mmm (rather than $2/m\ 2/m\ 2/m$) according to *Hermann-Mauguin*. The *Hermann-Mauguin* point group corresponding to the *Schönflies* C_{4v}, however, is not called $4m$, but $4mm$, as here two non-equivalent sets of vertical mirror planes (σ_v and σ_d according to *Schönflies*) are present. Except for the point groups 23 and 32 (T and D_3 according to *Schönflies*), all symbols may be unambiguously distinguished from each other.

11.3.2. Extension of Symmetry Concepts Caused by the Crystal Lattice

The *translation symmetry* in a crystal lattice on one hand drastically reduces the number of point groups. On the other hand, it permits the occurrence of further symmetry elements not found in simple molecules, but which can be present in polymers (assumed to be infinitely long) with one-dimensional periodicity (*Chapt. 7.2.2*):

- *screw axes*, by combination of a rotation with a translation parallel to the rotation axis,
- *glide planes*, by combination of a reflection with a translation parallel to the mirror plane.

It can easily be seen that additional symmetry elements must be taken into account, as, for example, double reflection within a crystal – in contrast to a molecule – need not necessarily lead back to the original position, but may just as well translate the subject into an equivalent position. Hence, a molecule in a crystal (assumed to be infinite) is able to glide, metaphorically, along the plane of reflection, as long as the glide component is so adjusted that a two-fold glide reflection corresponds to an allowed translation within the crystal (*Fig. 11.11*). The glide planes are denoted as a, b, c, n, or d, depending on the glide direction, where the two last-named glide planes have combined glide components along diagonals. Screw axes are designated by the symbol n_m; here m/n gives the translation component parallel to the axis. The following screw axes are allowed: 2_1, 3_1, 3_2, 4_1, 4_2, 4_3, 6_1, 6_2, 6_3, 6_4, and 6_5.

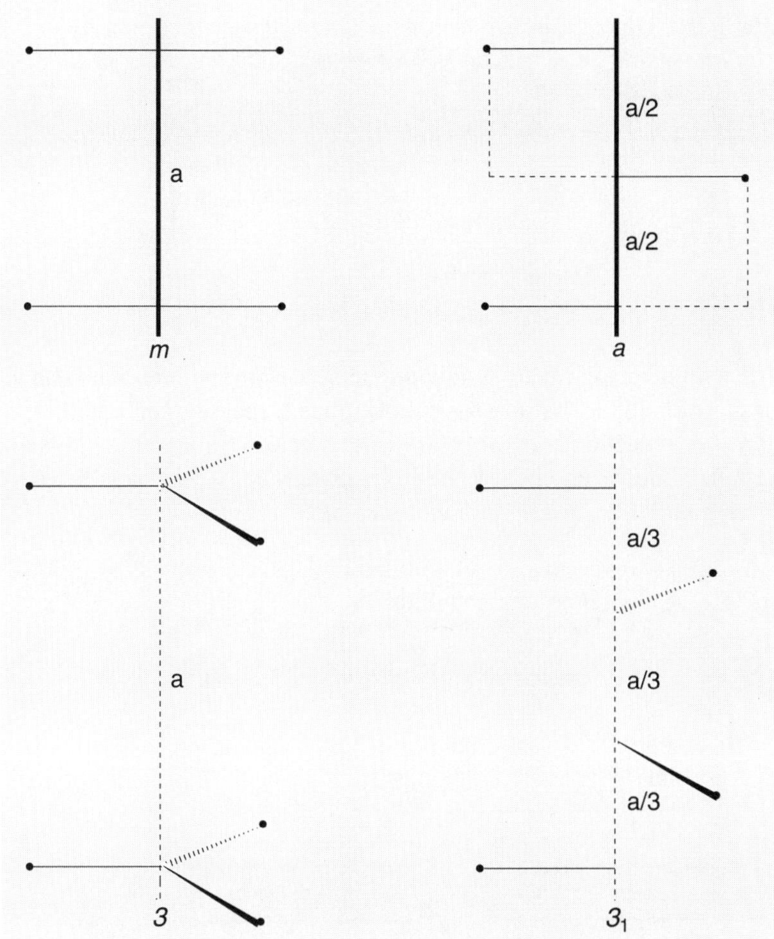

Fig. 11.11. *Comparison between a mirror plane (m) and a glide plane (a), and between a rotation axis (3) and a screw axis (3_1)*

Furthermore, besides the primitive unit cells (P), in which, in the simplest case, only the corners are occupied by atoms, centered unit cells are used for the description of the crystal lattice, the *Bravais* lattices A, B, C (each one-face centered), I (body centered), F (all-face centered), and R (rhombohedral).

There is a total of 230 combinations of symmetry elements within the crystal; these are known as *space groups*. The space-group symbol may be derived from the point-group symbol, by

- making appropriate modifications, if a screw axis instead of a rotation axis, or a glide plane instead of a mirror plane is present,
- giving the abbreviation for the *Bravais* lattice as a prefix.

As an example, the 13 space groups of the monoclinic crystal system, together with the point groups from which they originate, are summarized in *Table 11.2*. It can be seen that the *Schönflies* notation, although approved for molecules, has to capitulate before the variety of symmetry present in the crystal: *Schönflies* terminology for the space group $P2_1/c$ would be C_{2h}^5, but this gives no information about the *Bravais* lattice nor about the existence of a screw axis and a glide plane parallel to the c axis.

Table 11.2. *The 13 Monoclinic Space Groups and the Point Groups from Which They Derive*

Point group	*P* Lattices				*C* Lattices	
2	$P2$	$P2_1$			$C2$	
m	Pm		Pc		Cm	Cc
$2/m$	$P2/m$	$P2_1/m$	$P2/c$	$P2_1/c$	$C2/m$	$C2/c$

It would be logical to subdivide the space groups into achiral and chiral groups, as is the case for point groups, depending on whether or not rotary-reflection axes (or rotary-inversion axes) are present. Crystallographers, however, classify the space groups according to a different ordering principle, owing to practical considerations (*Fig. 11.12*).

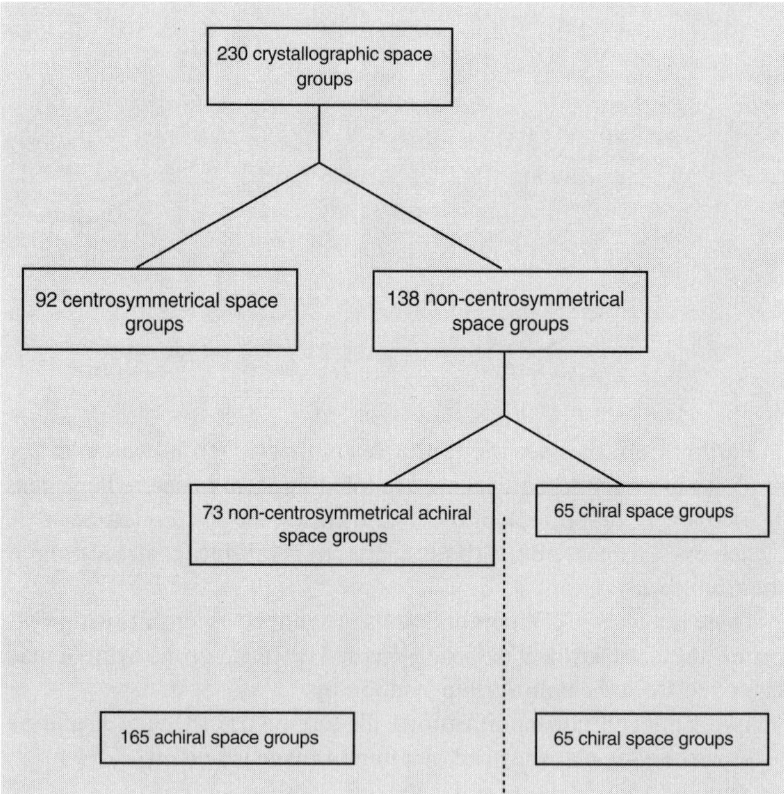

Fig. 11.12. *Subdivision of space groups*

In the course of an X-ray crystal-structure analysis, inversion centers have a special place among the symmetry elements. Their presence or absence is of exceptional importance regarding the difficulty of solving the 'phase problem' (see *Aspects of Organic Chemistry*, Volume 4: *Methods of Structure Determination*). Crystallographers, therefore, distinguish primarily between centrosymmetrical (with inversion centers) and non-centrosymmetrical (without inversion centers) space groups. Consequently, under non-centrosymmetrical space groups may be found not only chiral, but also achiral space groups (those with mirror planes or four-fold rotary-reflection axes). Every chiral space group possesses polar directions, the opposite ends of which cannot be transposed onto one another by any symmetry operation (except for translation): the crystal exhibits *one-dimensional chirality* along a polar direction. Those chiral space groups in which one of the three translation vectors (***a***, ***b***, or ***c***) coincides with a polar direction, and which, therefore, contain a polar axis, are of special importance.

11.4. Irreducible Representations

It was established in *Sect. 11.2.1* that, for a finite group of h elements to be unambiguously defined, all of the h^2 combinations must be known. These combinations are tabulated in the form of an $h \times h$ matrix; the so-called *group multiplication table*.

Every symmetry operation may be represented by a transformation matrix. This describes the coordinate changes experienced by the atoms in a molecule subjected to the symmetry operation; hence, for a molecule with n atoms, $3n \times 3n$ matrices are required, if the x-, y-, and z-coordinates are each considered separately in every case. The coordinate changes resulting from the combination of two symmetry operations are then given by multiplication of the corresponding transformation matrices. A set of matrices of this kind (one matrix for each symmetry operation of the group) thus behaves as dictated by the group multiplication table; this is known as a *representation* of the group.

For every group, there can be any number of different representations, but only a limited number with fundamental significance. By using proper mathematical operations, it is possible to split a $3n \times 3n$ matrix into smaller matrices, which still obey the group multiplication table. The smallest matrices to satisfy this condition comprise the irreducible representations of a group. In many cases, it is not necessary to have the matrices themselves, but only the respective sum of their principal diagonal elements (the 'character' of a matrix). Therefore, these sums are tabulated in *character tables* for each point group.

The character table for the point group C_{2v} is given as an example (*Table 11.3*). The *Schönflies* symbol for the symmetry point group is to be found in the upper left-hand corner. The individual symmetry operations are listed next to it, in the top row. Beneath these are tabulated the characters corresponding to the respective irreducible representations. Read horizontally, these give, for each irreducible representation, the transformation behavior (1 = symmetrical, −1 = antisymmetrical) relative to each symmetry operation. Beneath the point group symbol are listed the descriptors devised by *Robert S. Mulliken* for the individual irreducible representations. Depending on whether the original matrices are one-, two-, or three-dimensional, the representations are assigned the symbols A (or B), E, or T (or F). The choice of symbol A or B is based on the behavior relative to rotation about the

Table 11.3. *Character Table for the Point Group* C_{2v}

C_{2v}	E	$C_2(z)$	$\sigma_v(xz)$	$\sigma_v(yz)$
A_1	1	1	1	1
A_2	1	1	−1	−1
B_1	1	−1	1	−1
B_2	1	−1	−1	1

principal axis: A for symmetrical, B for antisymmetrical behavior. As *Table 11.3* shows, different representations of the same type are differentiated using numerical indices (A_1/A_2 or B_1/B_2). If an inversion center is present, symmetrical or antisymmetrical behavior relative to inversion is expressed by the indices g (from the German 'gerade' = even) or u (from 'ungerade' = odd). Analogous behavior relative to a horizontal mirror plane is indicated with the superscripts ′ or ″.

Further Reading

E. Heilbronner, J. D. Dunitz, *Reflections on Symmetry in Chemistry and Elsewhere,* Verlag Helv. Chim. Acta, Basel, and VCH Verlagsgesellschaft, Weinheim, 1993.

F. A. Cotton, *Chemical Applications of Group Theory,* Wiley-Interscience, New York, 1971.

H. H. Jaffé, M. Orchin, *Symmetry in Chemistry,* John Wiley, New York, 1965.

D. Wald, *Gruppentheorie für Chemiker,* VCH Verlagsgesellschaft, Weinheim, 1985.

I. and M. Hargittai, *Symmetry through the Eyes of a Chemist,* VCH Verlagsgesellschaft, Weinheim, 1986.

C. H. MacGillavry, *Symmetry Aspects of M. C. Escher's Periodic Drawings,* Bohn, Scheltema & Holkema, Utrecht, 1976.

K. Mathiak, P. Stingl, *Gruppentheorie,* Vieweg-Akademische Verlagsgesellschaft, Braunschweig, 1969.

12 Determination of the Absolute Configuration

When *Emil Fischer* arbitrarily assigned the D-configuration to the dextrorotatory enantiomer of glyceraldehyde at the end of the last century, it was clear to him that all of the absolute configurations derived from this might ultimately turn out to be wrong (*Chapt. 10.2.1.1*). More than 50 years were to elapse, until the absolute configuration of a chiral compound could experimentally be determined. Fortunately, *Fischer*, purely by chance, had hit upon the correct solution.

12.1. Determination of the Absolute Configuration by Means of Anomalous X-Ray Diffraction

The technique of X-ray crystal-structure analysis is founded upon the scattering of X-rays by electrons (*Fig. 12.1*). This phenomenon may be described in terms of electrodynamic scattering theory, based on *Maxwell*'s equations. An electromagnetic wave of wavelength λ causes an electron to undergo an acceleration, inducing a harmonic vibration about its resting state. In consequence, the electron emits a secondary, scattered wave of the same wavelength: a process known as coherent scattering. The intensity of the scattered wave is greatest in the direction of the incident wave, declining with increasing angles of 2θ. The phase difference between the incident and the scattered wave is, to the first approximation, equal to 180°. The precise value varies from element to element, and is also dependent on the wavelength used. If the phase

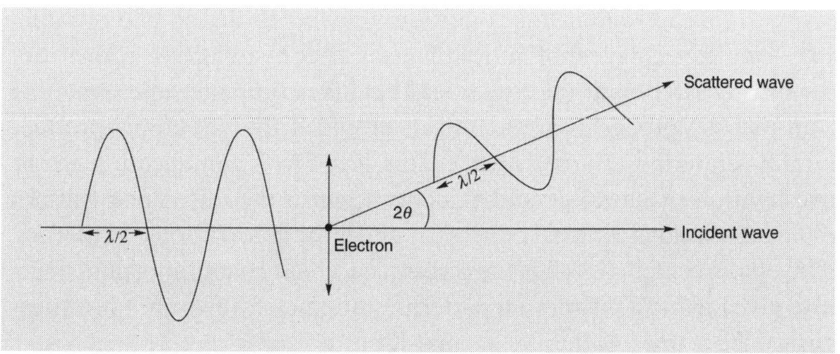

Fig. 12.1. *Schematic representation of the scattering of X-rays by electrons*

difference deviates markedly from this value, however, then the scattering is said to be *anomalous*.

The degree of scattering of any atom, the so-called *atomic scattering factor*, may be calculated from the degree of scattering of a single electron and tabulated electron densities of atoms. From this, in turn, the degree of scattering of all of the atoms in the unit cell of a single crystal can be worked out, provided that their positions relative to each other are known. Or, conversely, as the square of this quantity (the *structure factor*) can be determined experimentally, it is possible, in principle, to back-calculate the positions of atoms in an unknown structure from it. The structure factor is, therefore, of key significance in crystal-structure determination.

In calculating the degree of scattering of the whole crystal, it is necessary to consider the phase differences between scattered waves arising from the individual unit cells. Most scattered waves are quenched by the periodicity of the crystal, so that only very particularly defined values of 2θ lead to measurable intensities. A periodic diffraction diagram is obtained, the maxima of which are increasingly sharp, the more unit cells contribute to the diffraction. Diffraction only occurs, however, if the wavelength of the irradiation used is of the same order of magnitude as the dimensions within the lattice. As the lattice dimensions of crystals are approximately equivalent to interatomic distances, irradiation in the range of *ca.* 100 pm (*i.e.*, X-ray, neutron, or electron irradiation) is required for diffraction in crystal lattices.

W. Lawrence Bragg developed a graphic, clear interpretation of the diffraction process (*Fig. 12.2*). A crystal may be thought of as consisting of multitudes of layers, densely occupied by atoms, each separated from its neighbors by a distance *d*. Assuming that X-ray irradiation encountering such a 'lattice plane' is partially reflected, like at a semi-transparent mirror, a measurable intensity of reflected irradiation is only obtained, if no phase difference exists between the secondary (reflected) waves. The path difference of two beams, reflected from neighboring lattice planes, is dependent on the incident and reflected angles θ and amounts to $2d \sin\theta$. In order not to give rise to a phase difference, this path difference must correspond to an integer multiple of the wavelength λ. Thus, we arrive at *Bragg*'s equation:

$$2d \sin\theta = n\lambda$$

Fig. 12.2 shows that this condition may be fulfilled in two different ways for a particular set of lattice planes: either by reflection at the front side or by reflection at the rear side. The diffraction intensities resulting from the two processes are identical, provided that all atoms produce normal scattering (*Friedel*'s law). This leads to a complication in the investigation of chiral crystal structures; since the diffraction pattern obtained is always centrosymmetrical under these conditions. 'Addition' of inversion centers thus has the consequence that enantiomorphic structures give identical diffraction patterns, and, therefore, cannot be distinguished from one another by normal X-ray crystal-structure analysis.

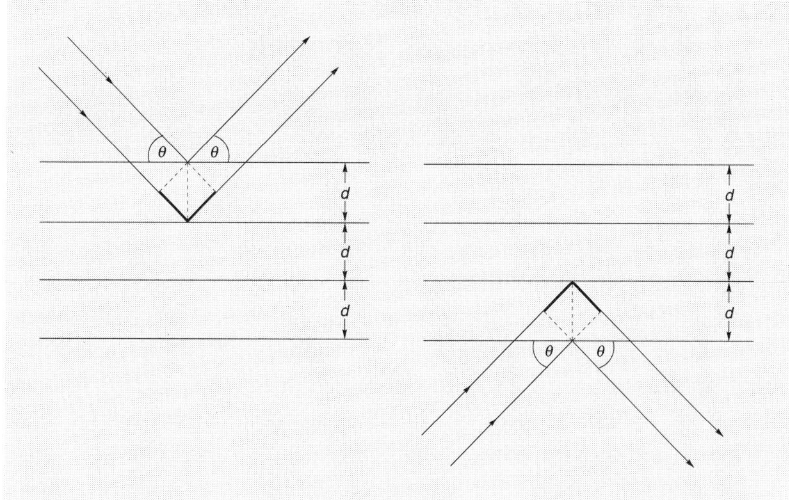

Fig. 12.2. *Reflection of X-rays at the front side* (left) *and the rear side* (right) *of a set of lattice planes*

If some atoms produce anomalous scattering, however, then measurable differences in intensity between X-rays reflected from the front and rear sides are the result. From this, it is possible to distinguish between two enantiomorphic structures, a technique first used by *Johannes Martin Bijvoet*. In 1951, he examined the crystal structure of Na-Rb (+)-tartrate, using a specially constructed X-ray tube, whose anode consisted of zirconium. The X-rays produced by this tube induced anomalous scattering from the Rb^{+} ions, so that the absolute configuration of the anion could be unequivocally determined from the differences in intensities mentioned.

$COO^{\ominus}$ $Na^{\oplus}$
H—C—OH
HO—C—H
$COO^{\ominus}$ $Rb^{\oplus}$

Improved measurement techniques now mean that this procedure is routine for crystallographers. The Cu X-rays most often used in the investigation of crystalline organic compounds require for that the presence of at least one atom of atomic number greater than 14, such as phosphorus, sulfur, or heavier atoms. The absolute configuration of a compound has, in certain cases, been successfully determined only from the relatively small degree of anomalous scattering of O-atoms present in the compound, but the certainty of the determination is improved substantially, if heavier atoms are present. Good experimental data are essential in any case.

In addition to this much-used method, a procedure developed by *Meir Lahav* and colleagues is applicable in special cases. It too requires crystals, but manages without X-rays. It is based on molecular recognition of 'tailor-made' additives at crystal surfaces and is briefly described below.

12.2. Determination of the Absolute Configuation by Molecular Recognition at Crystal Surfaces

12.2.1. On Crystal Faces

Interest in crystal structures has been promoted not only by X-ray crystallography, but also by the attention recently bestowed upon crystalline solids by organic chemists. If one regards crystals as solid materials, bounded by planar surfaces in accordance with the principles of crystallography, then the desire to classify symmetry-related crystal faces arises more or less of itself. Pairwise designation of crystal faces as equivalent (homotopic), enantiotopic, or diastereotopic seems appropriate, depending on whether the two half-spaces from which they are faced are equivalent, enantiomorphic, or diastereomorphic (see *Chapt. 4*, particularly *Chapt. 4.2* and *4.3*).

Centrosymmetrical crystals have, on opposite sides, parallel enantiotopic faces. They contain either achiral molecules or enantiomeric molecules, which may be transposed into one another by inversion. The orientation of the two enantiomers relative to a right-handed coordinate system may be unambiguously ascertained by normal X-ray diffraction; the crystal faces are also definitely assigned in the process. It is appropriate to regard a racemic crystal consisting purely of chiral molecules as *enantiopolar*, that is to handle it as though it were composed of two mutually enantiomorphic, chiral crystal structures, related to each other through an inversion center (or a glide plane).

The thermodynamically stable α-form of glycine, obtained by crystallization from water, is an example of a centrosymmetrical crystal structure. Four molecular conformations (labeled by the numbers **1** to **4** in *Fig. 12.3*) are accommodated in the crystal structure of α-glycine. One pair of conformers **1** and **2** are related to each other through a two-fold screw axis. The $C{-}H_{pro\text{-}R}$ bonds each point in the direction $+b$ and protrude out of the f1 face. The conformers **3** and **4** similarly form a molecule pair of this type, and are related to **1** and **2** through an inversion center. The $C{-}H_{pro\text{-}S}$ bonds each point in the direction $-b$ and protrude out of the f$\bar{1}$ face. The centrosymmetrical α-glycine, viewed microscopically, shows two enantiopolar sets of enantiomorphic glycine conformations of type **1**/**2** or **3**/**4**, respectively, and, viewed macroscopically, two enantiotopic crystal faces f1 and f$\bar{1}$.

Chiral crystals differ from centrosymmetrical crystals, in that their faces do not have to occur pairwise on opposite sides. *Fig. 12.4* shows the model of such a crystal with a polar axis. A special complication arises here in the structure determination using normal X-ray diffraction: because of the centrosymmetry of the diffraction pattern, the absolute orientation of a chiral molecule relative to the crystal axes and faces cannot be unambiguously established.

No inversion centers or symmetry planes, but only screw axes, may be discerned in the β- and γ-forms of glycine: each crystal contains only one of the enantiomorphic conformations.

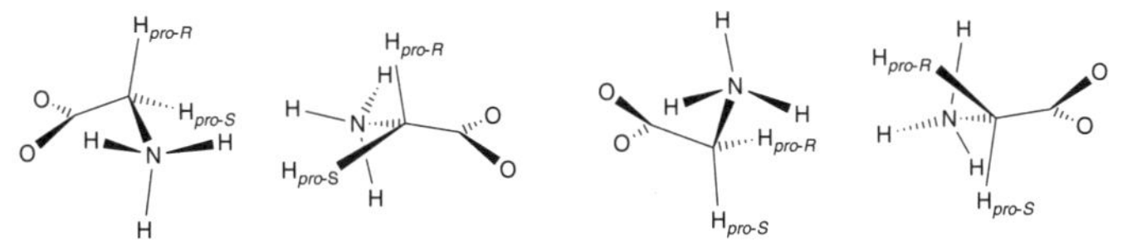

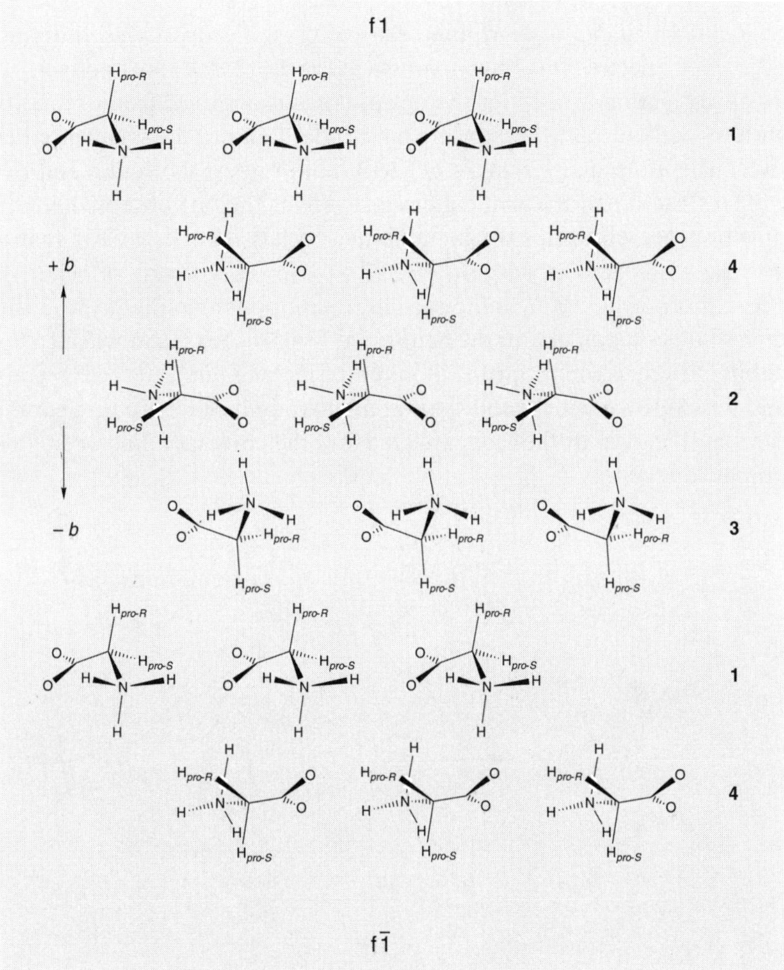

Fig. 12.3. *α-Form of glycine* (space group $P2_1/c$). The adjacent layers, which contain **1** and **4**, or **2** and **3**, are related through inversion centers, the adjacent layers, which contain **1** and **3**, or **2** and **4**, through a glide plane.

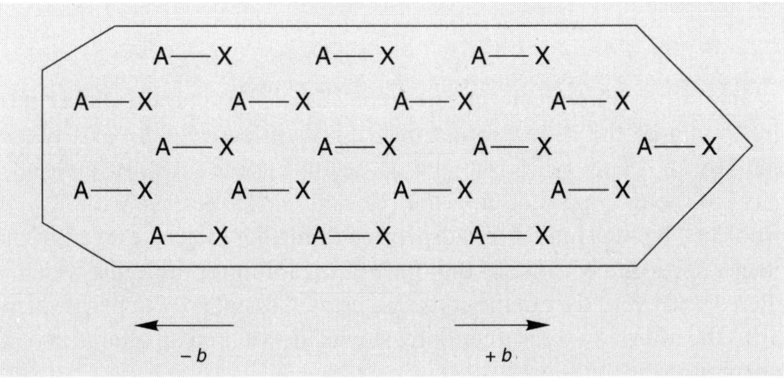

Fig. 12.4. *Model of a non-centrosymmetrical crystal with one-dimensional chirality along the polar axis* b

12.2.2. Targeted Influencing of Crystal Morphology

It has long been known that impurities in solution may drastically influence the morphology of growing crystals. Crystal growth occurs at crystal surfaces, and crystal surfaces have the ability to recognize individual host and guest molecules in a solution and to selectively attach them. Therefore, it is conceivable that the two crystal faces of a symmetry-related pair may adsorb inhibitors of crystal growth with differing degrees of selectivity: with consequences for the crystal morphology.

Investigation of the morphology of crystals obtained from solutions with or without added inhibitors have indeed shown a dependence between the molecular structure of the inhibitor guest molecules and the crystal structure of the host material, as well as the rate of growth in the direction perpendicular to the inhibited crystal face. It is clear that a crystal, with faces A and B (*Fig. 12.5*) will, as a result of selective occupation of face B by inhibitor molecules, preferentially grow in the direction perpendicular to the unaffected face A: the crystal will alter its habit. Differences in the proportions of the faces of uninfluenced crystals and those grown under conditions of additive-induced inhibition permit the identification of the faces affected and the corresponding crystallographic directions.

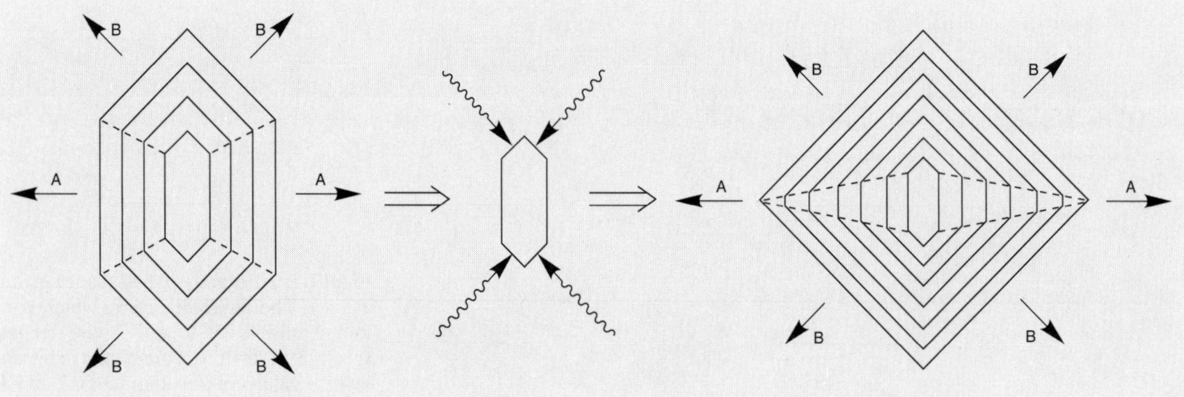

Fig. 12.5. *Influencing of crystal habit by selective inhibition of crystal growth perpendicular to the faces B*

Inhibitors whose molecules are only slightly structurally different to those ones of the substrate are particularly effective. The part of the inhibitor molecule which is unaltered, relative to the substrate molecule, may still be incorporated into the relevant crystal surface without any problems, but will hinder the adsorption of further molecule layers of the proper substrate because of that part of the inhibitor molecule which is altered and averted from the crystal interior. Crystal growth perpendicular to the affected face is, therefore, slowed down, and an altered crystal morphology is the result.

The deliberate disruption of crystal growth by selective inhibitors (*crystal engineering*) has made it possible to directly determine the absolute configuration of chiral molecules in chiral crystals, or, with the help of centrosymmetrical crystals, the absolute configuration of chiral inhibitors, by using 'tailor-made' additives in either case. We will explain the latter procedure in more detail.

12.2.3. Direct Determination of the Absolute Configuration of Chiral Additive Molecules with the Aid of Enantiopolar Crystals

Enantiopolar crystals are only suitable for the determination of configuration if a functional group W is oriented towards face f1, and not towards face f$\bar{1}$, in the (*R*)-enantiomers, and thus towards face f$\bar{1}$, and not towards face f1, in the (*S*)-enantiomers (*Fig. 12.6*). Because of the orientation of the group W, the additive W–S′–A can only be incorporated into the f1 and f2 faces, but not into the f$\bar{1}$ and f$\bar{2}$ faces. A similar consideration applies, on symmetry grounds, for the enantiomorphic additive A–R′–W.

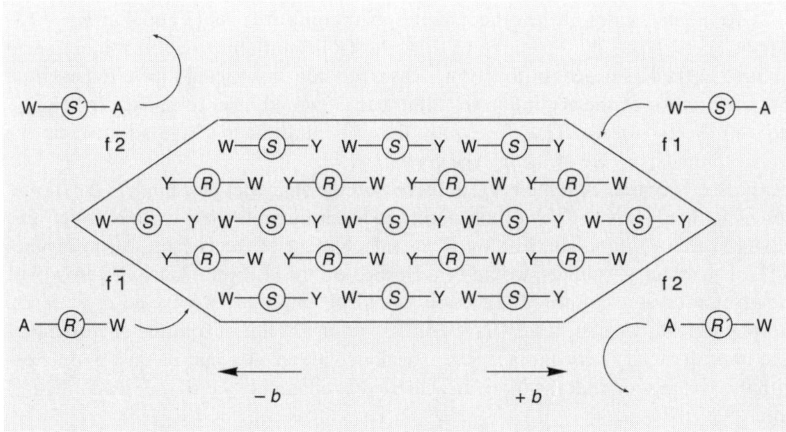

Fig. 12.6. *Model of an enantiopolar crystal consisting of the enantiomorphic crystal components W–S–Y and Y–R–W, and schematic representation of the attachment of the enantiomorphic additives W–S′–A and A–R′–W on the crystal surfaces f1, f2, f$\bar{1}$, and f$\bar{2}$*

This prerequisite is fulfilled in the already-mentioned α-glycine crystal (*Fig. 12.3*), as the individual glycine molecules are all present in chiral conformations. There are a total of two enantiopolar sets of glycine conformers (conformers **1** and **2**, and conformers **3** and **4** in *Fig. 12.3*). The f1 surface of a glycine crystal may be formed either by molecules in the conformations **1** and **2**, or the conformations **3** and **4**.

In one set of conformers (conformers **1** and **2**), the C–$H_{pro\text{-}R}$ bond of the CH_2 group is perpendicular, the C–$H_{pro\text{-}S}$ bond almost parallel to the f1 face. If an α-amino acid is introduced as an additive to the glycine solution, only its (*R*)-enantiomer is able to replace glycine molecules of this conformer set in the f1 surface. It is only here that the α-amino-acid side chain, which takes the place of the glycine $H_{pro\text{-}R}$-atom, protrudes out of the crystal surface (*Fig. 12.7*). The original position of the replaced glycine molecule cannot be taken by the (*S*)-enantiomer of an α-amino acid, as the side chain would have to replace the $H_{pro\text{-}S}$-atom, and would, therefore, have to be placed almost within the crystal surface.

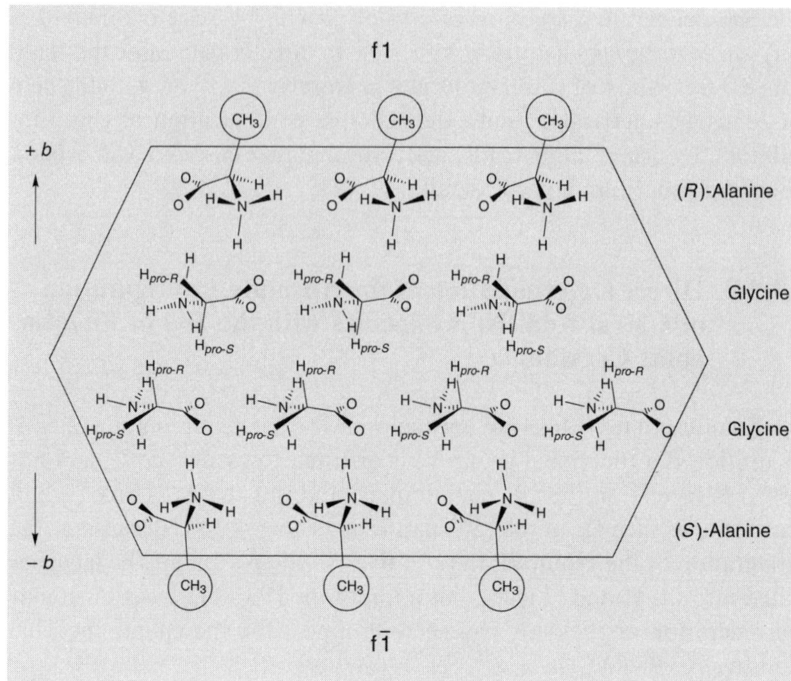

Fig. 12.7. *Section of the model of an α-glycine crystal, in whose f1 surface (R)-alanine, and in whose f1̄ surface (S)-alanine is incorporated*

Glycine molecules of the other enantiopolar conformer set (**3** and **4** in *Fig. 12.3*) cannot be replaced by either the (*S*)- or the (*R*)-enantiomer of the α-amino-acid additive in the f1 surface. In the former case, the side chain would have to penetrate into the interior of the crystal, in the latter case it would have to replace the $H_{pro\text{-}R}$-atom in the f1 surface. Therefore, only the (*R*)-enantiomer of the additive can be bound to the f1 surface of the glycine crystal.

Each (*R*)-amino-acid molecule adsorbed on the f1 crystal face hinders the formation of further layers of glycine molecules, and hence disrupts the growth of the glycine crystal in the $+b$ direction, leading to an enlargement of the f1 face. An (*S*)-amino acid, on symmetry grounds, would be adsorbed on the f1̄ crystal face, and growth of the glycine crystal in the $-b$ direction would be impeded. α-Glycine crystals are morphologically altered, if they have been grown in a solution containing a D-amino-acid impurity: the f1 crystal face is drastically enlarged. A result of this kind agrees with the assumption that the D-amino acid used as an impurity is of the (*R*)-configuration.

Further Reading

To Sect. 12.1.

M. J. Buerger, *Contemporary Crystallography*, McGraw-Hill, New York, 1970.

P. Luger, *Modern X-Ray Analysis on Single Crystals,* Walter de Gruyter, Berlin, 1980.

E. R. Wölfel, *Theorie und Praxis der Röntgenstrukturanalyse,* Vieweg, Braunschweig, 1987.

J. D. Dunitz, *X-Ray Analysis and the Structure of Organic Molecules,* 2nd Corr. Reprint, Verlag Helvetica Chimica Acta, Basel, and VCH Verlagsgesellschaft, Weinheim, 1995.

J. M. Bijvoet, A. F. Peerdeman, A. J. van Bommel, *Determination of the Absolute Configuration of Optically Active Compounds by Means of X-Rays, Nature* **1951,** *168,* 271.

To Sect. 12.2.

12.2.1.

D.Y. Curtin, I.C. Paul, *Chemical Consequences of the Polar Axis in Organic Solid-State Chemistry, Chem. Rev.* **1981,** *81,* 525.

12.2.2.

Z. Berkovitch-Yellin, L. Addadi, M. Idelson, L. Leiserowitz, M. Lahav, *Absolute Configuration of Chiral Polar Crystals, Nature* **1982,** *296,* 27.

L. Addadi, Z. Berkovitch-Yellin, I. Weissbuch, J. van Mil, L.J.W. Shimon, M. Lahav, L. Leiserowitz, *Growth and Dissolution of Organic Crystals with Taylor-Made Inhibitors – Implications in Stereochemistry and Materials Science, Angew. Chem. Int. Ed.* **1985,** *24,* 466.

I. Weissbuch, L. Addadi, M. Lahav, L. Leiserowitz, *Molecular Recognition at Crystal Interfaces, Science* **1991,** *253,* 637.

J.M. McBride, *Symmetry reduction in solid solutions: a new method for material design, Angew. Chem. Int. Ed.* **1989,** *28,* 377.

B. Kahr, J.M. McBride, *Optically anomalous crystals, Angew. Chem. Int. Ed.* **1992,** *31,* 1.

12.2.3.

I. Weissbuch, L. Addadi, Z. Berkovitch-Yellin, E. Gati, S. Weinstein, M. Lahav, L. Leiserowitz, *Centrosymmetric Crystals for the Direct Assignment of the Absolute Configuration of Chiral Molecules. Application to the α-Amino Acids by Their Effect on Glycine Crystals, J. Am. Chem. Soc.* **1983,** *105,* 6615.

13 NMR Spectroscopy

13.1. Physical Foundation

NMR (*Nuclear Magnetic Resonance*) spectroscopy is the indispensable method of current organic chemistry, employed for the determination of constitution, configuration, and conformation of organic compounds in solution. As atoms, or, more precisely, atomic nuclei in a molecule give rise to NMR signals, NMR spectroscopy makes available exceptionally well situated sensors with a range of a few femtometers. They act as molecular informants, since the spectroscopic properties of the nucleus are heavily dependent on its molecular environment. The interactions of nuclei with one another, through space or through the bonds connecting them, permit the measurement of internuclear distances or torsion angles.

For a relatively long time after its discovery by *Edward Purcell* and *Felix Bloch* (*Nobel* Prize 1952), the unique information provided by NMR, when compared to other spectroscopic methods, was difficult to exploit because of lack of sensitivity. NMR Signals are intrinsically weak, and were hard to detect. The introduction of the so-called pulse *Fourier* technique (1964) solved this problem at a stroke. The coming of multidimensional *Fourier* techniques (1974) enabled internuclear interactions to be determined even for macromolecules with several hundred nuclei. The development and application of these methods still remains an active area of research. Awarding the *Nobel* Prize for chemistry to *Richard R. Ernst* in 1991 underlined the recognition of the two fundamental achievements: *pulse Fourier spectroscopy* and *multidimensional NMR spectroscopy*.

13.1.1. The Make-up of the Atomic Nucleus

The nucleus of an atom is composed of *protons* and *neutrons*, with the exception of the ^{1}H-nucleus, which consists solely of one proton. The proton carries a single positive charge, the neutron, however, is uncharged. Common to both nuclear particles is their *spin*, which can be imagined pictorially as a rotation of each nuclear particle about its central axis. It is comparable with the rotation of the earth. With the rotation of an atomic nucleus, the charged elementary particles (*quarks*) inside that nucleus also rotate about the axis, so that a circular current flows. Such a current induces a magnetic moment along the rotation axis. The magnetic moment interacts with a magnetic field in the same way as a magnetic needle does with the earth's magnetic field. The most stable (lowest-energy) orientation of a charged particle is attained when its magnetic moment is oriented parallel to the external magnetic field, the least stable (highest-energy) orientation when the magnetic moment is aligned antiparallel to the external magnetic field.

If several elementary particles belong to one nucleus, the individual rotations must be added or subtracted to arrive at an overall rotation. We are now going to quantify this relationship.

13.1.2. Quantization of Spin

While macroscopic objects (the earth, for example) may rotate at whichever speed they wish about their center axis, this is not possible for microscopic objects (such as nuclear particles). The spin of a proton or neutron is constant: $I = h/2$. *Planck*'s constant h ($h = 6.6 \cdot 10^{-34}$ Js) is the unit for this rotation (*angular momentum*). It corresponds to the angular momentum of a miniaturized billiard ball of 1 ng mass and a radius of 1 nm, rotating about its own axis once every two hours. The atomic nucleus is, in general, composed of several neutrons and protons. The spin of an atomic nucleus consequently results from the spins of its individual components. As the spin of each individual nuclear component is 1/2 (in units of h), the total spin of an atomic nucleus is always a multiple of 1/2. Hence, the possible values of the spin of an atomic nucleus are 0, 1/2, 1, 3/2, ... While the charge of an atomic nucleus may be calculated by a simple addition of the charges of the individual nuclear components, this is not the case for the total spin. Therefore, the spins I for the atomic nuclei most frequently encountered in organic chemistry are listed in *Table 13.1*. Only the so-called *gg-nuclei*, with even numbers of protons and of neutrons (such as ^{4}He, ^{12}C, ^{16}O) have no nuclear spin.

According to the principles of quantum mechanics, a nucleus with a given spin I can only adopt $2I + 1$ different rotational states, relative to, for example, the z axis (of a *Cartesian* coordinate system). These $2I + 1$ states may be expressed by the relationship $m_z = -I, -(I-1), ..., (I-1), I$, where m_z is the expectation value of the z component of the spin I. For the case $I = 1/2$, two states are obtained: $m_z = +1/2$ and $m_z = -1/2$. One state corresponds to the parallel orientation of the z component of the spin to the z axis, the other one corresponds to the antiparallel orientation. This is reminiscent of the two different spin states of electrons. The latter may only be 'filled into' orbitals in pairs, and then only with opposite spins (*Chapt. 8.3.2.2*).

Just as the spin of a nuclear component produces a magnetic moment, the spin of an entire nucleus is assigned to the total moment $|\mu| = \gamma \cdot \hbar \cdot \sqrt{I(I+1)}$ ($\hbar = h/2\pi$). The proportionality between the spin I and the magnetic moment μ is described by the *gyromagnetic ratio* γ. The gyromagnetic ratio is dependent on the structure of the nucleus and, like the nuclear spin, cannot be derived from the number of nuclear particles by simple equations. Gyromagnetic ratios, relative to that of the hydrogen isotope ^{1}H, are reproduced in *Table 13.1*. The natural abundances of nuclear isotopes important in organic chemistry may also be taken from *Table 13.1*.

Table 13.1. *Natural Abundance, Spin, and Gyromagnetic Ratio* (relative to that one of ^{1}H) *of Nuclear Isotopes Relevant to Organic Compounds*

Isotope X	Natural abundance [%]	Spin	γ_X/γ_H
^{1}H	99.98	1/2	1
^{3}H	0	1/2	1.06663
^{13}C	1.108	1/2	0.25144
^{15}N	0.37	1/2	0.10133
^{19}F	100	1/2	0.94077
^{29}Si	4.7	1/2	0.19865
^{31}P	100	1/2	0.40481
^{57}Fe	2.19	1/2	0.03231
^{103}Rh	100	1/2	0.03147
^{2}H	0.0115	1	0.15351
^{6}Li	7.42	1	0.14716
^{14}N	99.63	1	0.07224
^{7}Li	92.58	3/2	0.38863
^{11}B	80.42	3/2	0.32084
^{23}Na	100	3/2	0.26451

13.1.3. Energy States of a Spin in an External Magnetic Field

The energy of a magnetic moment in an external magnetic field B_0, conventionally aligned along the z axis (unit vector $\vec{e}_z$), is given by:

$$E = -\vec{\mu}\, B_0\, \vec{e}_z = -\mu_z\, B_0 = -\gamma\, I_z\, \hbar\, B_0$$

Hence, for a nucleus with a spin $I = 1/2$, two levels, with the energies $E(m_z = \pm 1/2) = -\gamma\hbar\,(\pm B_0/2)$, are generated. In general, $2I + 1$ states with energetically equidistant separations $\gamma\hbar B_0$ are obtained. The tendency of a magnetized needle to orient itself parallel, and not antiparallel, to an external magnetic field provides an illustration of this phenomenon. The strength with which it tries to align itself with the field is dependent on the strength of the field.

13.1.4. NMR-Spectroscopic Transitions

Spectroscopy is the recording of the frequency-specific response of matter to electromagnetic radiation. If radiation is absorbed by matter (such as a molecule), then a transition from a lower-energy state to a higher-energy state takes place. In nuclear magnetic resonance, an alternating magnetic field, aligned transverse to the external B_0 field, and of frequency ν, induces transitions between energetically separated spin states, provided the resonance condition $h\nu = \gamma\hbar B_0 = \Delta E$ or $\omega = 2\pi\nu = \gamma B_0$ is fulfilled. The latter equation is the usual formulation of the resonance condition of NMR spectroscopy.

When this is the case, electromagnetic radiation is absorbed and, by measurement of this frequency-dependent absorption, an NMR spectrum is obtained. An NMR spectrum mirrors the distribution of energy differences of the various spin states. Owing to the finite lifetime (*relaxation*) T_2 of the spin state, electromagnetic radiation is also absorbed in the vicinity of the resonance $\omega = \gamma B_0$. This leads to a distinct shape of the NMR line (the *Lorentz* line). The shape of the *Lorentz* line, which consists of a real component and an imaginary component, is shown in *Fig. 13.1*.

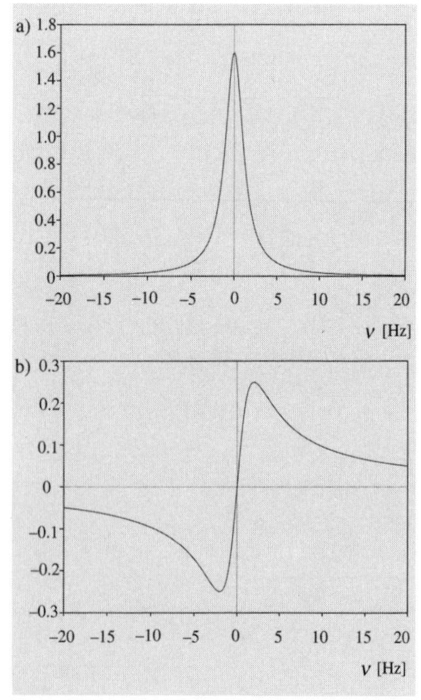

Fig. 13.1. a) *Absorption signal and* b) *dispersion signal of a* Lorentz *line.* The relaxation time T_2 is 0.5 s.

The real component (the *absorption signal*) is represented by $A(\omega)$, the imaginary component (the *dispersion signal*) by $D(\omega)$. The width of the absorption line at half of the total amplitude of the signal amounts to $1/\pi T_2$. The absorption line is narrower than the dispersion line. Normally, only the absorption component is displayed in the spectrum, as it reaches its maximum at the point where the resonance condition is exactly fulfilled, unlike the imaginary component, and converges more quickly to zero when the resonance condition $\omega = \gamma B_0$ is no longer exactly met.

$$A(\omega) = T_2/[1 + (\omega T_2)^2], \quad D(\omega) = \omega T_2^2/[1 + (\omega T_2)^2]$$

The integral under the absorption line is independent of the linewidth of the signal. Therefore, it is possible to directly determine the number of NMR-active nuclei in a compound, simply by integration.

13.2. Chemical Shift

13.2.1. Shielding and Deshielding of Nuclei

The resonance condition $\omega = \gamma B_0$ would result in the same resonance frequency being obtained for every nuclear isotope, irrespective of its chemical environment. In reality, the electrons situated around the nucleus alter the magnetic field B_0 in the vicinity of the nucleus. The external field will either be intensified above the value B_0, or weakened below it. The resonance conditions for identical nuclear isotopes in different chemical environments will, therefore, be altered. The displacement of the resonance frequency of a nuclear isotope relative to others of the same type because of its chemical environment is known as *chemical shift*. Chemically non-equivalent nuclei are described in NMR terminology as *anisochronous* (see *Chapt. 4.1.2*).

The NMR spectrum of EtOH serves as an example. The H-nuclei of the Me group are in a different chemical environment compared to the H-nuclei of the CH_2 group and the H-nucleus at the O-atom. The integrals of the three signals are in the ratio, from right to left, of 3:2:1 (*Fig. 13.2*), and may hence be assigned to the Me, the CH_2, and the OH group, respectively.

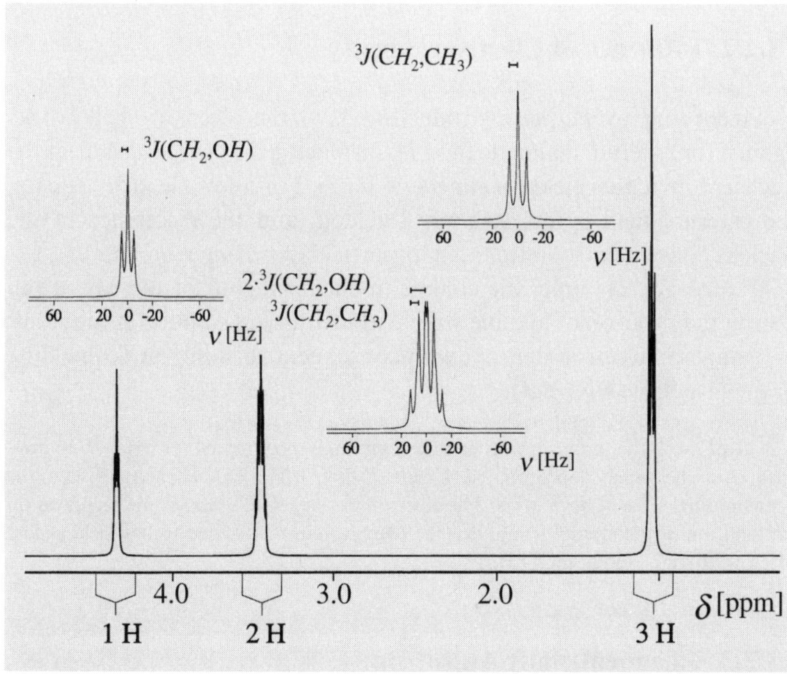

Fig. 13.2. *¹H-NMR spectrum of EtOH*

The chemical shift of the resonance frequency of the NMR signal $(-\sigma)$ may be obtained from the resonance condition $\omega = \gamma B_0 (1 - \sigma)$ as a dimensionless value. As it is much smaller than 1, it is multiplied by 10^6 and given in ppm (parts per million). The chemical shift of the sample under investigation is given relative to that of a reference substance (usually tetramethylsilane, TMS). This relative chemical shift is specified by δ. It, similarly, is given in ppm, and may be derived from the resonance frequency of the reference substance ν_{ref}, and that of the NMR line of interest ν, thus:

$$\delta[\text{ppm}] = 10^6 (\nu - \nu_{ref})/\nu_{ref}.$$

The resonance frequencies of the H-atoms of EtOH are $\delta(CH_3) = 1.05$ ppm, $\delta(CH_2) = 3.45$ ppm, and $\delta(OH) = 4.35$ ppm. To characterize chemical shifts, the expressions 'high field' and 'low field' are used for small δ values and large δ values, respectively. The Me group is said to resonate at high field, the OH group at low field. NMR Spectra are normally recorded with the chemical shift increasing from right to left.

Like the spin of a nucleus, or its gyromagnetic ratio, the chemical shift of nuclei in a molecule cannot be simply calculated. Moreover, it is dependent on the orientation of the molecule relative to the external magnetic field. In solution, varying orientations of the molecule adopted relative to the external field contribute to the chemical shift. A quantitative interpretation of the chemical shift for answering structural questions is only recently emerging. It is sensible, though, to discuss the chemical shifts of many related compounds relative to each other. We shall, therefore, qualitatively discuss some effects which influence chemical shift.

13.2.2. Influence of Electron Density

According to *Faraday*'s induction law, the electron shell works against the external magnetic field B_0, displacing the chemical shift of the nucleus to a higher field (*diamagnetic shift*). For a low electron density, the external field is less strongly shielded, and the resonance of the nucleus consequently shifted to a lower field (*paramagnetic shift*).

Hence, for example, the change in chemical shift of the Me group during the titration of alanine with acid in aqueous solution is the result of changes in electron density, with consequent shielding or deshielding of the H-nuclei (*Fig. 13.3*).

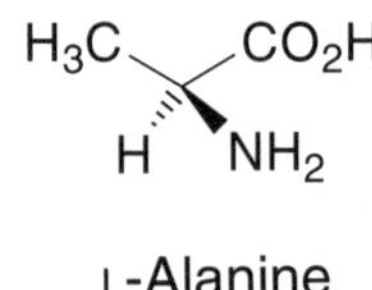

L-Alanine

At pH < 9, the amino group is overwhelmingly protonated, at pH > 9, in contrast, overwhelmingly deprotonated (*Chapt. 10.5.4.1*). The COOH group exists in the deprotonated state in both cases. The electron density on the Me group decreases on protonation of the amino group, *i.e.*, the Me resonance is shifted to low field at low pH, and to high field at high pH.

13.2.3. Chemical-Shift Anisotropy

The chemical shift of a nucleus is dependent on the orientation of the molecule relative to the external magnetic field B_0. This effect is particularly pronounced in molecules with conjugated π-electron systems. HMO Theory offers an explanation for their chemical-shift anisotropy. A planar, cyclic molecule with $4n + 2\,\pi$-electrons is 'aromatic' as defined by *E. Hückel* (*Chapt. 8.3.3.4.2* and *14.4.3*).

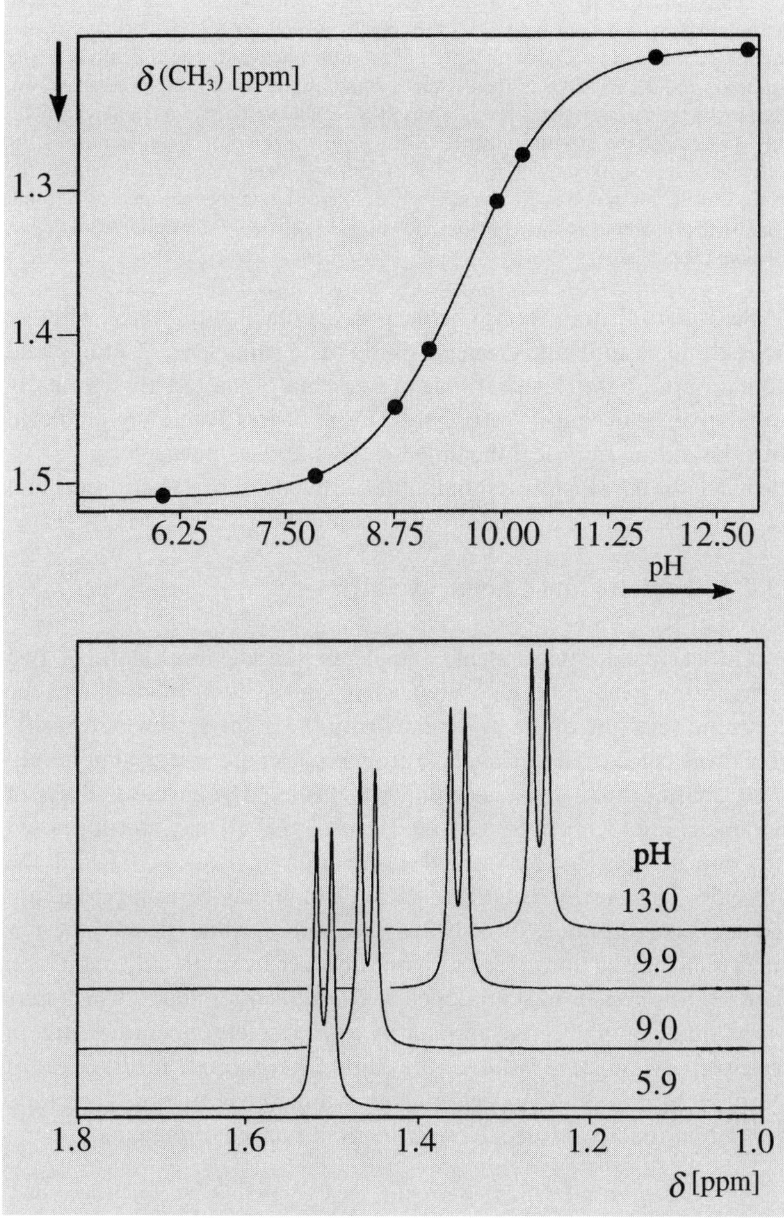

Fig. 13.3. *Chemical shift of the Me protons of alanine during titration of an aqueous solution.* The pK_2 of alanine, *ca.* 9.6, may be read off directly.

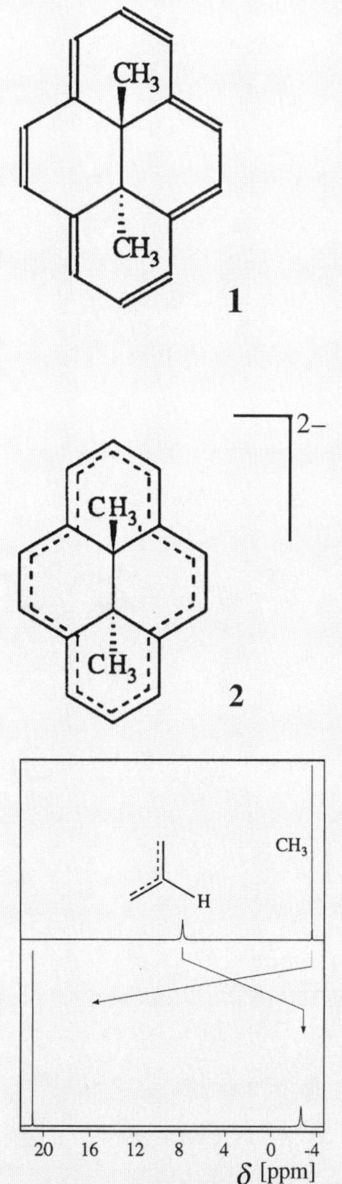

Fig. 13.4. *Chemical shifts of the 'exocyclic' protons and the angular Me-group protons in* **1** (above) *and* **2** (below)

Formula **1** (*Fig. 13.4*) represents such a compound with 14 π-electrons. If the plane of the molecule is oriented perpendicular to the external field B_0, a diamagnetic ring current is produced along the π-system. This ring current is comparable with a current produced by the introduction of a coil, perpendicularly, into the magnetic field. The additional magnetic field arising from the microscopic ring current weakens the external field in the interior of the 'coil', as required by *Lenz*'s law. The Me groups in the interior of **1**, the chemical shift of which would normally be expected to be at *ca.* 1 ppm, only attain resonance at a very high field caused by the diamagnetic ring current mentioned. By contrast, the magnetic field outside the 'coil' is strengthened, and the CH protons, whose chemical shift in olefins could be expected at 5–6 ppm, are in fact in case of **1** on resonance at 8 ppm.

Compound **1**, with $4n + 2$ π-electrons can be converted into the 'antiaromatic' compound **2**, with $4n + 4 = 4(n + 1)$ π-electrons (*Chapt. 14.4.2*) by chemical reduction. Since the electronic ground state of 'antiaromatic' compounds exhibits *triplet character*, and the HOMOs of the degenerate pair are each occupied by one electron, the spins of the single electrons align themselves parallel to the external magnetic field. This strengthens the external magnetic field B_0 in the interior of the 'antiaromatic' molecule, while weakening it outside. The paramagnetic ring current causes the protons of the Me group in **2** to be strongly deshielded, and they resonate at 22 ppm. The CH protons outside the π-system, however, are strongly shielded, resonating at -3 ppm (*Fig. 13.4*).

Because of their differing chemical environments, the nuclei within a molecule have different chemical shifts. The influences of atoms and atom groups on the chemical shift of a nucleus of interest are frequently cumulative, so chemical shifts can be more or less accurately predicted with the aid of empirical increments. This applies particularly to ^{13}C-chemical shifts, which, essentially, are influenced by constitution and configuration, and less by conformation, unlike ^{1}H-chemical shifts.

13.2.4. Topicity and Chemical Shift

The chemical environment, and hence the chemical shift, of two nuclei in the same molecule is dependent on the conformation and the electronic structure of the molecule. From the point of view of topicity, which was considered in *Chapt. 4*, atomic nuclei are arranged in homotopic, enantiotopic, and diastereotopic classes. The chemical shifts of two nuclei are identical (*isochronous nuclei*) if they may be transposed into one another by a symmetry operation (*Chapt. 11.1.1*) on the molecule. This is why *homotopic* nuclei, which may be transposed into one another by rotation, are always *isochronous*, while *enantiotopic* nuclei, which may be transposed into one another by rotary reflection, are, likewise, *isochronous* in achiral media. *Diastereotopic* nuclei, which may not be transposed into one another by any symmetry operation, are, in principle, always *anisochronous*. In chiral, non-racemic media, axes of rotary reflection cease to be valid as symmetry elements. Therefore, *enantiotopic* nuclei are also *anisochronous* in such environments.

The Me protons in EtOH are *homotopic*, the CH_2 protons are *enantiotopic*, and, therefore, chemically *equivalent* in achiral media. The chemical non-equivalence of enantiotopic protons in chiral, non-racemic media can be established using the example of benzyl alcohol ($PhCH_2OH$), by comparison of its spectra in an achiral solvent ($CDCl_3$) before and after the addition of the chiral ligand tris{3-[(trifluoromethyl)-hydroxymethylene]-*d*-camphorato}europium(III) ([Eu(tfc)$_3$]) (*Fig. 13.5*).

The enantiotopic CH_2 protons are chemically equivalent in $CDCl_3$, but, after addition of the chiral shift reagent, chiral complexes between [Eu(tfc)$_3$] and $PhCH_2OH$ are formed, in which the CH_2 protons are diastereotopic, and hence chemically non-equivalent. The two observed *doublets* illustrate this (*Fig. 13.5, c*). The shape of the *doublets* will be dealt with in detail in *Sect. 13.3.1*.

Topicity relationships are also helpful in the analysis of polymer tacticity (*Chapt. 7.2.1*). Despite their sometimes very high molecular mass, polymers of stereochemically uniform configuration have very

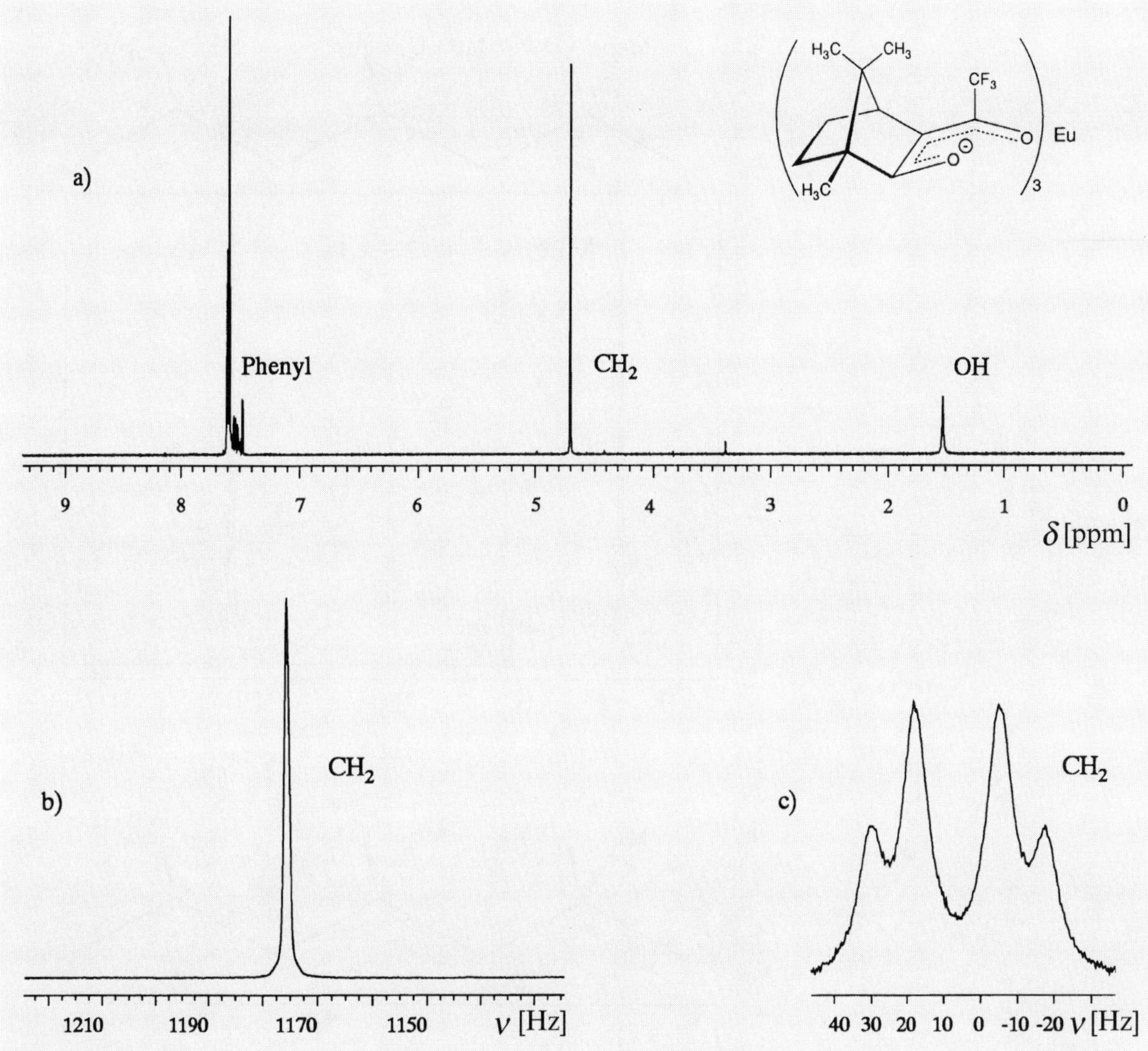

Fig. 13.5. *^{1}H-NMR Spectrum of benzyl alcohol in an achiral medium* ($CDCl_3$): a) *normal;* b) *section, high resolution;* c) *section after addition of [Eu(tfc)$_3$]*

simple spectra: a result of their translational symmetry, stemming from their constitutional repeating units. The tacticity of polymers may be clarified using NMR methods in such cases.

In *syndiotactic* poly(methyl methacrylate), the CH_2 protons are homotopic, as a vertical C_2 axis, located in the plane of the paper, runs through each methylene C-atom. Consequently, the two CH_2 protons appear as a *singlet* (*Fig. 13.6, a*).

In *isotactic* poly(methyl methacrylate), however, the CH_2 protons are diastereotopic, as none of the symmetry operations (translation about a constitutional repeating unit, reflection across a plane defined through CH_2 or $C(CH_3)$–$COOCH_3$) will transpose the two CH_2 protons into one another. Consequently, the two protons appear at different chemical shifts (*Fig. 13.6, b*).

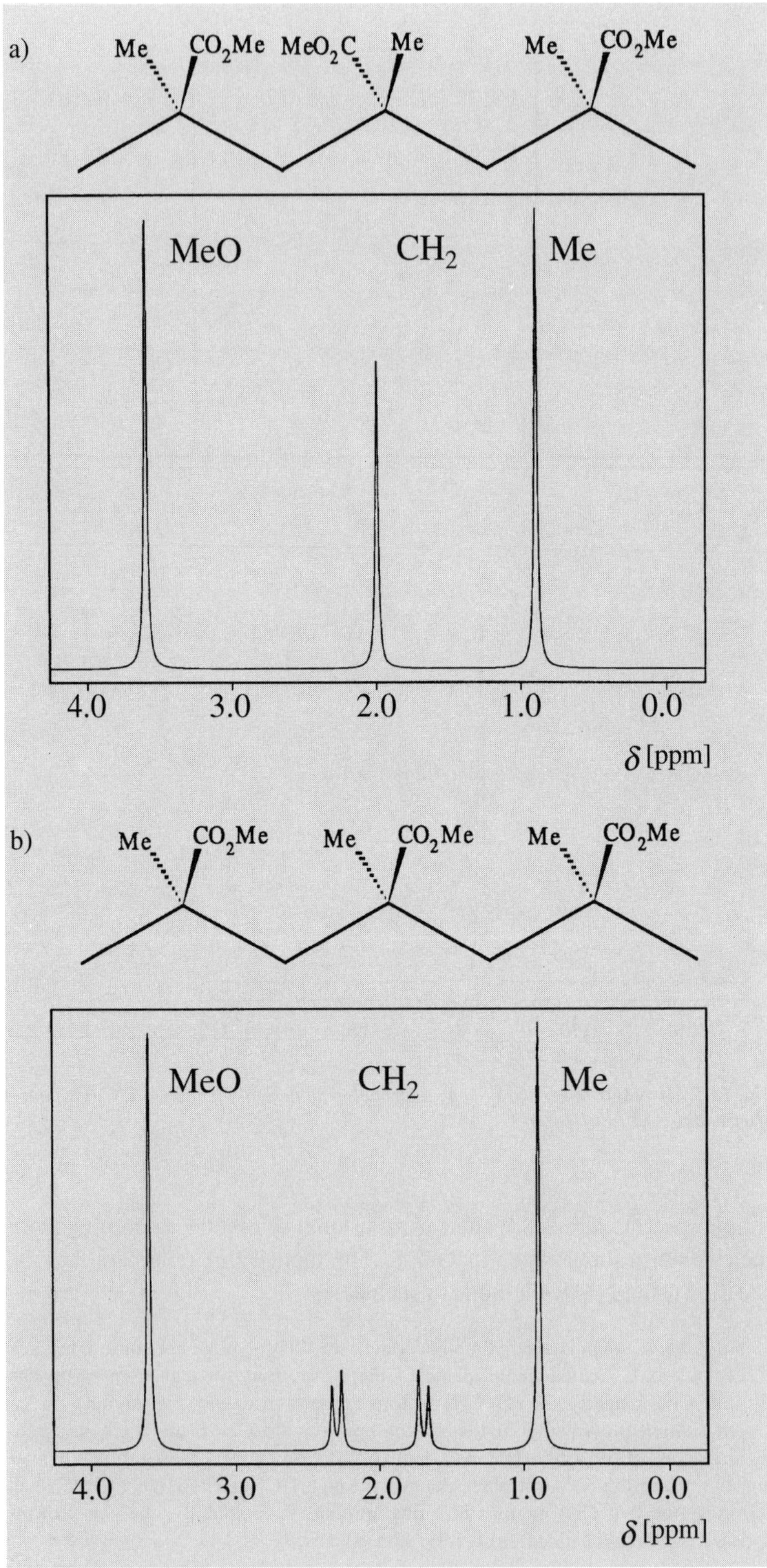

Fig. 13.6. a) *^{1}H-NMR Spectrum of syndiotactic poly(methyl methacrylate)*; b) *^{1}H-NMR Spectrum of isotactic poly(methyl methacrylate)*

13.2.5. Dynamic Effects

The chemical shift of a nucleus is dependent on the conformation of the molecule concerned. However, because of the difficulty in quantitatively predicting the chemical shifts of nuclei, it is not generally possible to determine the conformation of a molecule based solely on chemical shift. If a molecule in solution does exist in several conformations, then this fact is reflected in the spectrum. Interconversion of conformations of one molecule and chemical interconversion of molecules can be monitored by chemical exchange of nuclei with each other.

It is clear that a nucleus in a molecule in a particular conformation has a specific chemical shift. But what does the spectrum of a nucleus look like, when the latter is part of a molecule able to exist in two different conformations? If the positions adopted by the nucleus concerned in the two conformations are chemically non-equivalent, so that the nucleus in the two conformations gives rise to two anisochronous signals, of frequencies ν_1 and ν_2, and, moreover, if the two conformations exist in a chemical equilibrium, assuming equivalent populations, and converting into one another at a reaction rate k, then the real component of the spectrum obeys the following equation ($\omega = 2\pi\nu$):

$$A(\omega) = k(\omega_1 - \omega_2)^2/[(\omega - \omega_1)^2(\omega - \omega_2)^2 + 4k^2(\omega - (\omega_1 + \omega_2)/2)^2]$$

It is possible to distinguish between three cases (*Fig. 13.7*):

- slow chemical exchange of nuclei with each other: $k \ll |\omega_1 - \omega_2|$,
- rapid chemical exchange of nuclei with each other: $k \gg |\omega_1 - \omega_2|$,
- coalescence: $k = |\omega_1 - \omega_2|/8^{1/2}$.

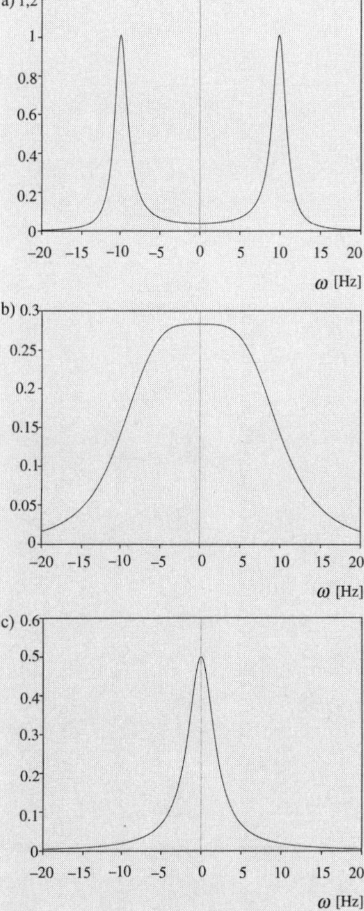

Fig. 13.7. *Dynamic NMR spectroscopy.* Two resonances at ω_1 and ω_2, with the difference in frequencies $\Delta\omega = \omega_1 - \omega_2$ (here $\Delta\omega = 20$ Hz), exchange in an equilibrium with identical forward and backward rate k: *a*) slow exchange; *b*) coalescence at $k = |\Delta\omega|/8^{1/2}$; *c*) rapid exchange.

13.2.5.1. Slow Chemical Exchange

From the point of view of NMR spectroscopy, two long-lived conformations behave like two different chemical compounds in a mixture. In this instance ($k \ll |\omega_1 - \omega_2|$), two undistorted *Lorentz* lines are obtained in the spectrum at ω_1 and ω_2 (*Fig. 13.7, a*). The width of the *Lorentz* lines at half amplitude is $2k$:

$$A(\omega) = k/[(\omega_1 - \omega)^2 + k^2] + k/[(\omega_2 - \omega)^2 + k^2]$$

For slow exchange, the number of conformations may be directly taken from the fine structure, and the rate of exchange of conformers from the line-width of the NMR signals. This only applies, however, as long as the intrinsic line-width of the signals due to spin relaxation is sufficiently small to be disregarded in comparison to k. The hindered rotation about the peptide bond is such an instance of a slow conformational exchange with respect to the NMR time scale.

13.2.5.2. Rapid Chemical Exchange

In the case of a rapid chemical exchange, only the arithmetic mean of the chemical shifts ω_1 and ω_2 of the two rapidly interchanging conformers is obtained. 'Rapid', in this sense, means that the difference of the resonance frequencies of the nuclei of interest in the conformations is small in comparison to the rate of conformational exchange ($k \gg |\omega_1 - \omega_2|$) (*Fig. 13.7, c*):

$$A(\omega) = 2\Delta_{1/2}/[(\omega_\Sigma - \omega)^2 + \Delta^2_{1/2}],\ \omega_\Sigma = (\omega_1 + \omega_2)/2,\ \Delta_{1/2} = (\omega_1 - \omega_2)^2/8k$$

The line-width is inversely proportional to the conformational-exchange rate, and proportional to the square of the difference in the chemical shifts of the nuclei concerned. Conformational exchanges which count as rapid on the NMR time scale are found in cases of rotation about a single bond, such as the pseudorotation in five-membered rings.

13.2.5.3. Coalescence

If the rate of conformational exchange is of the same order of magnitude as the difference in chemical shifts, then a very broad line is obtained. The minimum observed between the two signals for slow exchange disappears (*Fig. 13.7, b*). At the point of coalescence, the exchange rate is given directly by: $k = |\omega_1 - \omega_2|/8^{1/2}$.

13.2.5.4. Dynamic NMR Spectroscopy (Demonstrated on *cis*-Decalin)

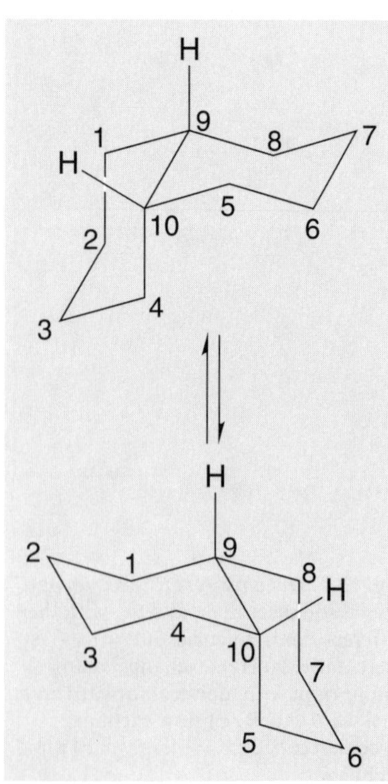

Using dynamic NMR spectroscopy, it is possible to study the rates of conformational exchange in equilibria. It is even possible to recognize degenerate reactions, in which educt and product are identical, or racemizations, in which educt and product are enantiomers, and hence give the same NMR spectra. The chair-chair inversion of *cis*-decalin is one such example.

cis-Decalin exists in solution in two enantiomorphic, C_2-symmetrical conformations (see *Chapt. 6.3.5.1*). The spectra of the two enantiomers are identical. Because of the C_2 axis bisecting the C(9)–C(10) bond, the nuclei C(1)/C(5), C(2)/C(6), C(3)/C(7), C(4)/C(8), and C(9)/C(10) are homotopic and their resonances isochronous. Hence, five different signals are to be expected in the case of slow conformational exchange. At low temperatures, the ^{13}C-NMR spectrum of *cis*-decalin does indeed show five signals (*Fig. 13.8*).

Ring inversion of *cis*-decalin converts the chemically non-equivalent resonance pairs C(1)/C(5) and C(4)/C(8), and C(2)/C(6) and C(3)/C(7) into one another. At 220 K, five ^{13}C-NMR signals are observed. The exchange of the enantiomorphic conformations is slow: C(1)/C(5)$\leftrightarrow$ C(4)/C(8); C(2)/C(6)$\leftrightarrow$C(3)/C(7). At higher temperature (280 K), however, the signals of these C-atoms coalesce. Because the difference in chemical shift of the exchanging pairs are, coincidentally, more or less

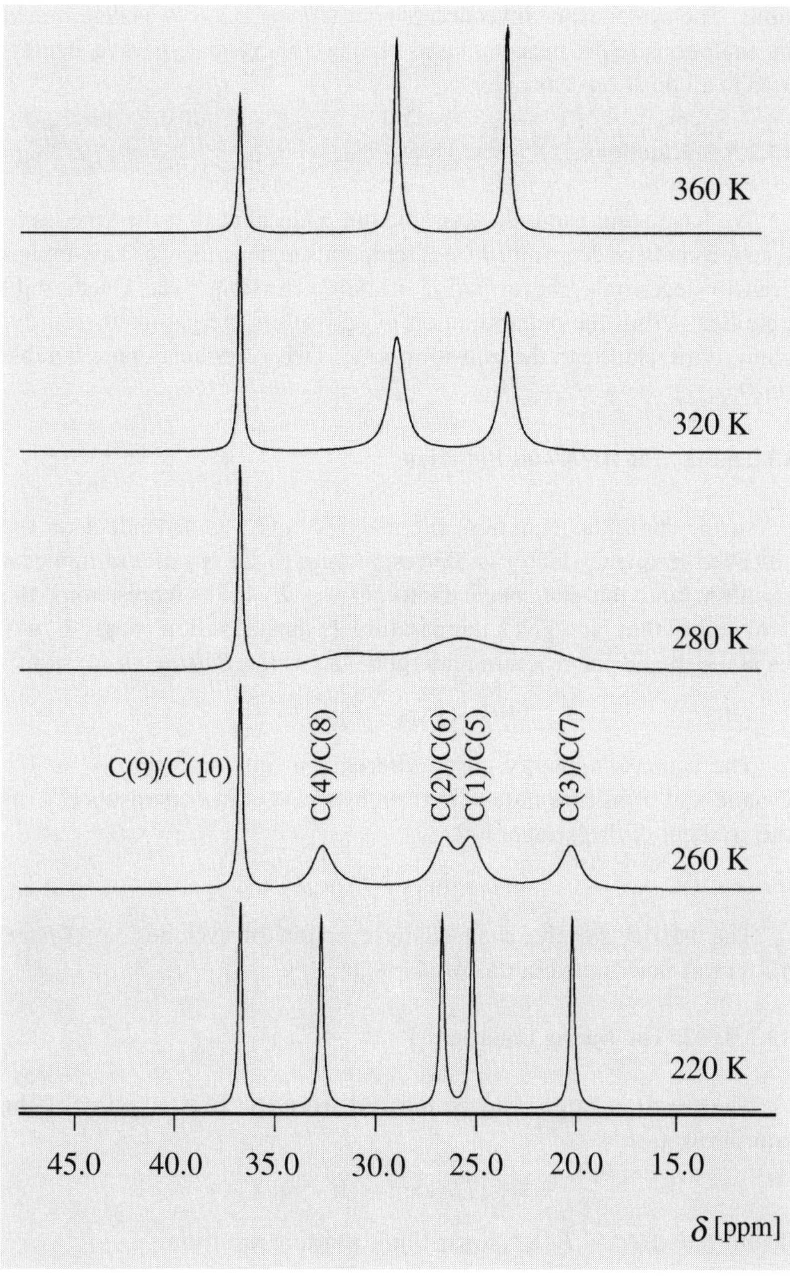

Fig. 13.8. *^{13}C-NMR Spectra of* cis-*decalin at different temperatures*

the same, all of the signals coalesce at this temperature. Upon increasing the temperature further, the NMR lines once again become sharp. Now, three sharp signals, for C(1)/C(4)/C(5)/C(8), C(2)/C(3)/C(6)/C(7), and C(9)/C(10), are obtained as a result of the time-averaged C_{2v} symmetry. The signal of C(9) and C(10) remains sharp over the whole temperature range, as these C-atoms are each located in chemically equivalent positions in the two enantiomorphic conformations.

The temperature-dependent measurement of reaction rates k permits the determination of energy barriers to conformational interconversions. The cyclohexane inversion barrier (*Chapt. 6.3.6*) was determined by analogous NMR measurements on a cyclohexane derivative, deuterated in all positions bar one.

13.2.5.5. Kinetics

We have seen, using the example of *cis*-decalin, that the kinetics of reactions may be determined from temperature dependence. This applies even to degenerate conformational interconversions. The kinetic data obtained permit the determination of activation energies or thermodynamic data relating to the transition state. Two approaches are available for this.

13.2.5.5.1. The *Arrhenius* Equation

In the *Arrhenius* equation, the reaction rate k is dependent on the so-called frequency factor A, corresponding to the maximum rate of a reaction, and the *Boltzmann* factor $\exp(-E_a/k_BT)$, representing the probability that, at a given temperature T, the activation energy E_a will be attained and the reaction undergone (k_B is the *Boltzmann* constant).

$$k = A\exp(-E_a/k_BT)$$

The activation energy – the difference in internal energies for the ground and transition states – is given by a plot of $\ln k$ against $(k_BT)^{-1}$ as the gradient of the straight line:

$$\ln k = \ln A - E_a(k_BT)^{-1}$$

The barrier for the chair-chair inversion of cyclohexane (*Chapt. 6.3.6*) was determined in this way.

13.2.5.5.2. The *Eyring* Equation

The reaction rate k can be derived from the free enthalpy of the transition state.

$$k = k_BT/h\exp(-\Delta G^{+}/k_BT)$$

As $\Delta G^{+} = \Delta H^{+} - T\Delta S^{+}$, logarithmic plotting results in:

$$\ln k = \ln(k_BT/h) + \Delta S^{+}/k_B - \Delta H^{+}/k_BT$$

Measurement of the temperature dependence of the reaction rate enables the thermodynamic functions ΔH^{+} and ΔS^{+} to be determined for the transition state. Analysis of this kind is not limited to the interpretation of NMR spectra. The temperature dependence of the rate of conversion of the cyclohexane twist-boat conformation into the chair conformation (*Chapt. 6.3.6*) was measured using IR spectroscopy at low temperature, allowing the energy barrier for this transition to be determined.

13.3. Scalar Coupling

13.3.1. Physical Foundation, Two-Spin Systems

In an NMR experiment, a nucleus is exposed to an external magnetic field, the strength of which is influenced not only by the electron shell of the nucleus, but also by other nuclei in the vicinity. If two nuclei, I_1 and I_2, both with a spin of 1/2, are considered, then there are two possible orientations for I_2: parallel ($m_z = 1/2$) or antiparallel ($m_z = -1/2$), relative to the external field B_0. Since a spin cannot adopt two states simultaneously, only one of these spin states can be adopted in each individual molecule. Measured over all of the molecules in the sample, the populations of the two states are practically the same. I_2 induces, dependent on its orientation relative to the external field B_0, an additional field in the vicinity of the nucleus I_1, altering the resonance frequency of this nucleus by a particular amount. This displacement is of the same magnitude for the two possible orientations of the spin I_2, but different in sign. For the parallel orientation of I_2, the resonance of nucleus I_1 is shifted by $J/2$, for the antiparallel orientation by $-J/2$. Since the two orientations of the spin I_2 are equal in probability, I_1 affords a spectrum consisting of two lines with half the expected intensity at $\nu_1 \pm J/2$. Such a pattern of two lines of equal intensity is called a *doublet* (*d*). The splitting of the frequency is called the *coupling constant J*. A reciprocal analysis applies to I_2. The splitting of the NMR resonance of I_2 is also J. I_1 and I_2 are said to couple with one another: they represent a two-spin system (*Fig. 13.9, a*).

Such a system is also known as an *AX* system, where the two different letters express the chemical non-equivalence of the nuclei I_1 and I_2. The degree of separation of the letters in the alphabet indicates the degree of separation of the resonance frequencies. The spectrum of the two CH_2 protons in $PhCH_2OH$ after the addition of $[Eu(tfc)_3]$ shows two resonances with a small difference in chemical shifts: here an *AB* system is under consideration (see *Fig. 13.5*). In an *AB* spectrum – unlike the spectrum of an *AX*-spin system – the line intensities of the *doublet* components are no longer the same. The four resonance lines form a 'roof' (*roof effect*), the inner line of each *doublet* component being more intense relative to the outer line.

13.3.2. Further Spin Systems, Their Specifications, and Spectra

If spin I_1 couples with two spins, I_2 and I_3, then these two spins I_2 and I_3 can assume a total of four spin orientations: both parallel (1/2,1/2), both antiparallel (−1/2,−1/2), and the two mixed states: (−1/2,1/2) and (1/2,−1/2). The coupling constant between I_1 and I_2 is called $J(1,2)$, that between I_1 and I_3, $J(1,3)$. Hence, the resonance of I_1 has four lines of equal intensity, with the frequencies $\nu_1 + (J(1,2) + J(1,3))/2$, $\nu_1 - (J(1,2) + J(1,3))/2$, $\nu_1 - (J(1,2) - J(1,3))/2$, and $\nu_1 + (J(1,2) - J(1,3))/2$. Such a pattern is called a *doublet* of *doublets* (*dd*) (*Fig. 13.9, b*). It is the spectrum of the *A* spin in an *AMX* system. Both couplings ($J(1,2)$ and $J(1,3)$) are to be found twice in the *multiplet*. If the couplings $J(1,2)$ and $J(1,3)$ are identical, then two lines of the *doublet* of *doublets* both occur at the chemical shift ν_1, giving a *triplet* (*t*) (*Fig. 13.9, c*). The Me group and the OH group of EtOH can serve as examples; both appear as *triplets* because of their coupling to the two enantiotopic CH_2 protons (*Fig. 13.2*). If I_1 has three coupling partners, then eight possible states exist, giving rise to a *doublet* of *doublets* of *doublets* (*ddd*) (*Fig. 13.9, d*), if all of the couplings of I_1 to the various spins are different. If two couplings are the same, then a *doublet* of *triplets* (*dt*)

(*Fig. 13.9, e*), and, if all three couplings are the same, then a *quadruplet* (*q*) (*Fig. 13.9, f*) is obtained. The two enantiotopic, chemically equivalent CH_2 protons in EtOH can serve as a further example. These couple with the three protons of the Me group and the proton of the OH group. A *doublet* of *quadruplets* (*dq*) is the result (*Fig. 13.2*). The spin system of EtOH is described as AM_2X_3.

Two chemically equivalent nuclei may still exhibit unequal coupling to a third spin. Nuclei of this kind are chemically, but not magnetically, equivalent. They are specified by means of the alphabet, describing

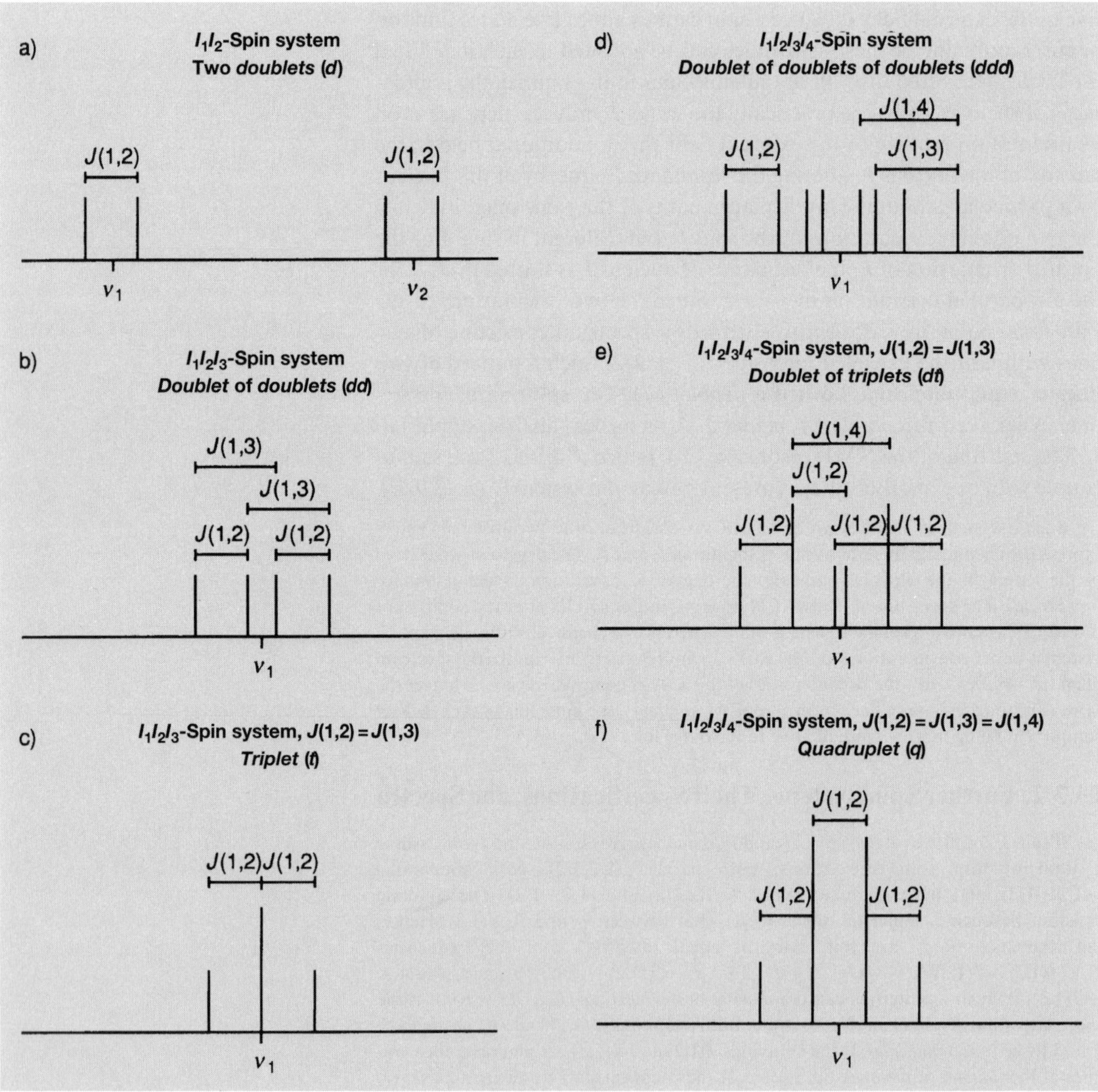

Fig. 13.9. *Coupling patterns for coupled spins of 1/2.* For *b)–f)* only spin I_1 is considered. *a)* Two-spin system with *doublets* of splitting $J(1,2)$. *b)* Three-spin system with different couplings $J(1,2)$ and $J(1,3)$ leads to a *doublet* of *doublets*. *c)* In the case of identical couplings, $J(1,2) = J(1,3)$, a *triplet* with line intensity ratio 1:2:1 is obtained. *d)* Four-spin system with three different couplings $J(1,2)$, $J(1,3)$, and $J(1,4)$. *e)* $J(1,2) = J(1,3) \neq J(1,4)$ gives a *doublet* of *triplets*. *f)* $J(1,2) = J(1,3) = J(1,4)$ gives a *quadruplet* with line intensities 1:3:3:1.

nuclei of this type with the same letter, but differentiating between them by primes. Hence, the spin system of 1,2-dibromobenzene is an *AA'BB'* system (*Table 14.1*), as the protons H–C(3)/H–C(6) and H–C(4)/H–C(5) are chemically equivalent. H–C(3) couples over three bonds with H–C(4), while the chemically equivalent H–C(6) couples with H–C(4) through four bonds. Therefore, the two chemically equivalent nuclei H–C(3) and H–C(6) are not magnetically equivalent. The same applies for H–C(4) and H–C(5). The *multiplet* structure in spectra of such spin systems is, in general, more complicated than that in spectra of spin systems lacking chemically equivalent, but magnetically non-equivalent, nuclei.

Simple inspection suffices to obtain coupling constants from *multiplets* with few lines or few different couplings. To obtain coupling constants from complex *multiplets*, two-dimensional NMR techniques must be employed (see *Aspects of Organic Chemistry*, Volume 4: *Methods of Structure Determination*).

13.3.3. Conformational Analysis with the Aid of Coupling

Coupling constants are very well suited for establishing local conformation. Couplings between nuclei which are connected to one another over three bonds (so-called ^3J or *'vicinal' coupling*) depend on the torsion angle in a manner determined by a *Karplus* curve (*Fig. 13.10*).

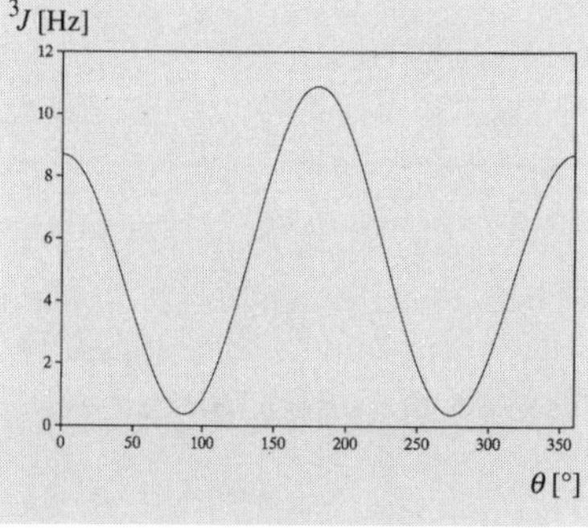

Fig. 13.10. *Typical* Karplus *curve* (^3J(H,H) = 9.4 cos$^2\theta$ – 1.1 cosθ + 0.4), *giving the relationship between torsion angle θ and ^{3}J*

Although the parameters specifying a given *Karplus* curve differ from case to case, the general tendency of the curve is fairly similar for almost all combinations of nuclei. The *Karplus* curve has a local maximum at $\theta = 0°$, a global maximum at $\theta = 180°$, and global minima at $\theta = 90°$ and 270°. Conformational analysis for compounds with atomic positions in the diamond lattice (*Chapt. 10.3.3*) is especially simple. Here, only torsion angles of ± 60° and 180° exist. The coupling for an *ap*-arrangement (θ = 180°) is easy to distinguish from that of an *sc*-arrangement (θ = ± 60°).

As an example, we will choose *β*-maltose octaacetate (*Fig. 13.11*). The following coupling constants may be determind: *J*(H–C(1'),H–C(2')) = 4.1 Hz; *J*(H–C(2'),H–C(3')) = 10.5 Hz; *J*(H–C(3'),H–C(4')) = 9.6 Hz; *J*(H–C(4'),H–C(5')) = 10.0 Hz; *J*(H–C(5'),H–C(6')) = 2.4 Hz; *J*(H–C(5'),H'–C(6')) = 3.5 Hz; *J*(H–C(1),H–C(2)) = 8.1 Hz; *J*(H–C(2),H–C(3)) = 9.2 Hz; *J*(H–C(3),H–C(4)) = 8.7 Hz; *J*(H–C(4),H–C(5)) = 9.6 Hz; *J*(H–C(5),H–C(6)) = 2.6 Hz; *J*(H–C(5),H'–C(6)) = 4.3 Hz. The diaxial couplings are above 8 Hz, the equatorial/axial and diequatorial couplings below 5 Hz. An unequivocal proof of the relative configurations of the stereocenters in the hexopyranose rings is, therefore, possible. The *α*-glycosidic connection between the two glucose units can immediately be recognized by the ^3J(H–C(1'),H–C(2')) coupling of 4.1 Hz. All other ^3J(H,H) couplings lie between 8 and 10 Hz, showing the diaxial positioning of the H-atoms. This is in agreement with the configuration of the glucose units in maltose, and with the assumption of the chair conformation for the two glucose residues with equatorial AcO groups (*Fig. 13.11*). The *β*-configuration is shown by the *J*(H–C(1),H–C(2)) coupling of 8.1 Hz.

The conformation about the C(1')–O(4) bond and the O(4)–C(4) bond cannot be established using ^3J(H,H) couplings, because of the lack of H-atoms at O(4). Methods

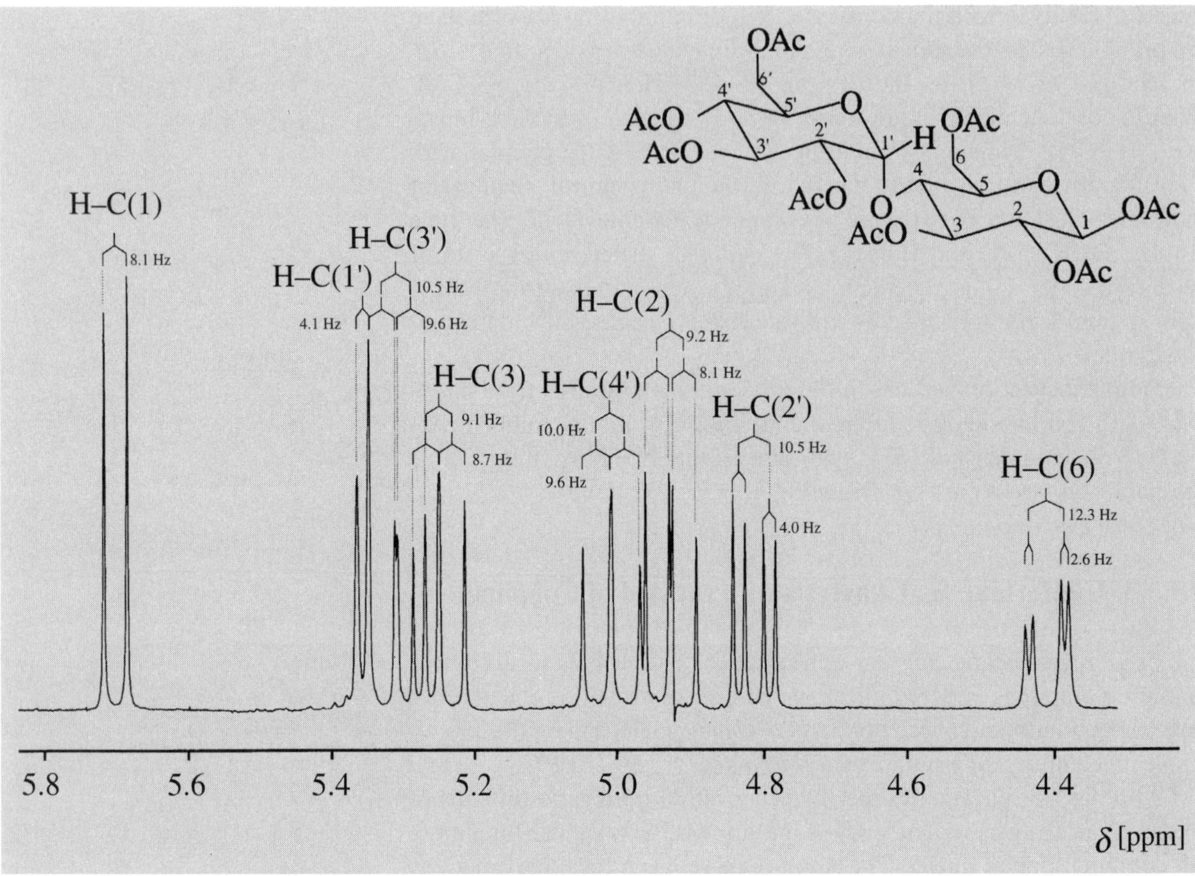

Fig. 13.11. *Section of a ^{1}H-NMR spectrum of β-maltose octaacetate*

for the determination of the conformation about such bonds, using coupling constants, have been reserved for *Aspects of Organic Chemistry*, Volume 4: *Methods of Structure Determination.*

13.4. Distance Measurements

NMR Spectroscopy is able to determine the spatial distance between NMR-active nuclei. This is possible by means of the quantitative determination of the so-called *NOE effect* (*nuclear* Overhauser *enhancement effect*). The NOE effect between two nuclei *i* and *j* is inversely proportional to the sixth power of the internuclear distance r_{ij} and directly proportional to the squares of the gyromagnetic ratios of the two nuclei *i* and *j*: NOE effect $\sim \gamma_i^2\gamma_j^2(r_{ij})^{-6}$.

Because of the size of the gyromagnetic ratio of H-atoms, in comparison with that one of other nuclei (*cf. Table 13.1*), NOE effects are mainly determined between protons. NOE Effects may be measured only up to a internuclear distance of about 500 pm, because of their rapid decrease with distance. However, distances through space are almost more important for the determination of the conformation of a compound than scalar couplings through bonds. The significance of NOE effects is explained below, with regard to the determination of secondary structure in peptides and proteins (*Chapt. 7.3.3.2.3*).

It is possible to distinguish between the various elements of secondary structure, as they show significant variations in the distance of protons. We will compare α-helix, β-pleated sheets, and β- and γ-loops with one another. The expected distances between H-atoms in the protein backbone for various elements of secondary structure are given in *Tables 13.2–13.5*. The structure of the idealized β- and γ-loops can be taken from *Fig. 13.12*.

Table 13.2. *NMR-Relevant Distances* [pm] *and Coupling Constants* [Hz] *for Characteristic Elements of Secondary Structure in Proteins*

Distance	α-Helix	3_{10}-Helix	Antiparallel β-pleated sheet	Parallel β-pleated sheet
$H_i^\alpha \cdots H_{i+1}$	350	340	220	220
$H_i^\alpha \cdots H_{i+2}$	440	380		
$H_i^\alpha \cdots H_{i+3}$	340	330		
$H_i^\alpha \cdots H_{i+4}$	420	330		
$H_i \cdots H_{i+1}$	280	260	430	420
$H_i \cdots H_{i+2}$	420	410		
$J(H_i,H_i^\alpha)$	4	4	9	9

Table 13.3. *Characteristic Distances* [pm] *Between Protons in a* β_I*-Loop*[a])

Distance	LL	LD	DL	DD
$H_2^\alpha \cdots H_2$	287	287	240	240
$H_2 \cdots H_3$	290	290	290	290
$H_2^\alpha \cdots H_3$	361	361	310	310
$H_3^\alpha \cdots H_3$	298	245	298	245
$H_3 \cdots H_4$	265	265	265	265
$H_2^\alpha \cdots H_4$	387	387	455	455
$H_3^\alpha \cdots H_4$	336	344	336	344

[a]) The distances have been taken from an idealized β_I-loop and are given for each of the four possible configurations (the first descriptor (D or L) corresponds to position 2, the latter one to position 3).

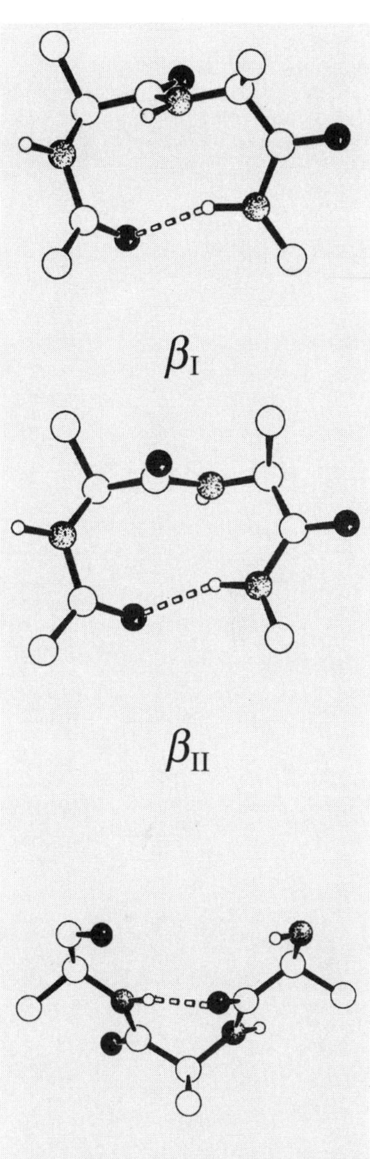

Fig. 13.12. β_{I}-, β_{II}-, *and* γ*-loops in proteins with* L*-amino acids in positions 2 and 3*

Table 13.4. *Characteristic Distances* [pm] *Between Protons in a β_{II}-Loop* [a])

Distance	LL	LD	DL	DD
$H_2^{\alpha}\cdots H_2$	287	287	240	240
$H_2\cdots H_3$	473	473	473	473
$H_2^{\alpha}\cdots H_3$	218	218	335	335
$H_3^{\alpha}\cdots H_3$	230	300	230	300
$H_3\cdots H_4$	279	279	279	279
$H_2^{\alpha}\cdots H_4$	352	352	445	445
$H_3^{\alpha}\cdots H_4$	336	344	336	344

[a]) The distances have been taken from an idealized β_{II}-loop and are given for each of the four possible configurations (the first descriptor (D or L) corresponds to position 2, the latter one to position 3).

Table 13.5. *Characteristic Distances* [pm] *Between Protons in a γ-Loop* [a])

Distance	L	D
$H_2^{\alpha}\cdots H_2$	229	296
$H_2^{\alpha}\cdots H_3$	371	270
$H_2\cdots H_3$	390	390

[a]) These distances have been taken from an idealized γ-loop (the descriptor (L or D) corresponds to position 2).

The enantiomorphic β_I'- and β_{II}'-loops may be obtained from reflection of β_I- and β_{II}-loops across the plane of the paper. It can clearly be seen, that β_I is sterically most favorable for L-,L-amino acids in positions 2 and 3 of the β-loop, while β_{II} is best fitted to L-,D-amino acids, as then the *Newman* strain between C_2^{β} and O_1 as well as C_3^{β} and O_2 is minimal (*Chapt. 7.3.3.5*). Consequently, β_I' is most favorable for D-,D- and β_{II}' for D-,L-amino acids in positions 2 and 3. The γ-loop preferentially incorporates D-amino acids, the enantiomorphic γ'-loop L-amino acids, in position 2.

If the distance between two H-atoms whose nuclei interact with one another through space, is inserted into the r^{-6} relationship, then a 'calculated' value for the magnitude of the NOE effect is obtained. In this way, it can quickly be seen that a distinction of secondary-structure elements by virtue of their varying $H\cdots H$ distances is very easy with the aid of NOE patterns.

Elements of secondary structure in proteins are often stabilized by hydrogen bonds. The amide H-atoms involved in hydrogen bonds exchange more slowly with the H-atoms of water than do amide H-atoms situated in unstructured regions of the protein. The time needed for exchange can reach many hours or even days. If, therefore, a protein is dissolved in D_2O, only the amide H-atoms of amino-acid residues located in the structured region of the protein will remain protonated, and hence visible in the ^1H-NMR spectrum, the other H-atoms exchange with deuterium and disappear in the ^1H-NMR spectrum. Recently, the formation of secondary structure during refolding of unstructured conformation of a denatured protein (with all amide protons exchanging rapidly with water or D_2O) into their structured, native conformation (a proportion of the amide protons is observable in the ^1H-NMR) has been an area of keen study. First results show that the elements of secondary structure (β-pleated sheets, α-helices) can assemble themselves within a few milliseconds.

The opportunity of identifying elements of secondary structure by means of distance measurements obtained from NOE spectroscopy is not limited to proteins. Distinction between A- and B-DNA may be made with the help of intramolecular H···H distances, just as the conformations about glycosidic bonds in oligosaccharides, or the pairing behavior of oligonucleotides with DNA duplexes in triple helices may be clarified.

NMR Spectroscopy is an indispensable method, able to provide information about the structure and dynamics of molecules and supramolecular systems. This chapter has only introduced NMR spectroscopy to the degree necessary for the discussion of structural problems mentioned in this volume. A thorough discussion has been reserved for *Aspects of Organic Chemistry*, Volume 4: *Methods of Structure Determination*, as the outline of the required techniques goes beyond the limits of this introductory consideration of NMR spectroscopy.

Further Reading

General

H. Friebolin, *Ein- und zweidimensionale NMR-Spektroskopie,* VCH Verlagsgesellschaft, Weinheim, 1988.

R.K. Harris, *Magnetic Resonance Spectroscopy,* Longman, New York, 1983.

M. Hesse, H. Meier, B. Zeeh, *Spektroskopische Methoden in der organischen Chemie,* Thieme Verlag, Stuttgart, 1984.

Pretsch, Clerc, Seibl, Simon, *Tabellen zur Strukturaufklärung organischer Verbindungen,* Springer-Verlag, Berlin, 1976.

H. Günther, *NMR-Spektroskopie,* Thieme Verlag, Stuttgart, 1983.

H.O. Kalinowski, S. Berger, S. Braun, *^{13}C-NMR-Spektroskopie,* Thieme Verlag, Stuttgart, 1984.

To Sect. 13.1.

E.M. Purcell, H.G. Torrey, R.V. Pound, *Resonance Absorption by Nuclear Magnetic Moments in a Solid, Phys. Rev.* **1946**, *69,* 37.

F. Bloch, W. Hansen, M.E. Packard, *Nuclear Induction, Phys. Rev.* **1946,** *69,* 127.

F. Bloch, *Nuclear Induction, Phys. Rev.* **1946,** *70,* 460.

R. R. Ernst, W. A. Anderson, *Application of Fourier-Transformation Spectroscopy to Magnetic Resonance, Rev. Sci. Instr.* **1966,** *37,* 93.

R. R. Ernst, *Sensitivity Enhancement in Magnetic Resonance, Adv. Magn. Reson.* **1966,** *2,* 1.

J. Jeener, *Ampère International Summer School* (Basko Polje, Jugoslawien) 1971.

W. P. Aue, E. Bartholdi, R. R. Ernst, *Two-Dimensional Spectroscopy. Application to Nuclear Magnetic Resonance, J. Chem. Phys.* **1976,** *64,* 2229.

R. Benn, H. Günther, *Modern Pulse Methods in High-Resolution NMR Spectroscopy, Angew. Chem. Int. Ed.* **1983,** *22,* 350.

K. Wüthrich, *NMR of Proteins and Nucleic Acids,* John Wiley & Sons, New York, 1986.

R. R. Ernst, G. Bodenhausen, A. Wokaun, *Principles of Nuclear Magnetic Resonance in One and Two Dimensions,* Clarendon, Oxford, 1987.

A. Bax, L. Lerner, *Two-Dimensional Nuclear Magnetic Resonance Spectroscopy, Science* **1986,** *232,* 960.

H. Kessler, M. Gehrke, C. Griesinger, *Two-Dimensional NMR Spectroscopy: Background and Overview of the Experiments, Angew. Chem. Int. Ed.* **1988,** *27,* 490.

R. R. Ernst, *Nuclear Magnetic Resonance Fourier Transform Spectroscopy, Angew. Chem. Int. Ed.* **1992,** *31,* 805.

C. Griesinger, O. W. Sørensen, R. R. Ernst, *A Practical Approach to Three-Dimensional NMR Spectroscopy, J. Magn. Reson.* **1987,** *73,* 574.

H. Oschkinat, C. Griesinger, P. J. Kraulis, O. W. Sørensen, R. R. Ernst, A. Gronenborn, G. M. Clore, *Three Dimensional NMR Spectroscopy of a Protein, Nature* **1988,** *332,* 374.

J. Mason, *Multinuclear NMR,* Plenum Press, New York, 1987.

A. E. Derome, *Modern NMR Techniques for Chemistry Research,* Pergamon Press, Oxford, 1987.

N. Chandrakumar, S. Subramanian, *Modern Techniques in High Resolution FT-NMR,* Springer-Verlag, New York, 1987.

J. K. M. Sanders, B. K. Hunter, *Modern NMR Spectroscopy,* University Press, Oxford, 1987.

Attar-ur-Rahman, *One and Two Dimensional NMR Spectroscopy,* Elsevier, Amsterdam, 1989.

A. Abragam, *The Principles of Nuclear Magnetism,* Clarendon Press, Oxford, 1978.

C. P. Slichter, *Principles of Magnetic Resonance,* Springer-Verlag, Berlin, 1978.

To Sect. 13.2.

L. M. Jackman, F. A. Cotton, *Dynamic Nuclear Magnetic Resonance Spectroscopy,* Academic Press, New York, 1975.

J. I. Kaplan, G. Fraenkel, *NMR of Chemically Exchanging Systems,* Academic Press, New York, 1980.

To Sect. 13.3.

H. Günther, *NMR-Spektroskopie,* Thieme Verlag, Stuttgart, 1983.

To Sect. 13.4.

J. H. Noggle, R. E. Shirmer, *The Nuclear Overhauser Effect, Chemical Applications,* Academic Press, New York, 1971.

K. Wüthrich, *NMR of Proteins and Nucleic Acids,* John Wiley & Sons, New York, 1986.

D. Neuhaus, M. Williamson, *The Nuclear Overhauser Effect in Structural and Conformational Analysis,* VCH Verlagsgesellschaft, New York, Weinheim, Cambridge, 1989.

J. B. Udgaonkar, R. L. Baldwin, *NMR evidence for an early framework intermediate on the folding pathways of ribonuclease A, Nature* **1988,** *335,* 694.

H. Roder, A. E. Gülnur, S. W. Englander, *Structural characterization of folding intermediate in cytochrome C by H-exchange labeling and proton NMR,* Nature **1988,** *335,* 700.

14 Benzene as a Special Case

14.1. *Kekulé*'s Proposed Constitution

In 1865, when *Kekulé* suggested a constitutional formula for benzene, in which C–C bonds alternated with C=C bonds in a six-membered ring, he also inserted the keystone for the constitutional theory of organic chemical compounds, founded only seven years before. Now it was possible, taking into account the tetravalency of the C-atom and the monovalency of the H-atom, to bind six CH groups together into a ring, using three *localized* C–C and three *localized* C=C bonds. The only procedure of critical testing available at the time, confirming whether or not the number of constitutional isomers predicted on the basis of *Kekulé*'s benzene and its derivatives agreed with those ones observed experimentally, led, however, into a dilemma.

Monosubstitution of benzene (or pentasubstitution with identical substituents) gives no constitutional isomers, and so raises no problems. Things are different, however, in the case of, for example, disubstitution. With two identical substituents, a total of four constitutional isomers would be expected; with two non-identical substituents even five. However, no more than three such constitutional isomers were ever found, in innumerable experimental studies.

Kekulé escaped this dilemma by developing a dynamic, rather than a static, concept of the constitution of the benzene ring: the C–C and C=C bonds are not localized, but change positions in the ring. The price he had to pay for this was inconsistency with *Butlerov*'s view, that every chemical compound can be described by *one* specific structural formula, and that every chemical structural formula corresponds uniquely to *one* particular chemical individual (*Chapt. 2.4*). *Kekulé*, who, besides *Loschmidt*, *Couper*, and *Butlerov*, had developed the constitutional model of the classical organic chemistry, also demonstrated, with his concept of the constitution of benzene and its derivatives, the limits of this model. A single *Kekulé* formula is not capable, by itself, of accurately representing the number of constitutional isomers of benzene and its derivatives. The other *Kekulé* formula, not generally given, must be inferred and a dynamic equilibrium between the two assumed. Hereby, benzene would average a representative of the D_{6h}, rather than the D_{3h} symmetry point group. *Robinson*'s symbol (for an electronic sextet) – a circle inside a regular hexagon – nicely fits with D_{6h} symmetry of benzene.

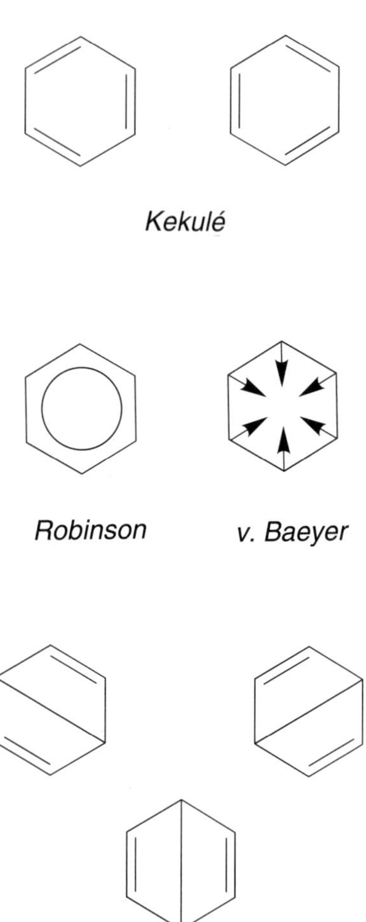

It would also be acceptable to use the symbol proposed by *A. v. Baeyer* (at the 1890 '*Benzolfest*'), indicating how '*the fourth valence, from our perspective, disappears*' from each C-atom of coordination number 3. This symbol for benzene had already been used by *Henry E. Armstrong* three years before. It fitted not only with the symmetry of benzene, but also with its thermodynamic stability (*Sect. 14.3*), which *Julius Thomsen* had determined experimentally.

It was possible to 'understand' the geometry and energy of benzene using two quantum-mechanical approaches, from which *Erich Hückel* had started out at the beginning of the thirties. One of these approaches leads to the *valence bond* (VB) theory, and to its qualitative subsidiary, the so-called *resonance theory*. For the quantitative theory, *Linus Pauling* had developed a simple mathematical method, with which it was possible to deal with the interaction between the two *limiting Kekulé formulae* of benzene in analogy with the resonance between two coupled pendulums of equal period. For precise calculation, however, three further functions must be taken into account for benzene: these correspond to the three descriptive *limiting Dewar formulae*. While the lucidity of the method is preserved by this amendment, the basis of the analogy with physical resonance is abolished. The hence inaccurately named *theory of resonance*, despite this fact, found favor with chemists as a source of inspiration, but, with time, was to lead to misuse and false interpretations. *Hückel*'s pointed comment that scientific progress had been held up by it for twenty years, is, therefore, very hard to deny.

The other quantum-theoretical approach was to lead to the general MO theory, to the special HMO theory, named after *Hückel*, and to its qualitative subsidiary: the qualitative HMO model. While the qualitative MO model is brought into service for those questions to which the structure model of the classical organic chemistry provides no answers, we do at least wish to have mentioned the concept of *resonance energy*, originating from the resonance theory and frequently employed in relevant literature (*vide infra*). However, we should refer in passing to the annotation of *Edgar Heilbronner* and *Hans Bock*, that resonance energies are not necessary for the interpretation of chemical and physical properties of compounds, and that their possible didactic advantages are canceled out by the confusion arising as a result of their uncritical application. First, though, we will look at ways in which the constitutional isomers to be found in one of the columns of *Fig. 14.1*, may be reliably distinguished from one another.

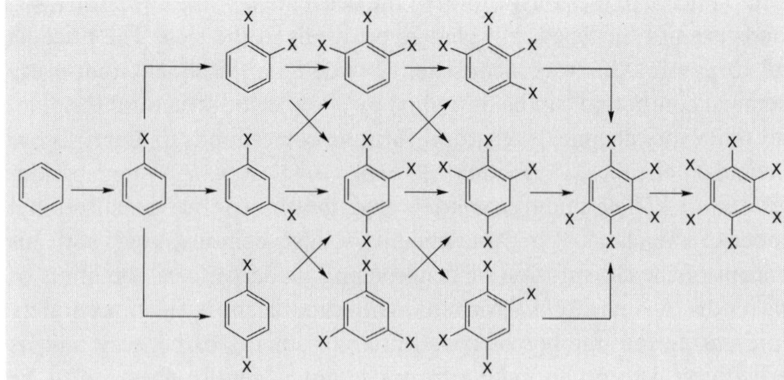

Fig. 14.1. *The sufficient set of constitutional formulae of benzene and its substitution products* (X for H). Constitutional isomers are grouped in columns.

14.2. Determination of Constitution in Benzene Derivatives

14.2.1. *Körner*'s Method

Wilhelm Körner had demonstrated as early as 1869 that the six benzene H-atoms are (in other words, of course) homotopic. His argumentation is essentially abstract, and dependent on a geometric concept in only one of his conclusions. *J. Michael McBride* has recently pointed out that *Körner*'s conclusions may be expressed in a fully abstract manner, using the language of group theory. Below we employ *Körner*'s method, but making use of the presupposed geometry of the benzene ring.

One way to distinguish between the three constitutional isomers of, for example, dibromobenzenes, is to determine the number of tribromobenzenes which can result from further substitution in each case (*Fig. 14.1*). From 1,4-dibromobenzene, only 1,2,4-tribromobenzene can be produced. From 1,2-dibromobenzene, two constitutional isomers, 1,2,3-tribromobenzene and 1,2,4-tribromobenzene, are obtained. Finally, 1,3-dibromobenzene affords three constitutional isomers: 1,2,3-, 1,2,4-, and 1,3,5-tribromobenzene. Similarly, 1,3,5-tribromobenzene correlates only with one tetrasubstituted product: 1,2,3,5-tetrabromobenzene. In contrast, 1,2,3-tribromobenzene leads to two constitutionally isomeric tetrabromobenzenes (1,2,3,4- and 1,2,3,5-tetrabromobenzene), while 1,2,4-tribromobenzene leads to three (1,2,3,4-, 1,2,3,5-, and 1,2,4,5-tetrabromobenzene).

The graph of *Fig. 14.1* shows the logic by which the structurally distinct representatives of an entire set are connected with one another. Use of *Körner*'s method – assessing how many constitutionally isomeric product components correlate with one educt of a given constitution – is not always so easy in practice as may appear at first glance: in some cases it is very difficult to verify the total number of possible isomers.

Independently of this, it would in any case be desirable to know the constitution of the individual product components: firstly to be able to identify them, and, not least, because in certain circumstances the constitution of just one product component is sufficient for determining that one of the educt. For example, if 1,3,5-tribromobenzene is found among the bromination products of one of the dibromobenzene isomers, then that educt must have been 1,3-dibromobenzene. A reliable method for distinguishing between the constitutional isomers in each column is offered by NMR spectroscopy.

14.2.2. NMR-Spectroscopic Assignment

Benzene itself shows only one signal in both its ^{1}H- and its ^{13}C-NMR spectra (at 7.26 and at 128.5 ppm, respectively): the six H-atoms and the six C-atoms are each homotopic.

Monosubstituted benzene derivatives exhibit a complex ^{1}H-NMR spectrum, with three groups of *multiplets* for the two *ortho*-protons, the two *meta*-protons, and the single *para*-proton. In the ^{13}C-NMR spectrum, four distinct signals (for C(1), C(2) and C(6), C(3) and C(5), and for C(4)) are observed for the six C-atoms.

The three constitutionally isomeric X_2-benzene derivatives are easy to identify NMR-spectroscopically. The 1,4-isomer has only homotopic protons, and so shows only a single signal in the ^{1}H-NMR spectrum, two in the ^{13}C-NMR spectrum. The 1,2-isomer displays a symmetrical ^{1}H-NMR spectrum of the *AA′BB′* type and a ^{13}C-NMR spectrum with three signals. Finally, the most complex spectra of the series are given by the 1,3-isomer: a ^{1}H-NMR spectrum of AB_2C type and a ^{13}C-NMR spectrum with four distinct signals. Similarly, it is possible to distinguish unambiguously between the three possible constitutional isomers of the tri- and tetrasubstituted benzene derivatives (*Table 14.1*).

Table 14.1. *NMR-Spectroscopic Determination of the Constitution of Benzene and Benzene Derivatives*

Compound	^{1}H	^{13}C
Benzene	A_6 *Singlet*	1 Signal
X-Benzene		
Monobromobenzene	*AA´BB´C Multiplet*	4 Signals
X_2-Benzene		
1,2-Dibromobenzene	*AA´BB´ Multiplet*	3 Signals
1,3-Dibromobenzene	AB_2C *Multiplet*	4 Signals
1,4-Dibromobenzene	*A A´A´´A´´´ Singlet*	2 Signals
X_3-Benzene		
1,2,3-Tribromobenzene	AB_2 *Multiplet*	4 Signals
1,2,4-Tribromobenzene	*ABC Multiplet*	6 Signals
1,3,5-Tribromobenzene	A_3 *Singlet*	2 Signals
X_4-Benzene		
1,2,3,4-Tetrabromobenzene	A_2 *Singlet*	3 Signals
1,2,3,5-Tetrabromobenzene	A_2 *Singlet*	4 Signals
1,2,4,5-Tetrabromobenzene	A_2 *Singlet*	2 Signals

14.3. Examining Energies

Structure and energy are the most important characteristics of a chemical compound. Consequently, the 'benzene case' has not only a structural aspect (D_{6h} symmetry rather than D_{3h}), but also an energetic one: *Thomsen*'s thermochemical examinations (*Sect. 14.1*) showed benzene to be much more stable than expected. This finding raises two problems:

- The statement 'more stable than' obviously relates to some reference substance.
- What does 'more stable than' mean to a chemist anyway?

For a molecule, the term 'stable' means a thermodynamic property, determinable by measurement of the relative molar *Gibbs* energy $\Delta_f G^\ominus$ at a given temperature ($^\ominus$ denotes the standard condition: a pressure of 0.1 MPa and the presence of the relevant stable phase). The change in the (molar) *Gibbs* energy in a chemical reaction r under standard conditions is given by the expression $\Delta_r G^\ominus = \Sigma_B \nu_B \Delta_f G^\ominus$ (B), where ν_B represents the stoichiometric coefficient of the reaction partner B. The relationship $\Delta_r G^\ominus = -RT \ln K$, between the change in *Gibbs* energy and the (dimensionless) thermodynamic equilibrium constant K, can be derived from the conditions of the chemical equilibrium.

A compound A is more stable than its isomer B, if the condition $\Delta_r G^\ominus > 0$ is fulfilled for the real or hypothetical reaction in which A is transformed into B. If, for the two reactions

$$P \rightarrow X + Y\ (\Delta_r G_1^\ominus);\ Q \rightarrow X + Z\ (\Delta_r G_2^\ominus)$$

the condition $\Delta_r G_1^\ominus > \Delta_r G_2^\ominus$ is met, then P, relative to Y, is more stable than Q, relative to Z. The expression 'more stable than' can apply both qualitatively and quantitatively in relation to either an expressly stated or a tacitly assumed standard.

The expressions 'stable' or 'more stable' should not be used synonymously with 'unreactive' or 'less reactive', as this would constitute mixing thermodynamic and kinetic concepts with one another. A relatively more stable chemical compound can by all means be more reactive towards a reaction partner than a reference compound.

14.3.1. Reaction Enthalpies, Enthalpies of Formation, and Bond Enthalpies

It is to be expected that one of the two diastereoisomers of (*E*)- and (*Z*)-but-2-ene is 'more stable' than the other. If both compounds are burned separately in pure O_2 (in a calorimetric bomb), then not only are the same product components (4 CO_2 and 4 H_2O) qualitatively and quantitatively found, but it can also be ascertained that, in the course of the exothermic reactions, a specific *reaction energy* $\Delta_c U$ (at constant volume) or *reaction enthalpy* $\Delta_c H$ (at constant pressure) has, in each case, been released into the surroundings:

$$\Delta_c H^\ominus\ ((E)\text{-But-2-ene}) = -2706.2\ \text{kJ/mol};$$
$$\Delta_c H^\ominus\ ((Z)\text{-But-2-ene}) = -2710.4\ \text{kJ/mol}.$$

Hence, the (*E*)-isomer, under standard conditions, is by 4.2 kJ/mol *thermodynamically more stable* than the (*Z*)-isomer (*Fig. 14.2*).

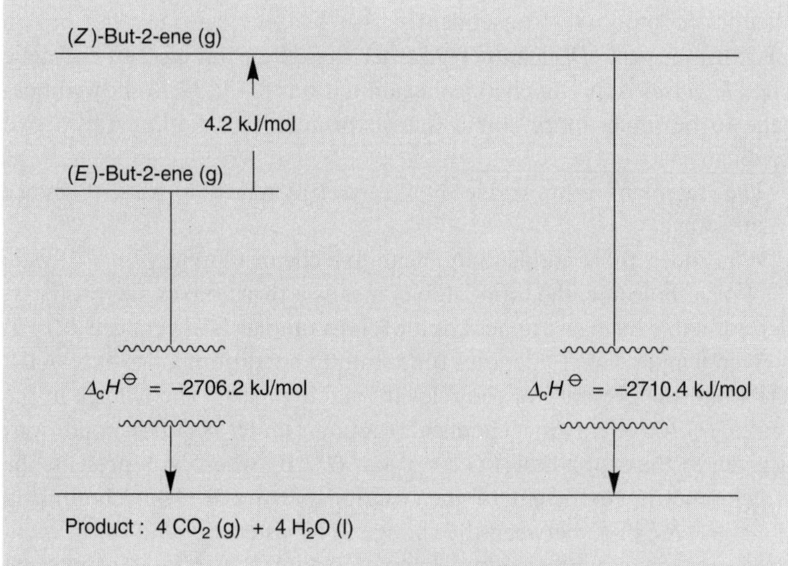

Fig. 14.2. *Graphic representation of calculation of the difference in thermodynamic stability between (*E*)- and (*Z*)-but-2-ene by means of their combustion enthalpies ($\Delta_c H^\ominus$)*

Instead of using the experimentally obtainable combustion enthalpies to calculate the difference in thermodynamic stability, it is also possible to commence from the experimentally determinable hydrogenation enthalpies (at 25 °C in glacial acetic acid):

$$\Delta_h H^\ominus \text{ ((}E\text{)-But-2-ene)} = -114.9 \text{ kJ/mol;}$$
$$\Delta_h H^\ominus \text{ ((}Z\text{)-But-2-ene)} = -119.1 \text{ kJ/mol.}$$

Hence, the difference in thermodynamic stabilities once more amounts to 4.2 kJ/mol (*Fig. 14.3*).

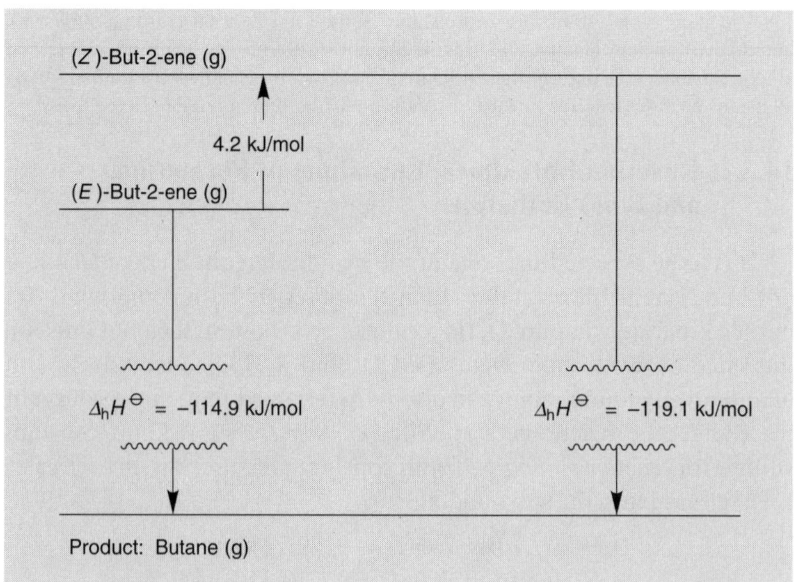

Fig. 14.3. *Graphic representation of calculation of the difference in thermodynamic stability between (*E*)- and (*Z*)-but-2-ene by means of their hydrogenation enthalpies ($\Delta_h H^\ominus$)*

If the combustion enthalpy of a hydrocarbon is known, together with those of carbon (graphite) and hydrogen (H_2(g)), then it is possible to calculate the so-called *enthalpies of formation* ($\Delta_f H^{\ominus}$) (*Fig. 14.4*).

$4\,CO_2 + 4\,H_2O \longrightarrow$	(Z)-$C_4H_8 + 6\,O_2$	$-\Delta_c H^{\ominus} = +2710.4$ kJ/mol
4 C (Graphite) + $4\,O_2 \longrightarrow$	$4\,CO_2$	$\Delta_c H^{\ominus} = -1574.0$ kJ/mol
$4\,H_2 + 2\,O_2 \longrightarrow$	$4\,H_2O$	$\Delta_c H^{\ominus} = -1143.3$ kJ/mol
4 C (Graphite) + $4\,H_2 \longrightarrow$	(Z)-C_4H_8	$\Delta_f H^{\ominus} = -6.9$ kJ/mol
$4\,CO_2 + 4\,H_2O \longrightarrow$	(E)-$C_4H_8 + 6\,O_2$	$-\Delta_c H^{\ominus} = +2706.2$ kJ/mol
4 C (Graphite) + $4\,O_2 \longrightarrow$	$4\,CO_2$	$\Delta_c H^{\ominus} = -1574.0$ kJ/mol
$4\,H_2 + 2\,O_2 \longrightarrow$	$4\,H_2O$	$\Delta_c H^{\ominus} = -1143.3$ kJ/mol
4 C (Graphite) + $4\,H_2 \longrightarrow$	(E)-C_4H_8	$\Delta_f H^{\ominus} = -11.1$ kJ/mol

Fig. 14.4. *Calculation of the enthalpies of formation $\Delta_f H^{\ominus}$ for (Z)- and (E)-but-2-ene*

This is possible because the reaction enthalpy is independent of the reaction pathway followed: whether an educt is transformed directly into a product or *via* an intermediate compound does not influence the reaction enthalpy (*Heß*' law of constant heat summation; *Fig. 14.5*).

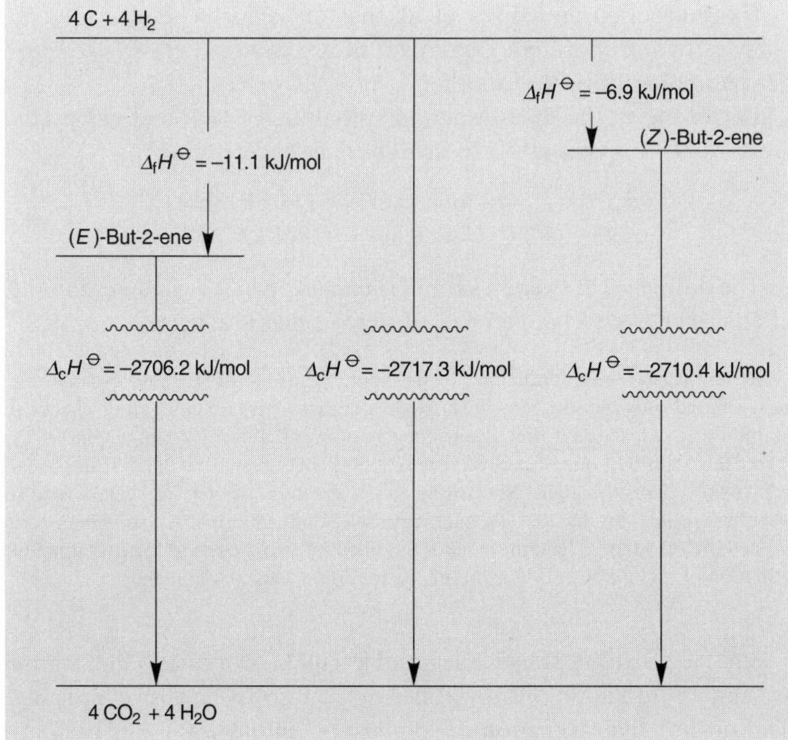

Fig. 14.5. *Graphic illustration of* Heβ' *law of constant heat summation for (E)- and (Z)-but-2-ene*

As the enthalpies of formation for many alkanes are known, it is possible to calculate those ones of the corresponding alkenes very simply, as soon as the relevant hydrogenation enthalpies are available (*Fig. 14.6*).

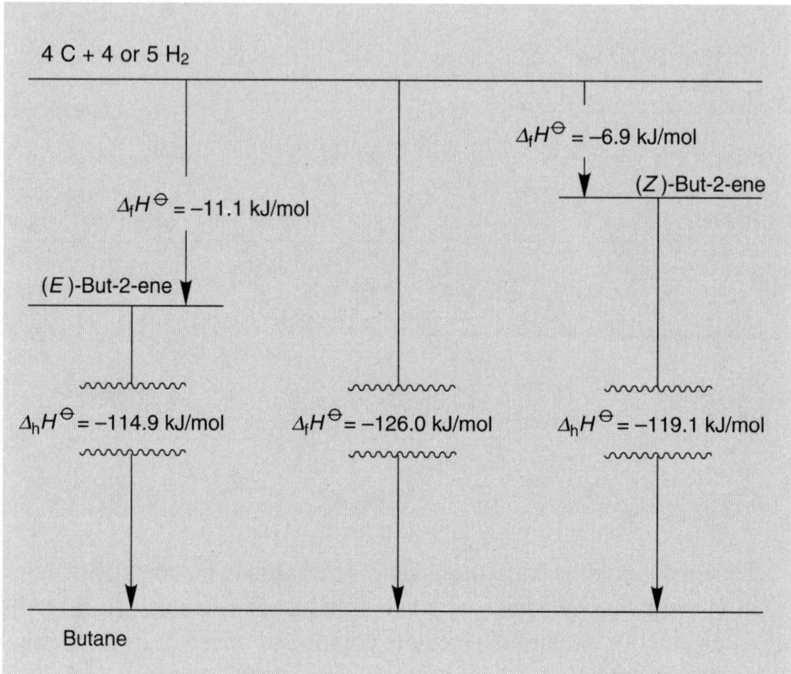

Fig. 14.6. *Graphic representation of calculation of the enthalpies of formation ($\Delta_f H^\ominus$) of (E)- and (Z)-but-2-ene from the enthalpy of formation of butane and the hydrogenation enthalpies of the two stereoisomers of but-2-ene*

Hydrogenation enthalpies of alkenes or cycloalkenes are used to compare two or more representatives of a compound class (stereoisomers, homologs) with one another.

For example, the hydrogenation enthalpies of (*E*)- and (*Z*)-cyclooctene (at 25 °C in glacial acetic acid) have been determined:

$$\Delta_h H^\ominus \text{ ((}E\text{)-Cyclooctene)} = -134.8 \text{ kJ/mol;}$$
$$\Delta_h H^\ominus \text{ ((}Z\text{)-Cyclooctene)} = -96.1 \text{ kJ/mol.}$$

The difference of over 38 kJ/mol is remarkable. It is greater than the difference between any other pair of stereoisomeric alkenes.

If the hydrogenation enthalpies of the homologs cyclopentene and cyclohexene are compared with one another (which would seem attractive for cultivating 'chemical intuition'), it can be seen that the hydrogenation enthalpy of cyclopentene is 7.1 kJ/mol less than that one of cyclohexene (catalytic hydrogenation at 25°C in glacial acetic acid). Here, the higher ring strain in cyclopentene relative to cyclohexene is overcompensated for by the lower energy of cyclohexane relative to the *Pitzer*-strained cyclopentane. The same result is obtained by comparison of the enthalpies of formation of cyclohexane, cyclopentane, cyclohexene, and cyclopentene.

Since it is scarcely practicable to subject all known compounds whose enthalpy of formation one might desire to know to expensive measurement of their hydrogenation or combustion enthalpies, it has been attempted – following the logic of the modular construction concept of the structure model of the classical organic chemistry – to interpret molecular enthalpies of formation as the sums of constant enthalpies of submolecular modules. *Table 14.2* contains abstract values for bond enthalpies, with the help of which it is possible to 'calculate' enthalpies of formation $\Delta_f H^\ominus$ for compounds under standard conditions in the gaseous state.

Table 14.2. *Values of Bond Enthalpies as Increments of Enthalpy of Formation.* C_d = C-atom with coordination number 3. For the calculation, the C=C and the C=O groups are both regarded as units with coordination number 4 or 2, respectively. (Source: *S.W. Benson*, Thermochemical Kinetics, 2nd Edn., Wiley, 1979, p. 25)

Bond	$\Delta_f H^\ominus$ [kJ/mol]	Bond	$\Delta_f H^\ominus$ [kJ/mol]
C–H	–16.02	S–H	–3.35
C–D	–19.79	S–S	–25.10
C–C	11.42	C_d–C	28.03
C–F	–219.66	C_d–H	13.39
C–Cl	–30.96	C_d–F	–163.18
C–Br	9.20	C_d–Cl	–20.92
C–I	58.99	C_d–Br	40.58
C–O	–50.21	C_d–I	90.79
O–H	–112.97	(O=)C–H	58.16
O–D	–116.73	(O=)C–C	–60.25
O–O	89.96	(O=)C–O	–211.29
O–Cl	38.07	(O=)C–F	–322.17
C–N	38.91	(O=)C–Cl	–112.97
N–H	–10.88	(NO_2)–O	–12.55
C–S	28.03	(NO)–O	37.66

The 'calculated' $\Delta_f H^\ominus$ value for ethyl acetate, as detailed in *Table 14.2*, is compared with the actual measured one below:

Ethyl acetate

$CH_3-C(=O)-O-CH_2-CH_3$

$\Delta_f H^\ominus$ ('calculated')

= 8 [C–H] + [(O=)C–C] + [(O=)C–O] + [C–O] + [C–C]

= (– 128.16 – 60.25 – 211.29 – 50.21 + 11.42) kJ/mol

= – 438.49 kJ/mol

$\Delta_f H^\ominus$ (measured)

= – 432.63 kJ/mol

For an organic compound X, the enthalpy of formation $\Delta_f H^\ominus$ in the gaseous (g), liquid (l), or solid (s) state is equivalent to the change in enthalpy arising from the formation of this compound from its elements (under standard conditions). At 298 K, the standard conditions of the elements are g for H_2, g for O_2, and s for C (graphite). The enthalpy of formation in the gas phase $\Delta_f H^\ominus$ (X(g)) differs from that one in condensed phases by the enthalpy of vaporization or sublimation.

Fig. 14.7 illustrates how $\Delta_f H^\ominus$ for methane, after dissociation or sublimation of the elements H (g, H_2) and C (s, graphite) and subsequent combination of the resulting atoms, is worked out.

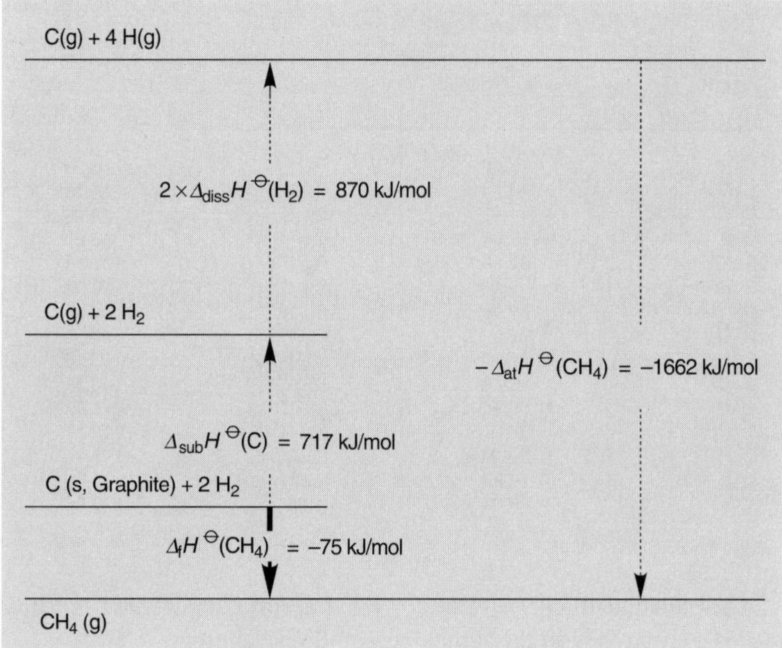

Fig. 14.7. *Graphic representation of the options for the formation of methane*

If, in a thought experiment, it is imagined that the formation of methane takes place by the synchronous movement of atoms, already in the correct relative orientations at large interatomic distances, towards their equilibrium distances, it is then possible to interpret the enthalpy of formation liberated as if each individual bond contributed a constant amount, the so-called *bond enthalpy* $\Delta_b H^\ominus$ (C–H).

The average C–H bond enthalpy of methane has been unambiguously determined from experimental values (enthalpies of formation of methane and its atoms from the elements). It is also used (not entirely without arbitrariness, but consistent with the modular construction concept) for the C–H bond enthalpy of ethane: to calculate the C–C bond enthalpy from the experimentally determined enthalpy of formation of this hydrocarbon.

Agreement between measured enthalpies of formation and those ones obtained by summation of enthalpy increments further improves when *enthalpy values for atom groups*, rather than those ones for bonds, are taken. *Table 14.3* contains those atom-group enthalpy increments

Table 14.3. *Enthalpy Increments for Atom Groups.* C_d = C-atom of coordination number 3, C_B = C-atom in benzene ring, C_t and C_a = C-atoms of coordination number 2 (C_a = allenic C-atom), C_{BF} = C-atom in a fused aromatic system such as naphthalene or anthracene. (Source: *S. W. Benson*, Thermochemical Kinetics, 2nd Edn., Wiley, 1979, p. 272f.)

Atom Group	$\Delta_f H^\ominus$ [kJ/mol]	Atom Group	$\Delta_f H^\ominus$ [kJ/mol]
$\mathrm{C-(H)_3(C)}$	−42.68	$\mathrm{O-(C_d)_2}$	138.07
$\mathrm{C-(H)_2(C)_2}$	−20.63	$\mathrm{O-(C_B)_2}$	−88.28
$\mathrm{C-(H)(C)_3}$	−7.95	$\mathrm{O-(C_d)(CO)}$	−189.12
$\mathrm{C-(C)_4}$	2.09	$\mathrm{O-(C_B)(CO)}$	−153.55
		$\mathrm{O-(O)(CO)}$	−79.50
$\mathrm{C_d-(H)_2}$	26.19	$\mathrm{O-(CO)_2}$	−194.56
$\mathrm{C_d-(H)(C)}$	35.94	$\mathrm{O-(O)_2}$	79.55
$\mathrm{C_d-(C)_2}$	43.26		
$\mathrm{C_d-(C_d)(H)}$	28.37	$\mathrm{CO-(H)_2}$	−108.78
$\mathrm{C_d-(C_d)(C)}$	37.15	$\mathrm{CO-(H)(C)}$	−121.75
$\mathrm{C_d-(C_B)(H)}$	28.37	$\mathrm{CO-(H)(C_B)}$	−121.75
$\mathrm{C_d-(C_B)(C)}$	36.15	$\mathrm{CO-(H)(C_d)}$	−121.75
$\mathrm{C_d-(C_t)(H)}$	28.37	$\mathrm{CO-(H)(C_t)}$	−121.75
$\mathrm{C_d-(C_B)_2}$	33.47	$\mathrm{CO-(H)(CO)}$	−105.86
$\mathrm{C_d-(C_d)_2}$	19.25	$\mathrm{CO-(H)(O)}$	−134.31
$\mathrm{C-(C_d)(C)(H)_2}$	−19.92	$\mathrm{CO-(C)_2}$	−131.38
$\mathrm{C-(C_d)_2(H)_2}$	−17.95	$\mathrm{CO-(C)(C_B)}$	−129.29
$\mathrm{C-(C_d)(C_B)(H)_2}$	−17.95	$\mathrm{CO-(C_B)_2}$	−107.95
$\mathrm{C-(C_t)(C)(H)_2}$	−19.79	$\mathrm{CO-(C)(O)}$	−146.86
$\mathrm{C-(C_B)(C)(H)_2}$	−20.33	$\mathrm{CO-(C)(CO)}$	−122.17
$\mathrm{C-(C_d)(C)_2(H)}$	−6.19	$\mathrm{CO-(C_d)(O)}$	−133.89
$\mathrm{C-(C_t)(C)_2(H)}$	−7.20	$\mathrm{CO-(C_B)(O)}$	−153.13
$\mathrm{C-(C_B)(C)_2(H)}$	−4.10	$\mathrm{CO-(C_B)(CO)}$	−112.13
		$\mathrm{CO-(O)_2}$	−125.10
$\mathrm{C-(C_d)(C)_3}$	7.03	$\mathrm{CO-(O)(CO)}$	−122.59
$\mathrm{C-(C_B)(C)_3}$	11.76		
		$\mathrm{C-(H)_3(O)}$	−42.17
$\mathrm{C_t-(H)}$	112.68	$\mathrm{C-(H)_2(O)(C)}$	−33.89
$\mathrm{C_t-(C)}$	115.27	$\mathrm{C-(H)_2(O)(C_d)}$	−27.20
$\mathrm{C_t-(C_d)}$	122.17	$\mathrm{C-(H)_2(O)(C_B)}$	−33.89
$\mathrm{C_t-(C_B)}$	122.17	$\mathrm{C-(H)_2(O)(C_t)}$	−27.20
		$\mathrm{C-(H)_2(O)_2}$	−67.36
$\mathrm{C_B-(H)}$	13.81	$\mathrm{C-(H)(O)(C)_2}$	−30.12
$\mathrm{C_B-(C)}$	23.05	$\mathrm{C-(H)(O)_2(C)}$	−68.20
$\mathrm{C_B-(C_d)}$	23.77		
$\mathrm{C_B-(C_t)}$	23.77	$\mathrm{C-(O)(C)_3}$	−27.61
$\mathrm{C_B-(C_B)}$	20.75	$\mathrm{C-(O)_2(C)_2}$	−77.82
		$\mathrm{C-(H)_3(CO)}$	−42.17
$\mathrm{C_a}$	143.09	$\mathrm{C-(H)_2(CO)(C)}$	−21.76
		$\mathrm{C-(H)_2(CO)(C_d)}$	−15.90
$\mathrm{C_{BF}-(C_B)_2(C_{BF})}$	20.08	$\mathrm{C-(H)_2(CO)(C_B)}$	−22.59
$\mathrm{C_{BF}-(C_B)(C_{BF})_2}$	15.48	$\mathrm{C-(H)_2(CO)(C_t)}$	−22.59
$\mathrm{C_{BF}-(C_{BF})_3}$	6.28	$\mathrm{C-(H)_2(CO)_2}$	−31.80
		$\mathrm{C-(H)(CO)(C)_2}$	−7.11
$\mathrm{O-(H)_2}$	−241.84	$\mathrm{C-(CO)(C)_3}$	5.86
$\mathrm{O-(H)(C)}$	−158.57	$\mathrm{C_d-(O)(H)}$	35.98
$\mathrm{O-(H)(C_B)}$	−158.57	$\mathrm{C_d-(O)(C)}$	43.10
$\mathrm{O-(H)(O)}$	−68.20	$\mathrm{C_d-(O)(C_d)}$	37.24
$\mathrm{O-(H)(CO)}$	−243.09	$\mathrm{C_d-(O)(CO)}$	48.53
		$\mathrm{C_d-(H)(CO)}$	20.92
$\mathrm{O-(C)_2}$	−97.07	$\mathrm{C_d-(CO)(C)}$	31.38
$\mathrm{O-(C)(C_d)}$	−127.61	$\mathrm{C_B-(O)}$	−3.77
$\mathrm{O-(C)(C_B)}$	−96.23	$\mathrm{C_B-(CO)}$	15.48
$\mathrm{O-(C)(O)}$	−18.83		
$\mathrm{O-(C)(CO)}$	−180.33		

with the aid of which $\Delta_f H^\ominus$ values for gas-phase compounds under standard conditions can be 'calculated'.

The 'calculated' $\Delta_f H^\ominus$ value for *tert*-butyl methyl ether (after *Table 14.3*) is compared with the measured value below:

$$CH_3-C(CH_3)_2-O-CH_3$$

***tert*-Butyl methyl ether**
$\Delta_f H^\ominus$ ('calculated')
$= 3\,[\,C\text{–}(H)_3(C)\,] + [\,C\text{–}(O)(C)_3\,] + [\,O\text{–}(C)_2\,] + [\,C\text{–}(H)_3(O)\,]$
$= (-128.04 - 27.61 - 97.07 - 42.17)$ kJ/mol
$= \underline{-294.89\ \text{kJ/mol}}$
$\Delta_f H^\ominus$ (measured)
$= \underline{-293 \pm 4\ \text{kJ/mol}}$

The *summation procedure* has proved its worth for acyclic molecules. If topological factors permit remote groups to approach each other (often (*Z*)- relative to (*E*)-alkenes, *sc*- relative to *ap*-conformations, or in cyclic systems), it is necessary to introduce correction factors: the existence of such correction factors makes it clear that exploiting additivity of increments of enthalpy of formation is only possible for particular compound classes. The need to take account of correction factors in the 'calculation' of enthalpies of formation from enthalpy increments is a useful sign that exceptional causes of stabilization or destabilization exist for particular compounds, leading to a divergence from the normal situation. Prototypes of such 'divergent cases' are cyclopropane and benzene.

While the 'calculated' value for $\Delta_f H^\ominus$ for cyclohexane agrees well with the measured one, there are considerable discrepancies in the cases of cyclopropane and benzene (*Table 14.4*).

'Real' cyclopropane is destabilized by 115 kJ/mol compared with 'calculated' (with the aid of the additivity relationship) cyclopropane. Intuition says that the appreciable angle deformation (*Baeyer* strain) is chiefly responsible for the destabilization.

According to *Baeyer*'s *strain theory*, cyclopropane is expected to display the greatest ring strain of all the homologs. However, the conventional ring strain of cyclopropane (115 kJ/mol) is only slightly greater than that of cyclobutane (110 kJ/mol). Conventional ring-strain energy is worked out in each case by subtracting 3·659 kJ/mol from $-\Delta_c H^\ominus$ for cyclopropane (2092 kJ/mol) and 4·659 kJ/mol from $-\Delta_c H^\ominus$ for cyclobutane (2746 kJ/mol). In any event, cyclopropane occupies a special place within the homologous series of cycloalkanes. This fact cannot be explained in the context of the structure model of the classical organic chemistry.

'Real' benzene, with D_{6h} symmetry, is stabilized by the magnitude of its *resonance energy*, in comparison with '*Kekulé* benzene', of D_{3h} symmetry and alternating C–C and C=C bonds.

Table 14.4. *Comparison Between 'Calculated' and Measured Standard Enthalpy of Formation* ($\Delta_f H^\ominus$ [kJ/mol]) *for Cyclopropane, Cyclohexane, and Benzene.* The values 'calculated' for benzene were obtained using the fictitious cyclohexa-1,3,5-triene as reference structure; this explains the glaring difference compared with the measured value.

Compound	Cyclopropane $(CH_2)_3$	Cyclohexane $(CH_2)_6$	Benzene $(CH{=}CH)_3$
$\Delta_f H^\ominus$ ('calculated' with values from *Table 14.2*)	−61.9	−123.8	+164.4
$\Delta_f H^\ominus$ ('calculated' with values from *Table 14.3*)	−61.9	−123.8	+170.2
$\Delta_f H^\ominus$ (measured)	+53.1	−123.4	+82.8
$\Delta_f H^\ominus$ (measured) − $\Delta_f H^\ominus$ ('calculated')	+115.0	+0.4	−81.6 −87.4

14.3.2. Calculation of Resonance Energy

The difference between measured and calculated (using increment sets) standard enthalpies of formation is an important measure of the validity of the structure model of the classical organic chemistry. For benzene, especially, there arises a substantial discrepancy between the values obtained using the two techniques. The resulting energy difference

$$RE = \Delta_f H^\ominus \text{(measured)} - \Delta_f H^\ominus \text{('calculated')}$$

is called the *resonance energy* (*RE*; *Sect. 14.1*). The definition of resonance energy raises a fundamental problem, however: resonance energy as a conceptual quantity is not measurable experimentally. It comes about as a result of comparing the experimentally determined enthalpy of formation of the compound in question with the enthalpy of formation calculated with the help of chemical models for a suitable reference structure.

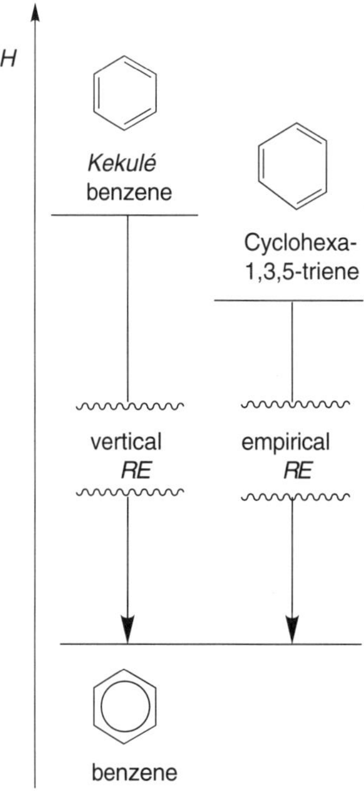

To determine the (vertical) resonance energy of benzene, the enthalpy of formation of *Kekulé* benzene is needed as a reference model. Attempts have been made to derive the hypothetical enthalpy of formation of the imaginary reference compound cyclohexa-1,3,5-triene with the aid of a suitable model compound (such as cyclohexene). In this sense, the difference between three times the hydrogenation enthalpy of cyclohexene and the hydrogenation enthalpy of benzene (−151 kJ/mol) can be regarded as the (empirical) resonance energy. This comparison, however, does not take account of the difference in strain energy between cyclohexatriene and cyclohexene (*Fig. 14.8*).

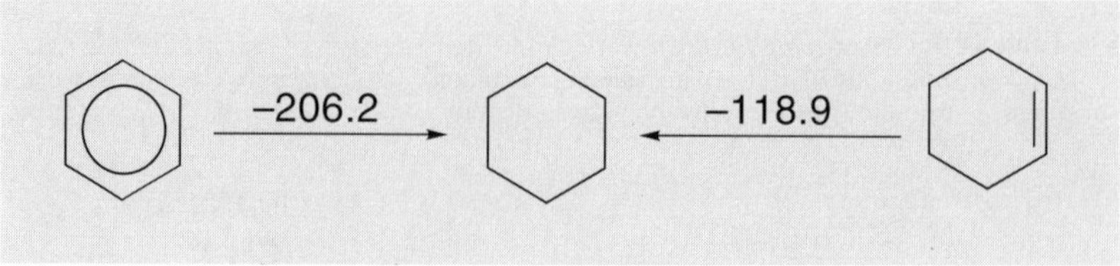

Fig. 14.8. *Comparison of the hydrogenation enthalpies* $\Delta_h H^\ominus$ [kJ/mol] *of benzene and cyclohexene*

In addition to the *chemical* approximation to the true structure of benzene, there is a *physical* approach, based on creating a force field acting on a reference structure (*Aspects of Organic Chemistry*, Volume 4: *Methods of Structure Determination*). Force-field techniques for calculating the structures and energies of organic molecules are based on an empirical strategy, corresponding to the modular construction approach already mentioned many times, but greatly refined. Every structural parameter (bond lengths, bond angles, torsion angles, distances between atoms not directly bound to one another) of a molecular model is varied using computer simulation, being iteratively refined, until the calculated values agree with the true ones (derived from experimental studies on a great many compounds) as closely as possible. This approach is analogous to using a very sophisticated mechanical model containing a great many springs of different lengths and strengths, representing the various intramolecular attractions and repulsions. Unlike the tabulated enthalpy increments (*Tables 14.2* and *14.3*), force-field calculations also take account of through-space interactions between atom groups, so that differences in, for example, the standard enthalpies of formation of (*Z*)- and (*E*)-alkenes are correctly reproduced.

To determine resonance energy, a modified version of *Norman L. Allinger*'s MM2 force field (*Aspects of Organic Chemistry*, Volume 4: *Methods of Structure Determination*) was used. To do justice to problems involving conjugated, unsaturated systems (*Chapt. 8.3.3.4.1*), *Wolfgang R. Roth*, acting on a suggestion made by *Michael J.S. Dewar*, not only distinguished roughly between C−C and C=C bonds, but introduced specific fine values, for example, for C−C bonds differing in the coordination number of the participating C-atoms (*Fig. 14.9*).

Roth let the resonance energy of buta-1,3-diene become zero and obtained for acyclic, conjugated polyenes a resonance energy, relative to buta-1,3-diene, that was also zero: experimental and calculated values of enthalpy of formation hence agreed with each other within the error limits of the methods. In the case of benzene, however, the resonance energy normalized to buta-1,3-diene attained a value of −108 kJ/mol. Benzene is exceptional then – whichever reference model may be used, and however its enthalpy of formation is determined. The structure model of the classical organic chemistry, taking no account of electronic structure, remains incapable of explaining benzene's remarkable thermodynamic stability. Here, the HMO model (see *Chapt. 8*) helps further.

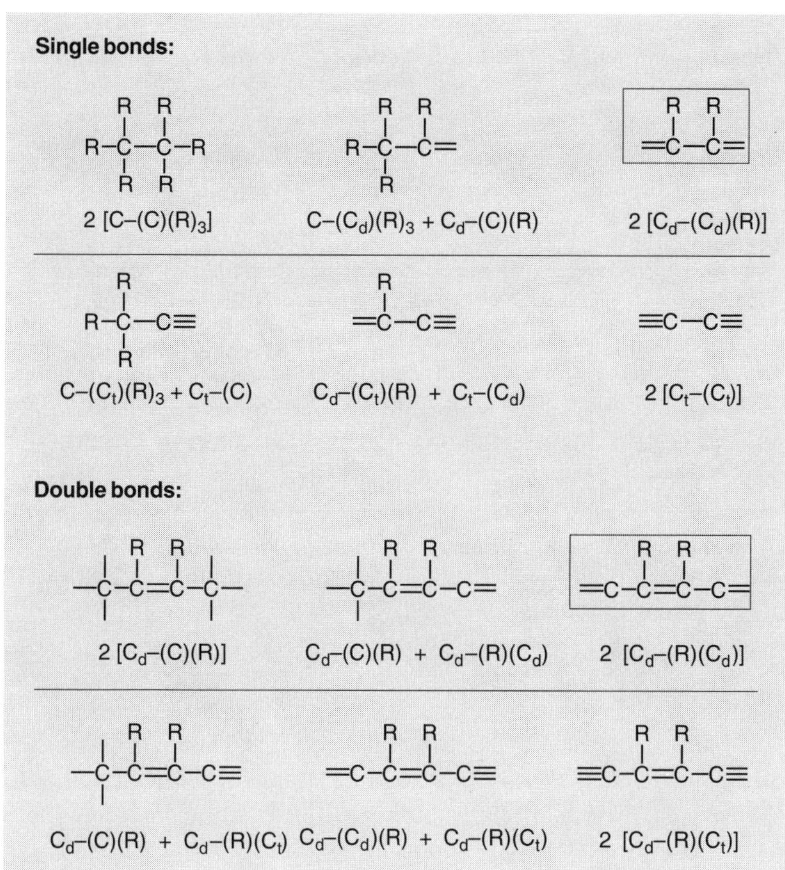

Fig. 14.9. *The different types of C−C and C=C bonds.* The framed examples apply for benzene; designations of structure fragments relate to *Table 14.3*.

14.4. Annulenes

14.4.1. Annulenes with $4n + 2$ π-Electrons: 'Aromatic' Molecules

Annulenes belonging to this subgroup are molecules in which the index m of the general molecular formula $(CH)_{2m}$ is an odd number. The prototype of this subgroup is benzene, known as [6]annulene in annulene terminology. The specific molecular formula is $(CH)_6$ ($m = 3$), and the number of π-electrons is 6 ($n = 1$ in the general expression $4n + 2$). Benzene (with an HMO π-electron energy of 8β) is *aromatic* (relative to hexa-1,3,5-triene with an HMO π-electron energy of 7β).

It should be noted that the six-membered carbocycles *benzene* and *cyclohexane* each play the roles of prototypes in their compound classes of annulenes and cycloalkanes (*Table 14.5*).

Table 14.5. *Structural Data for the Six-Membered Carbocycles Benzene and Cyclohexane*

Compound	Benzene	Cyclohexane
Symmetry point group	D_{6h}	D_{3d}
Topology	six-membered, planar ring	six-membered ring with two sets of C-atoms (with */without *), each in a plane
Valence angle (C–C–C)	idealized: 120° real: 120°	idealized: 109.5° real: 111.4°
Torsion angle (C–C–C–C)	idealized: 0° real: 0°	idealized: ± 60° real: ± 54.5°
	C_6 axis perpendicular to molecular plane	C_3 axis perpendicular to both planes

14.4.2. Annulenes with 4*n* π-Electrons: 'Antiaromatic' Molecules

In this subgroup of annulenes of general molecular formula $(CH)_{2m}$, m is an even number. Here are found both the lower and higher vinylogs of benzene: (*Z*,*Z*)-cyclobuta-1,3-diene (= [4]annulene, or simply cyclobutadiene) and (*Z*,*Z*,*Z*,*Z*)-cycloocta-1,3,5,7-tetraene (= [8]annulene, or simply cyclooctatetraene), with molecular formulae $(CH)_4$ and $(CH)_8$. For cyclobutadiene, $m = 2$ and $n = 1$; for cyclooctatetraene $m = 4$ and $n = 2$.

Unsubstituted cyclobutadiene can be trapped in an argon matrix at liquid helium temperatures, or even, as a *guest in a host macromolecule*, at room temperature. Spectroscopic examination, supported by supplementary calculation, has confirmed that cyclobutadiene in the ground state is of *singlet character* (*i.e.*, all of the bonding MOs are occupied by electron pairs) and rectangular geometry (D_{2h} symmetry), with topomerization of the rectangular isomers passing through a square transition structure (D_{4h} symmetry; *Fig. 14.10*).

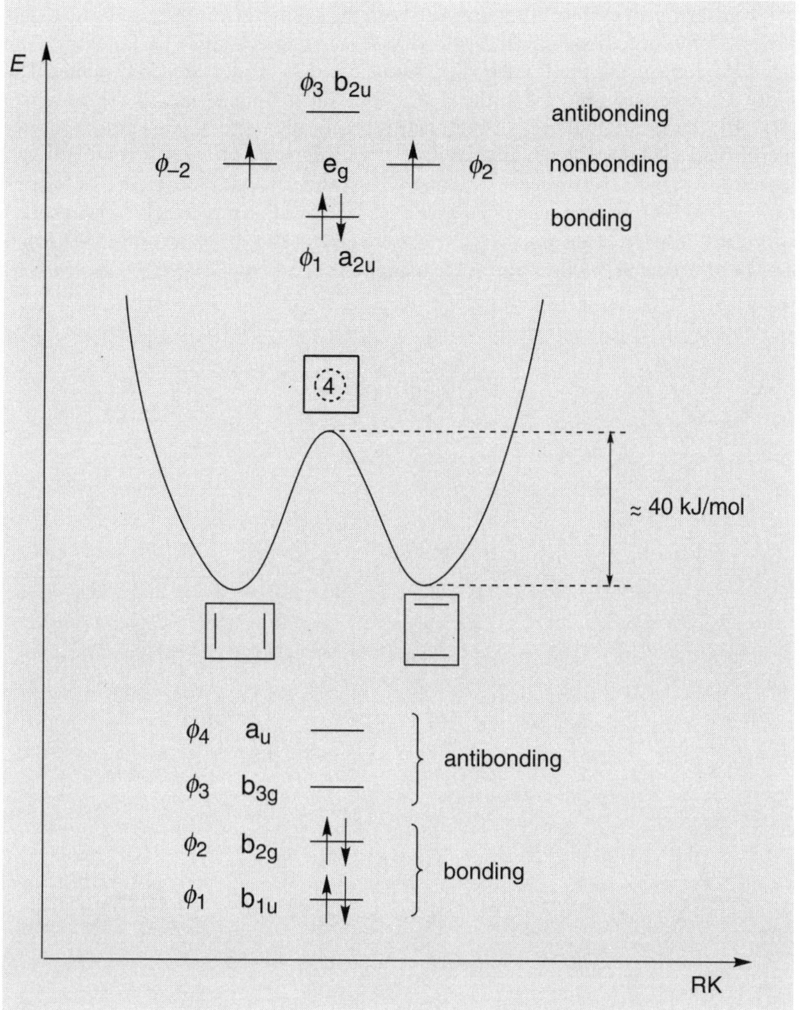

Fig. 14.10. *Isomerization of cyclobutadiene*

Cyclobutadiene (with an HMO π-electron energy of 4β) is *antiaromatic* (relative to buta-1,3-diene with an HMO π-electron energy of 4.5β).

The square cyclobutadiene with *delocalized* π-electrons, as required by the HMO model, has two degenerate MOs, each occupied by one electron, following *Hund*'s rule. In this case, the *Walsh* diagram shows that the degeneration of the two e_g-orbitals in the D_{4h} point group is removed by the lowering of symmetry to D_{2h}. Shortening of the one pair of C–C bonds and lengthening of the other pair alters the energy scheme in such a way that the lower-symmetry, rectangular molecule with four π-electrons is more stable. The square cyclobutadiene is the transition structure between the two molecules of D_{2h} symmetry. This effect has also been observed in other instances and is called the *Jahn-Teller effect*. For cyclobutadiene, the result is an electronic structure with *localized* C=C bonds (with no degenerate MOs), in which the bonding MOs are occupied by electron pairs and the antibonding MOs are unoccupied (*Fig. 14.10*).

Unsubstituted cyclooctatetraene has been the subject of considerable study. The preferred conformation has D_{2d} symmetry, resembling a boat in its topology. The bond lengths are 147 pm for the C–C bonds and 134 pm for the C=C bonds. The C=C–C angles are 126.1°, and the C=C–C=C torsion angles are all the same size (±57.9°): this means that cyclooctatetraene is practically free of strain. Two stereoisomerizations take place under normal conditions (*Fig. 14.11*): *ring inversion*, proceeding through a planar transition structure (D_{4h} symmetry) with localized C=C bonds ($\Delta G^{\ddagger}_{inv} = 39.7$ kJ/mol), and *bond shift via* a planar transition structure (D_{8h} symmetry) with delocalized π-electrons ($\Delta G^{\ddagger}_{bs} = 55.6$ kJ/mol). (The difference of 15.9 kJ/mol may be regarded, in this instance, as the magnitude of a *destabilization by conjugation*.)

Fig. 14.11. *Isomerization of cyclooctatetraene*

X-Ray crystal-structure analysis of $[K((CH_3OCH_2CH_2)_2O)]_2^+ [C_8H_4(CH_3)_4]^{2-}$ has shown that the cyclooctatetraene dianion is planar, the eight C-atoms of the ring forming a regular octagon. The average C–C bond length is 140.7 pm. With 10 π-electrons, the dianion meets the requirements of the *Hückel* rule and displays the structural characteristics of an aromatic compound in the *Hückel* sense.

Emanuel Vogel has demonstrated with creativity and technical dexterity how custom-built 10π-electron systems may be produced for the most diverse experimental examinations.

14.4.3. Some Remarks about the Term 'Aromatic'

The term 'aromatic' has so many vague meanings in organic chemistry that serious attempts have been made to abandon it entirely. At the beginning of the 19th century, pleasant smelling compounds were described as *aromatic* in the everyday sense. Some of these pleasant smelling compounds proved, on closer examination, to be derivatives of benzene: compounds in which one or more H-atoms of benzene had been substituted. For *Kekulé*, the word 'aromatic' had the meaning of a constitutional characteristic (*Sect. 14.1*). He was able to divide cyclic compounds precisely into *aromatic* and *non-aromatic* subgroups, even if these compounds only existed in the imagination of the chemist and not (yet) in reality.

Emil Erlenmeyer proposed using the word 'aromatic' for compounds exhibiting properties similar to benzene (in particular, the tendency to undergo substitution reactions rather than the addition reactions common with other unsaturated compounds). This was to give away the chance of obtaining an unambiguous classification, however. It could have been foreseen that, in the future, there would be compounds which could be described as 'fully aromatic', 'partially aromatic', or 'non-aromatic'. It could also have been predicted that people would look for different criteria for aromaticity in the *Erlenmeyer* sense, and that some compounds would then turn out to be aromatic viewed from one criterion and non-aromatic viewed from another one. Only benzene fully meets all of the criteria for *aromaticity*. To include other compounds, the relevant literature has been 'enriched' with words such as *homo-aromaticity*, *quasi-aromaticity*, and *pseudo-aromaticity*. The degree of conceptional confusion was made clear to observers when, on the occasion of an International Symposium on '*Aromaticity, Pseudo-Aromaticity, Anti-Aromaticity*' in Jerusalem in 1970, it was pointed out that, in the series of 'aromaticities', the term '*schizo-aromaticity*' was still absent. At any event, it was made apparent there that the term 'aromaticity' – if it should still be used at all – ought to be associated with a discontinuously changing structural characteristic and not with a continually varying property.

Such an (electronic) structural characteristic is offered by *Erich Hückel*'s HMO model, with its aromaticity criterion of $4n + 2$ π-electrons for monocyclic, periplanar, conjugated, unsaturated systems, in which the bonding MOs are fully occupied (*Chapt. 8.3.3.4.2*).

14.5. The Role of Models

In the context of the structure model of the classical organic chemistry, the term 'conjugation' merely implies that C–C and C=C bonds follow one another alternately in a polyene (*Chapt. 8.3.3.4.1*). The use of constitutional units like C–C and C=C as modules for the construction of organic molecules becomes problematic, when departing from the underlying structure-model framework, *i.e.*, when assigning two electrons which have combined into a pair to each straight-line bond symbol (*Chapt. 2.4*), one may wonder whether conjugation going beyond the formal description of constitution could also mean that overlapping orbitals might group together into sets of extended MOs, in which only their antibonding members are unoccupied by electrons. *Benzene* does not fit the constitutional model of the classical organic chemistry at all. The 'benzene story' with an air of mystery ends with the epistemological impression that (organic) chemistry, after all, is a theory-producing discipline. A somewhat similar role to that one given to benzene in *π-conjugation* has been attributed to cyclopropane in *σ-conjugation*. *σ-Aromaticity* was ascribed to the three-membered cycloalkane by *Dewar*.

If the three-membered ring compound *cyclopropane* is considered to be next neighbor to the 'two-membered ring compound' *ethene*, and ethene's well established separation into σ- and π-electrons (*Chapt. 8.3.3.4.1*) applied to cyclopropane, then, on constructing *Walsh* correlation diagrams, it is possible to get an intuitive awareness of the conventional ring-strain energy (*Sect. 14.3.1*), which, in cyclopropane (115 kJ/mol), is astonishingly low compared with cyclobutane (110 kJ/mol).

If two fragments of the *non-linear methylene* type are brought together in such a way that a two-membered ring is produced, with the σ-orbitals overlapping in the molecular plane and the p-orbitals perpendicular to it, then the MO set for the C=C region of ethene (*Fig. 14.12*), with clearly discernible MOs for σ- and π-electrons, is obtained.

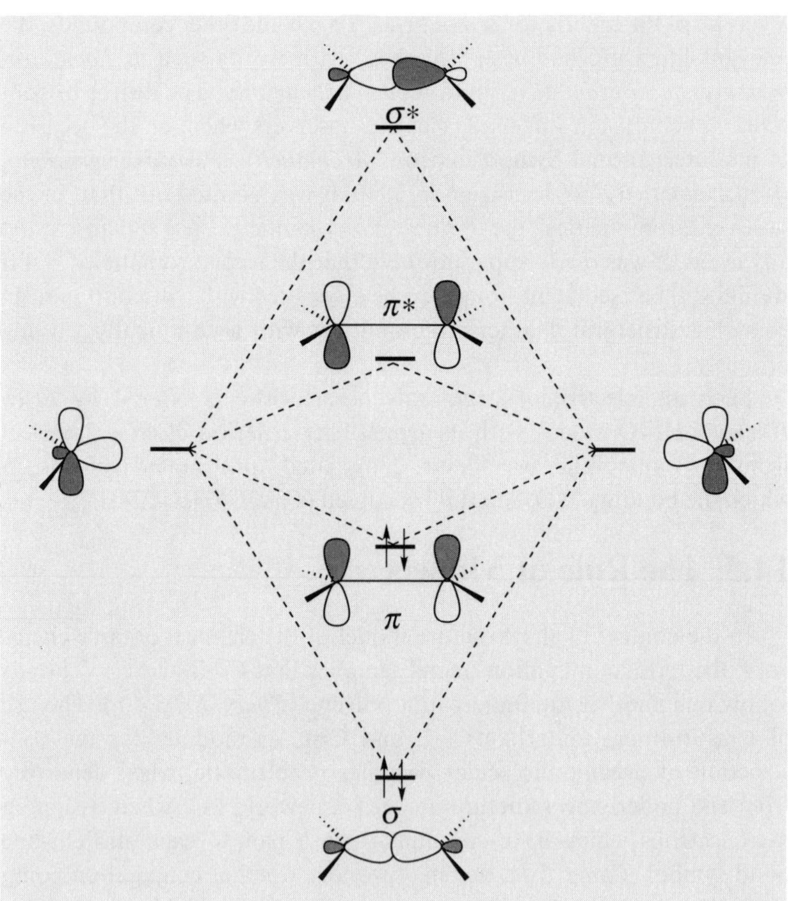

Fig. 14.12. *Linear combination of two fragments of the* non-linear methylene *type, gives the MOs for the C-atom skeleton of ethene* (D_{2h})

If three *non-linear methylene* fragments are brought together so as to form a triangle, with *radially* overlapping σ-orbitals and *tangentially* overlapping p-orbitals, then the MO set of cyclopropane, with the MO partial sets with radial and tangential group orbitals, is obtained (*Fig. 14.13*).

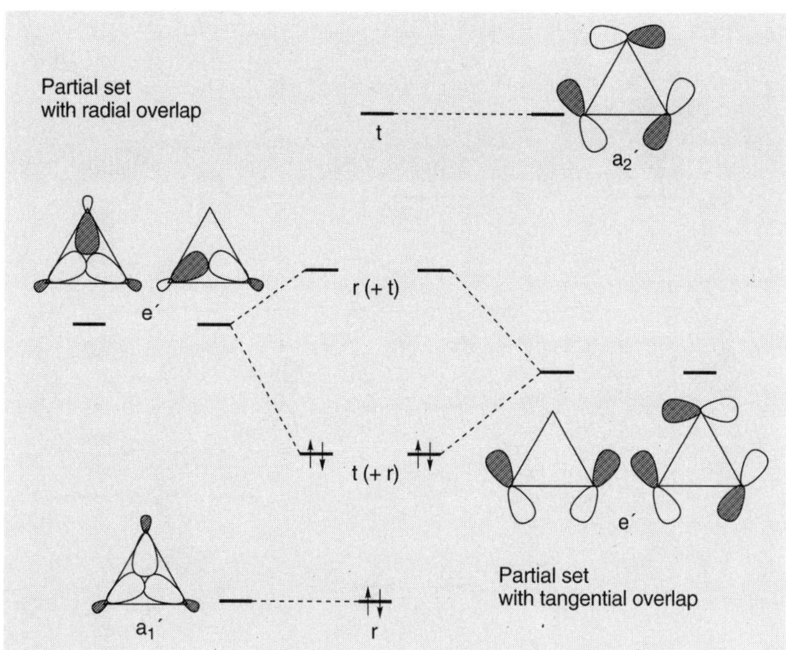

Fig. 14.13. *Linear combination of suitable group orbitals from the partial sets with radial and with tangential overlap gives the* Walsh *MOs for the ring skeleton of cyclopropane* (D_{3h})

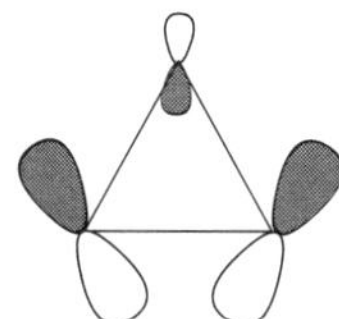

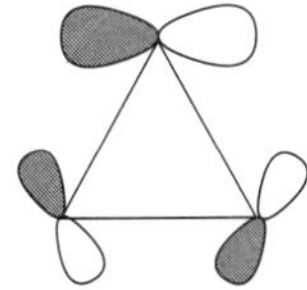

In the MO partial set with radial overlap, a strongly bonding a_1'-group orbital and a pair of antibonding e′-group orbitals can be seen. The three orbitals are ordered in a manner typical of a *Hückel* three-center cycle (*Chapt. 8.3.3.4.2*), and there is no nodal plane to be found in the bonding MO. In the MO partial set with tangential overlap, there are a pair of bonding e′-group orbitals and an antibonding a_2'-group orbital. These three orbitals are arranged in a manner typical of a *Möbius* three-center cycle (*Chapt. 8.3.3.4.2*), and no orbital without a nodal plane is to be found in the entire partial set. Linear combination of the e′-group orbitals from the two partial sets with tangential and radial overlap affords a bonding and an antibonding pair of degenerate e′-MOs in each case. The bonding e′-MOs of cyclopropane can be imagined as arising from the mixing of the antibonding e′-group orbitals with radial overlap and the bonding e′-group orbitals with tangential overlap. This overlapping is stronger than that between the p-orbitals of two CH_2 fragments and weaker than that between the σ-orbitals of two CH_2 fragments. Consequently, the bonding e′-MOs of cyclopropane lie at a higher energy than the σ-MOs of C–C bonds and at a lower energy than the π-MOs of C=C bonds.

If four fragments of the *non-linear methylene* type are brought together in such a way as to form a planar four-membered ring, with the σ-orbitals overlapping *radially*, the p-orbitals *tangentially*, then the MO set of an idealized cyclobutane with D_{4h} symmetry is obtained, *via* the MO partial sets with radial and tangential group orbitals (*Fig. 14.14*).

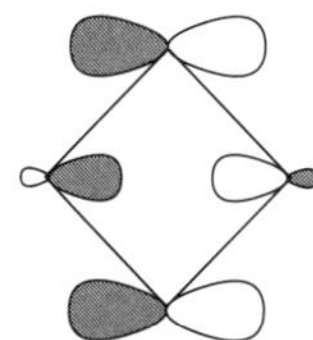

In the MO partial set with radial overlap a bonding a_{1g}-group orbital, a pair of nonbonding e_u-group orbitals, and an antibonding b_{2g}-group orbital can be seen. In the MO partial set with tangential overlap, there are a bonding b_{1g}-group orbital, a pair of nonbonding e_u-group orbitals, and an antibonding a_{2g}-group orbital.

The bonding e_u-MOs of planar cyclobutane may be described by imagining mixing between nonbonding e_u-group orbitals with radial overlapping and nonbonding e_u-group orbitals with tangential overlapping. It is only through this interaction that they become bonding orbitals at all; however, they remain very high in energy. They are σ-bonding between immediately adjacent atoms, but between C(1) and C(3), or between C(2) and C(4) they are π-antibonding.

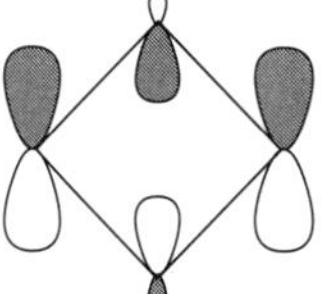

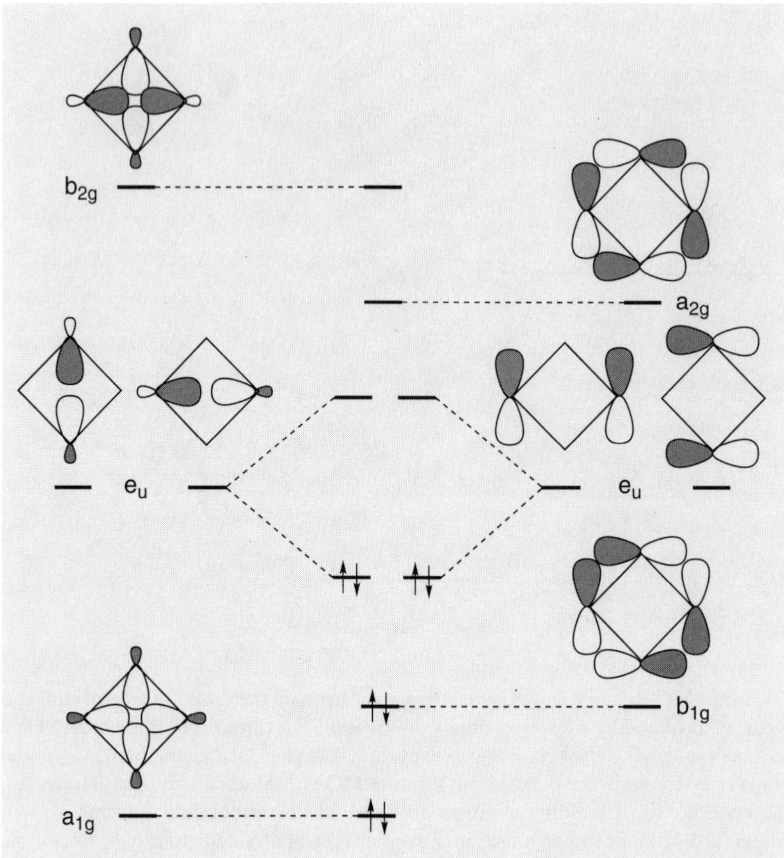

Fig. 14.14. *Linear combination of suitable group orbitals from the partial sets with radial* (left) *and tangential* (right) *overlap gives the MOs for the ring skeleton of cyclobutane* (D_{4h}) *of the* Walsh *type*

The almost identical texts in the descriptions of cyclopropane and cyclobutane, using the *Walsh* model, will not have remained unnoticed. Here, on the one hand, attention should be drawn to the formalism which has been followed in this procedure, regarding cycloalkanes as built up out of fragments of the *non-linear methylene* type. On the other hand, it should now be apparent that this model is capable of explaining the special situation of cyclopropane. There exist other models, not mentioned here, which also endeavor to interpret this phenomenon, assuming either a relatively low strain in cyclopropane or a relatively high strain in cyclobutane (or both).

The discussion of the special case of cyclopropane has not only demonstrated the greater flexibility and depth of the qualitative MO model in comparison with the structure model of the classical organic chemistry, but has once more illustrated the typical organic chemists' way of viewing things: the individual case of one chemical compound as a special case of a compound class, enabling structural cross-references to be made, leading to questions which may be answered experimentally with the aid of a suitable structure model. A structure model may be regarded as suitable if it is as simple and as capable of generalization as possible, and has survived the test of '*Ockham*'s razor' (*Entia non sunt multiplicanda praeter necessitatem*).

Further Reading

To Sect. 14.1.

A. Kekulé, *Über die Constitution der aromatischen Verbindungen, Annalen der Chemie und Pharmazie* **1866,** *137,* 129;
– *Theoretische Betrachtungen und historische Notizen über die Constitution des Benzols, Annalen der Chemie und Pharmazie* **1872,** *162,* 77.

G. Schultz, *Bericht über die Feier der Deutschen Chemischen Gesellschaft zu Ehren August Kekulés, Ber. Dtsch. Chem. Ges.* **1890,** *23,* 1265.

J. H. Wotiz (Ed.), *The Kekulé Riddle,* Cache River Press, Vienna, Il., 1993.

L. Pauling, *The Nature of the Chemical Bond,* Cornell University Press, Ithaca, 1960.

E. Hückel, *Ein Gelehrtenleben – Ernst und Satire,* Verlag Chemie, Weinheim, 1975.

To Sect. 14.2.

14.2.1.

G. Bruni, B. L. Vanzetti, *Vier Abhandlungen von Wilhelm Körner,* Ostwald's Klassiker der Exacten Wissenschaften, No. 174 (Ed. W. Engelmann), Leipzig, 1917.

J. M. McBride, *Completion of Koerner's Proof that the Hydrogens of Benzene are Homotopic. An Application of Group Theory, J. Am. Chem. Soc.* **1980,** *102,* 4134.

G. W. Wheland, *The Körner Absolute Method* in *Advanced Organic Chemistry,* 3rd Edn., Wiley & Sons, New York, 1960.

To Sect. 14.3.

14.3.1.

'IUPAC Recommendations 1994: Glossary of Terms used in Physical Organic Chemistry', Pure Appl. Chem. **1994,** *66,* 1077.

F. D. Rossini, K. S. Pitzer, W. J. Taylor, J. P. Ebert, J. E. Kilpatrick, C. W. Beckett, M. G. Williams, H. G. Werner, *Selected Values of Properties of Hydrocarbons,* United States Government Printing Office, Washington, DC, 1947.

K. B. Wiberg, *Thermochemistry* in *Determination of Organic Structures by Physical Methods,* Vol. 3 (Ed. F. C. Nachod, J. J. Zuckerman), Academic Press, New York, 1971.

G. B. Kistiakowsky, H. Romeyn, Jr., J. R. Ruhoff, H. A. Smith, W. E. Vaughan, *The Apparatus and the Heat of Hydrogenation of Ethylene, J. Am. Chem. Soc.* **1935,** *57,* 65.

G. B. Kistiakowsky, J. R. Ruhoff, H. A. Smith, W. E. Vaughan, *Hydrogenation of Some Simpler Olefinic Hydrocarbons, J. Am. Chem. Soc.* **1935,** *57,* 876;
– *Hydrogenation of Some Higher Olefins, J. Am. Chem. Soc.* **1936,** *58,* 137.

M. A. Dolliver, T. L. Gresham, G. B. Kistiakowsky, W. E. Vaughan, *Heats of Hydrogenation of Various Hydrocarbons, J. Am. Chem. Soc.* **1937,** *59,* 831.

R. B. Turner, W. R. Meador, R. E. Winkler, *Apparatus and the Heats of Hydrogenation of Bicyclo[2.2.1]heptene, Bicyclo[2.2.1]heptadiene, Bicyclo[2.2.2]octene and Bicyclo[2.2.2]octadiene, J. Am. Chem. Soc.* **1957,** *79,* 4116.

R. B. Turner, W. R. Meador, *Hydrogenation of Some cis- and trans-Cycloolefins, J. Am. Chem. Soc.* **1957,** *79,* 4133.

S. W. Benson, *Thermochemical Kinetics,* Wiley & Sons, New York, 1976.

O. V. Dorofeeva, L. V. Gurrich, V. S. Jorish, *Thermodynamic Properties of Twenty-One Monocyclic Hydrocarbons, J. Phys. Chem. Ref. Data* **1986,** *15,* 437.

14.3.2.

M. J. S. Dewar, H. N. Schmeising, *A Reevaluation of Conjugation and Hyperconjugation, Tetrahedron* **1959,** *5,* 166;
– *Resonance and Conjugation, Tetrahedron* **1960,** *11,* 96.

M. J. S. Dewar, *Resonance, Conjugation and Hyperconjugation, Chem. Eng. News,* **1965,** 86.

M. J. S. Dewar, C. de Llano, *Ground States of Conjugated Molecules, J. Am. Chem. Soc.* **1969,** *91,* 789.

W. R. Roth, O. Adamczak, R. Breackmann, H. W. Lennartz, R. Boese, *Die Berechnung von Resonanzenergien; das MM2ERW-Kraftfeld, Chem. Ber.* **1991,** *124,* 2499.

Selected Values of Properties of Hydrocarbons, Circular of the National Bureau of Standards, C 461, S. 141–143, Washington, DC, 1947.

To Sect. 14.4.

F. Sondheimer, *The Annulenes, Acc. Chem. Res.* **1972,** *5,* 81;
– *Recent Advances in the Chemistry of Large-Ring Conjugated Systems, Pure Appl. Chem.* **1963,** *7,* 363.

R. Breslow, *Antiaromaticity, Acc. Chem. Res.* **1973,** *6,* 393.

G. Maier, *Tetrahedrane and Cyclobutadiene, Angew. Chem. Int. Ed.* **1988,** *27,* 309.

D. J. Cram, M. F. Tanner, R. Thomas, *A cyclobutadiene that is stable at room temperature, Angew. Chem. Int. Ed.* **1991,** *30,* 1024.

J. F. M. Oth, *Conformational Mobility and Fast Bond Shift in the Annulenes, Pure Appl. Chem.* **1971,** *25,* 573.

F. A. L. Anet, A. J. R. Bourn, Y. S. Lin, *Ring Inversion and Bond Shift in Cyclooctatetraene Derivatives, J. Am. Chem. Soc.* **1964,** *86,* 3576.

G. Schröder, J. F. M. Oth, R. Merenyi, *Molecules Undergoing Fast, Reversible Valence-Bond Isomerization, Angew. Chem. Int. Ed.* **1965,** *4,* 752.

S. Z. Goldberg, K. N. Raymond, C. A. Harmon, D. H. Templeton, *Structure of the 10π-Electron Cyclooctatetraene Dianion in Potassium Diglyme 1,3,5,7-Tetramethylcyclooctatetraene Dianion, J. Am. Chem. Soc.* **1974,** *96,* 1348.

T. J. Katz, *The Cyclooctatetraenyl Dianion, J. Am. Chem. Soc.* **1960,** *82,* 3784.

E. Vogel, *Aromatic 10π-Electron Systems* in *Aromaticity,* Special Publ. No. 21, The Chemical Society, London, 1967.

14.4.3.

E. Heilbronner, in *Proceedings of an International Symposium 'Aromaticity, Pseudo-Aromaticity, Anti-Aromaticity'* (Ed. E. Bergmann, B. Pullman), The Israel Academy of Science and Humanities, Jerusalem, 1971.

D. Lloyd, D. R. Marshall, *An Alternative Approach to the Nomenclature of Cyclic Conjugated Polyolefins, together with some Observations on the Use of the Term 'Aromatic', Angew. Chem. Int. Ed.* **1972,** *11,* 404.

G. Binsch, *Aromaticity – An Exercise in Chemical Futility? Naturwissenschaften* **1973,** *60,* 369.

L. Salem, *The Molecular Orbital Theory of Conjugated Systems,* W. A. Benjamin, New York, 1966.

To Sect. 14.5.

A. D. Walsh, *The Structure of Ethylene Oxide, Cyclopropane, and Related Molecules, Trans. Farad. Soc.* **1949,** *45,* 179.

R. Hoffmann, R. B. Davidson, *The Valence Orbitals of Cyclobutane, J. Am Chem. Soc.* **1971,** *93,* 5699.

W. L. Jorgensen, L. Salem, *The Organic Chemist's Book of Orbitals,* Academic Press, New York, 1973.

E. Honegger, E. Heilbronner, A. Schmelzer, *Do Walsh-Orbitals 'Exist'? Nouv. J. Chim.* **1982,** *6,* 519.

M. J. S. Dewar, *Chemical Implications of σ-Conjugation, J. Am. Chem. Soc.* **1984,** *106,* 669.

D. Cremer, J. Gauss, *Reevaluation of the Strain Energies of Cyclopropane and Cyclobutane – CC and CH Bond Energies, 1,3-Interactions, and σ-Aromaticity, J. Am. Chem. Soc.* **1986,** *108,* 7467.

15 Hydrogen Bonds

15.1. Introduction

About 70% of the human body consists of H_2O. H_2O is of unique significance for life on Earth. It possesses a whole series of unusual properties: compared with other compounds of similar molecular mass, it has extraordinarily high melting and boiling points. It is liquid at room temperature (its heavier homolog, H_2S, is gaseous at room temperature, boiling at $-60\,°C$) and its high enthalpies of melting and vaporization (at 2.26 kJ/g the highest of all liquids) are of key importance for our climate. Equally unusual is the fact that the density of liquid H_2O at its melting point is higher than the density of ice. Because of its high dielectric constant, H_2O is also an excellent solvent for ions.

Wendell M. Latimer and *Worth H. Rodebush* postulated in 1920 that weak bonds between H_2O molecules (*hydrogen bonds*) lead to extended associations, and so were responsible for the unique properties of H_2O. An H-atom bound to an O-atom interacted with an additional O-atom in a neighboring molecule, thereby raising its coordination number from 1 to 2. Despite the damage inflicted on the theory of the monovalency of the H-atom, regarded as irrefutable until then, this concept won general acceptance, and an explosion of activity in the investigation of hydrogen bonds of the general type X–H···Y ensued. It was recognized that this kind of weak bond played an extremely important role, particularly in the formation of supermolecules. The most prominent examples are undoubtedly the α-helix in proteins (*Chapt. 7.3.3.4*) and the base pairs of the DNA double helix (*Chapt. 7.3.1.2.1*). In 1960, *George C. Pimentel* and *Aubrey L. McClellan* undertook the task of summarizing the state of knowledge of the time relating to this phenomenon. In their book '*The Hydrogen Bond*', which was for many years the standard work about hydrogen bonds, more than 2,000 relevant publications are cited; by 1985 this number had already grown to over 20,000.

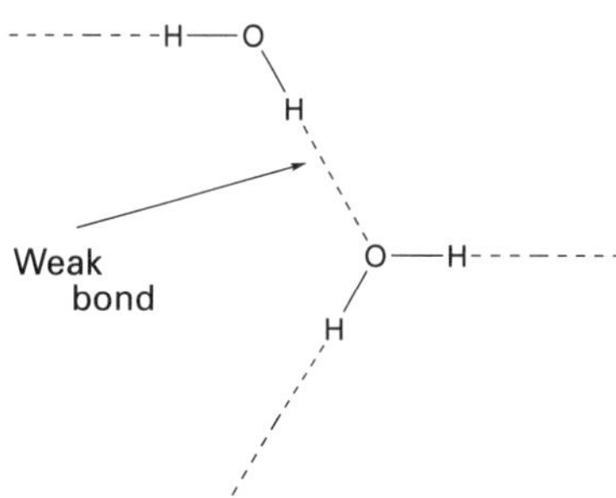

15.2. Properties of Hydrogen Bonds

The distinction between molecules, in which atoms are covalently bound to one another, and supermolecules, in which molecules are bound non-covalently to one another (*Chapt. 7.1*), implies that different forces for the existence of discrete chemical compounds may be anticipated. It is indeed appropriate to deal with strong and weak chemical bonds separately. Besides covalent (electron-pair) bonds, ionic (electrostatic) bonds are also included among strong bonds. Hydrogen bonds and *van der Waals* interactions are counted as weak bonds. In discussion of organic compounds, covalent bonds predominate for strong bonds, hydrogen bonds for weak bonds.

15.2.1. Experimental Evidence

The formation of supermolecules through hydrogen bonds alters the apparent mass, size, and form of the regarded molecule, and so influences its physical properties. Hydrogen bonds, therefore, may be investigated by a variety of methods. The effective molecular mass of compounds of this type, as determined by measurement of vapor density, depression of freezing point, or elevation of boiling point, is concentration-dependent, and in many cases permits definite conclusions to be drawn about the structure. Many carboxylic acids are dimers in the gas phase and in non-polar solvents, for example.

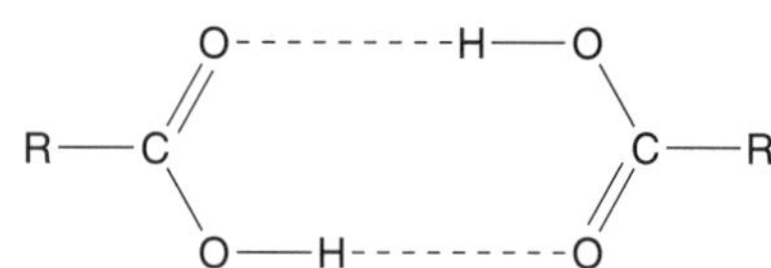

The entropy of vaporization of liquids with an extensive network of hydrogen bonds is significantly larger than predicted by *Trouton*'s rule. The reason for this is the partial dissociation of the supermolecules upon transition into the gas phase and the consequently heightened increase in disorder when compared with liquids, in which these interactions play no role. Changes in the physical values mentioned testify to the existence of association in a compound.

As hydrogen bonds influence the electronic structure of the functional groups participating in the interaction, they may be examined specifically using spectroscopic methods: primarily with IR and NMR spectroscopy. The X–H stretching vibration serves as the principal source of information in IR spectroscopy: the X–H bond is weakened by the H···Y interaction, and so the wave number and intensity of the corresponding absorption bands change in a characteristic way upon formation of hydrogen bonds. In the ^{1}H-NMR spectrum, the signal of the H-atom affected is shifted to significantly lower field.

Hydrogen bonds occur in all states of aggregation. They are encountered less frequently in vapor than in condensed phase, because of the higher temperatures and lower densities. Since powerful and reliable methods for the determination of crystal structures have been available for a long time, hydrogen bonds in crystals have been examined especially intensively. The study of hydrogen bonds with the aid of X-ray crystal-structure analysis, however, is made more difficult by the fact that H-atoms are more difficult to localize than any other atoms, because of their low electron density. Furthermore, X–H bond lengths determined by X-ray diffraction are about 10 pm too short, because H-atoms do not possess any core electrons. Neutron-diffraction experiments are, therefore, necessary for precise scrutiny; but the effort needed and the expense involved in this technique are far greater than for the use of X-rays. As a very large number of X-ray diffraction studies have been made, however, it is possible, in many cases, to obtain reliable information using statistical methods. *Robin Taylor* and *Olga Kennard* have shown, using the example of the N–H···O=C interaction, that X-ray-crystallographically determined crystal structures, in which the systematic shortening of the X–H bond is corrected, are, on average, just as reliable as those obtained by neutron-diffraction studies.

It is often difficult to produce conclusive proof that an H···Y bond does in fact exist. In many cases, one makes do with a structural criterion: a hydrogen bond is considered likely, if the distance between the two atoms H and Y is significantly smaller than the sum of their *van der Waals* radii. If the results used for this decision come from an X-ray crystal-structure analysis, account must be taken of the systematic error and uncertainty relating to the position of the H-atom. Therefore, the X···Y distance is often referred to, to demonstrate the existence of a hydrogen bond X–H···Y.

15.2.2. Structure and Energy

With a bond energy of *ca.* 10–40 kJ/mol, hydrogen bonds in general are an order of magnitude weaker than covalent bonds and an order of magnitude stronger than *van der Waals* interactions. They occupy a prominent position among non-covalent interactions. The strongest hydrogen bond known exists in the $[F{-}H{-}F]^-$ anion: it has a bond energy of over 100 kJ/mol. The distance of the two F-atoms is only 226 pm. The HF_2^- ion has $D_{\infty h}$ symmetry; this means it is linear with two equally long H–F bonds. Only a very few, very short hydrogen bonds show this structural feature. Hydrogen bonds are indeed preferentially linear, but in most cases the H···Y distance is appreciably larger than the X–H distance.

It is, therefore, possible to differentiate between symmetrical and unsymmetrical hydrogen bonds. The energy profile of a hydrogen bond X–H···X, in which the two bond partners of the H-atom are identical, may have two minima, each in the vicinity of one of the bond partners, or one minimum situated exactly between them (*Fig. 15.1*). Short (and hence strong) hydrogen bonds tend to represent the latter case, long (and hence weak) ones the former case. If the energy barrier between the two minima of an unsymmetrical hydrogen bond is so low that it can easily be surmounted at room temperature (**2** in *Fig. 15.1*), then this hydrogen bond can scarcely be distinguished from a symmetrical one, if X-ray crystal-structure analysis is being used. Only a structure determination at low temperature can clarify whether a distinct vibration about one minimum (**3** in *Fig. 15.1*) is occurring, or if two separate minima exist (**1** in *Fig. 15.1*): crystallographers use the term *static disorder* to describe the latter situation.

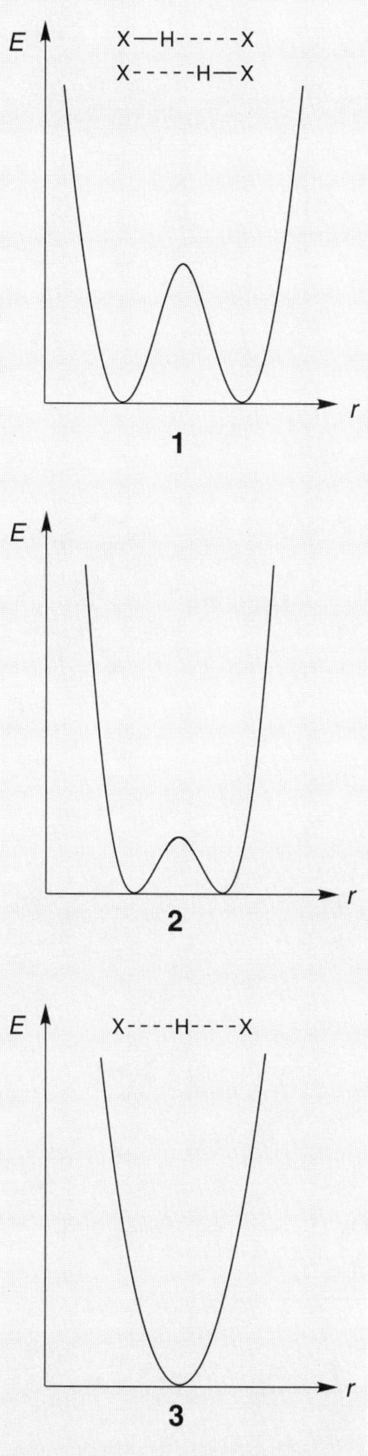

Fig. 15.1. *Energy profile of a hydrogen bond X–H···X with a strongly pronounced double minimum* (**1**), *weakly pronounced double minimum* (**2**), *or single minimum* (**3**)

Such a case is encountered in the crystal structure of ice (*Fig. 15.2*). Each O-atom is tetrahedrally surrounded by four other O-atoms ($r(O{\cdots}O) = 276$ pm), so that each H_2O molecule forms four almost linear hydrogen bonds and functions both as a double donor

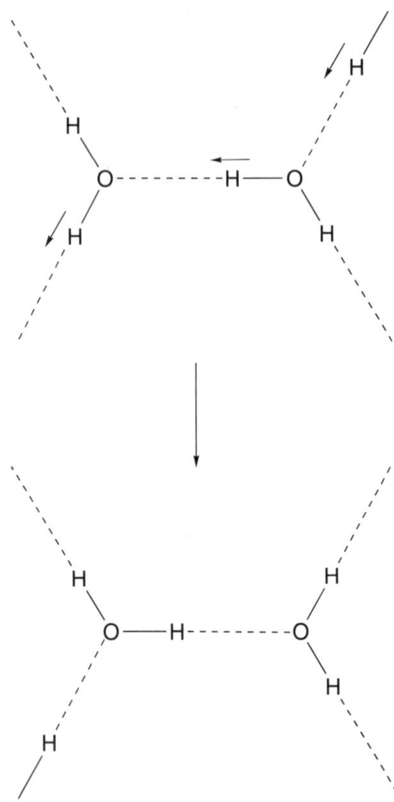

(O–H···O) as well as a double acceptor (O···H–O). The H-atom is to be found between two O-atoms in one of two locations situated close together, with equal probability of being in either. This is interpreted as a time-averaged (during the period of measurement) and space-averaged (over all unit cells) crystal structure, representing a summation of many structures, all with the same O-positions, but different H-positions. The individual structures can easily transmute into one another by synchronized change of positions by protons (or by rotation of H_2O molecules). They are fixed in place only at very low temperatures. So, ice displays a very high degree of static disorder. *Pauling* calculated the number of possible structures in 1935, and subsequently estimated a value for the zero-point entropy.

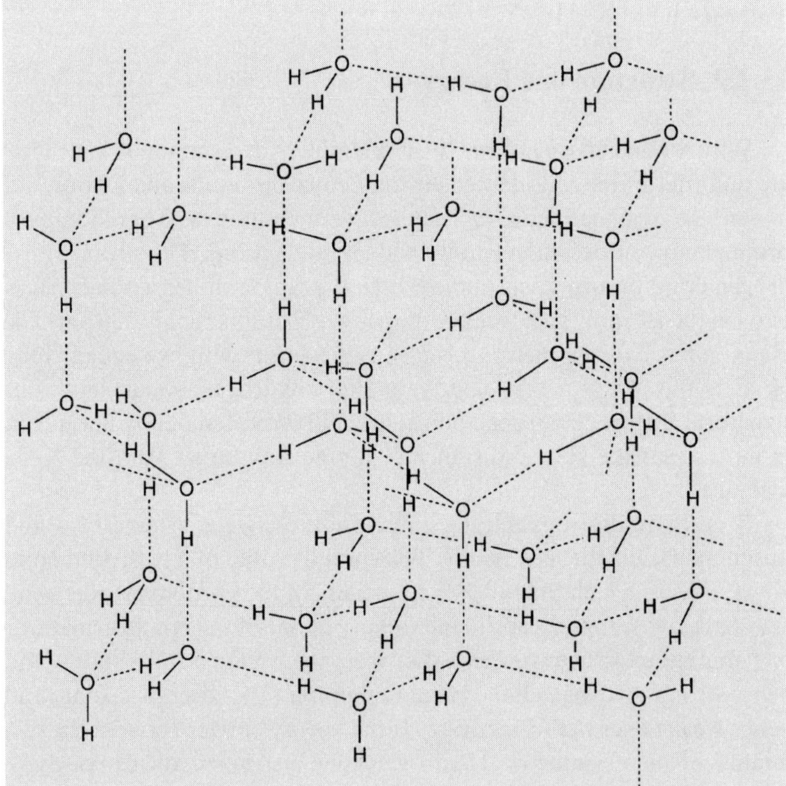

Fig. 15.2. *Section of the hexagonal crystal structure of ice* (ice Ih)

'Proton jumps' of this kind explain the high electrical conductivity of ice. The hexagonal structure of ice which is stable at 0 °C contains cage-like channels parallel to the six-fold axis. This 'loose' packing is partially destroyed on the melting of ice, so liquid H_2O, just above its melting point, has a higher density than ice. It can be estimated from the enthalpy of melting that only about 10% of the hydrogen bonds are abolished during the transition. Formation and destruction of hydrogen bonds proceeds at great speed in H_2O, with quasi-crystalline regions always present. The solubility of many substances (sugars, for example) in H_2O is founded on the formation of hydrogen bonds with the solvent.

The unusually high boiling points of HF and NH_3 may also be explained by the formation of hydrogen bonds. As HF has only one H-atom, long zig-zag chains with unsymmetrical, linear hydrogen bonds are to be found in liquid HF, unlike in H_2O (*Fig. 15.3*).

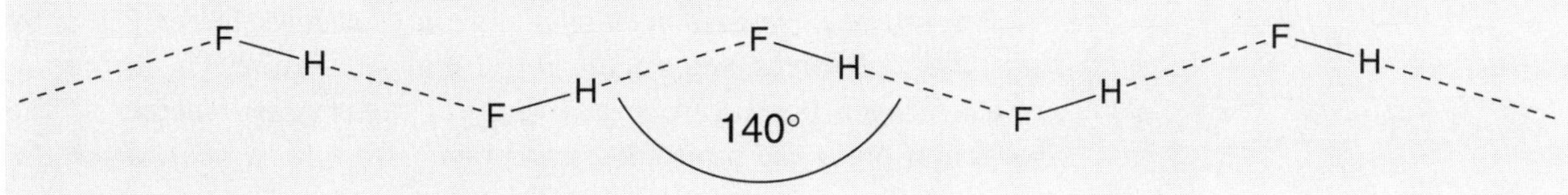

Fig. 15.3. *Section of the structure of liquid HF*

15.2.3. Donor and Acceptor Groups

A hydrogen bond may be interpreted as the result of an acid/base interaction in the *Brønsted* sense: a proton donor interacts with a proton acceptor. For strong hydrogen bonds, the following functional groups are the principal ones to be taken into account as donor and acceptor groups:

$$\overset{\delta-}{X}—\overset{\delta+}{H}\cdots\cdots\overset{\delta-}{|Y}$$

X-H: —COOH, —OH, —NH_2, —C(=O)—NH, H-Hal

Y: >C=O, –O–, —$\bar{N}R_2$, –S–, |(Hal)|$^\ominus$, –P=O

The organic chemist is naturally interested in the question of whether C—H bonds can also function as donor groups for hydrogen bonds. For C—H bonds which are activated by strongly electron-withdrawing groups, the answer has been known for a long time. $CHCl_3$ forms complexes with pyridine or acetone, and HCN exists in the crystalline state in the form of linear chains:

$$\cdots\cdots H—C\equiv N\cdots\cdots H—C\equiv N\cdots\cdots H—C\equiv N\cdots\cdots$$

To clarify whether less strongly activated C—H bonds also have a tendency towards the formation of C—H···Y hydrogen bonds, *Taylor* and *Kennard* studied 113 crystal structures determined by neutron diffraction. Their statistical analysis gave the result that C—H groups have a notable preference for forming intermolecular contacts to O-atoms (rather than to C- or H-atoms). The C—H···O angle generally amounts to 150–180°, and each C—H bond is directed towards one of the free electron pairs of the O-atom. There are similar suggestions also for C—H···N and C—H···Cl contacts, but the data available were insufficient for any statistically reliable conclusions to be drawn. Because of the

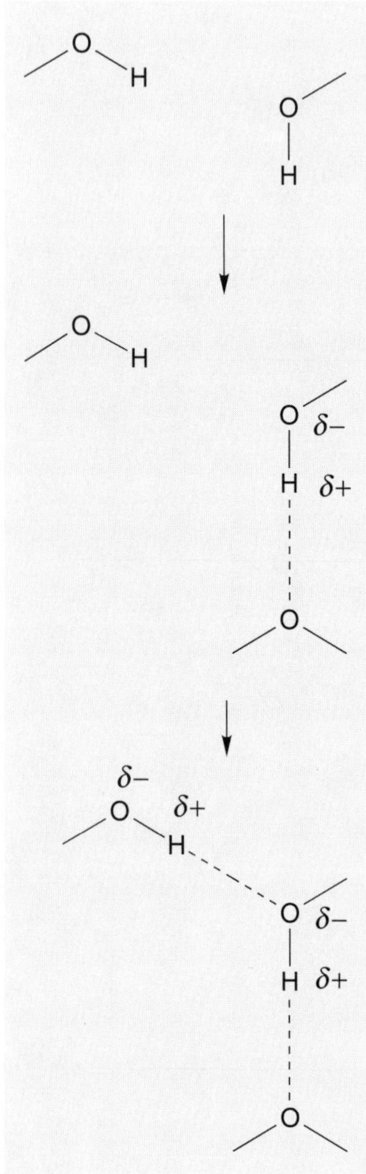

Fig. 15.4. *Favored formation of a second hydrogen bond subsequent to the formation of the first one* (cooperative effect)

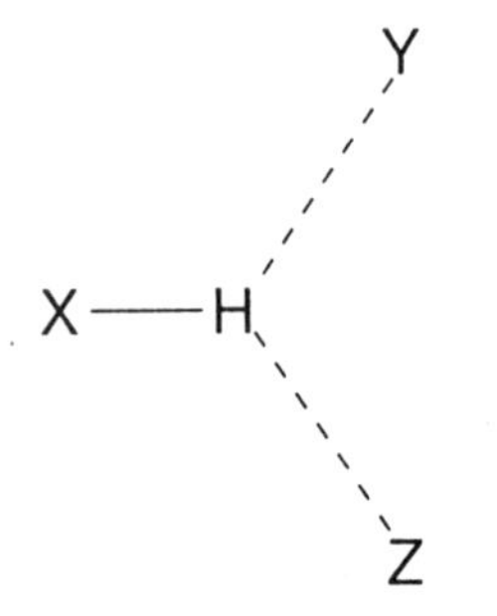

smaller donor strength of the C–H bond, C–H···Y hydrogen bonds are weaker than the hydrogen bonds mentioned up to now. They do play a significant role in the formation of crystals, however, if no better donors are present.

15.2.4. Hydrogen Bonds in Crystals

The lattice energies of crystals of organic molecules, held together by *van der Waals* forces, amount to *ca.* 1 kJ/mol per C-atom. The formation of a hydrogen bond lowers the energy of the crystal – merely by the reorientation of the molecules, and practically without loss of *van der Waals* energy – by several kJ/mol. The significance of hydrogen bonds in crystal formation can, therefore, scarcely be overestimated. Crystallographers know that, when suitable donor and acceptor groups are present, they will almost always form hydrogen bonds. As early as 1952 (when the positions of H-atoms in crystals could only be ascertained in occasional cases), *Jerry Donohue* pointed out that an OH or an NH group almost never adopts a position which would make the formation of a hydrogen bond impossible. H_2O Molecules are often found in the crystal structures of polar compounds (*water of crystallization*) for this reason.

Examination of hydrogen-bond patterns in crystals shows that long chains are favored over chelates. The cause of this is a *cooperative effect* (*Chapt. 7.3.1.2.1*). When, for example, an O–H···O hydrogen bond is formed, the O–H bond is weakened, so that the basicity of the O-atom is increased, making this atom a better hydrogen-bond acceptor (*Fig. 15.4*).

Besides simple hydrogen bonds, bifurcated ones often occur in crystals. In these, one H-atom interacts simultaneously with two acceptors, Y and Z. These hydrogen bonds are longer than those discussed previously.

The dominance of hydrogen bonds in crystal structures of compounds possessing suitable donor and acceptor groups will be demonstrated using a few examples.

Oxalic acid crystallizes in two forms, differing in their hydrogen-bond patterns. While the α-form exhibits a two-dimensional, infinite layer structure, chains resembling the previously mentioned carboxylic-acid dimers are found in the β-form (*Fig. 15.5*). The additional stabilization caused by the hydrogen bonds is shown by the fact that the respective enthalpies of sublimation of the two forms are 51 and 46 kJ/mol higher than that one of dimethyl oxalate.

In the crystal structure of *cyanuric acid*, all of the N–H bonds form linear N–H···O bonds, without the O-atoms of the C=O groups attaining their maximum coordination number of 3 (*Fig. 15.6*). Conversely, *melamine* has six donor groups available, but only three acceptor groups. It is, therefore, not possible to saturate all of the donors within the molecular plane, and a three-dimensional network of hydrogen bonds is the consequence. A 1:1 complex of cyanuric acid and melamine, however, possesses equal numbers of donor and acceptor groups, and enables a layer structure to exist, all N–H bonds participating in one hydrogen bond, and all C=O groups in two (*Fig. 15.7*).

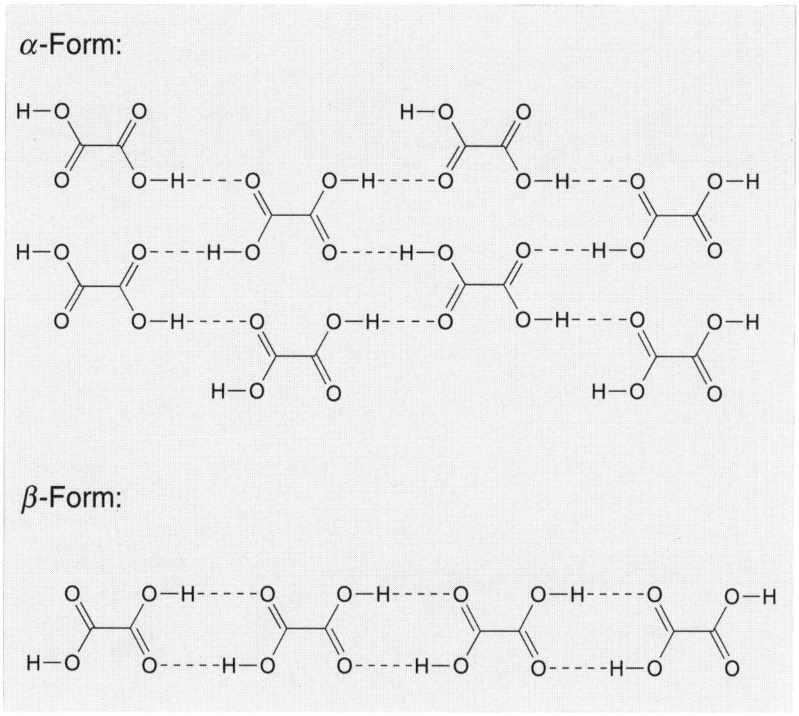

α-Form:

β-Form:

Fig. 15.5. *Sections of the crystal structures of α- and β-oxalic acid*

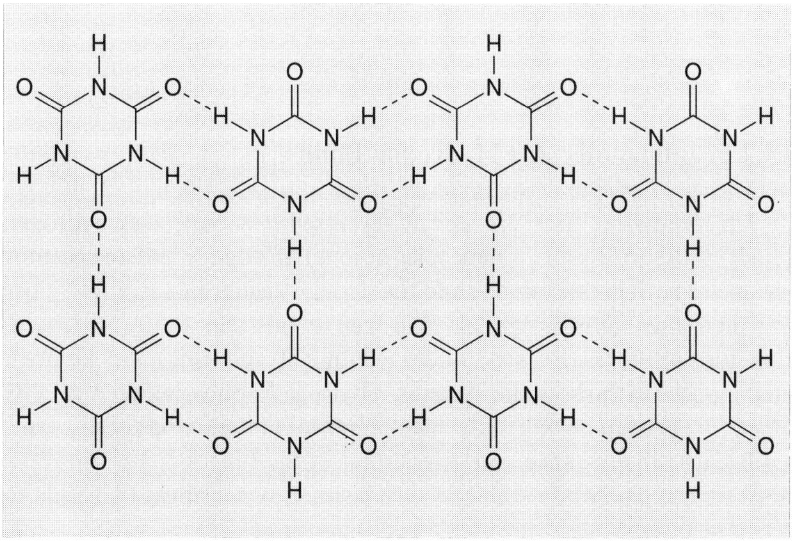

Fig. 15.6. *Section of the crystal structure of cyanuric acid*

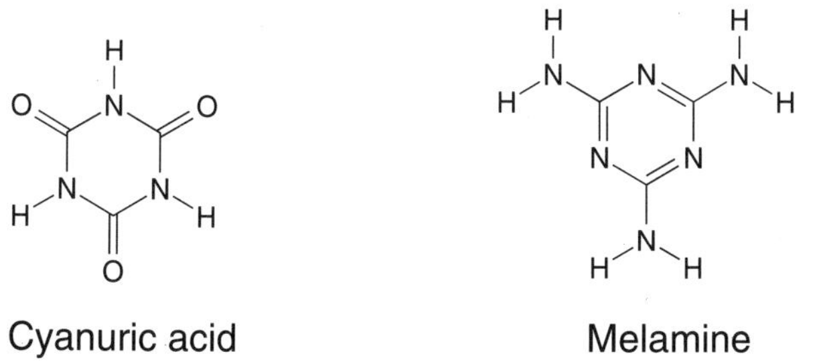

Cyanuric acid

Melamine

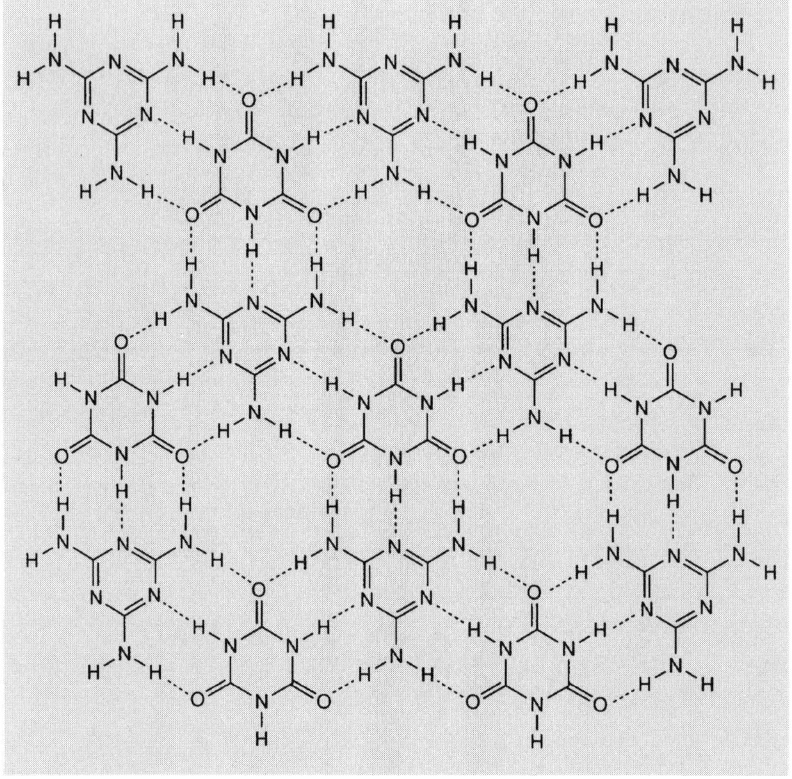

Fig. 15.7. *Section of the supposed structure of the cyanuric acid/melamine complex*

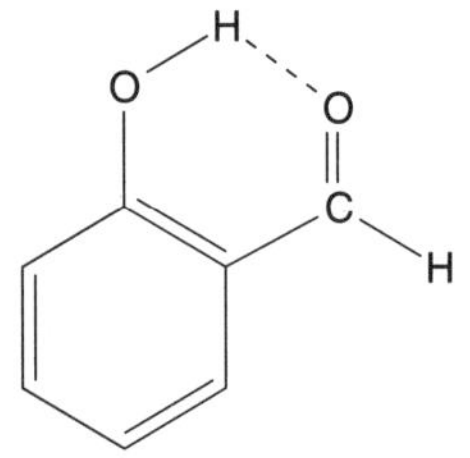

Salicylaldehyde

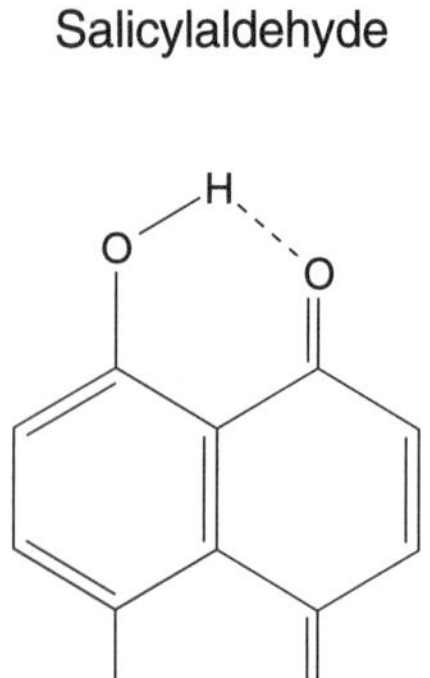

5,8-Dihydroxy-1,4-naphthoquinone

15.2.5. Intramolecular Hydrogen Bonds

Up to now, we have exclusively discussed *inter*molecular hydrogen bonds (those between two molecules or ions). If a donor and an acceptor group are both present in one and the same molecule and arranged close together, then *intra*molecular hydrogen bonds can be formed, with ring formation. Five-, six-, and seven-membered rings are favored, analogously to carbocyclic systems. Hydrogen bonds between an OH and a C=O group belong to the most important interactions of this type.

Because the distance and orientation of the functional groups may not be particularly adjustable in such cases, it is not always possible to form the optimal structure of the hydrogen bonds. Hence, intramolecular hydrogen bonds, unlike their linearly ordered intermolecular counterparts, are usually bent to some degree, and, as a rule, weaker than intermolecular hydrogen bonds. One exception is the monoanion of maleic acid: the crystal structure of the K-salt shows an unusually short intramolecular hydrogen bond, with $r(O \cdots O) = 240$ pm. It is linear and symmetrical, like in the HF_2^- ion (see opposite page).

Intramolecular hydrogen bonds are often the reason for substantial differences in boiling point, volatility, or solubility between benzene derivatives with two *ortho*-substituents and the corresponding *meta*- and *para*-substituted compounds. The enol isomer of β-dicarbonyl

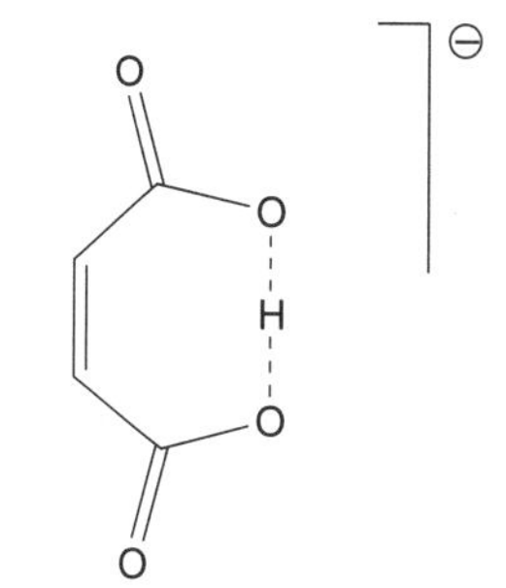

(= 1,3-dioxo) compounds is also stabilized to a significant degree by an intramolecular hydrogen bond. However, in solvents with hydrogen-bond donors, intermolecular hydrogen bonds between the C=O groups and the solvent molecules are preferentially formed. The rule '*similia similibus solvuntur*' would seem not to apply in this case, and the keto$\rightleftharpoons$enol equilibrium is shifted towards the keto isomer.

Keto isomer

Enol isomer

15.2.6. Hydrogen Bonds in Proteins and Nucleic Acids

Hydrogen bonds are of crucial importance for the structure and function of proteins (*Chapt. 7.3.3.2.3*) and nucleic acids (*Chapt. 7.3.1.2*). *Pauling*'s intensive study of the N–H···O=C interaction led to his proposal of the α-helix and the β-pleated sheet as important elements of secondary structure in proteins. Also *Watson* and *Crick*'s investigations into the structure of DNA remained without success until their recognition that base pairs (A=T and C≡G) could come into being through the formation of hydrogen bonds.

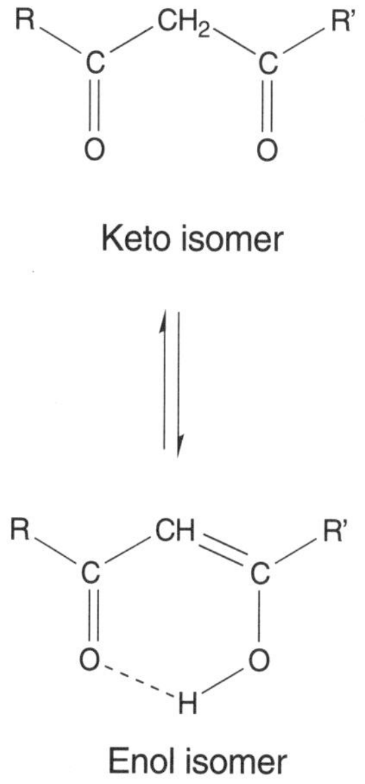

George A. Jeffrey examined the hydrogen bonds present in the crystal structures of nucleosides and nucleotides and arranged the donor and acceptor groups with regard to their strengths, using the atomic distances as a basis. This showed that H_2O molecules, because of the cooperative effect, are better acceptors than C=O groups, surpassed only by P=O groups in this property.

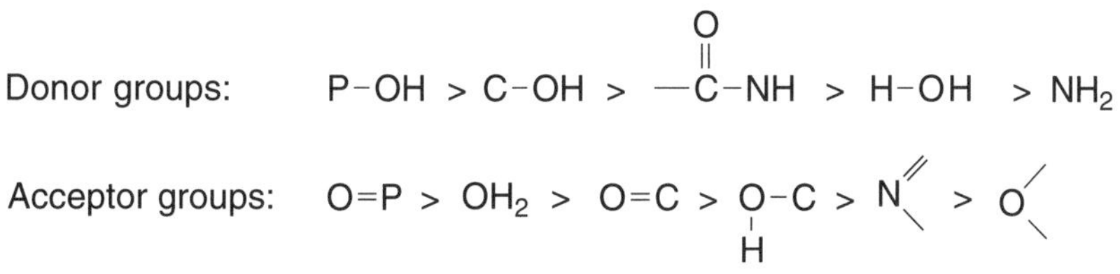

Donor groups: P–OH > C–OH > –C(=O)–NH > H–OH > NH_2

Acceptor groups: O=P > OH_2 > O=C > O(H)–C > N< > O<

15.3. Theoretical Approaches to the Description of Hydrogen Bonds

The first proposal for the theoretical description of hydrogen bonds came from *Pauling*. Supported by the stability of the HF_2^- ion and the strength of the O–H···O bridge, he postulated in 1928 that hydrogen bonds were ionic. The preference for linear hydrogen bonds X–H···Y could then be well explained with this simple model: the electrostatic repulsion between the atoms X and Y is at its least in a linear arrangement. Quantitative calculations remained unsatisfactory, though, as long as a spherically symmetrical distribution of electron density was assumed. Only the recently introduced description of electrostatic interaction with the aid of multipoles (charges, dipoles, and quadrupoles) was to lead to essentially better results.

As an alternative to the electrostatic model, *Pimentel* examined the stability of $[X{-}H{-}X]^-$ ions using an MO approach in 1951. Three MOs are formed by the interaction of the 1s-orbital of the H-atom with the $2p_X$-orbitals of the atoms X. According to the rules of play given in *Chapt. 8*, these must be occupied by four electrons (*Fig. 15.8*). Here, the middle MO comes into a decisive role. It is all the more stable, the more electronegative the atoms participating in the hydrogen bond. This explains the exceptional stability of the HF_2^- ion.

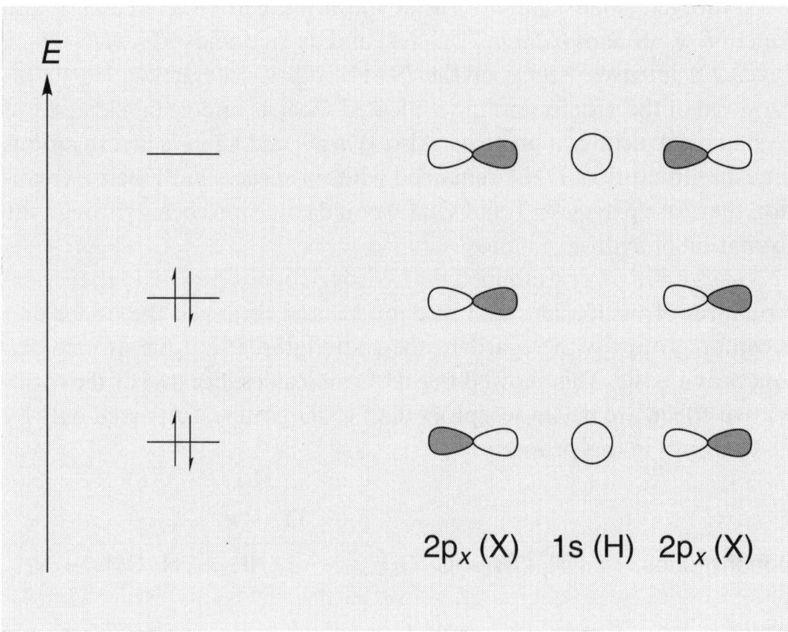

Fig. 15.8. *MO Scheme of [X–H–X]⁻*

More recent quantitative MO calculations have shown that the total energy of a supermolecule X–H···Y is made up of several energy components (electrostatic interaction, *van der Waals* repulsion, and charge transfer). The electrostatic energy is often the dominant component of these systems, and so the contemporary view of hydrogen bonding is not very far away from the original model.

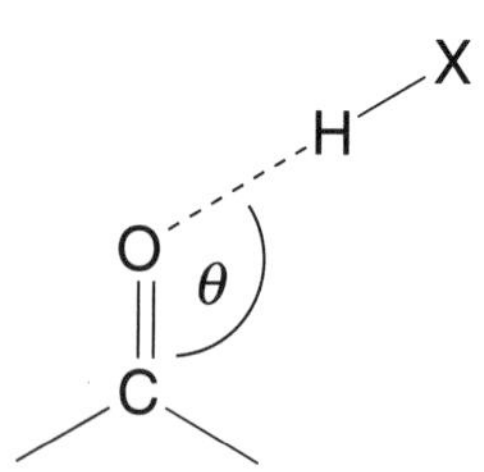

Exact quantum-chemical calculations (as well as the above-mentioned multipole models) agree well with experimental results. An example for that is the description of the X–H···O=C interaction. A search in the *Cambridge Structural Database*, which contains more than 100,000 crystal structures of organic and organometallic compounds, yielded altogether 181 crystal structures with intermolecular hydrogen bonds of this type between an NH or OH group and a ketone. The graph of the H-atoms participating in hydrogen bonds relative to the respective C=O group reveals an unequivocal preference of certain angular ranges (*Fig. 15.9*):

- practically all H-atoms are in the plane formed by the carbonyl C-atom and its ligands;

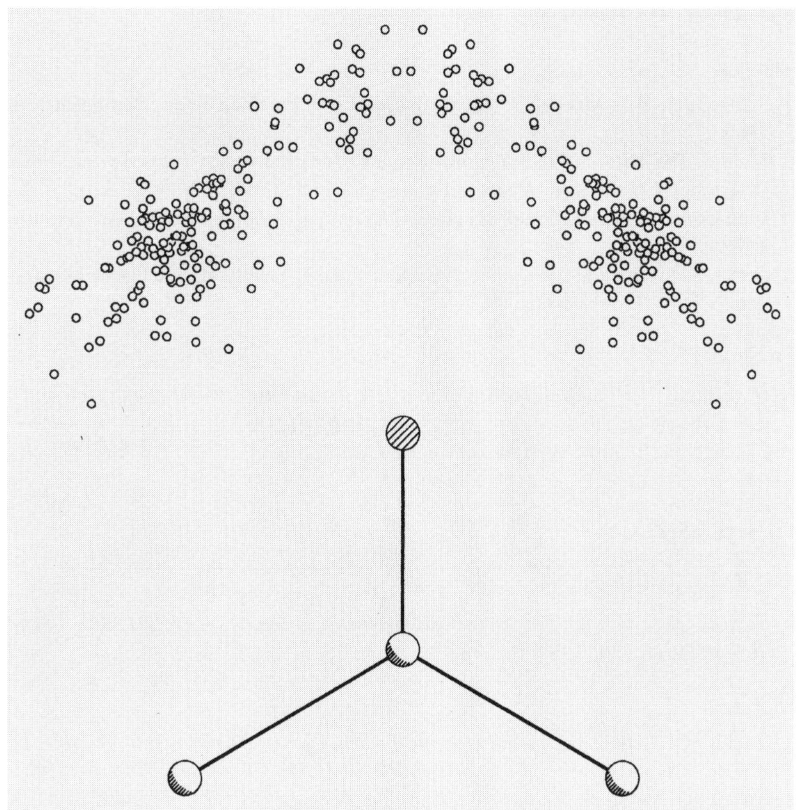

Fig. 15.9. C_{2v}-*Symmetrical graph of the intermolecular X–H···O=C bonds between an NH or OH group and a ketone as found in crystal structures.* For each hydrogen bond, the H-atom of the X–H group is shown relative to the C=O group.

- within this plane, the H-atoms are crowded along the direction of the lone-pairs of the O-atom (mean value: $\theta = 132°$). The distribution towards larger θ-values is rather broad, though, and also linear hydrogen bonds occur.

The preferred formation of X–H···O=C bonds which are linear at the H-atom, but angular at the O-atom was tacitly considered in previous sections (see *Fig. 15.7*). Also the structures of the nucleic-acid base pairs (*Chapt. 7.3.1.2*) and the elements of regular secondary structure of proteins (*Chapt. 7.3.3.3* and *7.3.3.4*) are in harmony with the observed distribution of hydrogen-bond donors. The experimental result shown in *Fig. 15.9* is not reproduced by a simple electrostatic model, but by the more recent approaches mentioned.

Further Reading

General

G. C. Pimentel, A. L. McClellan, *The Hydrogen Bond*, Freeman & Co., San Francisco, 1960.

L. Pauling, *The Nature of the Chemical Bond*, Cornell University Press, Ithaca, 1960.

M. D. Joesten, L. J. Schaad, *Hydrogen Bonding*, Marcel Dekker, New York, 1974.

P. Schuster, G. Zundel, C. Sandorfy (Ed.), *The Hydrogen Bond*, Elsevier, Amsterdam, 1976.

G. Geiseler, H. Seidel, *Die Wasserstoffbrückenbindung*, Vieweg-Akademische Verlagsgesellschaft, Braunschweig, 1977.

To Sect. 15.1.

W. M. Latimer, W. H. Rodebush, *Polarity and ionization from the standpoint of the Lewis theory of valence, J. Am. Chem. Soc.* **1920**, *42*, 1419.

M. L. Huggins, *50 Years of Hydrogen Bond Theory, Angew. Chem. Int. Ed.* **1971,** *10*, 147.

To Sect. 15.2.

15.2.1.

R. Taylor, O. Kennard, *Hydrogen-Bond Geometry in Organic Crystals, Acc. Chem. Res.* **1984**, *17*, 320.

15.2.3.

R. Taylor, O. Kennard, *Crystallographic Evidence for the Existence of C−H···O, C−H···N, and C−H···Cl Hydrogen Bonds, J. Am. Chem. Soc.* **1982**, *104*, 5063.

Z. Berkovitch-Yellin, L. Leiserowitz, *The Role Played by C−H···O and C−H···N Interactions in Determining Molecular Packing and Conformation, Acta Crystallogr.* **1984**, *B40*, 159.

15.2.4.

W.C. Hamilton, J.A. Ibers, *Hydrogen Bonding in Solids*, Benjamin, New York, 1968.

G.A. Jeffrey, S. Takagi, *Hydrogen-Bond Structure in Carbohydrate Crystals, Acc. Chem. Res.* **1978**, *11*, 264.

J.F. Griffin (Ed.), *The Hydrogen Bond: New Insights on an Old Story, Trans. Am. Cryst. Assoc.* **1986**, *22*.

15.2.6.

G.A. Jeffrey, W. Saenger, *Hydrogen Bonding in Biological Structures*, Springer-Verlag, Berlin, 1991.

M. Perutz, *The Significance of the Hydrogen Bond in Physiology* in *The Chemical Bond – Structure and Dynamics* (Ed. A. Zewail), Academic Press, San Diego, 1992.

To Sect. 15.3.

P.A. Kollman, L.C. Allen, *The Theory of the Hydrogen Bond, Chem. Rev.* **1972**, *72*, 283.

A.D. Buckingham, P.W. Fowler, *A model for the geometries of Van der Waals complexes, Can. J. Chem.* **1985**, *63*, 2018.

R. Taylor, O. Kennard, W. Versichel, *Geometry of the N−H···O=C Hydrogen Bond. 1. Lone-Pair Directionality, J. Am. Chem. Soc.* **1983,** *105*, 5761.

16 Base Pairing in Biology and Chemistry

The 1970 *National Academy of Sciences'* state-of-the-art report, entitled '*Biology and the Future of Man*', contains the statement: '*Those who are hopeful about synthesizing a cell in the foreseeable future have every reason to retain their optimism*'. This prediction was based on the assumption that the laws of physics and chemistry would be sufficient to enable comprehension of the functions performed within a living cell. The academic question of whether *chemistry* is or could become a branch of physics (*Chapt. 8.5*), however, cannot under any circumstances be extended to *biology*. Biological systems were not designed but rather evolved. They have their own irreducible history, being a link in a chain of events that came about with the aid of natural selection rather than as reducible progressions determined by the laws of physics and chemistry only.

Chemists may contribute to the explanation of the development of natural historical processes in various ways. They may track down relics of primitive molecules which formerly carried out, albeit in a simpler manner, the functions of contemporary biomolecules. Further, they may develop evolutionary models, in which the chemical reaction potential of those molecules which played a role in an earlier phase of evolution, later made redundant by natural selection, is explored. Important insights may be obtained through experiments implied by a suitably designed evolutionary model. For example, if the drama of evolution on the stage of the chosen model could be 'rewound' to go back to a particular point in time and then the 'tape would be allowed to run again', this might show whether an experimentally obtained result has arisen inevitably or coincidentally.

Max Delbrück and *Francis Crick*, two physicists who devoted themselves, early in their careers, to biological research, have described the mental shift which they believe necessary for non-biological scientists approaching biological problems.

'A trained physicist, becoming acquainted with biological problems for the first time, finds it incomprehensible that there are no 'absolute phenomena' in biology. Everything is dependent on its particular time and its particular place. The animal, plant, or microorganism worked with is nothing but one member in an evolutionary chain of changing forms, none of which can prevail indefinitely.'

'The laws of physics, it is believed, are the same everywhere in the universe. This is unlikely to be true in biology... While Occam's *razor is a useful tool in the physical*

sciences, it can be a very dangerous implement in biology... Biologists must constantly keep in mind that what they see was not designed, but rather evolved. It might be thought, therefore, that evolutionary arguments would play a large part in guiding biological research, but this is far from the case. It is difficult enough to study what is happening now. To try to figure out exactly what happened in evolution is even more difficult.'

16.1. Foundations

16.1.1. The Pathway to Molecular Genetics

The concept of the molecule was introduced as the reaction to the observed discreteness of chemical compounds (*Chapt. 2.1*). It became the basis of (molecular) genetics: a combinatorial discipline founded on the rules of heredity proposed by *Gregor Johann Mendel*. According to this, molecules, acting as carriers of discrete genetic information, are duplicated and, after cell division, passed on to the next generation. The pathway to molecular genetics was anything but straight; nonetheless, the most important ideas and observations on this route are worth discussing. In 1987, *Max Ferdinand Perutz*, in an article that he was invited to write on the effects of *Erwin Schrödinger*'s 1944 book '*What is Life?*', listed those pioneers who in his opinion had really opened up the pathway to molecular genetics.

He began with *Delbrück*, who, in 1935, first conceived of a gene as being a specific section of a linearly constituted macromolecule – the chromosome. Two years later, *John B.S. Haldane* made the assumption that the duplication of a gene might proceed by means of some form of 'negative' copy. In 1940, *Pauling* and *Delbrück* developed their ideas on '*the nature of the intermolecular forces operative in biological processes*', relating to the autocatalytic reproduction of molecules in general and to replication of genes in particular. They pointed out that complementarity between two molecules, arranged in a 'system side by side', should afford some priority over other supramolecular orientations (as we would say today), not forgetting to mention that the case in which the two molecules were not only complementary, but, at the same time, identical to one another, would merit particular attention. In 1944, *Oswald Theodore Avery* established that DNA molecules were *the* carriers of genetic characteristics. In 1953, *Watson* and *Crick* were able to deduce the mechanism of replication (by means of a 'negative' copy) from the double-helix structure of DNA (into which two mutually complementary polynucleotide single strands had assembled).

16.1.2. The *Darwin-Mendel* Theory of Evolution

Experience shows that characteristic features of a living individual recur in a more or less regular manner down through the generations: they are inherited. Intensive efforts are being made to establish which, out of the entire set of characteristics making up the *phenotype*, are causally connected with which out of the entire set of hereditary factors constituting the *genotype*. The laws of heredity drawn up by *Mendel*, which had escaped the notice of the scientific community for three decades, permit the discernment of a combinatorial regularity and a localizability of discrete hereditary factors on discrete cell particles, of which the latter required yet more decades before it was to find general recognition even among geneticists. To express *Mendel*'s laws in the language of today, which assumes knowledge of the molecular components of the cell, the following (italicized) terms are used.

Every *diploid cell* contains two sets of chromosomes: one from the mother and one from the father. The *chromosomes* consist of two core threads (*chromatides*). Each chromatide contains a *double-helical DNA molecule* as its fundamental constituent. DNA Sections responsible for particular (apparent or non-apparent) features of the organism concerned are called *genes*. The same gene is always to be found on the same section of a chromosome pair. Such genes exist either as pairs of *identical alleles* (homozygous) or as pairs of *non-identical alleles* (heterozygous). Germ cells contain only one set of chromosomes: these are called *haploid gametes*.

In the first example (*Fig. 16.1*), crossing is assumed between two homozygous parents, both endowed with identical alleles AA. Their gametes (G) are both identical (A). Members of the subsequent filial generations are genotypically and phenotypically identical to the parents.

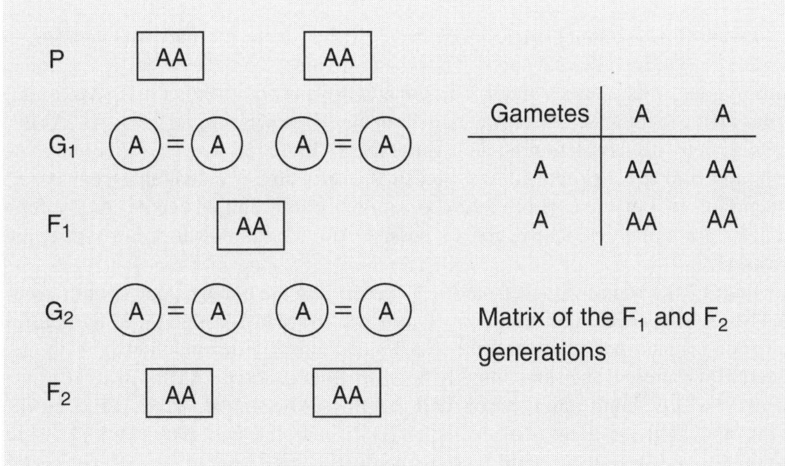

Fig. 16.1. *Genetic analysis by* Mendel*ian combinatorics: identical pairs of gametes G_1 are transmitted from homozygous parents* (P) *to progeny*. (The filial generations F_1 and F_2 represent the same genotype and the same phenotype.)

In the second example (*Fig. 16.2*), homozygous lines differing in only one gene are crossed with one another. One parent has the two identical alleles AA, the other the identical alleles aa. Their gametes both consist of only one allele: A in the former case, a in the latter. The heterozygous members of the first filial generation (F_1) represent the Aa genotype. If their gametes (both A or a, in either case) are combined with one another, the matrix of the F_2 generation – with three different genotypes (AA, Aa, and aa) in a 1:2:1 ratio – is obtained.

If one allele is completely dominant (A, say, so that the other allele, a, is wholly recessive), then, in the first filial generation (F_1), we observe not only uniform genotypes (Aa), but also uniform phenotypes (*Fig. 16.3*): the dominant allele A determines

P AA aa

G_1 (A)=(A) (a)=(a)

F_1 Aa

G_2 (A)≠(a) (A)≠(a)

F_2 AA Aa aa

Gametes	A	a
A	AA	Aa
a	aA	aa

Matrix of the F_2 generation

Fig. 16.2. *Genetic analysis by* Mendel*ian combinatorics: different pairs of gametes AA or aa are transmitted from homozygous parents* (P) *to progeny*. (Different from the second filial generation F_2, the homozygous members of the first filial generation F_1 represent the same genotype and phenotype.)

Fig. 16.3. *The stamp issued 1984 in honor of* Gregor Johann Mendel *representing schematically the outcome of dominant-recessive heredity* (here stand R for A and r for a)

that the phenotype of the homozygous parent type (AA) also holds for the members of F_1. Because the genotypes AA and Aa correspond to the dominant phenotype, and the genotype aa to the recessive phenotype, the phenotypes in the second filial generation (F_2) appear in the ratio 3:1. Two thirds of the dominant phenotypes are heterozygous: when these are crossed with one another they behave as representatives of F_1.

In the case of incomplete dominance of an allele, the phenotype of the heterozygote from F_1, with the uniform genotype Aa, lies somewhere between the phenotypes of the two homozygotes AA and aa. Members of the F_2 generation are distributed over three phenotypes, with a ratio between their genotypes of AA/Aa/aa = 1:2:1.

In the third example (*Fig. 16.4*), one individual from the parent generation is homozygous in two alleles A and B. The other individual from the parent generation is homozygous in the alleles a and b. The gametes are uniform in either case (AB or ab). From them, there arise uniform members of the F_1 generation, of genotype AaBb. Their gametes display all four possible genotypes (AB, Ab, aB, and ab) of the alleles A, a, B, and b in equal proportions. Upon further crossing of individuals from the F_1 generation, two of these gametes fuse at random into a diploid zygote. As the four gametes are equally probable, the 16 possible combinations will occur with equal frequency.

Four of the 16 individuals from the F_2 generation are homozygous in both genes (AABB, AAbb, aaBB, aabb), and only two of the 16 are identical to the homozygotes from the original parent generation (AABB and aabb). Four individuals of the F_2 generation represent the same, doubly heterozygous genotype AaBb. In total, there occur $3^2 = 9$ different genotypes (AABB, AABb, AAbb, AaBB, AaBb, Aabb, aaBB, aaBb, and aabb, in a ratio of 1:2:1:2:4:2:1:2:1). If the two alleles A and B are dominant, and the alleles a and b consequently recessive, the nine genotypes correspond to four phenotypes (AABB, AABb, AABb, AaBB, AaBb, AaBb, AaBB, AaBb, and AaBb; AAbb, Aabb, and Aabb; aaBb, aaBB, and aaBb, and, finally, aabb) in the ratio 9:3:3:1.

Designed crossing experiments with sexually reproducing organisms, combined with statistical analysis of the phenotype distribution of characteristic features of one or both parents in subsequent filial generations is the basis upon which the further development of *classical genetics* (in which undefined 'hereditary factors' played the key role) into *molecular genetics* (in which 'genes' as submolecular DNA sections transfer their structural information to phenotype functional molecules) has taken

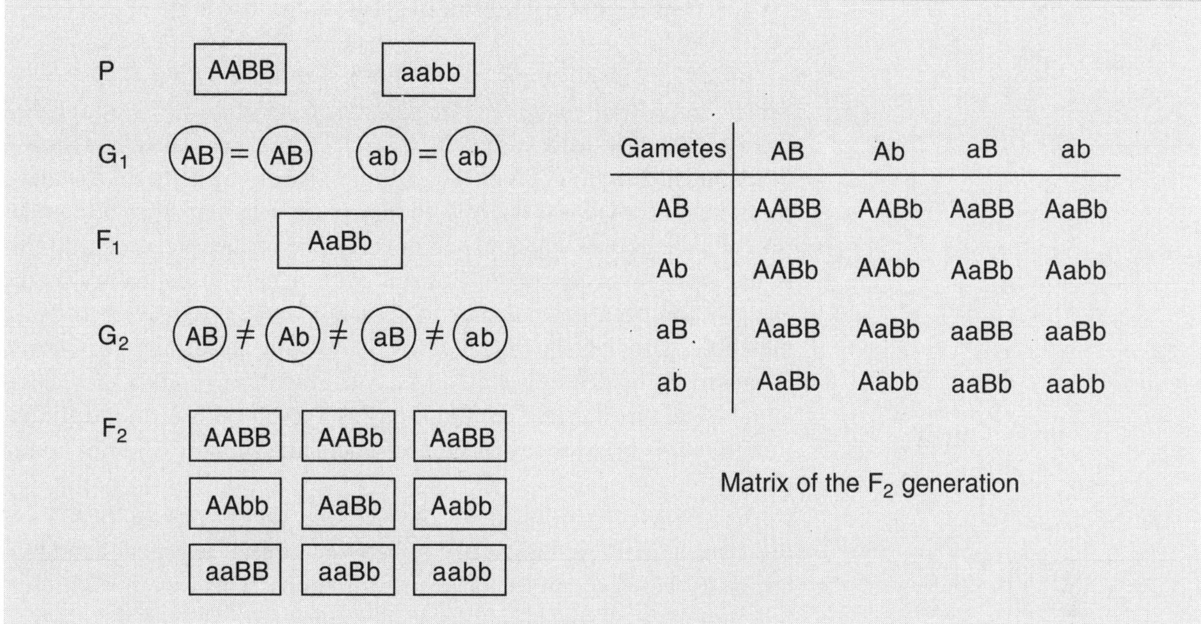

Fig. 16.4. *Genetic analysis by* Mendel*ian combinatorics: the allele pairs AABB and aabb of homozygous parents* (P) *are transmitted to progeny.* (Different from members of the second filial generation F_2, the members of the first filial generation F_1 have the same genotype AaBb.)

place. The foundation was laid by *Mendel.* With it, he performed a service to *Charles Robert Darwin*, albeit one neither recognized in their lifetimes.

In his 'theory of evolution by natural selection', *Darwin* had assumed that genetic characters or qualities were transmitted from parent to offspring by blending inheritance on crossing. This view was not compatible with the fact that particular phenotypical characteristics, which disappeared in a uniform F_1 generation, came to light once more in some individuals in the non-uniform F_2 generation. The obvious weakness in *Darwin*'s proposal, amounting to a disappearance of genetic variation, could be overcome using *Mendel*'s theory of particulate inheritance, in which this genetic diversity remained convincingly preserved. A 'synthesis' of *Darwin*'s theory of evolution by natural selection and *Mendel*'s theory of heredity was ultimately to result in *neo-Darwinism.*

The central element in *Darwin*'s theory of evolution is *natural selection.* Evolutionary biologist *Ernst Mayr* has advocated keeping in mind that natural selection, however, is only the second step in an overall two-stage process. The first, no less important step, is founded on new variation coming into being in every generation. To be able to alter the composition of a population – the total number of individual representatives of a given type of organism in a particular region – it is essential that these individuals are distinct from one another. Sexual reproduction ensures, above all, that geno- and phenotypical variation is guaranteed. New genotypes come into being in every generation, to be tested in subsequent generations for their suitability in the 'struggle for existence'.

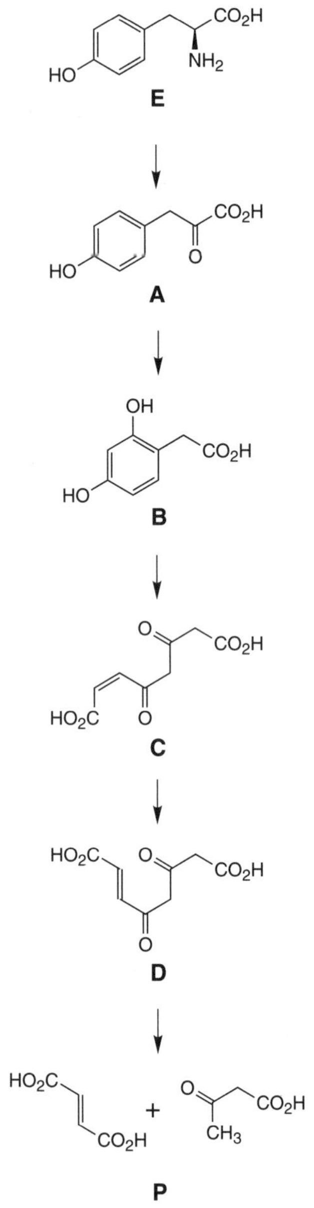

E

A

B

C

D

P

16.1.3. The Central Dogma of Molecular Biology

In 1923, there appeared a book, entitled '*Inborn Errors of Metabolism*', written by *Sir Archibald E. Garrod*. In it, he established links between, in each case, a gene and an enzyme which catalyzed a biochemical reaction. Were the enzyme absent, then the biochemical reaction concerned was not able to take place, resulting in an accumulation of a metabolite which would normally be further metabolized. This is the case, for example, in humans with a genetic deficiency in the enzyme homogentisate oxidase. This after *Mendel* recessive hereditary metabolic anomaly is called *alcaptonuria* and, according to *Garrod*, is caused by homogentisic acid (**A**), a degradation product of phenylalanine and tyrosine (**E**), accumulating as a result of its further degradation to (as we know today) acetoacetic acid and fumaric acid not taking place.

The concept developed by *Garrod*, *one gene – one enzyme – one biochemical reaction*, built a bridge between molecular genetics and biochemistry based on molecular transformations. This was to lead to intense analysis of which biochemical reactions are connected to which catalyzing enzymes, and which enzymes to which underlying genes.

In an analysis of this kind, it is attempted to identify anomalous individuals (among the *mutants* I–V), differing from the normal individual (the *wild type*) in lacking one of the five enzymes (1–5), and hence the associated gene, required for each of the five steps (**E** → **A**, **A** → **B**, **B** → **C**, **C** → **D**, and **D** → **P**) involved in the degradation of the educt **E** (tyrosine) into the product **P** (fumaric acid and acetoacetic acid) (*Fig. 16.5*).

Each mutant is characterized by having one of the five reaction steps blocked. Introduction of any of the compounds involved in the reaction chain from **E** to **P** after the blocked step, however, still results in the formation of product **P**. In this way, it is possible to deduce which gene, associated with the absent enzyme, is causing the phenotype to deviate from that of the wild type.

George Wells Beadle and *Edward Tatum* performed systematic experiments of this type upon the fungus *Neurospora*. Various mutants could be produced experimentally by the effect of X-rays on the wild type. In a medium providing minimum nutritional requirement for the wild type, the mutants ceased growing, until specific supplementary compounds were added. In each case, an hereditary alteration (*mutation*) of one particular gene had taken place. In 1942, *Beadle* and *Tatum* were able to prove an unequivocal relationship between a mutation and the lack of one particular enzyme. Thus they arrived at the result: *one gene – one enzyme*, or, expressed in terms of molecular structure, *one submolecular nucleic-acid segment – one polypeptide*.

The *genetic information* of a DNA molecule is translated in the living cell into the *functional information* of a protein. The cell's genetic information exists in the syntactic structure, *i.e.*, in the sequence of its nucleic acids (*Chapt. 7.3.1.2.1*). The functional information exists in the syntactic structure, *i.e.*, in the sequence of its proteins (*Chapt. 7.3.3.6*). To find

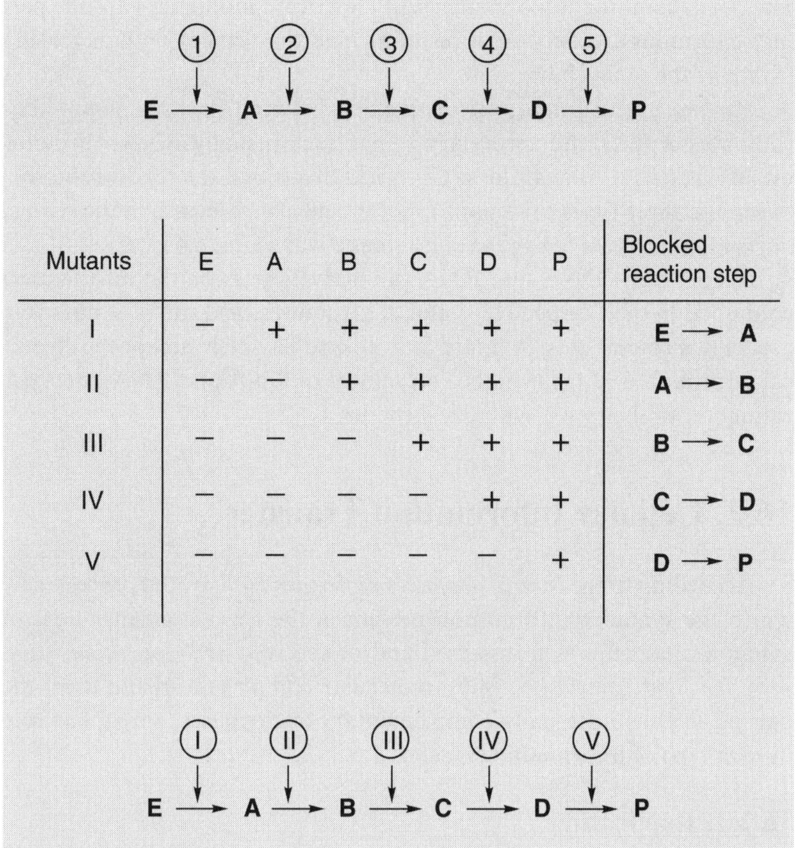

Mutants	E	A	B	C	D	P	Blocked reaction step
I	−	+	+	+	+	+	E → A
II	−	−	+	+	+	+	A → B
III	−	−	−	+	+	+	B → C
IV	−	−	−	−	+	+	C → D
V	−	−	−	−	−	+	D → P

Fig. 16.5. *Combinatorial analysis of the relationship: one gene – one enzyme – one biochemical reaction*

out how the syntactic structure of the nucleic acids can be 'translated' into the syntactic structure of the proteins should prove to become one of the most intractable problems of molecular biology.

Before particular details could be disclosed, it was already clear that the genetic information of nucleic acids is converted into the functional information of proteins. Conversely, no flow of information in the opposite direction takes place. This principle is to be found in the literature as the *central dogma of molecular biology*. To be precise, it was formulated rather more differentiated, not referring to nucleic acids as a whole, but distinguishing between the two types of naturally occurring nucleic acids: DNA and RNA (*Chapt. 10.5.2*). DNA and RNA both participate, actively as well as passively, in information transfer in the living cell.

Emile Zuckerkandl and *L. Pauling* have coined the terms *semantophoretic molecules* or *semantides*. Semantophoretic molecules are objects with a syntactic structure and a function intrinsic to that structure. Representatives of this class are linearly constituted macromolecules whose monomeric building blocks possess at least two functional groups, necessary for polymerization of the main chain, and whose (limited) variability is ensured by constitutionally distinct 'side chains'.

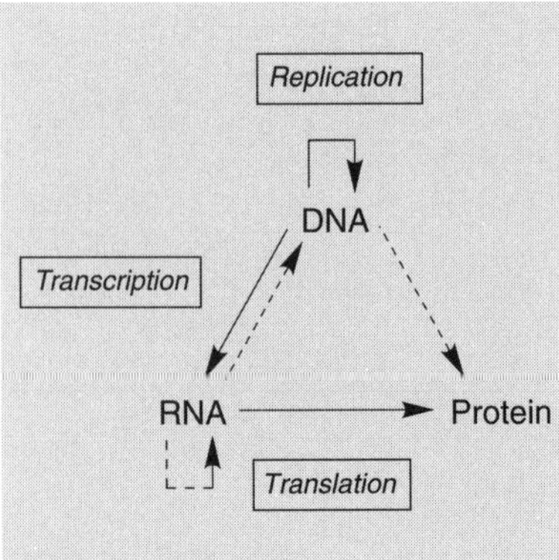

Fig. 16.6. *The 'central dogma of molecular biology', schematically sketched*

Statistical combination of numerically restricted monomer variants permits enormous diversity in the resulting macromolecules, with increasing length of the main chain.

Pauling and *Zuckerkandl* distinguish between primary semantides (DNA as a rule) and secondary semantides (usually RNA). Proteins qualify as tertiary semantides. *Crick* has illustrated the relationship between the three types of semantophoretic macromolecules in the *central dogma of molecular biology* in a schematic way (*Fig. 16.6*).

Accordingly, DNA and RNA can actively pass on the information contained in their respective syntactic structures, and can also passively accept it. Proteins, however, are only able to passively accept the chemical information of the syntactic structures of DNA and RNA, incorporating it into their own syntactic structures.

16.2. Cellular Information Transfer

According to the *central dogma of molecular biology*, the processes by which the syntactic information present in the specific arrangement of submolecular units is transferred and/or received are *replication, transcription*, and *translation*. Supramolecular complexes, arising from the pairing of constitutionally complementary nucleobases, play a key role in each type of information transfer.

16.2.1. Replication

For molecular information transfer, the first candidate to merit consideration would be a linearly constituted macromolecule with at least two different atom groups per *submolecular information unit*. If the two submolecular information units occurred with roughly equal frequency, the dual molecular system might be capable of storing information. The specific sequence of no fewer than four nucleobases on the sugar-phosphate backbone of a polynucleotide chain, as a readable sequence of submolecular information units, represents a syntactic structure.

Of the minimum requirements that a system must fulfill in order to be described as living, one is the ability to *self-replicate* the molecular information carrier comprising its syntactic structure. Replication in living systems (or in those ones which evolve into living systems), however, may not be completely free of error. Error-free replicating systems would produce no change in variation and so would not provide any basis for natural selection (*Sect. 16.1.2*). Mistakes (*mutations*) in replication lead to new sequences, and hence to an altered syntactic structure, *i.e.*, to a new source of information. The causal sequence, self-replication → mutation → selection → evolution, must, as *Manfred Eigen* showed, be fulfilled before non-living systems may evolve into living ones.

Along with the structure of double-helical DNA, *Watson* and *Crick* were also able to rationalize the genetic implications arising from it (*Fig. 7.30*). In their second publication (*Fig. 16.7*), the authors pointed out that the structure of DNA allows for recognition of the molecular template, required for the synthesis of each complementary single strand, and sought after for verification of molecular replication.

The genetic information may be recognized (read) in the supramolecular functional unit of a duplex. The duplex is held together by stabilizing interactions between the nucleobases. As well as the horizontal interactions between bases of the two complementary single strands, which are traced back to *hydrogen bonds* (*Chapt. 15*), vertical interactions take place between adjacent bases usually of the same single strand (see, however, *Chapt. 7.3.1.2.3*). These can be ascribed to *stacking forces* and – in aqueous solution – to so-called *hydrophobic forces*.

Stacking forces are non-covalent interactions between parallel oriented nucleobases. To what degree they are a consequence of *electron-donor/electron-acceptor interactions* (and thus stereoelectronic), and to what degree one of electrostatic (and thus electronic) interactions is not known. The term 'hydrophobic interactions', moreover, essentially covers the entropically assisted exclusion of water, from the region of the nucleobases in this instance.

Both single strands of a DNA duplex each serve as a template for enzymatic polymerization. At the single strand with *sense*-ordering of its bases (*Chapt. 7.3.1.2.1*), the complementary single strand with *antisense*-ordering, necessary for the new duplex, is synthesized. At the original single strand with *antisense*-ordering of its bases, the new strand with *sense*-ordering is synthesized. The two daughter duplexes arising as the products of a replication cycle each contain one parent strand and one newly synthesized single strand (*Fig. 16.8*): replication is said to be *semiconservative*.

Arthur Kornberg, recipient of the 1959 *Nobel* Prize for physiology or medicine, discovered that the synthesis of a DNA single strand proceeds enzymatically, with the involvement of *DNA polymerases*. DNA Polymerases can only ever enable the single strand being synthesized to grow in the $5' \rightarrow 3'$ direction, as the 2′-deoxyribonucleosides serving as building blocks each have a triphosphate group in the 5′-position. Consequently, synthesis at the two single strands of the original duplex is forced to proceed in different ways. Actually, forward replication can only proceed in a continuous manner at the *'leading' strand*. At the *'lagging' strand*, a reverse replication of segments of the complementary single strand takes place. These segments are later being connected to one another by *ligases*: replication is said to be *semidiscontinuous*.

Replication can only take place in the presence of a nucleic-acid template, activated building blocks (in the form of 2′-deoxynucleoside-5′-triphosphates), and a *'primer'*, as well as at least one DNA polymerase, and a whole range of DNA-binding proteins. A bacterial DNA polymerase is capable of synthesizing animal DNA, just as animal DNA polymerase is able to synthesize bacterial DNA. Information concerning the sequence of the DNA to be synthesized is contained entirely in the template. No DNA polymerase is able to give the signal for initiation of replication. That comes from the primer. The primer, as a rule, is a short RNA chain. Its function in DNA synthesis is to set up the scaffolding for a supramolecular machinery of synthesis, through base pairing with the

No. 4361 May 30, 1953 NATURE

GENETICAL IMPLICATIONS OF THE STRUCTURE OF DEOXYRIBONUCLEIC ACID

By J. D. WATSON and F. H. C. CRICK

Medical Research Council Unit for the Study of the Molecular Structure of Biological Systems, Cavendish Laboratory, Cambridge

THE importance of deoxyribonucleic acid (DNA) within living cells is undisputed. It is found in all dividing cells, largely if not entirely in the nucleus, where it is an essential constituent of the chromosomes. Many lines of evidence indicate that it is the carrier of a part of (if not all) the genetic specificity of the chromosomes and thus of the gene itself.

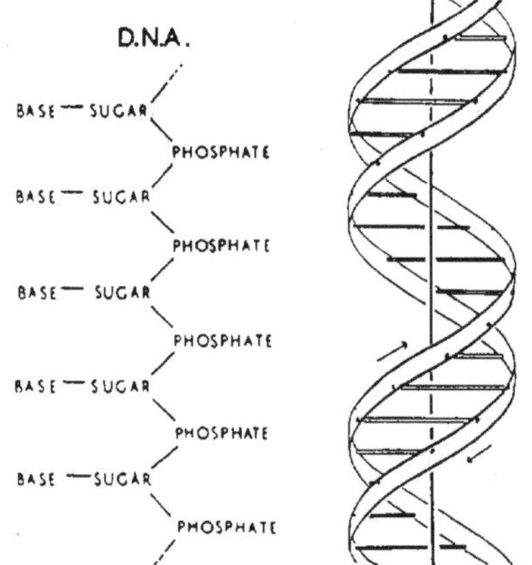

Fig. 1. Chemical formula of a single chain of deoxyribonucleic acid

Fig. 2. This figure is purely diagrammatic. The two ribbons symbolize the two phosphate-sugar chains, and the horizontal rods the pairs of bases holding the chains together. The vertical line marks the fibre axis

Until now, however, no evidence has been presented to show how it might carry out the essential operation required of a genetic material, that of exact self-duplication.

We have recently proposed a structure[1] for the salt of deoxyribonucleic acid which, if correct, immediately suggests a mechanism for its self-duplication. X-ray evidence obtained by the workers at King's College, London[2], and presented at the same time, gives qualitative support to our structure and is incompatible with all previously proposed structures[3]. Though the structure will not be completely proved until a more extensive comparison has been made with the X-ray data, we now feel sufficient confidence in its general correctness to discuss its genetical implications. In doing so we are assuming that fibres of the salt of deoxyribonucleic acid are not artefacts arising in the method of preparation, since it has been shown by Wilkins and his co-workers that similar X-ray patterns are obtained from both the isolated fibres and certain intact biological materials such as sperm head and bacteriophage particles[2,4].

The chemical formula of deoxyribonucleic acid is now well established. The molecule is a very long chain, the backbone of which consists of a regular alternation of sugar and phosphate groups, as shown in Fig. 1. To each sugar is attached a nitrogenous base, which can be of four different types. (We have considered 5-methyl cytosine to be equivalent to cytosine, since either can fit equally well into our structure.) Two of the possible bases—adenine and guanine—are purines, and the other two—thymine and cytosine—are pyrimidines. So far as is known, the sequence of bases along the chain is irregular. The monomer unit, consisting of phosphate, sugar and base, is known as a nucleotide.

The first feature of our structure which is of biological interest is that it consists not of one chain, but of two. These two chains are both coiled around a common fibre axis, as is shown diagrammatically in Fig. 2. It has often been assumed that since there was only one chain in the chemical formula there would only be one in the structural unit. However, the density, taken with the X-ray evidence[2], suggests very strongly that there are two.

The other biologically important feature is the manner in which the two chains are held together. This is done by hydrogen bonds between the bases, as shown schematically in Fig. 3. The bases are joined together in pairs, a single base from one chain being hydrogen-bonded to a single base from the other. The important point is that only certain pairs of bases will fit into the structure. One member of a pair must be a purine and the other a pyrimidine in order to bridge between the two chains. If a pair consisted of two purines, for example, there would not be room for it.

We believe that the bases will be present almost entirely in their most probable tautomeric forms. If this is true, the conditions for forming hydrogen bonds are more restrictive, and the only pairs of bases possible are:

adenine with thymine;
guanine with cytosine.

The way in which these are joined together is shown in Figs. 4 and 5. A given pair can be either way round. Adenine, for example, can occur on either chain; but when it does, its partner on the other chain must always be thymine.

This pairing is strongly supported by the recent analytical results[5], which show that for all sources of deoxyribonucleic acid examined the amount of adenine is close to the amount of thymine, and the amount of guanine close to the amount of cytosine, although the cross-ratio (the ratio of adenine to guanine) can vary from one source to another. Indeed, if the sequence of bases on one chain is irregular, it is difficult to explain these analytical results except by the sort of pairing we have suggested.

The phosphate-sugar backbone of our model is completely regular, but any sequence of the pairs of bases can fit into the structure. It follows that in a long molecule many different permutations are possible, and it therefore seems likely that the precise

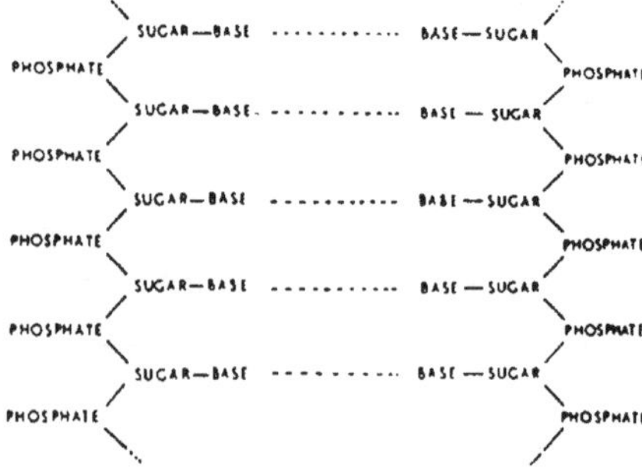

Fig. 3. Chemical formula of a pair of deoxyribonucleic acid chains. The hydrogen bonding is symbolized by dotted lines

Fig. 16.7. *Second original publication by* J.D. Watson *and* F.H.C. Crick *in* Nature **1953**

NATURE May 30, 1953 VOL. 171

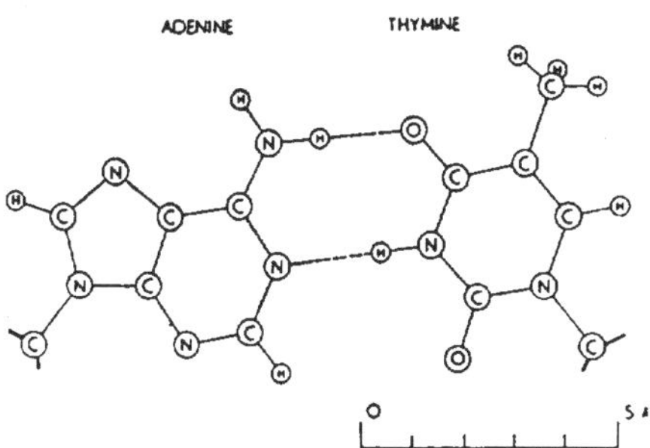

Fig. 4. Pairing of adenine and thymine. Hydrogen bonds are shown dotted. One carbon atom of each sugar is shown

GUANINE CYTOSINE

Fig. 5. Pairing of guanine and cytosine. Hydrogen bonds are shown dotted. One carbon atom of each sugar is shown

sequence of the bases is the code which carries the genetical information. If the actual order of the bases on one of the pair of chains were given, one could write down the exact order of the bases on the other one, because of the specific pairing. Thus one chain is, as it were, the complement of the other, and it is this feature which suggests how the deoxyribonucleic acid molecule might duplicate itself.

Previous discussions of self-duplication have usually involved the concept of a template, or mould. Either the template was supposed to copy itself directly or it was to produce a 'negative', which in its turn was to act as a template and produce the original 'positive' once again. In no case has it been explained in detail how it would do this in terms of atoms and molecules.

Now our model for deoxyribonucleic acid is, in effect, a *pair* of templates, each of which is complementary to the other. We imagine that prior to duplication the hydrogen bonds are broken, and the two chains unwind and separate. Each chain then acts as a template for the formation on to itself of a new companion chain, so that eventually we shall have *two* pairs of chains, where we only had one before. Moreover, the sequence of the pairs of bases will have been duplicated exactly.

A study of our model suggests that this duplication could be done most simply if the single chain (or the relevant portion of it) takes up the helical configuration. We imagine that at this stage in the life of the cell, free nucleotides, strictly polynucleotide precursors, are available in quantity. From time to time the base of a free nucleotide will join up by hydrogen bonds to one of the bases on the chain already formed. We now postulate that the polymerization of these monomers to form a new chain is only possible if the resulting chain can form the proposed structure. This is plausible, because steric reasons would not allow nucleotides 'crystallized' on to the first chain to approach one another in such a way that they could be joined together into a new chain, unless they were those nucleotides which were necessary to form our structure. Whether a special enzyme is required to carry out the polymerization, or whether the single helical chain already formed acts effectively as an enzyme, remains to be seen.

Since the two chains in our model are intertwined, it is essential for them to untwist if they are to separate. As they make one complete turn around each other in 34 A., there will be about 150 turns per million molecular weight, so that whatever the precise structure of the chromosome a considerable amount of uncoiling would be necessary. It is well known from microscopic observation that much coiling and uncoiling occurs during mitosis, and though this is on a much larger scale it probably reflects similar processes on a molecular level. Although it is difficult at the moment to see how these processes occur without everything getting tangled, we do not feel that this objection will be insuperable.

Our structure, as described[1], is an open one. There is room between the pair of polynucleotide chains (see Fig. 2) for a polypeptide chain to wind around the same helical axis. It may be significant that the distance between adjacent phosphorus atoms, 7·1 A., is close to the repeat of a fully extended polypeptide chain. We think it probable that in the sperm head, and in artificial nucleoproteins, the polypeptide chain occupies this position. The relative weakness of the second layer-line in the published X-ray pictures[3a,4] is crudely compatible with such an idea. The function of the protein might well be to control the coiling and uncoiling, to assist in holding a single polynucleotide chain in a helical configuration, or some other non-specific function.

Our model suggests possible explanations for a number of other phenomena. For example, spontaneous mutation may be due to a base occasionally occurring in one of its less likely tautomeric forms. Again, the pairing between homologous chromosomes at meiosis may depend on pairing between specific bases. We shall discuss these ideas in detail elsewhere.

For the moment, the general scheme we have proposed for the reproduction of deoxyribonucleic acid must be regarded as speculative. Even if it is correct, it is clear from what we have said that much remains to be discovered before the picture of genetic duplication can be described in detail. What are the polynucleotide precursors ? What makes the pair of chains unwind and separate ? What is the precise role of the protein ? Is the chromosome one long pair of deoxyribonucleic acid chains, or does it consist of patches of the acid joined together by protein ?

Despite these uncertainties we feel that our proposed structure for deoxyribonucleic acid may help to solve one of the fundamental biological problems—the molecular basis of the template needed for genetic replication. The hypothesis we are suggesting is that the template is the pattern of bases formed by one chain of the deoxyribonucleic acid and that the gene contains a complementary pair of such templates.

One of us (J. D. W.) has been aided by a fellowship from the National Foundation for Infantile Paralysis (U.S.A.).

[1] Watson, J. D., and Crick, F. H. C., *Nature*, 171, 737 (1953).

[2] Wilkins, M. H. F., Stokes, A. R., and Wilson, H. R., *Nature*, 171, 738 (1953). Franklin, R. E., and Gosling, R. G., *Nature*, 171, 740 (1953).

[3] (*a*) Astbury, W. T., Symp. No. 1 Soc. Exp. Biol., 66 (1947). (*b*) Furberg, S., *Acta Chem. Scand.*, 6, 634 (1952). (*c*) Pauling, L., and Corey, R. B., *Nature*, 171, 346 (1953); *Proc. U.S. Nat. Acad. Sci.*, 39, 84 (1953). (*d*) Fraser, R. D. B. (in preparation).

[4] Wilkins, M. H. F., and Randall, J. T., *Biochim. et Biophys. Acta*, 10, 192 (1953).

[5] Chargaff, E., for references see Zamenhof, S., Brawerman, G., and Chargaff, E., *Biochim. et Biophys. Acta*, 9, 402 (1952). Wyatt, G. R., *J. Gen. Physiol.*, 36, 201 (1952).

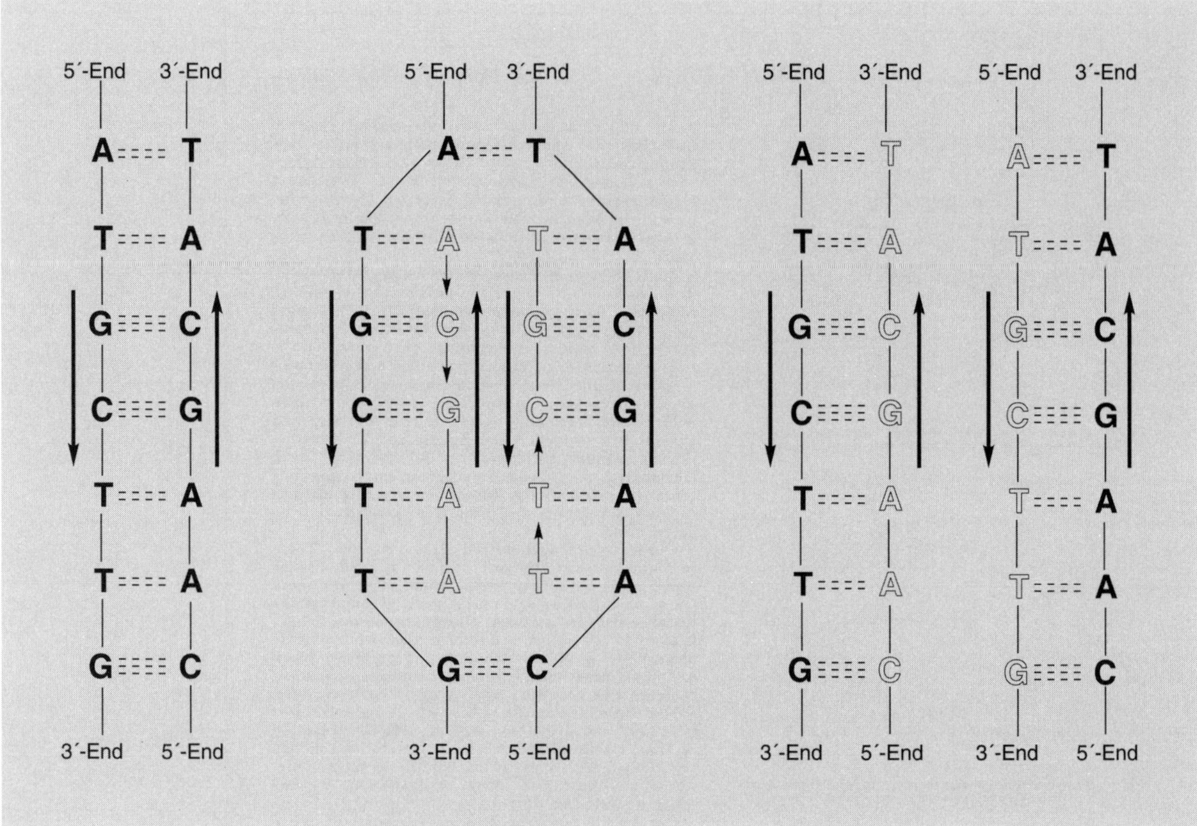

Fig. 16.8. *Schematic representation of the semiconservative DNA replication indicating the template to be copied* (bold letters) *and the single strand having been synthesized* (light letters)

template. Chain polymerization proceeds as a *quasi*-intramolecular process, then, and chain elongation takes place by means of nucleophilic attack of the 3′-OH group of the primer on the P(α)-atom of a nucleoside-triphosphate as a building block (*Fig. 16.9*).

Understanding the role played by *Watson-Crick* base pairing in semiconservative replication essentially solved the biological problem of how genetic information is passed on from cell to cell down through the generations, however complex the only partially understood chemical procedure of replication may be in reality (*Fig. 16.10*).

Detailed understanding of enzymatic DNA replication does not only benefit the knowledge of how genetic information is passed on *in vivo*, but is also made use of, in various ways, in an *in vitro* technique known as the *polymerase chain reaction* (PCR). Using PCR, a particular DNA fragment, found for example in a heterogenous DNA mixture, may be enriched a million-fold in a few hours, enabling handling of synthesized DNA samples using common laboratory methods. *Kary B. Mullis*, who developed this technique for specific *in vitro* synthesis of DNA, was awarded the 1993 *Nobel* Prize for chemistry in honor of this achievement. DNA Polymerase, required for enzymatic synthesis using PCR, had already been highlighted as 'Molecule of the Year' by the journal *Science*

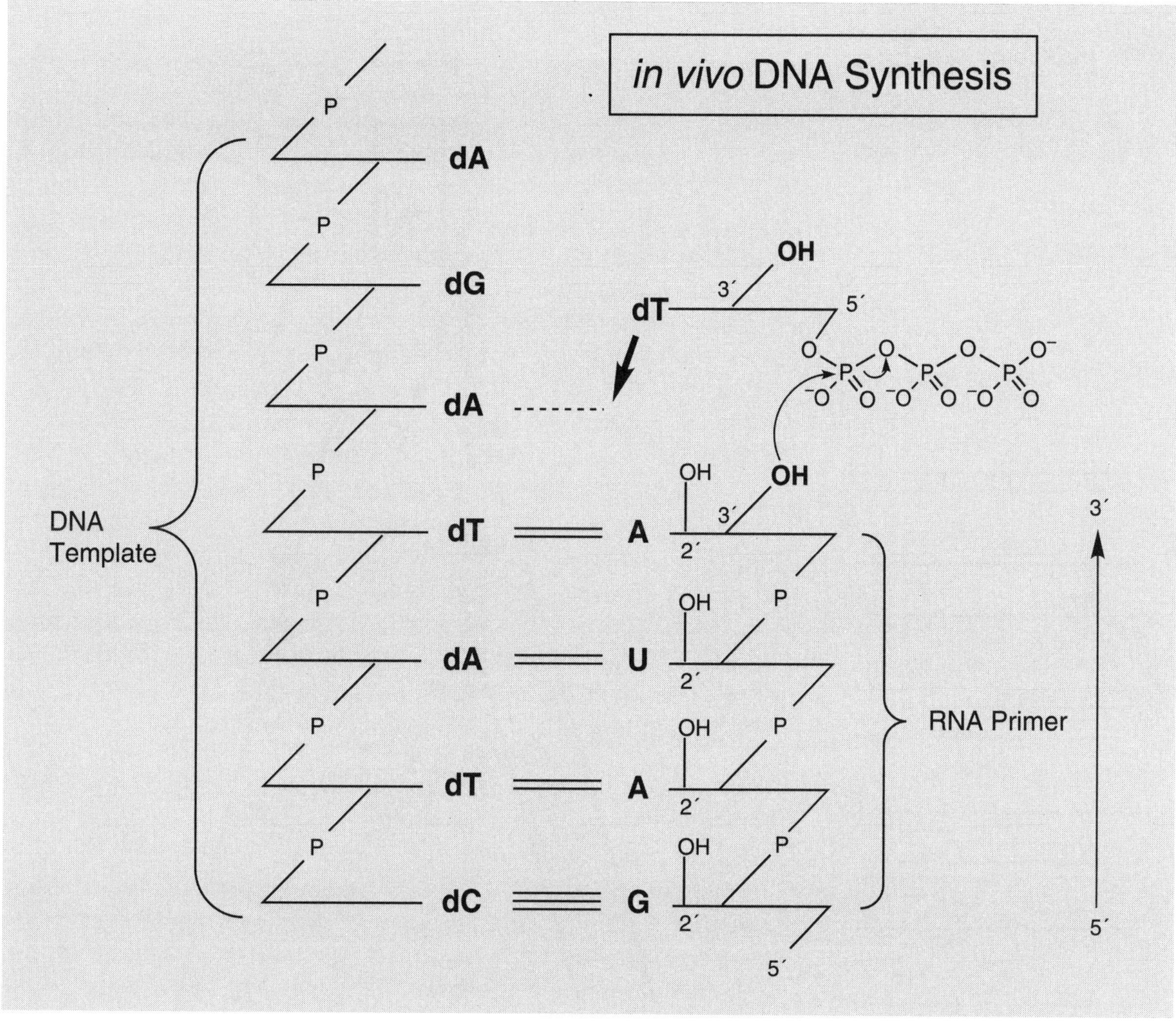

Fig. 16.9. *Growth of chain in 5′ →3′ direction during replication of a DNA template*

four years earlier. For the exponential multiplication of a segment of a DNA duplex, the sequences of both sides of the target sequence pair must be known exactly, so that the two oligodeoxynucleotides P_1 and P_2, intended for hybridization with them as primers, can be accurately made by chemical synthesis. The product of this molecular multiplication are discrete DNA duplexes, with 5′-ends corresponding to those of the oligodeoxynucleotides employed as the primers. PCR follows the procedure laid out in *Fig. 16.11*:

A DNA duplex, containing the target sequence to be propagated, is denatured by brief heating in the presence of P_1, P_2 (20–24mers), and the four synthetic building blocks dpppT, dpppC, dpppA, and dpppG, as well as a thermostable DNA polymerase (Taq-DNA polymerase, from the thermophilic bacterium *Thermophilus aquaticus*). After the two primers P_1 and P_2 have hybridized onto their respective single strands, the double-strand segments produced are elongated by the enzyme into

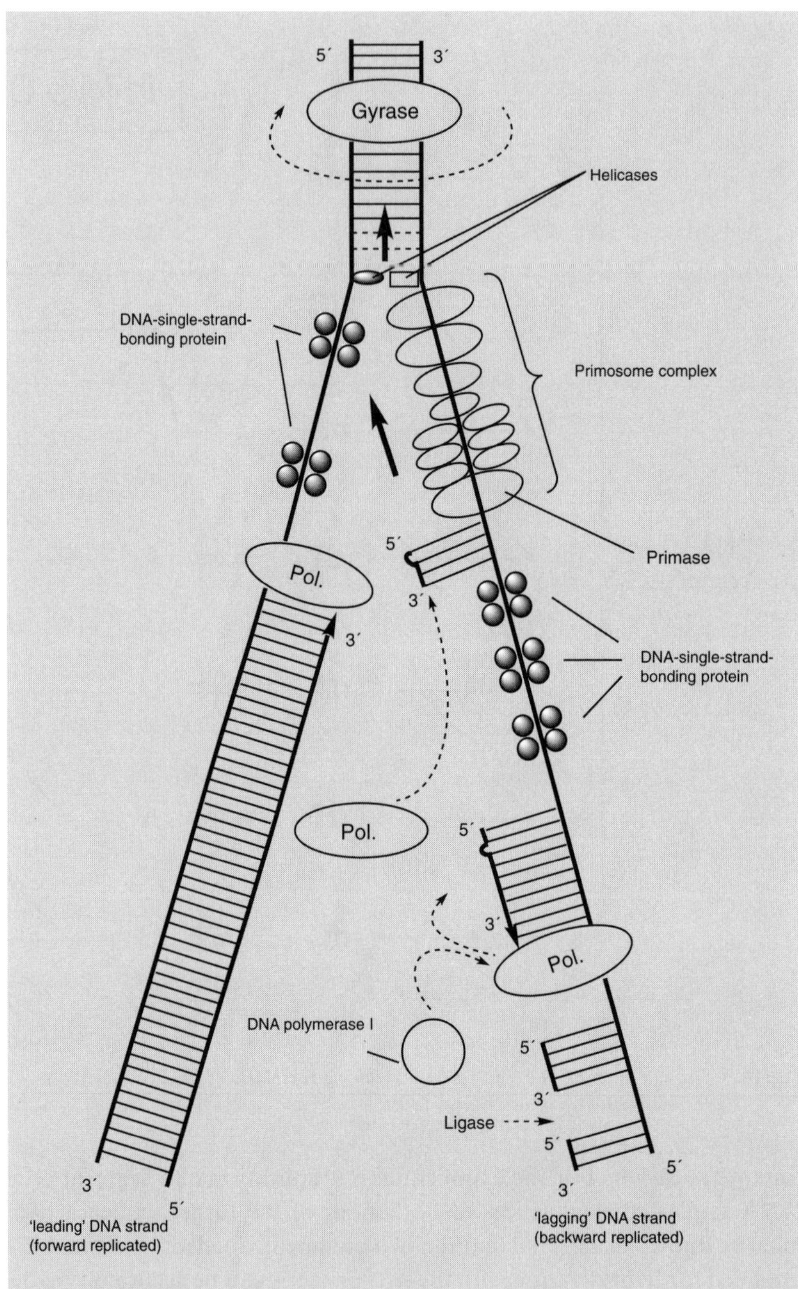

Fig. 16.10. *Schematic representation of the complex system with continuous forward and discontinuous backward replication* (Pol. = polymerase)

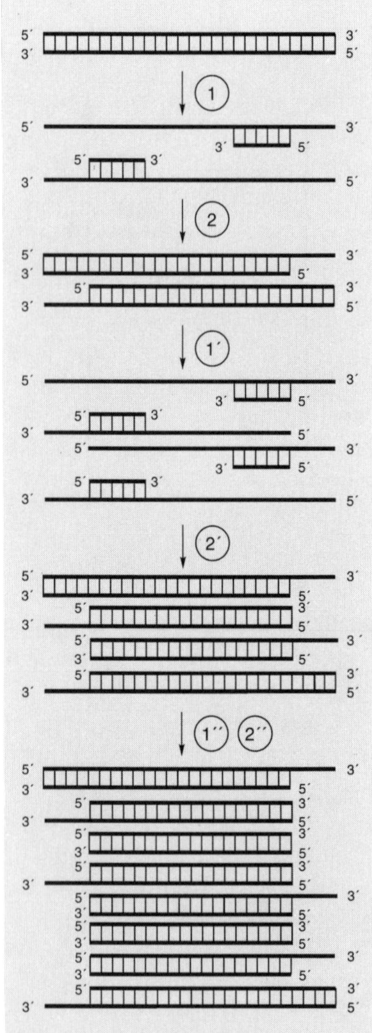

Fig. 16.11. *Schematic representation of three cycles of specific enzymatic amplification affording the target sequence increased eight-fold*

complete duplexes. This cycle of *1*) double-strand dissociation and hybridization of the two single strands with their corresponding primers, and *2*) enzymatic chain lengthening of each primer on the complementary DNA template is repeated until a sufficient quantity of the desired DNA duplex is present. In theory, 20 cycles will produce an approximately one million-fold enrichment, 30 cycles approximately a one billion-fold enrichment.

Is it possible to simulate the template-controlled replication of nucleic acids, proceeding with the *in vitro* or *in vivo* participation of suitable polymerases, in the absence of enzymes? DNA Single strands are not the only possible replicands: the proposed existence of a former RNA world (*Chapt. 7.3.1.1.2*) means that RNA single strands are also worthy of consideration. An oligonucleotide single strand, functioning as a template, is capable of complexing with complementary nucleoside monomers (*Fig. 16.12*). If the complex is sufficiently stable (below the 'melting temperature' of the corresponding duplex), and the nucleotide monomers are so oriented to one another that the 5′-triphosphate group of one nucleotide monomer is attacked by the nucleophilic 2′(3′)-OH group of its neighbor (*cf. Fig. 16.9*), then the minimum requirements for template-controlled synthesis of a complementary single strand are fulfilled.

Leslie E. Orgel's pioneering work on molecular replication has been a classic example of scientific method over decades in the current spectacle of the 'Origins of Life' debate. It identifies minimal conditions which must be fulfilled in order for template-controlled polymerization to be observed, and gives the impression that non-enzymatic replication with

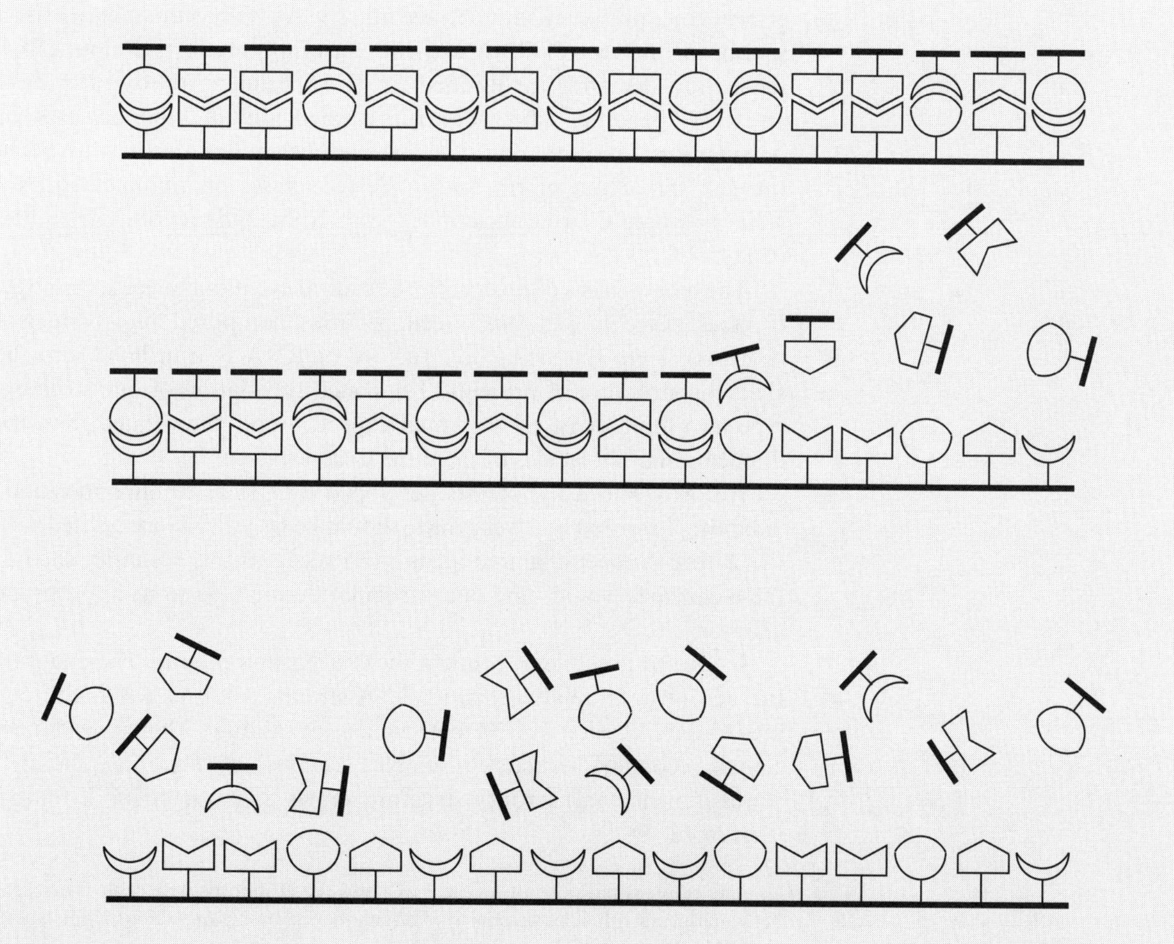

Fig. 16.12. *Schematic representation of matching the incoming nucleotide building blocks by base pairing to an oligonucleotide template*

suitably activated mono- and oligonucleotides can only be achieved in the laboratory in especially designed cases.

To invent reproducible experiments to find out how, in the *prebiotic phase* of evolution, the structure type of the nucleic acids could have arisen, and to provide a plausible explanation of how it actually happened, is a challenge directed at synthetic chemists. In *Aspects of Organic Chemistry*, Volume 3: *Synthesis*, we will look more closely at synthetic efforts to approach an RNA and a pre-RNA world.

16.2.2. Translation

In a bygone RNA world, without coded protein synthesis, information storage, transfer, and execution must have taken place through a functioning metabolism of substances still composed entirely of RNA molecules. In today's living world, in which, essentially, DNA acts as an information memory and proteins as functional mediators, RNA molecules are not only still in evidence, but also continue to play a central role, primarily in protein synthesis. As well as this, there exist a significant number of RNA activities carrying on old tradition: DNA replication demands involvement of RNA primers, retroviruses have an RNA genome, ribozymes act catalytically as ribonucleases or polymerases, and, finally, 2′-deoxyribonucleotides come into being through the action of ribonucleotide reductases on ribonucleotides – while nowhere do ribonucleotides arise from 2′-deoxyribonucleotides (*Chapt. 7.3.1.1.2*).

The biosynthesis of proteins is the central event of molecular biology. It takes place in *ribosomes* (cell particles composed of ribonucleoproteins). Here the syntactic structure of RNA is translated into the syntactic structure of proteins. The coded translation of one structure type into another is, as we would expect, more complicated than the duplication of molecules of the same structure type.

According to a hypothesis developed by *Crick*, 20 oligonucleotide 'adaptors', assisted by 20 enzymes, should be specifically associated with the 20 proteinogenic amino acids. According to this scenario, each of these enzymes would bind one particular amino acid to its appropriate adaptor.

It proved possible to confirm the adaptor hypothesis. The adaptors turned out to be relatively short RNA strands (*transfer RNA*, abbreviated to tRNA), each with 73 to 93 nucleotide residues. Their (two-dimensional) secondary structure resembles a cloverleaf (*Fig. 16.13, a*), their (three-dimensional) tertiary structure shows a characteristic L-folding (*Fig. 16.13, b*).

The close spatial position of the TψC- and the D-loops is noticeable. The stems between the individual leaves consist of double-helical base-pair sections of the type of an A-DNA duplex (*cf. Fig. 7.36*).

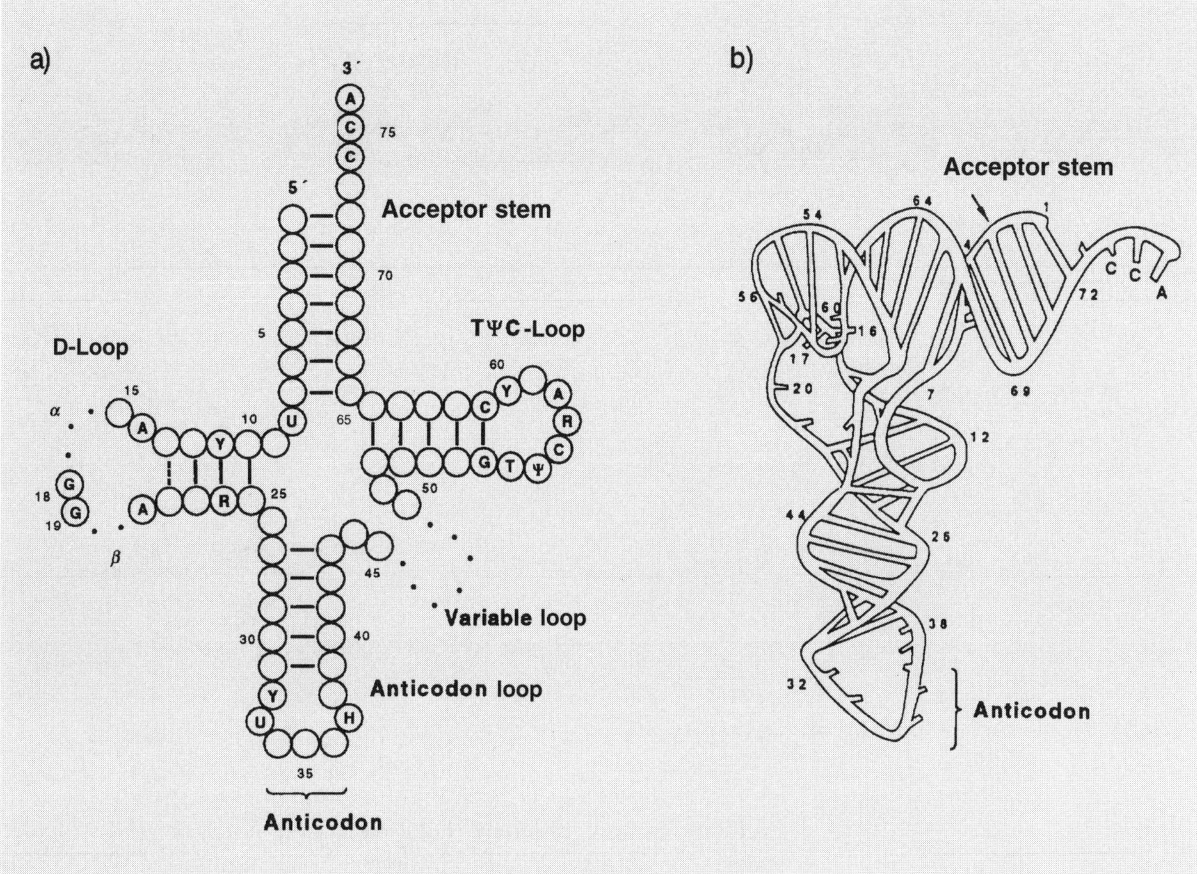

Fig. 16.13. *Schematic representations of* a) *the secondary (cloverleaf) structure and* b) *the (L-shaped) tertiary structure of a tRNA*

The posttranscriptional (cytidyl-cytidyl-adenosine) sequence is always found at the 3′-end of a tRNA. In a tRNA charged with its associated amino acid (*aminoacyl-tRNA*), the 3′-position (or sometimes the 2′-position) of the ribose in the terminal adenosyl building block carries the aminoacyl residue (*Fig. 16.14*).

Aminoacylation of a tRNA proceeds in two steps. In the first step, the amino acid concerned is converted into an activated aminoacyladenylic acid by interaction with adenosine triphosphate (ATP). In the second step, transfer of the aminoacyl residue to the terminal nucleotide of the tRNA takes place. Both processes are catalyzed by the same appropriate aminoacyl-tRNA synthetase. The cognate aminoacyl-tRNA synthetase, therefore, converts the correct amino acid into the *activated amino acid*, and is able to differentiate very selectively between *Phe* and *Tyr*, *Asn* and *Asp*, *Cys* and *His*, or *Lys* and *Trp*, for example, in the process. It emerged that, for each amino acid, there is at least one particular aminoacyl-tRNA synthetase present and either one single tRNA or a set of varied tRNAs.

The nucleosides of the anticodon and of the far removed acceptor stem (*Fig. 16.13*) determine, in general, whether a tRNA can be recognized by an aminoacyl-tRNA synthetase. In particular cases, the relative importance of the two RNA sections for the observed recognition

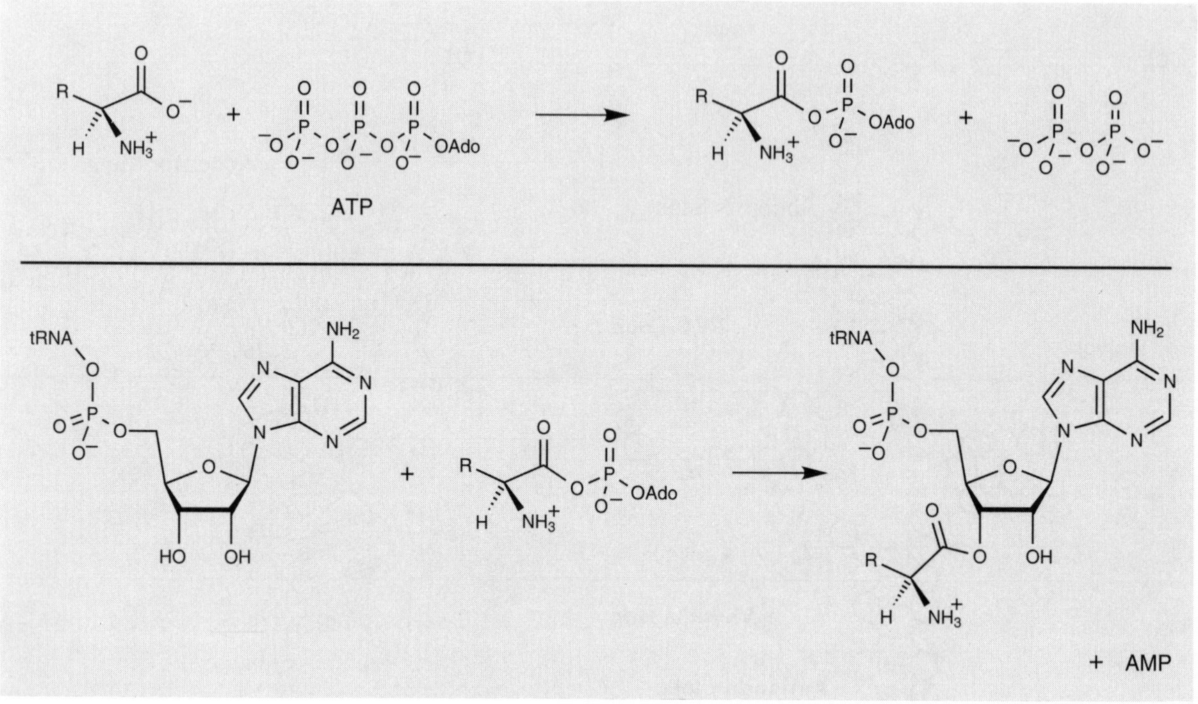

Fig. 16.14. *Aminoacylation of tRNA by aminoacyl-tRNA synthetases*

specificity is very different. High-resolution X-ray crystal-structure analyses on supermolecules consisting of an aminoacyl-tRNA synthetase and an aminoacyl-tRNA will probably provide information about the structural causes of the specificity of recognition. It would not be surprising, should it prove necessary to examine every individual case, as, in particular instances, characteristic conformational changes upon complex formation might well arise in either component.

A tRNA, charged specifically with the correct amino acid, is able, by base pairing, to recognize when its turn comes to act as a complementary 'building-block deliverer' during template-directed protein synthesis. It is informed so by the template RNA, described in this case as *messenger RNA* (mRNA). The acronym mRNA denotes the function of this RNA type. Protein synthesis takes place in eukaryots (and hence in cells with nuclei) in the cytoplasma at ribosomes. As the genetic information is stored in the DNA in the cell nucleus, it must be transferred to the ribosomes by a messenger: the mRNA. A ribosome, both structurally and functionally, is a highly complex synthesis machinery. About 50 proteins assemble around a core made up out of *ribosomal ribonucleic acids* (rRNAs). A ribosome couples with an mRNA single strand, travels the length of the mRNA chain, and, in so doing, induces base pairing between the given mRNA and a tRNA charged with the amino acid corresponding to its anticodon. Information transfer between the mRNA and the charged tRNA takes place by means of *codon-anticodon interaction* (*Fig. 16.15*).

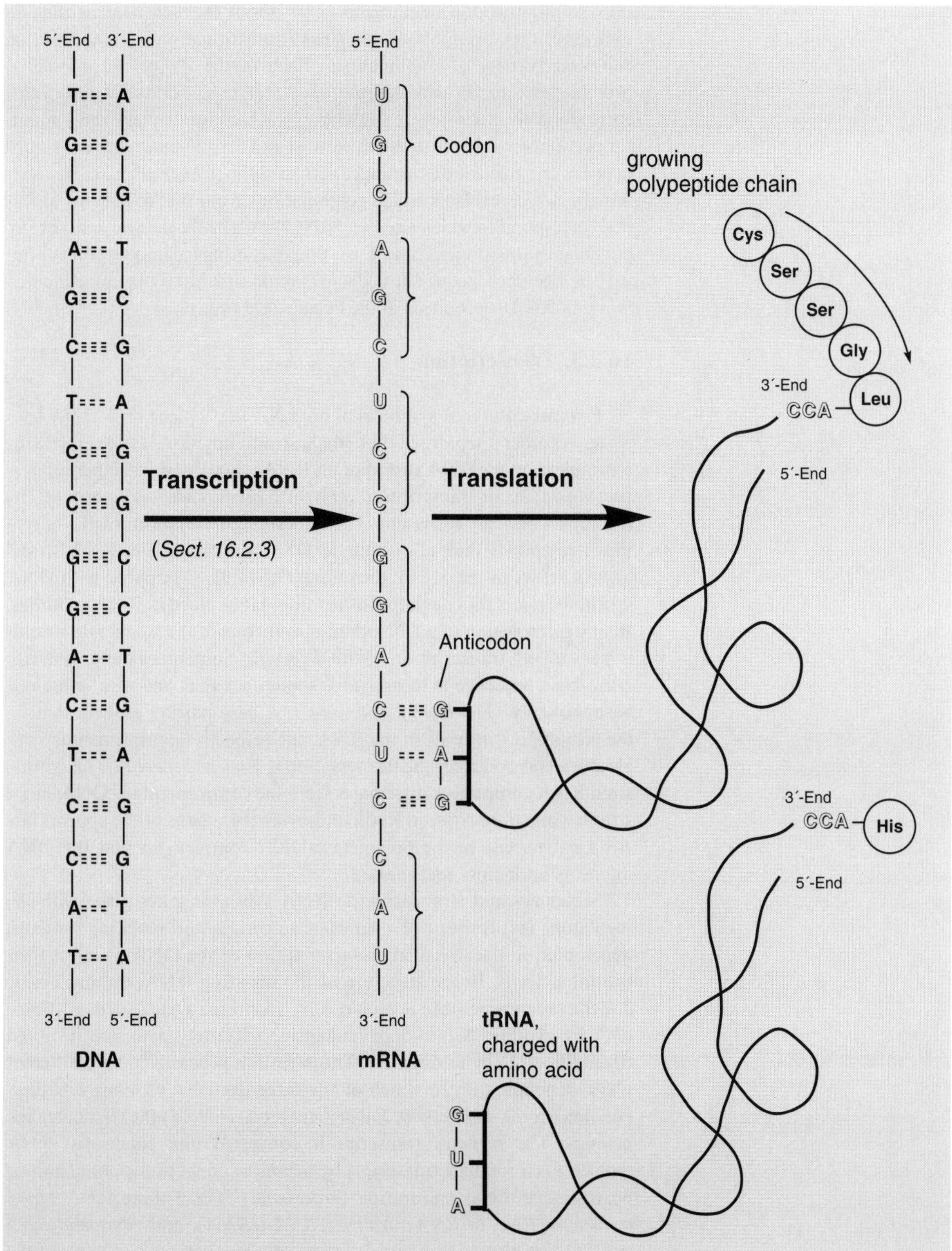

Fig. 16.15. *Transfer of inherent information from DNA* via *mRNA to charged tRNA*

Codon-anticodon interactions come about through base pairing. In each case, they bring about one single transformation selected from an enormous variety of combinations. Each of the 20 (or 21, *vide infra*) proteinogenic amino acids has one (or several) nucleoside sequences which correspond to it. Usually, in describing such codon triplets, the symbols for phosphoric-acid residues are omitted and $3' \rightarrow 5'$ connections assumed between the nucleotides arranged left to right (*Chapt. 10.5.2*). The 'dictionary' which explains which combinations from the 'four-letter alphabet' correspond to which amino acids is known as the *genetic code*. Before pursuing the question of how it was possible, using techniques from chemistry, to decipher the genetic code, we should first briefly examine the way in which RNA is produced in the living world today.

16.2.3. Transcription

Enzyme-catalyzed synthesis of an RNA takes place on a DNA template. Whether a separate DNA single strand functions as a template for a complementary DNA strand or an RNA strand – *i.e.*, whether replication *via* duplex or transcription *via* hybrid takes place – is determined by a complex regulatory system. The underlying mechanism of RNA synthesis resembles that of enzymatic DNA synthesis. The RNA strand always grows in the $5' \rightarrow 3'$ direction (*Fig. 16.9*). Compared with DNA synthesis, a less thorough 'proof reading' takes place in RNA synthesis. In any given region of a DNA duplex, only one of the two single strands is transcribed: transcription is *strand-specific*. Sometimes the strand with *sense*-base sequence is transcribed, sometimes that one with *antisense*-base sequence (*Fig. 16.16*). *Watson-Crick* base pairing ensures that, in the enzymatic synthesis of the RNA, the respective complementary single strand is produced: the transcription is *sequence-specific*. The synthesized RNA component dissociates from the complementary DNA single strand while transcription is still underway (by means of the appropriate *RNA polymerase* in the polymerase-DNA complex) so that the DNA duplex is once more regenerated.

Sequence- and strand-specific RNA synthesis takes place with the regulatory involvement of a series of activating and blocking proteins, which bind in the so-called *promoter* region of the DNA to effect their specific actions. In the structure of the resulting RNA, the nucleoside 2′-deoxyadenosine (dA) is replaced by adenosine (A), 2′-deoxycytidine (dC) by cytidine (C), 2′-deoxyguanosine (dG) by guanosine (G), and thymidine (dT) by uridine (U). Transcription proceeds in very different ways, depending on to which of the three domains of living creatures (*Archebacteria*, *Eubacteria*, *Eukaryots*; see *Sect. 16.4*) the particular cell belongs. The primary transcript is converted into functional RNA molecules (secondary transcript) by means of co-transcriptional and/or posttranscriptional maturation (*processing*). These three RNA types, *messenger RNA* (*mRNA*), *transfer RNA* (*tRNA*), and *ribosomal RNA* (*rRNA*), are always responsible for protein synthesis, irrespective of the type of cell in which RNA synthesis takes place.

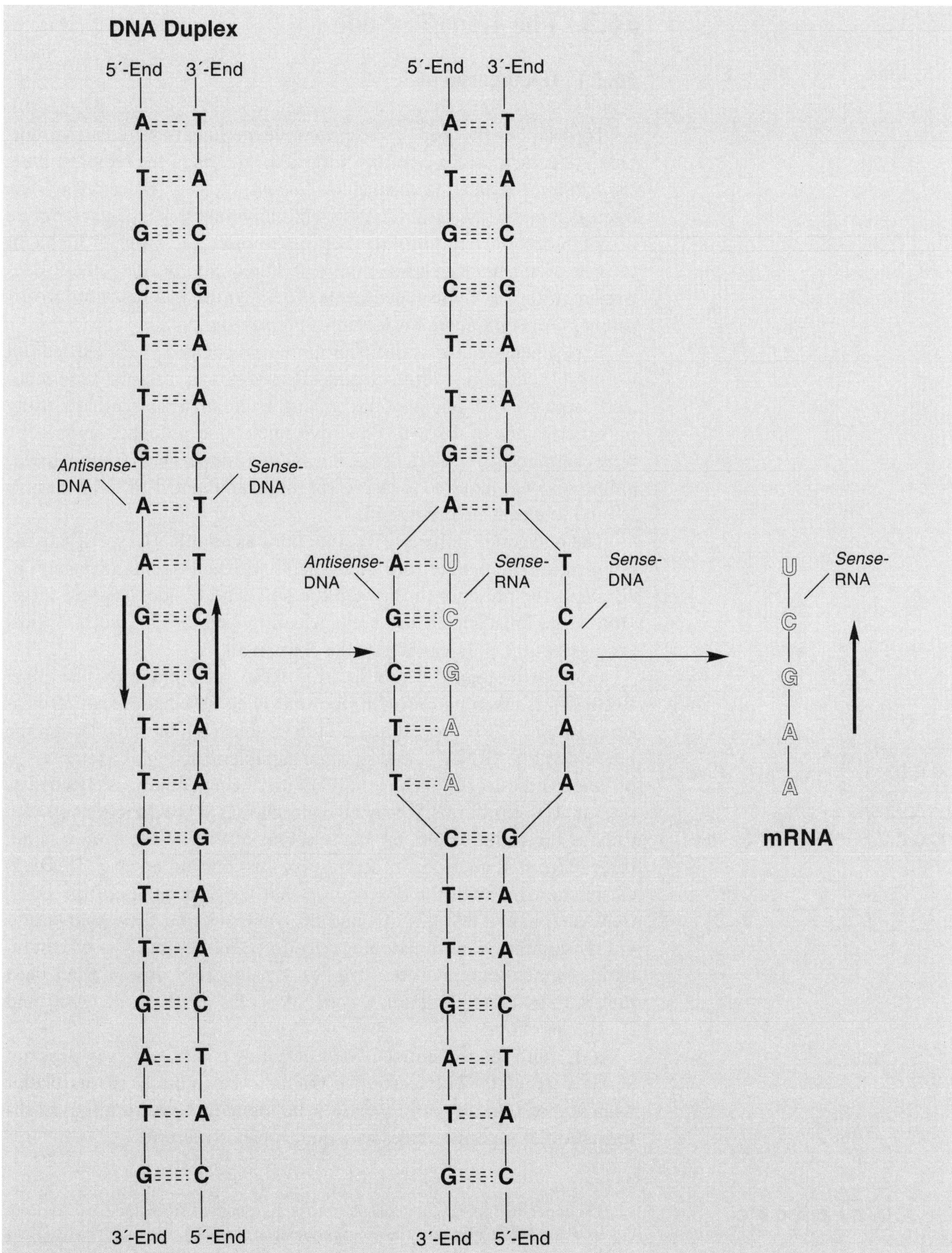

Fig. 16.16. *Schematic representation of the sequence- and strand-specific RNA synthesis emphasizing the DNA template to be copied* (bold letters) *and the mRNA strand having been synthesized* (light letters)

16.3. The Genetic Code

16.3.1. Decipherment

To decipher the genetic code, many experiments were carried out. Decisive contributions came from the laboratories of the biologist *Marshall Nirenberg* and the chemist *H. Gobind Khorana*. Since we are less interested in the historical succession of individual steps than in the logic of combinatorial relationships between molecules, we shall highlight those experiments which serve this end. Oligo- and polyribonucleotides were required for these experiments. Their synthesis is described in *Aspects of Organic Chemistry*, Volume 3: *Synthesis*.

Experiments aimed at deciphering the genetic code were carried out using synthetic poly- or trinucleotides. In the former instance, the amino-acid sequences of polypeptides arising from *in vitro* synthesis using polynucleotides of known, repetitive nucleotide sequence as mRNA were examined. In the latter case, it was determined which (radiolabeled) amino acid was required to be present for an aminoacyl-tRNA to be able to bind to a particular ribosome.

The nucleotide polymer $(U)_m$ functions as an mRNA, specifying the polypeptide $(Phe)_n$ in *in vitro* experiments with cell-free homogenates of *E. coli* in the presence of Phe. On the polypeptide side, *n* is very large. How large must *m*, on the polynucleotide side, be for the biological synthesis of the polypeptide to take place at all?

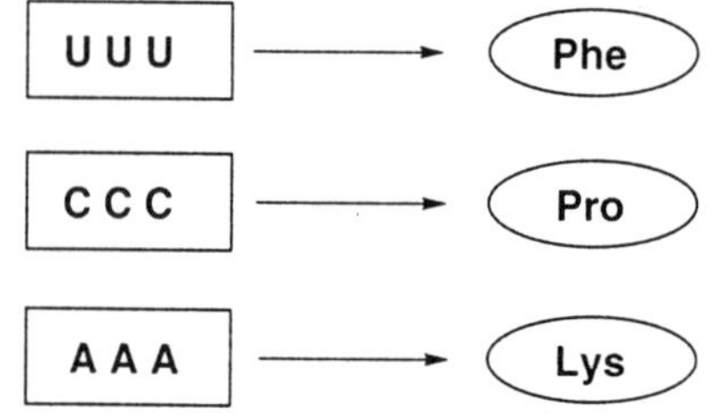

Prior to biological synthesis, the tRNA, charged with Phe (Phe-$tRNA^{Phe}$), is taken up onto the ribosome. This association can be measured using [^{14}C]Phe, revealing that the doublet $(U)_2$ has no effect on the binding of Phe-$tRNA^{Phe}$, and so is not capable of acting as a template in protein synthesis. $(U)_3$ behaves differently, being equally as effective as $(U)_4$ or $(U)_5$. From this, it may be deduced that UUU functions specifically as the codon triplet for the selection of Phe and that, in general, three sequential nucleobases act in specifying one amino acid. This conclusion concurs with the observation that the four letters of the DNA (RNA) alphabet, dA, dT, dG, and dC (A, U, G, and C), would permit $4^2 = 16$ combinations functioning as codon doublets, but $4^3 = 64$ combinations functioning as codon triplets. Analogously, it was established that $(C)_m$ and $(A)_m$ function as mRNA in the synthesis of $(Pro)_n$ and $(Lys)_n$, respectively.

$(G)_m$, with the sequential GGG nucleotide triplet, however, does not act as a template. The reason for this is to be found in the particular tendency of Gua to form hydrogen bonds with Gua, resulting in the formation of supermolecules with quadruplex structure.

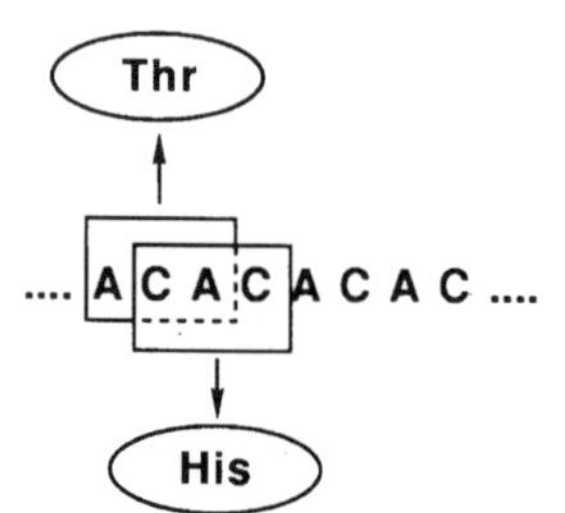

The copolymer $(A\text{-}C)_m$ contains the overlapping codons ACA and CAC, depending on where reading begins. The polypeptide determined by this copolymer consists of Thr and His in equal proportions. It is, therefore, clear that ACA and CAC are codons for Thr and His. But which codon specifies Thr and which His? This question may be answered experimentally from the stimulation of binding of [^{14}C]Thr-$tRNA^{Thr}$

to the ribosome, either by the codon triplet ACA or by the codon triplet CAC. Since ACA acts to stimulate binding, this codon specifies Thr. A similar experiment confirms the CAC triplet as a codon specifying His.

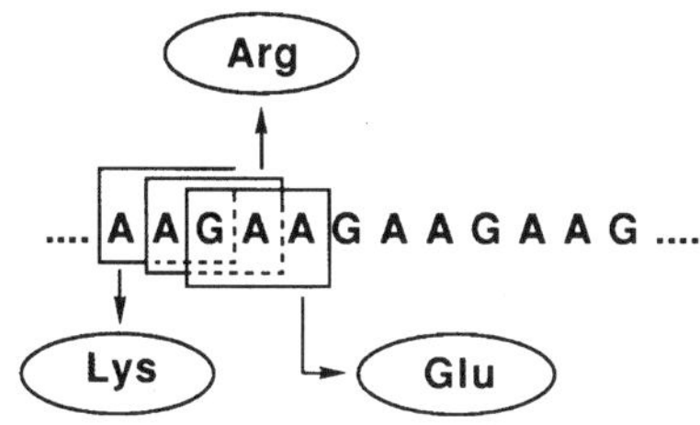

The copolymer $(A\text{-}A\text{-}G)_m$, depending on where reading begins, contains the overlapping triplets AAG, AGA, and GAA. In the presence of the codon triplet AAG, the binding of $[^{14}C]$Lys-$tRNA^{Lys}$ to *E. coli* ribosomes is stimulated, whereas the codon triplet AGA stimulates the binding of $[^{14}C]$Arg-$tRNA^{Arg}$, and the codon triplet GAA that of $[^{14}C]$Glu-$tRNA^{Glu}$. The codons AAG, AGA, and GAA, therefore, specify Lys, Arg, and Glu, respectively.

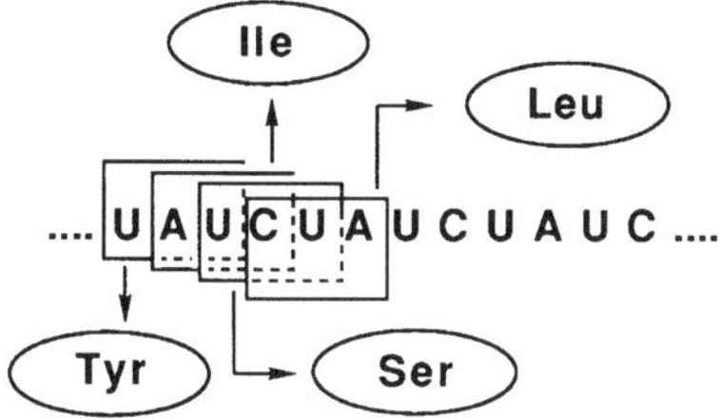

Depending on the starting point from where the sequence of the bases are read, the copolymer $(U\text{-}A\text{-}U\text{-}C)_m$ contains the codon triplets UAU, AUC, UCU, and CUA. It acts as mRNA for the synthesis of the polypeptide with the repetitive amino-acid pattern $(\text{Tyr-Leu-Ser-Ile})_n$. In the presence of the codon triplet UAU, the binding of $[^{14}C]$Tyr-$tRNA^{Tyr}$ to *E. coli* ribosomes is stimulated, whereas AUC promotes that of $[^{14}C]$Ile-$tRNA^{Ile}$, UCU that of $[^{14}C]$Ser-$tRNA^{Ser}$, and CUA that of $[^{14}C]$Leu-$tRNA^{Leu}$. The codons UAU, AUC, UCU, and CUA specify, therefore, Tyr, Ile, Ser, and Leu, respectively.

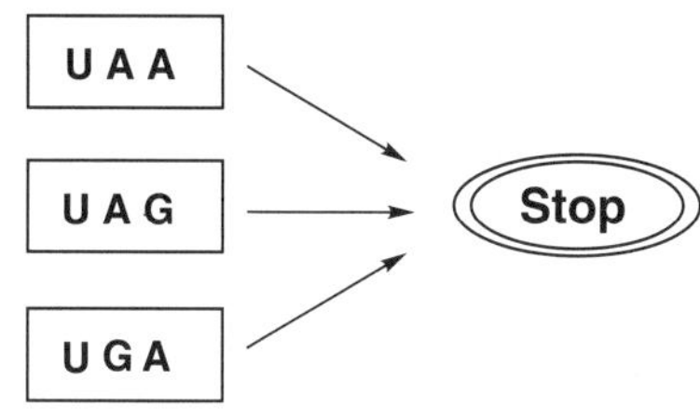

Analysis of the sequence of those polypeptides synthesized on an mRNA template with the polynucleoside sequence $(G\text{-}U\text{-}A\text{-}A)_m$ or $(A\text{-}U\text{-}A\text{-}G)_m$ led to surprises, as no periodic tetrapeptide sequences could be identified in these cases. A mixture of di- and tripeptides was observed instead. The reason for this is that the codon triplets UAA and UAG function as termination codons (*stop codons*), suppressing any continuation of protein synthesis. The codon triplet UGA is also a stop codon.

Among the 64 possible codon triplets there exist three termination codons. By experimentation, it has been possible to link the residual 61 codon triplets with the 20 proteinogenic amino acids formerly (*vide infra*) of molecular biologists' convention. The 'dictionary' with which these results were established is known for short as the *genetic code*, and is pictured on the inner back cover of this volume. Complementing the three termination codons is the initiation codon AUG. If this is located within a polynucleotide chain, rather than at the beginning, it specifies Met. The codon triplet UGA possesses a similar double function. Not only does it control the termination of protein synthesis, but also occasionally acts as the codon triplet specifying the amino acid selenocysteine.

Selenocysteine has been observed in a range of proteins from a great number of organisms. The genes responsible for the production of the enzymes glutathione peroxidase and formate dehydrogenase in the mouse or in *E. coli*, respectively, contain the codon triplet dTdGdA. Selenocysteine, therefore, belongs in the genetic code as the 21st proteinogenic amino acid. In *E. coli*, the codon triplet UGA of an mRNA is read for a Ser-$tRNA^{Ser}$, with serine being converted into selenocysteine before the aminoacyl-tRNA reaches the ribosome.

The genetic code determines the amino acid into which the codon triplet of an mRNA should be translated. In some mRNAs, there exists additional information, stored in the mRNA itself, which decides which of several possible readings of the genetic code should be applied. The term *recoding*, conveying an alteration of the meaning of code words during protein synthesis at the ribosome, is employed.

16.3.2. Degeneracy

Since the syntactic information of 62 codon triplets is translated, during protein synthesis, into the syntactic information of 21 amino acids, there must be 41 codons which either do not specify any amino acid, or which are synonymous with one of the 21 triplets already allowed for. The latter case is the true one (see *Table 10.18*): the code is said to be *degenerate*.

The genetic code may be represented in the form of a matrix with 16 equally large fields (*Fig. 16.17*). Half of these (boldface-framed) fields are characterized by the same amino acid being determined in all four codon triplets, solely by the first two codon nucleotides. The other (lightface-framed) half contain codon triplets which either assign one of two amino acids, or which assign one or two amino acids and give a stop command.

U U U Phe	U C U Ser	U A U Tyr	U G U Cys
U U C Phe	U C C Ser	U A C Tyr	U G C Cys
U U A Leu	U C A Ser	U A A Stop	U G A Stop
U U G Leu	U C G Ser	U A G Stop	U G G Trp
C U U Leu	C C U Pro	C A U His	C G U Arg
C U C Leu	C C C Pro	C A C His	C G C Arg
C U A Leu	C C A Pro	C A A Gln	C G A Arg
C U G Leu	C C G Pro	C A G Gln	C G G Arg
A U U Ile	A C U Thr	A A U Asn	A G U Ser
A U C Ile	A C C Thr	A A C Asn	A G C Ser
A U A Ile	A C A Thr	A A A Lys	A G A Arg
A U G Met	A C G Thr	A A G Lys	A G G Arg
G U U Val	G C U Ala	G A U Asp	G G U Gly
G U C Val	G C C Ala	G A C Asp	G G C Gly
G U A Val	G C A Ala	G A A Glu	G G A Gly
G U G Val	G C G Ala	G A G Glu	G G G Gly

Fig. 16.17. *The genetic code with its two subgroups in which either only the first and the second codon nucleotides* (boldface-framed), *or all three codon nucleotides* (lightface-framed) *determine the corresponding amino acid or termination of synthesis*

In the lightface-framed half of fields, it is crucial that successful distinction be made between the nucleotides in the third position of each codon, so as to avoid making mistakes in protein synthesis. For this, purine and pyrimidine bases must not in any circumstances be substituted for one another, and, in the codon-triplet pairs AUA/AUG and UGA/UGG, even A and G must not be interchanged with one another.

16.3.3. Codon-Anticodon Pairing

The anticodon region of an aminoacylated tRNA is capable of recognizing the codon region of an mRNA (*Fig. 16.18*).

Fewer than 62 tRNAs, and hence fewer than 62 anticodon regions, exist to interact with the 62 different codon triplets, corresponding to 21 amino acids as determined by the genetic code. One must, therefore, conclude that some tRNAs can introduce *their* amino acid into the protein which is going to be synthesized as a response to more than one codon. This is possible because the base pairing between codon and anticodon exhibits a certain flexibility. Only the first two bases of a codon triplet, and the last two bases of its complementary, antiparallel ordered anticodon triplet, enter into a stable *Watson-Crick* pair. The last base of the codon and the first base of the anticodon are connected very *loosely* with one another. *Crick* coined the term '*wobble rules*' to describe the rules governing the room for maneuver that this flexibility permits.

Of all nucleic acids, tRNAs encounter the greatest incidence of modified nucleosides. The enzymatically induced alterations, which, with few exceptions (inosine, for example), occur posttranscriptionally, may lead to conformational peculiarities, influencing supramolecular interactions with proteins or with other nucleic acids. Over 50 molecule types deviating from the canonical nucleosides A, G, U, or C are now known. Inosine (I) is a purine nucleoside composed of hypoxanthine (Hyp) and β-D-ribofuranose.

16.3.4. Universality?

The genetic code has a very wide range of validity. Exceptions may be found in, for example, the DNA of non-plant mitochondria (mitochondrial DNA, abbreviated as mtDNA), *i.e.* semiautonomous organelles in the cytoplasm of eukaryotic cells. Modern, highly compartmentalized eukaryotic cells are unanimously held to have come into being by symbiosis between one of their less advanced, organelle-free precursors and the precursor of a prokaryotic cell as an *endosymbiont* (*endosymbiont hypothesis*). If these endosymbionts originally behaved like cells within cells, then they could soon have adapted to meet the requirements of the host cells, through reciprocal division of labor. The organelles avail themselves of the entire mechanism of protein synthesis. Most of the mitochondrial proteins, however, are synthesized in the cell cytoplasm, following instructions from the genes in the cell nucleus. Only a few proteins are synthesized in the organelles under the command of their own genes.

The *genetic code* stands at the center of biology in a way similar to that in which the *periodic table of the elements* (see inner front cover of this volume) stands at the center of chemistry. There is one essential difference, however: since the periodic system of chemical elements is founded upon the specific electronic structure of atoms (*Chapt. 8.3*), it may reasonably be assumed that it is universally valid in the truest sense of the word. By contrast, the genetic code cannot be counted as univer-

If the gray-shaded nucleosides occupy the 1st position (wobble position) of the anticodon, then the tRNA

C	A	G	U	I
G	U	C	A	C
		U	G	A
				U

can recognize those codons in which one of the listed nucleosides occupies the 3rd position.

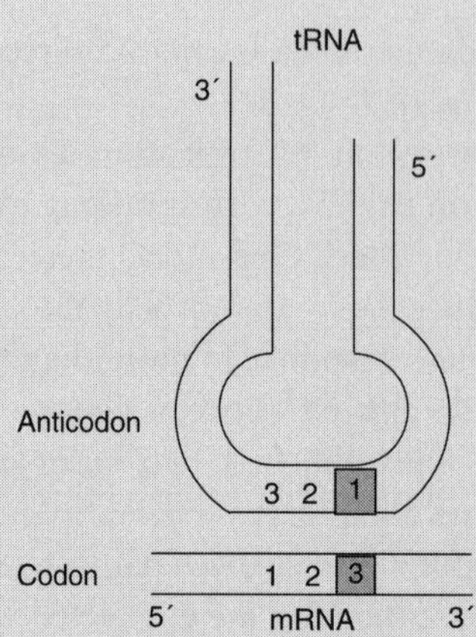

If the gray-shaded nucleosides occupy the 3rd position (wobble position) of the codon, then the mRNA

C	A	G	U
G	U	C	A
I	I	U	G
			I

can be recognized by a tRNA containing one of the listed nucleosides in the 1st position of the anticodon.

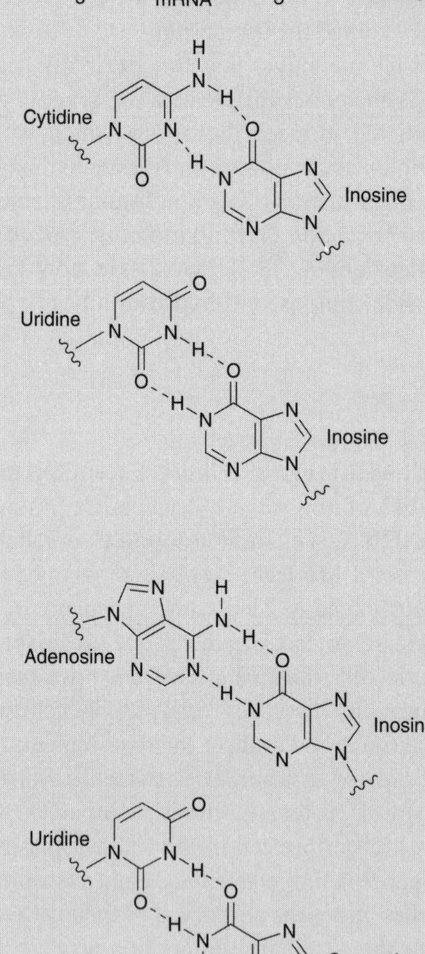

Fig. 16.18. *Formation of 'wobble pairs' between the third position of a codon triplet and the first position of an anticodon triplet leading to C/I, U/I, A/I, and U/G pairing*

sal, neither over time nor over space. It cannot, from our current viewpoint, be regarded as physically founded, but is a chance product of evolution: a chance product optimized by information-storing molecular replicators, resulting in teleonomic convergence towards an end appropriate for the present conditions. The trend of evolution expressed in that sentence may be illustrated by an analogous scenario: the fourth move of a chess player, who checkmates his or her opponent after the eighteenth move, seems, in retrospect, to have been specifically intended, and probably was indeed so, albeit in a way which could not have been unambiguously defined beforehand, and which was influenced by the opponent.

16.4. Documentation of Evolution Using Semantophoretic Molecules

The evolutionary biologist *Theodosius Dobzhansky*'s observation '*Nothing in biology makes sense except in the light of evolution*' would be ideologically tinged, were it not possible to test the theory of evolution by observation and experiment. It is only necessary, because of the relatively short human lifespan, to select a suitable object of observation, so that evolution may be unambiguously detected in the changes undergone by a natural population. By the way, there is no acceptable alternative to the theory of evolution.

The validity of the *genetic code* for everything that we acknowledge today as viable is a powerful evidence that all living beings share the same origin. It is possible to order them into phylogenetic families. Innumerable properties may be considered as classification criteria. Over the years, a shift of accent from morphological to molecular criteria has taken place. Sequences of proteins or nucleic acids have advanced more and more into the foreground.

After *Crick* had formulated the 'central dogma of molecular biology', *Zuckerkandl* and *Pauling* developed the concept of 'semantophoretic molecules as documents of evolution'. These semantophoretic molecules merit consideration as a basis for a molecular phylogeny. Non-semantophoretic molecules arising from enzyme-catalyzed reactions, such as, for example, the enzyme cytochrome C, used earlier for phylogenetic families, are designated *episemantic*, and molecules not synthesized by living cells, like butadiene, for example, as *asemantic*. Hence, at one end of a scale there stand semantophoretic molecules, especially the nucleic acids, at the other end are found the asemantic molecules. Asemantic molecules are not a source of phylogenetic information, episemantic molecules only an unreliable one. If it is desired to use semantophoretic molecules as indicators for the locating of a species (relative to other species) on the phylogenetic family tree, it must be taken into account that different proteins and different nucleic acids all have different rates of mutation. DNA from mitochondria, for example, evolves at a higher speed than DNA from the cell nucleus: mitochondrial DNA has no repair enzymes, unlike its counterpart in the nucleus. Ribosomal RNA evolves at a very slow rate. Therefore, rapidly evolving nucleic acids tend to be selected for examining species which have only recently diverged from a common ancestor, and slowly evolving nucleic acids for those species whose divergence took place in an early phase of evolution. In either case, if a pair of phylogenetically related species both possess the same nucleotide sequence (or amino-acid sequence), then this sequence will also be found in their common ancestors.

Using rRNA, *Woese* has recently established the evolutionary relationships of bacteria and hereby extended the phylogenetic family tree to every form of life. As a result, a new classification of all living things into three taxonomic domains – the eubacteria (*bacteria*), the archebacteria (*archaea*) and the eukaryots (*eucaria*) – could be made.

Eigen, using tRNA, has pursued the question of how old the genetic code is, a question important with regard to the origin of life.

M. Eigen, *Ruthild Winkler-Oswatitsch*, and *Andreas Dress* have established that the genetic code is younger than our planet, and cannot possibly be more than $3.8\,(\pm 0.6) \cdot 10^9$ years old. The conclusion was drawn with the aid of *statistical geometry* and representation of the results in *sequence space*. The process is essentially a procedure of establishing structural relationships between various sequences of proteins or nucleic acids. For nucleic acids, it must be taken into account that the sequences to be compared with one another are not of equal length, if particular nucleotides have been added (*insertion*) or removed (*deletion*), so it is necessary to concentrate on *homologies*, which may be distributed over the entire chain. Using computer-assisted techniques, it is possible to obtain the so-called *consensus sequence* out of a large quantity of homologous sequences. In this, the symbol (for the corresponding nucleoside) occurring most frequently in each position is listed. Each pair of sequences may be characterized by a divergence corresponding to the sum of differently occupied positions. For nucleic acids, with their four different nucleotide units, it is necessary to take account of a large number of divergence categories, and these are represented by complex geometric figures as multidimensional hypercubes in sequence space. It is then possible to deduce particular patterns of relationships from these geometric figures and, especially, to establish to what degree the 'memory' of the common ancestor has 'faded', or to what extent the original information has changed.

The question of the evolution of the genetic code has also been investigated by the *Eigen* school. *Table 16.1* offers information about current ideas.

Table 16.2 shows how, and on what hierarchical level, structural units of the semantophoretic biomacromolecular types relate to one another, and, additionally, to terms used in linguistics.

Analogies between human language and the chemical diversity of the immune system have been pointed out by *Niels K. Jerne*. In his *Nobel* Lecture of 1984, *Jerne* discusses the 'generative grammar of the immune system', and like *Noam Chomsky* takes a generative grammar as a basis for the ability of man to translate between orders of words and combinations of thought, assuming the existence of a generally applicable system of rules governing which syntactic structures are permissible and which are not. It is, therefore, not surprising when *Jerne* asks '*how is it possible that the immune system can produce antibodies capable of recognizing even those molecules which did not exist in the world until synthesized in a chemist's laboratory?*', while *Chomsky* proposes an answer to the ques-

Table 16.1. *Possible Evolution of the Genetic Code According to* M.Eigen et al. (1981). For the meaning N, R, and Y, see *Chapt. 10.5.2.*

1st Position — 2nd Position — 3rd Position

N Code in GNC, GNY frames

G	C	A	U
Gly	Ala	Asp	Val

RN Code in RNY frame

	G	C	A	U
G	Gly	Ala	Asp	Val
A	Ser	Thr	Asn?	Ile

RNN Code

	G	C	A	U	
G	Gly	Ala	Asp	Val	Y
	Gly	Ala	Glu	Val	R
A	Ser	Thr	Asn	Ile	Y
	Arg	Thr	Lys	Ile/Met	R

NNN Code

	G	C	A	U	
G	Gly	Ala	Asp	Val	C
	Gly	Ala	Asp	Val	U
	Gly	Ala	Glu	Val	G
	Gly	Ala	Glu	Val	A
A	Ser	Thr	Asn	Ile	C
	Ser	Thr	Asn	Ile	U
	Arg	Thr	Lys	Met ('START')	G
	Arg	Thr	Lys	Ile	A
C	Arg	Pro	His	Leu	C
	Arg	Pro	His	Leu	U
	Arg	Pro	Gln	Leu	G
	Arg	Pro	Gln	Leu	A
U	Cys	Ser	Tyr	Phe	C
	Cys	Ser	Tyr	Phe	U
	Trp	Ser	'STOP'	Leu	G
	'STOP'	Ser	'STOP'	Leu	A

Table 16.2. *Meaning of Linguistic Terms in the Languages of Semantophoretic Molecules*

Linguistic terms	Language of proteins	Language of Nucleic acids
Sentences	Polypeptides	Polynucleotides
Words	Amino acids	Nucleotide triplets
Letters	C^{α} with side chain	Nucleobases
Translation rule	The genetic code	

tion posed earlier by *Bertrand Russell* '*how is it that man, whose contact with the world is so brief, personal, and limited, can nonetheless know so much?*'

In this volume about the structure of molecules and supermolecules of organic chemistry, discussion has more than once been made of language as a vehicle of communication. The consistent use of symbols, which play an important role in chemical 'iconography' (*Chapt. 10*) for communication, not only between chemists, was emphasized repeatedly. Linguists have no doubt as to the function played by language in the development of cognitive faculties. *Wilhelm von Humboldt* saw clearly that linguistics has to amount to more than a mere demonstration of the history of language, but should also devote itself to the study of the human faculty of language as a biological survival trait. He rejects the view of language as being a preliminary accumulation of words which later had structure imposed upon them. In his opinion, languages are not only '*means to represent already perceived truth, but, much more, to discover things which were previously unknown*'. Different languages differ from one another in that each one is especially well suited to exposing particular facets of reality: each language affords a particular perspective on the world. *Arthur Kornberg* has described the contribution made by chemists, with their language, to the world-view of modern man: '*Much of life can be understood in rational terms if expressed in the language of chemistry. It is an international language, a language for all of time, and a language that explains where we came from, what we are, and where the physical world will allow us to go. Chemical language has great aesthetic beauty and links the physical sciences to the biological sciences*'.

Further Reading

General

P. Handler (Ed.), *Biology and the Future of Man,* Oxford University Press, New York, 1970.

S. J. Gould, *Wonderful Life,* W. W. Norton & Comp., New York, 1989.

W. Fontana, L. W. Buss, *What would be conserved if "the tape were played twice"? Proc. Natl. Acad. Sci. USA* **1994,** *91,* 757.

M. Delbrück, *A physicist looks at biology, Trans. Coun. Acad. Art. Sci.* **1949,** *38,* 173.

F. Crick, *What Mad Pursuit,* Basic Books, New York, 1988.

To Sect. 16.1.

16.1.1.

M. F. Perutz, *Physics and the riddle of life, Nature* **1987,** *326,* 555.

E. Schrödinger, *What is Life? & Mind and Matter,* Cambridge University Press, Cambridge, 1967.

N. W. Timofèeff-Ressovsky, K. G. Zimmer, M. Delbrück, *Über die Natur der Genmutation und der Genstruktur, Nachrichten aus der Biologie von der Gesellschaft der Wissenschaften zu Göttingen,* Mathematisch-Physikalische Klasse, Neue Folge, Fachgruppe VI, **1935,** *1,* 189.

J. B. S. Haldane in *Perspectives of Biochemistry* (Ed. J. Needham, D. E. Green), Cambridge University Press, Cambridge, 1937.

L. Pauling, M. Delbrück, *The Nature of the Intermolecular Forces Operative in Biological Processes, Science* **1940,** *92,* 77.

16.1.2.

C. Bresch, *Klassische und molekulare Genetik,* Springer-Verlag, Berlin, 1965.

A. J. F. Griffiths, J. H. Miller, D. T. Suzuki, R. C. Lewontin, W. M. Gelbart, *An Introduction to Genetic Analysis,* 5th Edn., W. H. Freeman & Comp., New York, 1993.

B. Lewin, *Genes V,* Oxford University Press, New York, 1994.

E. Mayr, *Eine neue Philosophie der Biologie,* Piper, München, 1988.

16.1.3.

A. E. Garrod, *Inborn Errors of Metabolism,* Oxford University Press, Oxford, 1923.

G. W. Beadle, *Nobel Lecture: Genes and chemical reactions in Neurospora* in *Nobel Lecture Physiology or Medicine 1942–1962,* Elsevier Publ. Comp., Amsterdam, 1964.

E. L. Tatum, *Nobel Lecture: A case history in biological research* in *Nobel Lecture Physiology or Medicine 1942–1962,* Elsevier Publ. Comp., Amsterdam, 1964.

F. H. C. Crick, *On Protein Synthesis, Symp. Soc. Exp. Biol.* **1958,** *12,* 548;
– *Central Dogma of Molecular Biology, Nature* **1970,** *227,* 561.

E. Zuckerkandl, L. Pauling, *Documents of Evolutionary History, J. Theoret. Biol.* **1965,** *8,* 357.

To Sect. 16.2.

L. Stryer, *Biochemistry,* 3rd Edn., W. H. Freeman & Comp., New York, 1988.

D. Voet, J. G. Voet, *Biochemistry,* Wiley, New York, 1990.

B. Alberts, D. Bray, J. Lewis, M. Raff, K. Roberts, J. D. Watson, *Molecular Biology of the Cell,* 2nd Edn., Garland, New York, 1989.

J. Darnell, H. Lodish, D. Baltimore, *Molecular Cell Biology,* 2nd Edn., Scientific American Books, New York, 1990.

J. D. Watson, N. H. Hopkins, J. W. Roberts, J. A. Steitz, A. M. Weiner, *Molecular Biology of the Gene,* 4th Edn., Vol. I and II, Benjamin/Cunnings Publ., Menlo Park, 1987.

M. Singer, P. Berg, *Genes & Genomes,* University Science Books, Mill Valley, California, 1991.

A.L. Lehninger, D.L. Nelson, M.M. Cox, *Principles of Biochemistry,* 2nd Edn., Worth Publ., New York, 1993.

H.F. Judson, *The Eighth Day of Creation,* Simon & Schuster, New York, 1979.

R. Dulbecco, *The Design of Life,* Yale University Press, New Haven, 1987.

16.2.1.

M. Eigen, *Steps toward life: A perspective on evolution,* Oxford University Press, New York, 1992.

J.D. Watson, F.H.C. Crick, *Genetical Implications of the Structure of Deoxyribonucleic Acid, Nature* **1953,** *171,* 964.

A. Kornberg, T.A. Baker, *DNA Replication,* W.H. Freeman & Comp., New York, 1992.

C.A. Hunter, *Interactions between aromatic systems: do they depend on electrostatic forces or charge-transfer transitions, Angew. Chem. Int. Ed.* **1993,** *32,* 1584.

A. Kornberg, *Nobel Lecture: The biologic synthesis of deoxyribonucleic acid* in *Nobel Lectures Physiology or Medicine 1942–1962,* Elsevier Publ. Comp., Amsterdam, 1964;
– *For the Love of Enzymes,* Harvard University Press, Cambridge, Mass., 1989.

R.L. Guyer, D.E. Koshland, Jr., *The Molecule of the Year, Science* **1989,** *246,* 1543.

K. Mullis, F. Faloona, S. Scharf, R. Saiki, G. Horn, H. Erlich, *Specific Enzymatic Amplification of DNA In Vitro: The Polymerase Chain Reaction, Cold Spring Harbor Symposia on Quantitative Biology,* **1986,** *L1,* 263.

K.B. Mullis, F.A. Faloona, *Specific Synthesis of DNA In Vitro via a Polymerase-Catalyzed Chain Reaction, Methods in Enzymol.* **1987,** *155,* 335.

K.B. Mullis, *The Unusual Origin of the Polymerase Chain Reaction, Scientific American* **1990,** *April,* 56.

K.B. Mullis, *Nobel Lecture: Polymerase chain reaction, Angew. Chem. Int. Ed.* **1994,** *33,* 1209.

K.B. Mullis, F. Ferré, R.A. Gibbs (Ed.), *The Polymerase Chain Reaction,* Birkhäuser, Boston, 1994.

L.E. Orgel, *RNA Catalysis and the Origin of Life, J. Theor. Biol.* **1986,** *123,* 127;
– *Evolution of the Genetic Apparatus: A Review, Cold Spring Harbor Symposia on Quantitative Biology,* **1987,** *L11,* 9;
– *Molecular replication, Nature* **1992,** *358,* 203.

S.L. Miller, L.E. Orgel, *The Origins of Life on the Earth,* Prentice-Hall, Englewood Cliffa, 1974.

G.F. Joyce, *Nonenzymatic Template-directed Synthesis of Informational Macromolecules, Cold Spring Harbor Symposia on Quantitative Biology,* **1987,** *L11,* 41.

G.F. Joyce, L.E. Orgel, *Prospects for Understanding the Origin of the RNA World* in *The RNA World* (Ed. R.F. Gasteland, J.F. Atkins), Cold Spring Harbor Laboratory Press, S. 1, Plainview, New York, 1993.

G. von Kiedrowski, *A self-replicating hexadeoxynucleotide, Angew. Chem. Int. Ed.* **1986,** *25,* 932.

A. Eschenmoser, *Toward a Chemical Etiology of the Natural Nucleic Acids' Structure,* The Robert A. Welch Foundation 37th Conference on Chemical Research, 1993;
– *Chemistry of Potentially Prebiological Natural Products, Origins of Life and Evolution of the Biosphere* **1994,** *24,* 389.

16.2.2.

D.L. Hatfield, B.J. Lee, R.M. Pirtle (Ed.), *Transfer RNA in Protein Synthesis,* CRC Press, Boca Raton, 1992.

A. Rich, *The structure and biology of transfer RNA, Current Biology* **1990,** Introductory Issue.

D. Moras, *Transfer RNA, Current Biology* **1991,** *1,* 410.

To Sect. 16.3.

H. G. Khorana, *Nobel Lecture: Nucleic acid synthesis in the study of the genetic code* in *Nobel Lectures Physiology or Medicine 1963–1970*, Elsevier, Amsterdam, 1972.

M. Nirenberg, *Nobel Lecture: The genetic code* in *Nobel Lectures Physiology or Medicine 1963–1970*, Elsevier, Amsterdam, 1972.

L. Maréchal-Dronard, A. Dietrich, J. H. Weil, *Adaption of tRNA Population to Codon Usage in Cellular Organelles* in *Transfer RNA in Protein Synthesis* (Ed. D. L. Hatfield, B. J. Lee, R. M. Pirtle), CRC Press, Boca Raton, 1992.

D. L. Hatfield, I. S. Choi, B. J. Lee, J.-E. Jung, *Selenocysteine, a New Addition to the Universal Genetic Code* in *Transfer RNA in Protein Synthesis* (Ed. D. L. Hatfield, B. J. Lee, R. M. Pirtle), CRC Press, Boca Raton, 1992.

F. H. C. Crick, *Codon-Anticodon-Pairing, The Wobble Hypothesis, J. Mol. Biol.* **1966,** *19,* 548.

To Sect. 16.4.

T. Dobzhansky, *Nothing in biology makes sense except in the light of evolution, Amer. Biol. Teacher* **1973,** *35,* 125.

M. Ridley, *Evolution,* Blackwell Scientific Publications, Cambridge, Mass., 1993.

E. Zuckerkandl, L. Pauling, *Documents of Evolutionary History, J. Theoret. Biol.* **1965,** *8,* 357.

C. R. Woese, *Bacterial Evolution, Microbiol. Rev.* **1987,** *51,* 221.

C. R. Woese, O. Kandler, M. L. Whellis, *Towards a natural system of organisms: Proposal for the domains Archaea Bacteria, and Eucarya, Proc. Natl. Acad. Sci. USA* **1990,** *87,* 4576.

C. R. Woese, *The use of ribosomal RNA in reconstructing evolutionary relationships among bacteria* in *Evolution at the Molecular Level* (Ed. R. K. Selander, A. G. Clark, T. S. Whittam), Sinauer Associates, Sunderland, Mass., 1991.

M. Eigen, R. Winkler-Oswatitsch, A. Dress, *Statistical geometry in sequence space: A method of quantitative comparative sequence analysis, Proc. Natl. Acad. Sci. USA* **1988,** *85,* 5913.

M. Eigen, R. Winkler-Oswatitsch, *Statistical Geometry in Sequence Space, Methods in Enzymol.* **1990,** *183,* 505.

M. Eigen, W. Gardiner, P. Schuster, R. Winkler-Oswatitsch, *The Origin of Genetic Information, Scientific American* **1981,** *244(3),* 78.

M. Eigen, B. F. Lindemann, M. Tietze, R. Winkler-Oswatitsch, A. Dress, A. von Haeseler, *How Old is the Genetic Code? Statistical Geometry of tRNA Provides an Answer, Science* **1989,** *244,* 673.

M. Eigen, *Viral Quasispecies, Scientific American* **1993,** *269(1),* 32.

N. K. Jerne, *Nobel Lecture: The generative grammar of the immune system* in *Nobel Lectures Physiology or Medicine 1981–1990*, World Scientific Publ., Singapore, 1993.

N. Chomsky, *Knowledge of Language,* Praeger, New York, 1986;
– *Reflections on Language,* Pantheon, New York, 1975.

B. Russell, *Human Knowledge. Its Scope and Limits,* Simon & Schuster, New York, 1948.

A. Kornberg, *The Two Cultures: Chemistry and Biology, Biochem.* **1987,** *26,* 6888.

Author Index

Subject Index

D

E

F

G

J

K

L

M

N

O

P

Q

R

T

U

V

W

X

Y

Z